Lecture Notes in Artificial Intelligence   5144

Edited by R. Goebel, J. Siekmann, and W. Wahlster

Subseries of Lecture Notes in Computer Science

Serge Autexier John Campbell
Julio Rubio Volker Sorge
Masakazu Suzuki Freek Wiedijk (Eds.)

Intelligent Computer Mathematics

9th International Conference, AISC 2008
15th Symposium, Calculemus 2008
7th International Conference, MKM 2008
Birmingham, UK, July 28 – August 1, 2008
Proceedings

 Springer

Series Editors

Randy Goebel, University of Alberta, Edmonton, Canada
Jörg Siekmann, University of Saarland, Saarbrücken, Germany
Wolfgang Wahlster, DFKI and University of Saarland, Saarbrücken, Germany

Volume Editors

Serge Autexier
DFKI, Stuhlsatzenhausweg 3, 66123 Saarbrücken, Germany
E-mail: autexier@dfki.de

John Campbell
University College London, Gower Street, London WC1E 6BT, UK
E-mail: j.campbell@cs.ucl.ac.uk

Julio Rubio
Universidad de La Rioja, 26004 Logroño (La Rioja), Spain
E-mail: julio.rubio@unirioja.es

Volker Sorge
University of Birmingham, Edgbaston,Birmingham, B15 2TT, UK
E-mail: V.Sorge@cs.bham.ac.uk

Masakazu Suzuki
Kyushu University, Hakozaki 6-10-1, Higashi-ku, Fukuoka, 812-8581, Japan
E-mail: suzuki@math.kyushu-u.ac.jp

Freek Wiedijk
Radboud University Nijmegen, Toernooiveld 1, 6525 ED Nijmegen, The Netherlands
E-mail: freek@cs.ru.nl

Library of Congress Control Number: Applied for

CR Subject Classification (1998): I.2.2, I.1-2, H.3, H.2.8, F.4.1, C.2.4, G.2, G.4

LNCS Sublibrary: SL 7 – Artificial Intelligence

ISSN   0302-9743

ISBN   978-3-540-85109-7 Springer Berlin Heidelberg New York

Springer is a part of Springer Science+Business Media

springer.com

© Springer-Verlag Berlin Heidelberg 2008

Typesetting: Camera-ready by author, data conversion by Scientific Publishing Services, Chennai, India
Printed on acid-free paper   SPIN: 12441570   06/3180   5 4 3 2 1 0

Preface

This volume contains the collected contributions of three conferences, AISC 2008, Calculemus 2008, and MKM 2008. AISC 2008 was the 9th International Conference on Artificial Intelligence and Symbolic Computation that is concerned with the use of AI techniques within symbolic computation as well as the application of symbolic computation to AI problem solving. Calculemus 2008 was the 15th Symposium on the Integration of Symbolic Computation and Mechanized Reasoning dedicated to the combination of computer algebra systems and automated deduction systems. MKM 2008 was the 7th International Conference on Mathematical Knowledge Management, an emerging interdisciplinary field of research in the intersection of mathematics, computer science, library science, and scientific publishing. All three conferences are thus in general concerned with providing intelligent computer mathematics. Although the conferences have separate communities and separate foci, there is a significant overlap in the interests in building systems for intelligent computer mathematics. For this reason it was decided to collocate the three events in 2008, at the University of Birmingham, UK. While the proceedings are shared, the submission process was separate. The responsibility for acceptance/rejection rests completely with the three Programme Committees. By this collocation we made a contribution against the fragmentation of communities which work on different aspects of different independent branches, traditional branches (e.g., computer algebra, theorem proving and artificial intelligence in general), as well as newly emerging ones (on user interfaces, knowledge management, theory exploration, etc.). This will also facilitate the development of systems for intelligent computer mathematics that will be routinely used by mathematicians, computer scientists, and engineers in their every-day business. In total, 37 papers were submitted to AISC. For each paper there were up to four reviews, out of which 14 papers were accepted for publication in these proceedings. Calculemus received 10 submissions, which were all reviewed by three Programme Committee members. This number was quite low, but fortunately all submissions were of good quality and on topic for the Calculemus conference. For this reason all papers were accepted for presentation at the conference and for publication in this proceedings volume. MKM received 34 submissions. For each paper there were at least two reviews; if the evaluation was not uniform there was a third review. After discussions, 18 papers were accepted for these proceedings. In addition to the contributed papers, the proceedings include the contributions of five invited speakers of AISC, Calculemus, and MKM. In the preparation of these proceedings and in managing the whole discussion process, Andrei Voronkov's EasyChair conference management system proved itself an excellent tool.

We would like to gratefully acknowledge the support of the School of Computer Science of the University of Birmingham and of the British Society for the Study of Artificial Intelligence and the Simulation of Behaviour (SSAISB).

May 2008   Serge Autexier
                                                  John Campbell
                                                    Julio Rubio
                                                  Volker Sorge
                                              Masakazu Suzuki
                                                Freek Wiedijk

AISC, Calculemus, and MKM Organization

Conference and Local Chair	Volker Sorge (University of Birmingham, UK)
Local Organization	Josef Baker (University of Birmingham, UK)
	Noureddin Sadawi (University of Birmingham, UK)
	Alan Sexton (University of Birmingham, UK)
WWW	http://events.cs.bham.ac.uk/cicm08/

AISC 2008 Organization

Programme Chair

Volker Sorge	University of Birmingham

Steering Committee

Jacques Calmet	University of Karlsruhe, Germany
John Campbell	University College London, UK
Eugenio Roanes-Lozano	Universidad Complutense de Madrid, Spain

Programme Committee

Alessandro Armando	University of Genoa, Italy
Christoph Benzmüller	Universität des Saarlandes, Germany
Russell Bradford	University of Bath, UK
Bruno Buchberger	RISC, Austria
Jacques Calmet	University of Karlsruhe, Germany
John Campbell	University College London, UK
Jacques Carette	McMaster University, Canada
Arjeh Cohen	Eindhoven University of Technology, The Netherlands
Simon Colton	Imperial College London, UK
Timothy Daly	Carnegie Mellon, USA
Lucas Dixon	University of Edinburgh, UK
William M. Farmer	McMaster University, Canada
Martin Charles Golumbic	University of Haifa, Israel
Hoon Hong	North Carolina State University, USA
Tetsuo Ida	University of Tsukuba, Japan
Tom Kelsey	University of St. Andrews, UK
George Labahn	University of Waterloo, Canada

Petr Lisonek	Simon Fraser University, Canada
Renaud Rioboo	ENSIIE, France
Eugenio Roanes-Lozano	Universidad Complutense de Madrid, Spain
Karem Sakallah	University of Michigan, USA
Jörg Siekmann	Universität des Saarlandes, DFKI, Germany
Elena Smirnova	Texas Instruments, USA
Dongming Wang	Beihang University, China and UPMC-CNRS, France
Stephen M. Watt	University of Western Ontario, Canada
Wolfgang Windsteiger	RISC, Austria

External Reviewers

John Abbott	Martin Brain	John Bullinaria
Marina De Vos	Louise Dennis	Jacques Fleuriot
Thomas French	Murdoch Gabbay	Enrico Giunchiglia
M. Skander Hannachi	Shan He	Nao Hirokawa
Heinz Kredel	Michael Maher	Jacopo Mantovani
Brendan McKay	Serge Mechveliani	Julian Padget
John Power	Jon Rowe	Thorsten Schnier
Joerg Siekmann	Alan Smaill	Luca Vigano
Marc Wagner	Johannes Waldmann	Jinzhao Wu
Hans Zantema	Jian Zhang	Hantao Zhang

Calculemus 2008 Organization

Programme Chairs

Julio Rubio	Universidad de La Rioja, Spain
Freek Wiedijk	Radboud University Nijmegen, The Netherlands

Programme Committee

Michael Beeson	San Jose State University, USA
Christoph Benzmüller	University of Cambridge, UK
Anna Bigatti	Università di Genova, Italy
Wieb Bosma	Radboud University Nijmegen, The Netherlands
Olga Caprotti	University of Helsinki, Finland
Jacques Carette	McMaster University, Canada
Thierry Coquand	Göteborg University, Sweden
James Davenport	University of Bath, UK
William Farmer	McMaster University, Canada
John Harrison	Intel Corporation, USA
Deepak Kapur	University of New Mexico, USA

Henri Lombardi	Université de Franche-Comte, France
Tomás Recio	Universidad de Cantabria, Spain
Renaud Rioboo	Université Pierre et Marie Curie, France
Volker Sorge	University of Birmingham, UK
Makarius Wenzel	Technische Universität München, Germany
Wolfgang Windsteiger	RISC-Linz, Austria

External Reviewers

Serge Autexier	Jeremy Avigad	Milad Niqui

MKM 2008 Organization

Programme Chairs

Serge Autexier	DFKI Saarbrücken & Saarland University, Germany
Masakazu Suzuki	Kyushu University, Japan

Programme Committee

Andrea Asperti	University of Bologna, Italy
Laurent Bernardin	Maplesoft, Canada
Thierry Bouche	Université de Grenoble I, France
Paul Cairns	University College London, UK
Olga Caprotti	University of Helsinki, Finland
Simon Colton	Imperial College, UK
Mike Dewar	NAG Ltd., UK
William Farmer	McMaster University, Canada
Herman Geuvers	Radboud University Nijmegen, The Netherlands
Eberhard Hilf	ISN Oldenburg, Germany
Tetsuo Ida	University of Tsukuba, Japan
Mateja Jamnik	University of Cambridge, UK
Fairouz Kamareddine	Heriot-Watt University, UK
Manfred Kerber	University of Birmingham, UK
Michael Kohlhase	Jacobs University Bremen, Germany
Paul Libbrecht	DFKI Saarbrücken, Germany
Bruce Miller	NIST, USA
Robert Miner	Design Science, Inc., USA
Bengt Nordström	Chalmers University of Technology, Sweden
Eugénio Rocha	University of Aveiro, Portugal
Alan Sexton	University of Birmingham, UK
Petr Sojka	Masaryk University, Czech Republic
Volker Sorge	University of Birmingham, UK

Andrzej Trybulec   University of Bialystok, Poland
Stephen Watt   The University of Western Ontario, Canada
Abdou Youssef   George Washington University, USA

External Reviewers

Josef Baker   David Carlisle   Martin Homik
Cezary Kaliszyk   Matti Pauna   Wilmer Ricciotti
Claudio Sacerdoti Coen   Christoph Zengler

Table of Contents

Contributions to AISC 2008

Invited Talks

Contributed Papers

Contributions to Calculemus 2008

Invited Talk

Contributed Papers

Contributions to MKM 2008

Invited Talks

Contributed Papers

Symmetry and Search – A Survey

Steve Linton

Searches of one form or another are a central unifying theme in Artificial Intelligence and effective management of symmetry is a key practical consideration in the practical application of a huge range of search techniques. It is almost never desirable to search, but frequently hard to avoid searching, very large numbers of equivalent sections of the search tree. Likewise many searches, if symmetry is not accounted for, will return very large numbers of unwanted equivalent solutions.

Symmetry, is, of course, a phenomenon much studied in pure mathematics, where group theory provides both a language for discussing it and a large body of knowledge about it. Computational group theory is a well-developed area within symbolic computation.

In this talk I will survey efforts to apply group theory and especially computational group theory to the management of symmetry in a range of AI search problems.

S. Autexier et al. (Eds.): AISC/Calculemus/MKM 2008, LNAI 5144, p. 1, 2008.

On a Hybrid Symbolic-Connectionist Approach for Modeling the Kinematic Robot Map - and Benchmarks for Computer Algebra

Jochen Pfalzgraf

Department of Computer Sciences, University of Salzburg, Austria
jochen.pfalzgraf@sbg.ac.at

Dedicated to Professor Jacques Calmet

Keywords: Robotic Kinematics, Computer Algebra (CA), Connectionism, Inverse Kinematics Problem, Artificial Neural Networks, Simulation, Benchmarks for CA.

1 Introduction

The kinematics model of a robot arm (we are considering open kinematic chains) is described by a corresponding robot map having the configuration space as its domain and the workspace as codomain. In other words, the robot map assigns to every configuration of the joint parameters a unique point of the workspace of the robot arm. We briefly discuss the general introduction of the robot map where the parameters of a translational joint are represented by points of the real line and the parameters of a rotational joint by points of the unit circle in the real plane, respectively. Thus, in general, a concrete joint configuration (point of the configuration space) is an element of an abelian Lie group being a direct product of some copies of the real line and the unit circle. The position and orientation of the endeffector of a robot arm is represented by an element of the euclidean motion group of real 3-space. The standard problems like the direct kinematics problem, the inverse kinematics problem and the singularity problem can easily be defined.

A classical method to establish the robot map is the approach by Denavit-Hartenberg. It leads to a completely symbolic description of the direct kinematics model of an arm and forms the basis for the treatment of the inverse kinematics problem. In order to represent an entire robot arm class it is of basic interest to find a completely symbolic closed form solution of the inverse kinematics problem. Using a two joint robot arm, B.Buchberger demonstrated the principle how to solve this problem with the help of a computer algebra (CA) system applying his Gröbner bases method (Buchberger Algorithm) - cf. [1], [2], [3], [4].

Later we made own investigations and constructed a more complex test example (cf. [5]). We observed that very hard performance problems arose, in the corresponding CA applications, when the degree of freedom of a robot arm increases. An interesting aspect is the fact that these investigations show a natural way how to construct benchmarks for CA.

S. Autexier et al. (Eds.): AISC/Calculemus/MKM 2008, LNAI 5144, pp. 2–16, 2008.

A completely different method to represent the kinematic model of a robot arm is a Connectionist Network approach. The idea is to learn a robot map with the help of a suitably chosen Artificial Neural Network (ANN) using a powerful ANN simulator. The training data for learning consists of selected input/output pairs generated by the given robot map. This approach also reaches soon its limits when the number of joints increases.

The experiences which we made with both approaches, symbolic and connectionist, formed the basis of the idea to combine both - constructing a hybrid system - trying to cope with the increasing complexity. Special investigations of a sample robot arm showed that it is worthwhile considering such a hybrid system for use in concrete applications.

In a short section we give a brief report on a former project dealing with the design and implementation of an ANN simulation tool (FlexSimTool). Its theoretical fundament was a general mathematical network structure model which we have developed based on notions from category theory and noncommutative geometry leading to the category of geometric networks (GeoNET) where learning can be interpreted in terms of morphisms ([6], [7]).

Some remarks on Benchmarks for CA conclude the article. We briefly describe two classes, one class is constructed with the help of the Denavit-Hartenberg approach for establishing the robot map of a robot arm, as considered in the corresponding section below. Another class of benchmarks for CA coming from Noncommutative Geometry (NCG) is briefly discussed.

2   The ROBOT Map and Computer Algebra Applications

A classical case of modeling a Knowledge-based System (KBS) in robotics is the development of the **Robot Map** leading to the mathematical model of a robot arm (open kinematic chain) which forms the basis for the formal definition of the problems in direct kinematics, inverse kinematics, singularity detection, path control and their computational treatment. A crucial point is the exploitation of the **geometry of a robot arm**. The robot arms we are considering are open kinematic chains with two types of joints, **translational joints** and **rotational joints**. For more details on the topics presented in this section we refer to [5] and the references cited there.

The *Base* of a robot is that part where a robot is fixed to the ground or where the wheels are attached, for example, in the case of a mobile robot. The first sub-arm is then linked with the base. The end of the last sub-arm of the whole robot (in that order) is mostly used for attaching the corresponding tools like screw driver, welding device, etc., it is called the *End Effector*, abbreviated by EE, for short.

Let t denote the number of translational joints and r the number of rotational joints. The robot map will be considered as a mathematical map $R : \mathcal{C} \to \mathcal{W}$ from joint space (configuration space) $\mathcal{C}$ to work space $\mathcal{W}$, this is the space where the movement of the end effector (EE) is described (cf. [5], section 2).

The *Degree of Freedom (DOF)* is defined by $\text{DOF} = r + t$ (this is the total number of joints), i.e. the "joint space manifold" $\mathcal{C}$ has dimension $r + t$.

Now we are going to consider the largest possible *joint space*.

The parameter of a translational joint runs through a real interval, we model this more generally by the whole real line $\mathbb{R}$. Rotational joint parameters (angles) can be represented by corresponding points of the unit circle S^1 in the real plane; we recall that $S^1 = \{e^{i\phi} | 0 < \phi \le 2\pi\}$ (for the algebraic description we prefer here the complex number representation of points of the real plane). A rotation through angle ϕ corresponds to a multiplication of a 2-dimensional vector by $e^{i\phi}$ in the corresponding "rotation plane". The complex numbers are represented as 2-dimensional vectors (and vice versa) with real and imaginary parts as x- and y- components in the real plane - we recall that $e^{i\phi} = \cos(\phi) + i\sin(\phi)$. The **configuration space (joint space) manifold** $\mathcal{C}$ can be described in the general case by $(\mathbb{R} \times \ldots \times \mathbb{R}) \times (S^1 \times \ldots \times S^1) = \mathbb{R}^t \times (S^1)^r$.

The cartesian product $(S^1)^r$ is also called *r-dimensional torus* $\mathbb{T}^r$ with $\mathbb{T} = S^1$, the 1-dimensional torus. This big configuration space is a nice differentiable manifold, actually it is an abelian Lie group.

Concerning the **work space** the question arises what is the complete information necessary to describe the location of the end effector (EE) of a robot? The complete information about the EE location is described by a *position vector* $a = (a_1, a_2, a_3) \in \mathbb{R}^3$ and the *orientation* of the EE. Orientation in Euclidean 3-space can be uniquely described by an orthonormal frame (representable by an orthogonal 3×3—matrix having determinant $+1$). The variety of all these frames can be described by the group of all proper rotations, the *Special Orthogonal Group*: $SO_3(\mathbb{R}) = \{A | A^t = A^{-1}, det(A) = +1\}$.

The **work space manifold** describing all possible EE locations in the most general case is $\mathbb{R}^3 \times SO_3(\mathbb{R})$, the *Euclidean Motion Group*. Again, this is a nice differentiable manifold and a Lie group (semidirect product of groups). It is of dimension 6, this is why 6-DOF robots are so widespread in practical applications. More precisely, a manipulator with the capability that the EE can reach every point in the complete work space must at least have six DOF, i.e. it needs six joints and at least three of them have to be rotational (to provide all possible orientations).

Thus, in the most general case, a robot can be mathematically described by a (differentiable) map, the **Robot Map** $R : \mathbb{R}^t \times (S^1)^r \longrightarrow \mathbb{R}^3 \times SO_3(\mathbb{R})$.

Let $R : \mathcal{C} \to \mathcal{W}$, in general, denote the robot map describing a selected robot model with joint space $\mathcal{C}$ and work space $\mathcal{W}$. Let $q \in \mathcal{C}$ denote the joint coordinate vector (n-dimensional) and $x \in \mathcal{W}$ the corresponding image $R(q)$ (m-dimensional) in the work space.

The degree of freedom (DOF) of the manipulator system is defined by the dimension of the configuration space manifold, i.e. $\text{DOF} = \dim(\mathcal{C}) = n$.

A *Redundant Manipulator* is defined as a robot arm having more DOF than is necessary in comparison with the dimension of the workspace, i.e.: $\dim(\mathcal{C}) = n > m = \dim(\mathcal{W})$. Redundant manipulators are of interest for complex industrial applications where difficult EE movements have to be performed and where redundancy is exploited practically.

With the help of the robot map it is now straight forward to formulate the two typical problems in robot kinematics (the direct or forward kinematics problem and the inverse kinematics problem). Let $x = (a, A) \in \mathcal{W}$ denote a point in work space specifying an EE location and $q = (q_1, \ldots, q_n) \in \mathcal{C}$ a point in the configuration space with joint coordinates (parameters) $q_1, \ldots, q_n$.

Forward (direct) kinematics problem: Given a joint configuration q, determine the EE location corresponding to these joint parameter values. That means in terms of the robot map for the given input q determine $\mathsf{R}(q) = x$.

Inverse kinematics problem: Given a point $x = (a, A)$ which specifies an EE location, then determine a suitable joint configuration $q = (q_1, \ldots, q_n)$ such that the robot points to the given EE location x. In terms of the robot map: Given x find a suitable q (pre-image) such that $x = \mathsf{R}(q)$.

When dealing with the general kinematics model a main problem is to express the location of the end effector EE in terms of the world coordinate system (WCS). This latter reference frame is frequently fixed in the robot base or in a selected point of the assembly hall where the robot is at work. To this end, local coordinate frames are attached to each sub-arm (link) and then the EE coordinates are successively transformed backwards by (local) transformations from the last sub-arm where the EE is situated to its predecessor and so on until the robot base is reached.

There are four invariants implicitly given by the **geometry of the robot**. In the figure below (showing the situation of two neighboring local coordinate systems) a_{i-1}, d_{i-1} are distance parameters and $\alpha_{i-1}, \Theta_{i-1}$ are angle parameters.

Let $\mathbf{B}_i$ denote the $i-$th coordinate frame and $\mathbf{O}_i$ its origin. The *Denavit and Hartenberg* approach describes the convention how to fix the local coordinate systems and to establish the corrresponding local transformations $\mathbf{B}_{i-1} \to \mathbf{B}_i$ using the parameters $\mathsf{a}_{i-1}, \alpha_{i-1}, \mathsf{d}_{i-1}, \Theta_{i-1}$ (cf. [5] and references there for more details - we mention here the books [8], [9], [10]).

The local transformation can then be expressed by the composition of the following operations: - $\mathrm{Rot}(z_{i-1}, \Theta_{i-1})$, - translation along z_i by distance d_{i-1}, - translation along new x_i-axis by distance a_{i-1}, - $\mathrm{Rot}(x_i, \alpha_{i-1})$. Where $\mathrm{Rot}(x, \Theta)$ means rotation around an axis x by an angle Θ.

In this way we obtain the corresponding local transformation matrix T_i^{i-1}. Thus, each pair of neighbored links contributes a local transformation T_i^{i-1}. The product T_n^0 of these successive local transformations $T_n^0 = T_1^0 \circ T_2^1 \ldots \circ T_n^{n-1}$ establishes a relationship between the coordinate system of the robot EE (system (n)) and the world coordinate system (system (0)) – involving all the "local contributions" (cf. the picture of the rtrr-robot for a special illustration).

In the article [5] I devised an own robot arm example with 4-DOF of type **rtrr** (cf. the corresponding figure below, the type notation is explained in a later section) and used it to apply the pure symbolic approach for computing the inverse kinematic solutions with the help of a computer algebra system and

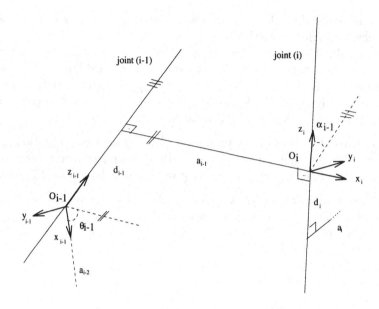

Fig. 1. Denavit-Hartenberg parameters

Buchberger's Algorithm (method of Gröbner bases) for solving the complex multivariate polynomial equations.

Actually, it was Bruno Buchberger who had the idea of such a symbolic modeling approach applying computer algebra. He demonstrated it with a 2-DOF rotational robot arm and found the closed form solution to the inverse kinematics problem. His example is included in [5], we point here to his publications [1], [2], [3].

The investigations of the rtrr-robot arm demonstrated the complexity that arises when many variables are involved in the corresponding multivariate polynomials and coefficient fields. Thus we found a rich source for constructing hard **benchmarks for computer algebra** (for details cf. [5] and the last section of this article).

3   Connectionist Approach: Learning the Robot Map

A completely different way to represent the kinematic model of a robot arm is a Connectionist Network approach. The idea is to learn a robot map with the help of a suitably chosen artificial neural network (ANN) using a powerful ANN simulator. The training data for learning consists of selected input-output pairs generated by a given robot map $R : C \rightarrow W$. We generate a set of pairs $Td = \{(q, x)\}$ where $q \in C$ and $x = R(q) \in W$ is the unique EE-location belonging to the input-configuration $q = (q_1, \ldots, q_n)$, we recall that $x = (a, A) \in \mathbb{R}^3 \times SO_3(\mathbb{R})$. Training a selected ANN using the first component q as input and the second component x as desired output of the ANN, where (q, x) runs through all pairs of Td, means that we do learn the forward kinematic model

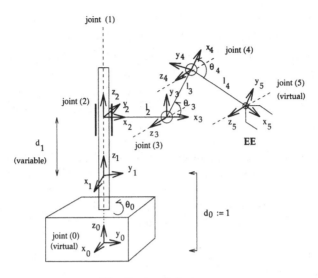

Fig. 2. rtrr-Robot

of the robot arm represented by its map **R**. This is the method of supervised learning in ANN theory.

Remark: We have to point out here that in the case where we do not have an explicit description of the robot map available, for a given robot arm, we use a corresponding robot simulation (RobSim) as a representation of the direct kinematic model. That means that we use the simulator RobSim for producing corresponding input-output pairs as previously described.

From the point of view of Inverse Kinematics (InvKin, for short) we can say that for the component x of a pair $(q, x) \in Td$ the component q is a suitable joint configuration fulfilling $R(q) = x$, i.e. it is a solution to the corresponding InvKin problem: "Given x find a suitable q such that the EE reaches the correct location x with the joint parameters in q".

Having produced a set of pairs $Td = \{(q, x)\}$, the idea now is to use this set for training a selected ANN in the following way. We take the second component x of a pair (q, x) as input and the first component q as desired output and apply, for example, supervised learning with the aim (and hope) that the training procedure yields a trained network (shortly denoted by $\mathcal{N}$) which represents the inverse kinematic (InvKin) model of the given robot arm.

Now it is necessary to test the trained network $\mathcal{N}$, i.e. to check the "InvKin-quality" of $\mathcal{N}$. To this end we select a set of pairs (q', x') which have not been used for training the net.

<u>TEST of $\mathcal{N}$:</u> Let (q', x') be such a pair of the test-set. Taking x' as input to $\mathcal{N}$ an actual output $q_{x'}$ is produced by the network. We have to check whether $q_{x'}$ is an acceptable configuration in comparison with the correct joint parameter configuration q'. This means we have to compare $R(q_{x'})$ with $R(q')$ using

a suitable measure for doing this. The result of this test should give insight how good the net $\mathcal{N}$ can function as an InvKin problem solving unit.

<u>Visualization of the TEST</u>: with the help of a Robot Simulation Tool (RobSim). We construct a Path Tracking Task by selecting a curve (path) γ in 3-space, $\gamma : [a, b] \longrightarrow \mathbb{R}^3$, where the EE has to move along. We produce a sequence of consecutive points on the graph of the curve ($graph(\gamma)$, for short), for example $\gamma(t_0), \gamma(t_1), \ldots, \gamma(t_i), \ldots, \gamma(t_N)$, where $t_0 < t_1 < \ldots < t_i < \ldots < t_N$ are elements of the real interval $[a, b]$. Usually we set $t_0 = a$ and $t_N = b$.

We use the points $\gamma(t_i)$ as input to the trained net $\mathcal{N}$ and, for the sake of simplicity, we select a suitable orientation $A \in SO_3(\mathbb{R})$ as second component of the input $x_i = (\gamma(t_i), A)$. There are cases where the handling of A can be simplified, for example if we consider movements in a plane (cf. the example in the figure below) where the axes of the rotational joints are perpendicular to the plane.

The network produces a sequence of corresponding actual outputs denoted by $q_0, q_1, \ldots, q_i, \ldots, q_N$.
This produced sequence of joint parameter configurations will be used as input to a robot simulator (RobSim). For the input q_i RobSim finds $R(q_i)$ and displays it graphically. Note that RobSim represents the direct kinematic model of the robot. We assume that we can also visualize with the same graphical system the given curve $graph(\gamma)$. Thus we can achieve a kind of an "optical comparison" of the produced sequence of results $R(q_0), \ldots, R(q_N)$ with the points $\gamma(t_0), \gamma(t_1), \ldots, \gamma(t_i), \ldots, \gamma(t_N)$ of the given curve.

A former diploma student (J.Sixt) in my working group ANNig at RISC-Linz did first experiments of this kind. For a first demonstration of the principle he used a simple robot arm with two rotational joints moving in the real plane and trained an ANN in the way as previously explained to generate a network $\mathcal{N}$ which represented the learned InvKin model. He selected as test curve γ a circle in the plane, this is simple enough, but not linear. More explicitly, he used for example $\gamma : [0, 1] \longrightarrow S^1$, $t \mapsto (cos 2\pi t, sin 2\pi t)$, for the path tracking test of $\mathcal{N}$. The visualization of the test ("optical comparison"), as desribed above, showed an interesting result. The curve produced by $\mathcal{N}$ and the circle were displayed with a simple computer graphics system and one could see a superposition of the circle by that produced curve. Its shape was a topological approximation in the sense that it was homeomorphic to the circle but not a circle itself. It looked like a closed chain of a few consecutive sine curve pieces ("waves") lying over the circle (cf. the picture). One might interpret this outcome as qualitatively acceptable or not.

Actually, then at RISC-Linz, I had the possibility to show the whole experiment to an engineer, an expert in robotics. He liked the approach and he found the result useful from the viewpoint of practical applications. He gave me the following explanation. In real robotics applications (where forward and inverse kinematics problems are treated) it can already be useful to have a reasonable InvKin solution which is a qualitatively acceptable approximation, as previously described in that experiment. The reason for this is the fact that in modern

Fig. 3. TEST: Path Tracking visualized with RobSim

robotics high level *Sensor Systems* will be applied for the fine tuning of a path tracking task. Thus, the approximate curve generated with the deployment of the network $\mathcal{N}$ is the rough guideline for concrete path tracking and on this basis the sensors achieve the fine tuning rather easily.

In particular cases this method can lead to much faster solutions to a given InvKin problem in comparison to try to find exact mathematical solutions and, consequently, from a practical (possibly industrial) point of view it could be the case that it is cheaper in a concrete project to apply such a connectionist approach in combination with sensors.

Final Remark: As expected, the method of learning the InvKin model on basis of the robot map also reaches soon its limits when the number of joints increases.

Network Structure Modeling and ANN Simulation
In the beginning of my time as a faculty at RISC-Linz (1990-1996), University of Linz, Austria, I established a neural network interest goup called ANNig. Rather soon I came into contact with the ANN Section of the big European project JESSI (Joint European Submicron Silicon Initiative). I was invited to apply for becoming an associated member of the project group JESSI-ANN (the project title of that group was: "Advanced Neural Circuits and Networks on Silicon"). The main topic of my proposal was the previously described idea and approach to learn the robot map and the InvKin model of a given robot arm using methods from connectionism. I had a cooperation contact with IMS, the Institute for Microelectronics Stuttgart (Germany), and it was the aim to apply our approach to support IMS in the development and test of ANN hardware (microchips) for fast robotics applications. Actually, with my group ANNig I became an associated member of JESSI-ANN and was included in the JESSI Blue Book in October 1990.

A few years ago, project work started in my working group (in Salzburg) to establish a flexible simulation tool (FlexSimTool) for simulation of a large class of ANN-types that have neurophysiological roots.

An own mathematical approach for modeling network structures that uses methods from Category Theory and Noncommutative Geometry (cf. [6], [7]) provided a precise formal guideline for the generic implementation of networks in that simulation tool. The first version of FlexSimTool was implemented in C++. The project terminated and it was not possible to continue it due to lack of manpower. There are still ideas to extend the simulator by further modules in order to have tools for fuzzy techniques, sensor data processing, logical reasoning (logical fibering module) and interfaces to: CA system, robot simulation, multiagent system simulation, heuristic optimization system (and others).

In my research work on ANN structure modeling, I introduced the category GeoNET where the instance of learning can be modeled by morphisms - a learning step is a corresponding morphism (cf. [7]). It can be shown theoretically with an argument from category theory that the mathematical network structure model can help to reduce complexity of learning in concrete ANN simulations. A final version of the FlexSimTool would have been well suited for the treatment of industrial application problems. Comparable, in some respect, to the power of the ANN simulator of a former cooperation partner, H.Geiger, who had developed a neurophysiologically inspired own ANN paradigm and implemented it in his ANN simulator which he applied in many industrial projects (a variety of them were on optical quality control). The former cooperation with him showed me the way to my work on GeoNET. It was H.Geiger who demonstrated the "economic effect" of the new categorical and geometric ANN structure modeling approach: In a concrete industrial project, where he exploited the new model in his ANN simulator for the first time, it could be observed that the effect of reduction of complexity of learning (as mentioned before) resulted in considerable reduction of project costs (cf. [6]) due to the fact that "learning became cheaper" than in former ANN simulations.

4 A Hybrid Symbolic-Connectionist Approach

The experiences with the symbolic and connectionist approaches (both have limitations) formed the basis of the idea to combine both methods - constructing a hybrid system - with the objective to exploit the advantages of each method and to try to cope with the increasing complexity when the degree of freedom grows. The simple idea is to decompose the joints of a robot arm into disjoint subsets of consecutive, neighbored joints and then model some subsets of consecutive joints symbolically, the other subsets with the connectionist method and finally compose the parts to represent the complete robot map.

Let us consider the general case of a robot arm R using the same notation as in the previous discussion of the robot map. The DOF (degree of freedom) is $n = r + t$, where r is the number of rotational joints and t the number of translational joints. Let $J_1, \ldots, J_n$ be the sequence of consecutive joints (cf. the rtrr-example above showing the local coordinates). Again, $q_1, \ldots, q_n$ denote the corresponding joint parameters. We do not consider the virtual joints in the base and the EE as depicted in the figure of the rtrr-robot example.

Now we describe the most general case of a subdivision of the set of joints into subsets of consecutive joints: $R_1, R_2, \ldots, R_k$, where every subset R_i contains at least two joints, we use the notation $\mid R_i \mid = \rho_i$, with $\rho_i > 1$, for $i = 1, \ldots, k$, and $\rho_1 + \ldots + \rho_k = n$. More explicitly, for n sufficiently large, we have the following decomposition of the whole set of joints

$$R_1 = \{J_1, \ldots, J_{\rho_1}\}, \ R_2 = \{J_{\rho_1+1}, \ldots, J_{\rho_1+\rho_2}\}, \ R_k = \{J_{\rho_1+\ldots+\rho_{k-1}+1}, \ldots, J_n\}.$$

But this is still not yet the complete description because we have to specify for each R_i its type. We point out that every R_i represents a possibly very small robot arm itself (we call it a "local part" or "sub-arm" of the given robot R). Consequently, such a sub-arm R_i has a local robot base, called $Base_i$, and a local end effector denoted by EE_i.

The letters R and R_i are also used as names of the corresponding robot maps, respectively. We explain the use of our notation by an example, for short. Let us take R_1, for example, and assume that $\rho_1 = 3$ and that J_1 is a translational joint and J_2, J_3 are rotational joints, then R_1 is of type trr. Corresponding to this, we denote R_1 by $< trr >_1$. Following this notation, the sub-arm R_i might have the type description $< rrtrt >_i$, for example.

The original objective was to combine symbolic and connectionist methods in a hybrid approach. To this end we have to determine which sub-arm is modeled symbolically and which one is represented by a trained ANN. We introduce the following notation:

Symbolic: $R_i CA_i$ (CA for computer algebra).

Connectionist: $R_i NN_i$ (NN for neural network).

Again, let us illustrate this notation by giving the following example of a robot of type $rtrrrtr$. We choose a decomposition into three sub-arms R_1, R_2, R_3 selecting the types and modeling methods as follows
$< rtr >_1 CA_1 < rr >_2 NN_2 < tr >_3 CA_3$.
The corresponding parameter tupels are $(q_1, q_2, q_3)_1, (q_4, q_5)_2, (q_6, q_7)_3$.

Sometimes it is technically more convenient to use a special notation for the local parts (sub-arms), introduced as follows. For R_i let $J_1^i, \ldots, J_{\rho_i}^i$ denote the sequence of consecutive local joints of that sub-arm. For a joint J_l^i the corresponding local coordinate frame is denoted by B_l^i with origin O_l^i.

Principle of Composing Robot Arms. If there is no reason to do it differently, we introduce the following convention to compose a pair of consecutive, neighbored robot sub-arms R_i and R_{i+1}. We fix the reference point of $Base_i$ in the first joint of R_i, more precisely in the origin O_1^i of the corresponding frame B_1^i. Since we want to combine R_i with R_{i+1}, at first sight it is plausible to fix the reference point of EE_i in the origin O_1^{i+1} of the frame B_1^{i+1} of the first joint of R_{i+1}, this is the joint J_1^{i+1}. The combination (composition) of both sub-arms will be denoted by $R_{i+1} \circ R_i$ according to the notation used for the composition of the corresponding local robot maps ("first R_i then R_{i+1}", from right to left).

Concerning the treatment of the InvKin problem with respect to the composition $R_{i+1} \circ R_i$ we introduce the following convention. Let (a, A) be a given position and orientation of EE_{i+1} (cf. the notation in the section on the robot map), this is the input to the InvKin model of the robot arm $R_{i+1} \circ R_i$. In the first step we apply the InvKin model of sub-arm R_{i+1} to this input obtaining corresponding joint parameters for the joints of R_{i+1}. For the second step, where we apply the InvKin model of R_i, we use as input (b, A), the position vector b of EE_i and the same orientation if it is possible (e.g. considering robot movement in a plane - otherwise we have to evaluate a new orientation depending on the geometry of R_{i+1}). The position vector b of EE_i is easy to compute, it is the sum of the position vector a and the vector pointing from EE_{i+1} to the origin O_1^{i+1} in the first joint of R_{i+1}, this is the position of EE_i.

rtrr-Robot as Demo-example of the Hybrid Approach. First Steps

A former diploma student (B.Jerabek) in my group here in Salzburg worked on my idea of a hybrid symbolic-connectionist approach in his diploma thesis. With a few selected robot arm types he made some ANN applications (learning a robot model), especially some experiments with rr- and rrr-arms moving in a plane perpendicular to the rotational joint axes which are pairwise parallel. His investigations showed advantages and limitations of the connectionist approach. Then he worked on the CA-ANN hybrid approach. It turned out that the rtrr-robot arm example is a good sample demonstrator for making first steps. We briefly describe the investigations and experiences of this work. The rtrr-robot was decomposed into the following parts $< rt >_1 CA_1$ and $< rr >_2 NN_2$. Based on the experiences gained so far it was plausible to model the first part symbolically (its workspace is 3-dimensional), the second part is represented by a trained network, NN_2, the $< rr >_2$ sub-arm moves in the plane perpendicular to the two parallel axes of the rotational joints.

The results of the experiments and tests showed a rather well functioning combination where the advantage of each method (symbolic CA_1, connectionist NN_2) could be deployed. One could say that these first experiences with the Hybrid CA-ANN Approach look promising and that it is worthwhile considering such a hybrid system for use in concrete applications. As previously remarked, we point to the importance of corresponding **sensor applications** for fine tuning in a path tracking task. Shortly speaking a general strategy could be: Exact Positioning and Control of the End Effector is the combination of the Mathematical Robot Model and Sensors. There remain interesting topics of future work in this direction.

Alternative Aspect and Question. Concerning the approach to model parts of a robot arm individually another aspect arises. Let us assume that we consider a sequence of small sub-arms where each R_i can be easily modeled with one and the same method, i.e. every local robot arm R_i has a well functioning CA_i- or NN_i-representation. Then it would be natural to think about the combination of these local representations without mixing different models and the basic question arises how to describe the composition (interaction) of neighboring subarms R_i and R_{i+1} (cf. the convention considered previously). In the special case of the $rtrr$-robot an example corresponding to such a question could be $< rt >_1 CA_1$ put together with $< rr >_2 CA_2$, both symbolically modeled separately. Question: Is there a possibility to compose these two parts in such a way that the treatment of the InvKin problem will be less complex than the computation of the purely symbolic closed form solution considered above?

In the field of hybrid symbolic-connectionist methods and numerical-symbolic scientific computing approaches there are many interesting and challenging questions and topics for future work on robotics. So far we have not done yet a literature search and review in my working group, this is also of future interest.

5 Conclusions: Benchmarks for Computer Algebra (CA)

As previously mentioned, the Denavit-Hartenberg approach for establishing the robot map (kinematic model of the robot arm) provides a systematic way to construct benchmark examples for computer algebra. This observation is based on experiences in work on finding a closed form, purely symbolic solution to the invese kinematics problem of a robot arm using the Buchberger Algorithm to compute corresponding Gröbner Bases with the help of a computer algebra system. The concrete example of the rtrr-robot arm which I constructed in [5] showed some hard performance problems in the CA applications and the main reason for this seems to be the fact that we have to deal with many parameters and variables. The following polynomial ring was used

$$\mathbb{Q}(l_2, l_3, l_4, p_x, p_y, r_{11}, r_{31})[c_0, c_3, c_4, s_0, s_3, s_4, d_1, p_z, r_{12}, r_{13}, r_{21}, r_{22}, r_{23}, r_{32}, r_{33}]$$

and the following lexical ordering for the polynomial variables $c_0 \prec c_3 \prec c_4 \prec s_0 \prec s_3 \prec s_4 \prec d_1 \prec p_z \prec r_{12} \prec r_{13} \prec r_{21} \prec r_{22} \prec r_{23} \prec r_{32} \prec r_{33}$.

The experience with this example gave the motivation to propose this symbolic approach to construct benchmarks for computer algebra. Obviously, it is easy to increase the degree of freedom of a robot arm arbitrarily by adding further joints (e.g. rtrrtrr-robot arm, etc.) and thus increase the degree of complexity in the computations. Thus we obtain a concrete nice class of benchmarks for CA from robotics.

Benchmarks from Noncommutative Geometry. Concluding, another class of benchmarks for CA is briefly introduced. Noncommutative Geometry (NCG) is a rather new field in geometry, it was introduced by Johannes André about 35 years ago as a generalization of classical affine geometry (selected references: [11,12,13,14], [15]). Elementary geometric configurations are the basis of the axioms which determine the structure of a geometry.

In the beginning of my own work on NCG, I introduced a new model to represent a geometric space $(X, <, >, R)$ by a so-called parallel map $<, >: X^2 \to R$, this led to the introduction of the category of Noncommutative Geometric Spaces **NCG** (selected references: [16,17,18], a brief summary of the main notions is contained in [19], for a summary with more details cf. [20]). Geometric axioms can be expressed by corresponding equations in the $<, >$-model (thus leading to a certain algebraization). The verification of an axiom amounts to verify the solvability of the corresponding system of equations and that means to verify the validity of corresponding geometric constraints.

This aspect suggested to think about computer applications for such automated deduction problems in NCG and I started to work on some selected examples of geometric spaces and concrete verification of geometric conditions (constraints) involving corresponding geometric configurations. At first sight the most interesting class of examples are "polynomial geometric spaces" being defined by polynomial parallel maps, the reason for that is the intention to apply well established methods for solving polynomial systems with computer algebra, especially the Buchberger Algorithm (Gröbner Bases method, [1], [2], [3], [4]), Cylindrical Algebraic Decomposition, c.a.d. (G.Collins, H.Hong, [21], [22]), Characteristic Sets (W.T.Wu, D.Wang, [23], [24]).

Considering the example of a circle space in the real plane, where a noncommutative line $x \sqcup y$, joining the points x and y, is a circle with center x and radius $\|x - y\|$, it is a special case ($n = 2$) of the more general sphere space:

Let $X := \mathbb{R}^n$ and $R := \mathbb{R}_+$, then the $(n-1)$ - *sphere space* is defined by the polynomial parallel map $< x, y > := \sum_{i=1}^{n} (x_i - y_i)^2$ for $x = (x_1, ..., x_n), y = (y_1, ..., y_n) \in X$. The value $< x, y >$ is called direction of the line $x \sqcup y$. Parallel lines have the same value ("direction").

We recall that *lines* are defined by $x \square y := x \sqcup y \cup \{< x, y >\}$, (for points $x, y \in X$) where $< x, y >$ is called *ideal point or direction (or color)* of the line $x \square y$ and $x \sqcup y := \{x\} \cup \{\zeta | < x, \zeta > = < x, y >\}$ is called the set of *proper points* of the line - it is the variety of solutions of the equation $< x, \zeta > = < x, y >$ together with the base point x.

Noncommutativity of the join operation $\sqcup$ means, that in general $x \sqcup y \neq y \sqcup x$ holds for lines.

A main axiom in NCG is called Tamaschke Condition (Tam), it deals with parallel shifting of triangels (formed by lines) over a space subject of corresponding constraints. In the case of a sphere space (more general polynomial geometric space) this is expressed by corresponding polynomial equations. The verification problem is equivalent to find solutions to these polynomial equations. This problem can be interpreted as a quantifier elimination problem and therefore it is amenable to apply the Cylindrical Algebraic Decomposition (c.a.d.) method.

In the article [25] the circle space (i.e. sphere space in the real plane) is discussed. The verification of (Tam) corresponds to the solution of the following quantifier elimination problem (we include the corresponding material of the article).

The general form of the (Tam) axiom is as follows

$$\forall x, y, z, x', y' \in X \quad \text{with} \quad < x, y > = < x', y' > \quad (I)$$

$$\exists \zeta \in X \qquad \text{s.th.} \quad < x', \zeta > = < x, z > \quad (II)$$
$$\text{and} \qquad\qquad\qquad < y', \zeta > = < y, z > \quad (III)$$

Translated into the corresponding polynomial equations we obtain the following system:

$$(x_1 - y_1)^2 + (x_2 - y_2)^2 - (x_1' - y_1')^2 - (x_2' - y_2')^2 = 0 \qquad (I)$$
(initial constraint)
$$(x_1' - \zeta_1)^2 + (x_2' - \zeta_2)^2 - (x_1 - z_1)^2 - (x_2 - z_2)^2 = 0 \qquad (II)$$
$$(y_1' - \zeta_1)^2 + (y_2' - \zeta_2)^2 - (y_1 - z_1)^2 - (y_2 - z_2)^2 = 0 \qquad (III)$$

These polynomials are considered in $\mathbb{R}(x_1, x_2, y_1, \ldots, y_2')[\zeta_1, \zeta_2]$.

Problem: Existence of solutions (zeros) in ζ_1, ζ_2 has to be verified.

This is the most general formulation of the verification problem (for arbitrary points) in a pure algebraic setting.

In its formulation as a quantifier elimination problem, we gave the above equations and the corresponding variables as input to the c.a.d. system QEPCAD implented and used by my former colleague at RISC-Linz, Hoon Hong.

Result: Since 12 variables (two variables for each of the points x, y, z, x', y', ζ) are involved this problem then, when I was at RISC-Linz, could not be treated - too many variables are involved. This was the comment by Hoon Hong when we discussed about this special c.a.d. application with his system QEPCAD.

In addition to the previous considerations, in [25] the above (Tam)-verification problem was treated with the Gröbner Bases approach using a CA implementation of Buchberger's Algorithm in Mathematica. Doing this, once more we could make the observation that these types of verification problems lead to a class of benchmarks for CA. At this point we should note that the circle space example is even the simplest case of a sphere space (for $n = 2$) and the degree of complexity can simply be increased by considering sphere spaces in higher dimensions.

The article [19] is the result of a short period of joint work with my former PhD student Wolfgang Gehrke who spent three months with me in Salzburg as a guest researcher in 1998. He is a very experienced programmer knowing several programming languages. In our cooperation he motivated why he used specific declarative programming languages (just as it was convenient) for all the implementations and computer applications.

To make it short, I just mention that constructing finite geometric spaces fulfilling given geometric configurations (axioms) can lead to very hard (constraint satisfaction) problems - it provides a rich source of benchmarks for computer applications and many interesting topics for future work.

References

1. Buchberger, B.: Ein algorithmisches Kriterium für die Lösbarkeit eines algebraischen Gleichungssystems (An Algorithmical Criterion for the Solvability of Algebraic Systems of Equations). Aequationes mathematicae 4/3, 374–383 (1970); English translation In: Buchberger, B., Winkler, F.(eds.) Gröbner Bases and Applications, Proceedings of the International Conference '33 Years of Gröbner Bases', RISC, Austria, London Mathematical Society Lecture Note Series, vol. 251, pp. 535–545. Cambridge University Press, Cambridge (1998)
2. Buchberger, B.: Gröbner bases: An algorithmic method in polynomial ideal theory. In: Bose, N.K. (ed.) Multidimensional Systems Theory, pp. 184–232. D.Reidel Publ. Comp., Dordrecht-Boston-Lancaster (1985)
3. Buchberger, B.: Applications of Gröbner bases in non-linear computational geometry. Rice, J.R.(ed.) Mathematical Aspects of Scientific Software 14, 59–87 (1987)
4. Buchberger, B., Collins, G., Kutzler, B.: Algebraic methods for geometric reasoning. Ann. Rev. Comput. Sci. 3, 85–119 (1988)
5. Pfalzgraf, J.: On geometric and topological reasoning in robotics. Annals of Mathematics and Artificial Intelligence 19, 279–318 (1997)
6. Pfalzgraf, J.: Modeling connectionist network structures: Some geometric and categorical aspects. Annals of Mathematics and AI 36, 279–301 (2002)

7. Pfalzgraf, J.: Modeling connectionist networks: categorical, geometric aspects (towards 'homomorphic learning'). In: Dubois, D.M. (ed.) Proceedings CASYS 2003. American Institute of Physics, AIP Conference Proceedings, Liège, Belgium, August 11-16, 2003, vol. 718 (2004) (Received a Best Paper Award)
8. Angeles, J.: Rational Kinematics. Springer, New York-Berlin (1988)
9. Craig, J.: Introduction to Robotics. Addison-Wesley Publ.Co, Reading (1986)
10. Paul, R.: Robot Manipulators. MIT Press, Cambridge Massachusetts (1982)
11. André, J.: On finite noncommutative affine spaces. In: Hall Jr., M., van Lint, J.H. (eds.) Combinatorics, 2nd edn. Proceedings of an Advanced Inst. on Combinat (Breukelen), pp. 65–113, Amsterdam Math. Centrum. (1974)
12. André, J.: Coherent configurations and noncommutative spaces. Geometriae Dedicata 13, 351–360 (1983)
13. André, J.: Endliche nichtkommutative Geometrie (Lecture Notes). Annales Universitatis Saraviensis (Ser. Math.) 2, 1–136 (1988)
14. André, J.: Configurational conditions and digraphs. Journal of Geometry 43, 22–29 (1992)
15. André, J., Ney, H.: On Anshel-Clay-Nearrings. In: Proc. Intern. Conf. Nearrings and Nearfields, Oberwolfach (1991)
16. Pfalzgraf, J.: On a model for noncommutative geometric spaces. Journal of Geometry 25, 147–163 (1985)
17. Pfalzgraf, J.: On geometries associated with group operations. Geometriae Dedicata 21, 193–203 (1986)
18. Pfalzgraf, J.: Sobre la existencia de espacios semiafines y finitos. In: Archivos de Investigación, vol. 3(1). (Proceedings) Instituto Profesional de Chillán, Chile (1985)
19. Gehrke, W., Pfalzgraf, J.: Computer-aided construction of finite geometric spaces: Automated verification of geometric constraints. Journal of Automated Reasoning 26, 139–160 (2001)
20. Pfalzgraf, J.: On a category of geometric spaces and geometries induced by group actions. Ukrainian Jour. Physics 43(7), 847–856 (1998)
21. Collins, G.E.: Quantifier elimination for the elementary theory of real closed fields by cylindrical algebraic decomposition. In: Brakhage, H. (ed.) GI-Fachtagung 1975. LNCS, vol. 33, pp. 134–183. Springer, Heidelberg (1975)
22. Hong, H.: Improvements in c.a.d.–based quantifier elimination (PhD thesis). Technical report, The Ohio State University, Columbus, Ohio (1990)
23. Wu, W.: Basic principles of mechanical theorem proving in elementary geometries. J. Automated Reasoning 2, 221–252 (1986)
24. Wang, D.: Elimination procedures for mechanical theorem proving in geometry. Annals of Mathematics and Artificial Intelligence 13, 1–24 (1995)
25. Pfalzgraf, J.: A category of geometric spaces: Some computational aspects. Annals of Mathematics and Artificial Intelligence 13, 173–193 (1995)

Applying Link Grammar Formalism
in the Development of
English-Indonesian Machine Translation System

Teguh Bharata Adji*, Baharum Baharudin, and Norshuhani Zamin

Department of Computer and Information Science,
Universiti Teknologi PETRONAS, Malaysia
adji@mti.ugm.ac.id,
{baharb,norshuhani}@petronas.com.my
http://www.utp.edu.my

Abstract. In this paper, we present a Machine Translation (MT) system from English to Indonesian by applying Link Grammar (LG) formalism. The Annotated Disjunct (ADJ) technique available in the LG formalism is utilized to map English sentences into equivalent Indonesian sentences. The ADJ is a promising technique to deal with target languages that do not have grammar formalism, parser, and corpus available like Indonesian language. An experimental evaluation shows that the applicability of LG for Indonesian language worked as expected. We have also discussed some significant issues to be considered in future development.

Keywords: Annotated Disjunct, Link Grammar, Parsing Algorithm, Natural Language Processing.

1 Introduction

Generally, Indonesia is a country in which English is not the first language. As such, the level of English competency among Indonesians is considered low. Due to the vast amount of available digital information nowadays is in English, there is a need for a means to translate this information into the Indonesian language.

Machine Translation (MT) is a study to automate the translation process of one natural language into other natural languages. A notable MT activity for Indonesian language is the Multilingual Machine Translation System (MMTS) project [7]. This project was conducted by the Agency for Assessment and Application of Technology (BPPT) as part of a multi-national research project between China, Indonesia, Malaysia, Thailand, and led by Japan. A team of NLP students from Gadjah Mada University in Indonesia [8] has built an English-Indonesian MT application using a direct method in Visual Basic. This method

---

* Teguh Bharata Adji is a lecturer at the Department of Electrical Engineering, Gadjah Mada University.

S. Autexier et al. (Eds.): AISC/Calculemus/MKM 2008, LNAI 5144, pp. 17–23, 2008.
© Springer-Verlag Berlin Heidelberg 2008

gives advantages when combined with the rule-based method using ADJ technique as explained in [1].

Currently, some English-Indonesian MT software systems have become available - Rekso Translator [5], Translator XP [9], and Kataku^TM.[1] There is a significant need to study the English-Malay MT research as Indonesian language shares many aspects with the Malay language. Close similarities are found in phonetic, morphology, semantics and syntax of both languages. An intensive research in English-Malay MT was conducted by Malaysia University of Science (USM) using example-based methods [4]. A technique to construct the Structured String-Tree Correspondence (SSTC) for Malay sentences by means of a synchronous parsing technique was introduced. The technique used synchronous parsing technique to parse the Malay sentences based on the English sentence parse tree together with the alignment result obtained from the alignment algorithms. The advantage is this technique can solve non-projective cases. The limitations to the approach are the extra work required to annotate all constituent levels and the effort to formalize the English and Malay grammars. In our research, the ADJ technique [2] is proposed since it has some advantages over the SSTC:

1. ADJ uses LG formalism which is more closely related to human intuition,
2. ADJ only needs parsing for the source language (English),
3. ADJ is suitable to MT where the target language does not have corpus and parser, such as Indonesian language.

However, the limitations to ADJ are it is unable to solve non-projective cases and bi-directional translation [1]. Other research [12] was also using LG to develop an MT system. The work used bilingual corpus to build a bilingual statistical parsing system that can infer a structural relationship between two languages. This model included syntax, but did not involve word-segmentation, morphology and phonology. As there is no available bilingual English-Indonesian corpus at present, building statistical parsing may not be possible in this study. Therefore, we merely utilize the link parser in [1] and [2].

2   Implementation

The architecture of the English to Indonesian MT system is given in Fig. 1. This system consists of four main modules: (1) pruning algorithm module, (2) parsing algorithm module, (3) ADJ algorithm module, (4) transfer rules algorithm module. The pruning algorithm and parsing algorithm modules have been introduced by Grinberg et. al. as part of the LG formalism [5]. After the pruning algorithm module prunes an English sentence, the result is parsed into a linkage. A linkage contains a sequence of words and all links that connect each word satisfying the linking requirement, as proposed by Sleator et. al. [10]. Fig. 2 illustrates the

---

[1] Kataku^TM is available at http://www.toggletext.com/kataku_trial.php but no detail is found on the underlying algorithm.

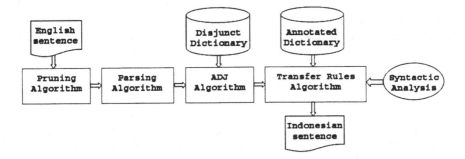

Fig. 1. Diagram of the developed MT system

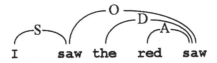

Fig. 2. An example of a linkage

linkage of an English sentence "I saw the red saw". In this example, S link connects subject-nouns to finite verbs, O link connects transitive verbs to objects, D link connects determiners to nouns, and A link connects adjectives to nouns.

Based on each word's links, a word disjunct generator is developed to get the disjunct of each word in a sentence. A disjunct is a list of left and right links of a word in a linkage. The proper disjunct for each word in the sentence "I saw the red saw" could be written in a disjunctive form [6] as follows:

- disjunct of "I": $(( )(S))$,
- disjunct of "saw": $((S)(O))$,
- disjunct of "the": $(( )(D))$,
- disjunct of "red": $(( )(A))$,
- disjunct of "saw": $((O,D,A)( ))$.

The two appearances of "saw" in Fig. 2 signify that this word is recognised as ambiquous. It is thus represented as two syntactically different words in the LG formalism. The first "saw" is a verb which has the disjunct of $d = ((S)(O))$, and the second "saw" is a noun of the disjunct of $d = ((O,D,A)( ))$. The first "saw" is translated into "melihat" in Indonesian and the second "saw" is translated into "gergaji".

The original parsing algorithm in LG generates a list of linkages at their own cost [10]. The ADJ algorithm module in this developed system is designed to consider only the first linkage in the generated list as it holds the lowest cost. Consequently, only single set of word disjuncts are obtained. These disjuncts are components used for producing ADJ. Processing disjunct annotations was explained in detail in [2]. The following lines describe the annotation briefly.

The single set of disjuncts of an English sentence that are obtained by ADJ algorithm are of the form $\{d_i \mid 1 \leq i \leq n, n =$ the total number of words in an English sentence$\}$. Thus, the words "I", "saw", "the", "red", and "saw" in the previous example have the disjuncts of d_1, d_2, d_3, d_4, and d_5 respectively sorted in sequence.

We developed the ADJ algorithm module to process the first linkage passed from the parsing algorithm module in obtaining the ADJ. ADJ is a set of source words (English), target words (Indonesian), and their associated disjuncts. It is represented in the following structure: $\{(W_i, W_i', \ d_i) \mid 1 \leq i \leq n\}$. For example, the sentence "I saw the red saw" is literally translated into the Indonesian sentence "Saya melihat itu merah gergaji" giving the following ADJ: $(I, Saya, ((\ )(S)))$, $(saw, melihat, ((S)(O)))$, $(the, itu, ((\ )(D)))$, $(red, merah, ((\ )(A)))$, $(saw, gergaji, ((O,D,A)(\ )))$. The ADJ algorithm is as follows:

```
0.  ADJ_algo(first_linkage)
1.      char** d ← get_words_disjuncts(sentence, first_linkage);
2.      for 1 <= i <= n
3.          char* d_i ← d[i];
4.          char* W_i ← sentence[i];
5.          W_i' ← insert_word_translation(W_i, d_i);
6.          ADJ ← annotate_disjuncts(W_i, W_i', d_i);
7.      return ADJ;
```

The variables used in the above algorithm are as follows:

- *first_linkage* is a struct data type variable,
- *sentence* (struct data type) is an English source sentence (SS) that consists of n words,
- i is for identifying the ith position of a word in an English sentence,
- $d_i = d[i]$ is the disjunct of W_i,
- W_i is the source word located in the ith position in the SS,
- W_i' is the translation of W_i, inserted by referring to its proper disjunct,
- ADJ is a struct data type variable consisting of W_i, W_i', and d_i. This ADJ is stored in an Annotated Dictionary (see Fig. 1).

Based on this set of ADJ, a transfer rules algorithm module is developed to arrange all the target words in a correct target sentence (TS) structure by referring to the syntactic analysis (see Fig. 1) of Indonesian language structure. For example, in English, adjective and determiner always precede the noun. This structure is against the structure in Indonesian language where the noun always precedes the adjective and determiner. Illustration in Fig. 3 shows how the above mapping problem can be solved using ADJ. When the sentence "I saw the red saw" is applied to the transfer rules algorithm module, the result produced is "Saya melihat gergaji merah itu". Now, the word "gergaji" (noun) precedes "merah" (adjective). The produced result also shows that the word "itu" (determiner) is located after "merah". This is due to the result of the word "the" in the 3rd position of the SS being mapped into "itu" in the 5th position of the

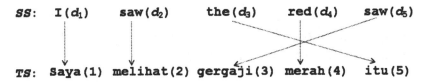

Fig. 3. Illustration of the English-Indonesian words mapping using ADJ

TS, while the word "saw" in the 5th position of *SS* is mapped into "gergaji" in the 3rd position of *TS*.

The transfer rules algorithm for implementing the illustration in Fig. 3 is:

```
0. char** transfer_rules_algo( )
1.     ADJ_algo(first_linkage);
2.     if (ADJ.disjunct[1] = d₁ & ADJ.disjunct[2] = d₂ & ...
           & ADJ.disjunct[n] = dₙ)
3.         for 1 ≤ i ≤ n
4.             temporary_word[i] ← ADJ.Wᵢ′;
5.             wordᵢ ← temporary_word[xᵢ];
6.     return word;
```

The variables used in the above algorithm are explained as follows:

- $disjunct[i]$ is a temporary variable for checking the disjunct of W_i,
- $temporary\_word[i]$ is a temporary buffer to store W_i',
- x_i is for identifying the position of W_i' before shifting to the correct position, where x_i has a value from 1 to n,
- $word_i$ is for locating the target word in the ith position of *TS*.

In the above transfer rule, line 2 checks whether all disjuncts of an input sentence equal certain disjuncts. If the input sentence is "I saw the red saw", and if the disjuncts matches with the condition, then all target words are stored in a *temporary_word* variable. For example, line 4 will store the Indonesian word "itu" (English: "the") in the 3rd position of *temporary_word*. Assigning $x_i = 3$ in line 5 will then shift the word "itu" from the 3rd to the 5th position in *TS*.

3   Results and Discussions

The performance of the system is discussed in this section. 150 English sentences that have been translated by three Indonesian linguists were used to generate the transfer rules. An unseen 150 English sentences were used to test the developed system. The same sentences were also used as input to the proprietary MT systems. The results from all tested systems were evaluated by the Indonesian linguists for accuracy. The accuracy of all the tested software is shown below:

- Translator XP = 52.00%,
- Rekso Translator = 56.78%,
- Kataku^{TM} = 61.00%,
- Link Grammar based MT system = 71.17%.

At present, we can consider the result produced by our system to be better than the result of the other software. Unfortunately, the error rate is still higher (28.83%). However, it seems quite possible that this will give a significant improvement in the accuracy if the amount of training data is large [11]. Hence, the accuracy is also very much reflected by the generation of more transfer rules. We have also analyzed the possible causes of the high error rate based on the data in Table 1.

Table 1. Performance analysis data

Category	Linguists' Score	Total Sentences	Total Sentences(%)
I	$5 \leq$ score < 6	2	1.33%
II	$6 \leq$ score < 7	55	36.67%
III	$7 \leq$ score < 8	58	38.67%
IV	$8 \leq$ score < 9	30	20.00%
V	$9 \leq$ score ≤ 10	5	3.33%

We categorized the score assigned by the Indonesian linguists into 5 categories together with the number of sentences (from the total of 150 tested sentences) fall within each category. It was found that 13.33% of the tested English sentences were not in LG formalism, hence some of their words have no disjuncts which have made the system fail to produce correct translations. This has contributed to the lower scores given by the linguists for sentences in Category I-IV. For example, all sentences in Category I dealt with idiomatic phrases or sayings such as "What a wonderful day!". Meanwhile Category V shows that the linguists were happy with the translated results which provide the best level of system's performance.

It was also found that Category II, III, and IV contributed to the most error i.e. 95.34%, which prompted us to further explore the causes. The sentences in those categories contained certain noun phrases (e.g. "the next morning", "Footballer Beckham"), verb phrases with prepositions (e.g. "look for", "dream of"), and translations that require morphological analysis. Further morphological studies on Indonesian language are vital since the language employs affixes more heavily than English [3]. Some of the tested sentences were in the English interrogative forms and negative forms which needs morphological analysis. For example, "paid" in the negative sentence "You will be paid" was translated into an Indonesian inflectional verb "dibayar". Meanwhile, "paid" is frequently translated into the inflectional verb "membayar" in affirmative sentences like "I paid you". For this example, surprisingly the ADJ algorithm can generate different disjuncts for the word "paid" in both forms. Two opportunities exist to improve the work i.e. adding a word stemmer and a morphological analyzer for the ADJ approach, which is likely to solve the mentioned problem.

4   Conclusions and Future Researches

The research contributes to a hybrid transfer approach using ADJ technique in LG formalism. In summary, the performance can be improved not only by considering the word stemmer and morphological analyzer but also the use of part of speech and ontology. Assuming that the system is already completed, the development of LG formalism for the Indonesian language should be possible provided that all source and target words are completely annotated with the ADJ. A new parser can be automatically induced for the language, without the aid of the language experts.

Acknowledgments. Thanks to Dr. Mohd Fadzil Hassan and Dr. Mohd Nordin Zakaria from Universiti Teknologi PETRONAS for their valuable helps.

References

1. Adji, T.B., Baharudin, B., Zamin, N.: Annotated Disjunct in Link Grammar for Machine Translation. In: 2007 ICIAS (International Conference on Intelligent & Advanced Systems), KL Convention Centre, Kuala Lumpur, November 25-28 (2007)
2. Adji, T.B., Baharudin, B., Zamin, N.: Building Transfer Rules using Annotated Disjunct: An Approach for Machine Translation. In: 2007 5th SCOReD (Student Conference on Research and Development), Malaysia, December 11-12 (2007)
3. Adriani, M., Asian, J., Nazief, B., Tahaghoghi, S.M.M.: Stemming Indonesian: A Confix-Stripping Approach. ACM Transactions on Asian Language Information Processing 6(4), Article 13 (December 2007)
4. Al-Adhaileh, M.H., Kong, T.E.: Synchronous Structured String-Tree Correspondence (S-SSTC). In: 20th IASTED 2002 International Conference, Innsbruck, Austria (February 2002)
5. Anonymous: Rekso Translator. Indonesia Commerce, Indonesia (2007), http://reksotranslator.com
6. Grinberg, D., Lafferty, J., Sleator, D.: A Robust Parsing Algorithm for A Link Grammar. In: 1995 4th International Workshop on Parsing Technologies, Prague (1995)
7. Nazief, B.: Development of Computational Linguistics Research: a Challenge for Indonesia. Faculty of Computer Science, University of Indonesia (internal publication, 1996)
8. Novento, F.: Perangkat Lunak Penerjemah Kalimat Inggris-Indonesia Menggunakan Metode Loading Data Sementara. Undergraduate final project, Electrical Engineering Department, Gadjah Mada University (2003)
9. Poulsen, C.S.: Translator XP. CV Media Internusa Enterprice, Yogyakarta, Indonesia (2008), http://translatorxp.com
10. Sleator, D.D., Temperley, D.: Parsing English with A Link Grammars. In: 3rd International Workshop on Parsing Technologies. ACL - SIGPARSE conference. University of Tilburg, The Netherlands (1993)
11. Tanaka, Y.: Example data for machine translation systems. In: IEEE International Conference on Systems, Man, and Cybernetics, October 7-10, vol. 2, pp. 915–920 (2001)
12. Venable, P.: Modelling Syntax for Parsing and and Translation. Ph.D. Thesis, Carnegie Mellon University (2003)

Case Studies in Model Manipulation for Scientific Computing

J. Carette, S. Smith, J. McCutchan, C. Anand, and A. Korobkine

Computing and Software Department, McMaster University, Hamilton, ON,
CANADA
{carette, smiths, mccutcjs, anandc, korobkao}@mcmaster.ca

Abstract. The same methodology is used to develop 3 different applications. We begin by using a very expressive, appropriate Domain Specific Language, to write down precise problem definitions, using their most natural formulation. Once defined, the problems form an implicit definition of a unique solution. From the problem statement, our model, we use mathematical transformations to make the problem simpler to solve computationally. We call this crucial step "model manipulation." With the model rephrased in more computational terms, we can also derive various quantities directly from this model, which greatly simplify traditional numeric solutions, our eventual goal. From all this data, we then use standard code generation and code transformation techniques to generate lower-level code to perform the final numerical steps. This methodology is very flexible, generates faster code, and generates code that would have been all but impossible for a human programmer to get correct.

1   Introduction

Collectively, the authors have been developing various scientific applications for several decades. Over time, we have independently drifted towards the same development methodology. The basic ingredients involve a (declarative) domain-specific language (DSL) in which to express our model(s)[1], model transformations, code generation and program transformation. The steps involved are shown in Fig. 1. Through 3 case studies, we show that the methodology is flexible, generates faster code, and generates code that would have been all but impossible for a human programmer to get correct.

For scientific applications, the most appropriate DSL is well-known: *mathematics*. More difficult is finding computer-based tools that can easily deal with the kinds of mathematics involved in typical scientific applications. Furthermore, not only does this language need to be "declarative," it should also allow direct manipulation, in and by the language itself, of mathematical expressions (and more generally of mathematical specifications). The only languages that currently combine the necessary richness and ease of manipulation are the languages of *Computer Algebra Systems*. In our case, because it is the system we are

---

[1] Note that where we use "model," mathematicians would use "problem" instead.

S. Autexier et al. (Eds.): AISC/Calculemus/MKM 2008, LNAI 5144, pp. 24–37, 2008.

1. *Express the Model* - the model is declaratively expressed in a DSL,
2. *Transform the Model* - transform the initial model into a form more suitable for computational solutions,
3. *Extract Structure* - structure and properties are directly extracted from the model,
4. *Optimize the Computation* - the structure is used to optimize the computational "solution" of the model,
5. *Generate the Code* - low-level code is generated for the solution.

Fig. 1. Typical model manipulation steps

(by far) the most familiar with, we have used Maple. It is then straightforward to directly phrase the kinds of models we are most interested in: (solutions of) differential equations, and (solutions of) continuous optimization equations.

We will show that given *explicit* representations of equations whose solution we seek, the *intentional* structure of those equations can be mined to obtain a wealth of information about the structure of the solution. This, in turn, allows one to make better choices about (numerical) solution methods. We call this step "model manipulation." This is the step where human creativity and ingenuity is most needed. This is also the step where the domain expert can bring important insights. We recommend spending relatively more time on model manipulation because an investment of time here makes subsequent steps much simpler to automate.

With a model rephrased in more computational terms, we can apply well-known techniques (like symbolic differentiation, common subexpression elimination, finding of differential or recurrence relations, etc.) to further optimize the computational structure of the model. At this point, classical code generation techniques can be applied to generate C code with embedded calls to optimized numerical libraries.

In scientific computation, there are at least two circumstances in which code generation has proven to be quite effective:

1. when complex program transformations are needed [11,22],
2. when a program can be expressed succinctly in a domain-specific language, but requires lengthy and complex code in a mainstream language. [6,7]

The first situation occurs most famously when *automatic differentiation* [13] is required and applicable. There is ample literature (from [25] onwards) that shows that smooth optimization problems are incomparably easier to solve when Jacobians and Hessians are available. Computing derivatives numerically is well-known to be a futile task, and computing them by hand (symbolically) is so fraught with error as to be deemed impossible. On the other hand, differentiation is a simple (symbolic) program.

The second situation from the above list is now emerging as rather common as well, which has caused the growing popularity of GUI-builders, lexer and parser generators, Java-from-DTD builders, etc. This trend is also present in the scientific computation community [8,9].

The problems well-suited to our approach are those which:

1. can be succinctly described using mathematics as the "domain language,"
2. needs information, like derivatives, easily obtained from the model, and
3. requires experimentation and manipulation at the "model" level.

The downsides of using a DSL, as given by [7], are not relevant when using a mathematical programming language (such as Maple).

It is worth repeating that the most important step is that of "model manipulation." Our aim is to automate *every other step of the problem-solving process* to ensure that a designer's time is spent thinking about the semantics and the structure of the problem to solve, and not wasted on mundane computational tasks. Eventually, we would hope to provide higher-level abstractions for this step as well. Several reviewers have, naïvely in our opinion, asked why we do not provide more automation for this step. The answer is simple: we only know how to automate very particular cases, and in fact believe that there are no general recipes to follow. This is not to say that particular cases cannot be fully automated, nor even that these particular cases are "rare" – quite the contrary.

Part of our aim is to free our own time, so that we can concentrate on the pure problem-solving parts of scientific software development. Once we have reliably achieved that, we will then strive to develop automated tools, using our scientific and engineering knowledge of typical solutions, to as many problem classes as possible.

We will present 3 applications developed using this methodology: real-time visual tracking of a target, data fitting in model-based time series and material behaviour modelling. To highlight the similarities between the examples, in each case reference will be made to the model manipulation steps shown in Fig. 1. Significantly more details on these examples can be found in [3].

2   Visual Tracking

This example is based on [1]. In visual tracking applications, a series of images captured from CCD (Charge-Coupled Device) cameras must be processed in real-time to extract information about spatial positioning. This information can be used for target identification, object measurement, and closed-loop target acquisition. Here we will focus on recognition of radially-symmetric, essentially compact targets, which we will call *spots*.

2.1   Model of Spot Fitting

We can *Express the Model* of spot recognition as the least squares fit between actual light intensity (ϕ_p) and the equation ($v_1 f(p) + v_0$) describing the spot:

$$\min_{\mathcal{U}} F = \sum_{p \in \Omega} (\phi_p - (v_1 f(p) + v_0))^2 \tag{1}$$

where v_0 is the background illumination, v_1 is the brightness of the centre of the spot, p is a pixel, Ω is a region of pixels and $f(p)$ is a polynomial. The function $f(p)$ depends on k_1, k_2, which determine the radial profile of the spot; b_x, b_y, which define the coordinates of the ellipse centre; and finally a_1, a_2, a_3, which define the shape of the elliptical boundary. We minimize F over $\mathcal{U}$, where $\mathcal{U} \subseteq \{v_0, v_1, k_1, k_2, b_x, b_y, a_1, a_2, a_3\}$. Figure 2 shows an example.

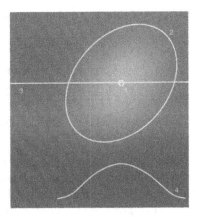

Fig. 2. Actual image of a gray-scale target, showing the spot's centre (1), shape (2) and cross-section (3, 4)

Equation 1 actually represents a family of models, distinguished by the choice of $\mathcal{U}$. Domain-specific information allows us to *Extract Structure* via appropriate choices of $\mathcal{U}$. Note that a naive implementation where we simultaneously optimize all variables will fail because the target recognition problem is not convex, forcing us into multiple solver stages.

2.2 Transformed Model (Newton's Method Solver)

The *Transformed Model* for finding the minimum in Eq. 1 consists of searching for a common zero of all the partial derivatives with respect to all the parameters of $\mathcal{U}$, using Newton's method. Denoting by $\mathbf{J}_{\mathcal{U}}$ the Jacobian of F and $\mathbf{H}_{\mathcal{U}}$ the Hessian of F with respect to the variables $\mathcal{U}$, Newton iteration for an iterative solution vector $\mathbf{u}_n$ is defined by:

$$\mathbf{u}_{n+1} = \mathbf{u}_n - \mathbf{H}_{\mathcal{U}}(\mathbf{u}_n)^{-1}\mathbf{J}_{\mathcal{U}}(\mathbf{u}_n) \tag{2}$$

2.3 Extracting Structure and Generating Code

To improve performance, we can *Extract Structure* (again) from the transformed model. In particular, we know that large arrays are needed to store the captured images and that computing the sum over each elements in those arrays is expensive. Efficient use of cache would help reduce execution time (as this is bounded by memory accesses). This is most easily done by localizing computations within a solver iteration. Furthermore, from their definitions, we know that Jacobian and Hessian matrices will contain many common subexpressions; therefore, optimization on the "the inner sum" is crucial. We also know that since Hessian matrices are symmetric, we only need to calculate their upper triangular portion. Using this information suggests that for the Jacobian and the Hessian we should jointly *Optimize the Code*. Measurements of floating point operations in code generated using these optimization strategies confirms our expectations (see Table 1).

Table 1. Number of flops per pixel in generated solvers

	jointly optimized +tryhard		separately optimized +tryhard	
b	78	112	97	152
a	88	135	117	176
a,b	205	325	220	396
a,b,v	230	394	284	461

There is an advantage to the joint optimization of the Jacobian and the Hessian matrices, which would not be feasible without *Optimizing the Code*. Maple's `codegen[optimize]` function, especially with the `tryhard` option, eliminates common subexpressions very effectively when these matrices are generated together. If the optimization of code is performed separately and the results are concatenated (which is closer to the code that would be obtained without using the model manipulation process), both the length of the solver and the number of flops per pixel are roughly doubled. This does not reflect the equally important reduction in memory traffic and reduction in local variables by jointly calculating the Jacobian and Hessian in one loop.

3 Parameter Estimation in Model-Based Time Series

This example of parameter estimation from time-series data is extracted from [2]. Parameter estimation is important in many problem domains including determination of rate constants in pharmaceutical drug transport, decomposing audio signals and voice recognition, and measurement of metabolite levels in Magnetic Resonance Spectroscopy (MRS) and Relaxometry. Figure 3 shows an example from MRS of the decomposition of a measured magnetic resonance spectrum for soya bean oil.

3.1 Expressing the Mathematical Model

Expressing the Model for parameter estimation shows that we have a more general version of the least squares fitting example presented in the previous section. A common method of parameter estimation for time series data involves modelling signal sources, $f(x_1, x_2, \ldots, x_n, t)$, (where the x_i are the model parameters and f is in general a vector-valued function) and fitting a superposition of the

Fig. 3. Soya bean oil spectrum (maroon) and component estimates

various sources to the measured data. Through minimization of an objective function F, an optimal set of parameters may be determined:

$$\min_{x_1^1, x_2^1, \ldots, x_n^1, \ldots, x_n^s} \sum_t \left\| y(t) - \sum_{s \in \{\text{sources}\}} a_s f_s(x_1^s, x_2^s, \ldots, x_n^s, t) \right\|^2. \tag{3}$$

where x_j^s denotes the x_j'th parameter of peak s. Equation 3 *Expresses the Model* for parameter estimation of a time series. An important part of the mathematical modelling step is to explicitly declare the class of functions f to consider, which parameters to optimize for, and how many superpositions of the basis function should be used for fitting. This gives important structural information from which we can *Extract Structure*.

As our objective functions will all be analytic, we can safely use Newton's method to solve the minimization problem; therefore, Newton's method forms the *Transformed Model*, as it did for the last example (see subsection 2.2).

3.2   Extracting Structure

Extracting Structure from the model shows the frequent occurrence of recurrence relations, since in many time-series models, a simple time evolution exists. This allows the use of recurrence relations instead of explicit calculations of the model function. This greatly increases the efficiency of objective function evaluations, as well as the calculation of the Jacobian and Hessian on each solver iteration. For instance, in the case of an exponentially damped oscillatory signal, $ae^{-(d+if)t}$ of frequency f, amplitude a, and damping coefficient d, the sequence $a, ae^{-(d+if)}, ae^{-2(d+if)}, \ldots$ can be calculated using the recursion $z_0 = a$, $z_{j+1} = kz_j$, $k = e^{-(d+if)}$.

We symbolically obtain the recurrence equation satisfied by the model f with respect to the main variable t via the `IsHypergeometricTerm` function from the `RationalNormalForms` Maple package. This function uses advanced symbolic techniques to decide if a given term $f(t)$ is such that $\frac{f(t+1)}{f(t)}$ is a rational function of t, and returns this rational function if this is the case.

Further efforts to *Extract Structure* show that if the model happens to have a simple dependence on the parameters, then it is usually the case that the derivatives that appear in the Jacobian and Hessian are simply expressible in terms of the model itself. Considering a simple model with first-order dependence on a parameter b,

$$f(b) = ae^{bp(x)} \text{ and } \frac{\partial f}{\partial b} = p(x)ae^{bp(x)} = p(x)f(b) \tag{4}$$

which shows that the derivative can be expressed in terms of f. If the dependence is algebraic, which can be considered to be a zeroth order differential equation, this can also be used for simplifications. As such dependencies are sources of redundant computations, it is important to factor them out.

gfun[holexprtodiffeq] is used to determine the differential equation(s) satisfied by the model f. The abbreviations stand respectively for *generating function* and *holonomic expression to differential equation*. The package gfun and the theory of holonomic (or D-finite) functions are described in [24] and [4], respectively.

3.3   Code Generation

We need to *Generate Code* that computes F, its Jacobian and Hessian, taking full advantage of the fact that F is a sum, and that all of its sub-terms satisfy a recurrence. Using this structure allows us to *Optimize the Computation*.

The Jacobian with respect to the parameters α is computed symbolically, using the previously computed differential relations. If the differential equation technique fails for any $a \in \alpha$, that partial derivative is computed by direct symbolic differentiation. Direct symbolic differentiation is then used on the Jacobian to get the Hessian. Any occurrence of $\frac{\partial f}{\partial a}$ in the Hessian is replaced by the Jacobian entry.

If f is a complex (vector) function, then f, and the Jacobian and Hessian of f are separated into real and imaginary parts at this point. We must eventually convert all our computations to real computations only, and this point in the algorithm is where we gain the most benefit: previous computations are simpler on the complex function, while more common sub-expressions can be pulled out from the expanded version.

The code to calculate F, **J** and **H** is combined with the code to calculate successive terms of f. This then makes up the body of a loop on the main variable t. Common sub-expression elimination is used on the loop body via codegen[optimize] with the tryhard option, and the optimized code is wrapped in a loop on t from 0 to $n - 1$, where n (number of data points) is an argument of the generated function. The loop is then spliced with the previous code and transformed into a C function.

The generation algorithm can be explained more specifically as

1. get recurrence relation for f on t (via IsHypergeometricTerm),
2. construct the Jacobian and the Hessian for the model function f in terms of f,
3. if f is a complex function, split the above into real and imaginary parts,
4. generate code to calculate the initial value of f, the recurrence ratio h, as well as code to calculate successive terms using h and the last calculated term; do this for each superposition of f;
5. generate code to calculate, by summing in a loop, F, Jacobian(F), Hessian(F); use previously computed relations on derivatives of f (from step (2)), as well as re-using the recurrence for f;
6. the above code uses local variables (in the generated code) to store the Jacobian and Hessian, to enable common-sub-expression elimination (as it cannot be done on Matrix/Vector entries).
7. generate "cleanup" code to assign locally stored Jacob(F) and Hess(F) to arrays that are "returned"

8. wrap F, Jacob(F), Hess(F) and recurrence code in a loop on t and apply sub-expression elimination optimization

9. "paste" code together and transform to C code

This is about 500 lines (counting comments and blank lines) of very clear Maple code. The core ideas fit in about 50 lines, with the rest needed to get around various idiosyncracies, keep the code modular and clean, and simply further automate the process.

Using model manipulation we have measured a 120-fold reduction in execution time for real valued exponential models when compared to a "vanilla" implementation, and a 540-fold reduction for complex valued exponential models. Although we would not expect this to be the case for all applications, we certainly expect significant gains for many applications.

4   Material Behaviour Modelling

Modelling the response of materials under loading is of critical importance to scientists and engineers. To model the deformation and stress within a solid body, we turn to the constitutive equation, which postulates a dependence of the stress on the history of deformation. A wide range of varied and complex constitutive equations are used in practise. Although the behaviour of these models can vary greatly, the underlying mathematics is very similar. Using the correct abstraction, a wide range of material behaviours form a family of material models. Using model manipulation we can quickly generate code for a specific member of this family.

4.1   The Mathematical Model Relating Stress and Deformation

The goal of material modelling is to find the stress ($\boldsymbol{\sigma} : \mathbb{R}^6$) as a function of time ($t : \mathbb{R}$). That is, to return the function $\boldsymbol{\sigma}(t) : \{t : \mathbb{R} | t_{beg} \leq t \leq t_{end}\} \rightarrow \mathbb{R}^6$, where t_{beg} and t_{end} delimit the duration of the simulation. The stress can be found by solving the constitutive equation, which in rate form is:

$$\dot{\boldsymbol{\sigma}} = \mathbf{D} \left(\dot{\boldsymbol{\epsilon}} - \gamma < \phi(F(\boldsymbol{\sigma}, \kappa)) > \frac{\partial Q(\boldsymbol{\sigma})}{\partial \boldsymbol{\sigma}} \right) \text{ and } \boldsymbol{\sigma}(t_{beg}) = \boldsymbol{\sigma}_0 \qquad (5)$$

where $< \phi(F) >= \phi(F)$, if $F > 0$, and 0 otherwise. This equation is based on the viscoplastic constitutive equation presented by Perzyna [19], which depends on the elastic constitutive matrix ($\mathbf{D} : \mathbb{R}^{6 \times 6}$), the fluidity parameter ($\gamma : \mathbb{R}$), the function ϕ ($\phi : \mathbb{R} \rightarrow \mathbb{R}$), the yield function ($F(\boldsymbol{\sigma}, \kappa) : \mathbb{R}^6 \times \mathbb{R} \rightarrow \mathbb{R}$), the plastic potential function ($Q(\boldsymbol{\sigma}) : \mathbb{R}^6 \rightarrow \mathbb{R}$), the stress tensor ($\boldsymbol{\sigma} : \mathbb{R}^6$), the strain rate tensor ($\dot{\boldsymbol{\epsilon}} : \mathbb{R}^6$) and the hardening parameter ($\kappa : \mathbb{R}^6 \rightarrow \mathbb{R}$), which measures the accumulated strain. In Eq. 5, the condition $F = 0$ defines a surface in 6 dimensional stress space, which can be visualized as in Fig. 4. Inside the surface ($F < 0$) the material response will be purely elastic, and outside the response is viscoplastic. When the material has yielded, which occurs when the stress path

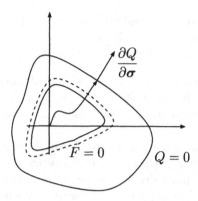

Fig. 4. Yield Function, Hardening and the Plastic Potential in Stress Space

reaches the yield surface, this surface may change shape, as shown in Fig. 4 by the dashed line. Details on material behaviour modelling can be found in [16].

The above constitutive equation, together with the equilibrium equation, are the *Expression of the Model*. The model is very similar between different problems. The only variabilities, which need to be set for a specific material before solving a given problem, are the following: F, Q, ϕ, κ, γ and the property vector, where the property vector consists of the material properties. These variabilities can be explicitly specified in a DSL that describes (declaratively) a particular material model from the family.

4.2   Transformed Model (Finite Element Algorithm)

The second step in the model manipulation process is *Transforming the Model*. In this example, the common parts of the model are transformed into their finite element (FE) method [27] equivalents. This step leaves the variabilities as unspecified; therefore, the algorithm will remain generic and thus be applicable to any material in the family. At the moment, there is no clear algorithm to automatically transform Eq. 5 to an FE equivalent. Currently the transformation seems to require human insight and expert knowledge of the available family of algorithms. However, by keeping the algorithm generic, multiple instances that apply to a variety of materials can quickly be generated.

The FE algorithm selected is a fully implicit time-stepping algorithm that includes a correction back to the yield surface when this is required. The algorithm involves vector and matrix operations and the calculation of the gradients of F and Q with respect to σ [3]. The FE equation to solve for the displacement degrees of freedom (**a**) is as follows:

$$\mathbf{Ka} = \mathbf{F} \tag{6}$$

where $\mathbf{K}$ is known as the stiffness matrix and $\mathbf{F}$ as the load vector. Neither of these quantities depends on **a**, which makes this a linear system of equations.

For the first iteration of the algorithm, the values of $\mathbf{K}$ and $\mathbf{F}$ are as follows:

$$\mathbf{K} = \int_V \mathbf{B}^T \mathbf{D}^{vp} \mathbf{B} dV; \mathbf{F} = \mathbf{R}_i - \int_V \mathbf{B}^T \boldsymbol{\sigma}_i dV + \int_V \mathbf{B}^T \Delta \boldsymbol{\sigma}^{vp} dV \qquad (7)$$

with

$$\mathbf{D}_{vp} = \mathbf{D} \left[\mathbf{I} - \Delta t C_1 \lambda' \frac{\partial Q}{\partial \boldsymbol{\sigma}} \left(\frac{\partial F}{\partial \boldsymbol{\sigma}} \right)^T \mathbf{D} \right], \lambda' = \frac{d\lambda}{dF} \qquad (8)$$

$$\Delta \boldsymbol{\sigma}^{vp} = \Delta t C_1 \lambda \mathbf{D} \frac{\partial Q}{\partial \boldsymbol{\sigma}} \qquad (9)$$

$$C_1 = [1 + \lambda' \Delta t (H_e + H_p)]^{-1} \qquad (10)$$

$$H_e = \left(\frac{\partial F}{\partial \boldsymbol{\sigma}} \right)^T \mathbf{D} (\frac{\partial Q}{\partial \boldsymbol{\sigma}}) \qquad (11)$$

$$H_p = -\frac{\partial F}{\partial \kappa} \left(\frac{\partial \kappa}{\partial \boldsymbol{\epsilon}^{vp}} \right)^T \frac{\partial Q}{\partial \boldsymbol{\sigma}} \qquad (12)$$

where $\mathbf{I}$ is the identity matrix.

For subsequent passes within an equilibrium iteration loop, the FE equations, which provide a correction $\Delta \mathbf{a}_i$ for $\mathbf{a}_i$, simplify to

$$\mathbf{K} = \int_V \mathbf{B}^T \mathbf{D} \mathbf{B} dV; \mathbf{F} = \mathbf{R}_i - \int_V \mathbf{B}^T \boldsymbol{\sigma}_i dV \qquad (13)$$

The equilibrium iteration loops ceases when the convergence criteria satisfies a given tolerance (toler) as follows:

$$\frac{\|\Delta \mathbf{a}\|}{\|\mathbf{a}\|} \leq \text{toler} \qquad (14)$$

where $\|\mathbf{a}\|$ represents the Euclidean norm of the vector $\mathbf{a}$. After solving for the displacements for a given time step the local stresses and strains are updated using a return map algorithm [26], which is described in [17].

4.3 Extracting Structure and Code Generation

After *Extracting the Structure*, which consists of the terms involving F, Q etc., from the *Transformed Model*, the next step is *Generating Code* using a DSL specification to replace the generic parts with material specific code. A program called MatGen [17] was developed to do this. MatGen needs to calculate the required derivatives and output source code for terms such as H_e (Eq. 11). Like the other examples in this paper, Maple was used to do this. Maple performs the necessary symbolic computations and is then used to convert from mathematical expressions into C expressions using the "CodeGeneration" package. These C expressions are inlined into a C++ class defining the material model. This class can then be used by an FE analysis program.

Note that the step *Optimize the Computation* was not emphasized in this example. Instead the goal was to automatically generate code for new constitutive equations in a manner that is simpler, less time consuming and less error prone than using hand calculations. We can illustrate that we reached this goal by considering the calculation of the example term, H_e. Comparing the symbolic output from Maple to a hand derived versions of H_e for a viscoelastic fluid shows the same result that $H_e = 3G$, where G is the shear modulus [17]. However, the hand derivation was complicated, took a nontrivial amount of time, and required expert knowledge. In particular, the hand derivation took 5 pages of equations and explanation [17, pages 77–81]. The derivation used the chain rule of calculus, several stress invariants, the Einstein index notation, vector calculus, and knowledge from continuum mechanics, such as the fact that the trace of the deviatoric stress tensor is zero. The MatGen version, on the other hand, only required using the DSL to specify the model for a viscous fluid, as follows: $F = Q = q; \phi = F; \kappa = 0; \gamma = 1/2\eta$, where q is the effective stress, which is provided by a macro in MatGen, and η is the material property of viscosity. The calculation of other terms in the FE algorithm are at least as complex, time consuming and error prone, as the calculation of H_e. In these other cases MatGen was just as simple and effective, although Maple was unable to simplify these other expressions to be identical to the hand derived versions. In these cases though the expressions were found to be equivalent by verifying their numerical agreement.

Although the *Optimize the Computation* step was not emphasized for the current example, the possibility certainly exists that the FE algorithm can be mined for structure in a manner similar to what was done for the previous two examples. Although the code generated by Maple is not currently efficient, an expert could potentially further *Extract Structure* to improve the efficiency. This possibility illustrates the current need for human insight in the model manipulation process. Additional human creativity and ingenuity at the initial stages of the model manipulation process can facilitate the subsequent automated steps and result in much more efficient code. Further investigation of material behaviour modelling is left as future work.

5   Related Work

The many people working on *Problem Solving Languages* [25] and *Problem Solving Environments* [10,12,14,18,20,21,23] (to cite just a few) implicitly believe in our thesis. By and large, they are however working at creating environments for solving particular problems. For each problem class, the solving methodology is well-enough understood that most of the process can be encapsulated in one piece of software.

Take one of the most impressive examples: SPIRAL [21]. They are essentially following the same approach that we are, but they have concentrated on documenting different aspects of their work. We have concentrated on pulling out the process, and making sure that we automate all that we can. In the domain of signal processing, they have achieved a high level of automation by doing exactly as we preach: automating the "rest" of the process, and then concentrating on the part where new mathematical insight makes a real difference. We believe that this can be done in general by using "mathematics" as the DSL.

Another approach is code extraction from constructive proofs, most notably from Coq proofs [15]. This is an extremely exciting prospect, but is too far on the leading edge of current research to be properly evaluated at this time. Certainly there are issues [5] where the style of one's proof has dramatic effects on the quality of the extracted software! Of course, there is also the issue that many parts of advanced mathematics (like holonomy) are not yet implemented in Coq, but have quite mature implementations in CASes.

6   Conclusion

We have demonstrated that the model manipulation development methodology for generation of (numeric) solution to scientific computation problems has several advantages.

1. The conventional approach, for example where the various gradients are worked out by hand in advance of implementation, is difficult and error prone. Replacing this step by symbolic processing reduces the workload, allows non-experts to deal with new problems, and increases reliability.
2. Although the generated code is for a particular numerical algorithm, given the existing framework, it is straightforward to generate new programs that meet the needs of other algorithms.
3. Any additional information available at the symbolic processing stage can be used to improve performance. For instance, if there is a known differential or recurrence relation in the model, this can be used for optimizing the code.
4. In certain situations, the performance gains from taking advantage of the problem structure can be impressive.

We have chosen to be pragmatic and reuse a well-known existing tool: Maple. We are well aware that this is a far from optimal choice. A better approach would require the use of a semantically richer tool (as provided by many theorem proving environments); but none of these tools have existing libraries as rich as Maple's. Certainly none of them, to our knowledge, contain tools for dealing with holonomic functions. We look forward to the day where semantically richer environments are as computationally capable as today's CASes.

We believe that we are discovering a new development methodology for high-level scientific applications that leverages DSLs, model transformations and program transformation to yield a process that is friendlier to the domain expert, provides insights into the original problem, and produces faster and more reliable code. We believe that tool developers who keep this process firmly in mind when they design new tools (or improve old ones) can produce environments which will improve the productivity of scientific software developers.

Acknowledgements

The financial support of the Natural Sciences and Engineering Research Council (NSERC) is gratefully acknowledged.

References

1. Anand, C., Carette, J., Korobkine, A.: Target recognition algorithm employing Maple code generation. In: Maple Summer Workshop (2004)
2. Anand, C.K., Carette, J., Curtis, A., Miller, D.: COG-PETS: Code generation for parameter estimation in time series. In: Maple Conference 2005 Proceedings, Maplesoft, pp. 198–212 (2005)
3. Carette, J., Smith, S., McCutchan, J., Anand, C., Korobkine, A.: Model manipulation as part of a better development process for scientific computing code. Technical Report 48, Software Quality Research Laboratory, McMaster University (2007)
4. Chyzak, F., Salvy, B.: Non-commutative elimination in Ore algebras proves multivariate holonomic identities. Journal of Symbolic Computation 26(2), 187–227 (1998)
5. Cruz-Filipe, L., Letouzey, P.: A Large-Scale Experiment in Executing Extracted Programs. In: 12th Symposium on the Integration of Symbolic Computation and Mechanized Reasoning, Calculemus 2005 (2005)
6. van Deursen, A., Klint, P.: Little languages: Little maintenance? Journal of Software Maintenance 10, 75–92 (1998)
7. van Deursen, A., Klint, P., Visser, J.: Domain-specific languages: An annotated bibliography. ACM SIGPLAN Notices 35(6), 26–36 (2000)
8. Dongarra, J., Eijkhout, V.: Self-adapting numerical software for next generation applications. Int. J. High Perf. Comput. Appl. 17, 125–131 (2003); also Lapack Working Note 157, ICL-UT-02-07
9. Fotinatos, J., Deak, R., Ellman, T.: Automated synthesis of numerical programs for simulation of rigid mechanical systems in physics-based animation. Automated Software Engineering 10(4), 367–398 (2003)
10. Gaffney, P.W., Houstis, E.N. (eds.): Programming Environments for High-Level Scientific Problem Solving, Proceedings of the IFIP TC2/WG 2.5 Working Conference on Programming Environments for High-Level Scientific Problem Solving, Karlsruhe, Germany, September 23-27, 1991. IFIP Transactions, vol. A-2. North-Holland, Amsterdam (1992)
11. Griewank, A.: Evaluating Derivatives: Principles and Techniques of Algorithmic Differentiation. In: Frontiers in Appl. Math., vol. 19. SIAM, Philadelphia (2000)
12. Hunt, K., Cremer, J.: Refiner: a problem solving environment for ode/dae simulations. SIGSAM Bull. 31(3), 42–43 (1997)
13. Kahrimanian, H.G.: Analytical differentiation by a digital computer. Master's thesis, Temple University (May 1953)
14. Kennedy, K., Broom, B., Cooper, K.D., Dongarra, J., Fowler, R.J., Gannon, D., Johnsson, S.L., Mellor-Crummey, J.M., Torczon, L.: Telescoping languages: A strategy for automatic generation of scientific problem-solving systems from annotated libraries. J. Parallel Distrib. Comput. 61(12), 1803–1826 (2001)
15. Letouzey, P.: A New Extraction for Coq. In: Geuvers, H., Wiedijk, F. (eds.) TYPES 2002. LNCS, vol. 2646, pp. 200–219. Springer, Heidelberg (2003)

16. Malvern, L.E.: Introduction to the Mechanics of Continuous Medium. Prentice-Hall, Englewood Cliffs (1969)
17. McCutchan, J.: A generative approach to a virtual material testing laboratory. Master's thesis, McMaster University (2007)
18. Parker, S.G., Miller, M., Hansen, C.D., Johnson, C.R.: An integrated problem solving environment: the SCIRun computational steering system. In: 31st Hawaii International Conference on System Sciences (HICSS-31) (1998)
19. Perzyna, P.: Fundamental problems in viscoplasticity. In: Advances in Applied Mechanics, pp. 243–377 (1966)
20. Pound, G.E., Eres, M.H., Wason, J.L., Jiao, Z., Keane, A.J., Cox, S.J.: A grid-enabled problem solving environment (pse) for design optimisation within matlab. In: IPDPS 2003: Proceedings of the 17th International Symposium on Parallel and Distributed Processing, p. 50.1. IEEE Computer Society, Washington (2003)
21. Püschel, M., Moura, J.M.F., Johnson, J., Padua, D., Veloso, M., Singer, B.W., Xiong, J., Franchetti, F., Gačić, A., Voronenko, Y., Chen, K., Johnson, R.W., Rizzolo, N.: SPIRAL: Code generation for DSP transforms. Proceedings of the IEEE, special issue on Program Generation, Optimization, and Adaptation 93(2) (2005)
22. Rall, L.B.: Automatic Differentiation: Techniques and Applications. LNCS, vol. 120. Springer, Berlin (1981)
23. Rice, J.R., Boisvert, R.F.: From scientific software libraries to problem-solving environments. IEEE Computational Science & Engineering 3(3), 44–53 Fall (1996)
24. Salvy, B., Zimmermann, P.: Gfun: a Maple package for the manipulation of generating and holonomic functions in one variable. ACM Transactions on Mathematical Software 20(2), 163–177 (1994)
25. Thames, J.M.: SLANG, a problem-solving language for continuous-model simulation and optimization. In: Proceedings of the ACM 24th National Conf. ACM, New York (1969)
26. Zienkiewicz, O.C., Taylor, R.L.: The Finite Element Method For Solid and Structural Mechanics, 6th edn. Elsevier Butterworth-Heinemann, Amsterdam (2005)
27. Zienkiewicz, O.C., Taylor, R.L., Zhu, J.Z.: The Finite Element Method Its Basis and Fundamentals, 6th edn. Elsevier Butterworth-Heinemann, Amsterdam (2005)

Mechanising a Proof of Craig's Interpolation Theorem for Intuitionistic Logic in Nominal Isabelle

Peter Chapman[1], James McKinna[2], and Christian Urban[3]

[1] School of Computer Science, University of St Andrews, Scotland
pc@cs.st-and.ac.uk
[2] Department of Computing Science, Radboud University, Nijmegen, Netherlands
james.mckinna@cs.ru.nl
[3] Institute for Computer Science, Technical University of Munich, Germany
urbanc@in.tum.de

Abstract. Craig's Interpolation Theorem is an important meta-theoretical result for several logics. Here we describe a formalisation of the result for first-order intuitionistic logic without function symbols or equality, with the intention of giving insight into how other such results in proof theory might be mechanically verified, notable cut-admissibility. We use the package *Nominal Isabelle*, which easily deals with the binding issues in the quantifier cases of the proof.

Keywords: Formal mathematics, Nominal Isabelle, automated reasoning, logic.

1 Introduction

Maehara's proof [11] of the Craig Interpolation Theorem stands as one of the more beautiful and intricate consequences of cut-elimination in the sequent calculus [13] for first-order logic. Properly stated, it involves both the polarity of subformulae, *and* the first-order language of terms that may occur in the interpolant formula. The aim of this paper is to present a formalised proof adapted for an intuitionistic G3-like system, in which routine informal considerations of free and bound variables in the language of first-order logic are rendered tractable by the use of Nominal Isabelle.

We build on the work of Ridge [10] and Boulmé [1] and have formalised the result in *Nominal Isabelle* [12]. The work of [10] is incomplete; there is a condition missing in the statement of the theorem. This condition constrains the interpolant formula F for a sequent $\Gamma_1, \Gamma_2 \Rightarrow \Delta_1, \Delta_2$, where the two computed sequents are $\Gamma_1 \Rightarrow \Delta_1, F$ and $\Gamma_2, F \Rightarrow \Delta_2$, as follows: the language of F, denoted $\mathcal{L}(F)$, should be *common* to the languages $\mathcal{L}(\Gamma_1, \Delta_1)$ and $\mathcal{L}(\Gamma_2, \Delta_2)$. The definition of common language in [10] accounts for the predicate constants (with their polarities), but *not* the individual constants (zero-arity functions and free variables). This condition is the most difficult to formalise, but is likewise an important part of the correct statement of the theorem; the additional complexity in enforcing the condition arises inductively when one considers the rules for the quantifiers.

S. Autexier et al. (Eds.): AISC/Calculemus/MKM 2008, LNAI 5144, pp. 38–52, 2008.

The work of Boulmé was undertaken in Coq [2], and used the locally-named syntax of McKinna and Pollack [5] to attempt to formalise Craig's theorem. This syntax was developed to deal with variable binding and reasoning up to α-conversion, as an alternative to de Bruijn notation. Substitution in a reduction rule was formalised by substituting a suitably fresh parameter (i.e. one that occured nowhere else in a particular derivation) for the bound variable, performing whichever reduction rule was needed, and then rebinding the original name. The method is constructive; a new fresh name need not actually be supplied, it is enough that such a fresh name actually exists, but the syntax stipulates that an explicit new parameter be given. Boulmé focused on a restricted language for classical first-order logic consisting of only $NAND$ and $\forall$ (hence expressively complete). The interpolant for a particular case was shown to be valid for that case in individual lemmata, rather than as inductive cases of one theorem.

Here, we use instead the *Nominal Isabelle* system, with the intention of handling variable binding more cleanly, and show that the proof of the theorem for full first-order intuitionistic logic can be formalised, including the tricky details that arise in the quantifier rule cases. The choice of intuitionistic logic simplifies the analysis of sequents $\Gamma \Rightarrow \Delta$, as Δ then consists of at most one formula. The nominal approach goes back to the work of Gabbay and Pitts [3].

Parts of this paper are the actual checked proof script; every result has been verified as correct by the *Isabelle* system. The type-setting facilities which come with *Isabelle* allow us to suppress the output of parts of the theory file for the sake of readability [7]. We intersperse the formal proofs with the proofs that one would normally see in a text book on proof theory. The informal proofs are displayed in a natural deduction style, and use the abbreviation Γ, Δ for $\Gamma \cup \Delta$.

2   The Development

2.1   A Brief Introduction to *Nominal Isabelle*

Nominal Isabelle is a package in the *Isabelle* proof assistant, specifically a package within *Isabelle-HOL*. A comprehensive introduction to the package is given in [12]. We can declare that certain types are *atoms*. This means that we can bind objects of this type within a datatype, and the system will automatically prove results about α-equivalence for the datatype. It is this mechanism which greatly reduces the number of lemmata which the user needs to prove when reasoning about variable binding.

We have formalised first-order intuitionistic logic as an object logic within *Isabelle*. As such, we cannot reuse the reserved symbols and words of *Isabelle*. For this reason, we have that object-level connectives are suffixed by $\star$. For instance, $\wedge\star$ represents conjunction within our object-logic, and $\wedge$ is conjunction in the meta-logic of *Isabelle*. We also cannot use $\perp$ to represent falsehood within the formalisation, so we have instead used *ff*.

Datatypes are introduced using the keyword **datatype**, followed by the name of the datatype and its constructors. After each constructor, we can introduce a notational abbreviation for that constructor in brackets, and augment this abbreviation with additional information about its precendence and associativity. Note that we have supressed this output in the paper. In the case where the datatype uses atoms, we use the keyword **nominal-datatype**. In clauses where binding actually occurs, the atom is enclosed within $\ll\gg$.

Functions are introduced introduced using the keyword **consts**, followed by its name, type, and any notational abbreviation for it in brackets. Note that *Isabelle* uses the syntax $A :: \tau$ for "A has type τ". If the function is defined by primitive recursion, then the keyword **primrec** is used, followed by the clauses for each constructor of the datatype over which recursive calls are made. There is an exception to this; when the the function is defined for a nominal datatype, we use the keyword **nominal-primrec**, and we are given a number of proof obligations to fulfill. Whilst these obligations are not trivial, the proofs of them are not the aim of the paper, and so we have suppressed the output in the document.

Rewriting forms the basis of the *Isabelle* system. Rules of inference and derived lemmata may be used as rewrite rules. The syntax for such rules in *Isabelle* is $[A_0; A_1; \ldots; A_n] \implies C$. Note here that *Isabelle* uses a semi-colon for the meta-level conjunction of premisses. In a natural deduction style, this rule would be represented as

$$\frac{A_0 \quad A_1 \quad \ldots \quad A_n}{C}$$

The proof is written using the *Isar* framework [8]. This allows us to write formalised proofs which are also human-readable. In particular, we can name statements so that we can refer to them by name later in a proof. For instance, "$a : A \in \Gamma_1 \cup \Gamma_2$" means "the statement $A \in \Gamma_1 \cup \Gamma_2$ has the name a." *Isar* proofs consist of a series of statements which are linked by various keywords, such as **from**, **have** and **by**. Whatever follows "**by**" is an *Isabelle* proof-tactic. Common ones are *auto*, *simp* and *blast*. As an example, the statement "**from** a **have** b **by** auto" can be read as "from the statement a we can derive the statement b using the auto tactic."

2.2 The Formalisation

2.2.1 (Formulae). We build formulae as follows. An *atomic formula* is a predicate applied to a list of terms. We formalise a logic without equality; terms are either variables or zero-arity function symbols, which can be simulated by variables. The result would have been obscured beneath a mass of technical details were we to consider a logic with equality, or terms constructed of non-nullary function symbols. The interested reader is directed to [13]. First-order formulae are built in the usual way; note that, since we are using intuitionistic logic, all of the logical connectives must be given as primitives. In the quantifier cases, the variable that is bound is more accurately called a representative of an α-equivalence class. This is the power of the nominal approach; it allows us to reason effectively about substitution and binding.

nominal-datatype *form* =
    Atom pred var list
    | *Conj form form* (- ∧* -)
    | *Disj form form* (- ∨* -)
    | *Impl form form* (- ⊃* -)
    | *ALL* «*var*»*form* (∀ * [-].-)
    | *EX* «*var*»*form* (∃ * [-].-)
    | *ff*

We have a notational shorthand for a quantification over a list of variables. These are "∀*s*" and "∃*s*" for the universal and existential quantifiers respectively. They are defined by primitive recursion on the list as follows

consts
    ALL-list :: *var list* ⇒ *form* ⇒ *form* (∀ *s* [-].-)
primrec
    ∀ *s* [*Nil*].*A* = *A*
    ∀ *s* [*x*#*xs*].*A* = ∀ * [*x*].(∀ *s* [*xs*].*A*)

consts
    EX-list :: *var list* ⇒ *form* ⇒ *form* (∃ *s* [-].-)
primrec
    ∃ *s* [*Nil*].*A* = *A*
    ∃ *s* [*x*#*xs*].*A* = ∃ * [*x*].(∃ *s* [*xs*].*A*)

2.2.2 (Free Variables and the Polarity of Predicates). We use an aspect of the *Nominal Isabelle* package when we consider the individual constants of a formula. We model the constants as the free variables of a formula.

nominal-primrec
    frees (*Atom n xs*) = *frees xs*
    frees (*A* ∧* *B*) = (*frees A*) ∪ (*frees B*)
    frees (*A* ∨* *B*) = (*frees A*) ∪ (*frees B*)
    frees (*A* ⊃* *B*) = (*frees A*) ∪ (*frees B*)
    frees (∀ * [*x*].*A*) = (*frees A*) − {*x*}
    frees (∃ * [*x*].*A*) = (*frees A*) − {*x*}
    frees (*ff*) = {}

The presence of implication as a primitive connective means that the positivity and negativity of a predicate must be defined simultaneously. We use a pair of lists, the first list containing the positive predicates, and the second containing the negative predicates, as follows

nominal-primrec
    pn (*Atom n xs*) = ([*n*],[])
    pn (*A* ∧* *B*) = (*let* (*pA,nA*) = (*pn A*) *in* (*let* (*pB,nB*) = (*pn B*) *in* (*pA*@*pB,nA*@*nB*)))
    pn (*A* ∨* *B*) = (*let* (*pA,nA*) = (*pn A*) *in* (*let* (*pB,nB*) = (*pn B*) *in* (*pA*@*pB,nA*@*nB*)))
    pn (*A* ⊃* *B*) = (*let* (*pA,nA*) = (*pn A*) *in* (*let* (*pB,nB*) = (*pn B*) *in* (*nA*@*pB,pA*@*nB*)))
    pn (∀ * [*x*].*A*) = *pn A*
    pn (∃ * [*x*].*A*) = *pn A*
    pn (*ff*) = ([],[])

We also need to define capture avoiding substitution. This is straightforward when using *Nominal Isabelle*; the package allows us to say when a variable is fresh for another, denoted $x \sharp y$. Here, we are really talking about α-equivalence classes, rather than individual variables. The notation $[t,x]A$ means we substitute the term t for the variable x in the formula A. Recall that our terms are simply variables.

nominal-primrec
$[z,y](Atom\ P\ xs) = Atom\ P\ ([z,y]xs)$
$[z,y](A \wedge_* B) = ([z,y]A) \wedge_* ([z,y]B)$
$[z,y](A \vee_* B) = ([z,y]A) \vee_* ([z,y]B)$
$[z,y](A \supset_* B) = ([z,y]A) \supset_* ([z,y]B)$
$x \sharp (z,y) \implies [z,y](\forall_* [x].A) = \forall_* [x].([z,y]A)$
$x \sharp (z,y) \implies [z,y](\exists_* [x].A) = \exists_* [x].([z,y]A)$
$[z,y]\!f\!f = f\!f$

2.2.3 (Rules of the Calculus). Rather than build derivations as explicit objects, we rather give an inductive definition of what it means to be a *provable sequent*, written $\Gamma \Rightarrow^* C$. A provable sequent is simply the root of a valid derivation; it is a straightforward induction to show that $\Gamma \Rightarrow^* C$ and $\exists d.d \vdash \Gamma \Rightarrow C$ are equivalent. Since we do not transform derivations (we are only interested in the roots of various derivations supplied by the induction hypothesis) this drastically simplifies the work. The rules we use are for a G3i calculus, for example given in [13], except that we have altered them slightly because we use sets and not multisets, and we have explicit weakening[1] . Note that we do not have the Cut rule; our provable sequents are by definition Cut-free.

inductive
   provable :: *form set* $\Rightarrow$ *form* $\Rightarrow$ *bool* (- $\Rightarrow_*$ -)
where
   Ax: $[\![finite\ \Gamma;\ C{\in}\Gamma]\!] \implies \Gamma \Rightarrow_* C$
   | $LBot$: $[\![finite\ \Gamma;\ f\!f{\in}\Gamma]\!] \implies \Gamma \Rightarrow_* C$
   | $ConjR$: $[\![\Gamma \Rightarrow_* A;\ \Gamma \Rightarrow_* B]\!] \implies \Gamma \Rightarrow_* A \wedge_* B$
   | $DisjR1$: $[\![\Gamma \Rightarrow_* A]\!] \implies \Gamma \Rightarrow_* A \vee_* B$
   | $DisjR2$: $[\![\Gamma \Rightarrow_* B]\!] \implies \Gamma \Rightarrow_* A \vee_* B$
   | $ImpR$: $[\![\{A\}{\cup}\Gamma \Rightarrow_* B]\!] \implies \Gamma \Rightarrow_* A \supset_* B$
   | $AllR$: $[\![x{\notin}frees\ \Gamma;\ \Gamma \Rightarrow_* A]\!] \implies \Gamma \Rightarrow_* \forall_* [x].A$
   | ExR: $[\![\Gamma \Rightarrow_* [y,x]A]\!] \implies \Gamma \Rightarrow_* \exists_* [x].A$
   | $ConjL$: $[\![(A \wedge_* B){\in}\Gamma;\ \{A,B\} \cup \Gamma \Rightarrow_* C]\!] \implies \Gamma \Rightarrow_* C$
   | $DisjL$: $[\![(A \vee_* B){\in}\Gamma;\ \{A\}{\cup}\Gamma \Rightarrow_* C;\ \{B\}{\cup}\Gamma \Rightarrow_* C]\!] \implies \Gamma \Rightarrow_* C$
   | $ImpL$: $[\![(A \supset_* B){\in}\Gamma;\ \Gamma \Rightarrow_* A;\ \{B\}{\cup}\Gamma \Rightarrow_* C]\!] \implies \Gamma \Rightarrow_* C$
   | $AllL$: $[\![(\forall_* [x].A){\in}\Gamma;\ \{[y,x]A\}{\cup}\Gamma \Rightarrow_* C]\!] \implies \Gamma \Rightarrow_* C$
   | ExL: $[\![(\exists_* [x].A){\in}\Gamma;\ x{\notin}frees\ (\Gamma,C);\ \{A\}{\cup}\Gamma \Rightarrow_* C]\!] \implies \Gamma \Rightarrow_* C$
   | wk: $[\![\Gamma \Rightarrow_* C]\!] \implies \{A\}{\cup}\Gamma \Rightarrow_* C$

Certain cases in the proof call for some derived rules. We have generalised weakening so that if $\Gamma \Rightarrow^* C$, and Γ is a subset of a finite set Γ', then $\Gamma' \Rightarrow^* C$. More

---

[1] The use of sets for contexts also means that contraction is admissible, since $\Gamma \cup A \cup A = \Gamma \cup A$.

importantly, we have derived four rules which perform the appropriate quantifier rule over all the variables in a given list. As an example, here is the derived rule corresponding to $R\forall$:

$$\frac{\Gamma \Rightarrow^\star C}{\Gamma \Rightarrow^\star \forall s\ L.C}\ R\forall s$$

where frees $L \cap$ frees $\Gamma = \emptyset$.

3 The Proof

We have formalised the following theorem

Theorem 3.1 (Craig's Interpolation Theorem). *Suppose that* $\Gamma \Rightarrow^\star C$. *Then, for any splitting of the context* $\Gamma \equiv \Gamma_1 \cup \Gamma_2$, *there exists an* E *such that*

1. $\Gamma_1 \Rightarrow^\star E$ *and* $\Gamma_2, E \Rightarrow^\star C$.
2. *Any predicate that occurs positively in* E *occurs positively in* Γ_1 *and* C *and negatively in* Γ_2.
3. *Any predicate that occurs negatively in* E *occurs negatively in* Γ_1 *and* C *and positively in* Γ_2.
4. *frees*$(E) \subseteq$ *frees*$(\Gamma_1) \cap$ *frees*(Γ_2, C)

We use the notation $\Gamma_1; \Gamma_2 \overset{E}{\Longrightarrow} C$ to represent that E is a suitable *interpolant* for a splitting $\Gamma_1 \cup \Gamma_2$. We have formalised this as

theorem *Craigs-Interpolation-Theorem*:
   assumes a: $\Gamma_1 \cup \Gamma_2 \Rightarrow\ast C$
   shows $\exists E.\ \Gamma_1 \Rightarrow\ast E \wedge \{E\} \cup \Gamma_2 \Rightarrow\ast C \wedge \Gamma_1, \Gamma_2, C \vdash E\ pnc$

where the notation "$\Gamma_1, \Gamma_2, C \vdash E$ pnc" is an abbreviation for E satisfying the conditions 2-4 with respect to Γ_1, Γ_2 and C. Normally, we would prove the theorem by induction on the height of the derivation of $\Gamma \Rightarrow C$, and then by case analysis on the last rule used in the derivation. This approach is possible in *Isabelle*, see [10] for instance. Here we prove the theorem by induction on the "provable sequent" definition. This means we show the theorem is valid for the conclusion of each rule given that it is valid for the premises. This induction scheme is derived and proved automatically by *Isabelle* when we write out inductive definitions.

Where a left rule was used to derive $\Gamma_1 \cup \Gamma_2 \Rightarrow^\star C$, there are two subcases: either the principal formula is in the left part of the split context, or it is in the right part. In other words, it is in Γ_1 or Γ_2. When we refer to "the left case" and "the right case", we really mean that "the principal formula is in the left part of the split context..." etc. Since we have a single succedent calculus, we have no such splitting when using right rules. This leads to a total of 22 subcases: 4 base cases, 10 cases from left rules, 5 cases from right rules, and 3 weakening cases.

In what follows, the names of the subsections refer to the rule(s) used in deriving the provable sequent. The use of variable binding is only evident in the first-order cases, therefore we give only sketches of the propositional cases, where there is no binding. They are still fully formalised, but the output is suppressed.

3.1 Axioms and $L\bot$

In the case where the derivation of $\Gamma_1, \Gamma_2 \Rightarrow^* C$ is an axiom, there are two cases. The left has $C \in \Gamma_1$ and the right has $C \in \Gamma_2$. In the former, we need to find a formula E such that

$$\Gamma_1 \Rightarrow^* E \text{ and } E, \Gamma_2 \Rightarrow^* C$$

A suitable candidate is $E \equiv C$, which would make both provable sequents instances of **Ax**. We must also check that $\Gamma, \Gamma', C \vdash C$ pnc, which is trivially true. For this case we can conclude $\Gamma_1; \Gamma_2 \overset{c}{\Longrightarrow} C$.

In the other case, we require an E so that

$$\Gamma_1 \Rightarrow^* E \text{ and } E, \Gamma_2 \Rightarrow^* C$$

which only hold for general Γ_1 if we have $E \equiv \bot \supset \bot$; in other words, $\top$. Since $\bot$ has neither free variables nor predicate symbols, the condition $\Gamma_1, \Gamma_2, C \vdash \bot \supset \bot$ pnc is trivially true for any Γ_1 and Γ_2. Therefore, for this case we have $\Gamma_1; \Gamma_2 \overset{\bot \supset \bot}{\Longrightarrow} C$.

Likewise, the provable sequents we require in the case where the rule used is $L\bot$ are straightforward. We have two subcases, where $\bot$ is either in Γ_1 or Γ_2. In the former, we have that the interpolant is $\bot$, and in the latter we have the interpolant is $\bot \supset \bot$:

```
case (LBot Γ C)
then have a1: finite Γ₁ ∧ finite Γ₂
     and a2: ff ∈ Γ₁∪Γ₂ by simp-all
have ff ∈ Γ₁ ∨ ff ∈ Γ₂ using a2 by blast
moreover
{assume ff ∈ Γ₁
 with a1 have ∃E. Γ₁ ⇒* E ∧ {E}∪Γ₂ ⇒* C ∧ Γ₁,Γ₂,C ⊢ E pnc by auto
}
moreover
{assume b2: ff ∈ Γ₂
 with a1 have Γ₁ ⇒* ff ⊃* ff
     and {ff ⊃* ff}∪Γ₂ ⇒* C
     and Γ₁,Γ₂,C ⊢ (ff ⊃* ff) pnc
   by (auto)
 then have ∃E. Γ₁ ⇒* E ∧ {E}∪Γ₂ ⇒* C ∧ Γ₁,Γ₂,C ⊢ E pnc by blast
}
ultimately show ∃E. Γ₁ ⇒* E ∧ {E}∪Γ₂ ⇒* C ∧ Γ₁,Γ₂,C ⊢ E pnc by blast
```

3.2 $R \supset$ and $L \supset$

For the right rule, there is only one possible way to split the antecedent, since there is no distinguished formula in it. We split the premiss on the right, and suppose the interpolant of the premiss is E. We therefore have the two sequents

$\Gamma_1 \Rightarrow^* E$ and $\Gamma_2, A, E \Rightarrow^* B$. Leaving the first alone, we have the simple deduction for the second

$$\frac{\Gamma_2, A, E \Rightarrow^* B}{\Gamma_2, E \Rightarrow^* A \supset B} \; R\supset$$

which gives us the interpolant for the whole as E:

$$\frac{\Gamma_1; A, \Gamma_2 \overset{E}{\Longrightarrow} B}{\Gamma_1; \Gamma_2 \overset{E}{\Longrightarrow} A \supset B}$$

There are two subcases for $L\supset$. The left subcase has $A \supset B \in \Gamma_1$ and is the most unusual of the propositional cases. Since the statement of the theorem says "for *any* splitting of the context", this means from our induction hypothesis we can choose whichever splitting we want. In this case, we choose a different splitting for the premisses than for the conclusion. Some brief experimentation reveals that we should split the first premiss as Γ_2 and Γ_1 and the second as $\Gamma_1 \cup B$ and Γ_2. In the formalisation below, we have instantiated the induction hypotheses to reflect this. This gives us four sequents: $\Gamma_2 \Rightarrow^* E_1$ and $\Gamma_1, E_1 \Rightarrow^* A$ and $\Gamma_1, B \Rightarrow^* E_2$ and $\Gamma_2, E_2 \Rightarrow^* C$.

Taking the first and fourth of these we can create the deduction

$$\frac{\dfrac{\Gamma_2 \Rightarrow^* E_1}{\Gamma_2, E_1 \supset E_2 \Rightarrow^* E_1} \; w \qquad \Gamma_2, E_2 \Rightarrow^* C}{\Gamma_2, E_1 \supset E_2 \Rightarrow^* C} \; L\supset$$

whereas using the second and third we can create the deduction

$$\frac{\Gamma_1, E_1 \Rightarrow^* A \quad \dfrac{\dfrac{\Gamma_1, B \Rightarrow^* E_2}{B, \Gamma_1, E_1 \Rightarrow^* E_2} \; w}{\dfrac{\Gamma_1, E_1 \Rightarrow^* E_2}{\Gamma_1 \Rightarrow^* E_1 \supset E_2} \; R\supset} \; L\supset}{}$$

This is precisely the form we need, with the interpolant being $E_1 \supset E_2$. We can therefore conclude that the following is a valid deduction for this case, recalling that $A \supset B \in \Gamma_1$

$$\frac{\Gamma_2; \Gamma_1 \overset{E_1}{\Longrightarrow} A \quad \Gamma_1, B; \Gamma_2 \overset{E_2}{\Longrightarrow} C}{\Gamma_1; \Gamma_2 \overset{E_1 \supset E_2}{\Longrightarrow} C}$$

We can see this formalised in the following fragment, where the language conditions are also verified

case $(ImpL\ A\ B\ \Gamma\ C\ \Gamma_1\ \Gamma_2)$
then have $(A \supset \! * B) \in \Gamma_1 \cup \Gamma_2$ **by** *simp*
then have $(A \supset \! * B) \in \Gamma_1 \vee (A \supset \! * B) \in \Gamma_2$ **by** *blast*
moreover

{ **assume** $b1$: $(A{\supset}*B){\in}\Gamma_1$

　have ihL: $\exists E.\ \Gamma_2 \Rightarrow* E \wedge (\{E\} \cup \Gamma_1) \Rightarrow* A \wedge \Gamma_2,\Gamma_1,A \vdash E\ pnc$ **by** $(simp)$

　have ihR: $\exists E.\ (\{B\}{\cup}\Gamma_1) \Rightarrow* E \wedge (\{E\} \cup \Gamma_2) \Rightarrow* C \wedge (\{B\}{\cup}\Gamma_1),\Gamma_2,C \vdash E\ pnc$ **by** $(simp)$

　from $ihL\ ihR$ **obtain** $E1\ E2$

　　where $c1$: $\Gamma_2 \Rightarrow* E1$ **and** $c2$: $\{E1\}{\cup}\Gamma_1 \Rightarrow* A$

　　and $d1$: $\{B\}{\cup}\Gamma_1 \Rightarrow* E2$ **and** $d2$: $\{E2\}{\cup}\Gamma_2 \Rightarrow* C$

　　and $c3$: $\Gamma_2,\Gamma_1,A \vdash E1\ pnc$ **and** $d3$: $(\{B\}{\cup}\Gamma_1),\Gamma_2,C \vdash E2\ pnc$ **by** $auto$

　from $d1$ **have** $\{B,E1\}{\cup}\Gamma_1 \Rightarrow* E2$ **using** $provable.wk$ **by** $(blast)$

　then have $\{E1\}{\cup}\Gamma_1 \Rightarrow* E2$ **using** $b1\ c2\ provable.ImpL$ **by** $(auto)$

　then have $\Gamma_1 \Rightarrow* E1 \supset E2$ **using** $provable.ImpR$ **by** $auto$

　moreover

　from $c1\ d2$ **have** $\{E1 \supset* E2\}{\cup}\ \Gamma_2 \Rightarrow* E1$

　　　　　and $\{E1 \supset* E2,E2\}{\cup}\Gamma_2 \Rightarrow* C$ **by** $(blast)+$

　then have $\{E1{\supset}*E2\}{\cup}\Gamma_2 \Rightarrow* C$ **using** $provable.ImpL$ **by** $(auto)$

　moreover

　from $c3\ d3$ **have** $\Gamma_1,\Gamma_2,C \vdash E1 \supset* E2\ pnc$ **using** $b1$ **by** $(auto)$

　ultimately have $\exists E.\ \Gamma_1 \Rightarrow* E \wedge \{E\}{\cup}\Gamma_2 \Rightarrow* C \wedge \Gamma_1,\Gamma_2,C \vdash E\ pnc$ **by** $blast$

}

In the right case $(A \supset B \in \Gamma_2)$, assuming via the induction hypothesis that the first premiss has interpolant E_1 and the second premiss interpolant E_2, we split both premisses on the right, so $\Gamma_1;\Gamma_2 \overset{E_1}{\Longrightarrow} A$ and $\Gamma_1;B,\Gamma_2 \overset{E_2}{\Longrightarrow} C$. We then obtain four sequents: $\Gamma_1 \Rightarrow^\star E_1$ and $\Gamma_2,E_1 \Rightarrow^\star A$ and $\Gamma_1 \Rightarrow^\star E_2$ and $\Gamma_2,B,E_2 \Rightarrow^\star C$. Naturally, we pair them up according to contexts. The first and third premisses therefore give

$$\frac{\Gamma_1 \Rightarrow^\star E_1 \quad \Gamma_1 \Rightarrow^\star E_2}{\Gamma_1 \Rightarrow^\star E_1 \wedge E_2}\ R\wedge$$

whereas the remaining two sequents give

$$\frac{\dfrac{\Gamma_2,E_1 \Rightarrow^\star A}{\Gamma_2,E_1,E_2 \Rightarrow^\star A}\ w \quad \dfrac{\Gamma_2,B,E_2 \Rightarrow^\star C}{\Gamma_2,B,E_1,E_2 \Rightarrow^\star C}\ w}{\dfrac{\dfrac{\Gamma_2,E_1,E_2 \Rightarrow^\star C}{}}{\Gamma_2,E_1 \wedge E_2 \Rightarrow^\star C}\ L\wedge}\ L{\supset}$$

which gives us the required interpolant as $E_1 \wedge E_2$:

$$\frac{\Gamma_1;\Gamma_2 \overset{E_1}{\Longrightarrow} A \quad \Gamma_1;B,\Gamma_2 \overset{E_2}{\Longrightarrow} C}{\Gamma_1;\Gamma_2 \overset{E_1 \wedge E_2}{\Longrightarrow} C}$$

3.3　$R\wedge$ and $L\wedge$

For $R\wedge$, we can only split the conclusion in one way, likewise we can only split the premisses in one way. Therefore, assuming that the interpolant for the first premiss is E_1 and the interpolant for the second premiss is E_2, we get four sequents: $\Gamma_1 \Rightarrow^* E_1$ and $\Gamma_2, E_1 \Rightarrow^* A$ and $\Gamma_1 \Rightarrow^* E_2$ and $\Gamma_2, E_2 \Rightarrow^* B$. Pairing them up by context, we get

$$\frac{\Gamma_1 \Rightarrow^* E_1 \quad \Gamma_1 \Rightarrow^* E_2}{\Gamma_1 \Rightarrow^* E_1 \wedge E_2} \; R\wedge$$

and

$$\frac{\dfrac{\Gamma_2, E_1 \Rightarrow^* A}{\Gamma_2, E_1, E_2 \Rightarrow^* A} \; w \quad \dfrac{\Gamma_2, E_2 \Rightarrow^* B}{\Gamma_2, E_1, E_2 \Rightarrow^* B} \; w}{\dfrac{\Gamma_2, E_1, E_2 \Rightarrow^* A \wedge B}{\Gamma_2, E_1 \wedge E_2 \Rightarrow^* A \wedge B} \; L\wedge} \; R\wedge$$

which means that $E_1 \wedge E_2$ is the interpolant:

$$\frac{\Gamma_1; \Gamma_2 \overset{E_1}{\Longrightarrow} A \quad \Gamma_1; \Gamma_2 \overset{E_2}{\Longrightarrow} B}{\Gamma_1; \Gamma_2 \overset{E_1 \wedge E_2}{\Longrightarrow} A \wedge B}$$

The two subcases for $L\wedge$ are simple. For the left case, assume that the interpolant is E, and split A, B likewise on the left, we get the two sequents $\Gamma_1, A, B \Rightarrow^* E$ and $\Gamma_2, E \Rightarrow^* C$. We leave the second of these alone, and taking the first apply $L\wedge$. We can then conclude that E is the interpolant:

$$\frac{\Gamma_1, A, B; \Gamma_2 \overset{E}{\Longrightarrow} C}{\Gamma_1; \Gamma_2 \overset{E}{\Longrightarrow} C}$$

The right case is symmetrical, therefore E, the interpolant supplied by the induction hypothesis, is also the interpolant for the conclusion:

$$\frac{\Gamma_1; A, B, \Gamma_2 \overset{E}{\Longrightarrow} C}{\Gamma_1; \Gamma_2 \overset{E}{\Longrightarrow} C}$$

3.4 $R\vee$ and $L\vee$

We have two rules for $R\vee$. However, the two cases are almost identical, so we will only show one. We can only split the conclusion in one way, and likewise the premiss. Suppose the interpolant from the induction hypothesis is E, and assume further that we used the rule $R\vee_1$. Then we have the sequents $\Gamma_1 \Rightarrow^* E$ and $\Gamma_2, E \Rightarrow^* A$. Using the rule $R\vee_1$ on the second, we obtain $\Gamma_2, E \Rightarrow^* A \vee B$. Therefore the interpolant in this case is E, and is given by the deduction

$$\frac{\Gamma_1; \Gamma_2 \overset{E}{\Longrightarrow} A}{\Gamma_1; \Gamma_2 \overset{E}{\Longrightarrow} A \vee B}$$

We get the same result if $R\vee_2$ was used in both situations.

Now we consider $L\lor$. In the left case assume that the interpolant for the first premiss is E_1, and the interpolant for the second premiss is E_2. Now, split both of the premisses on the left, to obtain the two sequents from the left premiss $\Gamma_1, A \Rightarrow^* E_1$ and $\Gamma_2, E_1 \Rightarrow^* C$, and the two sequents from the right premiss $\Gamma_1, B \Rightarrow^* E_2$ and $\Gamma_2, E_2 \Rightarrow^* C$. Again, pairing up by contexts, we have

$$\frac{\Gamma_2, E_1 \Rightarrow^* C \quad \Gamma_2, E_2 \Rightarrow^* C}{\Gamma_2, E_1 \lor E_2 \Rightarrow^* C} \; L\lor$$

and

$$\frac{\dfrac{\Gamma_1, A \Rightarrow^* E_1}{\Gamma_1, A \Rightarrow^* E_1 \lor E_2} \; R\lor \quad \dfrac{\Gamma_1, B \Rightarrow^* E_2}{\Gamma_1, B \Rightarrow^* E_1 \lor E_2} \; R\lor}{\Gamma_1 \Rightarrow^* E_1 \lor E_2} \; L\lor$$

This means the required interpolant is $E_1 \lor E_2$, giving a derivation:

$$\frac{\Gamma_1, A; \Gamma_2 \overset{E_1}{\Longrightarrow} C \quad \Gamma_1, B; \Gamma_2 \overset{E_2}{\Longrightarrow} C}{\Gamma_1; \Gamma_2 \overset{E_1 \lor E_2}{\Longrightarrow} C}$$

For the right case, we again split both premisses on the right, so the following deductions suffice, assuming that E_1 and E_2 are the interpolants,

$$\frac{\Gamma_1 \Rightarrow^* E_1 \quad \Gamma_1 \Rightarrow^* E_2}{\Gamma_1 \Rightarrow^* E_1 \land E_2} \; R\land$$

and

$$\frac{\dfrac{\dfrac{\Gamma_2, A, E_1 \Rightarrow^* C}{\Gamma_2, A, E_1, E_2 \Rightarrow^* C} \; w \quad \dfrac{\Gamma_2, B, E_2 \Rightarrow^* C}{\Gamma_2, B, E_1, E_2 \Rightarrow^* C} \; w}{\Gamma_2, E_1, E_2 \Rightarrow^* C} \; L\lor}{\Gamma_2, E_1 \land E_2 \Rightarrow^* E} \; L\land$$

meaning that $E_1 \land E_2$ is the interpolant:

$$\frac{\Gamma_1; A, \Gamma_2 \overset{E_1}{\Longrightarrow} C \quad \Gamma_1; B, \Gamma_2 \overset{E_2}{\Longrightarrow} C}{\Gamma_1; \Gamma_2 \overset{E_1 \land E_2}{\Longrightarrow} C}$$

3.5   R∃

A first attempt at finding an interpolant for this case would yield using the interpolant supplied by the induction hypothesis. Whilst it would give us the two provable sequents that we need for the theorem, this interpolant fails the language condition for the conclusion. Suppose the induction hypothesis gives us the two provable sequents $\Gamma_1 \Rightarrow^* E$ and $E, \Gamma_2 \Rightarrow^* [y, x]A$. The induction hypothesis will also gives us that the free variables of E are contained in the free variables of Γ_1 and the free variables of $\Gamma_2, [y, x]A$. Suppose that there were some free variables in E that were in the free variables of t, but *not* in the free variables of Γ_2 or A. These free variables will no longer appear in the conclusion, and so the language condition would fail when using E. This is the crucial difference between our definitions and formalisation and those in [10]: E would be a valid interpolant in that formalisation. We need to remove these free variables, which we do by quantification. In this case, we use existential quantification.

Let the set of such variables be L. Since they are finite, we can form a list from this set, which we will also call L, in a slight abuse of notation. We know that all the variables in this list will not appear in Γ_1, and hence every variable in L is not in the free variables of Γ_1, which means we can apply $R\exists$ for every variable in the list:

$$\frac{\Gamma_1 \Rightarrow^\star E}{\Gamma_1 \Rightarrow^\star \exists s\ L.E}\ R\exists s$$

On the second sequent, we can apply the derived rule $L\exists s$, *after* applying $R\exists$:

$$\frac{\dfrac{\Gamma_2, E \Rightarrow^\star [y, x]A}{\Gamma_2, E \Rightarrow^\star \exists x A}\ R\exists}{\Gamma_2, \exists s\ L.E \Rightarrow^\star \exists x A}\ L\exists s$$

We can see this argument formalised as follows

case (*ExR Γ y x A*)
then have *a1*: $\Gamma_1 \cup \Gamma_2 \Rightarrow* [y,x]A$
    and *ih*:$\exists E.\ \Gamma_1 \Rightarrow* E \wedge \{E\} \cup \Gamma_2 \Rightarrow* [y,x]A \wedge \Gamma_1, \Gamma_2, [y,x]A \vdash E\ pnc$ **by** *simp-all*
from *ih* **obtain** E **where** *b1*: $\Gamma_1 \Rightarrow* E$
            and *b2*: $\{E\} \cup \Gamma_2 \Rightarrow* [y,x]A$
            and *b3*: $\Gamma_1, \Gamma_2, [y,x]A \vdash E\ pnc$ **by** *blast*
have *finite* ((*frees E*) $-$ *frees* $(\Gamma_2, \exists * [x].A)$) **by** (*simp*)
then obtain L **where** *eq*: set $L = (frees\ E) - (frees\ (\Gamma_2, \exists * [x].A))$
  using *exists-list-for-finite-set* **by** *auto*
from *b1* **have** $\Gamma_1 \Rightarrow* \exists s\ [L].E$ **by** (*rule exists-right-intros*)
moreover
from *b2* **have** $\{E\} \cup \Gamma_2 \Rightarrow* \exists * [x].A$ **using** *provable.ExR* **by** *auto*
then have $\{\exists s\ [L].E\} \cup \Gamma_2 \Rightarrow* \exists * [x].A$ **using** *eq* **by** (*rule-tac exists-left-intros*)
moreover
from *b3* **have** $\Gamma_1, \Gamma_2, \exists * [x].A \vdash \exists s\ [L].E\ pnc$ **using** *eq* **by** (*auto*)
ultimately show $\exists E.\ \Gamma_1 \Rightarrow* E \wedge \{E\} \cup \Gamma_2 \Rightarrow* \exists * [x].A \wedge \Gamma_1, \Gamma_2, \exists * [x].A \vdash E$
pnc **by** *blast*

3.6 $L\forall$

We have the same problem in this case as in the case for $R\exists$. In the left subcase, we again define L as the list of variables which appear in t and nowhere else in the provable sequents supplied by the induction hypothesis, namely $\Gamma_1, [y, x]A \Rightarrow^\star E$ and $\Gamma_2, E \Rightarrow^\star C$, with $\forall x A \in \Gamma_1$. Using the former, we first apply $L\forall$, and then, since we know that the variables in L appear nowhere free in Γ_1 by construction, then we can apply one instance of $R\forall$ for every variable in L:

$$\frac{\dfrac{\Gamma_1, [y,x]A \Rightarrow^\star E}{\Gamma_1 \Rightarrow^\star E}\ L\forall}{\Gamma_1 \Rightarrow^\star \forall s\ L.E}\ R\forall s$$

Using the other sequent, we can simply apply our derived rule $L\forall s$, thus $\forall s\ L.E$ is the required interpolant.

In the right case, we again define L as above. The interesting sequent is now the second obtained from the induction hypothesis, $\Gamma_2, [y,x]A, E \Rightarrow^\star C$. We first apply $L\forall$, and then we know that the variables in L do not appear free in the new sequent, therefore we can apply $L\exists$ for each of the variables in L

$$\frac{\dfrac{\Gamma_2, [y,x]A, E \Rightarrow^\star C}{\Gamma_2, E \Rightarrow^\star C}\ L\forall}{\Gamma_2, \exists s\ L.E \Rightarrow^\star C}\ L\exists s$$

The proof for the left case is formalised as follows

case *(AllL x A Γ y C Γ_1 Γ_2)*
then have $\forall * [x].A \in (\Gamma_1 \cup \Gamma_2)$ **by** *simp*
then have $\forall * [x].A \in \Gamma_1 \vee \forall * [x].A \in \Gamma_2$ **by** *simp*
{ **assume** *b1*: $\forall * [x].A \in \Gamma_1$
  have *ih*: $\exists\ E.\ \{[y,x]A\} \cup \Gamma_1 \Rightarrow * E \wedge \{E\} \cup \Gamma_2 \Rightarrow * C \wedge \{[y,x]A\} \cup \Gamma_1, \Gamma_2, C \vdash E$ *pnc*
by *auto*
  from *ih* **obtain** *E* **where**
        c1: $\{[y,x]A\} \cup \Gamma_1 \Rightarrow * E$
    and *c2*: $\{E\} \cup \Gamma_2 \Rightarrow * C$
    and *c3*: $\{[y,x]A\} \cup \Gamma_1, \Gamma_2, C \vdash E$ *pnc* **by** *auto*
  have *finite* (*frees E − frees Γ_1*) **by** (*simp*)
  then obtain *L* **where**
     eq: *set L = frees E − frees Γ_1* **using** *exists-list-for-finite-set* **by** *auto*
  then have *set L $\cap$ frees Γ_1* = {} **by** *auto*
  from *c1* **have** $\Gamma_1 \Rightarrow * E$ **using** *provable.AllL* $\langle\forall * [x].A \in \Gamma_1\rangle$ **by** *auto*
  then have $\Gamma_1 \Rightarrow * \forall s\ [L].E$ **using** $\langle$*set L $\cap$ frees Γ_1* = {}$\rangle$ **by** (*rule forall-right-intros*)
  moreover
  from *c2* **have** $\{\forall s\ [L].E\} \cup \Gamma_2 \Rightarrow * C$ **by** (*rule forall-left-intros*)
  moreover
  from *c3* **have** $\Gamma_1, \Gamma_2, C \vdash \forall s\ [L].E$ *pnc* **using** *eq b1*
     by (*auto*)
  ultimately have $\exists\ E.\ \Gamma_1 \Rightarrow * E \wedge \{E\} \cup \Gamma_2 \Rightarrow * C \wedge \Gamma_1, \Gamma_2, C \vdash E$ *pnc* **by** *blast*
}

3.7 $R\forall$ and $L\exists$

In the case of $R\forall$, we can only split the premiss and conclusion in one way. Thus, we have the provable sequents, supplied by the induction hypothesis, $\Gamma_1 \Rightarrow^\star E$ and $\Gamma_2, E \Rightarrow^\star A$, and further that $x \notin \text{frees}(\Gamma_1, \Gamma_2, E)$. This means that we can simply apply $R\forall$ to the second of these two provable sequents, and then have the required sequents. Furthermore, since we know that $x \notin \text{frees}(E)$, we have that the free variables of E are contained in the free variables of $\Gamma_2, \forall x A$.

The two cases for $L\exists$ are symmetrical to that of $R\forall$; the induction hypothesis supplies that the quantified variable will be not be in the free variables of the interpolant, and so we can just apply the rule $L\exists$ to the appropriate sequent, leaving the other alone.

The weakening cases are uninteresting and so are not shown. All of the cases have now been shown, and the proof of the theorem is complete.

4 Mechanisation Statistics and Other Comments

The current work stands at 823 lines [2], including white space and comments. The first author adapted the proof in [10] for intuitionistic logic, and that comprised 1002 lines; moreover the theorem in that work was not as powerful as the one in the current work. We also use the more verbose *Isar* language, and not a tactic script, which necessarily adds to the length of our proof. The main insight that made this proof much shorter was the removal of explicitly mentioning derivations, in favour of the notion of a provable sequent. As an example, the first author attempted a proof of the theorem in the current work using explicit derivations, and the work was around 1200 lines. The use of *Nominal Isabelle* also greatly reduced the need for proving complicated lemmata about capture-avoiding substitution, which was required in this proof.

Weakening (**wk** from the definition in §2.2.3) is logically admissible because we have used a generalised axiom, **Ax**. We could prove **wk** as a lemma as part of the formalisation. We are interested in formalising a proof of Craig's Interpolation Theorem, rather than proving structural rules admissible for intutionistic first-order logic. Therefore, we have kept weakening as an explicit rule.

It would be a relatively straightforward to adapt this development and proof for classical logic without equality. We would need more subcases for each rule, since a sequent calculus for classical logic permits sets, or multisets, of formulae for succedents (see for instance [13]). However, we could also interdefine the connectives, meaning one needs to consider fewer rules. It would be possible, but by no means straightfoward, to extend the result to a logic with equality and non-nullary function symbols.

Our definitions and proof technique can also be used to formalise other meta-mathematical results from proof theory. Furthermore, the *Isar* language allows us to do this in a more human-readable way than other proof theory formalisations, such as [9].

Interpolation results form an important part of computer science. They can be applied to type-checking in C programs, as shown in [4], and also to model checking, as in [6], amongst other things.

---

[2] Available as part of the *Nominal Isabelle* distribution at "http://isabelle.in.tum.de/nominal/"

We believe that the current work is a clean and effective proof of a non-trivial result for first-order intuitionistic logic. It also shows how similar results for first-order logic could be mechanically verified, which in the abstract was stated as an aim of the paper. We have further shown that *Nominal Isabelle* is an effective and immensely useful tool when one deals with bound variables and substitution.

Acknowledgements. We would like to thank Roy Dyckhoff and the anonymous referees for their many helpful comments on an earlier version of this paper.

References

[1] Boulmé, S.: A Proof of Craig's Interpolation Theorem in Coq (1996), citeseer.ist.psu.edu/480840.html

[2] Coq Development Team. The Coq Proof Assistant Reference Manual Version 8.1 (2006), http://coq.inria.fr/V8.1/refman/index.html

[3] Gabbay, M.J., Pitts, A.M.: A new approach to abstract syntax involving binders. In: 14th Annual Symposium on Logic in Computer Science, pp. 214–224. IEEE Computer Society Press, Washington (1999)

[4] Jhala, R., Majumdar, R., Xu, R.-G.: State of the union: Type inference via craig interpolation. In: Grumberg, O., Huth, M. (eds.) TACAS 2007. LNCS, vol. 4424, pp. 553–567. Springer, Heidelberg (2007)

[5] McKinna, J., Pollack, R.: Pure type systems formalized. In: Bezem, M., Groote, J.F. (eds.) TLCA 1993. LNCS, vol. 664, pp. 289–305. Springer, Heidelberg (1993)

[6] McMillan, K.L.: Applications of craig interpolants in model checking. In: Halbwachs, N., Zuck, L.D. (eds.) TACAS 2005. LNCS, vol. 3440, pp. 1–12. Springer, Heidelberg (2005)

[7] Nipkow, T., Paulson, L., Wenzel, M.: A Proof Assistant for Higher-Order Logic. LNCS, vol. 2283. Springer, Heidelberg (2005)

[8] Nipkow, T.: Structured proofs in isar/hol. In: Geuvers, H., Wiedijk, F. (eds.) TYPES 2002. LNCS, vol. 2646, pp. 259–278. Springer, Heidelberg (2003)

[9] Pfenning, F.: Structural cut elimination i. intuitionistic and classical logic. Information and Computation (2000)

[10] Ridge, T.: Craig's interpolation theorem formalised and mechanised in Isabelle/HOL. Arxiv preprint cs.LO/0607058, 2006 - arxiv.org (2006)

[11] Takeuti, G.: Proof Theory. Studies in Logic and the Foundations of Mathematics, vol. 81. North-Holland Publishing Company, Amsterdam (1975)

[12] Tasson, C., Urban, C.: Nominal techniques in Isabelle/HOL. In: Nieuwenhuis, R. (ed.) CADE 2005. LNCS (LNAI), vol. 3632, pp. 38–53. Springer, Heidelberg (2005)

[13] Troelstra, A.S., Schwichtenberg, H.: Basic Proof Theory, 2nd edn. Cambridge Tracts in Computer Science, vol. 43. Cambridge University Press, Cambridge (2000)

AISC Meets Natural Typography

James H. Davenport

Department of Computer Science
University of Bath, Bath BA2 7AY
United Kingdom
J.H.Davenport@bath.ac.uk

Abstract. McDermott [12,13] introduced the concept "Artificial Intelligence meets Natural Stupidity". In this paper, we explore how Artificial Intelligence and Symbolic Computation can meet Natural Typography, and how the conventions for expressing mathematics that humans understand can cause us difficulties when designing mechanised systems.

1   Introduction

Notation exists to be abused[1]

> the abuses of language without which any mathematical text threatens to become pedantic and even unreadable. [3, pp. viii–ix]

but some abuse is more harmful than others, and may cause real problems in a mechanised context, or even to unwary human beings.

"Semantics", in a general context, has been defined as [14, 'semantic']

> Also, (the study or analysis of) the relationships between linguistic symbols and their meanings.

In the same vein, "Notation" has been defined as [14, 'notation' 6] (which goes on to give special meanings in mathematics, music, choreography and "in other disciplines, as chemistry, logic, chess, linguistics, etc.")

> The process or method of representing numbers, quantities, relations, etc., by a set or system of signs or symbols, for the purpose of record or analysis; (hence) any such system of signs or symbols.

Another way of looking at this paper is to ask how the semantics relate to the notation.

2   The Trivial Differences

While occasionally embarrassing, these are cases due to a difference in conventions and, at least in theory, could be avoided by a "sufficiently clever" context mechanism. Many of them are discussed in more detail in [8].

---

[1] On 10.6.2007, a quick use of Google demonstrated 783 uses of "abus de notation", roughly 10% of which were in english-language papers.

S. Autexier et al. (Eds.): AISC/Calculemus/MKM 2008, LNAI 5144, pp. 53–60, 2008.

Intervals. There are two well-known notations: the "Anglo-saxon" way $(0, 1]$ and the "French" way $]0, 1]$. The semantics are clear, though: in OpenMath they would be

```
<OMA>
  <OMS name="interval_oc" cd="interval1''"/>
  <OMI>0</OMI>
  <OMI>1</OMI>
</OMA>
```

Inverse functions. If f is a many–one function $\mathbf{C} \to \mathbf{C}$, its inverse[2], which will be denoted g, has two possible definitions: the one–one discontinuous one, and the one–many continuous one. It is usual in Anglo-saxon cultures to denote a[3] one–one function with a lower-case initial letter, as g, and the one–many one with an upper-case initial letter, as G. Regrettably, in France the convention is apparently reversed[4]. Here the situation is worse than in the previous example: the notations are not merely different but contradictory, and any attempt at understanding them will need to know the (linguistic, in this case) context. Attempting to understand precisely *which* one-to-one function is intended seems futile, and we also note (with regret) that there is no standard notation for distinguishing between functions which differ only in their branch cuts: the author and his colleagues have generally resorted to *ad hoc* subscripts or notation such as $\underbrace{\arctan}_{\text{Derive}}$. In terms of [13, p. 150], attempting to understand this mathematically is an example of the "unnatural language" fallacy: there is no internal way of deducing which function is meant (and indeed in *some* circumstances, *some* choices of the one-to-one function may not matter).

Metric tensor. It is possible to define the metric tensor for flat Minkowski space as $\begin{pmatrix} -1 & 0 & 0 & 0 \\ 0 & 1 & 0 & 0 \\ 0 & 0 & 1 & 0 \\ 0 & 0 & 0 & 1 \end{pmatrix}$ or its negative $\begin{pmatrix} 1 & 0 & 0 & 0 \\ 0 & -1 & 0 & 0 \\ 0 & 0 & -1 & 0 \\ 0 & 0 & 0 & -1 \end{pmatrix}$. Furthermore, one can

---

[2] We use a different letter, to avoid the usual problem of iterated functions versus inverse functions.

[3] It would be tempting, but wrong, to write "the one-one function". Since it is 'obvious' that the correct inverse of $x \mapsto x^2$ as $\mathbf{R} \to \mathbf{R}$ is the positive square root, we may be tempted to think there is an obvious inverse in other circumstances. While it is normal these days to define log to have imaginary part in $(-\pi, \pi]$, the author was initially taught to have the imaginary part in $[0, 2\pi)$. [1] changed the branch cut of arctan between printings, and systems have been known to be internally inconsistent [5].

[4] Various mathematical textbooks seem to indicate this. However [2, Arcsin] gives capitals to Arcsin, Arccos and Arctan, but not to the others. There is clearly an inconsistency here, as [2, Arctan] describes arctan as the inverse function, and makes no mention of Arctan. The other inverse functions seem to have no entries in [2].

decide that the temporal variable is the last, rather than the first, co-ordinate, giving $\begin{pmatrix} 1 & 0 & 0 & 0 \\ 0 & 1 & 0 & 0 \\ 0 & 0 & 1 & 0 \\ 0 & 0 & 0 & -1 \end{pmatrix}$ or its negative. Again, a human being generally has little difficulty with this, but it is hard to explain exactly why.

i **or** j This divergence of notation between electrical engineers and the rest of the world has been discussed before [7]: we merely note that we need a 'discipline' context as well as a 'linguistic' context to resolve ambiguities.

$0 \in \mathbf{N}$? This is discussed in [8]: we say here only that knowledge of linguistic context *may* help decide this question, but is far from certain.

3 Deep Semantics

Attempts to formalise mathematics often say "this has the usual mathematical meaning", or words to that effect. Let us look at **union** in this respect. There are three possible mathematical expressions, for which we give the MathML(-Content) and the OpenMath.

1. *n*-ary infix operator.
   Mathematics $a_1 \cup a_2 \cup a_3$
   LaTeX `a_1 \cup a_2 \cup a_3`
   OpenMath `<OMS name="union" cd="set1"/>`
   MathML `<apply> <union/> <i>a`$_1$`</i>...</apply>`
2. Acting on a set of arguments[5].
   Mathematics $\bigcup\{a_1, a_2, a_3\}$
   LaTeX `\bigcup \{a_1,a_2,a_3\}`
   OpenMath `<OMS name="big_union" cd="set3"/>`
   or `<OMS name="apply_to_list" cd="fns2"/>`
   MathML `<apply> <union/> <bvar>i</bvar> <domain ...> <set><i>`
   a_1 `</i>...</set>`
3. Iterating over a sequence.
   Mathematics $\bigcup_{i=1}^{3} a_i$
   LaTeX `\bigcup_{i=1}^3 a_i`
   OpenMath `big_union` on `make_list`
   MathML `<apply> <union/> <bvar>i</bvar> <lowlimit>...`

It could be argued that the OpenMath is trying to mimic the LaTeX too closely, and that MathML has the right idea, that there is only one concept of 'union'. This seems to the author as being analogous to the "wishful mnemonics" issue of [13]. $\cup$ and $\bigcup$ do *not* mean the same thing: we should note that $\cup\{\{a\}, \{b\}\} = \{\{a\}, \{b\}\}$, while $\bigcup\{\{a\}, \{b\}\} = \{a, b\}$.

In the analogy of [7], $\bigcup$ is a different part of the conjugation of $\cup$, and mathematics has more strikingly irregular verbs, so that Σ plays the same rôle to $+$ as $\bigcup$ does to $\cup$, and indeed as $\prod$ does to juxtaposition, or MathML's `&InvisibleTimes;`.

---

[5] And therefore clearly associative and commutative. Not quite so obviously however, it should also be idempotent, since $\{a, a\} = \{a\}$. This may explain why $\bigcup\{a_1, a_2, a_3\}$ looks natural, but $\Sigma\{a_1, a_2, a_3\}$ does not.

4 Plus or Minus

This is familiar to us all from the solution to the quadratic:

$$\frac{-b \pm \sqrt{b^2 - 4ac}}{2a}, \tag{1}$$

which can be seen as shorthand for

$$\left\{ \frac{-b - \sqrt{b^2 - 4ac}}{2a}, \frac{-b + \sqrt{b^2 - 4ac}}{2a} \right\}. \tag{2}$$

We are prepared to accept it in formulae such as [1, Equation 4.3.38]

$$\tan z_1 \pm \tan z_2 = \frac{\sin(z_1 \pm z_2)}{\cos z_1 \cos z_2}, \tag{3}$$

which we read as shorthand for two equations:

$$\tan z_1 - \tan z_2 = \frac{\sin(z_1 - z_2)}{\cos z_1 \cos z_2}$$

$$\tan z_1 + \tan z_2 = \frac{\sin(z_1 + z_2)}{\cos z_1 \cos z_2},$$

and the same is true of

$$\mathrm{Arctan}(z_1) \pm \mathrm{Arctan}(z_2) = \mathrm{Arctan}\left(\frac{z_1 \pm z_2}{1 \mp z_1 z_2} \right), \tag{4}$$

as meaning

$$\mathrm{Arctan}(z_1) + \mathrm{Arctan}(z_2) = \mathrm{Arctan}\left(\frac{z_1 + z_2}{1 - z_1 z_2} \right) \tag{5}$$

and

$$\mathrm{Arctan}(z_1) - \mathrm{Arctan}(z_2) = \mathrm{Arctan}\left(\frac{z_1 - z_2}{1 + z_1 z_2} \right). \tag{6}$$

But what of [1, Equations 4.6.26,27]

$$\mathrm{Arcsinh}\, z_1 \pm \mathrm{Arcsinh}\, z_2 = \mathrm{Arcsinh}\left(z_1 \sqrt{1 - z_2^2} \pm z_2 \sqrt{1 - z_1^2} \right) \tag{7}$$

$$\mathrm{Arccosh}\, z_1 \pm \mathrm{Arccosh}\, z_2 = \mathrm{Arccosh}\left(z_1 z_2 \pm \sqrt{(z_1^2 - 1)(z_2^2 - 1)} \right)? \tag{8}$$

As explained in [6], these have no such meaning, but are rather glosses on more complicated inclusions of the form $A = B \cup C$ or $A \subset B \cup C$ where A, B and C are multivalued expressions. In particular the $\pm$ on the left-hand side of (8) is redundant, since $\mathrm{Arccosh}(z) = -\mathrm{Arccosh}(z)$. (7) really means

$$\mathrm{Arcsinh}\, z_1 + \mathrm{Arcsinh}\, z_2 \subset \mathrm{Arcsinh}\left(z_1 \sqrt{1 - z_2^2} + z_2 \sqrt{1 - z_1^2} \right) \cup$$

$$\mathrm{Arcsinh}\left(z_1 \sqrt{1 - z_2^2} - z_2 \sqrt{1 - z_1^2} \right),$$

and the fact that the same equation holds for Arcsinh z_1 − Arcsinh z_2.

We are forced to conclude that ± has *no* definite meaning,

5   Pq

Chapter 16 of [1] is devoted to the elliptic functions such as sn, an area which has probably engendered more notational disputes and confusion as any other. [1, equation 16.25.1] defines

$$\mathrm{Pq}(u) = \int_0^u \mathrm{pq}^2(t)\mathrm{d}t \tag{9}$$

(where $\mathrm{pq}^2(t)$ means $\mathrm{pq}(t)^2$, and most certainly not $p \cdot q^2$ — see the next section). This is, of course, in defiance of (either of) the conventions of Arctan, but we are dealing with elliptic functions, not elementary ones. However, the joker here is that equation (9) applies whenever p and q are any of the letters s,c,n,d (note the order, which is traditional in the subject, and the implied assumption that $p \neq q$). Hence this equation is in fact shorthand for twelve equations of the form

$$\mathrm{Sn}(u) = \int_0^u \mathrm{sn}^2(t)\mathrm{d}t, \tag{10}$$

except that, when q is s, equation (9) should be read as

$$\mathrm{Pq}(u) = \int_0^u \left(\mathrm{pq}^2(t) - \frac{1}{t^2}\right)\mathrm{d}t - \frac{1}{u}, \tag{11}$$

where the changes are to remove the removable singularity at $t = 0$.

A similar equation, but this time with explanation, can be seen as

$$\mathrm{pq}(u) = \frac{\mathrm{pr}(u)}{\mathrm{qr}(u)} \qquad\qquad (\text{[1, Equation 16.3.4]})$$

(except that here there is no distinctness assumption, but pp is to be taken as the constant function 1).

To quote [1, coda to section 16.27]

> There is a bewildering variety of notations ... so that in consulting books caution should be used.

As an example of this, or showing that not all apparent misprints are such, we can see [1, Equation 17.2.8–10]

$$E(u|m) = \int_0^x (1 - t^2)^{-1/2}(1 - mt^2)^{1/2}\mathrm{d}t = \int_0^u \mathrm{dn}^2(w)\mathrm{d}w. \tag{12}$$

Does this tell us what $\mathrm{Dn}(u)$ is — indeed [1, Equation 16.26.3] has $\mathrm{Dn}(u) = E(u)$. However, the 'x' in equation (12) is not a misprint, and in fact [1, Equation 17.2.2] $x = \mathrm{sn}\, u$. So in Maple-speak

```
EllipticE(u,m)=int(sqrt((1-m*t^2)/(1-t^2)),t=0..JacobiSN(u,m))
              =int(JacobiDN(t,m)^2,t=0..u).
```

Quite how this is to be reconciled with [10, Equation 5.138(3)] —

$$\int \mathrm{dn}^2(u) = E(\mathrm{am}\,u, k)$$

— is not clear ($\mathrm{dn}(u)$ is really $\mathrm{dn}(u, m)$ of course, and $m = k^2$ here, and indeed throughout the theory, to the point where it appears to be improper to use any other letter).

6   The Meaning of Juxtaposition

Juxtaposition is a well-known trick notation in mathematics. It is normally believed to have two meanings (the first two listed below), but in fact has more. MathML-Presentation writes [4, 3.2.5.5] as follows.

> Certain operators that are "invisible" in traditional mathematical notation should be represented using specific entity references within mo elements, rather than simply by nothing.

Multiplication. A typical example would be ab, which could otherwise be rendered as $a \cdot b$. This is correctly encoded as ⁢ in MathML. We should note that this only applies to italic letters, juxtaposed roman letters are deemed to constitute a single lexeme, as in sin or pq (see (9)).

(Function) Application. A typical example would be $\sin x$, which could otherwise be rendered as $\sin(x)$, though even in this case there is ambiguity, since $\sin(x + y)$ is different from $2(x + y)$, and $f(x + y)$ is harder to understand. This is correctly encoded as ⁡ in MathML.

Concatenation. A typical example would be m_{12}, which could otherwise be rendered as $m_{1,2}$. This is correctly encoded as ⁣ in MathML. Even without this, MathML is less ambiguous than ordinary notation: m_{12} might equally be the twelfth item of a vector, but MathML would distinguish between

```
<msub>
  <mi> m </mi>
  <mrow>
    <mn> 1 </mn>
    <mn> 2 </mn>
  </mrow>
</msub>
```

and

```
<msub>
  <mi> m </mi>
```

```
  <mrow>
    <mn> 12 </mn>
  </mrow>
</msub>
```

(of course, the `<mrow>` is redundant in the latter case.

Addition. A typical example would be $4\frac{1}{2}$, which could otherwise be rendered as $4 + \frac{1}{2}$. This is correctly encoded as `&InvisiblePlus;` in MathML-3[6].

Summation. A typical example would be $a^i b_i$, which could otherwise be rendered as $\sum_i a^i b_i$. This is the summation convention, also called Einstein notation after its introducer [9, p. 781]. It has no MathML counterpart, nor would it be easy to see how to add one. This is, of course, not so much juxtaposition in the strict sense as 'proximity'.

7 Letters and Fonts

We have already seen that font can make a difference in what a compiler-writer would think of as the *lexing* of mathematics: thus 'pq' is a single token, whereas 'pq' is "p juxtaposed with q", which might become p `&InvisibleTimes;` q. Such lexing is already present in MathML(-Presentation) and (properly written) LaTeX: it only becomes an issue in areas such as the OCR of existing mathematics.

It is common to use fonts for semantic purposes, e.g. x might be the length of the vector **x** etc. Such conventions tend to be explained (in natural language) at the start of papers, but present a real problem to a parser, which would essentially have to convert x into $|\mathbf{x}|$ for internal purposes. Some authors also use change of case this way, as with $a = \det(A)$ etc. Fortunately for us, few authors go as far as [15]:

> Throughout this course, upper-case roman letters denote fields, and lower-case roman letters elements of the corresponding fields. Upper case fraktur letters denote the corresponding Galois groups, and lower case fraktur letters denote elements of the corresponding Galois groups.

8 Conclusion

We conclude that trying to make formal sense of natural typography can be *helped* by knowledge of both linguistic and discipline context, but that in many cases there is no obvious road to understanding questions such as whether $0 \in \mathbf{N}$ (the author probably knows), which branch cuts are intended (one hopes the author knows), or what is intended by $\pm$.

In the case of section 5 we clearly have a case of meta-notation. It is the author's contention that, in many cases (possibly even all), $\pm$ is really also

---

[6] Apparently added after several comments by this author.

metanotation, meaning "as appropriate with different choices of the signs". If the reader has a better theory, let it be explained!

Less this seems too pessimistic, we should state that, within a *given* corpus, it seems to be possible to do far better: authors generally do not change their minds wilfully during a paper ([11] is an unfortunate counter-example, dipping in and out of the summation convention several times in the course of one paper).

Acknowledgements. The author is grateful to many colleagues for their comments and suggestions. One of the referees pointed out the metric tensor, and the referees made many useful suggestions.

References

1. Abramowitz, M., Stegun, I.: Handbook of Mathematical Functions with Formulas, Graphs, and Mathematical Tables. US Government Printing Office (1964); (We quote from the tenth printing (1970): there are subtle, but occasionally significant, changes between printings)
2. Anonymous. Wikipedia, Français (2007), http://fr.wikipedia.org
3. Bourbaki, N.: Eléments de Mathématiques: Algèbre. C.C.L.S., Paris (1970)
4. World-Wide Web Consortium. Mathematical Markup Language (MathML) Version 3.0. W3C Working Draft (2008),
   http://www.w3.org/TR/2007/WD-MathML3-20080409
5. Corless, R.M., Davenport, J.H., Jeffrey, D.J., Watt, S.M.: According to Abramowitz and Stegun. SIGSAM Bulletin 2 34, 58–65 (2000)
6. Davenport, J.H.: MKM from book to computer: a case study. In: Proceedings Mathematical Knowledge Management 2003, pp. 17–29 (2003)
7. Davenport, J.H.: OpenMath in a Semantic Web,
   http://www.jem-thematic.net/node/592
8. Davenport, J.H., Libbrecht, P.: The Freedom to Extend OpenMath and its Utility. Mathematics in Computer Science (to appear, 2008)
9. Einstein, A.: Die Grundlage der allgemeinen Relativitaetstheorie (The Foundation of the General Theory of Relativity). Annalen der Physik Fourth Ser. 49, 284–339 (1916)
10. Gradshteyn, I.S., Ryzhik, I.M.: Table of Integrals, Series and Products, Jeffrey A. (ed.) 5th edn. Academic Press, London (1994)
11. Lawley, D.N.: A General Method for Approximating to the Distribution of Likelihood Ratio Criteria. Biometrika 43, 295–303 (2003)
12. McDermott, D.: Artificial Intelligence Meets Natural Stupidity. SIGART Newsletter 57, 4–9 (1976)
13. McDermott, D.: Artificial Intelligence Meets Natural Stupidity. Mind Design, 143–160 (1981)
14. Oxford University Press. Oxford English Dictionary (2008),
    http://dictionary.oed.com/entrance.dtl
15. Roseblade, J.E.: Galois Theory. Camridge University Lectures, Michaelmas Term (1973)

The Monoids of Order Eight and Nine

Andreas Distler[1] and Tom Kelsey[2]

[1] School of Mathematics and Statistics, Mathematical Institute,
North Haugh, St Andrews, KY16 9SS, UK
andreas@mcs.st-and.ac.uk
[2] School of Computer Science, Jack Cole Building,
North Haugh, St Andrews, KY16 9SX, UK
tom@cs.st-and.ac.uk

Abstract. We describe the use of symbolic algebraic computation allied with AI search techniques, applied to the problem of the identification, enumeration and storage of all monoids of order 9 or less. Our approach is novel, using computer algebra to break symmetry and constraint satisfaction search to find candidate solutions. We present new results in algebraic combinatorics: up to isomorphism and anti-isomorphism, there are 858,977 monoids of order 8 and 1,844,075,697 monoids of order 9.

1 Introduction

The aim of this paper is to find all solutions to a class of problems in algebraic combinatorics. This is a well-known research area: it is natural when discussing, for example, various types of latin squares to try to resolve the question of how many of each type exist. As well as obtaining the correct answer in terms of number of solutions, we aim to **store** each solution so that they can be analysed in terms of their structure by algebraists. This second aim means that we are not searching for a purely constructive solution to the enumeration problem; we generate and store a canonical example from each equivalence class of solutions.

The On-Line Encyclopedia of Integer Sequences [1] contains numerous examples of known initial sequences of enumerations of algebraic and combinatoric structures. The sequence that this paper extends is A058133: the numbers of monoids of order n, considered to be equivalent when they are isomorphic or anti-isomorphic. Currently values for $n \leq 7$ have been published.

Definition 1. *A* monoid *is an algebraic structure equipped with a closed and associative binary operator, and an identity element. More formally, a monoid is a tuple* $\langle S, *, e \rangle$ *where S is a set;* $* : S \times S \rightarrow S$ *satisfies* $x * (y * z) = (x * y) * z$ $\forall x, y, z \in S$*; and* $e \in S$ *satisfies* $x * e = x = e * x$ $\forall x \in S$.

If $\langle S, *, e \rangle$ *is a monoid, and* $|S| = n$*, then* $\langle S, *, e \rangle$ *has* order n.

Monoids can be thought of as semigroups having a multiplicative identity. Groups are special cases of monoids; each group element has a multiplicative inverse. Throughout this paper we consider only finite monoids, where S is a finite set.

S. Autexier et al. (Eds.): AISC/Calculemus/MKM 2008, LNAI 5144, pp. 61–76, 2008.

Definition 2. *Let* $\langle M_1, *_1, e_1 \rangle$ *and* $\langle M_2, *_2, e_2 \rangle$ *be two monoids of the same order. A bijection* $g : M_1 \to M_2$ *is an* isomorphism *if it respects the multiplication – i. e.* $(a *_1 b)^g = a^g *_2 b^g$ *– and an* anti-isomorphism *if it inverts the multiplication – i. e.* $(a *_1 b)^g = b^g *_2 a^g$. *If such a bijection exists the monoids are* isomorphic, *respectively* anti-isomorphic, *and are* equivalent *if either.*

In this paper we are interested in monoids up to equivalence. Thus we can choose the underlying set to be $\{1, 2, \ldots, n\}$. We then represent monoids by their multiplication tables, with rows and columns indexed from 1 up to n.

Table 1. Example multiplication table, and its image under $(1, 4)$

*	1 2 3 4
1	1 2 3 4
2	2 1 3 4
3	3 3 3 3
4	4 4 3 3

*	1 2 3 4
1	3 1 3 1
2	1 4 3 2
3	3 3 3 3
4	1 2 3 4

Isomorphism of monoids induces an action on tables. Given a permutation g of the members of S, we modify the table by permuting each row according to g, then each column, and finally permuting the values. An anti-isomorphism is the result of an isomorphism action followed by transposing the table. The effect of applying permutation $(1, 4)$ is shown in Table 1. Since permutations of a finite set, permutations of rows and columns of a multiplication table, and table transposition are all invertible, (anti-)isomorphism is a reflexive, symmetric and transitive relation on monoids, and is hence an equivalence relation.

In common with many algebraic enumeration problems, there is a combinatorial explosion as n increases. There are n choices for each of the n^2 positions in a multiplication table. For $n = 9$ and 10, the number of choices is approximately 1.5×10^{17} and 10^{20} respectively. This increase in problem size effectively rules out the obvious exhaustive search approach of generating each table, checking if the monoid axioms hold, then checking whether or not an (anti-)isomorphic version of the table has already been found.

Our approach is to develop algebraic results that allow us to devise algorithms for search-space reduction. We implement these algorithms in symbolic computational algebra; formulate the remaining problems in terms of constraints on solution tables; use advanced AI backtrack search techniques to find solutions; then (in some cases) use computational algebra to decide which canonical representive of (anti-)isomorphic equivalence class to accept as our unique solution.

In the remainder of this introduction we describe the computational algebra tools and techniques used, the basic principles of Constraint Satisfaction and the solver we use, and formalise notions regarding symmetry-breaking in Constraint Satisfaction Problems. In Section 2 we give a detailed derivation of the algebraic and AI search methods used, together with proofs of soundness and – where appropriate – completeness of the algorithms used. We summarise our results in Section 3, and provide concluding remarks and an indication of future avenues of research in Section 4.

1.1 GAP and Computational Algebra

Since our problem domain involves binary operators on finite sets, permutations, identity elements, action homomorphisms and symmetry groups, we use specialist software that provides robust, efficient and extensive implementations of algorithms in abstract algebra. GAP [2] (Groups, Algorithms and Programming) is a system for computational discrete algebra with particular emphasis on, but not restricted to, computational group theory. GAP provides a large library of functions which implement algebraic algorithms.

For our purposes, any advanced computational algebra system could be used. However, we rely heavily on efficient GAP code [3] that tests canonicity of an image of a set of points under the action of a permutation group. This allows isomorphic results to be eliminated efficiently.

1.2 Minion and Constraint Satisfaction

Definition 3. *A Constraint Satisfaction Problem (CSP) L is a set of constraints C acting on a finite set of variables $\Delta := \{A_1, A_2, \ldots, A_n\}$, each of which has a finite domain of possible values $D_i := D(A_i) \subseteq \Lambda$. A solution to L is an instantiation of all of the variables in Δ such that no constraint in C is violated.*

The class of CSPs is a generalisation of propositional satisfiability (SAT), and is therefore NP-complete. Solvers typically proceed by building a search tree, in which the nodes are assignments of values to variables and the edges lead to assignment choices for the next variable. If at any node a constraint is violated, then search backtracks. If a leaf is reached, then no constraints are violated, and the assignments provide a solution. Clearly these search trees are exponential, and for pathological cases each node may have to constructed. Heuristics exist for choices of variable and value for the next node, and again these need not lead to any reduction in search. The search tree can be pruned by enforcing levels of consistency: it is possible to check the effect of a variable-value instantiation on the domains of other variables. If such a check shows that a domain has become empty, it is safe to backtrack without exploring nodes that would otherwise be created. These checks have a computational cost, and the trade-off is between the effort of making checks – hopefully resulting in a pruned search tree – and the effort of searching a presumably larger tree with less expensive checks. The Handbook of Constraint Programming [4] provides details of CSP techniques.

A recent advance in Constraints is the "model and run" methodology, of users building constraint models and then executing them on a solver with few options. This methodology inspired the development of the constraint solver Minion [5]. A major feature of modern SAT solvers is their optimised use of modern computer architecture. Using this approach, Minion has been designed to minimise memory usage. The result of this is that Minion claims to offer *fast, scalable* constraint solving. Scalability as problem size increases is an important (and also neglected) factor in constraint solver construction. A key aim of our research is to test the claimed scalability of Minion.

1.3 Symmetry Breaking in CSPs

Constraint satisfaction problems (CSPs) are often highly symmetric. Given any solution, there can be others which are equivalent in terms of the underlying problem. Symmetries may be inherent in the problem, or be created in the process of representing the problem as a CSP. Without symmetry breaking (henceforth SB), many symmetrically equivalent solutions may be found and, in some ways more importantly, many symmetric equivalent parts of the search will be explored. An SB method aims to avoid both of these problems.

Permutation groups are the mathematical structures that best encapsulate symmetry. We describe the symmetries of a CSP as a permutation group of the literals (variable-value pairs) of the CSP, and obtain information regarding symmetric equivalence of search states from using GAP. Assignment of the form $(Var = val)$ are called *literals*, so a partial assignment is a conjunction of literals. We denote the set of all literals by χ, and adopt the convention of denoting variables by Roman capitals and values by lower case Greek letters.

Definition 4. *Given a CSP L, with a set of constraints C, and a set of literals χ, a symmetry of L is a bijection $f : \chi \to \chi$ such that a full assignment A of L satisfies all constraints in C if, and only if, $f(A)$ does.*

We denote the image of a literal $(X = \alpha)$ under a symmetry g by $(X = \alpha)^g$. The set of all symmetries of a CSP form a *group*: that is, they are a collection of bijections from the set of all literals to itself that is closed under composition of mappings and under inversion.

Definition 5. *Let G be a group acting on the set Ω. The stabiliser of an element $\omega \in \Omega$ is the set $g \in G$ such that $\omega^g = \omega$. This set is a subgroup of G. The orbit of an element $\omega \in \Omega$ is the set $\{\omega^g | g \in G\}$.*

The stabiliser of a literal $(X = \alpha)$ is the set of all symmetries in G that map $(X = \alpha)$ to itself. The orbit of a literal $(X = \alpha)$, denoted $(X = \alpha)^G$, is the set of all literals that can be mapped to $(X = \alpha)$ by a symmetry in G. That is

$$(X = \alpha)^G := \{(Y = \beta) : \exists g \in G \text{ s.t. } (Y = \beta)^g = (X = \alpha)\}.$$

Given a collection S of literals, the *pointwise* stabiliser of S is the subgroup of G which stabilises each element of S individually. The *setwise* stabiliser of S is the subgroup of G that consists of symmetries mapping the set S to itself.

There is a general technique, called "lex-leader", for generating constraints for any variable symmetry [6]. The idea is essentially simple: For each equivalence class of assignments under our symmetry group, we choose one to be canonical. We then add constraints before search starts which are satisfied only by canonical assignments. We generate canonical assignments by choosing an ordering of the variables and representing assignments as tuples under this variable ordering. Any permutation of variables g maps tuples to tuples, and the lexicographically least of these is our canonical assignment. This gives the set of constraints

$$\forall g \in G, \ V \preceq_{\text{lex}} V^g$$

where V is the vector of the variables of the CSP, $\preceq_{\mathrm{lex}}$ is the standard lexicographic ordering relation, defined by $AD\preceq_{\mathrm{lex}}BC$ iff either $A < B$ or $A = B$ and $D \leq C$, and V^g denotes the permutation of the variables by application of the group element.

Other SB techniques exist, but for this paper we will break symmetries either by lex-leader constraints, or by analysis of properties of monoids.

Definition 6. *Let S be a symmetry breaking technique for CSPs. S is sound if it does not rule out any valid solutions to a CSP. S is complete if it returns exactly one member of each equivalence class of solutions with respect to G.*

Obviously an unsound SB technique is worthless, but there is often a tradeoff in the computational costs involved with incomplete and complete techniques. In the event that no efficient SB method is available for certain symmetries, It may be desirable to break only a subset of the full group of symmetries of a problem. This leaves (partial) SB as a post process to be performed on the solutions obtained.

Remark 1. The above discussion applies to the situation where *all* solutions are required for a given CSP. If only the first solution (if any) is sought, then SB is not always an important consideration. In this paper we are always concerned with finding each symmetrically distinct solution to every CSP posed.

2 Methodology

Our underlying methodology is to use GAP to answer algebraic questions related to monoids. These answers allow us to generate suitable constraints for Minion programs, and eliminate (anti-)isomorphic solutions. In operational terms, GAP is the master process with Minion acting as a black box to provide solutions to carefully formulated CSPs. Our first task is to use the underlying algebra and symmetry of monoids to reduce the search space for the Minion programs.

There are two ways to achieve this. We can give restrictive constraints which will return an unchanged number of solution tables up to (anti-)isomorphism. Here we have to prove that for every monoid there is an equivalent one still in the search space. Secondly we can rule out a certain type of solution for which the number of equivalence classes is known. Here we have to show that the constraints remove all solutions of the specific type. We also have to provide the number of equivalence classes that we rule out, together with a proof where appropriate.

2.1 Reducing the Search Space 1: Identity Elements

The first simplification is to require that the first row and column of each table is the tuple $[1, 2, \ldots, n]$. This reduces the number of search variables from n^2 to $(n-1)^2$.

Proposition 1. *The above restriction is sound.*

Proof. Let M be a monoid on $\{1, 2, \ldots, n\}$ with identity $e \neq 1$. Then $M^{(1,e)}$ is a monoid isomorphic to M with 1 as identity. Thus it is sound to choose 1 to be the identity element in any solution. As $1 * x = x * 1 = x$ for all $x \in \{1, 2, \ldots, n\}$, the first row and column agree with our restriction, and hence a monoid equivalent to M is in the search space. □

The second simplification is that we can use existing results for semigroups of order $n - 1$ to obtain monoids of order n. The advantage of this approach is that the number of nonequivalent semigroups is known for $n = 1 \ldots 8$, as Integer Sequence A001423 in [1].

Proposition 2. *Let H be a semigroup with underlying set $\{2, \ldots, n\}$ and multiplication $*_H$. Define a multiplication $*$ on $\{1, \ldots, n\}$ by $1 * x = x * 1 = x$ $\forall x \in \{1, \ldots, n\}$, and $x *_H y$ otherwise. Then $\langle \{1, \ldots, n\}, *, 1 \rangle$ is a monoid.*

Proof. Element 1 is the required identity by definition of $*$. Consider the products $(a*b)*c$ and $a*(b*c)$. If any of a, b and c is 1, then $(a*b)*c = a*(b*c)$ immediately. Any product not involving 1 is associative, since semigroup multiplication is. Hence $\langle \{1, \ldots, n\}, *, 1 \rangle$ is a monoid. □

By construction, the table for any such monoid contains exactly one 1. Also, two non-equivalent semigroups will give two non-equivalent monoids.

Proposition 3. *Let $M = \langle \{1, \ldots, n\}, *, 1 \rangle$ be a monoid with fewer than two 1s in its table. Then $\{2, \ldots, n\}$ with multiplication $*$ forms a semigroup.*

Proof. Since $1 * 1 = 1$, the remaining values in the table for M are in $\{2, \ldots, n\}$. Therefore multiplication $*$ on $\{2, \ldots, n\}$ is closed and associative. □

Taken together, Propositions 2 and 3 show that the number of non-equivalent semigroups of order $n - 1$ is equal to the number of non-equivalent monoids having exactly one 1 in their table. It remains to search for tables of monoids that contain two or more 1s.

Remark 2. We have computed and stored all non-equivalent semigroups for order n up to 8 [7], using a similar combination of GAP and Minion. We do not report these calculations in detail, since the correct values have already been published.

2.2   Reducing the Search Space 2: Diagonals

To break more symmetries before the search we use an approach which was first introduced in computer search for semigroups [8] of order 6 (and subsequently used in the respective problem of order 8 [9]). The idea is to fix the diagonal entries first and consider no two equivalent diagonals. Observe that every diagonal entry is mapped to a diagonal entry under the action on the table described in Section 1. This yields an induced action on the diagonals and therefore induced equivalence classes of diagonals. As we made the restriction that 1 is to be the identity element of the monoid, the equivalence classes of diagonals will differ

---

Algorithm 1. Construct the connected digraphs with N vertices and a K cycle

---

Require: $K \leq N$ { cycle has not more than all vertices}
 1: $C \leftarrow \emptyset$
 2: **for all** p in PARTITIONS(N, K) **do** {the partition specifies the sizes of rooted trees at the vertices of the cycle}
 3:    $F \leftarrow$ FORESTS(p)
 4:    **for** $f \in F$ **do**
 5:       $\hat{f} \leftarrow$ set of all tuples of the elements of f
 6:       **for** *orbit* in ORBITS($C_K, \hat{f}$) **do** {the set of images forms orbits under the cyclic group $\langle (1, 2, \ldots, K) \rangle$}
 7:          *rep* $\leftarrow$ representative of *orbit* {arbitrary element in the orbit}
 8:          $D \leftarrow$ directed cycle of length K
 9:          **for** $i \in \{1, 2, \ldots, K\}$ **do**
10:             $D \leftarrow D$ merged with *rep*$_i$ joined at vertex i
11:          **end for**
12:          $C \leftarrow C \cap \{D\}$
13:       **end for**
14:    **end for**
15: **end for**
16: **return** C

---

from the ones in [8,9]. Our aim is to find or construct exactly one diagonal from each equivalence class.

As we describe in Section 3, this does not split the problem into more or less equivalent subproblems. It turns out that many diagonals give no solutions, and others very many. The aim of this approach, therefore, is not to parallelise the computation, but rather to safely reduce the number of diagonals to be tested from n^{n-1} to a more manageable number. Moreover, by fixing a diagonal we reduce the symmetry of the problem.

Proposition 4. *Let C be the family of directed, unlabelled graphs having $n - 1$ vertices, such that the outward degree of any vertex is less than or equal to 1 (and loops are allowed). Then C corresponds to the set of equivalence classes of diagonals with 1 in first position under the full symmetry group on $\{2, \ldots, n\}$.*

Proof. For any $c \in C$, every labelling of vertices from $\{2, \ldots, n\}$ leads to a diagonal in the following way: the first entry is 1, the entry in position $2 \leq k \leq n$ is the endpoint of the edge from vertex k, and 1 if no such edge exists. Two distinct labellings of c give two diagonals in the same equivalence class, since a re-labelling of c is simply a permutation of elements from $\{2, \ldots, n\}$, and hence is an element of the group acting on the diagonals. For any diagonal we can construct a labelled graph with vertices $\{2, \ldots, n\}$, and a directed edge from vertex $2 \leq k \leq n$ to the vertex given by the kth entry of the diagonal, unless the entry is 1. The unlabelled graph is in C.   □

Detailed information regarding the connected components of the members of C is given in [10, 3.4]. They can either be rooted trees with the direction of edges towards the root, or they can be a directed cycle (of length one or more) where every vertex in the cycle is the root of a tree. Algorithm 1 constructs and returns the latter. The former is constructed recursively using the one-one correspondence between rooted trees on n vertices, and forests of rooted trees on $n - 1$ vertices. All sets of forests whose tree sizes are specified by a partition p of $n - 1$ are returned by our function FORESTS(p).

2.3   Constraint Satisfaction Problems in Minion

We can now construct a CSP L for each diagonal D. The set of variables $\Delta :=$ $\{A_{1,1}, A_{1,2}, \ldots, A_{1,n}, A_{2,1}, \ldots, A_{2,n}, \ldots, A_{n,1}, \ldots, A_{n,n}\}$, consists of each entry in an $n \times n$ table, with each variable having domain $\{1, 2, \ldots, n\}$. The constraints are:

1. $a * (b * c) = (a * b) * c$ for all combinations of a, b and c;
2. the diagonal is fixed as D;
3. the first row and first column consist of $[1, 2, \ldots, n]$;
4. the table contains two or more 1 entries.

The first constraint is associativity, which in Minion is enforced using *element* constraints. The constraint *element*(*vector*, i, *val*) specifies that, in any solution, *vector*[i] = *val*. We add a new variable $A_{a,b,c}$ for each triple (a, b, c). The pair of constraints

$$element(row(a), b * c, A_{a,b,c}) \text{ and } element(column(c), a * b, A_{a,b,c})$$

then enforce associativity. Constraints 2 and 3 turn $n + 2(n - 1)$ of the n^2 variables into *ground variables*, having domain size exactly one. These implement the reduction in search space described in Sections 2.1 and 2.2, and reduce the symmetries remaining in each problem instance. Constraint 4 is a simple occurrence requirement.

Definition 7. *Define Y_D to be set of solutions to the CSP L defined by diagonal D, and Y_n to be the union of the Y_D for all D of length n. Let $\hat{Y}_D$ denote a set of representatives of non-equivalent solutions from Y; the union of these is $\hat{Y}_n$.*

2.4   Refinements and Optimisation

The CSPs described in Section 2.3 are sufficient to solve our identification and enumeration problem, up to any remaining symmetries. There are, however, a number of improvements that we can make. These involve further restricting the set of diagonals used, and imposing additional constraints for some of the remaining diagonals. We identify diagonals that cannot form part of a table with more than one 1 using the following proposition:

Proposition 5. *A monoid with table in Y_n either has an element $x \neq 1$ with $x^2 = 1$ or it contains a sequence of distinct elements $x_1, x_2, \ldots, x_k$ with $x_{i+1} = x^2_i$ for $i = 1, 2, \ldots, k - 1$, and $x_k^2 = x_1$.*

Proof. Let $\langle M, *, 1 \rangle$ be a monoid with table in Y_n . If M contains an element $x = 1$ with $x^2 = 1$ we are done. Otherwise there must be at least one pair of distinct elements $x_1, y_1 \in M \setminus \{1\}$ with $x_1 * y_1 = 1$. Define the two sequences of elements $(x_i)_{i=1,2,\ldots}, (y_i)_{i=1,2,\ldots}$ by

$$x_{i+1} = x_i^2 \text{ and } y_{i+1} = y_i^2, i = 1, 2, \ldots$$

Clearly $x_i * y_i = x_1^{2^{i-1}} * y_1^{2^{i-1}} = (x_1 * y_1)^{2^{i-1}} = 1$. As M contains only $n - 1$ elements not equal 1 there must be repetition in the sequence $(x_i)_{i=1,2,\ldots}$. Assume $x_{i+1} = x_i$ for some i. Then

$$1 = x_i * y_i = x_{i+1} * y_i = (x_i * x_i) * y_i = x_i * (x_i * y_i) = x_i * 1 = x_i$$

a contradiction. This shows that the period of the repetition is greater than 1 and completes the proof. $\qquad\square$

The conditions in Proposition 5 consider the squares of elements, i. e. the diagonal entries of the multiplication table. If diagonal D satisfies neither of the two conditions then Y_D will be empty, so we can exclude D.

For certain diagonals we can post *implied constraints*, which, although not strictly required, are likely to improve the performance of the solver. These are *all different* constraints which forbid equal values appearing in certain rows and columns. These constraints propagate very efficiently in CSP solvers, and often lead to early backtrack and hence pruning of the search tree.

Proposition 6. *Let a be an element of a monoid $\langle M, *, 1 \rangle$. If $a^k = 1$ for any k, then the values in row and column a of the table for M will be all different.*

Proof. Let $aM = \{a * m \mid m \in M\}$. The following inequalities hold in general

$$|a^k M| \leq |aM| \leq |M|.$$

By hypothesis, $a^k M = M$, so we have equality, and the values in row a of the table for M will be distinct. The case for right multiplication – hence column values – is similar. $\qquad\square$

From the diagonal we can compute $a^{2^{n-1}}$ for any element a. If this power equals 1, we add all different constraints on the row and column of a. In the event that diagonal D gives all different constraints on every row and column, then the solution set Y_D will consist entirely of groups. Since groups of small order are well known, we can safely reject D from our set of diagonals, provided that we add the number of groups not searched for to our final total. This number is obtained by inspection of the diagonals for groups.

2.5   (Anti-)Isomporph Rejection

We now address the problem of ruling out (anti-)isomporphs. There are two approaches available to us. We can either solve the Minion CSP instance, then

---

Algorithm 2. Enumerate and store Monoids of order n

Require: $n \leftarrow$ order of monoids
Require: $D \leftarrow$ set of inequivalent n-diagonals
Require: $S_N \times C_2 \leftarrow$ the symmetric group acting on n objects
1: **for** $d \in D$ **do**
2:      $stab \leftarrow$ the stabilizer of d in $S_n \times C_2$
3:      $P \leftarrow$ the Minion program for d
4:      **if** $stab$ is small **then**
5:          compute lex-leader constraints for d in GAP
6:          add these constraints to P
7:          obtain $\hat{Y}_d \leftarrow$ solutions of P from Minion
8:      **else**
9:          compute signature constraints for d in GAP
10:         add these constraints to P
11:         obtain $Y_d \leftarrow$ solutions of P from Minion
12:         $\hat{Y}_d \leftarrow$ (anti-)isomporph rejection of Y_d in GAP
13:     **end if**
14: **end for**
15: **return** $\hat{Y} = \cup_d \hat{Y}_d$

---

reject (anti-)isomorphs as a post process, or apply constraints which ensure that Minion only returns canonical solutions. Both methods involve GAP computation: the first requires an efficient minimal image test, the second requires the images of each literal $(X = \alpha)$ under the symmetry group of the CSP instance. Another key GAP calculation is the stabiliser of each diagonal in $S_n \times C_2$. This is the subgroup of $S_n \times C_2$ which fixes the diagonal entries of a table with respect to the isomorphism operation and transposition, and it represents the remaining symmetry to be broken. The minimum size of such a stabiliser is 2 (occurring whenever the diagonal permits no isomorphic solutions, but table transposition is still allowed), and the maximum size is $2(n-2)!$ (occurring when the diagonal consists of $[1, 1, 3, 4, \ldots, n-1, n]$, so that the only symmetry broken is the fixing of 1 at positions $(1,1)$ and $(2,2)$). We can apply lex-leader SB constraints for any diagonal, but each such constraint consists of a lexicographic requirement on two vectors – the first our canonical solution, the second its image under a group element – each containing n^3 Boolean variables (one for each literal). Posting a factorial number of such constraints is likely to slow Minion down; each constraint may be checked at each node in the search tree. However, most diagonal stabilisers are small, and posting lex-leader constraints is likely to be highly efficient.

Proposition 7. *Adding lex-leader SB constraints to a Minion CSP instance is both sound and complete.*

Proof. This is proved in [6].

The other approach, post-hoc isomorph rejection, is a simple concept. We store all solutions obtained by Minion, then check each solution to see if it is minimal in the stabiliser of the diagonal. If it is, we keep it; if not, we reject it. There are however some refinements and options to consider.

We define the *signature* of a solution table to be the n-tuple containing the number of occurrences of value k, in position k, for $k = 1, \ldots, n$. We use signatures to break more symmetries by posting constraints prior to search. We analyse the directed graph associated with each diagonal, as detailed in Section 2.2. There are two cases to consider. The first is to identify completely interchangeable vertices in rooted trees. We first look for roots of isomorphic trees, and then recurse down each tree. If we identify symmetric vertices, we post a linear ordering on the signature list of the labels of the vertices: we restrict the numbers of occurrences of these values in any solution. Each level in the rooted tree structure can provide zero or more such constraints. The second is to break any cyclic symmetry by fixing a minimal vertex in a cycle. Again, we recurse through the structure of the graph to identify any symmetry at a given level, posting linear ordering constraints whenever symmetries are found. We use the built in methods for Orbits and Stabilizers in GAP to find sets of equivalent values. At each level of the recursion we stabilise the vertices of the levels above.

Proposition 8. *Posting these ordering constraints on signatures is sound.*

Proof. If Y_D is empty there are no solutions to lose. Let $S \in Y_D$ be a solution of the original problem. If S violates a constraint on the first level then there is a symmetry $g \in Stab(D)$ – the stabiliser of the diagonal – such that S^g satisfies the constraint. Assume inductively that S satisfies all constraints up to level $l - 1$. If S violates a constraint on level l then there is a symmetry $g \in Stab(D)$ which fixes all the constraints of the levels above l such that S^g satisfies the constraint on level l. □

As examples, we first consider the diagonal $[1, 1, 3, 4, \ldots, n - 1, n]$. Its graph consists of 8 vertices labelled $2, \ldots, n$, with each vertex labelled 3 or higher forming a cycle of length one, with the empty rooted subtree. Since these vertices are indistinguishable when the labels are removed, it is safe to impose a linear order on the occurrences of values 3 through n in any solution table. We can then (anti-)isomorph reject in the stabilisers of the signatures of solutions: if g is a permutation with $k^g = l$ which maps solution S to solution T, then the signature of k has to equal the signature of l because of the linear ordering constraint posted. For an example that illustrates our extension of this simple linear ordering, consider diagonal $[1, 1, 3, 3, 4, 4, 3, 7, 7]$ which has graph with edges $4 \to 3, 7 \to 3, 5 \to 4, 6 \to 4, 8 \to 7, 9 \to 7$. We impose on the signatures the constraints $[4, 5, 6] \preceq_{\text{lex}} [7, 8, 9]$, $5 \leq 6$ and $8 \leq 9$. The first breaks the symmetry of the two equivalent sub-trees connected to vertex 3; the others break the symmetry in the equivalent leaves.

Posting signature constraints therefore has two advantages: we reduce the number of solutions returned by Minion, and we (anti-)isomorph reject in a smaller group. There is still the cost of computing the signature stabiliser, and this is non-trivial.

Table 2. Solutions & Timings

Method	n:	5	6	7	8	9
	Diagonals	27	81	242	699	2,026
Complete SB	Solutions	30	213	1,757	22,951	◇
	GAP cpu (s)	3	17	119	856	◇
	Minion cpu (s)	0	1	24	6,225	◇
	Total cpu (s)	3	18	143	7,081	◇
Isomorph rejection	Solutions	30	213	1,757	22,951	955,569
	GAP cpu (s)	2	13	96	989	197,587
	Minion cpu (s)	0	1	5	63	4,106
	Total cpu (s)	2	14	101	1,052	201,693
Combined 48	Solutions	30	213	1,757	22,951	955,569
	GAP cpu (s)	2	13	96	916	70,381
	Minion cpu (s)	0	1	5	68	3,150
	Total cpu (s)	2	14	101	984	73,531
Combined 240	Solutions	30	213	1,757	22,951	955,569
	GAP cpu (s)	2	13	96	720	24,386
	Minion cpu (s)	0	1	5	139	12,746
	Total cpu (s)	2	14	101	859	37,132

Legend: ◇ denotes timeout; *Combined 48* denotes complete SB only if the stabiliser of the diagonal in $S_n \times C_2$ has size ≤ 48 for $n = 9$, and ≤ 12 for $n = 5 \ldots 8$. *Combined 240* denotes complete SB only if the stabiliser of the diagonal in $S_n \times C_2$ has size ≤ 240 for $n > 8$, and ≤ 12 for $n = 5 \ldots 7$.

Our algorithm for (anti-)isomorph rejection is to order the returned Minion solutions lexicographically by signature, then only compute the stabiliser when the signature changes. This minimises the number of stabiliser calculations, but requires a potentially expensive sorting preprocess.

2.6   Enumeration and Storage of Monoids

Algorithm 2 describes our computational method. For each diagonal we generate a Minion instance that models associative multiplication tables having fixed first row, first column and diagonal values, and having at least two occurrences of value 1. The SB method used to break the remaining symmetries depends on the size of the stabiliser of the diagonal. A small stabiliser requires few complete SB constraints, whereas a large stabilizer indicates that signature constraints followed by (anti-)isomorph rejection may perform better. The definition of "small" is unclear *a priori* – we discuss suitable values obtained after experimentation in Section 3.

3   Results

Table 2 contains the timings for computations. We tested three approaches: only using SB constraints, only using post-solution (anti-)isomorph rejection, and the

Table 3. Pathological diagonals for $n = 9$

No.	DIAGONAL	STABILISER	$|Y_D|$	$|\hat{Y}_D|$
1	113456789	10,080	60,169	9,824
2	113333333	1,440	1,508,566	13,731
3	113344444	240	1,182,180	6,517
4	113335555	48	1,675,952	39,984
5	113333338	48	1,678,602	76,213
6	113333377	24	1,647,092	123,025
7	113333666	24	1,648,105	98,536
8	113344377	16	306,497	35,291
9	113343666	12	542,892	47,123
10	113335377	4	546,794	139,119

Legend: STABILISER is the size of the stabiliser of the diagonal in $S_n \times C_2$; $|Y_D|$ is the number of Minion solutions returned using signature constraints; $|\hat{Y}_D|$ is the number of solutions after (anti-)isomporph rejection.

combined approach set out in Algorithm 2. The first approach suffers badly in terms of Minion effort with increasing n. This is due to the large numbers of SB constraints posted for diagonals with large stabiliser. We were unable to obtain solutions for $n = 9$ in under two weeks, since Minion slowed markedly for certain diagonals. The number of diagonals shown is the number after applying the techniques described in Section 4; for $n = 9$ this reduced the number of diagonals from 2,598 to 2,026.

(Anti-)isomorph rejection is sufficiently efficient for $n = 9$. The combined approach works best, as expected. Smaller stabiliser diagonals mean faster Minion instances, with (anti-)isomorph rejection used when the stabiliser is too large. Both GAP and Minion times improve in the combined approach for $n = 9$. This is due to Minion returning fewer solutions for small stabiliser diagonals: the Minion search tree is pruned heavily by the SB constraints, and GAP has no (anti-)isomorph rejection to perform. We tested the combined approach with different cut-offs for the size of diagonal stabilizer, beyond which (anti-)isomporph rejection would be used.

The best trade-off between GAP and Minion computation was achieved when SB constraints were applied for diagonals having stabilisers with 240 or fewer elements. This roughly halved the total time of performing SB on stabilisers with 48 or fewer elements for $n = 9$. The optimal trade-off – with GAP performing roughly as much (anti-)isomporph rejection as Minion is performing complete symmetry-breaking – is not apparent before search starts.

Remark 3. Since GAP is the master process, the GAP times include all bookkeeping work such as generating Minion files, storing solutions, reporting statistics etc. This pads out the GAP times, but emphasises the speed of Minion.

The value for "small" used in practice was determined by analysis of results from (anti-)isomorph rejection. Table 3 lists several diagonals of order 9. Diagonal 1 is an obvious candidate for (anti-)isomorph rejection, since the stabiliser is large but only 60,169 solutions need to be tested for minimality. Diagonals 2 and 3 are a problem for both methods: the stabilisers are large and so are the numbers of solutions returned without SB constraints. This type of diagonal will make calculation for $n = 10$ much more difficult. However, diagonals such as 4 & 5 are better suited to SB constraints: we need only post 48 constraints to rule out over 1.5 million solutions. The remaining pathological diagonals require even fewer constraints. Hence, for $n = 9$ we first used SB constraints for stabilisers of size 48 and below. It transpires that – for this class of problems – using SB constraints for stabilisers of size 240 and below is a computational win, improving the balance between complete and incomplete symmetry-breaking. However, this does not appear to give an insight for the optimal choice of stabiliser size cutoff for higher order problems, since we need to know the size of Y_D and the size of the stabiliser before deciding upon an optimal strategy for obtaining the non-equivalent solutions $\hat{Y}_D$.

The numbers of monoids unique up to (anti-)isomorphism is given in Table 3. The values for $n = 8$ and $n = 9$ are new.

Table 4. Numbers of monoids up to (anti-)isomorphism

n	$|H_{n-1}|$	$|\hat{Y}_n|$	$|M_n|$
2	1	1	2
3	4	2	6
4	18	9	27
5	126	30	156
6	1,160	213	1,373
7	15,973	1,757	17,730
8	836,021	22,956	858,977
9	1,843,120,128	955,569	1,844,075,697

Legend: $|H_{n-1}|$ is the number of non-equivalent semigroups of order $n-1$; $|\hat{Y}_n|$ is the number of non-equivalent monoids with more than one 1 in their table; $|M_n|$ is the number of non-equivalent monoids, and is the sum of $|\hat{Y}_n|$ and $|\hat{Y}_n|$.

4   Discussion

We have analysed properties of monoids to obtain algorithms that permit the efficient enumeration and storage of monoids. Our implementation involves symbolic algebraic computation both as a pre-process and as a post-process. Our AI backtrack search tool, Minion, is fast – as claimed by its developers – but does not scale well on problems that grow with the factorial of the instance dimension. This is not surprising: there are no known solutions to the problem of combinatorial explosion.

As well as demonstrating the efficacy of combining symbolic computation with AI search, our results provide new numbers in algebraic combinatorics. We have stored each solution, and provide a library of monoids (and semigroups) [7] that can be accessed and analysed by the research community. We are confident in the accuracy of our results: we have used two different approaches to the same problem, and obtained identical answers in each case. In addition we have used similar algorithms to verify the enumeration of non-equivalent semigroups of order up to 8.

Future avenues of research involve solving similar problems having, as yet, unknown answers but being small enough to be tractable by our methods. In order to solve the monoid problem for $n = 10$, it is clear that our methods will work well for many diagonals. However, we have shown that pathological diagonals exist, for which neither complete symmetry breaking nor *post hoc* (anti-)isomorph rejection will be efficient. Moreover, the number of semigroups of order 9 is as yet unknown.

5   Revised Results and Conclusions

Since submitting this paper, we have re-run our calculations with a modified version of complete SB. When posting lex-leader constraints one has to choose a fixed order for the literals. It is known that this ordering plays a crucial role for the performance of the computation – although it is of no theoretical importance. Generally speaking, the ordering of the literals should agree with the search order of the constraint solver. This is to ensure that the violation of a SB constraint is discovered as early as possible during search. The implementation that provided the timings shown in Table 2 neglected this important fact. The timings for complete SB after correction of this mistake are given in Table 5. For these results we added a further optimisation, removing those literals from lex-ordering tests that cannot differ in a solution.

Table 5. Solutions & Revised Timings

Method	n:	5	6	7	8	9
	Diagonals	27	81	242	699	2,026
Revised Complete SB	Solutions	**30**	**213**	**1,757**	**22,951**	**955,569**
	GAP cpu (s)	2	13	91	686	10,557
	Minion cpu (s)	0	1	7	72	3,667
	Total cpu (s)	2	148	98	758	14,224

We can now revise our conclusions, based on these improved results. Minion can handle more symmetries than we envisaged without incurring a large computational penalty. The number of non-equivalent monoids of order less than 10 can be obtained using complete SB in a reasonable time. Our isomorph rejection method, as described in Section 2.5, is no longer crucial even for order 9. However, it remains an improvement for pathological diagonal no. 1, i.e. $[1, 1, 3, 4, 5, 6, 7, 8, 9]$, which has a stabiliser of size 10,080 in $S_9 \times C_2$.

Moreover, we are certain that the GAP times can be improved by implementing specialised stabiliser calculations for the action on tables and diagonals. This being the case, the problem of enumerating the monoids of order 10 is likely to be tractable using our methods.

Acknowledgements

Our work is supported by EPSRC grant EP/CS23229/1. We thank Ian Gent, Steve Linton, Victor Maltcev, James Mitchell and Nik Ruškuc for their help and comments.

References

1. Sloane, N.J.A.: The on-line encyclopedia of integer sequences (2008), http://www.research.att.com/~njas/sequences/Seis.html
2. GAP – Groups, Algorithms, and Programming, Version 4.4.10 (2007), http://www.gap-system.org
3. Linton, S.: Finding the smallest image of a set. In: ISSAC 2004: Proceedings of the 2004 International Symposium on Symbolic and Algebraic Computation, pp. 229–234. ACM, New York (2004)
4. Rossi, F., van Beek, P., Walsh, T. (eds.): The Handbook of Constraint Programming. Elsevier, Amsterdam (2006)
5. Gent, I.P., Jefferson, C., Miguel, I.: Minion: A fast scalable constraint solver. In: Brewka, G., Coradeschi, S., Perini, A., Traverso, P. (eds.) ECAI, pp. 98–102. IOS Press, Amsterdam (2006)
6. Crawford, J.M., Ginsberg, M.L., Luks, E.M., Roy, A.: Symmetry-breaking predicates for search problems. In: KR, pp. 148–159 (1996)
7. Distler, A., Mitchell, J.D.: smallsemi – a GAP package (2008), http://turnbull.mcs.st-and.ac.uk/~jamesm/smallsemi/
8. Plemmons, R.J.: There are 15973 semigroups of order 6. Math. Algorithms 2, 2–17 (1967)
9. Satoh, S., Yama, K., Tokizawa, M.: Semigroups of order 8. Semigroup Forum. 49, 7–29 (1994)
10. Harary, F., Palmer, E.M.: Graphical enumeration. Academic Press, New York (1973)

Extending Graphical Representations for Compact Closed Categories with Applications to Symbolic Quantum Computation

Lucas Dixon[1] and Ross Duncan[2]

[1] University of Edinburgh
l.dixon@ed.ac.uk
[2] University of Oxford
ross.duncan@comlab.ox.ac.uk

Abstract. Graph-based formalisms of quantum computation provide an abstract and symbolic way to represent and simulate computations. However, manual manipulation of such graphs is slow and error prone. We present a formalism, based on compact closed categories, that supports mechanised reasoning about such graphs. This gives a compositional account of graph rewriting that preserves the underlying categorical semantics. Using this representation, we describe a generic system with a fixed logical kernel that supports reasoning about models of compact closed category. A salient feature of the system is that it provides a formal and declarative account of derived results that can include 'ellipses'-style notation. We illustrate the framework by instantiating it for a graphical language of quantum computation and show how this can be used to perform symbolic computation.

Keywords: graph rewriting, quantum computing, categorical logic, interactive theorem proving, graphical calculi.

1 Introduction

Recent work in quantum computation has emphasised the use of graphical languages motivated by the underlying logical structure of quantum mechanics itself [1,15,3,5,6]. These techniques have a number of advantages over the conventional matrix-based approach to quantum mechanics:

- The visual representation abstracts over the values in the matrices. This removes detail that requires a lot of work for a human to interpret.
- Many properties have a natural graphical representation. For example, separability of quantum states can be inferred from disjoint subgraphs.
- The algebra of graphs generalises to domains other than vector spaces. In particular, it provides a representation for compact closed categories.

A major problem with these graphical representations is the lack of machinery for automating their manipulation. Unlike existing approaches to graph transformation, the graphs described here provide a representation of compact closed

S. Autexier et al. (Eds.): AISC/Calculemus/MKM 2008, LNAI 5144, pp. 77–92, 2008.

categories; hence we require a new approach to rewriting, which is sound with respect to the underlying semantics.

The main contribution of this paper is a graph-based formalism that is suitable for representing and evaluating quantum computations. We start by reviewing a formalism for graphical representations of compact closed categories. We also revisit a model of quantum computation based on this calculus. We then extend the graphical calculus in two significant ways driven by the need to express rules that could not be accounted for in the initial formalism. By combining these extensions, we provide a representation that captures an interesting and useful set of graph patterns. Notably, it can express the Spider Theorem which is normally written using informal ellipses notation (see Figure 4). We provide a semantics for our graph-based formalism in terms of the initial representation of compact closed categories.

Using our graph-based formalism as the representational foundation, we develop a simple logical framework for manipulating models of compact closed categories. This has a suitable rewriting mechanism where the axioms of the underlying object-formalism are expressed as equations between graphs. We then present a short case study that illustrates the framework by instantiating it for the introduced model of quantum computation. This shows how the framework can be used to symbolically perform simplifications of quantum programs as well as simulate computations.

2   Graphs and Compact Closed Categories

Graphs. A *directed graph*[1] consists of a 4-tuple (V, E, s, t) where V and E are sets, respectively of *vertices*[2] and *edges*, and s and t are maps

$$E \overset{s}{\underset{t}{\rightrightarrows}} V$$

which we call *source* and *target*. Let $\text{in}(v) := t^{-1}(v)$ and $\text{out}(v) := s^{-1}(v)$ denote the *incoming* and *outgoing* edges at a vertex v. The *degree* of a vertex v is $|\text{in}(v)| + |\text{out}(v)|$. To distinguish between elements of different graphs, we will use the subscript notation $G = (V_G, E_G, s_G, t_G)$.

Given graphs G and H, a *graph morphism* $f : G \to H$ consists of functions $f_E : E_G \to E_H$ and $f_V : V_G \to V_H$ such that:

$$s_H \circ f_E = f_V \circ s_G, \tag{1}$$
$$t_H \circ f_E = f_V \circ t_G. \tag{2}$$

These ensure that the structure of the graph is preserved by the morphism: an edge connected to a node gets mapped to a new edge that must be connected, in the same way, to the mapped node.

---

[1] Equivalently: a directed graph is a functor G from $\bullet \rightrightarrows \bullet$ to **Set**; a graph morphism is then a natural transformation $f : G \Rightarrow H$.

[2] We will use the words "vertex" and "node" interchangeably.

Definition 1. *Let $f : G \to H$ be a graph morphism and let $V' \subseteq V_G$. We say that f is strict for V' if $\forall e \in E_H$, if $s_H(e) \in f_V(V')$ or $t_H(e) \in f_V(V')$ then $\exists e' \in E_G$ such that $f_E(e') = e$.*

Strictness ensures that there are no additional edges connected to vertices in the image of V'.

Definition 2. *We call a graph morphism f an open embedding which is strict for V' if:*

1. *f_E is injective;*
2. *f_V restricted to V' is injective; and,*
3. *f is strict for V'.*

The intuition behind this definition is that the subgraph of G determined by V' should be preserved exactly by f; whereas other vertices may be identified and may contain additional incident edges.

Augmented by some additional structure, graphs form a *compact closed category*. In section 2.1 we will describe this structure, but first we review the basic properties of compact closed categories.

Compact Closed Categories

Definition 3. *A strict symmetric monoidal category [2] is called compact closed [10] when each object A has a chosen dual object A^*, and morphisms*

$$d_A : I \to A^* \otimes A \qquad e_A : A \otimes A^* \to I$$

where I is the tensor identity of the compact closed category, such that

$$A \cong A \otimes I \xrightarrow{\mathrm{id}_A \otimes d_A} A \otimes A^* \otimes A \xrightarrow{e_A \otimes \mathrm{id}_A} I \otimes A \cong A = \mathrm{id}_A \qquad (3)$$

$$A^* \cong I \otimes A^* \xrightarrow{d_A \otimes \mathrm{id}_{A^*}} A^* \otimes A \otimes A^* \xrightarrow{\mathrm{id}_{A^*} \otimes e_A} A^* \otimes I \cong A^* = \mathrm{id}_{A^*} \qquad (4)$$

Every arrow $f : A \to B$ in a compact closed category $\mathcal{C}$ has a *name* and *coname*:

$$\ulcorner f \urcorner : I \to A^* \otimes B, \qquad \llcorner f \lrcorner : A \otimes B^* \to I,$$

which are constructed as $\ulcorner f \urcorner = (\mathrm{id}_{A^*} \otimes f) \circ d_A$ and $\llcorner f \lrcorner = e_B \circ (f \otimes \mathrm{id}_{B^*})$. Hence there are natural isomorphisms $\mathcal{C}(A, B) \cong \mathcal{C}(I, A^* \otimes B) \cong \mathcal{C}(A \otimes B^*, I)$ making $\mathcal{C}$ monoidally closed[3]. Furthermore, f has a dual, $f^* : B^* \to A^*$, defined by

$$f^* = (\mathrm{id}_{A^*} \otimes e_B) \circ (\mathrm{id}_{A^*} \otimes f \otimes \mathrm{id}_{B^*}) \circ (d_A \otimes \mathrm{id}_{B^*})$$

By virtue of equations (3) and (4), $f^{**} = f$. Thus $(\cdot)^*$ lifts to an involutive functor $\mathcal{C}^{\mathrm{op}} \to \mathcal{C}$, making $\mathcal{C}$ equivalent to its opposite.

---

[3] In general compact closed categories are models of multiplicative linear logic where $A \multimap B$ is defined as $A^\perp \otimes B$.

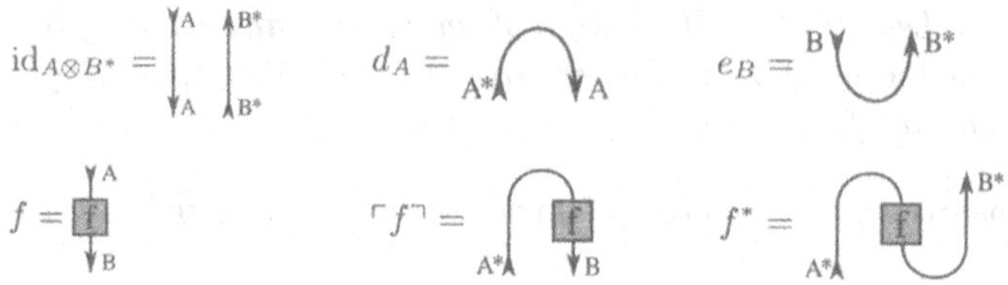

$$\mathrm{id}_{A \otimes B^*} = \quad\quad d_A = \quad\quad e_B =$$

Fig. 1. Compact Closed Structure as Graphs

2.1 Graph Representations for Compact Closed Categories

Graphs with certain additional structure give a representation for compact closed categories; we now give an overview of this construction. The details omitted here can be found in [8]. Pictorial representations are in Fig. 1. We make the convention that the domain of an arrow is at the top of the picture, and its codomain is at the bottom.

A *concrete graph* Γ is 5-tuple $(G, \mathrm{dom}\,\Gamma, \mathrm{cod}\,\Gamma, <_{\mathrm{in}(\cdot)}, <_{\mathrm{out}(\cdot)})$ where:

- $G = (V, E, s, t)$ is a graph;
- $\mathrm{dom}\,\Gamma$ and $\mathrm{cod}\,\Gamma$ are totally ordered disjoint sets of degree one vertices of G. The union of these sets is the *boundary* of Γ.
- $<_{\mathrm{in}(\cdot)}$ is a family of maps, indexed by V such that $<_{\mathrm{in}(v)}: \mathrm{in}(v) \xrightarrow{\cong} \mathbb{N}_k$ where $k = |\mathrm{in}(v)|$.
- $<_{\mathrm{out}(\cdot)}$ is a family of maps, indexed by V such that $<_{\mathrm{out}(v)}: \mathrm{out}(v) \xrightarrow{\cong} \mathbb{N}_{k'}$ where $k' = |\mathrm{out}(v)|$.

Since the sets $\mathrm{dom}\,\Gamma$ and $\mathrm{cod}\,\Gamma$ consist of vertices of degree one, we can assign a polarity to each one: $v \mapsto +$ if the edge incident at v is an incoming edge; $v \mapsto -$ otherwise. Hence $\mathrm{cod}\,\Gamma$ and $\mathrm{dom}\,\Gamma$ are *ordered signed sets*. Given any ordered signed set S we write S^* for the same ordered set with the opposite signing. Given two such sets we can define their disjoint union $R + S$ as the disjoint union of the underlying sets, inheriting the signing and the order from R and S, with the convention that $r < s$ for all $r \in R, s \in S$.

Proposition 1. *Concrete graphs form a compact closed category whose objects are ordered signed sets and whose arrows $f : A \to B$ are concrete graphs with $\mathrm{cod}\,f = B$ and $\mathrm{dom}\,f = A^*$.*

For each ordered signed set A, the identity map id_A has $\mathrm{dom}\,\mathrm{id}_A = A^*$ and $\mathrm{cod}\,\mathrm{id}_A = A$; its underlying graph has $E = A$ and $V = A^* + A$ with $t(a) = a$ and $s(a) = a^*$. Given a pair of concrete graphs $f : A \to B$ and $g : B \to C$ their composition $g \circ f : A \to C$ is constructed by merging the two graphs, erasing the vertices of $\mathrm{cod}\,f$ and $\mathrm{dom}\,g$ (called the *boundary vertices*), and identifying the edges previously incident at the deleted vertices. (Due to the opposite polarity of the domain and codomain the edges have compatible direction.) The tensor product on objects A, B is simply $A + B$; given $f : A \to B, g : C \to D$, the graph of $f \otimes g$ is the disjoint union of the graphs of f and g. The unit for the tensor is the empty set. The morphisms $d_A : I \rightarrow A^* \otimes A$, $e_A : A \otimes A^* \rightarrow I$ have the same underlying graph as id_A, but $\text{dom } d = \emptyset$, $\text{cod } d = A^* + A$, $\text{dom } e = A + A^*$ and $\text{cod } e = \emptyset$.

Remark 1. Given concrete graphs $f : A \rightarrow B$ and $g : B \rightarrow C$ there exists exactly one graph morphism $\tau : f \rightarrow (g \circ f)$ such that τ is an open embedding which is strict for the non-boundary nodes of f. Indeed the intuition behind an open embedding, is that any such map picks out a subgraph which forms a well defined arrow in its own right.

This category captures exactly the axioms for compact closed structure, in the sense that any freely generated compact closed category can be represented by concrete graphs. We will consider a collection of basic terms[4] F whose types are vectors of some set of basic types T. Then:

Definition 4. *A* T, F-*labelling* θ *for a concrete graph* Γ *is a pair of maps* $\theta_T :$ $E \rightarrow T$ *and* $\theta_F : (V - \text{cod } \Gamma - \text{dom } \Gamma) \rightarrow F$ *such that for each vertex* v, *if* $in(v) = \langle a_1, \ldots, a_n \rangle$ *and* $out(v) = \langle b_1, \ldots, b_m \rangle$ *then*

$$\theta v : \langle \theta a_1, \ldots, \theta a_n \rangle \rightarrow \langle \theta b_1, \ldots, \theta b_m \rangle$$

We say a concrete graph Γ *is* T, F-*labellable if there exists an* T, F-*labelling for it; and if* θ *is a labelling for* Γ, *then the pair* (Γ, θ) *is called a* T, F-*labelled graph.*

The T, F-labelled graphs form a compact closed category in the same way as the concrete graphs, subject to the further restriction that arrows are composable only when their labellings agree.

Theorem 1. *Let* C *be a compact closed category, freely generated by some set of arrows* F *and ground types* T; *then* C *is equivalent to the category of* T, F-*labelled graphs.*

Given a compact closed category C generated by some basic set of operations, the arrows of C have a canonical representation as labelled graphs. A consequence of the theorem is then that two arrows are equal by the equations of the compact closed structure if and only if their graph representations are equal.

As a final remark before moving on, note that the external structure of a vertex in a concrete graph is essentially the same as that of a complete graph; hence one can consistently view subgraphs as vertices, and abstract over the their internal structure.

3   Quantum Computations as Graphs

A substantial strand of work in quantum computation has involved the development of high-level models of quantum processes based on compact closed categories. In these formalisms, initiated in [1], quantum processes—such as quantum

---

[4] See [8] for a more thorough description of the nature of the terms.

logic gates, or the measurement of a qubit—correspond to arrows in a category, while the different quantum data types, usually just arrays of qubits, are the objects. The graphical representation described in the preceding section thus provides a very expressive notation for quantum processes.

While compact closed categories provide a suitable setting for reasoning about quantum computation, freely generated structure will not suffice: we need additional equations. These equations will be expressed as rewrites rules for graphs. In this section we will describe a set of generators and equations used to reason about quantum computation, and show how some of its formal properties lead to particular issues for the rewriting machinery.

Coecke and Duncan [4] propose a formal algebraic system for quantum computation built from the following collection of generators:

Objects: a single object, written Q.
Arrows: there are families of arrows X and Z:

$$\epsilon_Z : Q \to I, \qquad\qquad \epsilon_X : Q \to I,$$
$$\delta_Z : Q \to Q \otimes Q, \qquad\qquad \delta_X : Q \to Q \otimes Q,$$
$$\alpha_Z : Q \to Q \qquad\qquad \alpha_X : Q \to Q$$

where $\alpha \in [0, 2\pi)$, and in addition $H : Q \to Q$.

To each each arrow $f : A \to B$ we assign a formal adjoint $f^\dagger : B \to A$. Each arrow is represented as a small graph; its adjoint is the same graph written upside down by reflection in the x-axis. We use colours (light green and a darker red) to denote the two families:

The H arrow is denoted by and represents a Hadamard gate. The adjoints for H, α_X and α_Z will be defined equationally. The free compact closed category is then given by all graphs formed by composing and tensoring these basic graphs.

In terms of quantum processes, each edge in a graph represents a qubit, although several edges may represent the same physical qubit at different times. An edge may even represent a "virtual" qubit which stands for a correlation between different parts of the system. The maps δ_Z and ϵ_Z represent quantum operations which respectively copy and delete the eigenstates of the Pauli Z operator.[5] Notice that δ_Z has one edge in its domain representing the qubit to be copied, and two edges in its codomain for the two copies it produces. Similarly, ϵ_Z has one qubit as input and no outputs. The adjoints $\delta_Z^\dagger$ and $\epsilon_Z^\dagger$ correspond

---

[5] Uniform copying operations are forbidden by the no-cloning theorem [18], but such operations are possible if we demand only the eigenstates of some self-adjoint operator to be copied. Other states will not not copied. The same remarks hold true for erasing [11].

to an operation known as *fusion*, and to the operation of preparing a fresh qubit in a certain state. The α_Z corresponds to phase shift of angle α in the Z direction. The family of maps indexed by X are defined in exactly the same way, but relative to the Pauli X operator rather than the Z. All quantum operations may be defined by combining these simple operations—which are essentially classical—on two complementary observables.

We emphasise that this is a notation for representing quantum processes, rather than quantum states. In this setting a state is simply a process with no inputs; that is, a concrete graph with empty domain. Since our formalism is based on the underlying mathematical structure rather than any particular model of quantum computation, it is capable of representing quantum circuits, and measurement-based quantum computations, among other models. Indeed, an important application of this work is to show that states or computations implemented differently are equivalent.

To model the behaviour of quantum systems certain additional equations must be satisfied. At the level of objects, we ask that Q is self dual, i.e. that is $Q = Q^*$. Hence we use *undirected* graphs. The equations between arrows are discussed in detail in [4]; we present them in graphical form in Figure 2.

Consider the equations from Figure 2 which involve only one colour: these allow the remarkable *spider theorem*, first noted in [6], to be proved:

Theorem 2 (Spider Theorem). *Let G be a connected graph generated from δ_Z, ϵ_Z, α_Z and their adjoints; then G is totally determined by the number of inputs, the number of outputs, and the sum modulo 2π of the αs which occur in it.*

Hence any connected subgraph involving nodes of only one colour may be collapsed to a single vertex, with a single value α, giving a "spider". Informally, this can be depicted graphically as the equation in Figure 4-left. Conversely, a spider may be arbitrarily divided into sub-spiders, provided the total in- and out-degree is preserved, along with the sum of the αs. Furthermore, one can derive, from the Spider Theorem, n-fold versions of many of the other equations.

Spiders offer a very intuitive way to manipulate graphs, and are far more compact and convenient in calculations than the graphs built up naively from the generators. However, formalising spiders requires moving from finite graphs, where each vertex has bounded degree, and which are subject to a finite number of rewrite rules, to a system where nodes may have arbitrarily many edges, and there are infinitely many concrete rewrite rules. The desire to retain intuitive reasoning methods for these infinite families of rewrites motivates the extension from concrete graphs to *graph patterns*, the main subject of this paper.

4 Graphs with Variable Nodes

In the concrete representation, the following graphs represent different computations:

Comonoid Laws

Isometry, Frobenius, and Compact Structure

Abelian Unitary Group

$$\phi := \phi = |\qquad \left(\alpha\right)^{\dagger} = -\alpha \qquad \begin{smallmatrix}\alpha\\\beta\end{smallmatrix} = (\alpha+\beta) = \begin{smallmatrix}\beta\\\alpha\end{smallmatrix}$$

Bilinearity

Bialgebra Laws Let ◆ := ○ ; then:

Group Actions

H Properties

$$\left(H\right)^{\dagger} = H \qquad \begin{smallmatrix}H\\H\end{smallmatrix} = |$$

Colour Duality

Fig. 2. Graphical Equations for Quantum Systems. The $(\cdot)^{\dagger}$ functor gives a vertical symmetry to the category, hence for every equation we have a second equation obtained by flipping the diagram upside down. In addition, we have a "colour duality": each equation shown here gives rise to second, which is obtained by exchanging the two colours. The colour duality is derivable from the equations involving H.

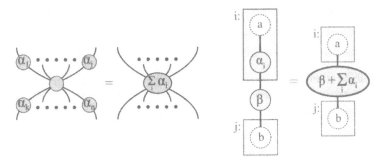

Fig. 3. [Left] An informal equation on graphs that expresses the Spider Theorem. [Right] The Spider Theorem expressed formally using graph patterns. The !-boxes are named i and j. The variable nodes are white and named a and b. The non-variable node data (the angle) is written inside the node when non-zero.

However, composition with semi-circles (d_Q and e_Q from Definition 3), on the co-domain or domain, allows an equation involving one of the above to easily lead to a derivation corresponding to either of the other two: given any one of the allows a trivial derivation of the others.

To address this, we formalise a representation that abstracts over the boundary nodes membership in the domain or co-domain. This gives rise to a *variable-node graphs*, in which boundary nodes have been generalised to *variable nodes*. The intuition of variable nodes is that they can replaced by concrete nodes in some graph in a process analogous to composition.

To formalise the semantics of variable-node graphs, we define *matching*, which captures the intuitive idea of a graph with variable nodes occurring within another graph:

Definition 5. *A variable-node (resp., concrete) graph G matches another graph H if there exists an open embedding $G \to H$ which is strict on the non-variable (resp., non-boundary) nodes. This open embedding is called a matching. The notation $G \leq_v H$ is used for G matches H.*

A graph with variable nodes, G, can be given a formal semantics by being interpreted as a set of concrete graphs, denoted by $[\![G]\!]_v$. The interpretation is simply the set of concrete graphs which the variable node graph matches.

Proposition 2. $G \leq_v H \Leftrightarrow [\![G]\!]_v \supseteq [\![H]\!]_v$.

Proof. The proof of the implication from left to right is a consequence of the fact that a composition of open embeddings is an open embedding. In particular, the embedding of $G \leq_v H$ composed with $[\![H]\!]_v$ is thus an open embedding of G into $[\![H]\!]_v$. Hence $[\![G]\!]_v \supseteq [\![H]\!]_v$. From right to left, we compose the open embedding $[\![G]\!]_v$, restricted to its subset $[\![H]\!]_v$, with the inverse of $[\![H]\!]_v$ to get $G \leq_v H$.

Fig. 4. An illustration of !-box graph matching using the !-box operations. This involves first copying !-box i_0 twice, then merging i_0 and i_1 and finally dropping i_3.

5   !-Boxes

If we wish formalise reasoning via spiders then an arbitrary number of repetitions of a subgraph need to be matched. To support this we introduce an operation, !-boxing (pronounced bang-boxing), on graph representations. Given a graph representation, this introduces a new notation by outlining a set of nodes. These nodes are said to be in a !-box. Intuitively, the resulting !-boxed graph can be thought of as representing a set of graphs with an arbitrary number of copies of the !-boxed nodes, where every copy connects, in the same way, to the nodes outside the !-box. The !-boxes can also be instantiated with zero copies. This erases all edges to and from the !-box.

More formally, a *!-box graph* is a pair $(G, \mathcal{B})$ where G is a graph and $\mathcal{B}$ is a set of disjoint subsets of V i.e. $b_1, b_2 \in \mathcal{B}$ then $b_1 \cap b_2 = \emptyset$.[6] To formalise the intuitive notion that a !-box represents arbitrary number of copies of the subgraph made from its nodes, we introduce *!-box matching*. This binary relation, written infix as $\leq_!$, is defined such that $(G, \mathcal{B}) \leq_! (H, \mathcal{C})$ whenever $(H, \mathcal{C})$ can be obtained from $(G, \mathcal{B})$ by a sequence the following operations:

copy: copies a !-box, b in a graph to produce a new graph with two copies of the !-boxed subgraph, b is the old one and b' is the new one. The set of !-boxes in the copied graph now also contains the new !-box b'. Any edges between a node, n, inside the !-box b, and a node, m, outside it, get copied so that there is a new edge from m to the new copy of n in b'.

drop: simply removes the !-box, but leaves its contents in the graph.

kill: removes from the graph all nodes in the !-box as well as any incident edges.

merge: combines two !-boxes, B_1 and B_2 into a single larger !-box $B_1 \cup B_2$. To ensure that copying after merging commutes with copying before, merging is restricted to !-boxes which do not have an edge between their nodes.

An example of matching with these operations is illustrated in Figure 5.

We give a formal semantics to !-box graphs by defining them in terms of a set of graphs in the underlying representation. In particular, we denote the interpretation of a !-box graph $(G, \mathcal{B})$ by $[\![(G, \mathcal{B})]\!]_!$ and say that its members are instances.

---

[6] One could consider more expressive notions of nested, or overlapping, node sets in the !-boxes. While such expressivity is interesting, it is not required for the system we formalise here.

Definition 6. *The* interpretation $[[(G, \mathcal{B})]]_!$ *is a set of concrete graphs defined by*

$$[[(G, \mathcal{B})]]_! = \{H \mid (G, \mathcal{B}) \leq_! (H, \emptyset)\}$$

i.e. the those graphs matched by the !-box graph that have no !-boxes.

Observe that every concrete instance of a !-box graph can be defined by pairing each !-box with the natural number that defines how many copies are made of it. Thus $[[G]]_!$ is isomorphic to the set of k-tuples of natural numbers, where k is the number of !-boxes. The need for the !-box matching operation, rather than using a direct k-tuple interpretation, is to allow matching between !-box graphs, and thus to provide a mechanism for derived rules.

Proposition 3. *!-Matching respects !-box semantics:* $G \leq_! H \Leftrightarrow [[G]]_! \supseteq [[H]]_!$. *The proof is a simple consequence from the definition of $[[G]]_!$ being a subset of the graphs that match G.*

For the purposes of this paper, the underlying graph representation of !-boxed graphs is variable-node graphs. This gives rise to *graph patterns* which we now discuss in more detail.

6   Graph Patterns

The representation of Compact Closed Categories as graphs, discussed in §2.1, is too restrictive for reasoning about quantum computation. In particular, graphical rules such the Spider Theorem, from Figure 4-left, are frequently needed, but not expressible. In this section we combine the *!-boxes* and *variable-node* extensions to graphs. We call this representation *graph patterns*. This forms a representation that allows us to express, in a finite way, certain infinite families of equations between concrete graphs. In particular, the Spider Theorem can now be represented as shown in Figure 4-right. We also extend the notion of matching for graph pattens. This provides the foundations for the rewriting machinery in §7 which can then be used to reason about quantum computation.

The semantics for a graph pattern G is a set of concrete graphs denoted by $[[G]]$ and define it as:

$$[[G]] = \{[[G']]_v \cdot G' \in [[G]]_!\}$$

this simply considers every interpretation of the !-boxes to give variable-node graphs for which we then appeal to their own semantics.

The specification for one pattern, G, to match another one, H is that it is more general with respect to the interpretation: $[[G]] \supseteq [[H]]$. However, graph patterns can correspond to a countably infinite number of concrete graphs. Thus matching between graph patterns cannot be implemented by simply unfolding all interpretations as concrete graphs and checking the membership relation.

Fortunately, it is quite easy to provide decidable matching: the size of the unfolding that needs to be considered can be bounded. The key observation is that a graph G_1 will never match a graph with fewer non-variable nodes.

Thus unfolding of G_1 can be bounded by the number of non-variable nodes in G_2. While this gives a generate and test style algorithm, it is not efficient. The intuition for an efficient algorithm is to search through one graph incrementally increasing the matched part.

7   Reasoning with Graph Patterns

In this section we describe how the graph pattern formalism can provide a *meta-level* framework for reasoning about models of compact closed categories. Following the terminology of logical frameworks such as Isabelle [12], we call the specification provided by the underlying model an *object-level* graph formalism. An object-level formalisation defines a set of rules which are treated as the axioms for the system; for instance, the equations from Figure 2. It also defines the data at the nodes and edges as well as corresponding data-matching behaviour. For its part, the meta-level provides generic machinery to manipulate graphs and derive new rules. We now describe the meta-level framework, noting the conditions for a rule to be valid, and prove the systems adequacy for rewriting. The resulting system forms the basis for an interactive proof assistant that supports reasoning about compact closed categories.

7.1   Equational Rules

In our framework, the axioms defined by an object-level model, as well as derived rules, are pairs of graph patterns. Such a pair represents the left and right hand sides of an equation. Rules are declarative in that they denote a set of concrete equational rules.

The intuitive idea of substitution with a rule is to replace a subgraph that matches the left hand side with the right hand side. However, not all pairs of rules make sense with respect to the underlying semantics. For an equation to be well defined with respect to the compact closed structure it must not be possible to change the type (the boundary nodes in the domain and co-domain) of a concrete graph by rewriting. Mapping this restriction back to pairs of graph patterns results in the following conditions on rules:

- There has to be a isomorphism between variable nodes in the left and right hand subgraphs. Given a matching against the left hand side of a rule, the target subgraph is replaced with the right hand side while keeping the same instantiations for the isomorphic variable nodes of the right hand side.
- Rules must also define a partial injective mapping between !-boxes on the left and right hand sides. The intuition for this mapping is that the unfolding used when matching a !-box on the left, is applied to the mapped !-box on the right before replacement.
- The interplay between !-boxes and variable nodes means that when a variable node appears within a !-box on one side of a rule, it must also appear under a mapped !-box on the other side.

Notationally, and implementationally, we annotate !-boxes and variable nodes in a graph with unique names. For example, Figure 4 shows the Spider Theorem. In this figure, the mapping between !-boxes is represented by !-boxes having the same name on the left and right of a rule. Similarly, the isomorphism between variable nodes is captured by the set of variable node names on the left and right hand side being equal.

7.2   Lifting Axioms and Adequacy

The axioms of an object formalism come from the semantics of the underlying system. For instance, the equations given in Figure 2 can be proved by matrix calculations in the underlying model. When such rules are expressed as graph patterns, we replace the concrete representation's boundary nodes with variable nodes. This operation is called *lifting*. An equation on graph patterns corresponds to an infinite (when their are variable nodes) set of equations between concrete graphs. Thus we might worry that the lifted equations express too much: they may allow rewrites which are not true.

We call the property that the lifted representation is a conservative extension of the initial theory *adequacy*. For models of compact closed categories, the proof of adequacy is quite simple: given an equation between concrete graphs, $G = H$, we observe that every instance of the lifted equation corresponds to the original equations composed with some graph. The graph is given by the unmapped subgraph using the open embedding from the lifted equations onto the considered instance. Thus if $G = H$ is true, then so is every instance of its lifting, and thus lifting produces an adequate representation.

7.3   Meta-level Logic and Derived Rules

Having defined what makes a valid rule, we now present the meta-logic of the framework. This is quite simple as it only involves dealing with object-level equations:

$$\frac{}{\Gamma \vdash A = A}\ \text{refl} \qquad \frac{\Gamma \vdash A = B}{\Gamma \vdash B = A}\ \text{sym} \qquad \frac{\Gamma \vdash A = B \qquad \Gamma \vdash C = D}{\Gamma \vdash C = (D[A/B])}\ \text{subst}$$

where Γ is the set of object-level axioms.

We assume that the axioms in Γ meet the validity conditions described earlier. These rules all preserve the validity conditions on equations and thus the system as a whole ensures only valid rules are derived. For the reflexivity rule (*refl*), we assumes that A is a well-formed pattern graph. This rule allows a new graph to be introduced. By then applying the *subst* rule, intermediate results are derived which can themselves be used to rewrite other rules and conjectures. In this way, the system allows derived rules to provide an abbreviation for a combination of steps.

A sets of rules can be applied automatically to simplify a graph or simulate computation in the object domain. For such rewriting to terminate, a suitable left-to-right ordering on rules needs to be observed, such as a decrease in the size of the subgraph. In §8 we illustrate simulating a quantum computation.

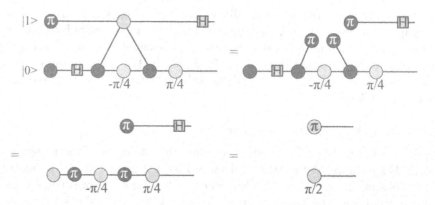

Fig. 5. An example computation of the Quantum Fourier Transform with inputs 1 and 0, performed symbolically by rewriting

8   A Case Study in Quantum Computation

The model of quantum computation introduced in §3 provides an object formalism for our meta-level framework. In particular, the object level axioms come from lifting the equations in Figure 2 and from the encoding of the Spider Theorem as shown in Figure 4.

Our model of quantum computation requires no data for the edges. The nodes on the other hand are either H (a Hadamard gate), with no additional data, or have an *angle* and a *colour*, expressing that they are defined in the X or Z basis. For their part, angles are expressed as rational numbers which correspond to the coefficient of π in the underlying matrix.

To allow composition of rules to compute the resulting angles we give the X and Z nodes an *angle expression*. When a node is within a !-box, the expression is a single *angle-variable* which gets instantiated to a new angle-variable in each of the unfoldings of the !-box. When a node is not within a !-box, the angle-expression is a mapping from a set of angle-variables to the corresponding rational coefficient. When an angle-expression contains an angle-variable within a !-box, this is interpreted as a sum of the variables that result from its unfolding. This rather simple expression language has a normal form by ordering the angle-variable by name. Matching then results in angle-variables being instantiated and the expressions in all affected nodes are then (re)normalised. An additional implementation detail must also be observed for the substitution rule: it must ensure that angle-variables in the rule being applied are distinct from those in the expression being rewritten.

The quantum Fourier transform is among the most important quantum algorithms, forming an essential part of Shor's algorithm [16], famous for providing polynomial factoring. In our graph pattern calculus this circuit becomes the top-left graph in Figure 8. This figure shows how computation can be symbolically performed by rewriting with the lifted equations from Figure 2 and the graph pattern version of the Spider Theorem.

9 Related Work

There are several foundational approaches to graph transformation, including algebraic approaches [7], node-label controlled [9], matrix based [17], and programmed graph replacement [14]. These provide general ways of understanding graph transformations which can then be implemented to provide machinery for a specific application. However, systems based on these theories do not provide machinery for the semantics of compact closed categories. Thus their notion of matching and replacement do not guarantee well-typed results. Furthermore, they do not provide machinery for rewriting of graphs with ellipses notation which is needed to represent the Spider Theorem.

The distinctive feature of our form of graph rewriting is that the graphs capture the structural properties of compact closed categories and rewriting is compositional: it preserves the type of the rewritten subgraph. However, our system can also be seen as an instantiation of a general graph rewriting system: matching provides the embedding information and the object-level node matching defines the application conditions and the attribute transfer function.

We note that our graphical notation has little connection to *graph states* as used in various approaches to measurement-based quantum computation [13]. In that approach the graph structure is used to provide a description of the entanglement in a state: it does not provide a complete description of a computation.

10 Conclusions and Further Work

We have introduced a representation for graphs which can formally characterise the ellipses notation used informally to represent certain infinite families of graph rewrites, such as the Spider Theorem. This representation provides the foundation for a simple meta-logic for reasoning about models of compact closed categories. We illustrated this by providing an account of quantum computation and showing how computation can be performed. Having developed the basic representational machinery and shown matching to be decidable, we are left with several exciting avenues for further research. The most immediate direction we are pursuing is to provide a full implementation - only a partial one is currently available[7]. Other areas of further work include considering confluence results for sets of rewrite rules, increasing the expressiveness of the representation for graph-patterns, and finding a complete set of rewrite rules for the considered model of quantum computation.

References

1. Abramsky, S., Coecke, B.: A categorical semantics of quantum protocols. In: LICS 2004, pp. 415–425. IEEE Computer Society, Los Alamitos (2004)
2. Asperti, A., Longo, G.: Categories, Types and Structures. MIT Press, Cambridge (1991)

---

[7] http://dream.inf.ed.ac.uk/projects/quantomatic

3. Coecke, B.: Kindergarten quantum mechanics. Lecture Notes (2005)
4. Coecke, B., Duncan, R.: Interacting quantum observables. In: Aceto, L., Damgård, I., Goldberg, L.A., Halldórsson, M.M., Ingólfsdóttir, A., Walukiewicz, I. (eds.) ICALP 2008. LNCS, vol. 5126, pp. 298–310. Springer, Heidelberg (2008)
5. Coecke, B., Paquette, E.O.: POVMs and Naimark's theorem without sums. In: Proc. of the 4th International Workshop on Quantum Programming Languages (2006)
6. Coecke, B., Pavlovic, D.: Quantum measurements without sums. In: The Mathematics of Quantum Computation and Technology. CRC Applied Mathematics & Nonlinear Science, Taylor and Francis (2007)
7. Corradini, A., Ehrig, H., Heckel, R., Korff, M., Löwe, M., Ribeiro, L., Wagner, A.: Algebraic approaches to graph transformation - part I: Single pushout approach and comparison with double pushout approach. In: Rozenberg, G. (ed.) Handbook of Graph Grammars and Computing by Graph Transformation. Foundations, vol. I, pp. 247–312. World Scientific, Singapore (1997)
8. Duncan, R.: Types for Quantum Computation. PhD thesis, Oxford University (2006)
9. Janssens, D., Rozenberg, G.: Graph grammars with node-label controlled rewriting and embedding. In: Proc. of the 2nd International Workshop on Graph-Grammars and Their Application to Computer Science, pp. 186–205. Springer, Heidelberg (1983)
10. Kelly, G.M., Laplaza, M.L.: Coherence for compact closed categories. Journal of Pure and Applied Algebra 19, 193–213 (1980)
11. Pati, A.K., Braunstein, S.L.: Impossibility of deleting an unknown quantum state. Nature 404, 164–165 (2000)
12. Paulson, L.C.: Isabelle: A generic theorem prover. Springer, Heidelberg (1994)
13. Raussendorf, R., Briegel, H.J.: A one-way quantum computer. Phys. Rev. Lett. 86, 5188–5191 (2001)
14. Schürr, A.: Programmed graph replacement systems, pp. 479–546. World Scientific Publishing Co, River Edge (1997)
15. Selinger, P.: Dagger compact closed categories and completely positive maps. In: Proc. of the 3rd International Workshop on Quantum Programming Languages (2005)
16. Shor, P.W.: Polynomial-time algorithms for prime factorization and discrete logarithms on a quantum computer. SIAM J.Sci.Statist.Comput. 26(5) (1997)
17. Velasco, P.P.P., de Lara, J.: Matrix approach to graph transformation: Matching and sequences. In: Corradini, A., Ehrig, H., Montanari, U., Ribeiro, L., Rozenberg, G. (eds.) ICGT 2006. LNCS, vol. 4178, pp. 122–137. Springer, Heidelberg (2006)
18. Wootters, W., Zurek, W.: A single quantum cannot be cloned. Nature 299, 802–803 (1982)

A Full First-Order Constraint Solver
for Decomposable Theories

Khalil Djelloul

Laboratoire d'Informatique Fondamentale d'Orléans
Bat. 3IA, rue Léonard de Vinci. 45067 Orléans, France

Abstract. Over the last decade, first-order constraints have been efficiently used in the artificial intelligence world to model many kinds
of complex problems such as: scheduling, resource allocation, computer
graphics and bio-informatics. Recently, a new property called *decomposability* has been introduced and many first-order theories have been
proved to be decomposable: finite or infinite trees, rational and real
numbers, linear dense order,...etc. A decision procedure in the form of
5 rewriting rules has also been developed. This latter can decide if a
first-order formula without free variables is true or not in any decomposable theory. Unfortunately, this decision procedure is not enough when
we want to express the solutions of a first-order constraint having free
variables. These kind of problems are generally known as *first-order constraint satisfaction problems*. We present in this paper, not only a decision procedure but a full first-order constraint solver for decomposable
theories. Our solver is given in the form of nine rewriting rules which
transform any first-order constraint φ (which can possibly contain free
variables) into an equivalent formula ϕ which is either the formula true,
or the formula false or a simple solved formula having at least one free
variable and being equivalent neither to true nor to false. We show the
efficiency of our solver by solving complex first-order constraints over finite or infinite trees containing a huge number of imbricated quantifiers
and negations and compare the performances with those obtained using
the most recent and efficient dedicated solver for finite or infinite trees.
This is the first full first-order constraint solver for any decomposable
theory.

1   Introduction

First-order constraints are first-order formulas built on a set of function and re-
lation symbols using the following logical symbols: $=, true, false, \neg, \wedge, \vee, \rightarrow, \leftrightarrow,$
$\forall, \exists, (,)$. Over the last decade, first-order constraints have been efficiently used
in the the artificial intelligence world to model many kinds of complex problems
such as: scheduling, resource allocation, configuration, temporal and spatial reasoning, computer graphics, bio-informatics [1,8]. However, in most of the cases,
the quantifiers are not used due to the inherent huge complexity in time and
space when solving first-order constraints with imbricated quantifiers, such as:

$$\exists x \forall y \begin{bmatrix} x = f(y, x) \wedge f(x, f(w, y)) = f(f(y, x), w) \wedge \\ \neg(\forall v \exists z \, (x = f(v, x) \rightarrow w = f(z, w))) \end{bmatrix},$$

S. Autexier et al. (Eds.): AISC/Calculemus/MKM 2008, LNAI 5144, pp. 93–108, 2008.

and if we use Maher's theory of finite or infinite trees [9,4] then solving such a constraint cannot be done with an algorithm of better complexity in time and space than a huge tower of powers of two, i.e. $2^{2^{2^{\cdots}}}$ whose depth is proportional to the number of imbricated quantifiers [3,12]. Due to this high complexity, only a few general first-order constraint solvers have been developed in the past and no one of them could solve complex first-order constraints with many imbricated quantifiers.

Recently, we showed that a lot of first-order theories such as: finite or infinite trees, real numbers, rational numbers, linear dense order without endpoints,...etc share a new property that we call *decomposability* [5]. We have then presented a decision procedure in the form of five rewriting rules which for any decomposable theory T can decide the satisfiability or unsatisfiability of any first-order proposition, i.e. any first-order constraint whose all variables are quantified, such as:

$$\exists u_2 \forall u_1 \exists u_3 \, \neg \begin{bmatrix} \exists v_1 \, v_1 = f(u_1, u_2) \wedge u_2 = g(u_1) \wedge \\ \neg(\exists w_1 \, v_1 = g(w_1)) \wedge \\ \neg(\exists w_2 \, u_2 = g(w_2) \wedge w_2 = g(u_3)) \end{bmatrix}.$$

However, even if this decision procedure [5] can be used for any complex proposition with any logical symbols, it suffers from three main problems:

(1) It can only decide if a proposition is true or not but cannot solve first-order constraints having at least one free variable (i.e a none quantified variable). In fact, if we model a given problem P by the following first-order constraint φ having u as free variable:

$$\neg \left[\exists v_1 \, u = f(v_1) \wedge \begin{bmatrix} \neg(\exists w_1 \, u = f(w_1) \wedge w_1 = v_1) \wedge \\ \neg(\exists w_2 \, y = f(v_1) \wedge w_2 = f(v_1)) \wedge \\ \neg(\exists w_3 \, u = f(v_1) \wedge v_1 = f(w_3)) \end{bmatrix} \right],$$

then solving the problem P means to transform φ into an equivalent simple formula - generally known as *solved formula* - from which we can easily extract the values of the free variable u such that the formula φ is true. Our decision procedure for decomposable theories [5] is not able to do produce such a solved formula and can only test if there exists at least one solution to our problem by solving the following proposition:

$$\exists u \, \neg \left[\exists v_1 \, u = f(v_1) \wedge \begin{bmatrix} \neg(\exists w_1 \, u = f(w_1) \wedge w_1 = v_1) \wedge \\ \neg(\exists w_2 \, y = f(v_1) \wedge w_2 = f(v_1)) \wedge \\ \neg(\exists w_3 \, u = f(v_1) \wedge v_1 = f(w_3)) \end{bmatrix} \right],$$

and the answer is either the formula *true* (i.e. the problem P has at least one solution) or the formula *false* (i.e. the problem P has no solutions). This is why the algorithm given in [5] is called *decision procedure* and not *full first-order constraint solver*.

(2) If we use our decision procedure on a formula φ which contains free variables then we can get an equivalent solved formula ϕ having free variables but being always false or always true. The appropriate solved formula of φ in this case should be the formula *false* or the formula *true* instead of ϕ. Let us take for instance the

theory Tr of finite or infinite trees (which was proved to be decomposable in [5]) and let us use our decision procedure on the following formula φ

$$\neg(\exists y\, x = f(y) \wedge \neg(\exists zw\, x = f(z) \wedge w = f(w))). \tag{1}$$

We get the following final formula ϕ

$$\neg(\exists y\, x = f(y) \wedge \neg(\exists z\, x = f(z))). \tag{2}$$

The problem is that this formula contains free variables but is always true in the theory Tr. In fact, it is equivalent to $\neg(\exists y\, x = f(y) \wedge \neg(\exists z\, x = f(y) \wedge x = f(z)))$, i.e. to $\neg(\exists y\, x = f(y) \wedge \neg(x = f(y) \wedge (\exists z\, z = y)))$, thus to $\neg(\exists y\, x = f(y) \wedge \neg(x = f(y)))$, which is finally equivalent to *true*. As a consequence, the solved formula of our initial formula (1) should be the formula *true* instead of (2). This is a good example which shows the limits of the decision procedures on first-order constraints having at least one free variable.

(3) The third problem is that the complexity of this decision procedure is exponential in time and space for most of the decomposable theories. As a consequence, the implementation of this decision procedure does not allow one to solve huge first-order constraints with many imbricated quantifiers. In fact, we succeeded in [5] to decide the validity of some very particular propositions (mainly formulas representing the solutions of a two player game) but never succeeded to decide the validity of randomly generated first-order constraints.

Much more elaborated algorithms are then needed when we want to induce solved formulas expressing solutions of complex first-order constraint satisfaction problems in decomposable theories. Of course, our goal in these kinds of problems is not only to know if there exist solutions or not, but to express these solutions in the form of a first-order formula φ which is either the formula *true* (i.e. the problem is satisfiable for all the values of the free variables) or the formula *false* (i.e. the problem is unsatisfiable for all the values of the free variables) or a simple solved formula having at least one free variable and being equivalent neither to true nor to false. Algorithms which are able to produce such a formula φ are called *first-order constraint solvers*.

Overview of the paper: we present in this paper not only a decision procedure, but *the first full first-order constraint solver for any decomposable theory*. This paper is organized in six sections followed by a conclusion. This introduction is the first section. Section 2 is dedicated to a brief review of first-order logic and decomposable theories. We present in section 3 the working formulas which are structured formulas having an important notion of *depth*. We also introduce the *box-checkers*. They are kind of black boxes that enable us to efficiently simplify some working formulas to *true* or *false* even if these latter contain free variables. In section 4, we present 9 rewriting rules which handle working formulas and transform a working formula of any depth d into an equivalent conjunction of final working formulas of depth 2 from which we can easily extract a solved formula written in a very simple form. The main idea behind these rules consists in: (1) a top-down propagation of constraints. In each level, conjunction of atomic formulas are propagated to the embedded sub-formulas. This step enables us to remove all inconsistent formulas which contradict their top-formulas using the box-checkers. (2) a bottom-up elimination of quantifiers using the property of decomposability followed by a very particular distribution. We present in section 5 our full first-order constraint solver in any decomposable theory T. This solver uses, among other things, the 9 rules of section 4. It transforms any first-order formula φ into a simple solved formula. Finally, we show in section 6 the efficiency of our algorithm and compare its performances with those of our decision procedure for decomposable theories [5], even if this latter can only answer by true or false. Among other things, our algorithm can solve formulas of a two player game involving more than 80 nested alternated quantifiers while the decision procedure overflows the memory starting from 40 nested alternated quantifiers. We also compare our performances with those of the most recent and efficient first-order constraint solver over finite or infinite trees which we have presented in [4] and show that even if our solver is general and can be used for any decomposable theory T, he gives very competitive results comparing with those of [4] which is specially optimized for the theory of finite or infinite trees and cannot be used for other decomposable theories such as rational or real numbers. The box-checkers, the 9 rewriting rules, the solver and the benchmarks are our new contributions in this paper.

2   Preliminaries

2.1   First-Order Formulas

Let V be an infinite set of variables. Let S be a set of symbols, called a signature and partitioned into two disjoint sub-sets: the set F of function symbols and the set R of relation symbols. To each function symbol and relation is linked a non-negative integer n called its *arity*. An n-ary symbol is a symbol of arity n. A first-order constraint or formula is an expression of the one of the eleven following forms:

$$s = t, \ r(t_1, \ldots, t_n), \ true, \ false,$$
$$\neg\varphi, \ (\varphi \wedge \psi), \ (\varphi \vee \psi), \ (\varphi \rightarrow \psi), \ (\varphi \leftrightarrow \psi), \tag{3}$$
$$(\forall x \, \varphi), \ (\exists x \, \varphi),$$

with $x \in V$, r an n-ary relation symbol taken from F, φ and ψ shorter formulas, s, t and the t_is terms, that are expressions of the one of the two following forms $x, f(t_1, ..., t_n)$, with x taken from V, f an n-ary function symbol taken from F and the t_i's shorter terms. The formulas of the first line of (3) are known as *atomic*, and *flat* if they are of one of the following forms:

$$true, \ false, \ x_0 = x_1, x_0 = f(x_1, ..., x_n), \ r(x_1, ..., x_n),$$

with the x_i's (possibly non-distinct) variables taken from V, $f \in F$ and $r \in R$. We denote by AT the set of the conjunctions of flat atomic formulas. A *proposition* is a formula without free variables.

A *model* is a couple $M = (D, F)$, where D is a non-empty set of individuals of M and F a set of functions and relations in D. We call *instantiation* of a formula φ by individuals of M, the formula obtained from φ by replacing each free occurrence of a free variable x in φ by the same individual i of D and by considering each element of D as 0-ary function symbol.

A *theory* T is a (possibly infinite) set of propositions. We say that the model M is a *model of* T, if for each element φ of T, $M \models \varphi$. If φ is a formula, we write $T \models \varphi$ if for each model M of T, $M \models \varphi$. A theory T is *complete* if for every proposition φ, one and only one of the following properties holds: $T \models \varphi$, $T \models \neg\varphi$.

Let M be a model and T a theory. Let $\bar{x} = x_1 \ldots x_n$ and $\bar{y} = y_1 \ldots y_n$ be two words on V of the same length. Let φ, and $\varphi(\bar{x})$ be formulas. We write

$$\exists \bar{x} \, \varphi \qquad \text{for } \exists x_1 ... \exists x_n \, \varphi,$$
$$\forall \bar{x} \, \varphi \qquad \text{for } \forall x_1 ... \forall x_n \, \varphi,$$
$$\exists? \bar{x} \, \varphi(\bar{x}) \text{ for } \forall \bar{x} \forall \bar{y} \, \varphi(\bar{x}) \wedge \varphi(\bar{y}) \rightarrow \bigwedge_{i \in \{1,...,n\}} x_i = y_i,$$
$$\exists! \bar{x} \, \varphi \qquad \text{for } (\exists \bar{x} \, \varphi) \wedge (\exists? \bar{x} \, \varphi).$$

The word $\bar{x}$, which can be the empty word ε, is called *vector of variables*. Note that semantically the new quantifiers $\exists?$ and $\exists!$ simply means "at most one" and "one and only one".

2.2 Decomposable Theories

We now recall the definition of *decomposable theories* [5]. Informally, this definition simply states that in every decomposable theory T each formula of the form $\exists \bar{x} \, \alpha$, with $\alpha \in AT$, is equivalent in T to a decomposed formula of the form $\exists \bar{x}' \, \alpha' \wedge (\exists \bar{x}'' \, \alpha'' \wedge (\exists \bar{x}''' \, \alpha'''))$ where the formulas $\exists \bar{x}' \, \alpha'$, $\exists \bar{x}'' \, \alpha''$, and $\exists \bar{x}''' \, \alpha'''$ have elegant properties which can be expressed using the following quantifiers: $\exists?$, $\exists!$ and $\exists_{\infty}^{\Psi(u)}$.

In all what follows, we will use the abbreviation wnfv for *"without new free variables "*. A formula φ is equivalent to a wnfv formula ψ in T means that $T \models \varphi \leftrightarrow \psi$ and ψ does not contain other free variables than those of φ.

Definition 2.2.1. *A theory T is called* decomposable *if there exists a set $\Psi(u)$ of formulas having at most u as free variable and three sets A', A'' and A''' of formulas of the form $\exists \bar{x} \, \alpha$ with $\alpha \in AT$ such that:*

1. *Every formula of the form $\exists \bar{x} \, \alpha \wedge \psi$, with $\alpha \in AT$ and ψ any formula, is equivalent in T to a wnfv decomposed formula of the form*

$$\exists \bar{x}' \, \alpha' \wedge (\exists \bar{x}'' \, \alpha'' \wedge (\exists \bar{x}''' \, \alpha''' \wedge \psi)),$$

with $\exists \bar{x}' \, \alpha' \in A'$, $\exists \bar{x}'' \, \alpha'' \in A''$ and $\exists \bar{x}''' \, \alpha''' \in A'''$.
2. *If $\exists \bar{x}' \alpha' \in A'$ then $T \models \exists? \bar{x}' \, \alpha'$ and for each free variable y in $\exists \bar{x}' \alpha'$, at least one of the following properties holds:*
    $- T \models \exists? y \bar{x}' \, \alpha',$

- *there exists $\psi(u) \in \Psi(u)$ such that $T \models \forall y\,(\exists \bar{x}'\,\alpha') \rightarrow \psi(y)$.*
3. *If $\exists \bar{x}''\alpha'' \in A''$ then for each x_i'' of $\bar{x}''$ we have $T \models \exists_\infty^{\Psi(u)} x_i''\,\alpha''$.*
4. *If $\exists \bar{x}'''\alpha''' \in A'''$ then $T \models \exists! \bar{x}'''\,\alpha'''$.*
5. *If the formula $\exists \bar{x}'\alpha'$ belongs to A' and has no free variables then this formula is either the formula $\exists \varepsilon\, true$ or $\exists \varepsilon\, false$.*

In [5] many first-order theories have been proved to be decomposable such as: theory of finite or infinite trees [9,4], Clark equational theories [2], rational and real numbers with addition and subtraction [7] and many combinations based on these theories [6]. From the proof of the decomposability of these theories we can deduce their completeness using a decision procedure which for every proposition produces either true or false [5]. However, this latter is not able to solve first-order constraints having free variables. To this end, we present in the next section some tools that will enable us to build a full first-order constraint solver for any decomposable theory T.

3   Working Formulas and Box-Checkers

Let T be a decomposable theory. The sets $\Psi(u)$, A, A', A'' and A''' are now known and fixed for all the following sections of this paper.

3.1   Normalized Formulas

Definition 3.1.1. *A normalized formula φ of depth $d \geq 1$ is a formula of the form*

$$\neg(\exists \bar{x}\, \alpha \wedge \bigwedge_{i \in I} \varphi_i), \tag{4}$$

with I a finite (possibly empty) set, $\alpha \in AT$ and the φ_is normalized formulas of depth d_i with $d = 1 + \max\{0, d_1, ..., d_n\}$ and all the quantified variables of φ have distinct names and different from the names of the free variables.

Property 3.1.2. *Every formula φ is equivalent in T to a wnfv normalized formula of depth $d \geq 1$.*

We have shown this property in detail in [5] where we gave a very simple algorithm which transforms any first-order formula into a normalized formula of depth $d \geq 1$.

3.2   Box-Checkers

We now introduce a property which uses two tools denoted by $BC1$ and $BC2$ (BC stands for box-checker) which will enable us to detect if a given normalized formula (which can possibly contain free variables) is equivalent to *true* or *false* in T. These checkers will be used in our solver to simplify some normalized formulas into *true* or *false* even if these latter have free variables.

Property 3.2.1. *Let φ be normalized formula of depth 2 of the form*

$$\neg(\exists \bar{x}\, \alpha \wedge \bigwedge_{i=1}^{n} \neg(\exists \bar{y}_i\, \beta_i)),$$

with $\alpha \in AT$ and $\beta_i \in AT$. Let us denote by $\bar{z}$ the vector of the free variables of φ. It is easy to check if φ is equivalent to true or to false even if it contains free variables. For that we proceed as follows:

- *Find the truth value of the proposition $\forall \bar{z}\, \varphi$. If the answer is true then φ is true in T. This step is made using a Box-checker that we denote by $BC1(\varphi)$.*
- *Find the truth value of the proposition $\forall \bar{z}\, \neg\varphi$. If the answer is true then φ is false in T. This step is made using a Box-checker that we denote by $BC2(\varphi)$. If $BC1(\varphi) = false$ and $BC2(\varphi) = false$ then φ is neither true nor false in T.*

We show that the complexity of such a checking is $n * Cpx$ if $n \neq 0$, and Cpx if $n = 0$, where Cpx is the complexity of the decomposability algorithm, i.e. the algorithm which can decompose each quantified conjunction of flat atomic formulas according to Definition 2.2.1.

The two checkers $BC1$ and $BC2$ are considered as two black-boxes: they are composed of two algorithms which are very technical and which use very complex properties of decomposable theories. As a consequence, we did not find it interesting to detail them in this paper. We prefer instead to detail the whole solver with many intuitive explanations and examples (see sections 4 and 5).

Example 1. Let Tr be the theory of finite or infinite trees [9,4] and let us check if the following normalized formula φ can be simplified in Tr into *true* or *false* even if it has x as free variable:

$$\neg(\exists y\, x = f(y) \wedge \neg(\exists z\, x = f(z)))$$

Let us compute $BC1(\varphi)$. According to Property 3.2.1, $BC1(\varphi)$ is true if the following formula is true in Tr

$$\forall x \neg(\exists y\, x = f(y) \wedge \neg(\exists z\, x = f(z))).$$

This latter is equivalent to

$$\forall x \neg(\exists y\, x = f(y) \wedge \neg(\exists z\, x = f(y) \wedge x = f(z))), \text{ i.e. to}$$

$$\forall x \neg(\exists y\, x = f(y) \wedge \neg(x = f(y) \wedge (\exists z\, z = y))) \text{ i.e. to}$$

$$\forall x \neg(\exists y\, x = f(y) \wedge \neg(x = f(y))), \text{ i.e. to}$$

$$\forall x \neg false,$$

which is finally equivalent to *true*. Thus, $BC1(\varphi) = true$ and thus according to Property 3.2.1, the formula φ is equivalent to *true* even if it contains one free variable. This example is the one given in the introduction of this paper (see the formula (1) page 3) and for which we have noted that the decision procedure of [5] cannot detect that φ is always true and cannot simplify the formula φ anymore. Our solver can completely solve φ and produce the solved formula *true* thanks to our box-checkers. We will also see in section 6 why these box-checkers greatly improve the efficiency of our solver comparing with the decision procedure of [5].

3.3   Working Formulas

We will now introduce the working formulas: they are normalized formulas having an integer over each negation which enables us to:

(1) link a semantic meaning to each sub-normalized formula of the form $\neg^k(\exists \bar{x}\, \alpha \wedge ...)$ according to the value of the integer k

(2) have a full control on the execution of the rewriting rules of our solver on the normalized formulas (cf. Section 4).

Definition 3.3.1. *A working formula is a normalized formula in which all the occurrences of $\neg$ are replaced by $\neg^k$ with $k \in \{0, ..., 4\}$ and such that each occurrence of a sub-formula of the form*

$$p = \neg^k(\exists \bar{x}\, \alpha \wedge q), \quad with \ \ k > 0, \tag{5}$$

satisfies the k first conditions of the condition list below. In (5) $\alpha \in AT$, q is a conjunction of working formulas of the form $\bigwedge_{i=1}^{n} \neg^{k_i}(\exists \bar{y}_i\, \beta_i \wedge q_i)$, with $n \geq 0$, $\beta_i \in AT$, q_i a conjunction of working formulas, and in the below condition list $\exists \bar{x}'\, \alpha'$ is the quantified conjunction of the flat atomic formulas of the immediate top-working formula[1] p' of p if it exists.

1. *If p' exists then $T \models \alpha \rightarrow \alpha'$.*
2. *$BC1(\neg(\exists \bar{x}\, \alpha)) = false$.*
3. *If p' exists then $BC1(\neg(\exists \bar{x}'\, \alpha' \wedge \neg(\exists \bar{x}\, \alpha))) = false$.*
4. *The formula $\exists \bar{x}\, \alpha$ belongs to A'.*

We strongly insist in the fact that $\neg^k$ does not mean that the normalized formula satisfies only the kth condition but all the conditions i with $1 \leq i \leq k$.

Example 2. The formula $\neg^3(\exists x\, y = f(x) \wedge \neg^2(\exists z\, z = f(y) \wedge y = f(x)))$ is a working formula of depth 2 in the theory Tr of finite or infinite trees. In fact:

- For $\neg^3$ we have $BC1(\neg(\exists x\, y = f(x))) = false$ because the formula $\forall y\, \neg(\exists x\, y = f(x))$ is false in Tr.
- For $\neg^2$ we have $Tr \models (z = f(y) \wedge y = f(x)) \rightarrow y = f(x)$ and $BC1(\neg(\exists z\, z = f(y) \wedge y = f(x))) = false$ because the formula $\forall xy\, \neg(\exists z\, z = f(y) \wedge y = f(x))$ is false in Tr.

Definition 3.3.2. *An* initial *working formula is a working formula which begins with $\neg^3$ and such that $k = 0$ for all the other occurrences of $\neg^k$. A* final *working formula is a working formula of depth less or equal to 2 with $k = 4$ for all the occurrences of $\neg^k$.*

---

[1] In other words, p' is of the form $\neg^{k'}(\exists \bar{x}'\, \alpha' \wedge p^* \wedge p)$ where p^* is a conjunction of working formulas and p is the formula (5).

4 Transformation of an Initial Working Formula into a Final Working Formula

In the following, we present nine rewriting rules which transform an initial working formula of any depth d into an equivalent conjunction of final working formulas. To apply the rule $p_1 \implies p_2$ to the working formula p means to replace in p a sub-formula p_1 by the formula p_2, by considering that the connector $\wedge$ is associative and commutative. In the following, α, β, λ represent conjunctions of flat atomic formulas, $\bar{x}$, $\bar{y}$ and $\bar{z}$ represent vectors of variables, q represents a conjunction of working formulas, r represents a conjunction of flat atomic formulas and working formulas. **All these letters can be subscripted or have primes**.

$$(1) \quad \neg^3(\exists \bar{x}\, \alpha \wedge q \wedge \neg^0(\exists \bar{y}\, r)) \implies \neg^3(\exists \bar{x}\, \alpha \wedge q \wedge \neg^1(\exists y\, \alpha \wedge r))$$

$$(2) \quad \neg^1(\exists \bar{x}\, \alpha \wedge q) \implies true$$

$$(3) \quad \neg^1(\exists \bar{x}\, \alpha \wedge q) \implies \neg^2(\exists \bar{x}\, \alpha \wedge q)$$

$$(4) \quad \neg^3(\exists \bar{x}\, \alpha \wedge q \wedge \neg^2(\exists \bar{y}\, \beta)) \implies true$$

$$(5) \quad \neg^3(\exists \bar{x}\, \alpha \wedge q \wedge \neg^2(\exists \bar{y}\, \beta \wedge q')) \implies \neg^3(\exists \bar{x}\, \alpha \wedge q \wedge \neg^3(\exists \bar{y}\, \beta \wedge q'))$$

$$(6) \quad \neg^3(\exists \bar{x}\, \alpha \wedge \bigwedge_{i \in I} \neg^4(\exists \bar{y}_i\, \beta_i)) \implies \neg^4(\exists \bar{x}'\, \alpha' \wedge \bigwedge_{i \in I'} \neg^4(\exists \bar{z}'\, \lambda_i'))$$

$$(7) \quad \neg^4(\exists \bar{x}\, \alpha \wedge \bigwedge_{i \in I} \neg^4(\exists \bar{y}_i\, \beta_i)) \implies true$$

$$(8) \quad \neg^4(\exists \bar{x}\, \alpha \wedge \bigwedge_{i \in I} \neg^4(\exists \bar{y}_i\, \beta_i)) \implies false$$

$$(9) \quad \neg^3 \left[\begin{array}{l} \exists \bar{x}\, \alpha \wedge q \wedge \\ \neg^4 \left[\begin{array}{l} \exists \bar{y}\, \beta \wedge \\ \bigwedge_{i \in I} \neg^4(\exists \bar{z}_i\, \lambda_i) \end{array} \right] \end{array} \right] \implies \left[\begin{array}{l} \neg^3(\exists \bar{x}\, \alpha \wedge q \wedge \neg^4(\exists \bar{y}\, \beta)) \wedge \\ \bigwedge_{i \in I} \neg^3(\exists \bar{x}\bar{y}\bar{z}_i\, \lambda_i \wedge q_0)^* \end{array} \right]$$

with I a finite possibly empty set. In rule (2), $BC1(\neg(\exists \bar{x}\, \alpha)) = true$. In rule (3), $BC1(\neg(\exists \bar{x}\, \alpha)) = false$. In rule (4), $BC1(\neg(\exists \bar{x}\, \alpha \wedge \neg(\exists \bar{y}\, \beta))) = true$. In rule (5), if $q' = true$ then $BC1(\neg(\exists \bar{x}\, \alpha \wedge \neg(\exists \bar{y}\, \beta))) = false$. In rule (6):

- The formula $\exists \bar{x}\, \alpha$ is equivalent in T to a decomposed formula of the form $(\exists \bar{x}'\, \alpha' \wedge (\exists \bar{x}''\, \alpha'' \wedge (\exists \bar{x}'''\, \alpha''')))$,
- The formula $\exists \bar{z}'\, \lambda'$ is the first part of the decomposition of the formula $\exists \bar{x}'''\bar{y}_i\, \alpha''' \wedge \beta_i$. In other words, the formula $\exists \bar{x}'''\bar{y}_i\, \alpha''' \wedge \beta_i$ is equivalent in T to a decomposed formula of the form $(\exists \bar{z}'\, \lambda' \wedge (\exists \bar{z}''\, \lambda'' \wedge (\exists \bar{z}'''\, \lambda''')))$. Moreover, the quantified variables of each formula $\exists \bar{z}_i'\, \lambda_i'$ are renamed by distinct names so that they respect the definition of the normalized formulas.
- I' is the set of the $i \in I$ such that $\exists \bar{y}_i'\beta_i'$ does not have free occurrences of any variable of $\bar{x}''$.

In rule (7), $BC1(\neg(\exists \bar{x}\, \alpha \wedge \bigwedge_{i \in I} \neg(\exists \bar{y}_i\, \beta_i))) = true$. In rule (8), $BC2(\neg(\exists \bar{x}\, \alpha \wedge \bigwedge_{i \in I} \neg(\exists \bar{y}_i\, \beta_i))) = true$. In rule (9), $BC1(\neg(\exists \bar{y}\, \beta \wedge \bigwedge_{i \in I} \neg(\exists \bar{z}_i\, \lambda_i))) = false$, $BC2(\neg(\exists \bar{y}\, \beta \wedge \bigwedge_{i \in I} \neg(\exists \bar{z}_i\, \lambda_i))) = false$, the set I is a none empty set and q_0 is the formula q in which all the occurrences of $\neg^k$ have been replaced by $\neg^0$. The formula $(\exists \bar{x}\bar{y}\bar{z}_i\, \lambda_i \wedge q_0)^*$ is the formula $(\exists \bar{x}\bar{y}\bar{z}_i\, \lambda_i \wedge q_0)$ in which we have renamed the variables of $\bar{x}$ and $\bar{y}'$ by distinct names and different from the names of the free variables.

How Does it work? The use of indices on the negations of the working formulas enables us to force the application of the rules to follow a clear strategy until reaching a conjunction of final working formulas. In fact, the algorithm follows two main steps:

(i) A top-down propagation of atomic formulas following the tree structure of the working formulas and using the rules (1),...,(5). In this step, atomic formulas are copied in all sub-working formulas by rule (1). Inconsistent sub-formulas as well as those which contradict their sub-formulas are removed by the rules (2) and (4).

(ii) A bottom-up elimination of quantifiers and depth reducing of the working formulas using the rules (6),...(9).

More precisely, starting from an initial working formula φ of the form $\neg^3(\exists\bar{x}\,\alpha \wedge \bigwedge_{i\in I} q_i)$, where all the q_i are working formulas whose negations are of the form $\neg^0$, rule (1) propagates the atomic formulas of α into a sub-formula q_i, with $i \in I$, and changes the first negation of q_i into $\neg^1$. The rules (2),...,(5) can now be applied. Inconsistent conjunction of flat atomic formulas that was created after propagation are removed by rule (2). Rule (3) is then applied and changes the first negation of q_i into $\neg^2$. The algorithm starts now a new phase which consists in removing the formulas which contradict their sub-formulas using Rule (4). Note that in rule (4), q is a conjunction of working formula of any depth d. This step is done using the box-checker $BC1$ and enables us to reduce directly the whole working formula to true without solving the sub-working formula q which can have a huge depth. The decision procedure given in [5] does not use this step and loses time and space by solving the formula q using a very costly rule which increases exponentially the size of the formula. This is why our new solver is much more efficient than this decision procedure. Once this step done, rule (5) is applied and changes the second negation into $\neg^3$. Rule (1) can now be applied again since all the nested negations are of the form $\neg^0$ and so on. This is the first step of our algorithm. Once the sub-working formulas of depth 1 are of the form $\neg^3(\exists\bar{y}_i\,\beta_i)$, the second step starts using rule (6) with $I = \emptyset$ on all these sub-working-formulas of depth 1 and transforms their negations into $\neg^4$. Rule (6) with $I \neq \emptyset$ is applied again on the sub-working-formulas of depth 2 of the form $\neg^3(\exists\bar{x}\,\alpha \wedge \bigwedge_{i\in I} \neg^4(\exists\bar{y}_i\,\beta_i))$ and produces working formulas of the form $\neg^4(\exists\bar{x}\,\alpha \wedge \bigwedge_{i\in I} \neg^4(\exists\bar{y}_i\,\beta_i))$. Inconsistent working formulas of depth 2 as well as those which are equivalent to true are then simplified to false or true by the rules (7) and (8). These rules are different from the rules (2) and (4). In fact, we can build many examples in which the rules (2) and (4) cannot be applied on a working formula φ but rule (7) can be applied. Once all these simplifications done, rule (9) can now be applied on the working formulas of depth $d > 2$ of the form $\neg^3(\exists\bar{x}\,\alpha \wedge q \wedge \neg^4(\exists\bar{y}\,\beta \wedge \bigwedge_{i\in I} \neg^4(\exists\bar{z}_i\,\lambda_i)))$. After each application of this rule, new working formulas containing negations of the form $\neg^0$ are created which implies the execution of the rules of the first step of our algorithm, starting by rule (1) and so on. After several applications of our rules, we get a conjunction of working formulas whose depth is less or equal to 2. The rules are then applied again until all the negations of these working formulas are of the form $\neg^4$. It is a conjunction of final working formulas.

Property 4.0.3. *Every repeated application of the preceding rewriting rules on an initial working formula p is terminating and producing a wnfv conjunction of final working formulas equivalent to p in T.*

Example 3. Let f and g be two distinct function symbols taken from F of respective arities 2, 1. Let w_1, w_2, v_1, u_1, u_2, u_3 be variables. Let us run our rules in the theory of finite or infinite trees on the following initial working formula

$$\neg^3 \begin{bmatrix} \exists v_1\, v_1 = f(u_1, u_2) \wedge u_2 = g(u_1) \wedge \\ \neg^0(\exists w_1\, v_1 = g(w_1)) \wedge \\ \neg^0(\exists w_2\, u_2 = g(w_2) \wedge w_2 = g(u_3)) \end{bmatrix}. \tag{6}$$

According to rule (1), the preceding formula is equivalent in T to

$$\neg^3 \begin{bmatrix} \exists v_1\, v_1 = f(u_1, u_2) \wedge u_2 = g(u_1) \wedge \\ \neg^1(\exists w_1\, v_1 = g(w_1) \wedge v_1 = f(u_1, u_2) \wedge u_2 = g(u_1)) \wedge \\ \neg^0(\exists w_2\, u_2 = g(w_2) \wedge w_2 = g(u_3)) \end{bmatrix}.$$

Rule (2) can be applied on the sub formula $\neg^1(\exists w_1\, v_1 = g(w_1) \wedge v_1 = f(u_1, u_2) \wedge u_2 = g(u_1))$. Thus, the preceding formula is equivalent in T to

$$\neg^3 \begin{bmatrix} \exists v_1\, v_1 = f(u_1, u_2) \wedge u_2 = g(u_1) \wedge \\ \neg^0(\exists w_2\, u_2 = g(w_2) \wedge w_2 = g(u_3)) \end{bmatrix},$$

which according to rule (1) is equivalent in T to

$$\neg^3 \begin{bmatrix} \exists v_1\, v_1 = f(u_1, u_2) \wedge u_2 = g(u_1) \wedge \\ \neg^1(\exists w_2\, v_1 = f(u_1, u_2) \wedge u_2 = g(u_1) \wedge u_2 = g(w_2) \wedge w_2 = g(u_3)) \end{bmatrix}.$$

Rule (3) followed by rule (5) is applied. Thus, the preceding formula is equivalent in T to

$$\neg^3 \begin{bmatrix} \exists v_1\, v_1 = f(u_1, u_2) \wedge u_2 = g(u_1) \wedge \\ \neg^3(\exists w_2\, v_1 = f(u_1, u_2) \wedge u_2 = g(u_1) \wedge u_2 = g(w_2) \wedge w_2 = g(u_3)) \end{bmatrix}.$$

Rule (6) with $I = \emptyset$ is applied and we get

$$\neg^3 \begin{bmatrix} \exists v_1\, v_1 = f(u_1, u_2) \wedge u_2 = g(u_1) \wedge \\ \neg^4(\exists \varepsilon\, v_1 = f(u_1, u_2) \wedge u_2 = g(u_1) \wedge u_1 = g(u_3)) \end{bmatrix}.$$

Once again rule (6) can be applied, with $I \neq \emptyset$ and we get the following final working formula

$$\neg^4 \begin{bmatrix} \exists \varepsilon\, u_2 = g(u_1) \wedge \\ \neg^4(\exists \varepsilon\, u_2 = g(u_1) \wedge u_1 = g(u_3)) \end{bmatrix}.$$

We have seen in the preceding example how the rules (1),...,(8) can be applied. Let us now see how rule (9) is applied.

Example 4. Let s and 0 be two function symbols taken from F of respective arities $1, 0$. Let w_1, w_2, u, v be variables. Let us run our rules in the theory of finite or infinite trees on the following working formula:

$$\neg^3\left[\exists\varepsilon\ true \wedge \left[\begin{array}{l}\neg^4(\exists\varepsilon\ u = s(v))\wedge\\ \neg^4(\exists w_1\ u = s(w_1) \wedge w_1 = s(v))\wedge\\ \neg^4(\exists\varepsilon\ v = u \wedge \neg^4(\exists\varepsilon\ v = u \wedge u = 0) \wedge \neg^4(\exists w_2\ v = u \wedge u = s(w_2)))\end{array}\right]\right].$$

By considering that

- $(\exists\bar{x}\ \alpha) = (\exists\varepsilon\ true)$
- $q = \left[\begin{array}{l}\neg^4(\exists\varepsilon\ u = s(v))\wedge\\ \neg^4(\exists w_1\ u = s(w_1) \wedge w_1 = s(v))\end{array}\right]$
- $(\exists\bar{y}\ \beta) = (\exists\varepsilon\ v = u)$
- $\bigwedge_{i\in I} \neg^4(\exists\bar{z}_i\ \lambda_i) = \left[\begin{array}{l}\neg^4(\exists\varepsilon\ v = u \wedge u = 0)\wedge\\ \neg^4(\exists w_2\ v = u \wedge u = s(w_2))\end{array}\right]$

rule (9) can be applied and produces the following formula

$$\left[\begin{array}{l}\neg^3(\exists\varepsilon\ true \wedge\neg^4(\exists\varepsilon u = s(v)) \wedge \neg^4(\exists w_1\ u = s(w_1) \wedge w_1 = s(v)) \wedge \neg^4(\exists\varepsilon\ v = u))\wedge\\ \neg^3(\exists\varepsilon\ v = u \wedge u = 0 \wedge \neg^0(\exists\varepsilon\ u = s(v)) \wedge \neg^0(\exists w_{11}\ u = s(w_{11}) \wedge w_{11} = s(v)))\wedge\\ \neg^3(\exists w_2\ v = u \wedge u = s(w_2) \wedge \neg^0(\exists\varepsilon u = s(v)) \wedge \neg^0(\exists w_{12}\ u = s(w_{12}) \wedge w_{12} = s(v)))\end{array}\right].$$

Now, only the rules (1),...,(8) will be applied until all the negations are of the form $\neg^4$. Rule (9) will not be applied anymore since there exists no working formulas of depth greater or equal to 3 and the rules (1),...,(8) never increase the depth of the working formulas.

5   A Full First-Order Constraint Solver for Decomposable Theories

Let p be a formula. Solving p in T proceeds as follows:

(1) Transform the formula $\neg p$ (the negation of p) into a wnfv normalized formula p_1 equivalent to $\neg p$ in T. For that we can refer to [5] where a simple algorithm was given.

(2) Transform p_1 into the following initial working formula p_2

$$p_2 = \neg^3(\exists\varepsilon\ true \wedge \neg^0(\exists\varepsilon\ true \wedge p_1)),$$

where all the occurrences of $\neg$ in p_1 are replaced by $\neg^0$.

(3) Apply our 9 rewriting rules on p_2 as many time as possible. According to Property 4.0.3 we obtain at the end a wnfv conjunction p_3 of final working formulas of the form

$$\bigwedge_{i=1}^{n} \neg^4(\exists\bar{x}_i\ \alpha_i \wedge \bigwedge_{j=1}^{n_i} \neg^4(\exists\bar{y}_{ij}\ \beta_{ij})).$$

Since p_3 is equivalent to $\neg p$ in T, then p is equivalent in T to

$$\neg \bigwedge_{i=1}^{n} \neg(\exists \bar{x}_i\, \alpha_i \wedge \bigwedge_{j=1}^{n_i} \neg(\exists \bar{y}_{ij}\, \beta_{ij})),$$

which is equivalent to the following disjunction p_4

$$\bigvee_{i=1}^{n} (\exists \bar{x}_i\, \alpha_i \wedge \bigwedge_{j=1}^{n_i} \neg(\exists \bar{y}_{ij}\, \beta_{ij})).$$

This is the final answer of our solver to the initial constraint p. Note that the negations which were at the beginning of each formula of p_3 have been removed and the top conjunction of p_3 has been replaced by a disjunction. As a consequence, the set of the solutions of the free variables of p_4 is nothing other than the union of the solutions of each simple formula of the form $\exists \bar{x}_i\, \alpha_i \wedge \bigwedge_{j=1}^{n_i} \neg(\exists \bar{y}_{ij}\, \beta_{ij})$. Thanks to using $BC1$, $BC2$ in our rules, we show easily that if p_4 contains at least one free variable then neither $T \models p_4$ nor $T \models \neg p_4$. The decision procedure given in [5] produces a final formula which is not in a so simple form than p_4 (it contains two levels of nested negations) and can even be simplified (in many cases) to true or to false (cf. the formula (2) in page 3).

6   Benchmarks

6.1   Two Partner Game

Let us consider the following two partner game: An ordered pair (i, j) is given, with i a non-negative (possibly null) integer and $j \in \{0, 1\}$. One after another, each player changes the values of i and j according to the following rules

- If $j = 0$ then the actual player should replace i by $i - 1$ in the pair (i, j).
- If $j = 1$ and i is odd then the actual player can either replace i by $i + 1$ or replace j by $j - 1$, in the pair (i, j).
- If $j = 1$ and i is even then the actual player can either replace i by $i + 1$ and j by $j - 1$ in the pair (i, j) or replace only i by $i + 1$ in the pair let (i, j)

The first player who cannot keep i non negative has lost. This game can be represented by the following directed infinite graph:

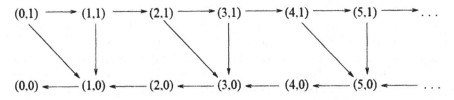

It is clear that the player which is at the position $(0, 0)$ and should play has lost. Suppose that it is the turn of player A to play. A position (n, m) is called *k-winning* if, no matter the way the other player B plays, it is always possible for A to win, after having made at most k moves. We have shown in [4] that for any integer k we can compute all the k-winning positions by solving a normalized formula of depth $2k$ in the theory of finite or infinite trees. For that, we have presented an efficient first-order constraint solver over finite or infinite trees. This latter uses very particular properties that hold only in finite or infinite trees and cannot be generalized to any decomposable theory T.

The times of execution (CPU time in milliseconds) of the formulas $winning_k(x)$ are given in the following table as well as a comparison with those obtained using our efficient first-order constraint solver over finite or infinite trees [4]. We also compare the performances of our 9 rules with those obtained using the decision procedure for decomposable theories [5] (even though the latter does not produce comprehensible results, i.e. explicit solved forms). The benchmarks are performed on a 2 Ghz Pentium IV processor, with 1024Mb of RAM. The symbol "-" below means exhausting memory.

k $(winning_k(x))$	1	2	4	5	7	10	20	40
C++ [5] (5 rules)	28	50	115	150	245	430	2115	–
C++ [4] (16 rules)	25	40	90	115	175	315	1490	15910
C++ (our 9 rules)	27	44	98	133	199	353	1688	17124

The decision procedure takes more time (until 25%), comparing with our 9 rules to solve the $winning_k(x)$ formulas of our game and overflows the memory for $k > 20$, i.e. 40 nested alternated quantifiers while our solver can solve formulas having more than 80 nested alternated quantifiers. On the other hand, we reach almost the same performances (in time and space) as those obtained using the most recent and efficient first-order constraint solver dedicated to finite or infinite trees [4].

We now discuss why our solver is much more efficient in time and space than the decision procedure of [5]. The latter uses many times a particular distribution (rule (5) in [5]) which decreases the depth of the normalized formulas but increases exponentially the number of conjunctions of the normalized formulas until overflowing the memory. Our solving algorithm uses a similar distribution (rule (9)) but only after a necessary propagation step which copies the flat atomic formulas into the sub-formulas and checks (using the box-checkers in the rules (2), (4), (7) and (8)) if there exists no normalized formulas which contradict their top-formula. Solving a $winning_k(x)$ formula in our game generates many huge formulas which contradict their top-formulas. Our algorithm removes directly these huge formulas after the first propagation step. The decision procedure [5] cannot detect this inconsistency and is obliged to apply a costly rule to decrease the size of these inconsistent formulas. At each application of this rule, the depth of the normalized formulas decreases but the number of conjunctions increase exponentially until overflowing the memory. This explains why for this game the decision procedure overflows the memory for $k > 20$ while our solver reaches $k = 40$.

Note also that the solved formulas obtained using our solver are neither equivalent to true nor to false (thanks to the use of $BC1$ and $BC2$ in our rules). The decision procedure of [5] does not warrant that any final formula with free variables cannot be simplified anymore to *true* or *false*. In fact, we have got a lot of final formulas with free variables on which we succeeded to use our new solver to simplify them into *true* or *false*.

6.2   Random Normalized Formulas

We have also tested our rules on randomly generated normalized formulas such that in each sub-normalized formula of the form $\neg(\exists \bar{x}\, \alpha \wedge \bigwedge_{i=1}^{n} \varphi_i)$, with the φ_i's normalized formulas and $n \geq 0$, we have:

- n is a positive integer randomly chosen between 0 and 4.
- The number of the atomic formulas in α is randomly chosen between 1 and 8. Moreover, the atomic formula *true* occurs at most once in α.
- The vector of variables and the atomic formulas of $\exists \bar{x}\, \alpha$ are randomly generated starting from a set containing 10 variables and 6 function symbols: $f_0, f_1, f_2, g_0, g_1, g_2$. Each function symbol f_j or g_j is of arity j with $0 \geq j \geq 2$.

The benchmarks were realized on a 2.5Ghz Pentium IV processor with 1024Mb of RAM as follows: For each integer $1 \geq d \geq 41$ we generated 10 random normalized formulas of depth d, we solved them and computed the average execution time (CPU time in milliseconds). Once again, the performances (time and space) of our 9 rules are impressive comparing with those of the decision procedure for decomposable theories and are very competitive comparing with our recent and efficient solver over finite or infinite trees [4]. The symbol "-" below means exhausting memory.

d	4	8	12	22	26	41
C++ [5] (5 rules)	108	375	1486	18973	–	–
C++ [4](16 rules)	88	202	504	3550	11662	2142824
C++ (our 9 rules)	94	221	612	4522	13654	2172632

7   Conclusion

We have presented in this paper the first full first-order constraint solver for any decomposable theory T. The decision procedure given in [5] is not able to reason on first-order constraints with free variables and can only answer by true or false to any proposition (formula without free variables). The solver was given in the form of nine rewriting rules which transform any first-order constraint φ into a solved formula ϕ so that ϕ is either the formula *true* or the formula *false* or a simple formula having at least one free variable and being equivalent neither to true nor to false.

The main idea behind our solver consists in (i) A top-down propagation of atomic formulas following the tree structure of the working formulas. (ii) A bottom-up elimination of quantifiers and depth reducing of the working formulas.

We have shown the efficiency of our algorithm by comparing its performances with those of our decision procedure for decomposable theories [5]. Among other things, our algorithm can solve formulas of a two player game involving more than 80 nested alternated quantifiers while the decision procedure overflows the memory starting from 40 nested alternated quantifiers. We have also compared our performance with those of the most recent and efficient first-order constraint solver over finite or infinite trees [4] and showed that even if our algorithm can be used for any decomposable theory T, it gives very competitive results comparing with a dedicated solver for one particular decomposable theory [4].

Currently, we are trying to find a more abstract characterization and/or a model theoretical characterization of the decomposable theories. The current definition gives only an algorithmic insight into what it means for a theory to be complete. We are also trying to test the performances of our solver on other kind of problems which can be modeled using other first-order theories such as: theory of lists [11], theory of queues [10] and the combination of real numbers with addition, subtraction, multiplication and a linear dense order relation without endpoints.

Acknowledgements. We thank Alain Colmerauer for our many discussions and his help in this work. We dedicate to him this paper.

References

1. Apt, K.: Principles of constraint programming. Cambridge University Press, Cambridge (2003)
2. Clark, K.L.: Negation as failure. In: Ed Gallaire, H., Minker, J. (eds.) Logic and Data bases. Plenum Pub. (1978)
3. Colmerauer, A., Dao, T.: Expressiveness of full first-order constraints in the algebra of finite or infinite trees. Journal of Constraints 8(3), 283–302 (2003)
4. Djelloul, K., Dao, T., Fruehwirth, T.: Theory of finite or infinite trees revisited. Theory and practice of logic programming (TPLP) (to appear, 2008)
5. Djelloul, K.: Decomposable theories. Theory and practice of logic programming (TPLP) 7(5), 583–632 (2007)
6. Djelloul, K., Dao, T.: Extension into trees of first-order theories. In: Calmet, J., Ida, T., Wang, D. (eds.) AISC 2006. LNCS (LNAI), vol. 4120, pp. 53–67. Springer, Heidelberg (2006)
7. Djelloul, K.: About the combination of trees and rational numbers in a complete first-order theory. In: Gramlich, B. (ed.) FroCos 2005. LNCS (LNAI), vol. 3717, pp. 106–121. Springer, Heidelberg (2005)
8. Fruehwirth, T., Abdennadher, S.: Essentials of Constraint Programming. Springer, Heidelberg (2003)
9. Maher, M.: Complete Axiomatizations of the Algebras of Finite, Rational and Infinite Trees. In: Proc. of LICS 1988, pp. 348–357 (1988)
10. Rybina, T., Voronkov, A.: A decision procedure for term algebras with queues. ACM transaction on computational logic 2(2), 155–181 (2001)
11. Spivey, J.: A Categorial Approch to the Theory of Lists. In: van de Snepscheut, J.L.A. (ed.) MPC 1989. LNCS, vol. 375, pp. 399–408. Springer, Heidelberg (1989)
12. Vorobyov, S.: An improved lower bound for the elementary theories of trees. In: McRobbie, M.A., Slaney, J.K. (eds.) CADE 1996. LNCS, vol. 1104, pp. 275–287. Springer, Heidelberg (1996)

Search Techniques for Rational Polynomial Orders[*]

Carsten Fuhs[1], Rafael Navarro-Marset[2], Carsten Otto[1], Jürgen Giesl[1],
Salvador Lucas[2], and Peter Schneider-Kamp[1]

[1] LuFG Informatik 2, RWTH Aachen University, Germany
{fuhs,giesl,psk}@informatik.rwth-aachen.de, carsten.otto@rwth-aachen.de
[2] DSIC, Universidad Politécnica de Valencia, Spain
{slucas,rnavarro}@dsic.upv.es

Abstract. Polynomial interpretations are a standard technique used in almost all tools for proving termination of term rewrite systems (TRSs) automatically. Traditionally, one applies interpretations with polynomials over the naturals. But recently, it was shown that interpretations with polynomials over the rationals can be significantly more powerful. However, searching for such interpretations is considerably more difficult than for natural polynomials. Moreover, while there exist highly efficient SAT-based techniques for finding natural polynomials, no such techniques had been developed for rational polynomials yet. In this paper, we tackle the two main problems when applying rational polynomial interpretations in practice: (1) We develop new criteria to decide when to use rational instead of natural polynomial interpretations. (2) Afterwards, we present SAT-based methods for finding rational polynomial interpretations and evaluate them empirically.

Topics. computer algebra systems and automated theorem provers, implementation and performance issues.

Keywords: termination, term rewriting, SAT solving, dependency pairs.

1 Introduction

Orders based on polynomial interpretations are essential for termination proofs. Recently, [16,17,18] showed that polynomial interpretations *over the rationals* are strictly more powerful for proving termination than those over the naturals.[1]

One of the most popular termination techniques that is implemented in virtually all current tools for termination analysis of TRSs is the *dependency pair* (DP) method, cf. e.g. [1,9,11,12,13]. In principle, rational polynomial interpretations can immediately be used in this method. In other words, the polynomial

---

[*] C. Fuhs, J. Giesl, C. Otto, and P. Schneider-Kamp were supported by the DAAD under grant D/06/12785 and by the DFG under grant GI 274/5-2. S. Lucas and R. Navarro-Marset were partially supported by the EU (FEDER) and the Spanish MEC, under grants TIN 2007-68093-C02-02 and HA 2006-0007. R. Navarro-Marset was partially supported by the Spanish MEC under FPU grant AP2006-026.
[1] Several such examples where this is *provably* the case are presented in Sect. 3.1.

S. Autexier et al. (Eds.): AISC/Calculemus/MKM 2008, LNAI 5144, pp. 109–124, 2008.
© Springer-Verlag Berlin Heidelberg 2008

constraints (over the rationals) which have to be generated are *the same* as those for polynomials with natural coefficients [16,18]. But as discussed in [18], the main problem when attempting to use rational polynomials in practice is that one needs *efficient and suitable methods* to find polynomial interpretations over the rationals automatically. Here, there are two main challenges:

Since searching for rational polynomial interpretations is much more time-consuming than for natural interpretations, one needs criteria to decide when to use rational interpretations. After recapitulating the necessary prerequisites on termination proving in Sect. 2, the first contribution of this paper are such criteria, presented in Sect. 3. Here, we first introduce *sufficient* criteria (i.e., criteria which state that the termination proof will fail when just using natural polynomials). Afterwards, we introduce *heuristics* to characterize the remaining termination problems where rational polynomials are "likely" to be needed.

The other challenge are efficient methods to search for rational interpretations. For interpretations over the naturals, until recently the best known techniques were dedicated constraint-based algorithms like [3]. However, recently a new approach was developed in [7] which proposes the use of SAT solvers for generating natural polynomial interpretations. This approach was implemented in the termination tool AProVE [10] and it leads to speed-ups in orders of magnitude over constraint-based algorithms. While there already exists a constraint-based algorithm for finding rational polynomial interpretations [18][2] (implemented in the tool MU-TERM [15]), a SAT-based approach similar to [7] could bring similar improvements when polynomials over the rationals are considered. The second contribution of this paper (in Sect. 4) is the development of two such SAT-based approaches. Finally, Sect. 5 contains an extensive experimental evaluation.

2 Termination Proving with Rational Polynomials

Definition 1 (Dependency Pairs). *For a TRS $\mathcal{R}$, the* defined symbols $\mathcal{D}$ *are the root symbols of left-hand sides of rules. All other function symbols are called* constructors. *For every defined symbol $f \in \mathcal{D}$, we introduce a fresh* tuple symbol $f^\sharp$ *with the same arity. To ease readability, we often write F instead of $f^\sharp$, etc. If $t = f(t_1, \ldots, t_n)$ with $f \in \mathcal{D}$, we write $t^\sharp$ for $f^\sharp(t_1, \ldots, t_n)$. If $\ell \to r \in \mathcal{R}$ and t is a subterm of r with defined root symbol, then the rule $\ell^\sharp \to t^\sharp$ is a* dependency pair *of $\mathcal{R}$. The set of all dependency pairs of $\mathcal{R}$ is denoted by $DP(\mathcal{R})$.*

Example 2. Consider the following TRS $\mathcal{R}$ from [20], where random(x) *computes a random number between 0 and x.*

$$\text{nonZero}(0) \to \text{false} \quad (1) \qquad \text{random}(x) \to \text{rand}(x, 0) \quad (3)$$

$$\text{nonZero}(\text{s}(x)) \to \text{true} \quad (2) \qquad \text{rand}(x, y) \to \text{if}(\text{nonZero}(x), x, y) \quad (4)$$

---

[2] [18] also presents an algorithm for *real* polynomial interpretations. Extending the results of the current paper to real interpretations is a topic for future work.

$$\mathsf{p}(0) \to 0 \qquad (5)$$
$$\mathsf{p}(\mathsf{s}(x)) \to x \qquad (6)$$
$$\mathsf{id\_inc}(x) \to x \qquad (7)$$

$$\mathsf{if}(\mathsf{false}, x, y) \to y \qquad (8)$$
$$\mathsf{if}(\mathsf{true}, x, y) \to \mathsf{rand}(\mathsf{p}(x), \mathsf{id\_inc}(y)) \qquad (9)$$
$$\mathsf{id\_inc}(x) \to \mathsf{s}(x) \qquad (10)$$

The defined symbols are nonZero, p, id_inc, random, rand, if, *and the DPs are*

$$\mathsf{RANDOM}(x) \to \mathsf{RAND}(x, 0) \qquad (11)$$
$$\mathsf{RAND}(x, y) \to \mathsf{IF}(\mathsf{nonZero}(x), x, y) \ (12)$$
$$\mathsf{RAND}(x, y) \to \mathsf{NONZERO}(x) \qquad (13)$$

$$\mathsf{IF}(\mathsf{true}, x, y) \to \mathsf{RAND}(\mathsf{p}(x), \mathsf{id\_inc}(y)) \ (14)$$
$$\mathsf{IF}(\mathsf{true}, x, y) \to \mathsf{P}(x) \qquad (15)$$
$$\mathsf{IF}(\mathsf{true}, x, y) \to \mathsf{ID\_INC}(y) \qquad (16)$$

The newset formulation of the DP method is the so-called *DP framework* [9,11]. In this framework, termination techniques operate on sets of dependency pairs instead of TRSs. We refer to such techniques as *DP processors*. Formally, a DP processor is a function *Proc* which takes a set of DPs as input and returns several new sets of DPs which then have to be solved instead. These DP processors are *sound*: if d is a set of DPs, $Proc(d) = \{d_1, \ldots, d_n\}$, and all $d_1, \ldots, d_n$ represent terminating problems, then the original problem d is also terminating.[3]

Termination proofs in the DP framework start with the initial set of DPs $DP(\mathcal{R})$. Then DP processors are applied repeatedly. If the final processors return empty sets, then termination is proved. In Thm. 5 and 6 we recapitulate the two most important DP processors. The first uses an *estimated dependency graph* to estimate which DPs (i.e., which "function calls") follow each other in evaluations.

Definition 3 (Estimated Dependency Graph). *Let $\mathcal{P}$ be a set of DPs. The nodes of the* estimated $\mathcal{P}$-dependency graph *are the pairs of $\mathcal{P}$ and there is an arc from $s \to t$ to $u \to v$ iff* REN(CAP(t)) *and u unify. Here,* CAP(t) *replaces all subterms of t with defined root symbol by fresh variables and* REN(t) *linearizes t by renaming all occurrences of variables into pairwise different fresh variables.*

Example 4. For the TRS in Ex. 2, we obtain the following estimated $DP(\mathcal{R})$-dependency graph.

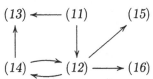

For example, the reason for the arc from (12) to (14) is that if t is the right-hand side of (12) and u is the left-hand side of (14), then REN(CAP(t)) = REN(IF(z, x, y)) = IF(z', x', y') and u = IF(true, x, y) clearly unify.

One can prove termination separately for each strongly connected component (SCC) of the estimated dependency graph. Therefore, the following processor modularizes termination proofs by decomposing the set of DPs.

---

[3] To ease readability we consider just sets of dependency pairs instead of *DP problems* [9,11]. This suffices for the presentation of the results of this paper. We also refer to [9,11] for a precise definition of "terminating" problems.

Theorem 5 (Dependency Graph Processor). *Let $\mathcal{P}$ be a set of DPs whose estimated dependency graph has n SCCs. For every $i \in \{1, \ldots, n\}$, let $\mathcal{P}_i$ be the set of DPs in the i-th SCC. Then the following DP processor is sound:*

$$Proc(\mathcal{P}) = \{\mathcal{P}_1, \ldots, \mathcal{P}_n\}$$

So in our example, the original set of DPs $DP(\mathcal{R}) = \{(11), \ldots, (16)\}$ is transformed to the subset $\mathcal{P}_1 = \{(12), (14)\}$, i.e., $Proc(DP(\mathcal{R})) = \{\mathcal{P}_1\}$.

The next processor is based on *reduction pairs* $(\succsim, \succ)$. Here, $\succsim$ is reflexive, transitive, monotonic (i.e., $s \succsim t$ implies $f(\ldots s \ldots) \succsim f(\ldots t \ldots)$ for all function symbols f), and stable (i.e., $s \succsim t$ implies $s\sigma \succsim t\sigma$ for all substitutions σ) and $\succ$ is a stable well-founded order compatible with $\succsim$ (i.e., $\succsim \circ \succ \,\subseteq\, \succ$ or $\succ \circ \succsim \,\subseteq\, \succ$).

The following processor generates inequality constraints which have to be satisfied by a reduction pair $(\succsim, \succ)$. The constraints require that all DPs in $\mathcal{P}$ are strictly or weakly decreasing (i.e., w.r.t. $\succ$ or $\succsim$) and all *usable rules* $\mathcal{U}(\mathcal{P})$ are weakly decreasing. Then one can delete all strictly decreasing DPs from $\mathcal{P}$.

The *usable rules* include all rules that can reduce the terms in right-hand sides of $\mathcal{P}$ when their variables are instantiated with normal forms. To ensure that it suffices to regard only the *usable* rules instead of *all* rules in the following processor, one has to demand that $\succsim$ is $\mathcal{C}_\varepsilon$-*compatible*, i.e., that $c(x, y) \succsim x$ and $c(x, y) \succsim y$ hold for a fresh function symbol c [11,13]. This requirement is satisfied by almost all quasi-orders used in practice.

Theorem 6 (Reduction Pair Processor). *Let $(\succsim, \succ)$ be a reduction pair where $\succsim$ is $\mathcal{C}_\varepsilon$-compatible. Then the following DP processor Proc is sound.*

$$Proc(\mathcal{P}) = \begin{cases} \mathcal{P} \setminus \succ & \text{if } \mathcal{P} \subseteq \succ \cup \succsim \text{ and } \mathcal{U}(\mathcal{P}) \subseteq \succsim \\ \mathcal{P} & \text{otherwise} \end{cases}$$

For any function symbol f, let $Rls(f) = \{\ell \to r \in \mathcal{R} \mid root(\ell) = f\}$. For any term t, the usable rules $\mathcal{U}(t)$ *are the smallest set such that*

$$\mathcal{U}(f(t_1, \ldots, t_n)) = Rls(f) \cup \bigcup_{\ell \to r \in Rls(f)} \mathcal{U}(r) \cup \bigcup_{i=1}^{n} \mathcal{U}(t_i)$$

For a set of dependency pairs $\mathcal{P}$, its usable rules are $\mathcal{U}(\mathcal{P}) = \bigcup_{s \to t \in \mathcal{P}} \mathcal{U}(t)$.

There are many techniques to search for reduction pairs automatically (recursive path orders, polynomial interpretations, etc. [4]). In this paper, we consider polynomial interpretations $\mathcal{P}ol$ which map every n-ary function symbol f to a polynomial $f_{\mathcal{P}ol} \in \mathbb{Q}_0[x_1, \ldots, x_n]$. So the coefficients of $f_{\mathcal{P}ol}$ are from $\mathbb{Q}_0 = \{\frac{p}{q} \mid p \in \mathbb{N}, q \in \mathbb{N} \setminus \{0\}\}$ and the variables $x_1, \ldots, x_n$ also range over $\mathbb{Q}_0$. This is in contrast to traditional polynomial interpretations where one uses $\mathbb{N} = \{0, 1, 2, \ldots\}$ instead of $\mathbb{Q}_0$. The mapping $\mathcal{P}ol$ is extended to terms by defining $[x]_{\mathcal{P}ol} = x$ for variables x and $[f(t_1, \ldots, t_n)]_{\mathcal{P}ol} = f_{\mathcal{P}ol}([t_1]_{\mathcal{P}ol}, \ldots, [t_n]_{\mathcal{P}ol})$. An interpretation $\mathcal{P}ol$ induces an order $\succ_{\mathcal{P}ol}$ and a quasi-order $\succsim_{\mathcal{P}ol}$ where $s \succsim_{\mathcal{P}ol} t$ iff $[s]_{\mathcal{P}ol} - [t]_{\mathcal{P}ol} \geq 0$ holds for all instantiations of the variables with numbers from $\mathbb{Q}_0$. To define $\succ_{\mathcal{P}ol}$ one needs a number $\delta > 0$ and then $s \succ_{\mathcal{P}ol} t$ iff $[s]_{\mathcal{P}ol} - [t]_{\mathcal{P}ol} \geq \delta$

holds for all instantiations of the variables with numbers from $\mathbb{Q}_0$. Then, $\succ_{Pol}$ is also well founded for rational polynomial interpretations [16,18].

Example 7. For the TRS of Ex. 2, the dependency graph processor reduced the set of DPs to $\mathcal{P}_1 = \{(12), (14)\}$. The rules for the defined symbols nonZero, p, and id_inc in the right-hand sides of (12) and (14) are usable, i.e., $\mathcal{U}(\mathcal{P}_1) = \{(1), (2), (5), (6), (7), (10)\}$. We have to find a reduction pair which makes the rules in $\mathcal{U}(\mathcal{P}_1)$ weakly decreasing and the DPs in $\mathcal{P}_1$ weakly or strictly decreasing. Then the strictly decreasing DPs are removed. We use $(\succsim_{Pol}, \succ_{Pol})$ with

$$
\begin{array}{ll}
0_{Pol} = 0 & \mathsf{p}_{Pol} = \tfrac{1}{2}\,x_1 \\[4pt]
\mathsf{s}_{Pol} = 2\,x_1 + 1 & \mathsf{id\_inc}_{Pol} = 2\,x_1 + 1 \\[4pt]
\mathsf{true}_{Pol} = 1 & \mathsf{RAND}_{Pol} = 2\,x_1 \\[4pt]
\mathsf{false}_{Pol} = 0 & \mathsf{IF}_{Pol} = x_1 + x_2 \\[4pt]
\mathsf{nonZero}_{Pol} = x_1 & \delta = 1
\end{array}
$$

Now all usable rules from $\mathcal{U}(\mathcal{P}_1)$ and all DPs from $\mathcal{P}_1$ are weakly decreasing. Moreover, the DP (14) is strictly decreasing since $[\mathsf{IF}(\mathsf{true}, x, y)]_{Pol} - [\mathsf{RAND}(\mathsf{p}(x), \mathsf{id\_inc}(y))]_{Pol} = 1 + x - 2 * \tfrac{1}{2}\,x \geq 1$. Thus, it is removed by Thm. 6 and the resulting set of DPs is $\{(12)\}$. Afterwards, another application of the dependency graph processor results in the empty set of DPs, since now the graph has no arcs anymore. Hence, termination of this example is proved.

To measure the performance of termination tools, there is an annual *International Termination Competition* [19] where the tools are applied to a large collection of TRSs (the so-called *Termination Problem Data Base* (TPDB)). The TRS of Ex. 2 comes from the TPDB (SchneiderKamp-trs-thiemann40), but none of the tools in the Termination Competition 2007 could show its termination.[4] Indeed, almost all termination tools use polynomial interpretations, but most of them are restricted to interpretations with natural or integer coefficients. If they were extended to rational coefficients, TRSs like Ex. 2 could easily be handled by virtually all existing tools. Thus, this TRS shows that rational polynomial interpretations indeed increase the power of termination proving substantially.

3   Criteria for Rational Polynomial Interpretations

In this section, we introduce criteria to decide when to use rational polynomial interpretations. In Sect. 3.1 we present sufficient criteria[5] which state that the

---

[4] [20] presents a (manual) termination proof for this TRS using an improved variant of predictive labeling, but their technique has not been implemented yet. In contrast, our proof is much easier and (apart from *rational* interpretations) it only uses standard methods that are already implemented in most termination provers.

[5] The criteria in Sect. 3.1 are restricted to *linear* polynomial interpretations which are used in the vast majority of automated termination proofs for TRSs, cf. [19]. All other results of the paper (i.e., the heuristics of Sect. 3.2 as well as the automation of Sect. 4) can be used for interpretations with polynomials of arbitrary degree.

termination proof will fail if one uses natural instead of rational interpretations. In particular, this proves that rational polynomials really increase power, i.e., that there are examples where termination can be proved with rational, but not with natural interpretations. Afterwards, Sect. 3.2 introduces heuristics to detect remaining cases where rational interpretations are also likely to be needed.

3.1   Sufficient Criteria for Rational Polynomial Interpretations

Our sufficient criteria are based on the following notions of monotonicity.

Definition 8 (Monotonicity). *Let $\mathcal{P}ol$ be a linear polynomial interpretation, let f be a function symbol with arity n, let $1 \leq i \leq n$, and let $f_{\mathcal{P}ol} = f_0 + f_1 x_1 + \ldots + f_n x_n$ with $f_0, \ldots, f_n \in \mathbb{Q}_0$. Then[6] f is monotonically increasing (MI) on i iff $f_i > 0$ and f is strongly monotonically increasing (SMI) on i iff $f_i \geq 1$. So if f is MI, but not SMI on i, then we have $0 < f_i < 1$, i.e., $f_i \notin \mathbb{N}$.*

Now we present sufficient criteria to detect when a function symbol must be MI but not SMI. This indicates that one has to use rational interpretations for the termination proof. We start with a criterion to detect that certain argument positions cannot be SMI. To this end, we check whether there are terms s and t where $s \succ_{\mathcal{P}ol} t$ must hold although s is embedded in t. To formalize the notion of embedding, we use the TRS $\mathcal{E}mb$ which consists of the rules $f(x_1, \ldots, x_n) \to x_i$ for all function symbols f and all $1 \leq i \leq n$ where n is the arity of f.

Theorem 9 (Sufficient Criterion for Non-SMI). *Let $\mathcal{P}ol$ be a linear polynomial interpretation. If $s \succ_{\mathcal{P}ol} t$ and $t \to_{\mathcal{E}}^* s$ for a set[7] of embedding rules $\mathcal{E} \subseteq \mathcal{E}mb$, then there is a rule $f(x_1, \ldots, x_n) \to x_i$ in $\mathcal{E}$ such that f is not SMI on i.*

Proof. Assume that for all $f(x_1, \ldots, x_n) \to x_i$ in $\mathcal{E}$, f is SMI on i. We show that $t \to_{\mathcal{E}}^m s$ implies $t \succsim_{\mathcal{P}ol} s$ by induction on m. This is a contradiction to $s \succ_{\mathcal{P}ol} t$.

Clearly, $t \to_{\mathcal{E}}^m s$ implies $t \succsim_{\mathcal{P}ol} s$ for $m = 0$. Now let $m > 0$, i.e., $t \to_{\mathcal{E}} t' \to_{\mathcal{E}}^* s$. So $t' \succsim_{\mathcal{P}ol} s$ by the induction hypothesis. Thus, it suffices to show $t \succsim_{\mathcal{P}ol} t'$.

As $t \to_{\mathcal{E}} t'$, we obtain $t = t[f(t_1, \ldots, t_i, \ldots, t_n)]_\pi$ and $t' = t[t_i]_\pi$ for some position π and some rule $f(x_1, \ldots, x_n) \to x_i$ in $\mathcal{E}$. Since $\mathcal{P}ol$ is linear, we have $f_{\mathcal{P}ol} = f_0 + f_1 x_1 + \ldots + f_n x_n$ for $f_0, \ldots, f_n \in \mathbb{Q}_0$ and as f is SMI on i, we have $f_i \geq 1$. Thus, $f(x_1, \ldots, x_n) \succsim_{\mathcal{P}ol} x_i$. As $\succsim_{\mathcal{P}ol}$ is monotonic and stable, this implies $t[f(t_1, \ldots, t_i, \ldots, t_n)]_\pi \succsim_{\mathcal{P}ol} t[t_i]_\pi$ and hence, $t \succsim_{\mathcal{P}ol} t'$ as desired.   □

---

[6] In general, a function $f_{\mathcal{P}ol}$ is *monotonically increasing* if $x_i - y_i > 0$ implies $f_{\mathcal{P}ol}(x_1, \ldots, x_i, \ldots, x_n) - f_{\mathcal{P}ol}(x_1, \ldots, y_i, \ldots, x_n) > 0$ for all numbers $x_1, \ldots, x_n, y_i$ and $f_{\mathcal{P}ol}$ is *strongly monotonically increasing* if $x_i - y_i \geq \delta$ implies $f_{\mathcal{P}ol}(x_1, \ldots, x_i, \ldots, x_n) - f_{\mathcal{P}ol}(x_1, \ldots, y_i, \ldots, x_n) \geq \delta$ for all numbers $x_1, \ldots, x_n, y_i$ and all $\delta > 0$. So obviously, $\frac{\partial f_{\mathcal{P}ol}}{\partial x_i} > 0$ implies that $f_{\mathcal{P}ol}$ is monotonically increasing and $\frac{\partial f_{\mathcal{P}ol}}{\partial x_i} \geq 1$ implies that $f_{\mathcal{P}ol}$ is strongly monotonically increasing.

[7] Explicitly considering the rules $\mathcal{E}$ which are needed to come from t to s (instead of considering $\mathcal{E}mb$) gives a better approximation of the "non-SMI" arguments.

Example 10. To illustrate the criterion of Thm. 9, we consider the following TRS from the TPDB (secret05-tpa2).

$$\mathsf{minus}(x, 0) \to x \qquad (17) \qquad \mathsf{f}(\mathsf{s}(x), y) \to \mathsf{f}(\mathsf{p}(\mathsf{minus}(\mathsf{s}(x), y)), \mathsf{p}(\mathsf{minus}(y, \mathsf{s}(x)))) \ (20)$$
$$\mathsf{minus}(\mathsf{s}(x), \mathsf{s}(y)) \to \mathsf{minus}(x, y) \ (18) \qquad \mathsf{f}(x, \mathsf{s}(y)) \to \mathsf{f}(\mathsf{p}(\mathsf{minus}(x, \mathsf{s}(y))), \mathsf{p}(\mathsf{minus}(\mathsf{s}(y), x))) \ (21)$$
$$\mathsf{p}(\mathsf{s}(x)) \to x \qquad (19)$$

This TRS has 11 DPs, but an application of the dependency graph processor yields the two subsets $\{(22)\}$ and $\{(23), (24)\}$, where

$$\mathsf{MINUS}(\mathsf{s}(x), \mathsf{s}(y)) \to \mathsf{MINUS}(x, y) \qquad (22)$$
$$\mathsf{F}(\mathsf{s}(x), y) \to \mathsf{F}(\mathsf{p}(\mathsf{minus}(\mathsf{s}(x), y)), \mathsf{p}(\mathsf{minus}(y, \mathsf{s}(x)))) \qquad (23)$$
$$\mathsf{F}(x, \mathsf{s}(y)) \to \mathsf{F}(\mathsf{p}(\mathsf{minus}(x, \mathsf{s}(y))), \mathsf{p}(\mathsf{minus}(\mathsf{s}(y), x))) \qquad (24)$$

The DP (22) can immediately be removed by the reduction pair processor. It remains to find a polynomial interpretation such that one of the DPs (23) and (24) is strictly decreasing and the other DP and the usable rules $\{(17), (18), (19)\}$ are weakly decreasing. For both DPs (23) and (24), the left-hand side is embedded in the right-hand side. For instance for (23), we have $\mathsf{F}(\mathsf{p}(\mathsf{minus}(\mathsf{s}(x), y)), \mathsf{p}(\mathsf{minus}(y, \mathsf{s}(x)))) \to_{\mathcal{E}}^{*} \mathsf{F}(\mathsf{s}(x), y)$ with $\mathcal{E} = \{\mathsf{p}(x_1) \to x_1, \mathsf{minus}(x_1, x_2) \to x_1\}$. So by Thm. 9, p or minus cannot be SMI on 1.

Now we present criteria for MI. Clearly, if one has to satisfy a *collapsing* inequality $s \succsim_{\mathcal{P}ol} x$ for a variable $x \in \mathcal{V}$, then the polynomial $[s]_{\mathcal{P}ol}$ must contain x. Hence, x is at a monotonically increasing position in s. For any position π in a term s, let $\mathrm{trace}(s, \pi)$ contain all pairs (f, i) such that π is below the i-th argument of the function symbol f. So $\mathrm{trace}(s, \varepsilon) = \varnothing$ and $\mathrm{trace}(f(s_1, \ldots, s_n), i\,\pi') = \{(f, i)\} \cup \mathrm{trace}(s_i, \pi')$. We omit the proof of Thm. 11, since it is obvious.

Theorem 11 (First Sufficient Criterion for MI). *Let $\mathcal{P}ol$ be a linear polynomial interpretation. If $s \succsim_{\mathcal{P}ol} x$ for $x \in \mathcal{V}$, then there exists a position π in s with $s|_\pi = x$ where f is MI on i for all $(f, i) \in \mathrm{trace}(s, \pi)$.*

Example 12. To illustrate the criterion from Thm. 11, we continue the example from Ex. 10. Since the rule (19) is usable, our polynomial interpretation has to satisfy $\mathsf{p}(\mathsf{s}(x)) \succsim_{\mathcal{P}ol} x$. We have $\mathsf{p}(\mathsf{s}(x))|_{11} = x$ and $\mathrm{trace}(\mathsf{p}(\mathsf{s}(x)), 11) = \{(\mathsf{p}, 1), (\mathsf{s}, 1)\}$. Hence, both p and s have to be MI on 1. Similarly, the rule (17) is also usable and therefore, we have to satisfy $\mathsf{minus}(x, 0) \succsim_{\mathcal{P}ol} x$. By Thm. 11 this implies that minus also has to be MI on 1.

As both p and minus are MI on 1 but at least one of them is not SMI on 1 (cf. Ex. 10), the constraints of the reduction pair processor are not satisfied by a linear polynomial interpretation over the naturals. More precisely, if $\mathsf{p}_{\mathcal{P}ol} = p_0 + p_1 x_1$ and $\mathsf{minus}_{\mathcal{P}ol} = m_0 + m_1 x_1 + m_2 x_2$ then $0 < p_1 < 1$ or $0 < m_1 < 1$.

Indeed, the following rational polynomial interpretation makes all usable rules weakly decreasing and both DPs (23) and (24) strictly decreasing. Hence, they can both be removed, which proves termination of this example.

$$0_{\mathcal{P}ol} = 0 \qquad\qquad \mathsf{minus}_{\mathcal{P}ol} = x_1$$
$$\mathsf{s}_{\mathcal{P}ol} = 2\,x_1 + 1 \qquad\qquad \mathsf{F}_{\mathcal{P}ol} = x_1 + x_2$$
$$\mathsf{p}_{\mathcal{P}ol} = \tfrac{1}{2}\,x_1 \qquad\qquad \delta = \tfrac{1}{2}$$

Example 13. The criteria presented so far can also detect the need for rational coefficients in the TRS of Ex. 2. As explained in Ex. 7, one has to find an interpretation such that one of the DPs (12) and (14) is strictly decreasing and the other DP and the usable rules $\{(1),(2),(5),(6),(7),(10)\}$ are weakly decreasing. So

$$\mathsf{RAND}(\mathsf{s}(x),y) \succsim_{\mathcal{P}ol} \mathsf{IF}(\mathsf{nonZero}(\mathsf{s}(x)),\mathsf{s}(x),y) \quad \textit{by weak decrease of (12)}$$
$$\succsim_{\mathcal{P}ol} \mathsf{IF}(\mathsf{true},\mathsf{s}(x),y) \qquad\qquad \textit{by weak decrease of (2)}$$
$$\succsim_{\mathcal{P}ol} \mathsf{RAND}(\mathsf{p}(\mathsf{s}(x)),\mathsf{id\_inc}(y)) \quad \textit{by weak decrease of (14)}$$

and as at least one of the DPs is strictly decreasing, we also have[8]

$$\mathsf{RAND}(\mathsf{s}(x),y) \succ_{\mathcal{P}ol} \mathsf{RAND}(\mathsf{p}(\mathsf{s}(x)),\mathsf{id\_inc}(y)).$$

Note that the term in the left-hand side is embedded in the right-hand side, i.e., $\mathsf{RAND}(\mathsf{p}(\mathsf{s}(x)),\mathsf{id\_inc}(y)) \to_{\mathcal{E}}^{*} \mathsf{RAND}(\mathsf{s}(x),y)$ *with* $\mathcal{E} = \{\mathsf{p}(x_1) \to x_1, \mathsf{id\_inc}(x_1) \to x_1\}$. *So by Thm. 9, one of the symbols* p *and* $\mathsf{id\_inc}$ *is not SMI on 1. But due to the usable rules (6) and (7), by Thm. 11 both* p *and* $\mathsf{id\_inc}$ *have to be MI on 1. Thus here we again need a rational polynomial interpretation. More precisely, if* $\mathsf{p}_{\mathcal{P}ol} = p_0 + p_1\,x_1$ *and* $\mathsf{id\_inc}_{\mathcal{P}ol} = i_0 + i_1\,x_1 + i_2\,x_2$, *then* $0 < p_1 < 1$ *or* $0 < i_1 < 1$.

Thm. 14 is a second criterion for MI which can be used instead of Thm. 11.

Theorem 14 (Second Sufficient Criterion for MI). *Let* $\mathcal{P}ol$ *be a linear polynomial interpretation. Let* $C[f(s_1,\ldots,s_n)] \succ_{\mathcal{P}ol} C[f(t_1,\ldots,t_n)]$ *and let there be an* $1 \le i \le n$ *such that* $s_j \in \mathcal{V}$ *for all* $j \neq i$. *Then* f *is MI on* i. *If moreover* t_i *is a variable that does not occur in* s_i, *then there must be an* $i' \neq i$ *with* $s_{i'} = t_i$ *and* f *is also MI on* i'.

Proof. Clearly, $C[f(s_1,\ldots,s_n)] \succ_{\mathcal{P}ol} C[f(t_1,\ldots,t_n)]$ for a context C implies $f(s_1,\ldots,s_n) \succ_{\mathcal{P}ol} f(t_1,\ldots,t_n)$. If $f_{\mathcal{P}ol} = f_0 + f_1\,x_1 + \ldots + f_n\,x_n$, then $pl = [f(s_1,\ldots,s_n)]_{\mathcal{P}ol} - [f(t_1,\ldots,t_n)]_{\mathcal{P}ol} = f_1\,([s_1]_{\mathcal{P}ol} - [t_1]_{\mathcal{P}ol}) + \ldots + f_n\,([s_n]_{\mathcal{P}ol} - [t_n]_{\mathcal{P}ol}) \ge \delta$. Thus we must have $f_i > 0$ (i.e., f is MI on i), because otherwise the polynomial pl is 0 or negative when instantiating all variables with 0.

Now let t_i be a variable that does not occur in s_i. If the variable t_i did not occur in s, then the coefficient for the variable t_i in the polynomial pl would be $-f_i$, i.e., pl would be negative if one instantiates t_i by a large enough number. Hence, there must be an $i' \neq i$ with $s_{i'} = t_i$ and $f_{i'} > 0$. $\qquad\Box$

Example 15. To illustrate the criterion of Thm. 14, we consider the following TRS from the TPDB (`Zantema-jw05`).

$$\mathsf{f}(\mathsf{f}(\mathsf{a},x),\mathsf{a}) \to \mathsf{f}(\mathsf{f}(x,\mathsf{f}(\mathsf{a},\mathsf{a})),\mathsf{a}) \qquad\qquad (25)$$

---

[8] To automate Thm. 9, one has to search for inequalities $s \succ_{\mathcal{P}ol} t$ where s is embedded in t. To this end, one could use *narrowing* on right-hand sides of DPs.

This TRS has 3 DPs:

$$\mathsf{F}(\mathsf{f}(\mathsf{a},x),\mathsf{a}) \to \mathsf{F}(\mathsf{f}(x,\mathsf{f}(\mathsf{a},\mathsf{a})),\mathsf{a}) \quad (26) \qquad\qquad \mathsf{F}(\mathsf{f}(\mathsf{a},x),\mathsf{a}) \to \mathsf{F}(\mathsf{a},\mathsf{a}) \quad (28)$$

$$\mathsf{F}(\mathsf{f}(\mathsf{a},x),\mathsf{a}) \to \mathsf{F}(x,\mathsf{f}(\mathsf{a},\mathsf{a})) \quad (27)$$

The dependency graph processor removes the DP (28). We first try to find a polynomial interpretation where the DP (27) is strictly decreasing and where the DP (26) and the usable rule (25) are weakly decreasing. This is easy by using $\mathsf{F}_{\mathcal{P}ol} = x_2$, $\mathsf{a}_{\mathcal{P}ol} = 1$, $\mathsf{f}_{\mathcal{P}ol} = 0$, and $\delta = 1$. Hence, (27) can be removed.

Finally, we have to find a polynomial interpretation where (26) is strictly decreasing and where the usable rule (25) is weakly decreasing. Now we can apply Thm. 14 by choosing "C", "$f(s_1,s_2)$", "i", and "$f(t_1,t_2)$" as follows: C is $\mathsf{F}(\square,\mathsf{a})$, $f(s_1,s_2)$ is $\mathsf{f}(\mathsf{a},x)$, i is 1, and $f(t_1,t_2)$ is $\mathsf{f}(x,\mathsf{f}(\mathsf{a},\mathsf{a}))$. So by Thm. 14, f is MI on 1 and as the variable t_1 does not occur in s_1, f is also MI on 2.

Moreover, strict decrease of (26) implies $\mathsf{F}(\mathsf{f}(\mathsf{a},\mathsf{a}),\mathsf{a}) \succ_{\mathcal{P}ol} \mathsf{F}(\mathsf{f}(\mathsf{a},\mathsf{f}(\mathsf{a},\mathsf{a})),\mathsf{a})$ where the left-hand side is embedded in the right-hand side, i.e., $\mathsf{F}(\mathsf{f}(\mathsf{a},\mathsf{f}(\mathsf{a},\mathsf{a})),\mathsf{a}) \to_{\mathcal{E}}^{*} \mathsf{F}(\mathsf{f}(\mathsf{a},\mathsf{a}),\mathsf{a})$ with $\mathcal{E} = \{\mathsf{f}(x_1,x_2) \to x_1\}$ or $\mathcal{E} = \{\mathsf{f}(x_1,x_2) \to x_2\}$. So by Thm. 9, f is neither SMI on 1 nor on 2. Hence if $\mathsf{f}_{\mathcal{P}ol} = f_0 + f_1\,x_1 + f_2\,x_2$, then both $0 < f_1 < 1$ and $0 < f_2 < 1$. Indeed, (26) is strictly decreasing and (25) is weakly decreasing if we use the following interpretation:

$$\mathsf{f}_{\mathcal{P}ol} = \tfrac{1}{4}x_1 + \tfrac{1}{4}x_2 \qquad \mathsf{F}_{\mathcal{P}ol} = 4\,x_1 \qquad \mathsf{a}_{\mathcal{P}ol} = 4 \qquad \delta = 2$$

3.2 Heuristics for Rational Polynomial Interpretations

The criteria from Sect. 3.1 are only sufficient, i.e., there are TRSs where rational interpretations are needed although the criteria are not fulfilled. Therefore, we now develop heuristics which indicate that rational polynomials are *likely* to be useful. So one should apply rational interpretations whenever one of the sufficient criteria of Sect. 3.1 or one of the following heuristical criteria is fulfilled.

The first heuristic suggests to apply rational interpretations whenever a destructor symbol occurs in the right-hand side of a DP. A *destructor* is a symbol which is the inverse function to a constructor. So if s is a constructor and we have a rule $\mathsf{p}(\mathsf{s}(x)) \to x$, then the symbol p is a destructor.

Heuristic 16 (Destructor Heuristic). *Let $\mathcal{P}$ be a set of DPs. If the TRS $\mathcal{R}$ contains $\mathsf{f}(\mathsf{c}(x_1,\ldots,x_n)) \to x_i$, c is a constructor, and f occurs in the right-hand side of a DP from $\mathcal{P}$, then apply rational polynomials in the processor of Thm. 6.*

For instance, in the TRS of Ex. 2, we indeed have the rule (6) for the destructor p and p occurs in the right-hand side of the DP (14). Hence, the above heuristic suggests to apply rational polynomial interpretations.

However, one can of course also formulate destructor rules in a different way. The next heuristic serves to detect such alternative formulations.

Heuristic 17 (Permutation Heuristic). *Let $\mathcal{R}$ be a TRS and $\mathcal{P}$ be a set of DPs. If $\mathcal{R} \cup \mathcal{P}$ contains a rule $C_1[t_1] \to C_2[t_2]$ where $t_1 = \mathsf{f}(\ldots,D_1[\mathsf{g}(\ldots)],\ldots)$*

and $t_2 = g(\ldots, D_2[f(\ldots)], \ldots)$ *and where at least one of the terms* t_1 *or* t_2 *con-tains two nested* f*-symbols or two nested* g*-symbols, then apply rational polyno-mials in the processor of Thm. 6. Here,* C_1, C_2, D_1, D_2 *are contexts and* f *and* g *may also be the same function symbol.*

As an example, we replace the rules $p(0) \to 0$ and $p(s(x)) \to x$ in the TRS of Ex. 2 by $p(s(0)) \to 0$ and $p(s(s(x))) \to s(p(s(x)))$. Now p still acts as a destructor and termination of the TRS can be proved almost[9] as before, but the destructor heuristic (Heuristic 16) fails. Instead, the permutation heuristic is applicable now.

Example 18. Another class of examples recognized by this heuristic are permu-tative TRSs like the following example Endrullis-pair3swap *from the TPDB.*

$$p(a(a(x_0)), p(x_1, p(a(x_2), x_3))) \to p(x_2, p(a(a(b(x_1))), p(a(a(x_0)), x_3)))$$

By two repeated applications of the dependency graph and the reduction pair processor, this example can easily be solved. However, in the reduction pair processor, one should use rational polynomial interpretations. This would be detected by the permutation heuristic above.[10]

Finally, the last heuristic detects rules where the same variable occurs twice in different arguments of a constructor on the right-hand side.

Heuristic 19 (Non-Linearity Heuristic). *Let* $\mathcal{R}$ *be a TRS and* $\mathcal{P}$ *be a set of DPs. If* $\mathcal{R} \cup \mathcal{P}$ *contains a rule* $\ell \to C[c(\ldots, t_1, \ldots, t_2, \ldots)]$ *where* $\mathcal{V}(t_1) \cap \mathcal{V}(t_2) \neq \varnothing$*, then apply rational polynomials in the processor of Thm. 6.*

Example 20. To illustrate this heuristic, consider the following example. Its be-havior is similar to Ex. 2, i.e., $f(s^n(0))$ *rewrites to* $f(s^m(0))$ *for any* $0 \le m < n$.

$$\begin{aligned}
f(s(x)) &\to f(\text{id\_inc}(c(x,x))) & \text{id\_inc}(s(x)) &\to s(\text{id\_inc}(x)) \\
f(c(s(x),y)) &\to g(c(x,y)) & \text{id\_inc}(c(x,y)) &\to c(\text{id\_inc}(x), \text{id\_inc}(y)) \\
g(c(s(x),y)) &\to g(c(y,x)) & \text{id\_inc}(0) &\to 0 \\
g(c(x,s(y))) &\to g(c(y,x)) & \text{id\_inc}(0) &\to s(0) \\
g(c(x,x)) &\to f(x)
\end{aligned}$$

When applying the dependency graph processor, the set of DPs can be split into the set of ID_INC-DPs (here the termination proof is trivial) and into the set with the F- and G-DPs. Due to the DP

$$F(s(x)) \ \to \ F(\text{id\_inc}(c(x,x))), \tag{29}$$

the non-linearity heuristic applies. One can use the rational polynomial in-terpretation with $F_{\mathcal{P}ol} = G_{\mathcal{P}ol} = x_1$, $0_{\mathcal{P}ol} = 0$, $s_{\mathcal{P}ol} = \text{id\_inc}_{\mathcal{P}ol} = x_1 + 1$,

---

[9] The only difference is that the polynomial interpretation of s must be modified. Instead of $s_{\mathcal{P}ol} = 2\,x_1 + 1$ we now use $s_{\mathcal{P}ol} = 2\,x_1^2 + 1$.

[10] For this example, a termination proof is also possible with matrix orders [6], but no tool found a proof with natural polynomial interpretations in the competitions.

$c_{\mathcal{P}ol} = \frac{1}{2}x_1 + \frac{1}{2}x_2$, and $\delta = \frac{1}{2}$ to remove all DPs with G on the right-hand side. Another application of the dependency graph processor removes the remaining DP with G on the left-hand side. To handle the last DP (29), we can use the interpretation $F_{\mathcal{P}ol} = \text{id\_inc}_{\mathcal{P}ol} = x_1$, $0_{\mathcal{P}ol} = s_{\mathcal{P}ol} = 1$, $c_{\mathcal{P}ol} = 0$, $\delta = 1$. In contrast, it is not clear how to prove termination of this system with natural polynomial interpretations.[11] For example, the tool AProVE [10] was the winner of the Termination Competition 2007 for TRSs, but the version of AProVE used at the competition fails on this example.

4   Generating Rational Interpretations by SAT Solving

In this section, we present two approaches to extend the SAT-based method of [7] in order to search for polynomial interpretations *over the rationals*. The approach of Sect. 4.1 transforms constraints over the rationals into constraints over the naturals which are then solved with the SAT-based technique of [7]. In contrast to that, Sect. 4.2 introduces a novel direct reduction of the search problem for rational polynomial interpretations into a SAT problem.

4.1   Transformation from Rationals to Naturals

To solve constraints over rational unknowns, one can reduce the problem to so-called *Diophantine* constraints where the unknowns are natural numbers. Subsequently, one can apply a Diophantine solver to solve the resulting constraints, cf. [16]. Such an approach was already implemented in the tool MU-TERM [15], but there the resulting Diophantine constraints were solved with the constraint-based solver CiME [2] instead of a more efficient approach using SAT solving. As shown in [18], this transformational approach in MU-TERM [15] is not competitive.[12]

We now illustrate our transformation in more detail. One starts with an *abstract* polynomial interpretation. It maps each function symbol to a polynomial with *abstract* coefficients. Thus, one has to determine the degree and the shape of the polynomial, but the actual coefficients are left open. For instance, for the TRS of Ex. 2 we could use an abstract polynomial interpretation $\mathcal{P}ol$ where $p_{\mathcal{P}ol} = p_0 + p_1 x_1$, $s_{\mathcal{P}ol} = s_0 + s_1 x_1$, etc. Here, p_0, p_1, s_0, s_1 are abstract coefficients.

To apply the reduction pair processor of Thm. 6, we obtain inequalities of the form $s \succ_{\mathcal{P}ol} t$ or $s \succsim_{\mathcal{P}ol} t$ that we would like to hold. These inequalities then lead to constraints on the abstract coefficients. To ensure $s \succsim_{\mathcal{P}ol} t$, it suffices to require that $[s]_{\mathcal{P}ol} - [t]_{\mathcal{P}ol}$ has only non-negative coefficients, cf. [14]. For $s \succ_{\mathcal{P}ol} t$, in addition we require that the constant coefficient of $[s]_{\mathcal{P}ol} - [t]_{\mathcal{P}ol}$ is > 0.[13] So

---

[11] However, one can prove termination using other techniques. For example, the tool Jambox [5] finds a proof using dependency pairs and matrix interpretations [6].

[12] It is much slower than MU-TERM's direct constraint-based approach [18] for finding rational polynomials. However, in Sect. 5 we show that our new SAT-based technique even significantly outperforms MU-TERM's direct constraint-based approach.

[13] This is sufficient, since we only regard finitely many inequalities of the form $s \succ_{\mathcal{P}ol} t$. Hence, δ can be defined to be the smallest constant coefficient of all these polynomials $[s]_{\mathcal{P}ol} - [t]_{\mathcal{P}ol}$, cf. [16,18].

to ensure $p(s(x)) \succ_{Pol} x$ with the abstract interpretation Pol above, we have to regard $[p(s(x))]_{Pol} - [x]_{Pol} = (p_0 + p_1 s_0) + (p_1 s_1 - 1) x$. Hence, we require

$$p_0 + p_1 s_0 > 0 \quad (30) \qquad\qquad p_1 s_1 - 1 \geq 0 \quad (31)$$

In this way, the search for a polynomial interpretation is transformed to the search for values of abstract coefficients satisfying certain inequalities.

In our setting, the values for the abstract coefficients may be numbers from $\mathbb{Q}_0$. To make this problem decidable, we restrict the possible values to numbers from a finite set $Dom = \{\frac{p}{q} \mid 0 \leq p \leq m \wedge 1 \leq q \leq n\}$. To transform this problem into a problem with abstract coefficients over the naturals instead of the rationals, we now apply the following transformation:

1. Replace all abstract variables a by fractions $\frac{a_N}{a_D}$ where a_N and a_D are new abstract variables. Here "N" stands for "numerator" and "D" stands for "denominator". The values for the abstract variables a_N and a_D are chosen from the domains $Dom_N = \{0, \ldots, m\}$ and $Dom_D = \{1, \ldots, n\}$, respectively. So in our example, the constraints (30) and (31) would be replaced by

$$\frac{p_{0_N}}{p_{0_D}} + \frac{p_{1_N}}{p_{1_D}} \frac{s_{0_N}}{s_{0_D}} > 0 \quad (32) \qquad\qquad \frac{p_{1_N}}{p_{1_D}} \frac{s_{1_N}}{s_{1_D}} - 1 \geq 0 \quad (33)$$

2. Multiply each constraint with the product of all its denominators. So (32) is multiplied by $p_{0_D} p_{1_D} s_{0_D}$ and (33) is multiplied by $p_{1_D} s_{1_D}$. This yields

$$p_{0_N} p_{1_D} s_{0_D} + p_{1_N} s_{0_N} p_{0_D} > 0 \quad (34) \qquad p_{1_N} s_{1_N} - p_{1_D} s_{1_D} \geq 0 \quad (35)$$

Now we obtained Diophantine constraints of the form $pl > 0$ or $pl \geq 0$ where pl is a (possibly non-linear) polynomial over abstract coefficients and where the values for the abstract coefficients are natural numbers.

3. Apply a Diophantine solver to search for suitable values for the abstract coefficients. In [7], it was shown how to translate Diophantine constraints into a satisfiability problem for propositional logic which can be handled by SAT solvers efficiently. In our example, the constraints (34) and (35) are for instance satisfied by $p_{0_N} = 0$, $p_{0_D} = 1$, $p_{1_N} = 1$, $p_{1_D} = 2$, $s_{0_N} = s_{0_D} = 1$, $s_{1_N} = 2$, $s_{1_D} = 1$. This corresponds to the values $p_0 = 0$, $p_1 = \frac{1}{2}$, $s_0 = 1$, $s_1 = 2$ for the original abstract coefficients. So with these values, the abstract interpretation with $p_{Pol} = p_0 + p_1 x_1$ and $s_{Pol} = s_0 + s_1 x_1$ is turned into the concrete interpretation with $p_{Pol} = \frac{1}{2} x_1$ and $s_{Pol} = 1 + 2 x_1$.

4.2 SAT Encoding for Searching Rational Interpretations

Next we present an alternative approach which encodes the search for rational polynomial interpretations *directly* into a SAT problem. One again starts with an abstract polynomial interpretation and thus, one obtains constraints like (30) and (31). In this approach, we follow a heuristic suggested in [18] and let the domains for the abstract variables have the form $Dom = \{2^{-k}, 2^{-k+1}, \ldots, 2^{\ell-1}, 2^{\ell}\} \cup \{0\}$ for $k, \ell \in \mathbb{N}$. The advantage of such domains is that they are particularly suitable for a SAT encoding. To encode constraints like (30) and (31) into a SAT problem, we now proceed as follows:

1. Up to now, the abstract coefficients like p_0, p_1, s_0, s_1 may take rational values from $\mathcal{D}om$. We now transform the constraints so that the abstract coefficients only take natural values from $\mathcal{D}om' = \{2^0, \ldots, 2^{k+\ell}\} \cup \{0\}$. To this end, every abstract coefficient a in the constraints is replaced by $\frac{1}{2^k} a'$ where a' is a fresh abstract coefficient. In our example, let $k = 1$ and $\ell = 2$, i.e., the values for the original abstract coefficients are from $\mathcal{D}om = \{2^{-1}, 2^0, 2^1, 2^2, 0\} = \{0, \frac{1}{2}, 1, 2, 4\}$. Then (30) and (31) are transformed into

$$\tfrac{1}{2} p_0' + \tfrac{1}{4} p_1' s_0' > 0 \quad (36) \qquad\qquad \tfrac{1}{4} p_1' s_1' - 1 \geq 0 \quad (37)$$

The values for p_0', p_1', s_0', s_1' are from $\mathcal{D}om' = \{2^0, 2^1, 2^2, 2^3, 0\}$.

2. To remove the rational numbers from the constraints, one now multiplies them with the least common multiple of all denominators occurring in the respective constraint. So (36) and (37) are both multiplied by 4 which yields

$$2 p_0' + p_1' s_0' > 0 \quad (38) \qquad\qquad p_1' s_1' - 4 \geq 0 \quad (39)$$

3. Now we have again obtained *Diophantine* constraints. The only difference to the Diophantine constraints handled in existing SAT encodings like [7] is that the domains used for the values of abstract coefficients are not intervals of natural numbers, but sets of powers of 2. In [7], one used a mapping $||.||$ from Diophantine constraints to propositional formulas such that a constraint α is satisfiable with values from a domain $\{0, 1, 2, 3, \ldots, 2^n - 1\}$ iff the propositional formula $||\alpha||$ is satisfiable. We now have to modify this mapping in order to handle domains of the form $\{2^0, 2^1, \ldots, 2^n\} \cup \{0\}$.

As usual, propositional formulas $\mathcal{F}$ are built from propositional variables $\mathcal{X}$, the constants 0 ("false") and 1 ("true"), and the usual Boolean connectives. Propositional interpretations are mappings $\mathfrak{I} : \mathcal{X} \to \{0, 1\}$ which can be extended to propositional formulas as usual (i.e., then we have $\mathfrak{I} : \mathcal{F} \to \{0, 1\}$). Moreover, one can extend $\mathfrak{I}$ further to *tuples* of formulas by defining

$$\mathfrak{I}(\langle \varphi_1, \ldots, \varphi_n \rangle) = 2^{n-1} * \mathfrak{I}(\varphi_1) + 2^{n-2} * \mathfrak{I}(\varphi_2) + \ldots + 2 * \mathfrak{I}(\varphi_{n-1}) + \mathfrak{I}(\varphi_n).$$

Hence, then $\mathfrak{I} : \mathcal{F}^n \to \mathbb{N}$. So if $b \in \mathcal{X}$ and $\mathfrak{I}(b) = 0$, then $\mathfrak{I}(\langle 1, b \vee \neg b, b \rangle) = 4 * \mathfrak{I}(1) + 2 * \mathfrak{I}(b \vee \neg b) + \mathfrak{I}(b) = 4 * 1 + 2 * 1 + 0 = 6$.

To determine $||.||$, one first defines the mapping of polynomials to *tuples* of propositional formulas. For numbers k, $||k||$ is the corresponding binary representation (e.g., $||6|| = \langle 1, 1, 0 \rangle$) and every abstract coefficient (i.e., Diophantine variable) a is mapped to an n-tuple of propositional variables (e.g., $||a|| = \langle a_1, a_2, a_3 \rangle$). Having defined $||pl_1||$ and $||pl_2||$ for polynomials pl_1 and pl_2, one can also define $||pl_1 + pl_2||$ and $||pl_1 * pl_2||$. Finally, one defines the mapping $||.||$ from Diophantine constraints like $pl > 0$ or $pl \geq 0$ to propositional formulas (not tuples of formulas). For details, we refer to [7].

To handle the new domains of the form $\{2^0, \ldots, 2^n\} \cup \{0\}$ we now extend propositional interpretations also to *pairs* of tuples of formulas. If Φ and Ψ are two tuples of propositional formulas, then we define

$$\mathfrak{I}(\ll \Phi, \Psi \gg) = \mathfrak{I}(\Phi) * 2^{\mathfrak{I}(\Psi)}$$

We now introduce a new mapping τ instead of $||.||$. For polynomials pl, $\tau(pl)$ is a *pair* of tuples of propositional formulas. For any number k, we define $\tau(k) = \ll ||m||, ||e|| \gg$ where $k = m * 2^e$ and m is an odd number (unless $k = m = 0$). So since $6 = 3 * 2^1$, we obtain $\tau(6) = \ll ||3||, ||1|| \gg$.

Every abstract coefficient (i.e., Diophantine variable) a is now mapped to a pair $\tau(a) = \ll a_0, \langle a_1, \ldots, a_{\lceil \log n \rceil} \rangle \gg$. Here, a_0 is just a single propositional variable (i.e., $\mathfrak{I}(a_0) \in \{0, 1\}$ for any interpretation $\mathfrak{I}$) and $\mathfrak{I}(\langle a_1, \ldots, a_{\lceil \log n \rceil} \rangle)$ can be any number between 0 and n. Hence, $\ll a_0, \langle a_1, \ldots, a_{\lceil \log n \rceil} \rangle \gg$ can indeed represent the numbers from $\{2^0, \ldots, 2^n\} \cup \{0\}$. Afterwards, one has to extend the mapping τ to more complex polynomials and to Diophantine constraints, similar to the mapping $||.||$ from [7].

In our example, we could finally obtain an interpretation with $\mathfrak{I}(\tau(p'_0)) = 0$, $\mathfrak{I}(\tau(p'_1)) = 1$, $\mathfrak{I}(\tau(s'_0)) = 2$, $\mathfrak{I}(\tau(s'_1)) = 4$. This would correspond to the solution $p_0 = \frac{1}{2} * p'_0 = 0$, $p_1 = \frac{1}{2} * p'_1 = \frac{1}{2}$, $s_0 = \frac{1}{2} * s'_0 = 1$, and $s_1 = \frac{1}{2} * s'_1 = 2$. With these values, the abstract interpretation with $\mathsf{p}_{\mathcal{P}ol} = p_0 + p_1 x_1$ and $\mathsf{s}_{\mathcal{P}ol} = s_0 + s_1 x_1$ is again turned into the concrete interpretation with $\mathsf{p}_{\mathcal{P}ol} = \frac{1}{2} x_1$ and $\mathsf{s}_{\mathcal{P}ol} = 1 + 2 x_1$.

5   Experiments and Conclusion

In Sect. 3, we developed new criteria to determine when to use rational interpretations in termination proofs. Moreover, in Sect. 4.1 and 4.2 we proposed two SAT-based approaches to automate the search for rational polynomials.

We implemented our contributions in the termination prover AProVE [10] and evaluated the performance of different variants of AProVE on all 2061 term and string rewrite systems from the TPDB. As in the Termination Competition 2007, we used a time limit of 120 seconds for each example.

In the following table, we only used the dependency graph and reduction pair processor, but no other termination techniques. In the first technique "Nat", we only searched for natural polynomials where the coefficients take values from $\{0, 1, 2, 3, 4\}$. In the technique "Rat + Sect. 4.1", we used rational coefficients from $\{\frac{p}{4} \mid 0 \leq p \leq 16\}$ instead[14] and applied the transformational technique of Sect. 4.1 to convert constraints over the rationals to constraints over the naturals. Here, we *always* search for rational polynomials, whereas in the technique "Rat + Sect. 4.1 + Sect. 3" we only search for rationals if this is suggested by the criteria from Sect. 3. Otherwise, we use natural polynomials with coefficients from $\{0, 1, 2, 3, 4\}$. Finally, in the technique "Rat + Sect. 4.2" we (always) use rational coefficients from $\{2^{-2}, 2^{-1}, 2^0, 2^1, 2^2, 0\}$ and apply the direct SAT-encoding from Sect. 4.2.[15] The column "Yes" shows the number of TRSs where the termination proof succeeds. "SucTime" gives the average runtime for successful examples and "FulTime" gives the average runtime for all examples.

---

[14] The idea of fixing the value of the denominator (e.g. to 4) and only to search for suitable values of the numerator was already proposed by [8].

[15] We also experimented with different ranges for the coefficients, but the above ranges gave the best results as far as power and runtimes are concerned.

Nat			Rat + Sect. 4.1			Rat + Sect. 4.1 + Sect. 3			Rat + Sect. 4.2		
Yes	SucTime	FulTime	Yes	SucTime	FulTime	Yes	SucTime	FulTime	Yes	SucTime	FulTime
606	1.9 s	2.9 s	742	3.1 s	15.4 s	685	2.6 s	11.0 s	696	6.1 s	29.2 s

Comparing "Nat" with the other setting shows that rational polynomials can significantly increase power, but they also increase runtimes. The comparison of "Rat + Sect. 4.1" with "Rat + Sect. 4.1 + Sect. 3" shows the usefulness of the criteria from Sect. 3: if one applies these criteria, then runtimes are not increased that much anymore, but (as long as one does not use any other termination techniques) one also loses several examples where rational interpretations were needed. Finally, the comparison with the last setting in the table shows that the method of Sect. 4.1 which transforms constraints over the rationals to constraints over the naturals is preferable to the direct SAT encoding from Sect. 4.2.

The next experiment compares "Rat + Sect. 4.1" with the existing constraint-based method [18] for generating rational interpretations, implemented in MU-TERM [15]. More precisely, we compare this version ("MU-TERM + [18]") with a version of MU-TERM where instead of [18] one calls AProVE (with the technique of "Rat + Sect. 4.1") externally. Since MU-TERM generates the polynomial constraints and it only calls AProVE with this set of constraints, the implementation of the criteria from Sect. 3 cannot be used here. In this table, we only ran MU-TERM on a collection of 79 TRSs from the TPDB. These are TRSs where MU-TERM needs rational polynomials in order to succeed with the proof. It turns out that in spite of the external calls, the new SAT-based implementation is indeed significantly faster than the previous non-SAT-based method of [18].

MU-TERM + [18]		MU-TERM + Rat + Sect. 4.1	
Yes	FulTime	Yes	FulTime
62	10.1 s	65	4.1 s

Finally, to measure the usefulness of our contributions in full termination provers, the next table compares the performance of *full* versions of AProVE on all 2061 examples. Here, many termination techniques are used in addition to the dependency graph and reduction pair processor. Moreover, there are also techniques to disprove termination (cf. column "No"). The next table shows that the results of the current paper are also useful when integrating them into such a powerful prover. AProVE-07 is the version which participated in the Termination Competition 2007 (and which won this competition in the category of TRSs). "AProVE-07 + Sect. 4.1" differs from AProVE-07 by using rational polynomials with the setting "Rat + Sect. 4.1" and "AProVE-07 + Sect. 4.1 + Sect. 3" uses "Rat + Sect. 4.1 + Sect. 3" instead. It is interesting to note that when integrating rational polynomials into this full version of AProVE, the criteria of Sect. 3 have quite positive effects. In other words, they reduce the runtimes and hardly affect the power. For details on our experiments (including details on runtimes and timeouts) and to run "AProVE-07 + Sect. 4.1 + Sect. 3" via a web-interface, we refer to http://aprove.informatik.rwth-aachen.de/eval/RATPOLO/.

APROVE-07				APROVE-07 + Sect. 4.1				APROVE-07 + Sect. 4.1 + Sect. 3			
Yes	No	SucTime	FulTime	Yes	No	SucTime	FulTime	Yes	No	SucTime	FulTime
1089	238	3.8 s	29.6 s	1119	238	5.2 s	30.4 s	1118	238	4.9 s	30.1 s

References

1. Arts, T., Giesl, J.: Termination of term rewriting using dependency pairs. Theoretical Computer Science 236, 133–178 (2000)
2. Contejean, E., Marché, C., Monate, B., Urbain, X.: CiME, http://cime.lri.fr
3. Contejean, E., Marché, C., Tomás, A.P., Urbain, X.: Mechanically proving termination using polynomial interpretations. Journal of Automated Reasoning 34(4), 325–363 (2005)
4. Dershowitz, N.: Termination of rewriting. Journal of Symbolic Computation 3, 69–116 (1987)
5. Endrullis, J.: Jambox, http://joerg.endrullis.de
6. Endrullis, J., Waldmann, J., Zantema, H.: Matrix interpretations for proving termination of term rewriting. In: Furbach, U., Shankar, N. (eds.) IJCAR 2006. LNCS (LNAI), vol. 4130, pp. 574–588. Springer, Heidelberg (2006)
7. Fuhs, C., Giesl, J., Middeldorp, A., Thiemann, R., Schneider-Kamp, P., Zankl, H.: SAT solving for termination analysis with polynomial interpretations. In: Marques-Silva, J., Sakallah, K.A. (eds.) SAT 2007. LNCS, vol. 4501, pp. 340–354. Springer, Heidelberg (2007)
8. Gebhardt, A., Hofbauer, D., Waldmann, J.: Matrix Evolutions. In: Proc. WST 2007 (2007)
9. Giesl, J., Thiemann, R., Schneider-Kamp, P.: The dependency pair framework: Combining techniques for automated termination proofs. In: Baader, F., Voronkov, A. (eds.) LPAR 2004. LNCS (LNAI), vol. 3452, pp. 301–331. Springer, Heidelberg (2005)
10. Giesl, J., Schneider-Kamp, P., Thiemann, R.: AProVE 1.2: Automatic termination proofs in the DP framework. In: Furbach, U., Shankar, N. (eds.) IJCAR 2006. LNCS (LNAI), vol. 4130, pp. 281–286. Springer, Heidelberg (2006)
11. Giesl, J., Thiemann, R., Schneider-Kamp, P., Falke, S.: Mechanizing and improving dependency pairs. Journal of Automated Reasoning 37(3), 155–203 (2006)
12. Hirokawa, N., Middeldorp, A.: Automating the dependency pair method. Information and Computation 199(1,2), 172–199 (2005)
13. Hirokawa, N., Middeldorp, A.: Tyrolean Termination Tool: Techniques and features. Information and Computation 205(4), 474–511 (2007)
14. Hong, H., Jakuš, D.: Testing positiveness of polynomials. Journal of Automated Reasoning 21(1), 23–38 (1998)
15. Lucas, S.: MU-TERM: a tool for proving termination of context-sensitive rewriting. In: van Oostrom, V. (ed.) RTA 2004. LNCS, vol. 3091, pp. 200–209. Springer, Heidelberg (2004)
16. Lucas, S.: Polynomials over the reals in proofs of termination: From theory to practice. RAIRO Theoretical Informatics and Applications 39(3), 547–586 (2005)
17. Lucas, S.: On the relative power of polynomials with real, rational, and integer coefficients in proofs of termination of rewriting. Applicable Algebra in Engineering, Communication and Computing 17(1), 49–73 (2006)
18. Lucas, S.: Practical use of polynomials over the reals in proofs of termination. In: Proc. PPDP 2007, pp. 39–50. ACM Press, New York (2007)
19. Marché, C., Zantema, H.: The termination competition. In: Baader, F. (ed.) RTA 2007. LNCS, vol. 4533, pp. 303–313. Springer, Heidelberg (2007)
20. Thiemann, R., Middeldorp, A.: Innermost termination of rewrite systems by labeling. In: Proc. WRS 2007. ENTCS 204, pp. 3–19 (2008)

Strategies for Solving SAT in Grids by Randomized Search

Antti E. J. Hyvärinen, Tommi Junttila, and Ilkka Niemelä

Helsinki University of Technology TKK
Department of Information and Computer Science
{Antti.Hyvarinen,Tommi.Junttila,Ilkka.Niemela}@tkk.fi

Abstract. Grid computing offers a promising approach to solving challenging computational problems in an environment consisting of a large number of easily accessible resources. In this paper we develop strategies for solving collections of hard instances of the propositional satisfiability problem (SAT) with a randomized SAT solver run in a Grid. We study alternative strategies by using a simulation framework which is composed of (i) a grid model capturing the communication and management delays, and (ii) run-time distributions of a randomized solver, obtained by running a state-of-the-art SAT solver on a collection of hard instances. The results are experimentally validated in a production level Grid. When solving a single hard SAT instance, the results show that in practice only a relatively small amount of parallelism can be efficiently used; the speedup obtained by increasing parallelism thereafter is negligible. This observation leads to a novel strategy of using grid to solve collections of hard instances. Instead of solving instances one-by-one, the strategy aims at decreasing the overall solution time by applying an alternating distribution schedule.

1 Introduction

This paper considers techniques for solving challenging instances of the *propositional satisfiability* (SAT) problem with the aid of computational *Grids*. Such techniques are of particular interest firstly due to the increasing use of SAT based technologies in computer aided verification and other application areas, and secondly since Grids are nowadays offering large quantities of affordable computing power. The first phenomenon is a consequence of recent developments in SAT solvers which have dramatically improved the computational power of the solvers, whereas the second seems to be a major trend in high-performance computing.

Our goal in this paper is to develop techniques for exploiting the parallel computing resources provided by a Grid in a way that allows us to use state-of-the-art SAT solvers with no or only minor modifications. To do this, we use the *Simple Distributed SAT* (SDSAT) framework, whose basic version consists of simply running N *randomized SAT solvers* in parallel until one of them finds the solution. We consider extensions of the basic version obtained by incorporating different *restart strategies* and study their effects in a specifically built simulation environment. The simulation environment comprises of (i) a Grid model taking into account the inherent communication and management delays, and (ii) run time distributions of a state-of-the-art randomized SAT

S. Autexier et al. (Eds.): AISC/Calculemus/MKM 2008, LNAI 5144, pp. 125–140, 2008.

solver when applied on several hard SAT instances. We also validate some of the results and parameters of our Grid model by using a production level Grid called NorduGrid (see http://www.nordugrid.org/),

The key idea we exploit is that a complete SAT solver can be turned into a *randomized search procedure* (RSP) in a natural way by slightly modifying the heuristic function used in the solver. For example, MiniSAT [1] 1.14 makes by default 2% of its heuristic choices pseudo-randomly; thus a natural modification to turn MiniSAT into a RSP is to seed its pseudo-random number generator differently for each run. Such a randomized search procedure, when provided with an input x, is guaranteed to give a correct result $\text{RSP}(x)$ when the computation of the procedure finishes. However, due to the randomization, the time required for computing $\text{RSP}(x)$ is not known in advance but is described by a random variable $T_{\text{RSP}(x)}$. The random variable $T_{\text{RSP}(x)}$, and thus the run time of $\text{RSP}(x)$, is completely characterized by its cumulative *run time distribution* function, $q_{\text{RSP}(x)}(t)$, giving the probability that the computation will terminate before or at time t. This randomization of a SAT solver may sound counter-intuitive as one usually tries to remove all non-determinism in order to make runs reproducible to ease benchmarking and debugging. However, in the SDSAT framework as well as when employing restart strategies to a RSP (discussed below), the goal is to exploit the *short runs* (if any) in the distribution to decrease the *expected run time* of the overall system.

The expected run time of a randomized search procedure can often be substantially reduced by periodically restarting the procedure [2]. For example, assume that $T_{\text{RSP}(x)} = 1\text{s}$ with probability 0.3 and $T_{\text{RSP}(x)} = 10\text{s}$ with probability 0.7. Then the expected run time $\mathbb{E}(T_{\text{RSP}(x)})$ is $0.3 \cdot 1\text{s} + 0.7 \cdot 10\text{s} = 7.3\text{s}$. If the RSP is modified so that it restarts itself immediately after time $t = 1\text{s}$, the expected run time becomes $\sum_{i=1}^{\infty} 0.7^{i-1} \cdot 0.3^{i} \cdot is \approx 3.3\text{s}$. Such a modification, where the procedure is forced to start from the beginning after running t_1 seconds, then after t_2 seconds and so forth, is called a *restart strategy* $S = (t_1, t_2, \ldots)$ and the time t_i the i:th *restart limit*. When a restart strategy is employed to an RSP, the result is a randomized *algorithm* that also has a run time distribution and an expected run time. The restart strategy employed in the previous example is a special case of a *fixed restart strategy* $S^t = (t, t, \ldots)$ and the algorithm corresponding to the fixed restart strategy S^t employed on RSP is denoted by $\text{FIXED}_{t,\text{RSP}}$ (or simply FIXED_t when RSP is implicitly known). Fixed restart strategies are important in our analysis, since if $q_{\text{RSP}(x)}(t)$ is known, then t can be chosen so that the expected run time of $\text{FIXED}_t(x)$ is the minimal among all the algorithms obtainable from $\text{RSP}(x)$ by employing *any* restart strategy [3]. However, in practice $q_{\text{RSP}(x)}(t)$ is not known: obtaining information about $q_{\text{RSP}(x)}(t)$ in general requires solving $\text{RSP}(x)$, which is the overall goal in many applications. To circumvent this problem, several *universal* restart strategies have been suggested [3,4]: they do not depend on the instance x and let the restart limits grow arbitrary large in order to preserve the completeness of the algorithm.

We first study the effect of applying several restart strategies on our benchmark set of hard SAT instances in the sequential setting. The results show that there are instances on which the optimal fixed restart strategy provides a substantial reduction in the expected run time. The two universal strategies considered can also reduce the expected run time on some instances but result in a bad performance on some others. The reason is that the universal strategies can spend too much time in trying to find a short run; when an instance has none, all that time is wasted.

Based on results in the sequential case, we consider ways to parallelize restart strategies in the SDSAT framework and use our simulation model to benchmark them. The results give rise to two major observations. First, parallelism seems to be an effective "luck enhancer"; when randomized solvers are run in parallel, the probability that one of them finds a short run grows quite quickly. This seems to render elaborate restart strategies practically useless in the parallel setting as the simple approach with no restarts tends to provide quite good results consistently. The second observation is that only a relatively small amount of parallelism seems to be effectively exploitable; after a certain amount, adding more parallel solvers does not seem to give any significant performance gain. There seems to be two reasons for this: (i) the probability that a short run is found is already quite high with a smallish number of parallel solvers, and (ii) the delays in the Grid environment reduce the effect of restart strategies.

The above results suggest that when solving a *set of instances*, a good speedup is not obtained by solving them one-by-one in a Grid. Instead, the instances should be solved in parallel by reserving a smallish amount of computing resources for each instance. We validate this idea in Sect. 6 both with the simulation model and by using a production level Grid.

Related Work. Techniques for learning or adapting restart strategies to improve the aggregate performance on a given collection of instances are studied, e.g., in [5,6,7,8,9]. A closely related topic is the use of algorithm portfolios [10,11]. The idea is combined with clause learning in [12]. Parallel restart strategies are studied in [13], without considering the practical limitations of a Grid. Guiding path [14,15] is a technique for distributed SAT solving based on dynamic partitioning of the problem with new assumptions. Such methods combine also with clause learning [16]. The techniques in grid-like environments have been investigated, for example, in [17,18,19,20]. The guiding path method is further developed in [21]. A different algorithm is presented in [22].

In this paper we extend previous work in three crucial respects: (i) We take into account the limitations of practical Grid environments which involve strict resource bounds and significant latencies due to communication and job management. (ii) We require minimal changes to the SAT solvers, and the changes are almost totally independent of the underlying solver technology. (iii) We use realistic run time distributions of the randomized search procedure obtained experimentally by running a state-of-the-art SAT solver on a representative collection of SAT instances from the application domain.

2   Grid Environment

The paper develops techniques for using loosely coupled, widely distributed Grid environments for solving challenging SAT problems. From an abstract point of view a Grid environment can be seen as consisting of a collection of computing resources called *primitive computing elements* (PCEs). A PCE can execute a sequential program given its input, hence, in practice corresponding to a CPU. A user can submit a *job* (a sequential program together with its input) to the Grid which executes it on one of its PCEs and gives results back to the user.

Next we briefly described three key characteristics which play an important role when developing Grid applications and the algorithms in this paper: (i) jobs in Grids experience significant delays but (ii) the run time of a job typically affects the effect of delays and (iii) communication between jobs is very limited when compared to traditional multi-processor environments such as clusters.

(i) The entry point of a Grid environment is a set of queues accepting jobs. Each queue is associated with a set of *computing elements* (CEs) corresponding to a set of CPUs. A job starts executing when the queue system assigns the job to a CE. Several causes of delays can be identified. Firstly, the time required for the job to reach a CE after submission to the corresponding queue depends on the amount and types of previously submitted jobs still in the queue, and the remaining run times of the jobs currently executing in the CEs. Secondly, if the submission of a job involves transmitting a large amount of data, the amount of network bandwidth may greatly affect the delays [23]. Thirdly, the run time of a job in a CE depends on the load potentially placed by other jobs on the neighboring CPUs, as well as the types of the CPUs in the CE. Finally, it is possible that jobs disappear due to maintenance breaks or various random faults. Efficient job management in Grids is a non-trivial task and is typically handled by special tools. In those experiments of this paper that are run in NorduGrid, we use a fault-tolerant and efficient job management system called the Grid Job Manager (GridJM) [24].

(ii) Note that the different delays above seem to suggest that a job with limited run time could experience shorter delays. For instance, most queue systems support a mechanism called *reservation*, where a complicated task requesting a CE of several CPUs will force the queue system to start to reserve CPUs. In this case, no new jobs requesting a CE will be assigned from the queue, unless the run time of the job is short enough to finish before the time expected for the requested CE of several CPUs to become available. On the other hand, since the delays are experienced by each job, it would be preferable to submit sufficiently long running jobs so that the delays do not dominate the total run time. As a reasonable compromise, in the experiments in NorduGrid we use jobs where the run time is limited to one hour.

(iii) Since a Grid can be formed by several independent but collaborating organizations which decide to share the computing resources, it is common that two jobs submitted to the Grid are not guaranteed to be able to communicate with each other at all. For example, such limitations are typically posed by the networks of the organizations in NorduGrid used in the experiments and, therefore, in the algorithms developed in the paper we assume that jobs cannot communicate directly with each other.

3 Simulation Environment

Realistic Grid systems pose certain challenges for exact algorithm benchmarking, since both the delays and the run times vary, rendering the reproduction of results difficult. To overcome these challenges, we construct a simple Grid model based on the following components:

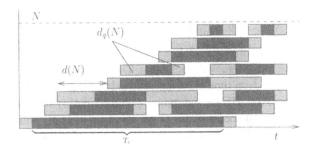

Fig. 1. A time line of an execution in Grid representing the number N of PCEs, queue delay $d_q(N)$, and the submit delay $d(N)$. In the example, the first job has executed the maximum allowed time T_c on a PCE.

(1) A unique central process M initiating new and monitoring old jobs, and a set of N PCEs receiving jobs from and reporting the results to M.
(2) An initiation delay describing the amount of time required to submit a job to the Grid. The delay $d(N)$ can be modeled as a random variable depending on the number of PCEs employed. The delay is executed by M and results in a bottleneck when initiating new computations.
(3) A queue delay is the sum of two components: the time spent queuing to the PCE, and the time spent receiving the results after the job has finished. The delay $d_q(N)$ can be modeled as a random variable depending on the number of PCEs employed. The delay is experienced by the job and does not form a bottleneck for submission.
(4) A maximum resource limit T_c describing the amount of time a PCE is allowed to execute before terminating a job and becoming ready to accept a new job.

We believe that this system provides a realistic model for distributed computing in Grids. (1) A central process managing jobs provides a natural synchronization mechanism. (2&3) Most such systems have a delay associated with the synchronization, and specifically shared distributed environments require certain communication in selecting the PCE to be employed. (4) Batch systems such as Grids usually limit the resources available to a single job, for example, to provide fairness in scheduling. The model does not directly consider the effect of various CPU models and the load on the CPUs on the run time. Such effects can be obtained by adjusting the queue delay and the resource limit accordingly.

We may study an application submitting jobs to the Grid through a central process M as a time line, illustrated in Fig. 1. The time advances to the right in the figure and the abstract PCEs can be seen as N bands placed on top of each other. The filled rectangles represent jobs, and the dark areas inside the jobs represents the CPU time, as opposed to the queuing delay. The time in the figure starts when the first job (the long rectangle at the bottom of the figure) is placed into a queue of a PCE. The second job is submitted immediately after this, and after the submission delay $d(N)$, reaches the queue. Meanwhile, the first job has reached the PCE, is executed in it, and finally the result is reported back to the central process after some queue delay.

When performing the actual simulations, we make the following simplifying assumptions on the model:

- submit delay $d(N) = d$ is constant for every PCE and does not depend on N, and
- queue delay $d_q(N) = d_q$ is constant for every PCE and does not depend on N.

If the effect of the number of PCEs is taken into account, the delays will increase since in practice the jobs will interfere with each other. This means that using the simplifying assumptions the resulting run time is underestimated and this error increases with the number of PCEs employed. Hence, the model with the simplifying assumptions gives overly optimistic results on speedups for larger numbers of PCEs which needs to be taken into consideration when evaluating the results. Nevertheless, these assumptions allow us to study the effect of delays in a simple yet reasonably realistic environment.

Run time distributions. As a representative collection of SAT instances we use a set of benchmarks from the SAT 2007 Competition (see http://www.satcompetition.org/2007/). The instances, with the full name, abbreviated name, and satisfiability, are listed below.

- mod2-rand3bip-sat-250-3.shuffled-as.sat05-2220, mod2-250, satisfiable.
- mod2-rand3bip-sat-280-1.sat05-2263.reshuffled-07, mod2-280, satisfiable.
- 9999990000001nc.shuffled-as.sat05-446, 99999900, unsatisfiable.
- clqcolor-10-07-09.shuffled-as.sat05-1258, clqcolor, unsatisfiable.
- cube-11-h14, cube, satisfiable.
- dated-10-13-s, dated, satisfiable.
- mizh-md5-48-5, mizh-md5, satisfiable.
- vmpc_28.shuffled-as.sat05-1957, vmpc_28, satisfiable.
- AProVE07-16, AProVE07, unsatisfiable.

The set covers both industrial and hand-crafted instances, having typical run time of thousands of seconds for a state-of-the-art SAT solver.

The SAT solver run time distributions are approximated by using a collection of samples for each instance. The samples are obtained by 100 separate randomized runs of a state-of-the-art SAT solver (MiniSAT version 1.14 with its pseudo-random number generator initialized differently for each run). Based on the randomized runs, we construct a distribution of run times with linear interpolation between the sample points, assuming probability 0 for runs shorter than the minimum sample and for runs longer than the maximum sample. We also studied the case with discrete distribution, but this did not significantly affect our results.

Table 1 documents for each instance the abbreviated names and the SAT solver run times for minimum, fifth percentile, median, 95th percentile and maximum of the samples. We also provide the average of the samples, i.e., an approximation of the expected run time of the solver on the instance, in the RSP column. The columns OPTIMUM, t^*, LUBY and WALSH will be explained in Sections 4 and 5. At this point, of particular interest are the large dynamics in certain distributions, such as vmpc_28 with over 19000-fold difference between minimum and maximum run time. We also provide the cumulative run time distributions for two of the test instances in Fig. 2. The distribution is the increasing graph $q(t)$. The horizontal lines in Figures 2(b) and 2(d) indicate the maximum and minimum run times of the instance and the vertical line indicates the maximum run time on the x-axis, i.e., the value of t where $q(t) = 1$. The remaining graphs will be explained in Sections 4 and 5.

Table 1. Characteristics of the run times for the test instances

Instance	Min	5%	Median	95%	Max	RSP	OPTIMUM	t^*	LUBY	WALSH
mod2-250	40.16	97.16	1210	2675	3088	1181	1181	∞	2715	1510
mod2-280	9.184	55.71	1732	6611	7775	2382	918.4	9.184	1274	1718
99999900	1072	1204	2056	3101	3725	2065	2065	∞	25070	4560
clqcolor	1198	1300	1922	2955	4329	1900	1900	∞	23060	4158
cube	2629	2896	4708	7936	10049	4832	4832	∞	106200	18500
dated	10.09	46.53	803.0	12550	37930	2279	716.1	29.08	901.5	993.3
mizh-md5	49.76	128.7	861.7	5784	9489	1660	1236	899.3	3403	1471
vmpc_28	0.1370	3.905	394.7	1730	2720	623.3	12.71	0.2560	137.4	279.6
AProVE07	879.4	1071	1471	2713	2855	1564	1564	∞	17330	3381

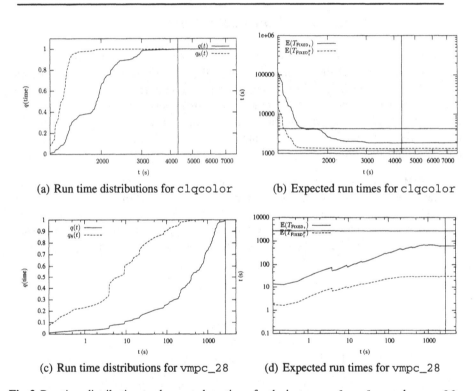

(a) Run time distributions for clqcolor

(b) Expected run times for clqcolor

(c) Run time distributions for vmpc_28

(d) Expected run times for vmpc_28

Fig. 2. Run time distributions and expected run times for the instances clqcolor and vmpc_28

It can be argued that 100 samples is not enough to give us a realistic view of the run time distribution of an instance. In order to estimate the magnitude of the error introduced to the finite distribution, we compare the distributions of cube with 100 samples and 1000 samples. The results are reported in the first two rows of Table 2. Even though the minimum run time decreases and the maximum run time increases, the distribution seems to remain relatively stable when increasing the number of samples. To have an impression on how, for example, a short run would affect the results, we

Table 2. Comparison of the distributions for cube with 100 samples (cube_{100}), 1000 samples (cube_{1000}), and a modified distribution with one artificial short run inserted (cube_{1001m})

Instance	Min	5%	Median	95%	Max	RSP	OPTIMUM	t^*	LUBY	WALSH
cube_{100}	2629	2896	4661	7617	8821	4832	4832	∞	106200	18500
cube_{1000}	1441	2990	4914	7664	14051	5067	5067	∞	97360	31510
cube_{1001m}	0.7352	2990	4914	7647	14051	5061	725.9	0.735	5101	30280

inserted an artificial short sample and constructed the corresponding distribution. The resulting distribution has the same dynamics as the distribution of vmpc_28.

4   Restart Strategies in a Sequential Setting

Given a randomized search procedure RSP and a problem instance x, it is possible to associate a run time distribution $q_{\mathrm{RSP}(x)}(t)$ with the run time of $\mathrm{RSP}(x)$. Employing a restart strategy S on RSP results in a new algorithm with a potentially different run time distribution. In this section we discuss the effect of using several such algorithms on our collection of SAT instances by comparing the run time distributions $q_{\mathrm{RSP}(x)}(t)$ with the run time distributions of the new algorithms. We use the following restart strategies and corresponding algorithms:

- OPTIMUM. The fixed restart strategy S^t and the corresponding algorithm FIXED_t mentioned in Sect. 1 have the property that there is a restart limit t^* which is optimal for a given RSP and instance x [3]. If the cumulative distribution function $q(t)$ of the instance is known, the optimal restart limit t^* may be determined by minimizing the expected run time $\mathbb{E}(T_{\mathrm{FIXED}_t(x)})$ as a function of the restart limit t,

$$\mathbb{E}(T_{\mathrm{FIXED}_t(x)}) = \frac{t - \int_{t'=0}^{t} q(t')dt'}{q(t)}, \tag{1}$$

  i.e., $t^* = \mathrm{argmin}(\mathbb{E}(T_{\mathrm{FIXED}_t(x)}))$. Determining t^* can be done in our simulation environment but not usually in practice as the distribution $q(t)$ is typically not known.
- LUBY. Luby et al. [3] define the universal strategy $S^L = (l(1), l(2), \ldots)$ where

$$l(i) = \begin{cases} 2^{k-1}, & \text{if } i = 2^k - 1, k \in \mathbb{N} \\ l(i - 2^{k-1} + 1), & \text{if } 2^{k-1} \leq i < 2^k - 1. \end{cases}$$

  When the strategy S^L is employed on a RSP, the corresponding algorithm is called LUBY. In [3] it is further shown that the expected run time of $\mathrm{LUBY}(x)$ is within a logarithmic factor from the expected run time of $\mathrm{OPTIMUM}(x)$ independently of x.
- WALSH. Another universal strategy is the strategy $S^W = (w(1), w(2), \ldots)$, where $w(i) = 2^{1.2i}$, presented in [4]. The strategy differs from S^L, for example, in the rate of growth. Clearly, the restart limits in S^W grow exponentially, whereas S^L grows only linearly with respect to i. The corresponding algorithm will be referred to as WALSH.

Table 1 compares the three algorithms against the run time of RSP. Column RSP reports the expected run time of $RSP(x)$ for different instances x. Using the run time distribution $q_{RSP(x)}(t)$, we computed the optimum restart limit t^* for each instance minimizing Eq. (1). The resulting expected run time is reported on column OPTIMUM and the corresponding restart limit in column t^*. The value ∞ is used to mark the cases when run times for OPTIMUM(x) and RSP(x) are equal. In this collection of instances, in five cases out of nine the expected run time of OPTIMUM(x) is equal to that of RSP(x). Some of the satisfiable instances, though not all, seem to profit from employing a fixed restart strategy with small restart limit. As an example, the expected run time for the algorithm FIXED$_t$ with input vmpc_28, is shown in Fig. 2(d) as a function of the restart limit t (graph labeled $\mathbb{E}(T_{FIXED_t})$). In other cases, the expected run times of algorithms with larger restart limits compare favorably to those with smaller restart limits. An example is shown in Fig. 2(b).

The results for the two universal strategies are shown in columns LUBY and WALSH of Table 1. Based on the results, it seems that in most cases the instances having $\mathbb{E}(T_{OPTIMUM(x)}) \neq \mathbb{E}(T_{RSP(x)})$ also profit of more complex strategies. We also note that LUBY performs very badly on many instances with a high minimum run time. This is a consequence of the slow growth of the restart limit in the strategy S^L. In general, the algorithm WALSH seems to offer a relatively robust approach, resulting in good speedup where such speedup would be obtainable with FIXED$_{t^*}$ given that t^* is known, and still performing usually well in cases where $\mathbb{E}(T_{OPTIMUM(x)}) = \mathbb{E}(T_{RSP(x)})$. This is a slightly surprising result, since to our knowledge no optimality result exists for the strategy S^W.

5   Parallel Solving of a Single Instance

In the previous section we discussed several restart strategies and resulting sequential algorithms when the strategies are employed to a RSP. In this section we develop a number of *parallel algorithms* for Grid environments based on the restart strategies. Here we consider a Grid environment as an efficient distributed system for running jobs. Hence, the algorithmic design boils down to approaches to constructing a sequence of jobs $j_1, j_2, \ldots$ to be submitted to the Grid for execution based on a RSP and a restart strategy. Since each job has a resource limit T_c limiting the execution time, we employ a *finite restart strategy* (discussed below) on the RSP which guarantees that the run time of the resulting algorithm is not more than T_c. Hence, each job j_i consists of the RSP, the input x to be solved and a finite restart strategy.

A *finite restart strategy* $S = (t_1, t_2, \ldots, t_n)$ is a finite sequence of restart limits which, when employed on a RSP, will terminate the resulting algorithm unless a solution is found by the end of the restart limit t_n. The *length* of the finite restart strategy S, denoted by $|S|$, is n. Given a restart strategy $S = (t_1, t_2, \ldots)$ and a resource limit T_c, we define an operator finite(S) for constructing finite restart strategies from S as

$$\text{finite}(S) = \begin{cases} (T_c) & \text{if } t_1 > T_c \\ (t_1, t_2, \ldots, t_m) & \text{where } m \text{ maximizes } \sum_{i=1}^{m} t_i \leq T_c \text{ otherwise.} \end{cases}$$

For any restart strategy S, the run time of the algorithm obtained by employing finite(S) on a RSP is less than or equal to T_c.

The most intuitive way of constructing jobs from a restart strategy $S = (t_1, t_2, \ldots)$ is to assign the job j_i the restart strategy (t_i) for $i = 1, 2, \ldots$. In practice this approach performs very badly due to the high delays in actual Grid environments. Therefore, the parallel algorithms we propose are based on two general *schemes* for constructing a sequence of jobs, given a restart strategy S.

- *Straightforward scheme.* Given a restart strategy S for constructing jobs we define a sequence of restart strategies $S_1, S_2, \ldots$ in the following way: let $S_1 = S$ and given a strategy S_i, the restart strategy S_{i+1} is constructed from S_i by removing the first $|\text{finite}(S_i)|$ restart limits from S_i. Given an environment with N PCEs, in the straightforward scheme jobs are constructed from the sequence $S_1, S_2, \ldots$ by assigning the restart strategy $\text{finite}(S_1)$ for the jobs $j_1, \ldots, j_N$, then $\text{finite}(S_2)$ for the jobs $j_{N+1}, \ldots, j_{2N}$ and so forth. This strategy is discussed in [13].
- *Faithful scheme.* In this scheme given a restart strategy S we construct the sequence $S_1, S_2, \ldots$ as above and then assign the job j_1 the restart strategy $\text{finite}(S_1)$, the job j_2 the restart strategy $\text{finite}(S_2)$, and so forth.

Parallel Algorithms. Given the randomized search procedure and the distributed environment, the parallel algorithm is uniquely determined by the used scheme (introduced above) and the restart strategy. Furthermore, for a fixed restart strategy, the straightforward and faithful schemes result in the same parallel restart strategy, and thus the same algorithm. We will discuss six parallel algorithms:

- The *maximum parallel algorithm* $\text{FIXED}^p_{T_c}$ is formed from the fixed restart strategy S^{T_c} and either straightforward or faithful scheme.
- The *optimal parallel algorithm* $\text{FIXED}^p_{t^*}$ is formed by finding a value t^* which minimizes the parallel run time distribution

$$\mathbb{E}(T_{\text{FIXED}^p_t(x)}) = \frac{t - \int_{t'=1}^{t}(1 - (1 - q(t'))^N)dt'}{1 - (1 - q(t))^N} \qquad (2)$$

for $\text{RSP}(x)$ with the run time distribution $q(t)$. Equation (2) is obtained from Eq. (1) by substituting $q(t)$ with the corresponding parallel distribution $1 - (1 - q(t))^N$. However, as shown in [13], there are run time distributions for which $\text{FIXED}^p_{t^*}$ does not result in minimum expected run time over all parallel algorithms.
- The *faithful parallel Luby and Walsh algorithms* LUBY-F^p and WALSH-F^p are constructed by using the faithful scheme on the strategies S^L and S^W, respectively.
- The *straightforward parallel Luby and Walsh algorithms* LUBY-S^p and WALSH-S^p are constructed by using the straightforward scheme on the strategies S^L and S^W, respectively.

Zero-Delay Parallel Environment. In this subsection we consider an idealized Grid environment captured by the Grid model, where we set the delays $d = d_q = 0$ and the resource limit $T_c = 3600s$. This provides us with a lower bound on the run times achievable in more realistic Grid environments.

Table 3. Results for different strategies and the zero-delay parallel environment

Instance	N	FIXED$_{t*}^p$	FIXED$_{T_c}^p$	LUBY-S^p	WALSH-S^p	LUBY-F^p	WALSH-F^p
mod2-250	16	105.7	116.2	334.2	177.5	171.8	114.0
	64	47.25	47.25	194.6	84.86	50.23	45.32
mod2-280	16	61.82	84.52	71.44	76.65	67.65	79.32
	64	19.36	21.55	22.29	25.69	21.44	24.58
99999900	16	1219	1219	14657	2910	1620	1238
	64	1097	1097	14530	2784	1213	1094
clqcolor	16	1293	1293	14730	2963	1553	1301
	64	1223	1223	14660	2899	1287	1224
cube	16	2891	2891	33600	6777	8105	2996
	64	2682	2682	33410	6570	3086	2687
dated	16	48.44	64.12	59.30	53.29	63.46	60.15
	64	15.89	16.33	15.92	16.05	14.69	19.26
mizh-md5	16	133.8	133.8	525.8	116.6	162.1	125.4
	64	73.23	73.23	259.2	126.1	84.53	81.76
vmpc_28	16	0.834	7.293	4.694	6.065	4.366	11.22
	64	0.251	0.539	0.6507	0.7994	0.6550	0.5003
AProVE07	16	1049	1049	11040	2285	1299	1064
	64	918.8	918.8	7823	1823	1056	915.4

We report the results for the maximum parallel algorithm in column FIXED$_{T_c}^p$ of Table 3 for 16 and 64 PCEs. For comparison, we also report on the column FIXED$_{t*}^p$ the results when using the optimal parallel algorithm, in which case we use $T_c = \infty$.

The speedup is in most cases linear with respect to the added resources, and for vmpc_28 even super-linear, for both FIXED$_{T_c}^p$ and FIXED$_{t*}^p$. For some instances, however, the speedup is negligible. It seems that there are certain distributions which do not allow for speedup when parallelized in this manner after a certain amount of PCEs has been reached. Two different examples of this phenomenon are closer studied in Figures 2(b) and 2(d) for $N = 1$ and $N = 8$. The graphs labeled $\mathbb{E}(T_{\text{FIXED}_t^p})$ in the figures are the expected run times of the algorithm FIXED$_t^p$ with the respective instance as a function of the restart limit t. In Fig 2(b), the run time of the algorithm FIXED$_t^p$ with large values of t is almost equal to that of the shortest sampled run (the lower horizontal line) which can also be seen from the run time distribution of the algorithm FIXED$_t^p$ when $N = 8$, $q_8(t)$, in Fig 2(a). The situation is different in Fig 2(d), where the shortest run is much shorter than the expected run also when $N = 8$.

We also note that the difference between FIXED$_{T_c}^p$ and FIXED$_{t*}^p$ becomes insignificant when N increases. The intuitive explanation for this is that the benefit of aggressive restarting can be obtained by running several solvers in parallel. The important consequence of the phenomenon is that with a large number of PCEs, the significance of the restart strategies decreases.

The remaining columns in Table 3 show the behavior of the strategies S^L and S^W. The results are obtained by simulating 100 runs of the parallel algorithms and reporting the mean time required to find the solution. The columns LUBY-S^p and WALSH-S^p correspond to the straightforward parallel restart strategy for S^L and S^W. This scheme

Table 4. Comparison of 64-PCE S^L and S^W with $f = 1.0$s, $f = 15.0$s, and $f = 100.0$s

Instance	LUBY-F^p			WALSH-F^p		
	$f = 1.0$	$f = 15.0$	$f = 100.0$	$f = 1.0$	$f = 15.0$	$f = 100.0$
mod2-250	68.09	50.23	47.19	46.54	48.85	48.55
mod2-280	34.71	21.44	20.16	23.82	21.69	18.55
99999900	1372	1213	1166	1093	1096	1105
clqcolor	1345	1287	1262	1220	1224	1222
cube	3950	3086	2977	2696	2688	2677
dated	28.43	14.69	18.71	18.74	15.85	18.57
mizh-md5	98.35	84.53	74.76	82.65	72.48	79.45
vmpc_28	0.5140	0.6550	0.6560	0.4717	0.5401	0.4870
AProVE07	1088	1056	992.2	930.2	936.2	914.76

has the benefit that small restart limits are attempted often. However, especially S^L suffers from the repeating of the short runs in cases where the smallest run time is high. The results corresponding to the faithful scheme are reported in columns LUBY-F^p and WALSH-F^p. In most cases the faithful scheme performs significantly better than the straightforward scheme, and when this is not the case, the difference is relatively small.

To further enhance the strategies S^L and S^W, we studied the effect of multiplying the restart limits of the strategies by a constant factor f in Table 4 for 64 PCEs. Based on these results, the factor does not seem to have a significant effect on the run times. The runs in Table 3 (as in Table 5) are measured with $f = 15.0$.

We study the effect of a larger sample base similar to the case in Table 2 in the zero-delay environment. The results are reported in Table 5. For this particular instance, the strategy FIXED$_{t*}^p$ is equal to the maximum strategy both when the amount of samples is 100 and 1000. In this case, when the number of samples is increased, the expected solving time decreases for most algorithms. There is no significant difference between WALSH-F^p and FIXED$_{T_c}^p$ whereas LUBY-F^p suffers from a larger number of short unsuccessful runs (even though not visible in Table 2, the distributions are significantly different when $t \leq T_c$; e.g. $q(3600s) \approx 0.24$ in the 100 samples distribution but only approximately 0.14 in the 1000 samples case). Since cube is a satisfiable instance, it is possible that there is a short run time for the randomized SAT solver. Since the 1000 samples did not reveal a short run time, it might be that the run is extremely improbable. To study the effect of such a short successful run we modify the distribution of cube to include a single short run. The resulting run times are given in the row

Table 5. Effect of additional samples on the zero-delay solving of cube with 64 PCEs

Instance	FIXED$_{t*}^p$	FIXED$_{T_c}^p$	LUBY-F^p	WALSH-F^p
cube$_{100}$	2682	2682	3086	2687
cube$_{1000}$	2364	2364	3760	2270
cube$_{1001m}$	11.86	2175	969.8	2185

labeled cube_{1001m}. In this case, LUBY-F^p is better than FIXED$^p_{T_c}$ because of the higher probability of finding the short run.

Non-Zero Delay Parallel Environment. The simulation results from the parallel environment with zero submission delay and zero queuing delay provide some insight to how the parallelization method based on randomizing algorithms can perform on the benchmark set. However, realistic parallel environments in general, and Grid environments in particular, always include some overhead related to initializing the computations. As described in Sect. 3, we divide the delays into two categories: submit delay d and queue delay d_q. Typical values in NorduGrid are $d = 12$s and $d_q = 125$s. However, the two values seem to vary strongly. The simulated experiments are presented in Table 6 under the title "large delay". All results are obtained by computing the mean run time over 100 samples using $T_c = 3600$s for the jobs.

The results show that almost always the maximum parallel algorithm FIXED$^p_{T_c}$ outperforms those based on universal restart strategies on these instances. It is worth noting that increasing the number of PCEs four-fold brings next to nothing in speedup, a consequence of the long queuing delays.

It is possible that the submission and queue delays are significantly shorter in, say, some other Grid environments. We simulate the effect of smaller delays by using submission delay $d = 5$s and queue delay $d_q = 30$s. The results are reported under the caption "small delay". Even though the strategies S^L and S^W are now more competitive, their effectiveness still suffers from the high delays and it can be argued that the maximum timeout is a sufficient approximation of the optimum. The super-linear speedup observed in zero-delay environment cannot be observed in either of the delayed environments. For certain instances, such as 99999900 and cube, already a smallish number of parallel runs suffices to find a short run from the samples. As a result, obtainable speedup is small.

We confirm these results by repeating them for two instances in the NorduGrid Grid environment. We select two instances which according to the simulated results are illustrative examples on the techniques used in parallel solving. The instance vmpc_28 shows super-linear speedup in simulations in zero-delay environments, but only a moderate speedup in delayed environments using the techniques we have studied. The instance AProVE07, on the other hand, has a less dynamic distribution in the simulations and yields no significant speedup at the transition from 16 to 64 PCEs even in the zero-delay environment. The results are presented in Table 7. The submission delays seem to be below the average delay of 12 seconds, but the results correspond approximately to the simulated results. No speedup seems to be achieved when the number of PCEs is increased.

6 Parallel Solving of a Set of Instances

In this section we propose an algorithm for solving a collection of SAT problems efficiently in a Grid environment based on the results on solving a single instance. The results indicate that (i) an increase in the number of PCEs does not result in a corresponding speedup when solving a single instance and (ii) for a large number of problems to solve, a good speedup is not obtained by using all the resources for solving

Table 6. Results for different strategies and delayed parallel environments. The two rows for each instance correspond to $N = 16$ (top) and $N = 64$ (bottom).

Instance	small delay				large delay			
	$\text{FIXED}^p_{t_*}$	$\text{FIXED}^p_{T_c}$	LUBY-F^p	WALSH-F^p	$\text{FIXED}^p_{t_*}$	$\text{FIXED}^p_{T_c}$	LUBY-F^p	WALSH-F^p
mod2-250	177.0	145.1	232.7	164.4	352.8	379.3	399.8	399.3
	161.5	157.7	182.7	133.5	364.4	355.7	422.7	350.4
mod2-280	125.8	159.0	137.4	150.7	306.8	331.0	321.4	350.0
	118.1	126.2	135.2	132.4	296.3	327.7	320.9	340.1
99999900	1242	1268	1672	1306	1431	1477	1984	1527
	1208	1246	1401	1253	1432	1485	1756	1490
clqcolor	1340	1353	1455	1378	1506	1525	1846	1577
	1328	1351	1448	1352	1508	1536	1777	1554
cube	2882	2960	9209	3067	3094	3117	9233	3195
	2792	2840	3489	2842	3050	3121	4159	3145
dated	112.1	140.5	138.2	126.1	272.2	323.8	281.6	312.1
	104.7	114.1	116.6	117.4	284.3	309.2	293.8	305.8
mizh-md5	181.4	190.4	268.7	199.4	352.6	391.2	445.0	395.3
	190.4	186.3	208.5	195.0	379.8	385.2	464.8	392.0
vmpc_28	43.27	67.35	62.70	65.49	155.7	206.7	198.4	214.0
	42.18	68.06	62.59	64.30	155.5	218.3	200.0	212.0
AProVE07	1073	1089	1313	1127	1262	1289	1569	1310
	1073	1065	1205	1061	1292	1299	1568	1300

Table 7. Experimental results in Grid for selected instances. Reported is the average over 10 runs using the strategy S^{T_c}.

Instance	PCEs	Time	d
vmpc_28	8	105.4	3.333
	16	125.7	7.668
	64	134.5	5.189

Instance	PCEs	Time	d
AProVE07	8	1624	5.917
	16	1574	9.714
	64	1271	8.555

a single problem at a time, but rather by dedicating only a certain amount of PCEs for a single problem and solving multiple problems simultaneously instead. These observations lead to the following *locally-aided fair-share algorithm*: Given a collection of instances, the instances are sent for solving in a round-robin manner by using the maximum parallel algorithm $\text{FIXED}^p_{T_c}$. In addition, the problems are also solved locally at the same time using an algorithm similar to LUBY with the modified strategy $S^{L,C} = (\min\{l(1), C\}, \min\{l(2), C\}, \ldots)$, where C is a maximum local run time constant, in a round-robin manner.

We provide experimental evidence that the proposed algorithm is efficient in a real Grid environment. For this experiment, we select 8 problems from our benchmark set of 9 problems and run them in parallel with 64 PCEs, reserving at most eight PCEs per problem. This enables us to compare the results of this experiment against a strategy where 64 PCEs are dedicated for a single instance at a time. We first exclude cube from the set of instances, since this problem is in the limit of solvable problems within

3600 seconds in our Grid environment, having expected run time of 4708 seconds in the simulation environment. The resulting run time for the full instance set is 1865 seconds. The sum of the simulated run times for these instances from Table 6 is 5916 seconds. This results in a speedup 3.17 compared to the strategy of using 64 PCEs per instance. When these results are compared against a simple strategy of running the problems on a single PCE with no delays, the speedup computed from the results of Table 1 is 7.32.

However, we note that the results can be significantly worse if a difficult instance, such as cube, is included in the set of problems to solve. We repeated the above experiment with 10 repetitions, now using 72 PCEs, resource limit $T_c = 7200$ seconds and including cube to the set of problems to solve. This resulted in a speedup of 1.76 with average solving time of 5136 seconds in the Grid environment compared to the expected solving time of 9037 seconds with long delays and 64 PCEs in Table 6. When these results are compared against a simple strategy of running the problems on a single PCE with no delays, the speedup is 3.60.

7   Conclusions

In this paper we have developed techniques for solving collections of hard SAT instance in a Grid using a randomized SAT solver. We have compared different approaches using a simulation framework consisting of a grid model capturing the communication and management delays, and a representative collection of run-time distributions of a randomized solver. The results are experimentally confirmed also in NorduGrid which is a European-wide distributed production level Grid. When solving a single hard SAT instance, the results show that in practice often (i) a relatively small number of parallel jobs suffices to increase the probability of finding a short run in the distribution to a significant level and (ii) the non-negligible delays in a Grid eliminate super linear speedups that could be obtained in an ideal environment without any delays. Hence, attempts to decrease the overall expected run time by using clever universal restart strategies or by finding optimal restart limits do not lead to significant improvements compared to using the resource limit implied by the Grid environment as the restart limit. These observations lead to a novel strategy of using Grid to solve collections of hard instances. Instead of solving instances one-by-one, the strategy aims at decreasing the overall solution time by applying an alternating distribution schedule.

Acknowledgments. The authors wish to thank the anonymous reviewers for their valuable comments. The financial support of the Academy of Finland (projects 122399 and 112016), Helsinki Graduate School in Computer Science and Engineering, and Jenny and Antti Wihuri Foundation is gratefully acknowledged.

References

1. Eén, N., Sörensson, N.: An extensible SAT-solver. In: Giunchiglia, E., Tacchella, A. (eds.) SAT 2003. LNCS, vol. 2919, pp. 502–518. Springer, Heidelberg (2004)
2. Gomes, C.P., Selman, B., Crato, N., Kautz, H.A.: Heavy-tailed phenomena in satisfiability and constraint satisfaction problems. J. Automated Reasoning 24(1/2), 67–100 (2000)

3. Luby, M., Sinclair, A., Zuckerman, D.: Optimal speedup of Las Vegas algorithms. Inf. Process. Lett. 47(4), 173–180 (1993)
4. Walsh, T.: Search in a small world. In: IJCAI, pp. 1172–1177. Morgan Kaufmann, San Francisco (1999)
5. Kautz, H.A., Horvitz, E., Ruan, Y., Gomes, C.P., Selman, B.: Dynamic restart policies. In: AAAI/IAAI, pp. 674–681 (2002)
6. Ruan, Y., Horvitz, E., Kautz, H.A.: Restart policies with dependence among runs: A dynamic programming approach. In: Van Hentenryck, P. (ed.) CP 2002. LNCS, vol. 2470, pp. 573–586. Springer, Heidelberg (2002)
7. Streeter, M., Golovin, D., Smith, S.F.: Restart schedules for ensembles of problem instances. In: AAAI, pp. 1204–1210. AAAI Press, Menlo Park (2007)
8. Huang, J.: The effect of restarts on the efficiency of clause learning. In: IJCAI, pp. 2318–2323 (2007)
9. Wu, H., van Beek, P.: On universal restart strategies for backtracking search. In: Bessière, C. (ed.) CP 2007. LNCS, vol. 4741. Springer, Heidelberg (2007)
10. Gomes, C.P., Selman, B.: Algorithm portfolios. Artificial Intelligence 126(1-2), 43–62 (2001)
11. Wu, H., van Beek, P.: On portfolios for backtracking search in the presence of deadlines. In: ICTAI, pp. 231–238 (2007)
12. Inoue, K., et al.: A competitive and cooperative approach to propositional satisfiability. Discrete Applied Mathematics 154(16), 2291–2306 (2006)
13. Luby, M., Ertel, W.: Optimal parallelization of Las Vegas algorithms. In: Enjalbert, P., Mayr, E.W., Wagner, K.W. (eds.) STACS 1994. LNCS, vol. 775, pp. 463–474. Springer, Heidelberg (1994)
14. Boehm, M., Speckenmeyer, E.: A fast parallel SAT-solver: Efficient workload balancing. Annals of Mathematics and Artificial Intelligence 17(4-3), 381–400 (1996)
15. Zhang, H., Bonacina, M., Hsiang, J.: PSATO: A distributed propositional prover and its application to quasigroup problems. J. Symbolic Computation 21(4), 543–560 (1996)
16. Feldman, Y., Dershowitz, N., Hanna, Z.: Parallel multithreaded satisfiability solver: Design and implementation. Electronic Notes in Theoretical Computer Science 128(3), 75–90 (2005)
17. Blochinger, W., Westje, W., Küchlin, W., Wedeniwski, S.: ZetaSAT – Boolean satisfiability solving on desktop grids. In: CCGrid 2005, pp. 1079–1086. IEEE, Los Alamitos (2005)
18. Jurkowiak, B., Li, C., Utard, G.: A parallelization scheme based on work stealing for a class of SAT solvers. Journal of Automated Reasoning 34(1), 73–101 (2005)
19. Sinz, C., Blochinger, W., Küchlin, W.: PaSAT — Parallel SAT-checking with lemma exchange: Implementation and applications. In: SAT 2001. Electronic Notes in Discrete Mathematics, vol. 9, pp. 12–13. Elsevier, Amsterdam (2001)
20. Chrabakh, W., Wolski, R.: GridSAT: A chaff-based distributed SAT solver for the grid. In: SC 2003. IEEE, Los Alamitos (2003)
21. Hyvärinen, A.E.J., Junttila, T., Niemelä, I.: A distribution method for solving SAT in grids. In: Biere, A., Gomes, C.P. (eds.) SAT 2006. LNCS, vol. 4121, pp. 430–435. Springer, Heidelberg (2006)
22. Forman, S., Segre, A.: NAGSAT: A randomized, complete, parallel solver for 3-SAT. In: SAT 2002, Proceedings (2002),
   http://gauss.ececs.uc.edu/Conferences/SAT2002/sat2002list.html
23. Pitkanen, M.J., et al.: Using the grid for enhancing the performance of a medical image search engine. In: CBMS 2008. IEEE, Los Alamitos (accepted for publication, 2008)
24. Hyvärinen, A.E.J.: GridJM a Computer Program,
   http://www.tcs.hut.fi/~aehyvari/gridjm/

Towards an Implementation of a Computer Algebra System in a Functional Language

Oleg Lobachev and Rita Loogen

Philipps–Universität Marburg, Fachbereich Mathematik und Informatik
Hans–Mehrwein–Straße, D–35032 Marburg, Germany
{lobachev,loogen}@informatik.uni-marburg.de

Abstract. This paper discusses the pros and cons of using a functional language for implementing a computer algebra system. The contributions of the paper are twofold. Firstly, we discuss some language–centered design aspects of a computer algebra system — the "language unity" concept. Secondly, we provide an implementation of a fast polynomial multiplication algorithm, which is one of the core elements of a computer algebra system. The goal of the paper is to test the feasibility of an implementation of (some elements of) a computer algebra system in a modern functional language.

Keywords: computer algebra, software technology, language and system design.

1 Introduction

With the flow of the history of computing, exact methods gained more and more importance. It was clear since almost the beginning, that imprecise, *numerical* operations may and will fail. The Wilkinson Monster $\prod_{j=1}^{20}(x-j)$ is a nice – and old! [45,46] – example for the thesis "the way we compute it matters". One of the crucial points of computer algebra systems (CAS) is the implementation of fast algorithms. One of the core algorithms is fast multiplication, be it of numbers or of polynomials. Current approaches include methods by Karatsuba, Toom and Cook [20,44,24] and Schönhage and Strassen [38,37]. An implementation of the latter in the functional language Haskell [33] is presented in this paper to test the suitability of functional languages for implementing computer algebra algorithms. Our vision is an open–source flexible computer algebra system, that can easily be maintained, extended and optimised by the computer algebra community. The mainstream computer algebra systems like Maple [34] or Mathematica [47] provide highly optimised routines with interesting but hidden implementation details. However, the closed–source nature of such systems does not enable us to analyse their internals. On the contrary, the following modern CAS are examples for systems with freely available source code: CoCoA [8], DoCon [27], GAP [11], and GiNaC [16,13]. Our approach follows the philosophy of the GiNaC library, which extends a given language (C++) by a set of algebraic capabilities, instead of inventing a separate interface language for that purpose.

S. Autexier et al. (Eds.): AISC/Calculemus/MKM 2008, LNAI 5144, pp. 141–154, 2008.

We plan to implement a computer algebra system in a modern functional language like `Haskell`. Several features of such languages, like lazy evaluation, improve numerical computations [5,6]. Lazy evaluation is also helpful for designing algorithms in scientific computing [22]. Other features could as well be useful for a CAS [28]. Incidentally, both functional programming languages [25,42,29,2,35] and computer algebra systems [15, 14, 18, 39] are present in the field of parallel and distributed computing.

Plan of the Paper

The second section discusses the benefits of functional languages for implementing computer algebra algorithms. Section 3 pushes the *language unity concept* for CAS, i.e. choosing the same language for implementing and using a CAS. Section 4 presents a few case studies. We

a) compare different `Haskell` implementations of polynomial multiplication,
b) compare `Haskell` and imperative implementations for computing factorials,
c) consider the FFT–based implementation of polynomial multiplication by Schönhage and Strassen.

Section 5 concludes the paper. Code samples are presented in Figures 3 and 4 in Section 4.

2   Advantages of Functional Languages

We consider `Haskell` [33] as a base of our thoughts. Some of the key features of most functional programming languages, all of them found in `Haskell`, are:

- *Lazy evaluation* means that no expression is evaluated if it is not required. This can be combined with *memorisation*, when no expression is evaluated more than once. We should think of lazy evaluation as of a double–edged sword. Indeed it reduces the amount of required computations and the end user of the CAS has the freedom of writing his/her own programs in a way more corresponding to standard mathematical nomenclature. However, worse performance will be observed, if lazy evaluation fails to outweigh its overhead by skipping evaluations. A detailed comparison is beyond the scope of this paper. However nice applications of lazy evaluation in the context of scientific computing can be found in papers by Jerzy Karczmarczuk [21, 22, 23]
- Functional languages provide *infinite data structures*, notably: lists. Such lists can be easily implemented with lazy evaluation. Infinite data structures enable "more mathematical" definitions of e. g. sequences and series. On the one hand, this means "more conforming to the current mathematical nomenclature" as in e.g. `factorial n = product [1..n]` and, on the other hand, "nice in describing typical mathematical concepts" including infinite sequences. A classical example for this is `fibs = 0 : 1 : zipWith (+) fibs (tail fibs)`[1].

---

[1] See `http://haskell.org/haskellwiki/The_Fibonacci_sequence` for a sublinear time implementation of the same sequence.

- *Referential transparency* enables a "more mathematical" semantics: for function f, f(5) has the same value, whenever it is evaluated, pretty much as $f(5)$ in a mathematical notation. Consider an example in C.

  ```
  int i = 5;
  i = ++i + i++;
  ```

  This example is rather unnatural, but the result value of i depends on the implementation – try it in any imperative language of your choice. In a pure functional language, such dubious definitions are not possible.
- In the context if a CAS *strong typing* gives some benefits. For example, it is possible to produce an error at compile time for a product of matrices of incompatible dimensions. On the other hand, *type inference* is possible. However there are some problems with Haskell type system in a computer algebra context. For instance, if you define a factor ring over a commutative ring, it may or may be not a field: it depends on the properties of the ideal. If rings, domains, etc. are defined as types, the Haskell type system would not be able to determine at compile time, whether this *instance* of type "factor ring" is a field or not. Papers by S. Mechveliani, for instance [26], discuss this problem and suggest an appropriate solution.
- Haskell's *hierarchical module system*, being a rather software engineering issue, provides the possibility to structure large programs efficiently.
- Another benefit of modern functional languages is the possibility to *prove* the correctness of implementations.

3 The Two Languages of a CAS

Computer algebra systems possess two different languages, we shall call them in this paper as follows. The *internal* language of a CAS is the language the system is written in, the implementation language. Since the end user of the CAS wants to perform some kind of *programming*, there is also a second language. The *external* language of a CAS is the language for user interaction, the interface language. The idea of "language unity" is to utilise the same language for both purposes, i.e. as internal and as external language.

It is desirable to write as much as possible of the CAS itself in its external language. This gives the user the opportunity to inspect and (if needed) to modify some external functions of the CAS. However, for several reasons, this is impossible in most CAS. Firstly, the external language of most CAS is "weaker" than their internal one in the sense that some technical things may be hard or even impossible. On the other hand, the external language is better suited for the typical computer algebra operations: we may expect, e. g. polynomials and matrices as native objects or an interesting handling of lists, non–existent in the internal language of the CAS if this language is imperative. Especially advanced features like type safety and generic programming are desired in the external language. A recent development is to utilise a general purpose dynamic language like Ruby [10], Groovy [3] or Python [43] for interconnecting different programs, building a composite computer algebra system [40].

Secondly, unfortunately, the external language of most CAS is not as fast as the internal one. The cause may be the interpreted origin of these languages or their very high level nature. This is often avoided by compiling the input files to some kind of byte code. Other speedup approaches compromise the extensibility. The implementation of the S programming language for statistical computations, GNU R, utilises a Scheme dialect as its external language. The whole R system could be implemented in Scheme. But because of performance lack in core operations, these are replaced with function calls from the bundled C library. These functions can still be overloaded and replaced by the user's own version, but one cannot simply look into the routines, which are sped up this way. There is also a third option: to use a functional language and to perform optimisations in the language compiler typical for a functional language. This way our external language could be feature–rich and reasonably fast, but it will have the price of writing a, say, LISP interpreter in an imperative language.

An interesting approach in this field was taken by Christian Bauer, Alexander Frink, Richard Kreckel et al., the developers of GiNaC [4, 13]. This computer algebra system was written in C++ and it maintains C++ as its main interface. It is made in a very simple way: GiNaC is rather a computer algebra *library*, than a complete system. So the primary use of GiNaC is to give one a possibility of writing his/her own C++ programs, while using arbitrary precision numbers, polynomials, matrices, expression evaluation and other nice and fast computer algebra functions, offered by the GiNaC library. As the authors of GiNaC state:

> Its design is revolutionary in a sense that contrary to other CAS it does not try to provide extensive algebraic capabilities and a simple programming language, but instead accepts a given language (C++) and extends it by a set of algebraic capabilities.

This approach is very interesting and powerful, but the interactive front end program of GiNaC, the ginsh, is less powerful due to a rather weak language. It was, however, never intended to be a complete GiNaC interface. The possibility to use all the GiNaC features at an interactive prompt requires a C++ *interpreter*. While interpreting C++ is not very nice (although possible: see e.g. [9])[2], it is much easier with Haskell: aside from the glorious Glasgow Haskell Compiler [12], we have Hugs, the Haskell interpreter. Also GHC itself offers an interactive version, GHCi. The latter is capable of loading pre-compiled object files into the interpreted environment. With this achievement one has the possibility to write a computer algebra system, whose *external* interface language equals its *internal* implementation language, and this language is a functional one.

The idea of GiNaC was not born in vain: most *long* CAS–supported computations are run in "batch mode", with no user interaction. It seems plausible not to wait in front of a command prompt for the result for hours, days or even

---

[2] There is also a third party GiNaC interface language project,
http://swiginac.berlios.de/

months.[3] On the other hand, most of CAS–based *development* is done in an interactive environment, in a "shell". If one could use the same language both for developing and for lengthy computations, this would be a major success in saving developers' work time [32] and gaining stability of computations.

Now why not just make both: a compiler *and* an interpreter of CAS' external language? The problem is, that despite many efforts, the external languages of computer algebra systems are slow. On the other hand, we already have a fast language in our CAS–developing project. This is the language, the CAS itself is written in, the *internal* language. One may oppose, however, the whole game with computer algebra system's *external* language was started, because the internal language was not high–level enough for vectors, matrices, polynomials and all the other expressions, which are eagerly wanted in a full–fledged CAS. Now we come back to the beginning of this paper. Functional languages *are* complicated and high–level enough to have all the aforementioned objects and properties [33,41,27,7,17]. Functional languages have very compact code size and rapid development times [32]. Most functional languages have very interesting data structures and language design features, which benefit both featuring them as an internal or as an external language, see [31] for details. And some modern functional languages already have an efficient compiler and an interpreter implemented, which leads us to the future goal of internal and external language fusion. Haskell is an example of a such language.

Concluding: an implementation of a CAS in a functional language utilising the above "language unity" concept will greatly reduce code size and improve readability, at the same time it shall not reduce the performance significantly. In order to test the feasibility of these assumptions we consider several case studies.

4 First Case Studies

Now as we have seen some *theoretical* reasons for a CAS to be implemented in Haskell, let's take a look at some examples. At first we shall examine the univariate polynomials. One can hardly imagine a computer algebra system without them, polynomials are used in thousands of higher–level algorithms and the operations with the polynomials should be fast. Unfortunately as of today neither of available Haskell software packages implementing univariate polynomials uses sub–quadratic algorithms like Karatsuba[4], Toom–Cook or Schönhage–Strassen algorithms. As one of the examples we demonstrate an implementation of Schönhage–Strassen algorithm in Haskell. But first we look at the schoolbook case.

---

[3] In this case one might think of porting his/her CAS–based program to some low-level language and let, say, FORTRAN run the number–crunching mills. However this is a highly interactive and bug–ridden process. And the FORTRAN program is to be tested for errors *again*, before the real computations may begin: the thoroughly tested CAS–routines are not enough!

[4] Although an implementation of this algorithm in Haskell was presented in [19,36].

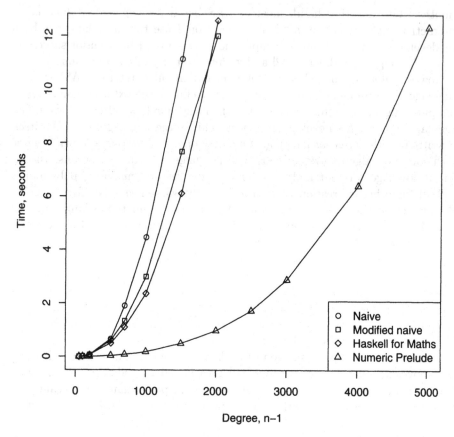

Fig. 1. Multiplication of univariate polynomials of degree $n - 1$. Runtime comparison of naive implementations.

All the tests were run on the same machine[5] with the same compiler – GHC 6.8.2. For the same n, each test was run ten times and the mean value of measured execution time has been determined. We utilise standard `Haskell` lists for representing the polynomials. The complete system would use some kind of generalisation layer, probably based on type classes, to abstract the implementation from the given representation. It would be sufficient to redefine the few standard functions on lists to obtain the implementation of the same algorithm for yet another data structure. No modification of the presented code would then be required.

4.1   Naive Polynomial Multiplication

We have tested four different $\mathcal{O}(n^2)$ implementations:

1. our own naive implementation with lists of integers
2. our naive implementation, modified à la Numeric Prelude,

---

[5] AMD Athlon 64 X2 4000+ CPU with 1 Gb RAM, running Gentoo Linux.

3. the implementation from *Haskell for Math* [1],
4. the implementation from *Numeric Prelude* [41].

We multiply two dense univariate $(n-1)$–grade polynomials with random co-efficients. The coefficients are random signed 32–bit integers: what we test here are the polynomial multiplication implementations, not the hardware multiplication of small integers, nor even different libraries for arbitrary precision integers. Nor do we test the quadratic algorithms – they all represent pretty the same "school" multiplication – or compiler options, but the impact of the particular implementation decisions on the performance. The naive implementation uses a "dumb" list of Ints, the other implementations build a chain of types similar to the algebraic objects. One can e.g. define addition and subtraction for elements of the additive group, multiplication for elements of this group embedded into a ring, and finding an inverse for invertible elements of this ring embedded into a field. An overview of test results is provided in Figure 1. Time is measured in seconds. The Numeric Prelude implementation is much better than the other implementations which show similar runtimes. Note that the simplest implementation is *not* the fastest one and that the type hierarchy enables optimisations. Nevertheless, we conclude the strong need for sub–quadratic implementations.

4.2 Computing Factorial

We would like to discuss briefly another example. We take a well–known and very quickly growing function on integers: the factorial. We have tested the famous Haskell one–liner factorial n = product [1..n], and two C++ implementations. Both C++ versions are based on the CLN [16] – the arbitrary precision library used in GiNaC. One implementation uses the built–in factorial function from the CLN. It makes use of table look–ups and computes some parts of the factorial value in divide and conquer fashion. The other C++ implementation is not optimised, but it still uses CLN built–in multiplication and large integers. We find this implementation comparable with the naive Haskell implementation. Arbitrary long integers are provided in Haskell out of the box. We are not willing to discuss the details of arbitrary precision arithmetic implementation in Haskell compiler runtime, our focus is to demonstrate how competitive the functional approach is. The graphical representation of the obtained results is shown in Figure 2. The timings of the Haskell version lie in between both C++ versions.

This small example shows that Haskell implementations, even in their simplest and primitive form are competitive with implementations in some industry–used programming language which are more sophisticated in programming effort. The optimised version outperforms both naive versions, thus motivating us to create implementations of fast algorithms in Haskell.

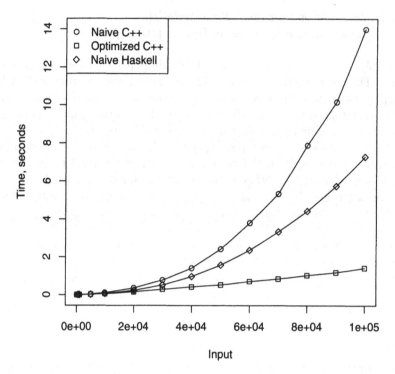

Fig. 2. Computing the factorial

4.3 Fast Polynomial Multiplication

The essence of Schönhage and Strassen's method for fast polynomial multiplication[6] is the way a *convolution* is performed. A convolution in $\mathbb{C}[x]$ corresponds to multiplication, as in "each with every". A convolution in Fourier–transformed space is just a component–wise multiplication. So if we want to compute a product of two polynomials, we compute their Fourier transformed (e. g. with the routine in Figure 3), then multiply the transformed functions component wise and then, with the inverse Fourier transformation, transform the product back to a polynomial (Figure 4). The presented version performs twice as well as the full version at the price of not computing the complete product. However, the current implementation for computing the full product can be easily obtained from this code. The functions `zipWith`, `splitAt`, `length`, `concat` and `transpose` are provided by `Haskell` standard libraries. `zipWith` "zips" two lists with a supplied binary function, e.g. `zipWith (+) [1,2,3] [4,5,6]` results in `[5,7,9]`. `splitAt` splits a list into two parts at the provided offset. `length` returns the length of a list. `concat` concatenates a list of lists to a list. `transpose`,

---

[6] ... over the domains supporting the fast Fourier transform, just like complex numbers $\mathbb{C}$. If the domain does not support FFT, the fast multiplication is still possible, through an implicit algebraic extension of the original domain. For details please refer to the original paper [38] or a standard book on this topic [44].

```
fft     :: [Complex Double] -> [Complex Double]
fft f   = mix [fft (l @+ r), fft ((l @- r)@* w)]
          where (l, r)  = splitAt (length f 'div' 2) f
                mix     = concat . transpose
                (@+) f g = zipWith (+) f g  -- @-, @* analog
                -- w is list of powers of an n-th primitive root of unity.
```

Fig. 3. Implementation of Cooley–Tukey algorithm in `Haskell`

```
(%*%)     :: (Num a) => [a] -> [a] -> [a]
(%*%) f g = unlift $ ifft ((fft $ lift f) @* (fft $ lift g))
            -- where lift   :: (Num a) => [a] -> [Complex Double]
            --       unlift :: [Complex Double] -> [Int]
            -- ifft is the inverse fft, basicly the same fft with
            -- different twiddle factors.
            -- And (@*) is still element-wise multiplication
```

Fig. 4. FFT–based multiplication modulo $x^n - 1$ in `Haskell`

as the name says, transposes a list of lists. The functions `lift`, `unlift`, `ifft` and `(@*)` are part of our implementation. The inverse Fourier transformation is nothing spectacular and is pretty much the forward Fourier transformation with different values. As the fast Fourier transform (FFT) for a polynomial in $\mathbb{C}[x]$ of degree $n - 1$ can be performed in $\mathcal{O}(n \log n)$ time and the component-wise multiplication in $\mathcal{O}(n)$, we can multiply two polynomials of degree $n - 1$ in $\mathbb{C}[x]$ in $\mathcal{O}(n \log n)$ time [44]. Due to limitations of the naive implementation we receive the remainder of the product after the division through $x^n - 1$. But it is still possible to compute the whole product without changing the asymptotic complexity, for example, applying one step of the Karatsuba algorithm first, or just padding both arguments to the length of the product.

The technical representation of a polynomial in our case is a list of coefficients. The Cooley–Tukey decimation in frequency algorithm is utilised, using a divide–and–conquer approach for computing the Fourier transform. This is the simplest FFT algorithm, there exist some more sophisticated variants [44, 30]. Figure 5 presents the results, we used the same kind of input as in Figure 1. A sub–quadratic method for polynomial multiplication is definitely superior. The bottom line is the FFT–based multiplication algorithm, we compute the whole product. It is clearly visible, that the current FFT algorithm is relying on the fact, that the length of its input is a power of two. The rapidly ascending lines correspond to the values shown in Figure 1. Unfortunately, we have no explanation for the decreasing values of the FFT-based algorithm for $n \in [16000..20000]$.

Now we have seen a fast polynomial multiplication in `Haskell`. By using advanced algorithms we significantly increase the performance, the implemented functions can be used by any other `Haskell` program, as we have not tweaked the compiler. The size of the code base is modest for the task it accomplishes. This case study shows that it is possible to extend `Haskell` with further implementations of fast computer algebra algorithms, obtaining in the end a computer

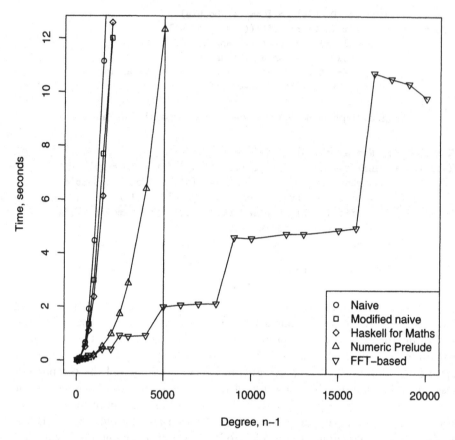

Fig. 5. Multiplication of dense univariate polynomials of degree $n - 1$ revised. Naive Implementations vs. FFT–based. The left side of the plot corresponds to the Figure 1.

algebra library. The main interface to this system is the language itself, direct interaction with the library is possible with an interpreter.

5   Related Work

Writing a computer algebra system in a functional programming language is not a really new idea. The first generation CAS named Macsyma was written in LISP 1.5 dialect called MACLISP, and LISP is considered to be the first functional language ever. Axiom CAS has some interesting aspects. It features an embedded (although detachable) functional programming language [7]. In addition, it uses a hierarchical structure of mathematical objects (like: monoid – group – ring – integrity domain – field) to specify and perform operations on them.

The DoCon computer algebra library [27] is at the first glance very similar to our intention. It utilises Haskell as implementation language. Being a library, it also has Haskell as an interface language. However, DoCon pursues a different goal. DoCon is an algebra *framework*, implementing different mathematical objects and their relations, thereby heavily dependent on `Haskell`'s type system. For instance, it is easy to define a residue domain modulo some polynomial ideal in DoCon. However, we focus on the computer algebra *algorithms*. We would like to have e.g. a fast polynomial multiplication, while representing the polynomials as simply as possible. Moreover, we are interested in *parallelising* our algorithm implementations. Because of high communication costs, we need to keep the underlying data structures as "dumb" as possible. It will be interesting to utilise the DoCon approach in our own work and to share our results with the current DoCon implementation.

6 Conclusions and Future Work

We propose to unify the internal implementation and the external interface language of computer algebra systems and to use a functional language to achieve this integration. The usage of a functional language in a computer algebra field drastically reduces the size of the source code. Secondly, it does not affect the performance. Hence, is not required to mix two different languages in an implementation of a CAS. We have shown that functional programs are competitive with mainstream imperative programs and significantly easier to develop.

Concerning the performed case studies, a possible direction of the future work would be the optimisation of the fast Fourier transform. Some practical tests in the parallel context indicate an optimisation potential in switching to decimation in time. From the theoretical viewpoint, it will be interesting to reconstruct the Fourier transformed values in special cases, the so–called pruned FFT algorithm. It would also be of interest to try other FFT algorithms, for example, the r–radix implementations.

Concerning the future goals of this work: Modern functional languages and computer algebra are two rapidly developing research areas, an intersection of these two areas is highly interesting. A third component to mix into this "cocktail" of computational algebra and functional programming topics is parallelism. Computer algebra applications tend to be quite resource hungry and functional languages have great potential in parallelism, which is being currently quite extensively investigated. With respect to our gradually evolving practical implementation, modern algorithms of computer algebra should be implemented in relevant `Haskell` software packages, as a naive implementation typically leads to asymptotically bad complexity. One should carefully design such implementations, as design choices play a significant role for the execution times in the same complexity class. Such choices gain even more on importance in the parallel setting. The aforementioned algorithms should provide

- fast polynomial multiplication – tackled in this paper,
- fast integer multiplication – our current approach is to use fast polynomial multiplication,
- efficient Euclid's algorithm for polynomials,

- efficient vector and matrix computations,
- framework for symbolic computation and object manipulation.

Such foundation will be a solid base for more complex research areas, including

- algorithms of numerical number theory,
- implementation of public key cryptography,
- algorithms of computational algebraic geometry, based on Gröbner bases,
- symbolic integration and summation,
- parallel computations.

As for FFT–based multiplication, we provide our `Haskell` implementation of polynomial multiplication, a multiplication routine for arbitrary long integers based on top of it and an interface script to SCSCP [39] on request.

Acknowledgement

We would like to thank to anonymous referees for their helpful and detailed comments.

References

1. Amos, D.: Haskell for Math program,
   http://www.polyomino.f2s.com/david/haskell/codeindex.html
2. Armstrong, J.: Programming Erlang. In: The Pragmatic Programmers, LLC (2007)
3. Barclay, K., Savage, J.: Groovy Programming: An Introduction for Java Developers. Morgan Kaufmann Publishers Inc, San Francisco (2006)
4. Bauer, C., Frink, A., Kreckel, R.: Introduction to the GiNaC Framework for Symbolic Computation within the C++ Programming Language. J. of Symbolic Computation 33, 1–12 (2002)
5. Benouamer, M.O., Michelucci, D., Peroche, B.: Error-free boundary evaluation based on a lazy rational arithmetic: a detailed implementation. Computer Aided Design 26(6), 403–416 (1994)
6. Bird, R.S., Jones, G., De Moor, O.: More haste, less speed: lazy versus eager evaluation. J. of Functional Programming 7(5), 541–547 (1997)
7. Bronstein, M., Davenport, J., Fortenbacher, A., et al.: AXIOM – the 30 year horizon (2003), http://portal.axiom-developer.org/public/book2.pdf
8. Capani, A., Niesi, G.: CoCoA 3.0 User's Manual. Dipartimento di Matematica, Università di Genova, Via Dodecaneso, Genova (Italy), vol. 35, I-16146 (1995)
9. Cint, the C/C++ interpreter, version 5.16.19,
   http://root.cern.ch/root/Cint.html
10. Flanagan, D., Matsumoto, Y.: The Ruby Programming Language. O'Reilly, Sebastopol (2008)
11. The GAP Group. GAP – Groups, Algorithms, and Programming, Version 4.4.10 (2008)
12. The Glorious Glasgow Haskell Compilation System User's Guide (February 2008), http://www.haskell.org/ghc/docs/latest/users_guide.pdf
13. GiNaC program, http://www.ginac.de

14. HPC-Grid for Maple program,
    http://www.maplesoft.com/products/toolboxes/HPCgrid/index.aspx
15. gridmathematica2 program,
    http://www.wolfram.com/products/gridmathematica/
16. Haible, B., Kreckel, R.: CLN, a class library for numbers manual (2005),
    http://www.ginac.de/CLN/cln.ps
17. Hall, C., Hammond, K., Jones, S.P., Wadler, P.: European Symposium On Programming. In: Sannella, D. (ed.) ESOP 1994. LNCS, vol. 788, pp. 241–256. Springer, Heidelberg (1994)
18. Hammond, K., Al Zain, A., Cooperman, G., Petcu, D., Trinder, P.: Symgrid: a framework for symbolic computation on the grid. In: Kermarrec, A.-M., Bougé, L., Priol, T. (eds.) Euro-Par 2007. LNCS, vol. 4641, Springer, Heidelberg (2007)
19. Herrmann, C.A., Lengauer, C.: HDC: A Higher–Order Language for Divide–and–Conquer. Parallel Processing Letters 10(22), 239–250 (2000)
20. Karatsuba, A., Ofman, Y.: Multiplication of many-digital numbers by automatic computers. Doklady Akad. Nauk SSSR 145, 293–294 (1962); Translation in Physics–Doklady 7, 595–596 (1963)
21. Karczmarczuk, J.: The most unreliable technique in the world to compute pi (1998)
22. Karczmarczuk, J.: Scientific computation and functional programming. Computing in Science & Engineering 1(3), 64–72 (1999)
23. Karczmarczuk, J.: Functional differentiation of computer programs. Higher–Order and Symbolic Computation 14(1), 35–57 (2001)
24. Knuth, D.E.: The Art of Computer Programming, 3rd edn., vol. 2. Addison–Wesley (1998)
25. Loogen, R., Ortega-Mallén, Y., Peña-Marí, R.: Parallel Functional Programming in Eden. Journal of Functional Programming 15(3), 431–475 (2005)
26. Mechveliani, S.D.: Haskell and computer algebra. Pereslavl-Zalessky, Russia (manuscript, 2000)
27. Mechveliani, S.D.: DoCon. The Algebraic Domain Constructor Manual. Program Systems Institute, Pereslavl–Zalessky, Russia, Version 2.11 (2007)
28. Milmeister, G.: Functional kernels with modules. Master's thesis, ETH Zürich (1995)
29. Nikhil, R.S., Arvind, L.A., Hicks, J., Aditya, S., Augustsson, L., Maessen, J., Zhou, Y.: pH Language Reference Manual, Version 1.0. Massachusetts Institute of Technology, Computation Structures Group Memo No. 396 (1995)
30. Nussbaumer, H.J.: Fast Fourier Transform and Convolution Algorithms. Springer, Berlin (1981)
31. Okasaki, C.: Purely Functional Data Structures. Cambridge University Press, Cambridge (1998)
32. Jones, M.P., Hudak, P.: Haskell vs. Ada vs. C++ vs. awk vs.... An experiment in software prototyping productivity, Yale University, Department of Computer Science (July 1994)
33. Jones, S.P. (ed.): Haskell 98 Language and Libraries: The Revised Report. Cambridge University Press, Cambridge (2003)
34. Redfern, D.: The Maple Handbook: Maple V Release 4. Springer, Heidelberg (1995)
35. Van Roy, P. (ed.): Multiparadigm Programming in Mozart/Oz. In: Van Roy, P. (ed.) MOZ 2004. LNCS, vol. 3389. Springer, Heidelberg (2005)
36. Schaller, C.: Elimination von Funktionen höherer Ordnung in Haskell–Programmen. Master's thesis, Universität Passau (September 1998)

37. Schönhage, A.: Asymptotically fast algorithms for the numerical multiplication and division of polynomials with complex coefficients. In: Calmet, J. (ed.) ISSAC 1982 and EUROCAM 1982. LNCS, vol. 144, pp. 3–15. Springer, Heidelberg (1982)
38. Schönhage, A., Strassen, V.: Schnelle Multiplikation großer Zahlen. Computing 7(3–4), 281–292 (1971)
39. Symbolic Computation Infrastructure for Europe project, http://www.symbolic-computation.org/
40. Stein, W.: Sage: Open Source Mathematical Software (Version 2.10.2) The Sage Group (2008), http://www.sagemath.org
41. Thurston, D., Thielemann, H.: Haskell Numeric Prelude program, http://darcs.haskell.org/numericprelude/
42. Trinder, P.W., Barry Jr., E., Davis, M.K., Hammond, K., Junaidu, S.B., Klusik, U., Loidl, H.-W., Jones, S.L.P.: GpH: An Architecture–Independent Functional Language. In: Glasgow Functional Programming Workshop, Pitlochry, Scotland (September 1998)
43. van Rossum, G.: The Python Language Reference Manual. Network Theory Ltd. (2006)
44. von zur Gathen, J., Gerhard, J.: Modern Computer Algebra, 2nd edn. Cambridge University Press, Cambridge (2003)
45. Wilkinson, J.H.: Rounding Errors in Algebraic Processes. Prentice Hall, Englewood Cliffs (1963)
46. Wilkinson, J.H.: The perfidious polynomial. In: Golub, G.H. (ed.) Studies in Numerical Analysis, Mathematical Association of America, Washington, D.C, vol. 24, pp. 1–28 (1984)
47. Wolfram, S.: Mathematica: a system for doing mathematics by computer. Wolfram Research, Inc. (1991)

Automated Model Building: From Finite to Infinite Models

Nicolas Peltier

LIG INPG/CNRS
46, avenue Félix Viallet
38031 Grenoble Cedex, France
Nicolas.Peltier@imag.fr

Abstract. We propose a method for using existing finite model builders for constructing infinite models of first-order formulae. The considered interpretations are represented by tree tuple automata. Our approach is based on formula transformation. It is proven to be sound (i.e. all the constructed interpretations are models of the original formula) and complete for the considered class of interpretations (i.e. a model is eventually built for any formula having a model representable by a tree automaton).

1 Introduction

Many works in Automated Theorem Proving aim at defining efficient algorithms for identifying valid (resp. unsatisfiable) formulae and constructing proofs (resp. refutations). However, at the beginning of the 90s, Automated Model Building emerged as a very important, – and complementary – trend of research. Constructing a counter-example (resp. model) of a formula is a very natural and convincing way of proving that it is not valid (resp. that it is satisfiable). This is very useful for some applications: for instance, in program verification, this feature is critical for detecting bugs and giving hints to correct them.

Many systems have been developed for the automated construction of models in first-order logic: Finder [13], Sem [14], FMC [11], Mace [10], Paradox [5] etc. Of course, since first-order logic is undecidable (semi-decidable) no complete algorithm exists for model construction.

Most existing automated model builders try to construct finite interpretations by enumeration, using sophisticated techniques for pruning the search space, in particular for detecting and discarding isomorphic interpretations. This is obviously a critical point, since the number of interpretations is exponential w.r.t. the cardinality of the domain (see for instance [2,1] for more details on this problem). However, many satisfiable formulae have no finite models, or have only models of very large size. Finite model builders are useless or inefficient on such formulae.

Thus, methods have also been developed to construct automatically infinite models, or more precisely to construct *finite representations* of infinite models. These approaches are mainly based on deduction and use refinements of

S. Autexier et al. (Eds.): AISC/Calculemus/MKM 2008, LNAI 5144, pp. 155–169, 2008.

deductive techniques such as resolution and tableaux for extracting models from satisfiable formulae. The reader can consult [3] for a complete synthesis of deduction-based approaches in automated model building. Of course, since the interpretations are infinite (non enumerable) objects, no computer program can generate them explicitly: what can be built is only a *finite* representation of these objects. Various formalisms have been used for this purpose, with various expressive powers and complexities: sets of atoms (see for instance [9]), equational formulae [4], contexts with exceptions [7] etc.

In this work we investigate another possibility, based on completely different principles. Rather than designing new algorithms and systems, we propose to use existing *finite* model builders for constructing *infinite* structures (instead of finite ones). For this purpose, starting from a formula ϕ, we apply a transformation algorithm on ϕ in order to obtain a formula $\psi = \Delta(\phi)$ with the following properties:

- ψ is satisfiable iff ϕ is satisfiable (satisfiability is preserved).
- Any **finite** model of ψ can be seen as a finite representation of an *infinite* model of ϕ.

In other words, rather than using the finite model builder for enumerating the (finite) interpretations themselves, we propose to use it for enumerating finite structures representing (possibly infinite) interpretations.

This principle can be used for various model representation formalisms. In this paper, we restrict ourselves to interpretations representable by finite tree automata on tuples of terms [6]. It is well known that this formalism is strictly more expressive than finite models and than finite sets of ground atoms [3]. We show how to define the above transformation appropriately for this class of interpretations. This is not easy since computing complements or projections of regular (term) languages necessarily involves some fixpoint computations (e.g. to check that a given final state is non reachable), which is in principle out of the scope of first-order logic. We show how to overcome this issue if the considered automaton is finite.

Our idea is well-illustrated by the schema below (see Figure 1).

The formula ϕ is transformed into a formula $\psi = \Delta(\phi)$, then a finite model $\mathcal{M}'$ of ψ is built. $\mathcal{M}'$ can be mapped to a finite tree automaton. The set of terms accepted by this automaton defines an interpretation $\mathcal{M}$ validating ϕ. $\mathcal{M}$ is infinite in general (since its domain is the set of ground terms).

The rest of the paper is structured as follows.

- In Section 2, we introduce some preliminary definitions (about first-order logic and tree automata). The notations are mainly standard, although we adapt some of the usual definitions to better suit our purposes.
- In Section 3, we describe the transformation algorithm.
- In Section 4, we give a simple example of application.
- In Section 5, we prove the soundness of our transformation algorithm and its completeness w.r.t. the class of representable models.
- Section 6 contains a short conclusion and some lines of future works.

Due to space restriction, the most complex proofs are omitted.

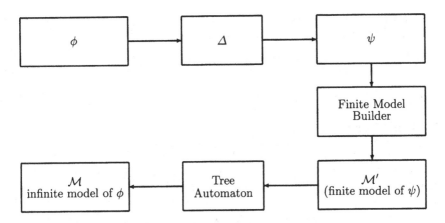

Fig. 1. From Finite to Infinite models

2 Preliminaries

2.1 Basic Definitions

In this section, we briefly review the definitions and notations that are necessary for the understanding of our work. We assume some familiarity with the usual notions in Logic (missing definitions can be found for instance in [8]).

We assume three disjoint sets of symbols are given: a set of *function symbols* Σ, a set of *predicate symbols* Ω and a set of *variables* $\mathcal{X}$. Let *ar* be a function mapping each symbol in $\Sigma \cup \Omega$ into a natural number. Constants are function symbols of arity 0. Ω contains a special symbol $\approx$ of arity 2 (in infix notation).

Terms and formulae are built inductively as usual, using a set of logical symbols $\neg, \vee, \wedge, \Rightarrow, \Leftrightarrow, \forall, \exists$. $FVar(\phi)$ denotes the set of free variables occurring in the formula ϕ (it is defined as usual). A term (formula) is said to be *closed* if it contains no (free) variables.

A *signature* is a subset of $\Sigma \cup \Omega$. An *interpretation* $\mathcal{I}$ *of* a signature $\mathcal{S}$ is defined by a nonempty set $D_{\mathcal{I}}$ called the domain of $\mathcal{I}$ and by a function mapping each function symbol f of arity n in $\mathcal{S}$ into a n-ary function $f_{\mathcal{I}}$ from D^n to D (in particular constant symbols are mapped to elements of D) and each predicate symbol p of arity n in $\mathcal{S}$ to a n-ary relation $p_{\mathcal{I}}$ on the domain D.

A *variable assignment* for a term or formula ϕ and an interpretation $\mathcal{I}$ is a function mapping each free variable in ϕ to an element in $D_{\mathcal{I}}$. The (truth) value of a term or formula ϕ in a pair $(\mathcal{I}, \sigma)$, where $\mathcal{I}$ is an interpretation and σ a variable assignment, is denoted by $[\phi]_{\mathcal{I}_\sigma}$ and defined inductively as usual.

A *position* is a finite sequence of natural numbers. ϵ denotes the empty sequence and $p.q$ denotes the concatenation of the sequences p and q. For any formula ϕ, the set of positions $Pos(\phi)$ occurring in ϕ and the formula $\phi_{|p}$ occurring at position p in ϕ are defined inductively as follows:

- $\epsilon \in Pos(\phi)$ and $\phi_{|\epsilon} \stackrel{def}{=} \phi$.
- If ϕ is of the form $\neg\psi$ (resp. $(\exists x)\psi$ or $\psi \wedge \psi'$) and p is a position in ψ, then $1.p \in Pos(\phi)$ and $\phi_{|1.p} \stackrel{def}{=} \psi_{|p}$.
- If ϕ is of the form $\psi' \wedge \psi$ and p is a position in ψ, then $2.p \in Pos(\phi)$ and $\phi_{|2.p} \stackrel{def}{=} \psi_{|p}$.

2.2 Tuple Term Automata

We recall some basic definitions on tuple tree automata (TTA for short). For technical convenience we adapt some of the definitions (see [6] for more details).

Let $\perp$ be a new symbol not occurring in $\Sigma, \Omega, \mathcal{X}$.

If t is a term, then $t_{|i}$ denotes the i-th argument of the term t (i.e. the term t_i if t is of the form $f(t_1,\ldots,t_n)$ with $i \in [1..n]$) and $\perp$ otherwise (for instance $a_{|1} = \perp$). If $\boldsymbol{t} = (t_1,\ldots,t_n)$ is a vector of terms, then $\boldsymbol{t}_{|i}$ denotes the vector $(t_{1|i},\ldots,t_{n|i})$. For any ground term t, we denote by $head(t)$ the head symbol of t i.e. the symbol f s.t. $t = f(\ldots)$. If $\boldsymbol{t} = (t_1,\ldots,t_n)$ then $head(\boldsymbol{t}) \stackrel{def}{=} (head(t_1),\ldots,head(t_n))$.

Definition 1. A (n,m)-TTA is a tuple (S, s_0, τ, S_F) where S is a finite set, $s_0 \in S$, $S_F \subseteq S$, τ is a function mapping each pair $(\boldsymbol{s}, \boldsymbol{f}) \in S^m \times (\Sigma \cup \{\perp\})^n$ to an element in S, s.t. $\tau(s_0^m, \perp^n) = s_0$.

S is a set of *states*, s_0 is the *initial state* of the automaton, S_F is a set of *final states* and τ is a *transition function*. Intuitively, a (n,m)-TTA recognizes n-tuples of terms, built on a set of function symbols of maximal arity m.

Let $\mathcal{A} = (S, s_0, \tau, S_F)$ be a TTA. If $\boldsymbol{t}$ is a n-tuple of terms, then $\mathcal{A}(\boldsymbol{t})$ is inductively defined as follows: if $\boldsymbol{t} = \perp^n$ then $\mathcal{A}(\boldsymbol{t}) \stackrel{def}{=} s_0$, otherwise $\mathcal{A}(\boldsymbol{t}) \stackrel{def}{=} \tau((\mathcal{A}(\boldsymbol{t}_{|1}),\ldots,\mathcal{A}(\boldsymbol{t}_{|m})), head(\boldsymbol{t}))$. Intuitively, $\mathcal{A}(\boldsymbol{t})$ is the state associated to the tuple $\boldsymbol{t}$. A tuple of terms is said to be *accepted* by $\mathcal{A}$ if $\mathcal{A}(\boldsymbol{t}) \in S_F$.

Example 1. We assume that 3 function symbols a, f, g of arity $0, 1, 2$ respectively are given. Let $\mathcal{A}$ be the $2, 2$-automaton defined as follows:

$$\mathcal{A} \stackrel{def}{=} (\{s_0, s_1, s_2, s_3, s_4\}, s_0, \tau, \{s_3\})$$

where the transition function τ is defined as follows:

S^2	Σ^2	S
s_0, s_0	$(\perp, \perp) \rightarrow$	s_0
s_0, s_0	$(\perp, a) \rightarrow$	s_1
s_1, s_0	$(\perp, f) \rightarrow$	s_2
s_1, s_2	$(a, g) \rightarrow$	s_3

and $\tau(\boldsymbol{s}, \boldsymbol{t}) \stackrel{def}{=} s_4$ otherwise (s_4 is a "deadlock" state, useful only to make the automaton deterministic).

The reader can easily check that the set of tuples accepted by $\mathcal{A}$ is $\{(a, g(a, f(a)))\}$.

The following definition will be useful in the following.

Let μ be a function symbol of arity m. If t, s are two terms built on $\perp, \mu$, we write $t \leq s$ iff either $t = s$ or $t = \perp$ or $t = \mu(t_1, \ldots, t_n), s = \mu(s_1, \ldots, s_m)$ and for any $i \in [1..m]$ we have $t_i \leq s_i$. We write $t < s$ if $t \leq s$ and $t \neq s$.

If t is a n-tuple of terms, then $l(t)$ denotes a term built on the signature $\perp, \mu$ inductively defined as follows: if $t = \perp^n$ then $l(t) \stackrel{def}{=} \perp$, otherwise $l(t) \stackrel{def}{=} \mu(l(t_{|1}), \ldots, l(t_{|m}))$. $l(t)$ encodes in some sense the "shape" of the vector t. For instance, $l(f(a), g(f(a), b)) = \mu(\mu(\mu(\perp, \perp)), \mu(\perp, \perp))$.

3 Transformation Algorithm

In this section we define an algorithm transforming any formula ϕ into a formula $\Delta(\phi)$ satisfying the properties informally described in the Introduction.

3.1 Restricting the Input Language

For technical convenience, we make some further assumptions about the considered formulae. These assumptions help to improve the readability of the forthcoming sections without entailing any loss of generality.

We assume that the formulae contain no function symbol and no occurrences of $\vee, \forall, \Rightarrow, \Leftrightarrow$. Obviously this does not restrict the expressive power of the language, since all these connectives can be expressed by using only $\neg, \wedge$ and $\exists$ (structural transformation algorithms can be used to avoid an explosion of the size of the formula). Moreover, functions may be denoted by predicate symbols, by *flattening* the terms as it is done for instance in Prolog, for instance $p(f(a))$ becomes $\neg a(x) \vee \neg f(x, y) \vee p(y)$, with the axioms $(\exists x)a(x)$ and $(\forall x)(\exists y)f(x, y)$. Moreover, we also assume that all the atoms occurring in the formulae are *linear*, i.e. contain at most one occurrence of each variable. Since the formulae contain no function symbol, this implies that any atom is of the form $p(x_1, \ldots, x_n)$ where the x_i's are *pairwise distinct variables*.

This is possible since if any atom of the form $p(x_1, \ldots, x_n)$ where $x_i = x_j$ for some $i, j \in [1..n]$, $i \neq j$ can be replaced by $(\exists x'_j)[(x'_j \approx x_i) \wedge p(x_1, \ldots, x_{j-1}, x'_j, x_{j+1}, \ldots, x_n)]$, where x'_j is a new (fresh) variable distinct from $x_1, \ldots, x_n$. It is clear that the last formula is equivalent to $p(x_1, \ldots, x_n)$. By repeating this process as many times as needed we eventually obtain a formula only containing linear atoms.

We assume a total (arbitrarily chosen) ordering on the variables $\prec$ is given. For any formula $(\exists x)\psi$, x must be greater than every free variable in $(\exists x)\psi$ according to $\prec$ (if it is not the case then we simply rename some of the variables accordingly).

3.2 A-Interpretations

Let us start by an informal summary of our approach. We assume a set of function symbols $\mathcal{S}_N$ is given ($\mathcal{S}_N$ is chosen arbitrarily, hence not related to the formula at hand, which, as explained in Section 3.1, contains no function symbol).

Our method tries to construct, for any (satisfiable) formula ϕ satisfying the condition of Section 3.1, a model of ϕ represented by a tree automaton on tuples of terms built on $\mathcal{S}_N$. This model has the following property: for any n-ary predicate symbol p, there exists a TTA $\mathcal{A}$ s.t. the interpretation of p is exactly the set of n-tuples accepted by $\mathcal{A}$.

Interpretations satisfying this requirement are called A-interpretations:

Definition 2. *An interpretation $\mathcal{I}$ of a set of predicate symbols S is called a A-interpretation (w.r.t. a set of function symbols $\mathcal{S}_N$) iff it satisfies the following conditions:*

- *The domain of $\mathcal{I}$ is the set of ground terms built on the set of function symbols $\mathcal{S}_N \cup \bot$.*
- *For any n-ary predicate symbol $q \in S$, there exists a (n, m)-automaton accepting exactly the set of tuples $\boldsymbol{t}$ s.t. $\boldsymbol{t} \in [q]_{\mathcal{I}}$.*

A formula ϕ is A-satisfiable iff there exists an A-interpretation $\mathcal{I}$ s.t. $\mathcal{I} \models \phi$. In this case, $\mathcal{I}$ is called a A-model of ϕ.

In order to construct the automaton corresponding to each predicate symbol, we construct a new formula ψ from ϕ, in such a way that any finite model of ψ can be associated (in a natural and "canonic" way) to the collection of automata denoting a model of ϕ. The transformation algorithm is parametrized by the set of symbols $\mathcal{S}_N$. The reader should note that the signature corresponding to the formula ψ is distinct from $\mathcal{S}_N$ and also distinct from the signature of the original formula ϕ.

3.3 Signatures

We actually use three distinct signatures:

1. The set of symbols occurring in the *original formula*. This set contains only predicate symbols, including the predicate $\approx$. It is denoted by $\mathcal{S}_I$.
2. The set of symbols $\mathcal{S}_N$ defining the *domain* on the constructed model. We assume that all the function symbols in $\mathcal{S}_N$ have the same arity (denoted by m in the following). The only constant symbol is the symbol $\bot$. If it is not the case then we can simply add the constant symbol $\bot$ as additional argument to some of the function symbols (for instance the term $g(f(a), b)$ becomes $g(f(a(\bot, \bot), \bot), b(\bot, \bot))$).
3. The signature of the *transformed* formula.

The first two signatures are already known. The purpose of this section is to define the constant and function symbols used in the transformed formulae.

Let ϕ be a first-order formula. We assign to ϕ a signature $\mathcal{S}_\phi$ containing the following symbols:

- The constant symbol $\bot$;
- A constant symbol v_p for each position p in ϕ;
- A constant symbol c_q for each function or predicate symbol q occurring in ϕ or in $\mathcal{S}_N$;
- A function symbol $\#^n$ of arity n for any $n \in \mathbb{N}$;
- 3 function symbols n, i, t of arity $1, 1, 3$ respectively;
- 5 predicate symbols $F, in, <, \leq, \mu_s$ of arity $1, 4, 2, 2, m + 1$ respectively.

We assume that all these symbols are *pairwise distinct*.

Intuitively, the intended meaning of the newly introduced symbols is as follows.

- $<$ simply denotes an (arbitrarily chosen) ordering.
- $\#^n(t_1, \ldots, t_n)$ can be seen as a n-tuple $(t_1, \ldots, t_n)$.
- n associates to each state a state corresponding in some sense to the "negation" of x.
- Each constant symbol of the form v_p (where p is a position) or c_q (where q is a predicate symbol) will be associated to an automaton. The automaton associated to c_q, where q is a n-ary predicate symbol, will recognize the set of tuples $(t_1, \ldots, t_n)$ s.t. $q(t_1, \ldots, t_n)$ holds, and the automaton associated to v_p will recognize the set of tuples $(t_1, \ldots, t_n)$ s.t. $\phi_{|p}\{x_i \to t_i \mid i \in [1..n]\}$ holds, where $x_1 \prec \ldots \prec x_n$ is the (ordered) set of variables in ϕ. In particular $c_\approx$ recognizes the set of tuples (t, t). The transition function, initial and final state of these automata are specified using the symbols t and F:
- For any (n, m)-automaton a, $t(a, \#^m(s), \#^n(f))$ denotes the image of the pair (s, f) by the transition function associated to a (s is a m-vector of states and f is a n-vector of function symbols). $i(a)$ denotes its initial state, and $F(x)$ indicates whether the state x is final or not. In particular, a state $\#^2(e_1, e_2)$ is final iff e_1, e_2 are final and $n(e)$ is final if e is not.
- In order to encode projections, some states will be associated to sets of states. Thus the formula $in(n, e', e, l)$ will hold if e' is a state in e. n denotes the automaton in which e occurs. l is an additional argument, which encodes in some sense, the shape of the *minimal* derivation yielding the state e'. This point will be explained in more details later, when handling existential quantifications.
- μ_s denotes a *partial* strictly increasing function (μ_s cannot be total otherwise there is no finite model).

The following section formalizes these ideas.

3.4 Computing $\Delta(\phi)$

We denote by $\Delta'(\phi)$ the *conjunction* of the following formulae (V_1), (V_2), (V_3), (I), (O_1), (O_2), (O_3), (E), (M_1), (M_2), (M_3), (T_ϕ).

The following formulae specify the meaning of the symbols, following the above intuitive explanations.

(V_1) $F(n(e)) \Leftrightarrow \neg F(e)$ ($n(e)$ is final iff e is not final).
(V_2) $F(\#^2(e_1, e_2)) \Leftrightarrow (F(e_1) \wedge F(e_2))$ ($\#^2(e_1, e_2)$ is final iff e_1, e_2 are both final).

(V_3) $F(e) \Leftrightarrow (\exists e', n, l)(F(e') \wedge in(n, e', e, l))$ (e is final iff it contains a final state).

(I) $(\forall v) \bigwedge_{n \leq N} t(v, \#^m(i(v), \ldots, i(v)), \#^n(\bot, \ldots, \bot)) \approx i(v)$ where N is the maximal number of free variables of a subformula in ϕ (by the definition of TTA's, the image of $(s_0^m, \bot^n)$ must be s_0).

(O_1) $(\forall l, l', l'')(l < l' \wedge l' < l'') \Rightarrow (l < l'')$ ($<$ is transitive).

(O_2) $(\forall l)(l \not< l)$ ($<$ is irreflexive).

(O_3) $(\forall l, l')(l \leq l') \Leftrightarrow (l < l' \vee l \approx l')$ (definition of $\leq$).

(E) $(\forall f, g)F(t(c_\approx, \#^m(e_1, \ldots, e_m), \#^2(f, g))) \Leftrightarrow (f \approx g \wedge \bigwedge_{i=1}^m F(e_i)$ (this defines the automaton associated to $c_\approx$ corresponding to syntactic equality between terms built on $\mathcal{S}_N$: $f(t_1, \ldots, t_m) \approx g(s_1, \ldots, s_m)$ holds iff $f = g$ and if $t_i \approx s_i$ holds for any $i \in [1..n]$).

The following formulae specify the interpretation of μ_s. As explained before, it can be seen as a partial function mapping some tuples $(x_1, \ldots, x_m)$ to an element y s.t. $\mu_s(y, x_1, \ldots, x_m)$ holds and $y > x_1, \ldots, x_m$.

(M_1) $(\forall x, y_1, \ldots, y_m, z_1, \ldots, z_m)[\mu_s(x, y_1, \ldots, y_m) \wedge \mu_s(x, z_1, \ldots, z_m)] \Rightarrow \bigwedge_{i=1}^m (y_i \approx z_i)$. This formula expresses the fact that this function is injective.

(M_2) $(\forall y_1, \ldots, y_m)(\neg\mu_s(\bot, y_1, \ldots, y_m))$. This ensures that $\bot$ is not in the co-domain of the partial function corresponding to μ_s.

(M_3) $(\forall y_1, \ldots, y_m, x)(\mu_s(x, y_1, \ldots, y_m)) \Rightarrow \bigwedge_{i=1}^m x > y_i$. This expresses the fact that the partial function corresponding to μ_s is strictly increasing.

We denote by (T_ϕ) the conjunction $\bigwedge_{p \in Pos(\phi)} \Gamma(p)$, where $\Gamma(p)$ is defined below and by $\Delta(\phi)$ the formula: $\Delta'(\phi) \wedge (T)$, where (T) is the formula $F(i(v_\epsilon))$.

Now, we define the formula $\Gamma(p)$. It depends on the form of the subformula occurring at position p in ϕ. Let p be a position in ϕ, let $\psi = \phi_{|p}$ and $v = v_p$. We distinguish several cases.

Atoms. Assume that ψ is an atom. By definition, ψ is of the form $q(\boldsymbol{x})$ for some predicate symbol q (possibly $\approx$) and some tuple of distinct variables $\boldsymbol{x} = x_1, \ldots, x_n$.

Let $\gamma(i)$ be the rank of the variable x_i in the ordered set of the free variables in $FVar(\psi)$. By definition γ is a bijective function from $[1..n]$ to $[1..n]$. $\gamma^{-1}(i)$ is the index of the i-th variable in the $\prec$-ordered set of the free variables in $FVar(\psi)$.

Let $\boldsymbol{t} = (t_1, \ldots, t_n)$ be a n-tuple. We assign to this tuple the n-tuple $\lambda_{\boldsymbol{x}}(\boldsymbol{t}) = (t_{\gamma^{-1}(1)}, \ldots, t_{\gamma^{-1}(n)})$ ($\lambda_{\boldsymbol{x}}(\boldsymbol{t})$ contains the same elements as the tuple $\boldsymbol{t}$ but in the order induced by the ordering $\prec$ on the components of $\boldsymbol{x}$).

Let e be a variable. Let $\boldsymbol{u}$ be an (arbitrarily chosen) n-tuple of pairwise distinct variables, distinct from the variable e. We define $\Gamma(p)$ as follows.

$$\Gamma(p) \overset{def}{=}$$
$$[(\forall e, \boldsymbol{u})t(v, e, \#^n(\lambda_{\boldsymbol{x}}(\boldsymbol{u}))) \approx t(c_q, e, \#^n(\boldsymbol{u}))]$$
$$\wedge i(v) \approx i(c_q).$$

Informally, this formula relates the transition function corresponding to the automaton associated to the position p to the one corresponding to the symbol q. The transition function for p is the same as the transition function for q, but applied to the tuple $\lambda_x(\boldsymbol{u})$ instead of $\boldsymbol{u}$. The initial states are the same.

Negations. Assume that ψ is of the form $\neg\psi'$. Let $v' = v_{p.1}$.
Let $e_1, \ldots, e_m, x$ be (arbitrary chosen) distinct variables. We define:

$$\Gamma(p) \stackrel{def}{=}$$

$$(\forall e_1, \ldots, e_m, x)[t(v, \#^m(n(e_1), \ldots, n(e_m)), x) \approx n(t(v', \#^m(e_1, \ldots, e_m), x))]$$

$$\wedge i(v) \approx n(i(v')).$$

This formula states that the automata corresponding to ψ is the same as the one corresponding to ψ', excepted that the states e are replaced by $n(e)$. Due to the formula (V_1), this means that final states become non final – which implies that accepted tuples become non accepted – and conversely (non accepted tuples become accepted).

Conjunctions. Assume that $\psi = (\psi_1 \wedge \psi_2)$. Let $v_i = v_{p.i}$ (v_i is the symbol corresponding to ψ_i).
Let $\boldsymbol{y}^i = y_1^i \prec \ldots \prec y_{n_i}^i$ be the ordered sequence of variables occurring in ϕ_i. Let $\boldsymbol{x} = x_1, \ldots, x_n$ be the ordered sequence of variables occurring in ϕ.
Let e_j^i ($i = 1, 2, j = 1, \ldots, m$) be distinct variables (distinct from the variables in $\boldsymbol{x}$).

$$\Gamma(p) \stackrel{def}{=}$$

$$[(\forall e_1^1, \ldots, e_m^1, e_1^2, \ldots, e_m^2, \boldsymbol{x})$$

$$t(v, \#^m(\#^2(e_1^1, e_1^2), \ldots, \#^2(e_m^1, e_m^2)), \#^n(\boldsymbol{x}))$$

$$\approx \#^2(t(v_1, \#^m(e_1^1, \ldots, e_m^1), \#^{n_1}(\boldsymbol{y}^1)), t(v_2, \#^m(e_1^2, \ldots, e_m^2), \#^{n_2}(\boldsymbol{y}^2)))]$$

$$\wedge i(v) \approx \#^2(i(v_1), i(v_2))$$

The state corresponding to a given vector $\boldsymbol{t}$ in the automaton associated to v is of the form $\#^2(e_1, e_2)$ where e_i is the state corresponding to $\boldsymbol{t}$ in the automaton associated to v_i. By the formula (V_2), $\#^2(e_1, e_2)$ is final iff e_1 and e_2 are final. Thus $\boldsymbol{t}$ is accepted iff it is accepted by the automata corresponding to v_1 and v_2.

Existential Quantifiers. This is the trickiest part since the computation of the automaton corresponding to an existential quantification (i.e. a projection) normally involves the computation of a fixpoint which is in principle beyond the scope of first-order logic.
Assume that $\psi = (\exists x)\psi'$. Let $v' = v_{p.1}$. Let $x_1 \prec \ldots \prec x_{n+1}$ be the set of free variables in ψ'. We assume that x occurs in $x_1, \ldots, x_{n+1}$ (otherwise we can simply remove the quantifier $\exists x$). By the restriction imposed in Section 3.1, we must have $x = x_{n+1}$ (otherwise the variables are renamed).

For any n-vector $\boldsymbol{t} = (t_1, \ldots, t_n)$ and for any term s, we denote by $\boldsymbol{t}.s$ the $(n+1)$-vector $t_1, \ldots, \ldots, t_n, x$ (s is inserted at position $n+1$ in $\boldsymbol{t}$). Note that we may have $n = 0$, in this case $\boldsymbol{t}$ is the empty vector and $\boldsymbol{t}.s$ is simply the term s.

Before giving the formal definition of the formula $\Gamma(p)$, we start by some informal explanation in order to help the reader to grasp the intuitive idea behind the definitions below. Let $\mathcal{A}'$ be the automaton corresponding to v'. As explained before, we want to construct an automaton $\mathcal{A}$ corresponding to v. $\mathcal{A}$ should accept the set of terms $\boldsymbol{t}$ s.t. there exists s s.t. $\boldsymbol{t}.s$ is accepted by $\mathcal{A}'$. To this purpose, we build an automaton in which any state e corresponds to a finite set of states occurring in $\mathcal{A}'$. $\mathcal{A}$ is defined in such a way that for any term $\boldsymbol{t}$, $\mathcal{A}(\boldsymbol{t})$ is the set of states $\mathcal{A}'(\boldsymbol{t}.s)$, where s is a ground term. The more natural way of defining this automaton is as follows: $e' \in t(v, \#^m(e_1, \ldots, e_m), \#^n(\boldsymbol{x}))$ iff there exists a symbol $g \in \mathcal{S}_N$ and $e_1', \ldots, e_m'$ occurring in $e_1, \ldots, e_m$ respectively s.t. $e' = t(v', \#^m(e_1', \ldots, e_m'), \#^n(\boldsymbol{x}.g))$. However, this definition is not well-founded (for instance e can occur in $e_1, \ldots, e_m$). To capture the intended idea, one would have to specify that the set of states $t(v', \#^m(e_1', \ldots, e_m'), \#^n(\boldsymbol{x}.g))$ is the *smallest* set having this property. However, this is beyond the scope of first-order logic.

In order to overcome this difficulty, we add an argument to the membership predicate. We specify, for each pair (e', e), s.t. $e' \in e$, the minimal (according to subterm ordering) term l s.t. there exists a vector $\boldsymbol{t}$ with $e' = \mathcal{A}'(\boldsymbol{t})$ and $l = l(\boldsymbol{t})$ (see Section 2.2 for the definition of $l(\boldsymbol{t})$). This avoids loops in the definition hence ensures that the set of states e is well-defined. Since the number of states is finite, the number of minimal terms is also finite (thus our definition does not threaten the existence of a finite model).

We use the predicate symbol in for this purpose. If $in(v, e', e, l)$ for some l, then e' belongs to e. According to the formula (V_3), this implies that e is final if it contains a final state.

Terms of the form $l(\boldsymbol{t})$ are encoded by repeated applications of the partial function μ (which is injective and increasing). Note that if μ_s would be assumed to be total, then no finite model would exist, which explains why we used a predicate symbol for encoding μ_s instead of a function symbol.

For any term t, we denote by t^n the term $\#^n(\underbrace{t, \ldots, t}_{n \text{ times}})$.

$\Gamma(p)$ is defined as the conjunction of the formulae $\Gamma_1(p), \Gamma_2(p), \Gamma_3(p)$ defined as follows.

Let $e, e', l, e_1, \ldots, e_m, e_1', \ldots, e_m', x_1, \ldots, x_n, u, l_1, \ldots, l_m$ be distinct variables. Let $\boldsymbol{x} = (x_1, \ldots, x_n)$.

$$\Gamma_1(p) \overset{def}{=} (\forall e, e') in(v, e', e, \bot) \Leftrightarrow (e' \approx i(v') \wedge e \approx i(v))$$

$$\Gamma_2(p) \overset{def}{=}$$
$$(\forall e, e', l, e_1, \ldots, e_m, \boldsymbol{x})(in(v, e', e, l) \wedge (e \approx t(v, \#^m(e_1, \ldots, e_m), \#^n(\boldsymbol{x}))) \Rightarrow$$
$$[(\exists l_1, \ldots, l_m, e_1', \ldots, e_m', u)([\bigvee_{g \in \mathcal{S}_N} u \approx c_g] \wedge e' \approx t(v', \#^m(e_1', \ldots, e_m'), \#^{n+1}(\boldsymbol{x}.u))$$
$$\wedge \mu_s(l, l_1, \ldots, l_m) \wedge \bigwedge_{j=1}^{m} in(v, e_j', e_j, l_j)]$$

$$\Gamma_3(p) \overset{def}{=}$$
$$(\forall e, e', e_1, \ldots, e_m, e'_1, \ldots, e'_m, l_1, \ldots, l_m, u)$$
$$(e \approx t(v, \#^m(e_1, \ldots, e_m), \#^n(\boldsymbol{x}))) \wedge$$
$$e' \approx t(v', \#^m(e'_1, \ldots, e'_m), \#^{n+1}(\boldsymbol{x}.u) \wedge \bigwedge_{i=1}^{m} in(v, e'_i, e_i, l_i)$$
$$\Rightarrow (\exists l) in(v, e', e, l) \wedge [\mu_s(l, l_1, \ldots, l_m) \vee \bigvee_{j=1}^{m}(l \leq l_j)]$$

Intuitively, if a state e is reachable from a given tuple of states $e_1, \ldots, e_m$ and a tuple of function symbols $\boldsymbol{x}$, then e' occurs in e for some term l iff there exist m elements $e'_1, \ldots, e'_m$ occurring in $e_1, \ldots, e_m$ for some term $l_1, \ldots, l_m$ respectively, a function symbol g s.t. e' is reachable from $e'_1, \ldots, e'_m$ and $\boldsymbol{x}.g$. The disjunction $\bigvee_{j=1}^{m}(l \leq l_j)$ in $\Gamma_3(p)$ is due to the fact that l must be minimal (thus either l is of the form $\mu(l_1, \ldots, l_m)$ or l occurs in one of the l_i's).

3.5   Complexity of the Transformation

It is obvious that the size of the obtained formula is simply *quadratic* w.r.t. the size of original one, provided that a structure-preserving transformation (see for instance [12]) is used for eliminating the connective $\Leftrightarrow$. The number of conjuncts is linear and the size of each conjunct is also linear.

4   Example

We provide an example of application. The formula is deliberately simple and is chosen only to illustrate the construction.

Let ϕ be the following formula:

$$\phi = (\exists x) \neg p(x, x)$$

First, we transform ϕ in order to satisfy the pre-conditions of Section 3.1.

$$\phi \equiv (\exists x, y)(\neg p(y, x) \wedge (x \approx y))$$

We assume that $\mathcal{S}_N$ contains two function symbols $0, s$ of arity $0, 1$ respectively. Let $u = c_p$ and $v = c_{\approx}$. We use the precedence $x \prec y$. For the sake of clarity we assume that $c_0 = 0, c_s = s$.

We denote by a, b, c, d, d', d'' the constant symbols v_p corresponding respectively to the positions $1111, 112, 111, 11, 1, \epsilon$, i.e. to the subformulae $p(y, x), x \approx y, \neg p(x, x), \neg p(y, x) \wedge (x \approx y), (\exists y)(\neg p(y, x) \wedge (x \approx y))$ and ϕ.

We have

$$\Gamma(1111) = (t(a, x, \#^2(y_1, y_2) \approx t(u, x, \#^2(y_2, y_1))) \wedge i(a) \approx i(u).$$

Similarly:

$$\Gamma(112) = (t(b, x, \#^2(y_1, y_2)) \approx t(v, x, \#^2(y_1, y_2))) \wedge (i(b) \approx i(v)).$$

Then $\Gamma(111) = (t(c, n(x), y) \approx n(t(a, x, y)) \wedge i(c) \approx n(i(a))$
The conjunction is handled as follows:

$$\Gamma(11) = t(d, \#^2(x_1, x_2), y) \approx \#^2(t(b, x_1, y), t(c, x_2, y))$$

$$\wedge i(d) \approx \#^2(i(b), i(c)).$$

Then comes the most difficult part, i.e. the existential quantifiers. The formula at position 1 contains only one free variable x. We have:

$$\Gamma_1(1) \stackrel{def}{=} (\forall e, e') in(d', e', e, \bot) \Leftrightarrow (e' \approx i(d) \wedge e \approx i(d'))$$

$$\Gamma_2(1) \stackrel{def}{=}$$
$$(\forall e, e', l, e_1, x)(in(d', e', e, l) \wedge (e \approx t(d', \#^1(e_1), \#^1(x)))) \Rightarrow$$
$$[(\exists l_1, e'_1, g)(g \approx 0 \vee g \approx s) \wedge e' \approx t(d, \#^1(e'_1), \#^2(x, g))$$
$$\wedge \mu_s(l, l_1) \wedge in(d', e'_1, e_1, l_1)]$$

$$\Gamma_3(1) \stackrel{def}{=}$$
$$(\forall e, e', e_1, e'_1, l_1, g)$$
$$(e \approx t(d', \#^1(e_1), \boldsymbol{x})) \wedge e' \approx t(d, \#^1(e'_1), \#^2(\boldsymbol{x}.g)) \wedge in(d', e'_1, e_1, l_1)$$
$$\Rightarrow (\exists l) in(d', e', e, l) \wedge [\mu_s(l, l_1) \vee (l \leq l_1)]$$

The formula ϕ contains no free variable. Thus:

$$\Gamma_1(\epsilon) \stackrel{def}{=} (\forall e, e') in(d'', e', e, \bot) \Leftrightarrow (e' \approx i(d') \wedge e \approx i(v))$$

$$\Gamma_2(p) \stackrel{def}{=}$$
$$(\forall e, e', l, e_1)(in(v, e', e, l) \wedge (e \approx t(v, \#^1(e_1), \#^0))) \Rightarrow$$
$$[(\exists l_1, e'_1, g)(g \approx 0 \vee g \approx f) \wedge e' \approx t(v', \#^1(e'_1), \#^1(c_g))$$
$$\wedge \mu_s(l, l_1) \wedge in(v, e'_1, e_1, l_1)]$$

$$\Gamma_3(p) \stackrel{def}{=}$$
$$(\forall e, e', e_1, e'_1, l_1, g)$$
$$(e \approx t(v, \#^1(e_1))) \wedge e' \approx t(v', \#^m(e'_1), \#^1(g) \wedge in(v, e'_1, e_1, l_1)$$
$$\Rightarrow (\exists l) in(v, e', e, l) \wedge [\mu_s(l, l_1) \vee (l \leq l_1)]$$

$\Delta(\phi)$ is the conjunction of $\Delta'(\phi)$ and the following formulae:

(T_0) $(\forall x)(\neg F(i(x)))$.
(V_1) $F(n(e)) \Leftrightarrow \neg F(e)$.
(V_2) $F(\#^2(e_1, e_2)) \Leftrightarrow (F(e_1) \wedge F(e_2))$.
(V_3) $F(e) \Leftrightarrow (\exists e', n, l)(F(e') \wedge in(n, e', e, l))$.
(I) $(\forall v) t(v, \#^1(i(v)), \#^0) \approx i(v) \wedge t(v, \#^1(i(v)), \#^1(\bot)) \approx i(v) \wedge t(v, \#^1(i(v)), \#^2(\bot, \bot)) \approx i(v)$.
(O_1) $(\forall l, l', l'')(l < l' \wedge l' < l'') \Rightarrow (l < l'')$

(O_2) $(\forall l)(l \not< l)$.
(O_3) $(\forall l, l')(l \leq l') \Leftrightarrow (l < l' \vee l \approx l')$.
(M_1) $(\forall x, y, z)[\mu_s(x, y) \wedge \mu_s(x, z)] \Rightarrow y \approx z$.
(M_2) $(\forall y)(\neg \mu_s(\bot, y))$.
(M_3) $(\forall y, x)(\mu_s(x, y)) \Rightarrow x > y$.
  (E) $(\forall f, g)F(t(u, \#^1(e), \#^2(f, g))) \Leftrightarrow (f \approx g \wedge F(e))$.
  $-$ $F(i(d''))$.

5   Soundness of the Transformation

We shall prove that the transformation algorithm specified in the previous section is sound, in the following sense: for any formula ϕ (satisfying the conditions in Section 3.1), $\Delta(\phi)$ has a finite model iff ϕ has a A-model, i.e. a model in which the interpretation of each predicate symbol is the accepted tuples of a automaton on the set of symbols S_N.

We first show how to "come back", i.e. how to construct the interpretation of the original formula ϕ from the model of $\Delta(\phi)$.

Let $\mathcal{I}$ be an interpretation of domain D validating the formula (I) (see Section 3.4 for the definition of (I)). From $\mathcal{I}$, we associate to any $v \in D$ a (m, n)-TTA denoted by $\mathcal{A}(\mathcal{I}, v)$ and defined as follows:

$$\mathcal{A}(\mathcal{I}, v) = (S, s_0, \tau, S_F) \text{ where:}$$

  $-$ $S \stackrel{def}{=} D$.
  $-$ $s_0 \stackrel{def}{=} i_{\mathcal{I}}(v)$.
  $-$ $S_F \stackrel{def}{=} F_{\mathcal{I}}$.
  $-$ $\tau((s_1, \ldots, s_m), (f_1, \ldots, f_n)) = t_{\mathcal{I}}(v, \#_{\mathcal{I}}^m(s_1, \ldots, s_m), \#_{\mathcal{I}}^n(c_{f_1 \mathcal{I}}, \ldots, c_{f_n \mathcal{I}}))$.

Note that τ is a transition function because, since (I) is satisfied, we have $\tau(s_0^m, \bot^n) = s_0$. Thus $\mathcal{A}(\mathcal{I}, v)$ is an (n, m)-automaton.

We denote by $\mathcal{I}^\star$ an A-interpretation of S_I s.t. for any n-ary predicate symbol $q \in S_I$: $q_{\mathcal{I}^\star}$ is the set of n-tuples accepted by $\mathcal{A}(\mathcal{I}, c_q)$.

Let ϕ be a formula of free variables $x_1 \prec \ldots \prec x_k$. We denote by $Sol_{\mathcal{I}}(\phi)$ the set of tuples $(v_1, \ldots, v_k) \in D$ s.t. $[\phi]_{\mathcal{I}\{x_i \to v_i | i \in [1..k]\}} = true$.

Lemma 1. *Let ϕ be a formula. Let $\psi = \Delta'(\phi)$. Let $\mathcal{I}$ be a finite model of ψ. For any position p in ϕ, the set of tuples accepted by $\mathcal{A}(\mathcal{I}, [v_p]_{\mathcal{I}})$ is exactly the set of tuples $Sol_{\mathcal{I}^\star}(\phi_{|p})$.*

Corollary 1. *Let ϕ be a formula. Let $\mathcal{I}$ be a finite model of $\Delta(\phi)$. $\mathcal{I}^\star \models \phi$.*

Proof. Let $\mathcal{A} = \mathcal{A}(\mathcal{I}, [v_\epsilon]_{\mathcal{I}})$. Due to the formula (T), we have $i(\epsilon) \in [F]_{\mathcal{I}}$. But since ϕ contains no free variable we have $\mathcal{A}(\emptyset) = i(\epsilon)$ (where $\emptyset$ denotes an empty vector of terms) hence $\emptyset$ is accepted by $\mathcal{A}$. By Lemma 1, this implies that $\emptyset \in Sol_{\mathcal{I}^\star}(\phi)$, i.e. $\mathcal{I}^\star \models \phi$.

Remark 1. If one wants to consider only terms built on $\mathcal{S}_N$ (without $\perp$) and/or handling function symbols of arity distinct than m, then this condition has to be added explicitly in the formula. More precisely, any quantification $(\exists x)\phi$ has to be replaced by $(\exists x)(D(x)\wedge\phi)$, where D is a special predicate symbol encoding to fact that x does not contain $\perp$. Obviously, the set of terms having this property can be defined be a TTA. This automaton can be specified by adding axioms into $\Delta(\phi)$:

$$(\forall x)\neg F(\tau(c_D, x, \perp))$$

$$(\forall x) \bigvee_{f\in\mathcal{S}_N} F(\tau(c_D, \#^m(x_1,\ldots,x_m), c_f)) \Leftrightarrow \bigwedge_{i=1}^{ar(f)} F(x_i) \wedge \bigwedge_{i=ar(f)+1}^{m} x_i \approx i(c_D)$$

The first formula states that a term of the form $\perp(\ldots)$ is not accepted and the second formula states that for any $f \in \mathcal{S}_N$, $f(t_1,\ldots,t_k)$ is accepted iff $k = ar(f)$ and if $t_1,\ldots,t_k$ are accepted.

Now we show the converse, i.e. that any A-model of ϕ can be associated to a finite model of $\Delta(\phi)$.

Lemma 2. *Let ϕ be a formula. Let $\mathcal{I}$ be a A-model of ϕ. There exists a finite interpretation $\mathcal{J}$ s.t. $\mathcal{J}^* = \mathcal{I}$ and $\mathcal{J} \models \Delta(\phi)$.*

Corollary 2. *Let ϕ be a formula. An A-interpretation $\mathcal{I}$ is a model of ϕ iff $\Delta(\mathcal{A})$ has a finite model $\mathcal{J}$ s.t. $\mathcal{J}^* = \mathcal{I}$.*

Proof. The proof follows immediately from Lemma 1 and 2.

6   Conclusion

We have defined a method for using any existing finite model builder (such as FINDER, SEM, MACE etc.) for constructing infinite models of first-order formulae. The constructed interpretations are represented by finite tree automata. We have shown that the method is correct and complete in the sense that a model can be obtained in finite time for any formula having a model representable in this formalism (obviously this is not the case for any satisfiable formula).

From a practical point of view, we remark that a large part of the obtained formula does not depend on the considered formula. Most conjuncts in $\Delta'(\phi)$ may be seen as the axioms of an "underlying theory", specifying some properties of the symbols used in the formula. In order to make the construction efficient, this theory could be made built-in in finite model builders. Moreover the interpretation of some of the symbols should be fixed a priori instead of being reconstructed at each time. For instance, the terms of the form $\#^n(t)$ are used in our construction in order to denote a tuple t. Affecting an interpretation to these symbols by enumeration is useless and time-consuming. Similarly, $<$ could be directly interpreted as a fixed ordering among the element of the domain. The inclusion and efficient handling of such theories in finite model builders deserve to be investigated in the future.

References

1. Audemard, G., Benhamou, B.: Reasoning by symmetry and function ordering in finite model generation. In: Voronkov, A. (ed.) CADE 2002. LNCS (LNAI), vol. 2392, pp. 226–240. Springer, Heidelberg (2002)
2. de la Tour, T.B.: A note on symmetry heuristics in SEM. In: Voronkov, A. (ed.) CADE 2002. LNCS (LNAI), vol. 2392, pp. 181–194. Springer, Heidelberg (2002)
3. Caferra, R., Leitsch, A., Peltier, N.: Automated Model Building. Applied Logic Series, vol. 31. Kluwer Academic Publishers, Dordrecht (2004)
4. Caferra, R., Zabel, N.: A method for simultaneous search for refutations and models by equational constraint solving. Journal of Symbolic Computation 13, 613–641 (1992)
5. Claessen, K., Sorensson, N.: New techniques that improve mace-style finite model finding. In: Proceedings of the CADE-19 Workshop: Model Computation - Principles, Algorithms, Applications (Miami, USA) (2003)
6. Comon, H., Dauchet, M., Gilleron, R., Jacquemard, F., Lugiez, D., Tison, S., Tommasi, M.: Tree automata techniques and applications (1997), http://www.grappa.univ-lille3.fr/tata
7. Fermueller, C., Pichler, R.: Model representation via contexts and implicit generalizations. In: Nieuwenhuis, R. (ed.) CADE 2005. LNCS (LNAI), vol. 3632, pp. 409–423. Springer, Heidelberg (2005)
8. Fitting, M.: First-Order Logic and Automated Theorem Proving. Texts and Monographs in Computer Science. Springer, Heidelberg (1990)
9. Gottlob, G., Pichler, R.: Working with ARMs: Complexity results on atomic representations of Herbrand models. Information and Computation 165, 183–207 (2001)
10. McCune, B.: MACE 2.0 reference Manual and Guide. Technical report, Argonne National Laboratory (2001)
11. Peltier, N.: A new method for automated finite model building exploiting failures and symmetries. Journal of Logic and Computation 8(4), 511–543 (1998)
12. Plaisted, D., Greenbaum, S.: A structure-preserving clause form translation. Journal of Symbolic Computation 2, 293–304 (1986)
13. Slaney, J.: Finder (FINite Domain EnumeratoR): Notes and guides. Technical report, Australian National University Automated Reasoning Project, Canberra (1992)
14. Zhang, J., Zhang, H.: SEM: a system for enumerating models. In: Proc. IJCAI 1995, vol. 1, pp. 298–303. Morgan Kaufmann, San Francisco (1995)

A Groebner Bases Based Many-Valued Modal Logic Implementation in Maple

Eugenio Roanes-Lozano[1], Luis M. Laita[2],
and Eugenio Roanes-Macías[1]

[1] Universidad Complutense de Madrid, Facultad de Educación,
Depto. de Algebra, c/ Rector Royo Villanova s/n, 28040-Madrid, Spain
{roanes,eroanes}@mat.ucm.es
[2] Universidad Politécnica de Madrid, Facultad de Informática,
Depto. de Inteligencia Artificial
Campus de Montegancedo, Boadilla del Monte, 28660-Madrid, Spain
laita@fi.upm.es

Abstract. The authors developed in the nineties a Groebner bases based polynomial model for classic Boolean algebra and many-valued modal logics and for rule based expert systems (RBES) based on these logics. Following this approach, they have designed and developed RBES in different fields. Now two *Maple* packages that can perform knowledge extraction and consistency checking in RBES which underlying logic is either classic Boolean or Kleene's or Lukasiewicz's many-valued modal have been developed and can be freely obtained from the authors. They extend the possibilities of Maple's built-in "Logic" package.

Keywords: Logic and Symbolic Computing, Groebner Bases, Rule Based Expert Systems.

1 Introduction

We developed in the nineties an algebraic model for classic Boolean logic and many-valued modal logics (with a prime number of truth values[1]) and for rule based experts systems (RBES) based on these logics [7,9,10].

These works extend previous works by Kapur-Narendran [6] and Hsiang [5] and Alonso et al. [1,4], where how to perform effective calculations in classic Boolean logic and many-valued modal logics using Groebner bases (GB) [2,3] is respectively treated.

Following this approach, we have designed and developed RBES in different fields. The implementations were written in the computer algebra system (CAS) *CoCoA* [15]. Most of them are medicine applications; some of the topics treated

---

[1] For the base field of the algebraic model, $\mathbb{Z}_p$ (where p is the number of truth values of the logic), to be a field. If the desired number of truth values was not prime, it would be sufficient to consider a prime number greater than the given number and not to use some truth values.

S. Autexier et al. (Eds.): AISC/Calculemus/MKM 2008, LNAI 5144, pp. 170–183, 2008.

are: coronary bypass clinical practice guidelines, anorexia [8] and migraine detection, evaluation and treatment...

We have always used *CoCoA* for RBES developing so far, because, although we are used to deal with *Maple* [16] (we have written a book [14], and have used *Maple* to develop applications in transportation engineering, pharmacokinetics, mechanical theorem proving in geometry...), the external packages of the old versions of *Maple* that could compute GB in finite characteristic presented difficulties, similar to those analyzed in sections 7 and 8, that we were not able to bypass.

The article is structured as follows. Section 2 provides some introductory notes about RBES and Section 3 gives the flavor of what a Groebner basis is and can do. Section 4 briefly describes polinomial models for classic Boolean and many-valued logics and RBES whose underlying logics are these ones. Section 5 presents an example of the polynomial translation of the logical connectives of Lukasiewicz's three-valued logic. Section 6 shows how this translation can be implemented in the CAS *CoCoA*. Sections 7 and 8 discuss the novelty of this work: two different implementations of these polynomial models for Boolean and modal many-valued logics and RBES written in *Maple 10 & 11* and *Maple 11*, that can be freely obtained from the authors. They can perform knowledge extraction and consistency checking in RBES which underlying logic is either classic Boolean or Kleene's or Lukasiewicz's many-valued modal. They use a GB-based inference engine and extend the possibilities of Maple's built-in `Logic` package. Finally, Section 9 contains a brief comparison of the performances of *CoCoA* and *Maple* for this purpose.

As *Maple* is one of the most widely used CAS, it is good to have an implementation of the polynomial model presented here available in this CAS. Moreover, *Maple* has some extra advantages, like plotting and GUI developing capability, a wide offer of specialized packages for different applications (e.g. graphs)... that can be used in a fruitful synergy with this new implementation.

2   Some Introductory Notes about RBES

A "rule based expert system" (to be hereinafter denoted as "RBES") has a "knowledge base" (hereinafter denoted as "KB") and an "inference engine" (hereinafter denoted as "IE"). A graphic user interface is frequently provided.

2.1   Literals and Rules

The KB consists of a certain number of logical formulae, of the form:

$$\wedge_{i=1}^{k} \maltese x[i] \longrightarrow \vee_{j=1}^{l} \maltese x[j]$$

where symbol $\maltese$ may mean:

- no symbol at all or "negation" (denoted $\neg$), if the underlying logic is classsic Boolean,

- no symbol at all, "negation", "possibility" (denoted $\Diamond$), "necessity" (denoted $\Box$), or a combination of these symbols, like "necessarily-not" (denoted $\Box\neg$), equivalent to $\neg\Diamond$, if the underlying logic is a modal many-valued one

(we shall use this notation hereinafter).

For example, formula

$$x[1] \wedge x[2] \wedge \neg x[3] \wedge \Diamond x[4] \wedge \neg x[5] \longrightarrow x[13] \vee \Box\neg x[17]$$

is read as: "IF $x[1]$ and $x[2]$ and *not* $x[3]$ and possibly $x[4]$ and *not* $x[5]$ HOLD, THEN $x[13]$ or necessarily-not $x[17]$ HOLDS".

The formulae of this type are known as "production rules" (and are usually refereed to as, simply, "rules").

The symbols $x[1], x[2], x[3]$... which appear in the production rules of our system, are called "propositional variables". If $x[k]$ is one of these symbols, any expression of the form $\maltese x[k]$ is called a "literal".

Therefore, the left hand side (or antecedent) of a production rule is a conjunction of literals and the right hand side (or consequent) of a production rule is a disjunction of literals. This is because the ocurrences of "$\wedge$" in the consequent and "$\vee$" in the antecedent of a rule can be avoided: if such symbols ocurred in a rule, the rule could be split into rules without them.

2.2   Potential Facts and Facts

In addition to production rules, the KB contains a set of "potential facts", which are the literals that do appear at the left hand side of at least one rule, but such that no literal containing the same propositional variable appears in any right hand side.

For instance, in case the underlying logic is classic Boolean, the whole set of potential facts can be something like

$$A = \{x[k] : k = 1, ..., 26\} \cup \{\neg x[k] : k = 1, ..., 26\} \ .$$

In case the underlying logic is classic Boolean, that a given set of facts (i.e., a subset of the set of potential facts that is stated as true) is "consistent" means that, from each pair formed by a potential fact and its contrary, only one literal is chosen. That it is "maximal" means that such a choice must be made for each pair.

In case the underlying logic is a modal many-valued one, that a given set of facts (i.e., a subset of the set of potential facts that is stated as true) is "consistent" means that, for each propositional variable appearing in the potential facts of the system, $x[k]$, at most one expression of the form $\maltese x[k]$, is chosen. That it is "maximal" means that such a choice must be made for each propositional variable appearing in the potential facts of the system.

2.3   The Inference Engine

The "IE" is a an automated tool (in our case, a program in the CAS), that verifies consistency (that is, checks that the system does not lead to contradictions) and draws consequences from the information contained in the KB. The latter corresponds to the logical concept of "tautological consequence" (a logical formula A_0 is a "tautological consequence" of the formulae $A_1, A_2, ..., A_m$, denoted $\{A_1, A_2, ..., A_m\} \models A_0$, if and only if, whenever $A_1, A_2, ..., A_m$ are true, then A_0 is true).

An inconsistency is found when all formulae can be obtained from a consistent set of facts, the rules and the integrity constraints (in case the logic is many-valued, two different types of consistency can be distinguished [11]).

3   A Brief Note about Groebner Bases

In the early '60s, both Heisuke Hironaka and Bruno Buchberger independently proved that, for each polynomial ideal, a basis completely identifying it always existed. They denoted their bases as "standard bases" and "Groebner bases" (GB), respectively. The latter's great advantage was that it provided a constructive method (Buchberger's algorithm).

Some GB are particularly important: we call them "reduced Groebner bases". We say that a Groebner basis is reduced if and only if the leading coefficient of all its polynomials is 1 and we can't "simplify" any of its polynomials by adding a linear algebraic combination of the rest of the polynomials in the basis.

The input to Buchberger's algorithm is a polynomial set, a term order (for instance, "total degree" or "pure lexicographical"), and a variable order (for instance, $x > y > z$) and its output is the ideal's reduced GB with respect to the specified term and variable orders.

The key point is that, once the term order and the variable order are fixed, such a reduced GB completely characterizes the ideal: any ideal has a unique reduced GB. As a consequence we have two most important results:

i) two sets of polynomials generate the same ideal if and only if their reduced GB are the same,

ii) $\{1\}$ is the only reduced Groebner basis for the ideal that is equal to the whole ring.

An elementary introduction to the topic can be found in [12,13].

4   Polynomial Models

Sections 4.1 and 4.2 summarize [7,9,10].

4.1   A Polynomial Model for Propositional Logic

Let $(\mathcal{C}, \vee, \wedge, \neg, \rightarrow)$ be a propositional Boolean algebra, with $\underline{0}$ and $\underline{1}$ respectively denoting contradiction and tautology and $X_1, X_2, ..., X_n$ denoting the propositional variables.

Let us consider the propositional residue class ring

$$\mathcal{A} = \mathbb{Z}_2[x_1, x_2, ..., x_n] / \langle x_1^2 - x_1, x_2^2 - x_2, ..., x_n^2 - x_n \rangle$$

and let us define the operation: $\forall a, b \in \mathcal{A}$, $a \tilde{+} b = a + b - ab$. Then $(\mathcal{A}, \tilde{+}, \cdot, 1+,$ *"is a multiple"*) is a Boolean algebra.

Moreover, the natural homomorphism of Boolean algebras, φ, from $\mathcal{C}$ into $\mathcal{A}$, where uppercase letters correspond to lowercase letters, turns out to be an ordering preserving isomorphism.

The principal ideal of the Boolean algebra $(\mathcal{C}, \vee, \wedge, \neg, \rightarrow)$ generated by $B \in \mathcal{C}$ is $E_B = \{X \in \mathcal{C} : X \rightarrow B\}$. That the ideals of $\mathcal{C}$ correspond by φ to the ideals of $\mathcal{A}$, and that these ideals are exactly the same as the ideals of the polynomial residue class ring $\mathcal{A}$, can be proven.

This model can be extended to p-valued modal logics (where p is a prime number) by considering the residue class ring

$$\mathcal{A} = \mathbb{Z}_p[x_1, x_2, ..., x_n] / \langle x_1^p - x_1, x_2^p - x_2, ..., x_n^p - x_n \rangle$$

and corresponding polynomial translations of the logic connectives.

4.2 A Polynomial Model for the Propositional Boolean Algebra Associated to a RBES

The Boolean algebra associated to a RBES whose underlying logic is classic Boolean is a structure $(\mathcal{C}^*, \vee, \wedge, \neg, \rightarrow)$ where $\rightarrow$ is the relation obtained applying the rules of logical deduction to the implications of $\mathcal{C}$ and the rules, facts and integrity constraints of the RBES (consequently, $\rightarrow$ is not the usual implication), and $\mathcal{C}^*$ is the set of equivalence classes defined by this enlarged equivalence relation $\leftrightarrow$ in $\mathcal{C}$.

If the rules $Rule_1, ..., Rule_v$, the facts $Fact_1, ..., Fact_m$ and the integrity constraints $IC_1, ..., IC_u$ of a RBES are added as true to the Boolean algebra $\mathcal{C}$ of the previous section, the structure obtained is isomorphic to the image of $\mathcal{A}$ in the natural surjective homomorphism

$$\psi : \mathcal{A} \longrightarrow \mathcal{A}/J$$

where J is the ideal

$$\langle \varphi(\neg Rule_1), ..., \varphi(\neg Rule_v), \varphi(\neg Fact_1), ..., \varphi(\neg Fact_m), \varphi(\neg IC_1), ..., \varphi(\neg IC_u) \rangle$$

That every formula can be obtained from the facts, rules and integrity constraints stated as true corresponds to the idea of RBES forward reasoning inconsistency.

In the polynomial model, the equality $J = \mathcal{A}$ translates the RBES concept of forward reasoning inconsistency. In such case of degeneracy, the whole residue class ring collapses to a single element.

The main theorem on which the inference engine is based is the following:

Theorem 1. *A formula A_0 is a tautological consequence of the formulae in the union of a consistent subset of the set of potential facts and the set of all production rules and integrity constraints of a RBES, if and only if, the polynomial translation of the negation of A_0 belongs to the ideal J of A/I, generated by the polynomial translations of the negations of the given potential facts and of all the production rules and the integrity constraints of the RBES[2].*

The key point is that this theorem can take advantage of the two important results enunciated at the end of the previous section.

This model and the mentioned results can also be extended to p-valued modal logics (where p is a prime number). A proof that the two structures are isomorphic and a general proof of Theorem 1 (in the many-valued case) can be found in [10].

5   Polynomial Translation of the Logical Connectives

For instance, the polynomial expressions corresponding to the basic logical formulae in Lukasiewicz's three-valued logic (if 2 is assigned to "true", 1 to "undetermined" and 0 to "false") are detailed afterwards.

- $\neg M$ is translated into the polynomial: $2 - m$
- $\Diamond M$ is translated into the polynomial: $2m^2$
- $\Box M$ is translated into the polynomial: $m^2 + 2m$
- $M \vee N$ is translated into the polynomial: $m^2n^2 + m^2n + mn^2 + 2mn + m + n$
- $M \wedge N$ is translated into the polynomial: $2m^2n^2 + 2m^2n + 2mn^2 + mn$
- $M \rightarrow N$ is translated into the polynomial: $2m^2n^2 + 2m^2n + 2mn^2 + mn + 2m + 2$
- $M \leftrightarrow N$ is translated into the polynomial: $m^2n^2 + m^2n + mn^2 + 2mn + 2m + 2n + 2$

(note that the coefficients of these polynomials belong to $\mathbb{Z}_3$).

For the different logics these polynomials can be obtained by solving an algebraic system (that is obtained from the truth tables).

6   CoCoA Implementation

These polynomial translations can be input almost directly in *CoCoA 4.3*, that provides a "Normal Form" command, NF(pol,I), that returns the reduction of polynomial *pol* modulo ideal I (the algorithm of NF is also included in the theory of GB).

---

[2] Observe that we could alternatively check that A_0 belongs to the ideal $I + J$ of A. We'll work this way in *CoCoA* to avoid working in a residue class ring.

We have to define the polynomial ring and ask *CoCoA* to use it:

```
USE Z/(3)[x[1..15]];
```

and then we can define the ideal MEMORY.I, generated by the expressions $x_i^3 - x_i$ (denoting it this way, it is a global variable, and it can be an input to NF):

```
MEMORY.I:=Ideal([x[K_]^3-x[K_] | K_ In 1..15]);
```

and introduce the polynomial expressions for the three-valued connectives of Lukasiewicz modal logic (the operators are prefix ones)[3]:

```
NEG(M):=NF(2-M,MEMORY.I);
POS(M):=NF(2*M^2,MEMORY.I);
NEC(M):=NF(M^2+2*M,MEMORY.I);
OR(M,N):=NF(M^2*N^2+M^2*N+M*N^2+2*M*N+M+N,MEMORY.I);
AND(M,N):=NF(2*M^2*N^2+2*M^2*N+2*M*N^2+M*N,MEMORY.I);
IMP(M,N):=NF(2*M^2*N^2+2*M^2*N+2*M*N^2+M*N+2*M+2,MEMORY.I);
IFF(M,N):=NF(M^2*N^2+M^2*N+M*N^2+2*M*N+2*M+2*N+2,MEMORY.I);
```

For instance, the *CoCoA* implementation of the polynomial expressions of a 7-valued modal logic can be found in [10].

With the notation of Theorem 1, whether A_0 is a consequence of a set of formulae or not, can be checked with *CoCoA* just typing:

```
NF(NEG(A[0]),MEMORY.I+J);
```

(where J is the polynomial ideal generated by the polynomial expressions of the negation of the formulae in the given set). If the output of the command is 0, the answer is "yes", otherwise the answer is "no".

It can also be directly checked using the Boolean command IsIn, that tests ideal memberships, by typing:

```
NEG(A[0]) IsIn MEMORY.I+J;
```

With the notation of Section 4.2, whether the RBES is inconsistent or not can also be checked using Groebner bases by typing in *CoCoA*:

```
GBasis(MEMORY.I+J);
```

If the output is 1, the RBES is inconsistent; otherwise (the output is normally a large set of polynomials), that set of facts doesn't lead to an inconsistency.

Again, using IsIn makes it even simpler, as it directly returns "true" or "false", just typing:

```
1 IsIn MEMORY.I+J;
```

---

[3] In *CoCoA 4.7* these operators would be defined using Define...EndDefine instead of directly :=.

7   Maple Implementation I (Using Groebner and Ore_algebra Packages)

Let us try to translate the model into *Maple*, as done above with *CoCoA*.

In *Maple 10 & 11* it is possible to define a polynomial ring over a finite field using *Maple*'s 'Ore_Algebra package. Either directly NormalForm command or Reduce command (using a trick) can be used.

We load the Groebner and Ore_algebra packages first, and then define the list of variables, the polynomial ring and the order that will be used by the GB-related commands:

```
> with(Groebner):
> with(Ore_algebra):
> SV:=x[1],x[2],x[3]:
> A:=poly_algebra(SV,characteristic=3):
> Orde:=MonomialOrder(A,'plex'(SV)):
```

and the ideal I (denoted iI, as I is a reserved word in *Maple*), using map in order to save time:

```
> fu:=v->v^3-v:
> iI:=map(fu,[SV]);
            3           3           3
  iI := [x[1]  - x[1], x[2]  - x[2], x[3]  - x[3]]
```

We can now define the functions that associate to the logical connectives their polynomial expressions, as done in *CoCoA* above.

7.1   Maple Implementation I: First Attempt (Fails)

According to the "help file" of Reduce and NormalForm commands in *Maple 11*:

> "The Reduce command reduces a polynomial by a list of polynomials with respect to a given monomial order. In other words, it returns the remainder of the full pseudo-division of a polynomial by a list of polynomials with respect to a given monomial order."

We can try to use this command to define the polynomial translation of the connectives[4]:

```
> 'NEG' :=(m::algebraic) -> Reduce(2-m,iI,Orde):
> '&AND':=(m::algebraic,n::algebraic) ->
>          Reduce(expand(2*m^2*n^2+2*m^2*n+2*m*n^2+m*n),iI,Orde):
. . .
```

---

[4] Note that in the definition of Orde in Section 7, plex appeared surrounded by right single quotes or apostrophes ("'"). Enclosing an expression or subexpression in apostrophes delays its evaluation by one level. Meanwhile, left single quotes ("'"), are used in this section: enclosing an expression in left single quotes converts the expression into a name.

(for instance, in the first of these lines we define NEG, that reduces "2 minus the input" w.r.t. the ideal iI using the order denoted Orde).

Although many formulae, like $x[1] \wedge x[2]$ or $x[1] \wedge x[1]$ are correctly translated into polynomials, others are not!:

```
> (2*x[1]) &AND (2*x[1]);
```
$$x[1]$$

(as idempotency holds, the output should be $2 \cdot x[1] = -x[1] \neq x[1]$).

This is an intriguing behavior, so we decided to construct truth tables from the polynomial expressions of the connectives to try to find out what is happening, and all the truth tables obtained are incorrect (no $2s$ can be found in the truth tables)! For instance, for &AND, we obtain:

```
> with(linalg);
> M:=Matrix(3,1):
> for i from 0 to 2 do
>     for j from 0 to 2 do M[i+1,j+1]:=i &AND j od;
> od;
> evalm(M);
```

$$\begin{pmatrix} 0 & 0 & 0 \\ 0 & 1 & 1 \\ 0 & 1 & 1 \end{pmatrix}$$

The problem is that, when dealing with ideals of a polynomial ring over a field, the coefficients of the generators don't care (in our particular case, all ideals are principal). But in this application we are substituting symbolic or numerical expressions in the reduced form of the polynomial, so the extra multiplicative coefficient introduced in the pseudo-division procedure should be taken into account.

7.2   Maple Implementation I: Second Attempt (Using Reduce)

Command Reduce has an optional fourth argument, 's', where the multiplicative coefficient is stored.

Surprisingly, it is enough to specify that coefficient in the implementation of the polynomial expressions of the logic connectives for command Reduce to change its behavior, and for everything to work correctly (we do not multiply by the multiplicative coefficient!). For example, if we type:

```
> '&AND':=(m::algebraic,n::algebraic)->
>         Reduce(expand(2*m^2*n^2+2*m^2*n+2*m*n^2+m*n),iI,Orde,
>              's'):
```

instead, then, e.g. (2*x[1]) &AND (2*x[1]); works as expected. Moreover, all truth tables obtained from these commands are also correct now.

7.3   Maple Implementation I: Third Attempt (Using `NormalForm`)

Another option would be to directly use the right command (`NormalForm`), instead of `Reduce`. If we define the commands as follows:

```
> NEG :=(m::algebraic) -> NormalForm(2-m,iI,Orde):
> NEC :=(m::algebraic) -> NormalForm(expand(m^2+2*m),iI,Orde):
> POS :=(m::algebraic) -> NormalForm(expand(2*m^2),iI,Orde):
> '&AND':=(m::algebraic,n::algebraic) ->
>           NormalForm(expand(2*m^2*n^2+2*m^2*n+2*m*n^2+m*n),
>                       iI,Orde):
> '&OR' :=(m::algebraic,n::algebraic) ->
>           NormalForm(expand(m^2*n^2+m^2*n+m*n^2+2*m*n+m+n),
>                       iI,Orde):
> '&IMP' :=(m::algebraic,n::algebraic) ->
>           NormalForm(expand(2*m^2*n^2+2*m^2*n+2*m*n^2+m*n+2*m+2),
>                       iI,Orde):
> '&IFF' :=(m::algebraic,n::algebraic) ->
>           NormalForm(expand(m^2*n^2+m^2*n+m*n^2+2*m*n+2*m+2*n+2),
>                       iI,Orde):
```

Then everything works fine. This implementation is available for *Maple 10 & 11*.

8   Maple Implementation II (Using Only Groebner Package)

In *Maple 11*, to first define a ring in order to perform GB-related commands when the base ring is a finite field is no longer needed. Now `NormalForm` and `Reduce` commands of *Maple*'s `Groebner` package already include an option for specifying the characteristic of the base field. Nevertheless, a strange behavior showed up in the first one, as the output sometimes includes fractions! Again, we could bypass this obstacle.

Now we only have to load the `Groebner` package and define the list of variables and the ideal `iI`:

```
> with(Groebner):
> SV:=x[1],x[2],x[3]:
> fu:=v->v^3-v:
> iI:=map(fu,[SV]):
```

8.1   Maple Implementation II: First Attempt (Fails)

That the characteristic is 3 can be specified as an input of `NormalForm` command. For instance:

```
> '&AND':=(m::algebraic,n::algebraic) ->
>           NormalForm(expand(2*m^2*n^2+2*m^2*n+2*m*n^2+m*n),
>                       iI,plex(SV),characteristic=3):
```

(we do not include the code of the other connectives for the sake of brevity). Surprisingly, there are fractions in the polynomial expressions:

```
> p &AND q;
```

$$1/2\ x[1]^2\ x[2]^2 + 1/2\ x[1]^2\ x[2] + 1/2\ x[1]x[2]^2 + x[1]x[2]$$

and in the truth tables reconstructed from the polynomial expressions, for instance, of $\wedge$:

$$\begin{pmatrix} 0 & 0 & 0 \\ 0 & 1 & 1 \\ 0 & 1 & \frac{1}{2} \end{pmatrix}$$

(we expected all these $\frac{1}{2}s$ to be $2s$). Therefore, the base field is not $\mathbb{Z}_3$, as expected. We have reported this behavior to *Maplesoft*.

8.2 Maple Implementation II: Second Attempt (Using Reduce)

Let us try Reduce command, instead of NormalForm.

According to the "help file" of these commands in *Maple 11*:

> "The Reduce command is similar to NormalForm, except that pseudo-division is performed. The result is a pseudo-remainder r, denominator s, and quotient Q such that r/s = Sum(Q[i]*G[i],i = 1 .. n) and r/s is the normal form. Note that NormalForm and Reduce return the same quotients Q (they are not scaled by s). In fact, both commands use the same underlying implementation."

So we use the Reduce command but multiply the output by the denominator s and include mod 3 afterwards. The reason for including mod 3 is that the output of the Reduce command is computed modulo 3, but if it is directly multiplied by s (in $\mathbb{Z}$) then, although the truth tables are correct, coefficients strictly greater than 2 can appear in the polynomial expressions of formulae. This is the case, for instance, for $x[1] \wedge x[2]$. Observe that, in the implementations of Section 7 and in the *CoCoA* implementation, we switch to a polynomial ring over a finite field, meanwhile in the implementations of Section 8 we have to simplify each calculation.

If we define:

```
> 'NEG' :=(m::algebraic) ->
>          Reduce(2-m,iI,plex(SV),'s',characteristic=3)*s mod 3:
> 'NEC' :=(m::algebraic) ->
>          Reduce(expand(m^2+2*m),
>                 iI,plex(SV),'s',characteristic=3)*s mod 3:
> 'POS' :=(m::algebraic) ->
>          Reduce(expand(2*m^2),
>                 iI,plex(SV),'s',characteristic=3)*s mod 3:
```

```
> '&AND':=(m::algebraic,n::algebraic) ->
>         Reduce(expand(2*m^2*n^2+2*m^2*n+2*m*n^2+m*n),
>              iI,plex(SV),'s',characteristic=3)*s mod 3:
> '&OR' :=(m::algebraic,n::algebraic) ->
>         Reduce(expand(m^2*n^2+m^2*n+m*n^2+2*m*n+m+n),
>              iI,plex(SV),'s',characteristic=3)*s mod 3:
> '&IMP' :=(m::algebraic,n::algebraic) ->
>         Reduce(expand(2*m^2*n^2+2*m^2*n+2*m*n^2+m*n+2*m+2),
>              iI,plex(SV),'s',characteristic=3)*s mod 3:
> '&IFF' :=(m::algebraic,n::algebraic) ->
>         Reduce(expand(m^2*n^2+m^2*n+m*n^2+2*m*n+2*m+2*n+2),
>              iI,plex(SV),'s',characteristic=3)*s mod 3:
```

everything works fine. For instance:

```
> x[1] &AND x[2];
          2    2           2                   2
     2 x[1]  x[2]  + 2 x[1]  x[2] + 2 x[1] x[2]  + x[1] x[2]
```

And all truth tables are correctly reconstructed from the truth tables. For instance, for ∧:

$$\begin{pmatrix} 0 & 0 & 0 \\ 0 & 1 & 1 \\ 0 & 1 & 2 \end{pmatrix}$$

is obtained.

9   Performance Comparison

The tests carried out with real RBES show very similar times for *WinCoCoA 4.3* and *Maple 11 Classic* running the implementation of Section 7.3, and much worse times for the implementation of Section 8.2. We believe that the difference is due to when the modular reductions take place.

For instance, checking the consistency of a real example from [8] of a RBES which underlying logic is classic Boolean, with 69 propositional variables and 182 rules, takes slightly more than 38 seconds in both *WinCoCoA 4.3* and *Maple 11 Classic* running the implementation of Section 7.3. Meanwhile, the computation hasn't yet finished after 20 minutes in *Maple 11 Classic* running the implementation of Section 8.2. Command GBasis was used in *WinCoCoA 4.3* and command Basis was used in *Maple 11 Classic*. The computer was a standard portable computer running the most common operating system.

10   Conclusions

We were finally able to implement our algebraic approach to RBES consistency checking and knowledge extraction in the CAS *Maple*.

If working with packages Ore_algebra and Groebner (of *Maple* version *10* or later) is chosen, it is simpler to use the implementation that uses command NormalForm than the one that uses command Reduce.

Meanwhile, if working only with the new version of package Groebner of *Maple* version *11* or later, then Reduce command must be used instead.

The implementation that uses both packages, Ore_algebra and Groebner, has turned out to be much faster than the other one, and very similar in speed to the one written in *WinCoCoA 4.3*.

We believe that, due to the size of *Maple*'s community of users and the wide variety of possibilities that this CAS offers, these implementations can be really useful both for RBES design and implementation and for logic and RBES teaching (that the inference engine is an algebraic one, based on an ideal membership, can be kept hidden to the user, that only deals with logic processes or knowledge extraction and consistency checking in a RBES).

The two packages, for *Maple 10 & 11* and *Maple 11*, dealing with classic Boolean and Lukasewicz's or Kleene's many-valued modal logics, can be freely obtained from the authors.

Acknowledgments

We thank the anonymous referees for their most valuable comments, that have really improved the article.

References

1. Alonso, J.A., Briales, E.: Lógicas Polivalentes y Bases de Gröbner. In: Martin, C. (ed.) Actas del V Congreso de Lenguajes Naturales y Lenguajes Formales. University of Seville, pp. 307–315 (1995)
2. Buchberger, B.: An Algorithm for Finding a Basis for the Residue Class Ring of a Zero-Dimensional Polynomial Ideal (Ph.D. Thesis in German). Math. Institute - University of Innsbruck (1965)
3. Buchberger, B.: Applications of Gröbner Bases in Non-Linear Computational Geometry. In: Rice, J.R. (ed.) Mathematical Aspects of Scientific Software. IMA, vol. 14, pp. 60–88. Springer, Heidelberg (1988)
4. Chazarain, J., Riscos, A., Alonso, J.A., Briales, E.: Multivalued Logic and Gröbner Bases with Applications to Modal Logic. Journal of Symbolic Computation 11, 181–194 (1991)
5. Hsiang, J.: Refutational Theorem Proving using Term-Rewriting Systems. Artificial Intelligence 25, 255–300 (1985)
6. Kapur, D., Narendran, P.: An Equational Approach to Theorem Proving in First-Order Predicate Calculus. In: Proceedings of the 9th International Joint Conference on Artificial Intelligence (IJCAI 1985), vol. 2, pp. 1146–1153 (1985)
7. Laita, L.M., de Ledesma, L., Roanes-Lozano, E., Roanes-Macías, E.: An Interpretation of the Propositional Boolean Algebra as a k-algebra. Effective Calculus. In: Calmet, J., Campbell, J.A. (eds.) AISMC 1994. LNCS, vol. 958, pp. 255–263. Springer, Heidelberg (1995)

8. Pérez-Carretero, C., Laita, L.M., Roanes-Lozano, E., Lázaro, L., González-Cajal, J., Laita, L.: A Logic and Computer Algebra-Based Expert System for Diagnosis of Anorexia. Mathematics and Computers in Simulation 58, 183–202 (2002)
9. Roanes-Lozano, E., Laita, L.M., Roanes-Macías, E.: Maple V in A.I.: The Boolean Algebra Associated to a KBS. CAN Nieuwsbrief 14, 65–70 (1995)
10. Roanes Lozano, E., Laita, L.M., Roanes-Macías, E.: A Polynomial Model for Multivalued Logics with a Touch of Algebraic Geometry and Computer Algebra. Mathematics and Computers in Simulation 45(1), 83–99 (1998)
11. Roanes-Lozano, E., Roanes-Macías, E., Laita, L.M.: Geometric Interpretation of Strong Inconsistency in Knowledge Based Systems. In: Ganzha, V.G., Mayr, E.W., Vorozhtsov, E.V. (eds.) Computer Algebra in Scientific Computing. Proceedings of CASC 1999, pp. 349–363. Springer, Heidelberg (1999)
12. Roanes-Lozano, E., Roanes-Macías, E., Laita, L.M.: The Geometry of Algebraic Systems and Their Exact Solving Using Groebner Bases. Computing in Science and Engineering 6(2), 76–79 (2004)
13. Roanes-Lozano, E., Roanes-Macías, E., Laita, L.M.: Some Applications of Gröbner Bases. Computing in Science and Engineering 6(3), 56–60 (2004)
14. Roanes-Macías, E., Roanes-Lozano, E.: Cálculos Matemáticos por Ordenador con Maple V.5, Ed. Rubiños-1890, Madrid (1999)
15. CoCoA: a system for doing Computations in Commutative Algebra, http://cocoa.dima.unige.it
16. URL, http://www.maplesoft.com/

On the Construction of Transformation Steps in the Category of Multiagent Systems

Thomas Soboll

University of Salzburg, Department of Computer Sciences, A-5020 Salzburg, Austria
tsoboll@cosy.sbg.ac.at

Abstract. Based on the category MAS of base diagrams of Multiagent Systems (MAS) and morphisms between them, a transformation system for MAS can be established using the well known Double Pushout (DPO) Approach. An important part in the DPO approach is to find a pushout complement for a given situation. This is usualy done by checking the so called "gluing condition". In this contribution a new approach for the pushout complement construction in the category MAS is introduced. For illustration a simple robot example is presented.

1 Introduction

In previous work we introduced a new method for MAS base diagram transformations [1] based on the Double Pushout Approach (DPO) introduced by Ehrig, Pfender, and Schneider [2]. Fundamental important for our work is the observation that the general communication and cooperation structure of a MAS can be represented by a corresponding arrow diagram, called base diagram of the MAS. The basic notion of base diagram has been introduced recently by J. Pfalzgraf [3]. To each MAS we associate such a base diagram, which represents the complete relational structure (i.e. communication in the general sense). The nodes of this arrow diagram represent agents, the arrows (and paths of arrows) hold the communication and cooperation information. This gives a category by its own right, more precisely a typed category [1].

In a MAS communication and cooperation (in general relations) between agents can change. This fact gives rise to the definition of the category MAS of all MAS where the objects are base diagrams of Multiagent Systems and the morphisms are MAS morphisms i.e. structure preserving maps between base diagrams. Based on this category MAS a transformation system for Multiagent Systems can be established by applying the double pushout approach to Multiagent Systems.

In this contribution we present a new approach to construct transformation steps. The main difficulty applying the DPO Approach is to construct the so called pushout complement. In [4] checking the so called "gluing condition" solves this problem, in this paper we introduce an alternative algorithm. The proposed

S. Autexier et al. (Eds.): AISC/Calculemus/MKM 2008, LNAI 5144, pp. 184–190, 2008.

concepts have been implemented in java and have been applied to a simple
cooperating robot scenario.

2 The Category MAS

Base diagrams of MAS and maps between those form the category MAS . Roughly
speaking a base diagram is a snapshot of the actual cooperation structure in a
MAS, it can be modeled by a typed category. We present an example of a MAS
object (see Fig. 1) and a MAS morphism (see Fig. 2). For more details regarding
the category MAS we refer to [1].

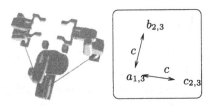

Fig. 1. A simple example of a base diagram (MAS -object). The left hand side shows
three robots (a,b and c), two robots (b,c) with a gripper (index 2) and an assembly
robot (a) equipped with a welding device (index 1). Index 3 indicates that the robot
"is not assigned to a task". Let $A = \{a, b, c\}$ be the set of agents and $C \subseteq A \times A$ be
the communication relation $C = \{(a, b), (a, c), (b, a), (c, a)\}$ i.e. the robots a and b as
well as a and c are able to communicate (arrow type c). The underlying base diagram
is depicted in the right hand side. Note that the base diagram is not only an object in
MAS but is a category as well. Therefore in the diagram exist the identity arrows, as
well as a composition operation on the morphisms, i.e. there is an arrow of arrow type
c from b to c and vice versa. To increase readability this arrows are not drawn in the
diagram.

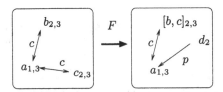

Fig. 2. Example of a MAS morphism. We observe that F maps the agents a to a, b
and c to $[b, c]$, the arrows between a and c as well as the arrows between a and b to
the arrows between a and $[b, c]$, the arrow type c to c and the object types 1 to 1, 2 to
2 and 3 to 3. We can see F is structure preserving.

A transformation system is defined by a concept introduced by Ehrig, Pfender,
and Schneider [2] the so called double pushout (DPO) approach, which is a far
developed concept in the field of algebraic graph transformations [4]. A transformation rule is defined by a production $p = (p_L, p_R)$, which is a pair of morphisms
with common domain. Given a MAS production $p = (p_L : I \to L, p_R : I \to R)$,

a MAS object MAS^L and a MAS morphism $m : L \to MAS^L$, called match, define a direct transformation step as follows: An object MAS^R is called direct derivable from an object MAS^L via p ($MAS^L \Rightarrow^p MAS^R$), iff there exists a context object MAS^C with corresponding MAS morphism $g : I \to MAS^C$, such that MAS^L and MAS^R are pushout objects in the following diagram.

$$
\begin{array}{ccccc}
L & \xleftarrow{\ p_L\ } & I & \xrightarrow{\ p_R\ } & R \\
{\scriptstyle m}\downarrow & & {\scriptstyle g}\downarrow & & {\scriptstyle g'}\downarrow \\
MAS^L & \xleftarrow{\ p'_L\ } & MAS^C & \xrightarrow{\ p'_R\ } & MAS^R
\end{array}
$$

This diagram illustrates a double pushout, for more details we refer to the book [4].

3   Pushout Complements

In the previous section the Double Pushout Approach as a tool for MAS transformations was discussed. It is essential to note that in general the application of a production or a rule to a given object, in the case of MAS transformations to a base diagram of a MAS, leads to the following situation. We have a production, a base diagram as well as the match which is a morphism from the left hand side of the production to the given base diagram. Now we need to construct a so called pushout complement [4].

Definition 1. *Given two* **C** *morphisms* $f : A \to B$ *and* $g : B \to C$ *a pushout complement is a* **C** *object PC together with two* **C** *morphisms* $f' : PC \to C$ *and* $g' : A \to PC$ *such that the following diagram yields a pushout square.*

$$
\begin{array}{ccc}
A & \xrightarrow{\ f\ } & B \\
{\scriptstyle g'}\downarrow & & {\scriptstyle g}\downarrow \\
PC & \xrightarrow{\ f'\ } & C
\end{array}
$$

Pushout complements need not exist in any case and if they exist they need not be unique. This is of course a problem that must be handled. In the field of graph transformations usually the so called "gluing condition" [4] is checked to find a pushout complement. In the next section we describe an alternative construction algorithm, that in the case that a PC exists, yields the minimal pushout complement. In the case that no pushout complement exists, the algorithm produces a production that transforms the base diagram in one for which the pushout complement construction works.

Recall that given two monomorphisms $f : A \to B$ and $g : B \to C$ in **SET** the pushout complement can be constructed in two steps. We construct the coproduct complement CC of g and then construct the coproduct of A and CC denoted by $A + CC$, this is the pushout complement we search. This suggests to construct pushout complements in MAS in a similar way. There are two problems with this construction: First is that we want to be able to construct pushout

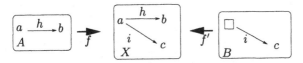

Fig. 3. Failed Coproduct Complement Construction in MAS. We observe that even though f is monic the coproduct complement B is no MAS object (the arrow i is dangling).

complements for all arrows not only for monomorphims. The second problem is that (unlike in **SET**) in general coproduct complements do not even exist for monics (see Fig. 3). The need for a new notion arises. We introduce quasi coproduct complements that fulfill our needs.

Definition 2. *Let* **C** *be a category that has coproducts. Given a morphism* $f :$ $A \to X$ *a quasi coproduct complement is a* **C** *object B together with a* **C** *morphism* $g : B \to X$ *such that the following diagram (1) commutes with e being an epimorphism and for each other* **C** *object B' with* **C** *morphism* $g' : B' \to X$ *and e' being an epimorphism there exists a (not necessarily unique) morphism* $k : B \to B'$ *such that* $g' \circ k = g$ *(diagram (2)).*

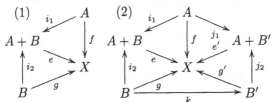

In the sequel we will restrict the morphims g and g' to monics, then k is the unique morphism that makes the diagram (2) commute.

Proof: If g' is monic and the diagram commutes for $k : B \to B'$ (i.e. $g' \circ k = g$), given another morphism $k' : B \to B'$ s.t. the diagram commutes (i.e. $g' \circ k' = g$), it follows immediately that $g' \circ k = g' \circ k'$, but g' is monic i.e. $k = k'$.

4   Construction of Pushout Complements

MAS is (Epi,Mono)-structured i.e. every MAS-morphism f can be factored into an epimorphism fe and a monomorphism fm s.t. $f = fm \circ fe$. In the sequel given an morphism $f : A \to B$ we will denote the quasi coproduct complement object as $B - A$. In every (Epi,Mono)-structured category that has pushouts and coproducts we construct the pushout complement of a given diagram

$$A \xrightarrow{f} B$$
$$m \downarrow$$
$$C$$

as follows (see Fig. 4):

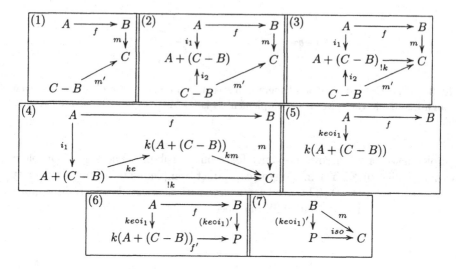

Fig. 4. Pushout Complement Construction

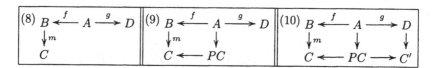

Fig. 5. Application to the DPO Approach

The first step is to compute the quasi coproduct complement $C - B$ (1). Next we construct the coproduct of A and $C - B$ (2). Due to the universal property of coproducts it follows that there is exactly one morphism $k : A + (C - B) \rightarrow C$ such that the diagram (3) commutes. Next we compute the image of $A + (C - B)$ over k via the (Epi,Mono)-factorization $k = km \circ ke$ and denote the image object as $k(A + (C - B))$ (4). Now we build the pushout of the diagram (5) and get the pushout square (6). $k(A + (C - B))$ together with the morphisms $ke \circ i_1$ and f' is the pushout complement for the diagram (1) if there exists an isomorphism $iso : P \rightarrow C$ such that the diagram (7) commutes.

In the case that there is no isomorphism from P to C s.t. the diagram (7) commutes i.e. the base diagrams are not structurally the same we interpret $(m, (ke \circ i_1)')$ as a production and try to apply it. This results in a base diagram for which the initial production is applicable.

We can apply this result to the DPO as follows (see Fig 5): Given a MAS -production $(f : A \rightarrow B, g : A \rightarrow D)$, a MAS object C and a match m (8). We construct the pushout complement PC as described above and get the diagram (9), the last step is to construct the right hand side pushout (10). This is a transformation step from C to C' ($C \Rightarrow C'$).

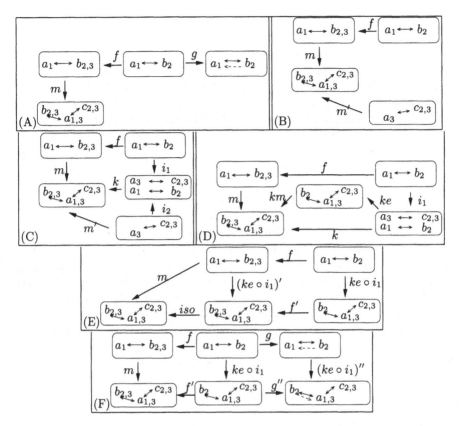

Fig. 6. Example: Given the production $p = (f, g)$ (upper part of (A), which advises an agent with a gripper to deliver a color cube to an agent with a welding device, indicated by the additional arrow $\dashrightarrow$), the initial base diagram from Fig. 1, together with a suitable match m, we apply the construction algorithm. We construct the quasi coproduct complement (step (1) above) and receive diagram (B). Next we build the coproduct and the unique arrow k (step (2),(3)), the result is depicted in (C). Now we factor k (step (4)) and get the diagram (D). The last steps (5),(6),(7) lead to the diagram (E). We observe that there is an isomorphism iso that makes the diagram (E) commute i.e. we found a pushout complement for the start situation (A). To finish the transformation we build the right hand side pushout (step (9),(10)) and get the diagram (F).

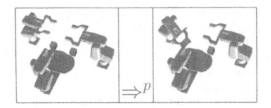

Fig. 7. In the robot simulation transforming the initial base diagram (see Fig. 1) via the production p from Fig. 6 into a new base diagram (down-right object in diagram(F) of Fig. 6) triggers the robot b to deliver a color cube to agent a

5   Conclusion

The concept of MAS transformations, which is an adaption of graph transformations [4] to typed categories, is a natural way to describe changes in the base diagram of Multiagent Systems. In this contribution we have pointed out a new approach for pushout complement construction based on the new notion of a quasi coproduct complement. This construction produces the minimal pushout complement. A special feature is that if the approach does not find a suitable pushout complement, its result can be regarded as a new production that transforms the given base diagram in a base diagram for which a pushout complement exists.

Future work will concern local-global modeling aspects. It turned out that logical fiberings [5] provide a concept to assign a system of distributed logics to a MAS in a natural way. The basic idea is to assign a logical fiber to every agent, this fiber models the local logical state space of an agent, the entire logical fiber bundle forms the global logical state space of the whole MAS. For more details we refer to [6]. This motivates the introduction of a 'Relational Fibering' with the aim to model local global interactions in the relational structure of a MAS. We assign a relational fiber to every agent, the fiber models the relational information attached to the agent. A first application of this approach is to compute subcategories of a MAS on demand, by taking the collection of the fibers over a defined set of agents as a starting point.

References

1. Pfalzgraf, J., Soboll, T.: On a General Notion of Transformation for Multiagent Systems. In: Proceedings of the Conference: Integrated Design and Process Technology, IDPT-2007. Society for Design and Process Science (2007), issn: 1090 9389
2. Ehrig, H., Pfender, M., Schneider, H.J.: Graph Grammars: an Algebraic Approach. In: Proceedings of FOCS, pp. 167–180. IEEE, Los Alamitos (1973)
3. Pfalzgraf, J.: On an Idea for Constructing Multiagent Systems (MAS) Scenarios. In: Lasker, G., Pfalzgraf, J. (eds.) Advances in Multiagent Systems, Robotics and Cybernetics: Theory and Practice. International Institute for Advanced Studies in Systems Research and Cybernetics, vol. 1 (2006)
4. Ehrig, H., Ehrig, K., Prange, U., Taentzer, G.: Fundamentals of Algebraic Graph Transformation, vol. 1. Springer, Heidelberg (2006)
5. Pfalzgraf, J.: On Logical Fiberings and Polycontextural Systems. In: Jorrand, P., Kelemen, J. (eds.) FAIR 1991. LNCS, vol. 535. Springer, Heidelberg (1991)
6. Pfalzgraf, J.: On Logical Fiberings and Automated Deduction in Many-valued Logics Using Gröbner Bases. In: Laita, L.M., Alonso, J.A., Roanes-Lozano, E. (eds.) Revista Real Academia de Ciencias, Serie A de Matemáticas (RACSAM), Special Issue on Symbolic Computation in Logic and Artificial Intelligence, vol. 98(1), Royal Academy of Sciences of Spain (2004)
7. Adámek, J., Herrlich, H., Strecker, G.E.: Abstract and Concrete Categories. The Joy of Cats. John Wiley and Sons Inc., Chichester (1990)
8. Wooldrige, M.J.: An Introduction to Multiagent Systems. and Sons LTD. John Wiley and Sons LTD, Chichester (2002)

Increasing Interpretations*

Harald Zankl and Aart Middeldorp

Institute of Computer Science
University of Innsbruck
6020 Innsbruck, Austria
{harald.zankl,aart.middeldorp}@uibk.ac.at

Abstract. The paper at hand introduces a refinement of interpretation based termination criteria for term rewrite systems in the dependency pair setting. Traditional methods share the property that—in order to be successful—all rewrite rules must (weakly) decrease with respect to some measure. The novelty of our approach is that we allow some rules to increase the interpreted value. These rules are found by simultaneously searching for adequate polynomial interpretations while considering the information of the dependency graph. We prove that our method extends the termination proving power of linear natural interpretations. Furthermore, this generalization perfectly fits the recursive SCC decomposition algorithm which is implemented in virtually every termination prover dealing with term rewrite systems.

Keywords: term rewriting, termination, polynomial interpretations.

Related Topics: implementations of symbolic computation systems, logic and symbolic computing.

1   Introduction

Termination of term rewriting systems (TRSs) has been a very active area of research for the last decades. In the early days many different (mostly non-modular) techniques have been developed based on syntactic and/or semantic aspects. In the recent past the demand for suitable ways for automating the methods grew. The international competition of termination tools[1] gave a strong stimulus in that direction. In this competition every tool can only spend a fixed amount of time on checking a rewrite system for (non-)termination. Since a vast number of termination criteria are known (and implemented), tool authors have to cleverly select a strategy which determines the order in which to apply the different methods and/or come up with fast implementations of termination criteria. In 2004 Kurihara and Kondo [17] were the first to encode a termination method in propositional logic. In 2006 for the first time termination analyzers

---

* This research is supported by FWF (Austrian Science Fund) project P18763.
[1] http://www.lri.fr/~marche/termination-competition/

S. Autexier et al. (Eds.): AISC/Calculemus/MKM 2008, LNAI 5144, pp. 191–205, 2008.

incorporated translations to SAT (Jambox [4] and Matchbox [20]) in the competition and astonished the termination community by the gains in power and speed. Another important issue of a termination method is locality which means that the method should fit the dependency pair method [1]. The technique we propose in this paper satisfies both demands, (a) it is modular and local in the sense that it perfectly fits the recursive SCC decomposition algorithm [12] and (b) it allows an efficient implementation using SAT solving.

The paper is organized as follows. In Section 2 the necessary definitions for graph reasoning, polynomial interpretations, and dependency pairs are given. Section 3 motivates our approach by means of an example and already suggests that special care is needed for generalizing the approach to the recursive SCC algorithm. Afterwards in Section 4 the main theorem is formally stated. Implementation details are presented in Section 5. An assessment of our contribution can be found in Section 6 before ideas for future work are addressed in Section 7.

2   Preliminaries

The termination method we present relies on (dependency) graph reasoning. The next subsection defines graphs and related concepts.

2.1   Graphs

Let N be a finite set. A *graph* $\mathcal{G} = (N, E)$ is a pair such that $E \subseteq N \times N$. Elements of N (E) are called *nodes* (*edges*). A *labeled graph* is a pair $(\mathcal{G}, \ell)$ consisting of a graph $\mathcal{G} = (N, E)$ and a labeling function $\ell \colon N \to \mathbb{Z}$ that assigns to every node an integer. A *path* from n_1 to n_m in a graph $\mathcal{G} = (N, E)$ is a finite sequence $[n_1, \ldots, n_m]$ of nodes such that $(n_i, n_{i+1}) \in E$ for all $1 \leqslant i < m$. A path is called *elementary* if all its nodes are distinct. The *length* (or *cost*) of a path $[n_1, \ldots, n_{m-1}, n_m]$ is $\ell(n_1) + \cdots + \ell(n_{m-1})$. The *distance* between two nodes a and b is the maximal length of an elementary path from a to b. A *cycle* $[n_1, \ldots, n_m]$ is a path with $m > 1$, $n_1 = n_m$, and $i \neq j$ implies $(n_i, n_{i+1}) \neq (n_j, n_{j+1})$ for all $0 \leqslant i, j < m$. A cycle $[n_1, \ldots, n_{m-1}, n_m]$ is called *elementary* if $n_1, \ldots, n_{m-1}$ are pairwise distinct. The definition of length carries over naturally from paths to cycles. Furthermore we define the *distance* $\mathsf{d}(n)$ for a single node n as the maximal length of an elementary cycle starting in n if such a cycle exists. A *strongly connected component* (SCC) is a maximal set of nodes such that there is a path from every node to every other node. Maximality means that the property of being an SCC is lost if a further node is added. For esthetic reasons, labels of nodes are associated to edges in graphical representations of graphs throughout the paper, where edges (n, m) are labeled with $\ell(n)$.

Example 1. In the labeled graph of Figure 2.1, $p_1 = [1, 2, 3, 4, 1]$ is an example of a (non-elementary) path and an elementary cycle. The (non-elementary) path $p_2 = [1, 4, 1, 4, 1]$ is no cycle since the edge $(1, 4)$ appears twice. We have length$(p_1) = 0$ and length$(p_2) = 2$. The distance of node 1 is 1 since it is the maximum length of the elementary cycles $[1, 4, 1]$ and $[1, 2, 3, 4, 1]$.

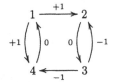

Fig. 1. A labeled graph

2.2 Polynomial Interpretations

For a signature $\mathcal{F}$ a polynomial interpretation $\mathcal{I}$ [18] maps each n-ary function symbol $f \in \mathcal{F}$ to a polynomial $f_{\mathcal{I}}$ over the natural numbers in n indeterminates. The induced mapping from terms to polynomials is denoted by $[\cdot]_{\mathcal{I}}$. For two terms s and t we have $s >_{\mathcal{I}} t$ if $[s]_{\mathcal{I}} > [t]_{\mathcal{I}}$ holds for all possible instantiations of variables by natural numbers. The comparison $s \geqslant_{\mathcal{I}} t$ is similarly defined. For polynomials with coefficients ranging over the natural numbers these problems are known to be undecidable (Hilbert's 10th problem). By fixing an upper bound for the coefficients the search space becomes finite. In typical implementations polynomials are ordered by absolute positiveness criteria [14]. Thus, in order to test whether $p > q$ holds for *linear* polynomials $p = c_0 x_0 + \cdots + c_n x_n + c_{n+1}$ and $q = d_0 x_0 + \cdots + d_n x_n + d_{n+1}$, a sufficient condition is $c_i \geqslant d_i$ for all $0 \leqslant i \leqslant n$ and $c_{n+1} > d_{n+1}$. The test $p \geqslant q$ is similar except for the constant case, i.e., $c_{n+1} \geqslant d_{n+1}$.

There already exist generalizations of polynomial interpretations, e.g., to rational and real coefficients [19] or to negative constants as well as coefficients [11]. Furthermore matrix [5], quasi-periodic [22], and arctic [15] interpretations do also extend the termination proving power significantly. All these extensions share the property that the rewrite rules under consideration must weakly decrease and at least one rule has to decrease strictly. Our approach differs from these ones in the sense that we allow a possible increase for some rules (under the side condition that some other rules eliminate that increase). In order to detect possible candidates where the interpreted value might increase when applying a rule, the dependency pair method in combination with the dependency graph (Definition 3) refinement is employed.

2.3 Dependency Pairs

We assume basic familiarity with term rewriting [2]. In the recent past there has been much research related to the dependency pair method [1] and its refinements. In this subsection we just recall the very basic definitions.

Definition 2. *Let $\mathcal{R}$ be a TRS over a signature $\mathcal{F}$. The defined symbols are the root symbols of the left-hand sides of the rewrite rules in $\mathcal{R}$. The original signature $\mathcal{F}$ is extended to a signature $\mathcal{F}^{\sharp}$ by adding for every defined symbol f a fresh symbol $f^{\sharp}$ with the same arity as f. For a term $t = f(t_1, \ldots, t_n)$*

with defined symbol f *we denote* $f^\sharp(t_1, \ldots, t_n)$ *by* $t^\sharp$. *In examples one often uses capitalization, i.e., one writes* F *for* $f^\sharp$. *If* $l \to r \in \mathcal{R}$ *and* t *is a subterm of* r *with defined root symbol, then the rule* $l^\sharp \to t^\sharp$ *is a dependency pair of* $\mathcal{R}$. *We write* $\mathrm{DP}(\mathcal{R})$ *for the set of all dependency pairs of* $\mathcal{R}$.

Dependency pairs correspond to recursive function calls. They are the basic ingredient for the dependency graph [1], which is kind of a call-graph that visualizes the order in which these recursive calls can be performed.

Definition 3. *Let* $\mathcal{R}$ *be a TRS. The nodes of the* dependency graph $\mathrm{DG}(\mathcal{R})$ *are the dependency pairs of* $\mathcal{R}$ *and there is an edge from node* $s \to t$ *to node* $u \to v$ *if there exist substitutions* σ *and* τ *such that* $t\sigma \to_{\mathcal{R}}^* u\tau$.

The dependency graph is not computable in general but sound approximations exist. Here soundness means that every edge in the original graph is also an edge in the estimated graph and hence it forms an over-approximation of the actual dependency graph.

Next, the notion of a reduction pair [1] is defined. We simplify the original definition by omitting argument filterings since they are automatically built in when dealing with polynomial interpretations (as zero coefficients correspond to deleting positions of an argument filtering).

Definition 4. *A* reduction pair $(\gtrsim, >)$ *consists of a rewrite pre-order* $\gtrsim$ *(a pre-order on terms that is closed under contexts and substitutions) and a well-founded order* $>$ *that is closed under substitutions such that the inclusion* $\gtrsim \cdot > \cdot \gtrsim \; \subseteq \; >$ *(compatibility) holds.*

The main theorem dealing with dependency pairs and including a dependency graph formulation is not given here but in Section 4 since then it is easier to see the differences between the usual theorem and our formulation.

3   A Simple Example

This section demonstrates the limitations of polynomial interpretations and suggests an improvement by additionally considering the order of recursive calls encoded in the dependency graph.

Example 5. Consider the TRS consisting of the following three rules:

$$f(0, x) \to f(1, g(x)) \tag{1}$$
$$f(1, g(g(x))) \to f(0, x) \tag{2}$$
$$g(1) \to g(0) \tag{3}$$

The dependency pairs

$$F(0, x) \to G(x) \tag{4}$$
$$F(0, x) \to F(1, g(x)) \tag{5}$$
$$F(1, g(g(x))) \to F(0, x) \tag{6}$$
$$G(1) \to G(0) \tag{7}$$

admit the following dependency graph:

$$(7) \longleftarrow (4) \longleftarrow (6) \underset{\longleftarrow}{\overset{\longrightarrow}{}} (5)$$

The idea in [12] is to find a reduction pair $(\gtrsim, >)$ for every SCC $\mathcal{S}$ such that all rules in $\mathcal{S} \cup \mathcal{R}$ decrease weakly and at least one rule in $\mathcal{S}$ decreases strictly. In the sequel we will show that the (only) SCC consisting of the nodes (5) and (6) cannot be handled by reduction pairs based on traditional implementations of linear polynomial interpretations. To be able to address all possible polynomial interpretations, we consider our problem as an abstract constraint satisfaction problem. Consequently the coefficients for the polynomials are variables whose values are natural numbers. Similarly to [6] a term $F(x, y)$ is transformed into an abstract linear polynomial $F_0 x + F_1 y + F_2$. Doing so for the SCC mentioned above results in the constraints

$$F_0 0_0 + F_1 x + F_2 \geqslant F_0 1_0 + F_1(g_0 x + g_1) + F_2$$
$$F_0 1_0 + F_1(g_0(g_0 x + g_1) + g_1) + F_2 \geqslant F_0 0_0 + F_1 x + F_2$$

where at least one inequality is strict. By simple mathematics the inequations simplify to

$$F_0 0_0 + F_1 x \geqslant F_0 1_0 + F_1 g_0 x + F_1 g_1 \qquad (8)$$
$$F_0 1_0 + F_1 g_0 g_0 x + F_1 g_0 g_1 + F_1 g_1 \geqslant F_0 0_0 + F_1 x \qquad (9)$$

From the fact that one of the above inequalities has to be strict it is obvious that $F_1 > 0$. The constraints for x in (8) demand $g_0 \leqslant 1$ and similarly (9) gives $g_0 \geqslant 1$. Hence the constraint problem is equivalent to

$$F_0 0_0 \geqslant F_0 1_0 + F_1 g_1 \qquad (10)$$
$$F_0 1_0 + F_1 g_1 + F_1 g_1 \geqslant F_0 0_0 \qquad (11)$$

which demands $g_1 > 0$ to make one inequation strict. The (simplified) constraint for rule (3) amounts to

$$1_0 \geqslant 0_0 \qquad (12)$$

The proof is concluded by the contradictory sequence

$$F_0 0_0 \geqslant F_0 1_0 + F_1 g_1 \geqslant F_0 0_0 + F_1 g_1$$

where the first inequality derives from (10), the second one from (12), and the contradiction from the fact that $F_1, g_1 > 0$ which we learned earlier.

Although we just proved that there is no termination proof for the system above with linear polynomials, we will present a termination proof right now. Assume the weakly monotone interpretation

$$F_N(x, y) = x + y \quad f_N(x, y) = 0 \quad g_N(x) = x + 1 \quad 0_N = 0 \quad 1_N = 0$$

Table 1. Rules with increasing interpretations

$$f(0, x) \rightarrow f(1, g(x)) \qquad\qquad 0 \geqslant 0 \tag{1}$$
$$f(1, g(g(x))) \rightarrow f(0, x) \qquad\qquad 0 \geqslant 0 \tag{2}$$
$$g(1) \rightarrow g(0) \qquad\qquad 1 \geqslant 1 \tag{3}$$
$$F(0, x) \rightarrow F(1, g(x)) \qquad\qquad x \geqslant x + 1 \tag{5}$$
$$F(1, g(g(x))) \rightarrow F(0, x) \qquad\qquad x + 2 \geqslant x \tag{6}$$

which orients almost all rules of interest correctly as can be seen in Table 1.

The idea to turn this interpretation into a valid termination proof is to combine the information of the dependency graph with the interpretation. From the (labeled) dependency graph

$$(7) \xleftarrow{\;0\;} (4) \xleftarrow{\;-2\;} (6) \underset{+1}{\overset{-2}{\rightleftarrows}} (5)$$

one infers that the two dependency pairs (5) and (6) are used alternately. The labels of the graph are computed as follows: From Table 1 one infers that an application of rule (6) *decreases* the interpreted value by the constant 2 (hence label -2) whereas rule (5) *increases* the value by the constant 1 (hence label $+1$). Consequently, after performing the cycle once the total value decreases by at least one. Therefore, the cycle cannot give rise to an infinite rewrite sequence.

3.1 From Cycles to SCCs

The above idea naturally extends from plain cycles to SCCs as described below. Nevertheless some care is needed when the dependency graph contains more complicated SCCs as the following example demonstrates. Consider the TRS $\mathcal{R}$ consisting of the five rules

$$f(0, 0, x, g(g(g(g(y))))) \rightarrow f(0, 1, g(g(x)), y)$$
$$f(0, 1, g(x), y) \rightarrow f(1, 1, x, g(g(y)))$$
$$f(1, 1, x, y) \rightarrow f(0, x, x, y)$$
$$g(0) \rightarrow g(1)$$
$$g(x) \rightarrow x$$

and the only SCC

$$F(0, 0, x, g(g(g(g(y))))) \rightarrow F(0, 1, g(g(x)), y) \tag{1}$$
$$F(0, 1, g(x), y) \rightarrow F(1, 1, x, g(g(y))) \tag{2}$$
$$F(1, 1, x, y) \rightarrow F(0, x, x, y) \tag{3}$$

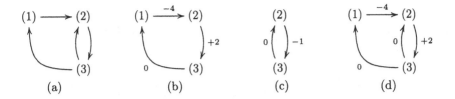

Fig. 2. Different parts of (labeled) dependency graphs

The corresponding SCC of the dependency graph depicted in Figure 2(a) contains the two cycles $[1, 2, 3, 1]$ and $[2, 3, 2]$. The first one is handled by the increasing interpretation

$$F_N(x, y, z, w) = w \quad f_N(x, y, z, w) = 0 \quad g_N(x) = x + 1 \quad 0_N = 0 \quad 1_N = 0$$

For the second we take the interpretation as above but with $F_N(x, y, z, w) = z$. Hence for the elementary cycle $[1, 2, 3, 1]$ the interpreted value decreases by 2 in every loop. Similarly there is a decrease of 1 for the elementary cycle $[2, 3, 2]$. The two labeled graphs in Figures 2(b) and 2(c) describe the symbiosis of the interpretations and the elementary cycles. The only problem is, that

$$
\begin{aligned}
f(0, 0, 0, g(g(g(g(g(y)))))) &\rightarrow f(0, 1, g(g(0)), y) \rightarrow f(1, 1, g(0), g(g(y))) \\
&\rightarrow f(0, g(0), g(0), g(g(y))) \rightarrow f(0, g(1), g(0), g(g(y))) \\
&\rightarrow f(0, 1, g(0), g(g(y))) \rightarrow f(1, 1, 0, g(g(g(g(g(y)))))) \\
&\rightarrow f(0, 0, 0, g(g(g(g(g(y)))))) \rightarrow \cdots
\end{aligned}
$$

constitutes a non-terminating sequence in this TRS. What exactly went wrong can be seen when considering the whole SCC of the labeled dependency graph (using the first interpretation, cf. Figure 2(d)). In the conventional setting it suffices to consider only the two cycles. This is the case because a strict decrease in every single cycle ensures a strict decrease in larger cycles by combining the partial proofs lexicographically. The example above shows that this is no longer true for increasing interpretations. The problematic non-terminating sequence corresponds to a run $[1, 2, 3, 2, 3, 1]$ where the interpreted value is increased in the elementary cycle $[2, 3, 2]$ and consequently the length of $[1, 2, 3, 2, 3, 1]$ is zero and there is no decrease. Considering (infinitely many!) possibly non-elementary cyclic paths is undoable. Hence the smart thing is to work with SCCs instead. To recognize dangerous runs, it suffices to compute the distance for every node. For the graph in Figure 2(d) we have $d(1) = -2$, $d(2) = 2$, and $d(3) = 2$. Only if for every node the distance is smaller than or equal to zero we know that problematic runs as demonstrated above cannot occur. Furthermore we know that in such a case we can delete nodes with negative distance because on every possible run the interpreted value decreases. If for the SCC under consideration one had managed to find a weakly monotone interpretation with labeled dependency graph like the one in Figure 3(a) (which is of course impossible since the system at hand is not

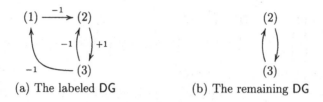

(a) The labeled DG   (b) The remaining DG

Fig. 3. A hypothetically labeled DG

terminating) then deleting node (1) would have been possible since $d(1) = -1$, $d(2) = 0$, and $d(3) = 0$. In such a situation node (1) could safely be removed and one could proceed with the simpler graph in Figure 3(b) with a possibly totally different interpretation.

4   Correctness of the Approach

The example in the preceding section shows that SCCs that consist of more than just one cycle need special attention. For usual reduction pairs it is sufficient to consider single cycles and hence in the literature theorems are usually dealing with cycles; theoretically there is no difference in power when considering cycles or SCCs but all fast implementations follow the recursive SCC approach [12]. The reason is that normally the formulation for cycles is a bit easier but in our setting it is essential to switch to an SCC treatment in order to avoid reasoning about an infinite number of possibly non-elementary cyclic paths as the example of the previous section demonstrates.

It is well known that (linear) weakly monotone polynomial interpretations over the naturals form a valid reduction pair. Note that there are strictly stronger formulations of the theorem since both restrictions—to polynomials and natural numbers—are severe.

Theorem 6. *Let $\mathcal{I}$ be a weakly monotone polynomial interpretation over the naturals. Then $(\geqslant_{\mathcal{I}}, >_{\mathcal{I}})$ is a reduction pair.*

Definition 7 ([12]). *Let $\mathcal{R}$ be a TRS, $\mathcal{S}$ a subset of the dependency pairs in $\mathsf{DG}(\mathcal{R})$, and $(\gtrsim, >)$ a reduction pair. The notation $(\gtrsim, >) \models_\exists \mathcal{R}, \mathcal{S}$ means that*

$$\mathcal{R} \subseteq \gtrsim \qquad \mathcal{S} \subseteq \gtrsim \cup > \qquad \mathcal{S} \cap > \neq \varnothing$$

In words the above definition says that all considered rules ($\mathcal{R}$ and $\mathcal{S}$) are weakly decreasing and at least one rule in $\mathcal{S}$ is strictly decreasing. The most basic theorem concerning dependency pairs (using the notation of [12]) and including the usage of the dependency graph is then formulated as follows.

Theorem 8 ([1]). *A TRS $\mathcal{R}$ is terminating if and only if for every cycle $\mathcal{C}$ in $\mathsf{DG}(\mathcal{R})$ there exists a reduction pair $(\gtrsim, >)$ such that $(\gtrsim, >) \models_\exists \mathcal{R}, \mathcal{C}$.*

There are many generalizations of the theorem above—usable rules [1,9,10], argument filterings [1], and reduction triples [13]—to name a few. To keep the presentation and discussion simple we present our work without these refinements (although our results directly generalize).

Definition 9 ([12]). *Let $\mathcal{R}$ be a TRS and $\mathcal{S}$ a subset of the dependency pairs in $\mathsf{DG}(\mathcal{R})$. We write $\models \mathcal{R}, \mathcal{S}$ if there exists a reduction pair $(\gtrsim, >)$ such that $(\gtrsim, >) \models_\exists \mathcal{R}, \mathcal{S}$ and $\models \mathcal{R}, \mathcal{S}'$ for all SCCs $\mathcal{S}'$ of the subgraph of $\mathsf{DG}(\mathcal{R})$ induced by the pairs $l \to r \in \mathcal{S}$ such that $l \not> r$.*

The theorem below states that concerning termination proving power it makes no difference if one considers cycles or performs a recursive SCC computation. The latter has the advantage that the number of SCCs is linear in the number of nodes in the dependency graph whereas the former might be exponential.

Theorem 10 ([12]). *Let $\mathcal{R}$ be a TRS. The following conditions are equivalent:*

- *$\models \mathcal{R}, \mathcal{S}$ for every SCC $\mathcal{S}$ in $\mathsf{DG}(\mathcal{R})$*
- *$\models_\exists \mathcal{R}, \mathcal{C}$ for every cycle $\mathcal{C}$ in $\mathsf{DG}(\mathcal{R})$*

We now show how to label the dependency graph by a given interpretation $\mathcal{I}$. When considering a root rewrite step which applies a rule $l \to r$, the change of the interpreted value is $[r]_\mathcal{I} - [l]_\mathcal{I}$. The idea is to label every edge by the constant part of that difference.

Definition 11. *For a polynomial p we denote the constant (non-constant) part of p by $\mathsf{cp}(p)$ ($\mathsf{ncp}(p)$). For a term t and a polynomial interpretation $\mathcal{I}$ we abbreviate $\mathsf{ncp}([t]_\mathcal{I})$ by $\mathsf{ncp}_\mathcal{I}(t)$. This notation naturally extends to rules and TRSs, e.g., $\mathsf{ncp}_\mathcal{I}(l \to r) = \mathsf{ncp}_\mathcal{I}(l) \to \mathsf{ncp}_\mathcal{I}(r)$ and $\mathsf{ncp}_\mathcal{I}(\mathcal{R}) = \{\mathsf{ncp}_\mathcal{I}(l \to r) \mid l \to r \in \mathcal{R}\}$. The same notation is freely used for $\mathsf{cp}_\mathcal{I}$.*

Definition 12. *Let $\mathcal{I}$ be an interpretation and DG a dependency graph. The labeled dependency graph $\mathsf{DG}_\mathcal{I}$ is defined as (DG, ℓ) with $\ell(l \to r) = \mathsf{cp}([r]_\mathcal{I} - [l]_\mathcal{I})$ for every node $l \to r$ in DG. By $\mathsf{d}_\mathcal{I}(n)$ we denote the distance of a node $n \in \mathsf{DG}_\mathcal{I}$.*

The next definition presents analogous versions of Definitions 7 and 9 in the setting of increasing interpretations.

Definition 13. *Let $\mathcal{R}$ be a TRS and $\mathcal{S}$ a subset of the dependency pairs in $\mathsf{DG}(\mathcal{R})$. We write $\models_\exists^\mathcal{I} \mathcal{R}, \mathcal{S}$ if $\mathcal{I}$ is an interpretation over the naturals and*

$$\mathcal{R} \cup \mathsf{ncp}_\mathcal{I}(\mathcal{S}) \subseteq \geqslant_\mathcal{I} \quad \mathsf{d}_\mathcal{I}(\mathcal{S}) \subseteq \mathbb{Z}^{\leqslant 0} \quad \mathsf{d}_\mathcal{I}(\mathcal{S}) \cap \mathbb{Z}^{< 0} \neq \varnothing$$

Consequently $\models^\mathcal{I} \mathcal{R}, \mathcal{S}$ if $\models_\exists^\mathcal{I} \mathcal{R}, \mathcal{S}$ and for all SCCs $\mathcal{S}'$ of the subgraph of $\mathsf{DG}_\mathcal{I}(\mathcal{R})$ induced by the pairs $l \to r \in \mathcal{S}$ such that $\mathsf{d}_\mathcal{I}(l \to r) \not< 0$ there exists an interpretation $\mathcal{I}'$ such that $\models^{\mathcal{I}'} \mathcal{R}, \mathcal{S}'$.

Now we are ready to present the main theorem. In Section 3 we already showed that this extends the termination proving power of natural linear interpretations.

Theorem 14. *A TRS $\mathcal{R}$ is terminating if for every SCC S in the dependency graph $\mathsf{DG}(\mathcal{R})$ there exists a weakly monotone polynomial interpretation $\mathcal{I}$ over the naturals such that $\models^{\mathcal{I}} \mathcal{R}, S$ holds.*

Proof. We show that under the assumption $\models^{\mathcal{I}}_{\exists} \mathcal{R}, S$ with $s \to t \in S$ satisfying $\mathsf{d}_{\mathcal{I}}(s \to t) < 0$ there cannot be a non-terminating rewrite sequence that applies $s \to t$ indefinitely. The theorem follows immediately from that property. For a proof by contradiction assume the existence of such a sequence:

$$s_0 \to_{s \to t} t_0 \to^{*}_{S \cup \mathcal{R}} s_1 \to_{s \to t} t_1 \to^{*}_{S \cup \mathcal{R}} s_2 \to_{s \to t} t_2 \to^{*}_{S \cup \mathcal{R}} s_3 \to \cdots$$

Since $\geqslant_{\mathcal{I}}$ is closed under contexts and substitutions, for all terms u, v, and all rules $l \to r \in \mathcal{R} \cup S$ with $u \to_{l \to r} v$ we get $\mathsf{ncp}_{\mathcal{I}}(u) \geqslant \mathsf{ncp}_{\mathcal{I}}(v)$. Because the infinite sequence was chosen such that the rule $s \to t$ is used infinitely often it is obvious that when starting from term s_0 one must cycle in the dependency graph in order to reach s_1. The fact that $\mathsf{d}_{\mathcal{I}}(s \to t) < 0$ together with $\mathsf{d}_{\mathcal{I}}(S) \subseteq \mathbb{Z}^{\leqslant 0}$ ensures that every cycle containing the node $s \to t$ decreases the constant part of the interpretation strictly (note that $\mathsf{cp}_{\mathcal{I}}(\mathcal{R}) \subseteq \geqslant$ by definition). Hence, $\mathsf{cp}_{\mathcal{I}}(s_0) > \mathsf{cp}_{\mathcal{I}}(s_1)$. Repeating this argument gives rise to the sequence

$$\mathsf{cp}_{\mathcal{I}}(s_0) > \mathsf{cp}_{\mathcal{I}}(s_1) > \mathsf{cp}_{\mathcal{I}}(s_2) > \mathsf{cp}_{\mathcal{I}}(s_3) > \cdots$$

which contradicts the well-foundedness of $>$ over the natural numbers. $\square$

5   Implementation

Almost all fast implementations of polynomial interpretations are based on a transformation to a SAT problem. Also many other termination criteria are very suitable for a SAT encoding as can be seen by the vast amount of literature. The major drawback is that one has to work with abstract encodings all the time. Hence when labeling the dependency graph one does not have concrete integers at hand but some propositional formulas which abstractly encode the range of all possible values. Since encoding polynomials in SAT has already been described in detail [6], in this paper we refrain from giving all implementation issues. The only encoding which is discussed here is how to compute the distance between two (not necessarily distinct) nodes within a labeled graph.

5.1   General Algorithm

The idea is to compute the distance of a node by means of a transitivity closure. The integer variable R_{abi} is $-\infty$ if b is not reachable in at most 2^i steps from a and otherwise this variable keeps the (currently known) distance from a to b. It is obvious that in a graph (N, E) an elementary cycle contains at most $|N|$ edges and hence for $k' \geqslant k := \lceil log_2(|N|) \rceil$ one has surely reached a fixed point, i.e., $R_{abk'} = R_{abk}$ for all a, $b \in N$.

More precisely, the variables R_{ab0} reflect the edges of the graph and hence b is reachable from a with a cost of $\ell(a)$ if $(a, b) \in E$ and it is unreachable if $(a, b) \notin E$. Thus we initialize these variables as follows:

$$R_{ab0} = \begin{cases} \ell(a) & \text{if } (a, b) \in E \\ -\infty & \text{otherwise} \end{cases}$$

Since R_{abi} might be $-\infty$, addition and maximum operation are extended naturally, i.e., $n + -\infty = -\infty + n = -\infty$ and $\max(n, -\infty) = \max(-\infty, n) = n$ for all $n \in \mathbb{Z} \cup \{-\infty\}$. For $0 \leqslant i < k$ we define

$$R_{ab(i+1)} = \max(R_{abi}, \max_{m \in N}\{R_{ami} + R_{mbi}\})$$

If one first forgets about the max then the above formula expresses that b is reachable from a in at most 2^{i+1} steps with a cost of $R_{ab(i+1)}$ if it is already reachable within 2^i steps with that cost or there is a mid-point[2] m and the cost from a to m and the one from m to b just sum up. Taking the maximum of all possible costs ensures that we consider a worst case scenario. In the end we want to test if $R_{nnk} \leqslant 0$ for all $n \in N$. Note that it might happen that the value R_{nnk} does not emerge from an elementary cycle (because it might happen that one cycles more than once). Nevertheless the idea remains sound because if the length of a maximal elementary cycle is smaller than zero, then the length remains smaller than zero if we go along that cycle more often. Dually this property holds for distances greater than zero. For a demonstration consider the following example.

Example 15. In the labeled graph from Example 2.1 we have $k = \lceil log_2(4) \rceil = 2$ and

$d(1) = 1$	$d(2) = 0$	$d(3) = 0$	$d(4) = 1$
$R_{112} = 2$	$R_{222} = 0$	$R_{332} = 0$	$R_{442} = 2$

The reason for the different values is that R_{112} does not correspond to an elementary cycle; we have $d(1) = 1$ (see Example 1) but $R_{112} = 2$ since it derives from the cyclic path $[1, 4, 1, 4, 1]$. A similar argument explains the discrepancy of $d(4)$ and R_{442}.

5.2   Special Algorithms

The encoding for computing maximal paths in SAT from the previous subsection has complexity $\mathcal{O}(n^2 log(n))$ where n is the number of nodes in the underlying SCC of the labeled DG. To get a faster implementation we specialize the algorithm for SCCs that have a special shape:

---

[2] Fortunately Zeno of Elea was wrong and this approach constitutes a valid method for computing reachability.

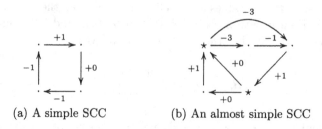

(a) A simple SCC   (b) An almost simple SCC

Fig. 4. Two special shapes of SCCs

(a) Simple SCCs: An SCC is called *simple* if it contains exactly one cycle, i.e., omitting any edge would destroy the property of being an SCC. An example of this shape is depicted in Figure 4(a). Linear time suffices to decide if a given SCC S is simple (the number of edges equals the number of nodes). In such a case the encoding specializes to

$$\sum_{n \in S} \ell(n) < 0$$

which expresses that the constant part of the interpretation $\mathcal{I}$ decreases when cycling. The encoding is linear in the size of the nodes.

(b) Almost simple SCCs: An SCC is called *almost simple* if it is not simple and there exists a node n (called *selected node*) such that after deleting *all* outgoing edges of n there is no non-empty sub-SCC left. Here we will exploit the fact that in every cycle within this SCC we pass the node n. The nodes indicated with $\star$ in Figure 4(b) satisfy this property. In the encoding we demand that $-\ell(n) > \ell(m_1) + \cdots + \ell(m_p)$ holds where n is the selected node and $m_1, \ldots, m_p$ are the nodes in the SCC that have a positive label. The underlying idea is that node n decreases the interpretation more than all other rules together might increase it and since that node n must be passed in every cyclic run there cannot be infinite reductions. For every selected node n the encoding is of linear size.

Note that the specialization for case (a) is exact whereas (b) is an approximation.

6   Assessment

In this paper we showed that increasing interpretations are strictly more powerful than standard linear interpretations over the naturals. Clearly for SCCs consisting of just a single rule they are of equal power.

The reason why the TRS of Example 5 cannot be proved terminating by means of linear polynomials is that we cannot differentiate constant 0 from 1 by the interpretation. Hence it is not so astonishing that the problematic SCC can be handled by matrix interpretations [5] of dimension two. Actually all of the tools (dedicated to proving termination) participating in the TRS category of the 2007 edition of the international termination competition can handle this system. All the proofs rely on matrix interpretations with dimension two. As a pre-processing step AProVE [8] and T_TT_2[3] use dependency pair analysis whereas Jambox [4] performs a reduction of right-hand sides [21].

It is an easy exercise to construct (larger) TRSs than Example 5 such that all tools of the termination competition fail. To disallow Jambox the rewriting of right-hand sides we introduce overlaps. To knock-out the matrix method just increasing the size of the system suffices. Since T_TT_2 can still prove these examples by bounds [7,16] we ensure the TRS to be not left-linear which makes increasing interpretations the only successful method.

Example 16. For the TRS where $g^8(x)$ is a shortcut for $g(g(g(g(g(g(g(g(x))))))))$

$$f(0,0,0,x) \rightarrow f(0,0,1,g(x)) \qquad\qquad f(0,0,1,x) \rightarrow f(0,1,0,g(x))$$
$$f(0,1,0,x) \rightarrow f(0,1,1,g(x)) \qquad\qquad f(0,1,1,x) \rightarrow f(1,0,0,g(x))$$
$$f(1,0,0,x) \rightarrow f(1,0,1,g(x)) \qquad\qquad f(1,0,1,x) \rightarrow f(1,1,0,g(x))$$
$$f(1,1,0,x) \rightarrow f(1,1,1,g(x)) \qquad\qquad f(y,y,y,g^8(x)) \rightarrow f(0,0,0,x)$$
$$g(g(0)) \rightarrow 1 \qquad\qquad g(g(1)) \rightarrow g(g(0))$$

none of the existing termination tools succeeds in proving termination within a 60 seconds time limit. Increasing interpretations produce a successful—and very intuitive—proof for the challenging SCC. It considers the changes of F's fourth argument. Both the general approach described in Section 5.1 and the specialization (b) from Section 5.2 yield the increasing interpretation $F_N(x,y,z,w) = w$, $g_N(x) = x + 1$, $f_N(x,y,z,w) = 0_N = 1_N = 0$ which ensures that all nodes have a negative distance and hence the whole problematic SCC can be removed. The only difference between the two is that it takes the first method almost half a minute whereas the optimized encoding succeeds within a fraction of a second.

The theory of increasing interpretations as described above directly applies to the matrix method [5] as well. Note that when interpreting dependency pairs the constant part amounts to a natural number and hence the dependency graph is labeled in exactly the same fashion.

7   Future Work

Generalizing the approach in such a way that not only the constant part of the interpretation is used as additional information in the dependency graph but also the non-constant part, is highly desirable. We anticipate that this would make the approach significantly more powerful. The only drawback is that probably this generalization applies to a very restricted class of TRSs only. To get a feeling for the problems that arise consider the non-terminating system

$$f(s(x)) \rightarrow g(s(x)) \qquad\qquad g(x) \rightarrow f(x)$$

---

[3] http://colo6-c703.uibk.ac.at/ttt2

which admits the dependency pairs (1) $F(s(x)) \to G(s(x))$ and (2) $G(x) \to F(x)$. The increasing interpretation $F_N(x) = 2x$, $G_N(x) = x$, $f_N(x) = 0$, $s_N(x) = x + 1$ would remove both dependency pairs since there is a strict decrease for every cycle in the labeled dependency graph, which looks like

$$(1) \overset{-x-1}{\underset{+x}{\rightleftarrows}} (2)$$

The problem in this example is that in the two dependency pairs the variable x does not correspond to the same term. For this example it is obvious that in any minimally non-terminating sequence, $s(x)$ is substituted for the variable x in the second rule. Hence, one should not consider the original system but immediately change the variable x in the second rule on both sides to $s(x)$. Then increasing interpretations are no longer successful. However such a transformation is not always possible. In the example above for every minimally non-terminating sequence there are no $\mathcal{R}$-steps and hence one can compute the substitution for x in the second rule by unification. Similar cases can be dealt with narrowing [1].

To conclude, we summarize that increasing interpretations can be extended to allow an increase also in the variable part if the TRS under consideration satisfies two properties: (a) all dependency pairs are variable disjoint (this can always be achieved by renaming) and (b) for every minimally non-terminating sequence

$$s_0 \to_{\mathsf{DP}(\mathcal{R})} t_0 \to^*_{\mathcal{R}} s_1 \to_{\mathsf{DP}(\mathcal{R})} t_1 \to^*_{\mathcal{R}} s_2 \to_{\mathsf{DP}(\mathcal{R})} t_2 \to^*_{\mathcal{R}} \cdots$$

the $\mathcal{R}$-sequences are empty (and hence the values for variables can possibly be computed by unification). Note that one sufficient condition for (b) is that the set of usable rules is empty.

Acknowledgments. We thank Niklas Eén and Niklas Sörensson for developing and providing MiniSat [3]. We thank Sarah Winkler for writing a suitable OCaml interface for MiniSat that paved the way for an integration of increasing interpretations into the termination prover T_TT_2.

References

1. Arts, T., Giesl, J.: Termination of term rewriting using dependency pairs. Theoretical Computer Science 236, 133–178 (2000)
2. Baader, F., Nipkow, T.: Term Rewriting and All That. Cambridge University Press, Cambridge (1998)
3. Eén, N., Sörensson, N.: An extensible SAT-solver. In: Giunchiglia, E., Tacchella, A. (eds.) SAT 2003. LNCS, vol. 2919, pp. 502–518. Springer, Heidelberg (2004)
4. Endrullis, J.: Jambox (2007), http://joerg.endrullis.de
5. Endrullis, J., Waldmann, J., Zantema, H.: Matrix interpretations for proving termination of term rewriting. Journal of Automated Reasoning 40(2-3), 195–220 (2008)

6. Fuhs, C., Giesl, J., Middeldorp, A., Schneider-Kamp, P., Thiemann, R., Zankl, H.: SAT solving for termination analysis with polynomial interpretations. In: Marques-Silva, J., Sakallah, K.A. (eds.) SAT 2007. LNCS, vol. 4501, pp. 340–354. Springer, Heidelberg (2007)

7. Geser, A., Hofbauer, D., Waldmann, J., Zantema, H.: On tree automata that certify termination of left-linear term rewriting systems. Information and Computation 205(4), 512–534 (2007)

8. Giesl, J., Schneider-Kamp, P., Thiemann, R.: AProVE 1.2: Automatic termination proofs in the dependency pair framework. In: Furbach, U., Shankar, N. (eds.) IJCAR 2006. LNCS (LNAI), vol. 4130, pp. 281–286. Springer, Heidelberg (2006)

9. Giesl, J., Thiemann, R., Schneider-Kamp, P., Falke, S.: Mechanizing and improving dependency pairs. Journal of Automated Reasoning 37(3), 155–203 (2006)

10. Hirokawa, N., Middeldorp, A.: Dependency pairs revisited. In: van Oostrom, V. (ed.) RTA 2004. LNCS, vol. 3091, pp. 249–268. Springer, Heidelberg (2004)

11. Hirokawa, N., Middeldorp, A.: Polynomial interpretations with negative coefficients. In: Buchberger, B., Campbell, J.A. (eds.) AISC 2004. LNCS (LNAI), vol. 3249, pp. 185–198. Springer, Heidelberg (2004)

12. Hirokawa, N., Middeldorp, A.: Automating the dependency pair method. Information and Computation 199(1,2), 172–199 (2005)

13. Hirokawa, N., Middeldorp, A.: Tyrolean termination tool: Techniques and features. Information and Computation 205(4), 474–511 (2007)

14. Hong, H., Jakuš, D.: Testing positiveness of polynomials. Journal of Automated Reasoning 21(1), 23–38 (1998)

15. Koprowski, A., Waldmann, J.: Arctic termination ... below zero. In: Voronkov, A. (ed.) RTA 2008. LNCS, vol. 5117. Springer, Heidelberg (to appear, 2008)

16. Korp, M., Middeldorp, A.: Proving termination of rewrite systems using bounds. In: Baader, F. (ed.) RTA 2007. LNCS, vol. 4533, pp. 273–287. Springer, Heidelberg (2007)

17. Kurihara, M., Kondo, H.: Efficient BDD encodings for partial order constraints with application to expert systems in software verification. In: Orchard, B., Yang, C., Ali, M. (eds.) IEA/AIE 2004. LNCS (LNAI), vol. 3029, pp. 827–837. Springer, Heidelberg (2004)

18. Lankford, D.: On proving term rewrite systems are noetherian. Technical Report MTP-3, Louisiana Technical University, Ruston, LA, USA (1979)

19. Lucas, S.: On the relative power of polynomials with real, rational, and integer coefficients in proofs of termination of rewriting. Applicable Algebra in Engineering, Communication and Computing 17(1), 49–73 (2006)

20. Waldmann, J.: Matchbox: A tool for match-bounded string rewriting. In: van Oostrom, V. (ed.) RTA 2004. LNCS, vol. 3091, pp. 85–94. Springer, Heidelberg (2004)

21. Zantema, H.: Reducing right-hand sides for termination. In: Middeldorp, A., van Oostrom, V., van Raamsdonk, F., de Vrijer, R. (eds.) Processes, Terms and Cycles: Steps on the Road to Infinity. LNCS, vol. 3838, pp. 173–197. Springer, Heidelberg (2005)

22. Zantema, H., Waldmann, J.: Termination by quasi-periodic interpretations. In: Baader, F. (ed.) RTA 2007. LNCS, vol. 4533, pp. 404–418. Springer, Heidelberg (2007)

Validated Evaluation of Special Mathematical Functions

Franky Backeljauw, Stefan Becuwe, and Annie Cuyt

Universiteit Antwerpen, Department of Mathematics and Computer Science,
Middelheimlaan 1, B-2020 Antwerpen, Belgium
{franky.backeljauw, stefan.becuwe, annie.cuyt}@ua.ac.be

Abstract. Because of the importance of special functions, several books and a large collection of papers have been devoted to the numerical computation of these functions, the most well-known being the Abramowitz and Stegun handbook [1]. But up to this date, no environment offers routines for the provable correct evaluation of these special functions.

We point out how series and limit-periodic continued fraction representations of the functions can be helpful in this respect. Our scalable precision technique is mainly based on the use of sharpened a priori truncation and round-off error upper bounds, in case of real arguments. The implementation is validated in the sense that it returns a sharp interval enclosure for the requested function evaluation, at the same cost as the evaluation.

1 Introduction

Special functions are pervasive in all fields of science and industry. The most well-known application areas are in physics, engineering, chemistry, computer science and statistics. Often encountered functions are the Gauss hypergeometric function $_2F_1(a, b; c; x)$, the Bessel functions of integer and half-integer order, the (complementary) error function to name just a few. Because of their importance, several books and a large collection of papers have been devoted to algorithms for the numerical computation of these functions.

Virtually all present-day computer systems, from personal computers to the largest supercomputers, implement the IEEE floating-point arithmetic standard, which provides 53 binary or approximately 16 decimal digits accuracy. For most scientific applications, this is more than sufficient. For instance, in electromagnetic simulation models the final required accuracy is usually in the order of only 2 to 3 significant digits.

However, for a rapidly expanding body of applications, 64-bit IEEE arithmetic is no longer sufficient. These range from some exploratory mathematical investigations to large-scale physical simulations performed on highly parallel supercomputers. In these applications, portions of the code typically involve numerically sensitive calculations, which produce results of questionable accuracy using conventional arithmetic. These inaccurate results may in turn induce other errors, such as taking the wrong path in a conditional branch.

S. Autexier et al. (Eds.): AISC/Calculemus/MKM 2008, LNAI 5144, pp. 206–216, 2008.

Such blocks of code benefit enormously from validated numerical techniques, possibly in combination with high-precision arithmetic. Indeed, a reliable numeric technique delivers a floating-point enclosure for the exact result rather than a computed estimate.

Up to this date, even environments such as Maple, Mathematica, MATLAB and libraries such as IMSL, CERN and NAG offer no routines for the provable correct evaluation of special functions. The following quotes concisely express the need for new developments in the evaluation of special functions:

- *"Algorithms with strict bounds on truncation and rounding errors are not generally available for special functions. These obstacles provide an opportunity for creative mathematicians and computer scientists."* Dan Lozier, general director of the DLMF project, and Frank Olver [2,3].
- *"The decisions that go into these algorithm designs — the choice of reduction formulae and interval, the nature and derivation of the approximations — involve skills that few have mastered. The algorithms that MATLAB uses for gamma functions, Bessel functions, error functions, Airy functions, and the like are based on Fortran codes written 20 or 30 years ago."* Cleve Moler, founder of MATLAB [4].

2   Validated Function Evaluation

Let us assume to have at our disposal a scalable precision IEEE 754-854 compliant [5] floating-point implementation of the basic operations, comparisons, base and type conversions, in the rounding modes upward, downward, truncation and round-to-nearest. Such an implementation is characterized by four parameters: the internal base β, the precision t and the exponent range $[L, U]$. Here we aim at least at implementations for $\beta = 2$ at precisions $t \geq 53$, and at implementations for use with $\beta = 2^i$ or $\beta = 10^i$ where $i > 1$.

We denote by $\oplus, \ominus, \otimes, \oslash$ the exactly rounded (to the nearest) floating-point implementation of the basic operations $+, -, \times, \div$ in the chosen base β and precision t. Hence these basic operations are carried out with a relative error of at most $1/2\, \beta^{-t+1}$ which is also called $1/2$ ulp in precision t:

$$\left| \frac{(x \circledast y) - (x * y)}{x * y} \right| \leq \frac{1}{2}\beta^{-t+1}, \qquad * \in \{+, -, \times, \div\}.$$

The realization of a machine implementation of a function $f(x)$ in that floating-point environment is essentially a three-step procedure:

1. For a given argument x, the evaluation $f(x)$ is often reduced to the evaluation of f for another argument $\tilde{x}$ lying within specified bounds and for which there exists an easy relationship between $f(x)$ and $f(\tilde{x})$. The issue of argument reduction is a topic in its own right and mostly applies to only the simplest transcendental functions [6]. In the sequel we skip the issue of argument reduction and assume for simplicity that $x = \tilde{x}$.

2. After determining the argument, a mathematical model F for f is constructed and a truncation error

$$\frac{|f(x) - F(x)|}{|f(x)|} \tag{1}$$

comes into play, which needs to be bounded. In the sequel we systematically denote the approximation $F(x) \approx f(x)$ by a capital italic letter.

3. When implemented, in other words evaluated as $\mathrm{F}(x)$, this mathematical model is also subject to a round-off error

$$\frac{|F(x) - \mathrm{F}(x)|}{|f(x)|} \tag{2}$$

which needs to be controlled. We systematically denote the implementation $\mathrm{F}(x)$ of $F(x)$ in capital typewriter font.

The technique to provide a mathematical model $F(x)$ of a function $f(x)$ differs substantially when going from a fixed finite precision context to a finite scalable precision context. In the former, the aim is to provide one optimal mathematical model, requiring as few operations as possible. Here optimal means that the model's complexity is minimal with respect to the truncation error bound imposed by the fixed finite precision. In the latter, the goal is to provide a generic technique, from which a mathematical model yielding the imposed accuracy, is deduced at runtime. Hence best approximants are not an option since these models have to be recomputed every time the precision is altered and a function evaluation is requested. At the same time the generic technique should generate an approximant of as low complexity as possible.

We aim, on the one hand, at a generic technique suitable for use in a multi-precision context, which on the other hand, is efficient enough to compete with the traditional hardware algorithms when the base β is set to 2 and the precision t to 53. We also want our implementation to be reliable, meaning that a sharp interval enclosure for the requested function evaluation is returned without any additional cost.

Besides series representations, as presented in Section 3, continued fraction representations of functions can be very helpful in the multiprecision context. A lot of well-known constants in mathematics, physics and engineering, as well as elementary and special functions enjoy very nice and rapidly converging continued fraction representations. In addition, many of these fractions are limit-periodic. Both, series and continued fraction representations, are classical techniques to approximate functions and there's a lot of literature describing implementations that make use of them [7]. But so far, no attempt is made at an efficient yet provable correct implementation.

It is well-known that the tail or remainder term of a convergent Taylor series expansion converges to zero. It is less well-known that the tail of a convergent continued fraction representation does not necessarily converge to zero. It does not even need to converge at all. A suitable approximation of the usually disregarded continued fraction tail may speed up the convergence of the continued fraction approximants. This idea is elaborated in Section 4.

3   Taylor Series Development

For simplicity, but without loss of generality, we assume that the Taylor series of $f(x)$ is given at the origin:

$$f(x) = \sum_{n=0}^{\infty} a_n x^n. \tag{3}$$

If we want the total error $|f(x) - \mathbf{F}(x)|/|f(x)|$ to be bounded by $\alpha \beta^{-t+1}$ we must determine N such that for $\mathbf{F}(x) = p_N(x)$, the partial sum of degree N of (3), the truncation error

$$\left| \frac{f(x) - p_N(x)}{f(x)} \right| \le \frac{\alpha}{2} \beta^{-t+1}$$

and evaluate $p_N(x)$, possibly in a working precision s slightly larger than the user precision t, such that the computed value $\mathbf{F}(x) = \mathbf{pN}(x)$ satisfies

$$\left| \frac{p_N(x) - \mathbf{pN}(x)}{f(x)} \right| \le \frac{\alpha}{2} \beta^{-t+1}.$$

An upper bound for the error $|f(x) - p_N(x)|$ is obtained from the sequence of coefficients $\{a_n\}_n$. If this sequence is strictly decreasing with all $a_n > 0$ and with

$$r_n = a_n/a_{n-1} \le R < 1, \qquad n \ge 1,$$

then

$$\sum_{n=N+1}^{\infty} a_n \le a_N \sum_{n=0}^{\infty} R^n = \frac{a_N}{1 - R}.$$

If the sequence is alternating with $\{(-1)^n a_n\}_n$ positive and decreasing and if N is odd, then

$$\sum_{n=N+1}^{\infty} a_n \le a_{N+1}.$$

Furthermore in both cases $|f(x)| \ge |p_N(x)|$ and hence

$$\left| \frac{p_N(x) - \mathbf{pN}(x)}{f(x)} \right| \le \left| \frac{p_N(x) - \mathbf{pN}(x)}{p_N(x)} \right|.$$

A standard method for the evaluation of the polynomial $p_N(x)$ is Horner's scheme, namely

$$p_N(x) = a_0 + x(a_1 + x(a_2 + x(\ldots + x a_N))). \tag{4}$$

Since the coefficients a_n of $p_N(x)$ are often related by a simple ratio $r_n = a_n/a_{n-1}$, Horner's scheme can be rewritten as

$$p_N(x) = a_0(1 + x r_1(1 + x r_2(1 + x r_3(\ldots + x r_N)))). \tag{5}$$

Let $\tilde{a}_0$ and $\tilde{r}_n$ denote the machine representations available for a_0 and r_n respectively. Let $\delta(\cdot)$ denote an upper bound for the relative error (expressed as a multiple of the working precision $\frac{1}{2}$ ulp) due to replacing the expression $(\cdot)$ by its floating-point counterpart. Hence

$$\tilde{a}_0 = a_0(1 + \delta_0), \qquad |\delta_0| \le \frac{1}{2}\delta(a_0)\beta^{-s+1},$$

$$\tilde{r}_n = r_n(1 + \delta_n), \qquad |\delta_n| \le \frac{1}{2}\delta(r_n)\beta^{-s+1}, \qquad n = 1, \dots, N.$$

A round-off error analysis of the nested scheme

$$\begin{aligned}
q_N(x) &= 1, \\
q_n(x) &= 1 \oplus x \otimes \tilde{r}_{n+1} \otimes q_{n+1}(x), \qquad n = N-1, \dots, 0, \\
\text{p}_N(x) &= \tilde{a}_0 \otimes q_0(x)
\end{aligned}$$

provides the bound [8, pp. 69]

$$\left| \frac{p_N(x) - \text{p}_N(x)}{p_N(x)} \right| \le \left| \frac{\epsilon(N)}{1 - \epsilon(N)} \right| \frac{p_N^+(|x|)}{|p_N(x)|},$$

where

$$p_N^+(x) = \sum_{n=0}^{N} |a_n| x^n,$$

$$\epsilon(N) = \frac{1}{2}\left(\delta(a_0) + N\left(3 + \delta(x) + \max_{n=1,\dots,N} \delta(r_n) \right) \right) \beta^{-s+1}.$$

Note that the factor

$$\frac{p_N^+(|x|)}{|p_N(x)|} \ge 1. \tag{6}$$

It equals 1 if $a_n \ge 0$ for all n and $x \ge 0$, or if $(-1)^n a_n \ge 0$ for all n and $x \le 0$. Otherwise this factor can sometimes be arbitrarily large.

4   Continued Fraction Representation

Let us consider a continued fraction representation of the form

$$f = \cfrac{a_1}{1 + \cfrac{a_2}{1 + \dots}} = \frac{a_1|}{|1} + \frac{a_2|}{|1} + \dots = \sum_{n=1}^{\infty} \frac{a_n|}{|1}, \qquad a_n := a_n(x), \quad f := f(x) \tag{7}$$

with $a_n \ge -\frac{1}{4}$. Here a_n is called the n-th partial numerator. The continued fraction is said to be limit-periodic if the limit $\lim_{n \to \infty} a_n$ exists (it is allowed to be $+\infty$). We respectively denote by the N-th approximant $f_N(x; w_N)$ and N-th tail $t_N(x)$ of (7), the values

$$f_N(x; w_N) = \sum_{n=1}^{N-1} \frac{a_n}{\mid 1} + \frac{a_N}{\mid 1+w},$$

$$t_N(x) = \sum_{n=N+1}^{\infty} \frac{a_n}{\mid 1}.$$

We restrict ourselves to the case where a sequence $\{w_n\}_n, w_n \neq 0$ can be chosen such that $\lim_{n\to\infty} f_n(x; w_n) = \lim_{n\to\infty} f_n(x; 0)$.

The tails $t_N(x)$ of a convergent continued fraction can behave quite differently compared to the tails of a convergent series which always go to zero. We illustrate the different cases with an example. Take for instance the continued fraction expansion

$$\frac{\sqrt{1+4x} - 1}{2} = \sum_{n=1}^{\infty} \frac{x}{\mid 1}, \qquad x \geq -\frac{1}{4}.$$

Each tail $t_N(x)$ converges to the value $1/2(\sqrt{1+4x} - 1)$ as well and hence the sequence of tails is a constant sequence. More remarkable is that the even-numbered tails of the convergent continued fraction

$$\sqrt{2} - 1 = \sum_{n=1}^{\infty} \left(\frac{(3 + (-1)^n)/2}{\mid 1} \right) = \frac{1}{\mid 1} + \frac{2}{\mid 1} + \frac{1}{\mid 1} + \frac{2}{\mid 1} + \cdots$$

converge to $\sqrt{2} - 1$ while the odd-numbered tails converge to $\sqrt{2}$ (hence the sequence of tails does not converge), and that the sequence of tails $\{t_N(x)\}_N = \{N+1\}_N$ of

$$1 = \sum_{n=1}^{\infty} \frac{n(n+2)}{\mid 1}$$

converges to $+\infty$. When carefully monitoring the behaviour of these continued fraction tails, very accurate approximants $f_N(x; w_N)$ for f can be computed by making an appropriate choice for w_N. For instance, when $\lim_{n\to\infty} a_n = a < +\infty$ then an estimate of the N-th tail is given by $(\sqrt{1+4a} - 1)/2$. The appearance of the square root explains the condition $a_n \geq -1/4$.

The relative truncation error $|f(x) - f_N(x; w_N)|/|f(x)|$ is bounded by the so-called interval sequence theorem [9]. Let the sequence of intervals $\{[L_n, R_n]\}_n$ with $-1/2 \leq L_n \leq R_n < \infty$ be given such that we have for

$$b_n := (1 + \text{sign}(L_n) \max(|L_n|, |R_n|))L_{n-1},$$
$$c_n := (1 + \text{sign}(L_n) \min(|L_n|, |R_n|))R_{n-1},$$

that

$$b_n \leq a_n \leq c_n, \qquad 0 \leq b_n c_n.$$

Then

$$\left| \frac{f(x) - f_N(x; w_N)}{f(x)} \right| \leq \frac{R_N - L_N}{1 + L_N} \prod_{n=1}^{N-1} M_n,$$

$$L_N \leq w_N \leq R_N, \qquad M_n = \max\left\{ \left| \frac{L_n}{1 + L_n} \right|, \left| \frac{R_n}{1 + R_n} \right| \right\}.$$

The L_n and R_n are tails of continued fractions constructed with the entries b_n and c_n [9] which actually bound the floating-point uncertainty on a_n. If the partial numerators a_n of the continued fraction (7) satisfy $a_n \geq -1/4$, then we know that:

- in case all $a_n > 0$ and $w_N \leq t_N$, the even approximants satisfy $f_N(x; w_N) \leq f(x)$,
- in case all $a_n < 0$ and $w_N \leq t_N$, all approximants satisfy $f_N(x; w_N) \leq f(x)$.

Hence we obtain for the round-off error on the computed value $F(x) = \texttt{fN}(x; w_N)$:

$$\left| \frac{f_N(x; w_N) - \texttt{fN}(x; w_N)}{f_N(x; w_N)} \right| \leq \left| \frac{f_N(x; w_N) - \texttt{fN}(x; w_N)}{\texttt{fN}(x; w_N)} \right|.$$

If the machine representation $\tilde{a}_n = a_n(1 + \delta_n)$ with $|\delta_n| \leq 1/2\, \delta(a_n)\beta^{-s+1}$ then [10, pp. 156–158] [11]

$$\left| \frac{f_N(x; w_N) - \texttt{fN}(x; w_N)}{\texttt{fN}(x; w_N)} \right| \leq \frac{1}{2}\left(4 + \Delta\right)\left(1 + M + \ldots + M^{N-1}\right)\beta^{-s+1},$$

$$\Delta = \max_{n=1,\ldots,N} \delta(a_n), \qquad M = \max_{n=1,\ldots,N} M_n,$$

where $s \geq t$ is the working precision.

5   Example: The Error Function

We consider the error function and the complementary error function

$$\mathrm{erf}(x) = \frac{2}{\sqrt{x}} \int_0^x e^{-t^2}\, dt,$$

$$\mathrm{erfc}(x) = \frac{2}{\sqrt{x}} \int_x^\infty e^{-t^2}\, dt$$

for $x \in \mathbb{R}$. These functions are closely related to one another through

$$\mathrm{erfc}(x) = 1 - \mathrm{erf}(x). \tag{8}$$

Furthermore, we can limit the discussion to $x > 0$ since

$$\mathrm{erf}(0) = 0,$$
$$\mathrm{erf}(-x) = -\mathrm{erf}(x),$$
$$\mathrm{erfc}(-x) = 2 - \mathrm{erfc}(x).$$

5.1 Series Implementation $0 < x \leq 1$

The Maclaurin series of $\mathrm{erf}(x)$ is defined by

$$\frac{\mathrm{erf}(x)}{2/\sqrt{\pi}} = \sum_{n=0}^{\infty} \frac{(-1)^n x^{2n+1}}{(2n+1)n!}. \tag{9}$$

Its coefficients are related by the ratio

$$r_n = -\frac{2n-1}{n(2n+1)},$$

which can be computed using one floating-point division, if we assume that N is such that $N(2N+1)$ remains exactly representable in the base β precision s floating-point system in use. Then $\max_{n=1,\dots,N} \delta(r_n) = 1$. A sufficient lower bound for $\mathrm{erf}(x)$ is given by

$$e(x) = x - \frac{x^3}{3}$$

for which $\mathrm{erf}(x)/e(x) \leq 1.121$. The factor (6) is bounded by 2. To compute the series using (5) we replace x by x^2 and a_0 by x, with $\delta(x^2) = 1$ and $\delta(a_0) = 0$ given that x is a floating-point number.

In *Table 1* we display the evaluation of $\mathrm{erf}(x)$ in a scalable precision floating-point system with $\beta = 2$ and $t = 125$ for a number of x-values. We also list the degree N of the partial sum and the working precision s.

Table 1.

x	$\mathrm{erf}(x)$	N	s
0.125	0.14031620480 ...	15	139
0.250	0.27632639016 ...	19	139
0.375	0.40411690943 ...	21	139
0.500	0.52049987781 ...	25	139
0.625	0.62324088218 ...	27	140
0.750	0.71115563365 ...	29	140
0.875	0.78407506105 ...	31	140
1.000	0.84270079294 ...	35	140

5.2 Continued Fraction Implementation on $1 < x$

Using (8) in combination with

$$\mathrm{erfc}(x) = \frac{e^{-x^2}}{\sqrt{\pi}} \left(\frac{2x/(2x^2+1)}{1} \Big| + \sum_{n=2}^{\infty} \frac{\frac{-(2n-3)(2n-2)}{(2x^2+4n-7)(2x^2+4n-3)}}{1} \Big| \right)$$

the values in *Table 2* are obtained. Again $\beta = 2, t = 125$ and the approximant number N and the working precision s are listed.

Here for $n \geq 2$ all $a_n < 0$ with

$$\delta(a_n) = 7, \qquad M = 0.85.$$

We can safely put that the integers $4N - 3$ and $(2N - 3)(2N - 2)$ can be represented exactly and that $\delta(x^2) = 1$ since $\delta(x) = 0$. For the additional factors $\exp(-x^2)/\sqrt{\pi}$ in combination with $2x/(2x^2 + 1)$ a separate error analysis is made.

Table 2.

x	erfc(x)	N	s
1.750	$1.3328328780\ldots e{-}2$	77	143
2.500	$4.0695201744\ldots e{-}4$	41	142
3.250	$4.3027794636\ldots e{-}6$	27	143
4.000	$1.5417257900\ldots e{-}8$	20	142
4.750	$1.8485047721\ldots e{-}11$	16	142
5.500	$7.3578479179\ldots e{-}15$	14	143
6.250	$9.6722041318\ldots e{-}19$	12	142
7.000	$4.1838256077\ldots e{-}23$	11	144

6   Special Function Support

In Table 3 we indicate which functions and which argument ranges (on the real line) are covered by our implementation. For the definition of the special functions we refer to [10].

Table 3.

special function	series	continued fraction
$\gamma(a,x)$		$a > 0, x \neq 0$[1]
$\Gamma(a,x)$		$a \in \mathbb{R}, x \geq 0$
erf(x)	$|x| \leq 1$	identity via erfc(x)
erfc(x)	identity via erf(x)	$|x| > 1$
dawson(x)	$|x| \leq 1$	$|x| > 1$

---

[1] For $a > 0, a > x$ a faster implementation making use of series is under development.

Table 3. (*continued*)

special function	series	continued fraction
Fresnel $S(x)$	$x \in \mathbb{R}^2$	
Fresnel $C(x)$	$x \in \mathbb{R}^2$	
$E_n(x),\ n > 0$		$n \in \mathbb{N}, x > 0^3$
$_2F_1(a, n; c; x)$		$a \in \mathbb{R}, n \in \mathbb{Z},$ $c \in \mathbb{R} \setminus \mathbb{Z}_0^-, x < 1$
$_1F_1(n; c; x)$		$n \in \mathbb{Z},$ $c \in \mathbb{R} \setminus \mathbb{Z}_0^-, x \in \mathbb{R}$
$I_n(x)$	$n = 0, x \in \mathbb{R}$	$n \in \mathbb{N}, x \in \mathbb{R}$
$J_n(x)$	$n = 0, x \in \mathbb{R}$	$n \in \mathbb{N}, x \in \mathbb{R}$
$I_{n+1/2}(x)$	$n = 0, x \in \mathbb{R}$	$n \in \mathbb{N}, x \in \mathbb{R}$
$J_{n+1/2}(x)$	$n = 0, x \in \mathbb{R}$	$n \in \mathbb{N}, x \in \mathbb{R}$

A similar implementation in the complex plane is the subject of future research.

References

1. Abramowitz, M., Stegun, I.: Handbook of mathematical functions with formulas, graphs and mathematical tables. U.S. Government Printing Office, NBS, Washington (1964)
2. Cipra, B.A.: A new testament for special functions. SIAM News 31(2) (March 1998)
3. Lozier, D.: NIST Digital Library of Mathematical Functions. Annals of Mathematics and Artificial Intelligence 38(1–3) (May 2003)
4. Moler, C.B.: Cleve's corner: The tetragamma function and numerical craftsmanship: MATLAB's special mathematical functions rely on skills from another era. Technical note, The MathWorks, Inc (2002)
5. Floating-Point Working Group: IEEE standard for binary floating-point arithmetic. SIGPLAN 22, 9–25 (1987)
6. Muller, J.M.: Elementary functions: Algorithms and implementation. Birkhäuser, Basel (1997)
7. Lozier, D., Olver, F.: Numerical Evaluation of Special Functions. In: Gautschi, W. (ed.) AMS Proceedings of Symposia in Applied Mathematics, vol. 48, pp. 79–125 (1994); Updated version (December 2000), http://math.nist.gov/nesf/
8. Higham, N.: Accuracy and stability of numerical algorithms. SIAM, Philadelphia (1996)

---

[2] When $|x| > 1$ the implementation is slow. A faster series version is planned in the near future.

[3] When $0 < x \le 1$ the implementation is slow. A faster series version is planned in the near future.

9. Cuyt, A., Verdonk, B., Waadeland, H.: Efficient and reliable multiprecision implementation of elementary and special functions. SIAM J. Sci. Comput. 28, 1437–
   1462 (2006)
10. Cuyt, A., Brevik Petersen, V., Verdonk, B., Waadeland, H., Jones, W.: Handbook
   of Continued Fractions for Special Functions. Springer, Heidelberg (2008)
11. Jones, W., Thron, W.: Numerical stability in evaluating continued fractions. Math.
   Comp. 28, 795–810 (1974)

MetiTarski: An Automatic Prover for the Elementary Functions

Behzad Akbarpour and Lawrence C. Paulson

Computer Laboratory, University of Cambridge, England
{ba265,lcp}@cl.cam.ac.uk

Abstract. Many inequalities involving the functions ln, exp, sin, cos, etc., can be proved automatically by MetiTarski: a resolution theorem prover (Metis) modified to call a decision procedure (QEPCAD) for the theory of real closed fields. The decision procedure simplifies clauses by deleting literals that are inconsistent with other algebraic facts, while deleting as redundant clauses that follow algebraically from other clauses. MetiTarski includes special code to simplify arithmetic expressions.

1 Introduction

Many branches of mathematics, engineering and science require reasoning about the *elementary functions*: logarithms, sines, cosines and so forth. Few techniques are known for automatically proving statements involving such functions. We have been working on an approach that involves replacing functions by polynomial upper or lower bounds, attempting to reduce the problem to the theory of real closed fields (RCF), and then applying a suitable decision procedure.

The theory of *real closed fields* (RCF) concerns equalities and inequalities involving addition, subtraction and multiplication. (We call logical formulas in this theory *algebraic*.) *Real closed* means every positive number has a square root. The decision procedure works by eliminating quantifiers from the supplied formula; for example, $\exists x.\, ax^2 + bx + c = 0$ reduces to $(a \neq 0 \wedge b^2 - 4ac \geq 0) \vee (a = 0 \wedge b \neq 0) \vee (a = b = c = 0)$. Both universal and existential quantifiers can be eliminated, but our current experiments involve refuting purely existential formulas.

Tarski proved the decidability of RCF in the 1930s, but his procedure was impractical [10]. McLaughlin and Harrison [17] recently implemented a more efficient procedure credited to Hörmander [13] and Cohen. We used it in earlier work [1,2], but unfortunately it fails to terminate if applied to a polynomial of degree greater than six. QEPCAD-B [6,12] is an advanced implementation of *cylindrical algebraic decomposition* (CAD), which is the best available decision procedure for the complete theory of RCF [10]. CAD is still doubly exponential in the number of variables, but it is polynomial in other parameters such as size of the input. In our experience, QEPCAD usually returns quickly. Its main drawback is that we have to run it as a separate process, while Harrison's ML code could be integrated with that of Metis.

S. Autexier et al. (Eds.): AISC/Calculemus/MKM 2008, LNAI 5144, pp. 217–231, 2008.

Our approach to proving inequalities involving elementary functions is to replace function occurrences one by one with appropriate upper or lower bounds. Once we have also eliminated occurrences of division, we can call QEPCAD. Daumas et al. [9] present families of upper and lower bounds for square roots, trigonometric functions, logarithms and exponentials. The approach can obviously be generalized to handle a wide variety of well-behaved functions.

Our approach requires a full theorem prover even to prove simple inequalities. The bounds typically have side conditions that must be proved. Case analysis is necessary when eliminating division and often when substituting bounds. We chose to modify a resolution theorem prover, rather than implementing a theorem prover from scratch. Impressive examples of the latter approach include Analytica [8] and Weierstrass [5]. However, we felt that writing an entire prover would require more effort than modifying a resolution prover, while delivering inferior results. We were also inspired by SPASS+T [18], which effectively combines the resolution theorem prover SPASS with various SMT solvers. For the resolution prover, we chose Joe Hurd's Metis [14]. Compared with leading provers, it is slow (being coded in Standard ML rather than C) and it lacks many refinements (such as advanced data structures for indexing). However, it implements the superposition calculus [4] and its code is extremely clear.

Paper outline. We begin by describing (§2) how we modified the resolution prover Metis. We then discuss (§3) the upper and lower bounds we use. We present a table of new results (§4) along with brief conclusions (§5).

2   Modifications to the Resolution Prover

In order to make this paper self contained, we briefly outline the general approach, which was described in our previous paper [1].

A resolution prover accepts its problems in *conjunctive normal form*. Seeking a contradiction, the conjecture to be proved is negated and conjoined with any axioms. The entire problem is then transformed into a conjunction of disjunctions. Each disjunction is called a *clause*. Each member of a disjunction is an atomic formula or its negation, and is called a *literal*. Each resolution inference

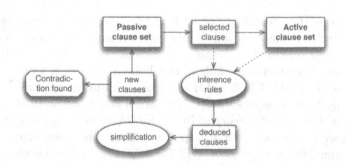

Fig. 1. The main loop of resolution

combines two clauses and yields new clauses. If the empty clause emerges, the proof is finished because the desired contradiction has been found.

A resolution prover's main loop (Fig. 1) manages two sets of clauses, *Active* and *Passive* [16]. At the start, all clauses belong to Passive. At each iteration, a Passive clause is selected and moved to Active. New clauses are inferred by resolving this clause with each of the other Active clauses. The new clauses are simplified and then added to Passive.

MetiTarski is a version of the Metis prover modified in several respects:

- Its implementation of the Knuth-Bendix ordering supports *subterm coefficients* [15]. This encourages the replacement of functions by bounds, even if the bounds superficially appear to be more complex.
- The integer constants are available. Note that all variables are assumed to range over the real numbers.
- A list of all ground algebraic clauses is maintained. They may contain only constants (including Skolem constants) and the functions $+$, $-$ and $\times$.
- Arithmetic expressions are simplified in order to identify redundant forms and to isolate the elementary functions.
- Ground algebraic literals that are inconsistent with existing algebraic facts are deleted from every new clause. This brings us closer to the empty clause.
- New clauses in that follow in RCF from existing algebraic facts are regarded as redundant and deleted. This reduces the use of space and time.

2.1 Arithmetic Simplification

MetiTarski uses a sparse recursive representation of polynomials. Originally, our sole objective was to map all variants of an expression to a unique canonical form [1]. We have added the objective of supporting the proof search by identifying occurrences of division and functions.

Horner normal form. The idea behind our representation is that any polynomial in x can be rewritten in recursive form as

$$p(x) = a_n x^n + \cdots + a_1 x + a_0 = a_0 + x(a_1 + x(a_2 + \cdots x(a_{n-1} + x a_n))).$$

We can express a multivariate polynomial as a polynomial in one variable whose coefficients are polynomials in other variables. For example, we can represent the polynomial $3xy^2 + 2x^2yz + zx + 3yz$ as

$$[y(z3)] + x([z1 + y(y3)] + x[y(z2)]),$$

where the terms in square brackets are considered as coefficients.

The Treatment of Division. Our normal form supports the operations of addition, subtraction, and multiplication. Division by an integer (or a rational number) does not present a problem, since a coefficient can be a rational number: the divisor is recursively supplied to the normal form conversion.

General nested divisions can occur in expressions as a proof develops. Without special treatment, they will be difficult to eliminate and the RCF decision procedure will not be applicable. We transform an expression containing division into a rational function according to the following rules. (We identify E with $\frac{E}{1}$ if necessary.)

$$\frac{x_1}{y_1} + \frac{x_2}{y_2} = \frac{x_1 y_2 + x_2 y_1}{y_1 y_2} \qquad\qquad \frac{x_1}{y_1} \times \frac{x_2}{y_2} = \frac{x_1 x_2}{y_1 y_2}$$

$$\frac{x_1}{y_1} - \frac{x_2}{y_2} = \frac{x_1 y_2 - x_2 y_1}{y_1 y_2} \qquad\qquad \frac{x_1}{y_1} \div \frac{x_2}{y_2} = \frac{x_1 y_2}{y_1 x_2}$$

The effect is to replace nested divisions by one single division, which as the outermost symbol can be eliminated by one proof step using an appropriate division axiom. In this example, three divisions are replaced by one.

$$\left(\frac{x}{y}\right) \frac{1}{\left(x + \frac{1}{x}\right)} = \frac{x^2}{y(x^2 + 1)}$$

We add literals to the resulting clause to account for the possibility of division by zero. In particular, if we simplify $x_1/y_1 + x_2/y_2$ then we make the resulting clause conditional on $y_1 \neq 0$ and $y_2 \neq 0$. However, for $(x_1/y_1) \times (x_2/y_2)$ and $(x_1/y_1) \div (x_2/y_2)$, no such conditions are necessary. That is because we define $x/0 = 0$. It is trivial to see that $(x_1/y_1)(x_2/y_2) = 0$ if and only if any of x_1, x_2, y_1, y_2, are zero, and in this they agree with the corresponding right-hand side. The alternative of introducing an error value and defining $x/0 = \infty$ would introduce great complications.[1]

Isolating Function Occurrences. We attempt to restore inequalities to a natural form. For example, we simplify $X \leq Y$ by normalizing $X - Y$, yielding an equivalent form $X' \leq Y'$ where X' and Y' both have positive coefficients. We have strengthened this process to isolate occurrences of elementary functions.

In the Horner normal form transformation, we regard any non-algebraic term (typically a function occurrence) as a variable. We order the variables, taken in this general sense, using Metis's built-in Knuth-Bendix ordering. This ensures that one of the function occurrences will appear as the outermost "variable." If we detect this situation, we leave this term by itself on one side of the inequality, for example as $\ln(t) \leq u$. We even divide both sides by any constant coefficient, but at present we have no way of isolating the function in cases such as $x \ln(t) \leq u$. This challenge is a focus of our current work.

2.2   Algebraic Literal Deletion

Literal deletion [1] simplifies new clauses that emerge from inference rules. For each ground algebraic literal in such a clause, we conjoin it with the negations of all ground algebraic literals in that clause (its context) and with all ground algebraic clauses known to the prover. We then form the existential closure of this formula, taking as variables all Skolem constants present in that formula. If the

---

[1] Harrison [11, §2.5] discusses various approaches to formalizing "undefined", taking $1/0 = 0$ as an example.

RCF solver (QEPCAD) reduces this formula to **false**, then that literal is deleted. This is the primary mechanism by which the decision procedure contributes to deduction.

As a small example, suppose we are trying to prove

$$\forall x. -3 < x < 1 \rightarrow \ln(1 - x) \leq -x$$

with the help of a range-restricted polynomial upper bound f_2,

$$\forall x. 2 \leq x \leq 4 \rightarrow \ln(x) \leq f_2(x).$$

Skolemization of the conjecture will yield three clauses, with u a Skolem constant:

$$-3 < u \qquad u < 1 \qquad \neg[\ln(1 - u) \leq -u].$$

At the start of the proof, $-3 < u$ and $u < 1$ will be the only elements of our list of ground algebraic clauses. As the proof proceeds, using axioms to be described later, a resolution step will eventually substitute our upper bound, yielding the following unsimplified clause:

$$f_2(1 - u) \leq -u \lor 2 > 1 - u \lor 1 - u > 4.$$

Ordinary arithmetic simplification can reduce $2 > 1 - u$ to $u > -1$, and $1 - u > 4$ to $-3 > u$, but if $f_2(1 - u)$ is a complicated polynomial, then only QEPCAD can achieve a real simplification: we give it the formula

$$\exists u.\, f_2(1 - u) \leq -u \land \underbrace{u \leq -1 \land -3 \leq u}_{\text{negated literals}} \land \underbrace{-3 < u \land u < 1}_{\text{algebraic clauses}}.$$

Provided f_2 is a sufficiently tight bound, the result will be **false** and the literal can be deleted. The literal $u < -1$ turns out to be consistent with its context. Then we call QEPCAD for $-3 > u$:

$$\exists u. -3 > u \land u \leq -1 \land -3 < u \land u < 1.$$

This again is **false**, and the final simplified clause is

$$u > -1.$$

It is a ground algebraic clause and will be added to our list. We have tightened the range of u to $-1 < u < 1$; if it becomes empty, then we have reached a contradiction.

In this example, the constraints that accumulate are essentially linear. More generally, the use of rational function upper and lower bounds causes an accumulation of algebraic constraints, which eventually turn out to be inconsistent.

2.3   Algebraic Subsumption

Resolution theorem provers generate many redundant clauses. To conserve space, they typically delete any clause that is a syntactic instance of another. We generalize this redundancy criterion, known as *subsumption*, by performing an analogous redundancy check in the RCF theory.

When a new clause is generated, we identify its ground algebraic literals and form their disjunction. If this disjunction is an algebraic consequence of existing algebraic facts, then we ignore the clause. Thus the formulas given to QEP-CAD do not contain redundant conjuncts. This technique can even improve the performance of some failing proofs so that they fail finitely rather than running forever. The resulting performance improvement depends on other aspects of the formalization; at present only four percent of our problems are proved significantly faster when this technique is enabled.

Recall our previous example, where the ground algebraic clauses included $-1 < u$ and $u < 1$. Suppose that a resolution step yields the following clause:

$$\ln(1 - u) \leq u^2 \vee u^2 < 2 \vee 4u > 3.$$

Algebraic subsumption will call QEPCAD with the formula

$$\exists u. \underbrace{u^2 \geq 2 \wedge 4u \leq 3}_{\text{negated literals}} \wedge \underbrace{-1 < u \wedge u < 1}_{\text{algebraic clauses}}.$$

QEPCAD returns **false**, indicating that the algebraic part of the clause follows from $-1 < u < 1$. The clause is discarded.

2.4   Modified Knuth-Bendix Ordering

The execution of a modern resolution prover is governed by an ordering [4]. This ordering serves a twofold purpose: first, to eliminate redundant combinations of inferences that would lead to identical results; second, to draw the prover's attention to literals of high priority. For us, high priority literals are those involving the functions we wish to eliminate.

Ordered resolution frequently employs a heuristic entitled *negative selection*: a literal's sign is taken into account, in addition to its rank in the ordering. Specifically, only maximal negative literals can be selected for resolution. Metis employs negative selection by default but also offers (via a simple change to its source code) unsigned literal selection. With this option, 67 percent of our problems are proved; with negative selection, only eight percent are proved; with no ordering whatever, 54 percent are proved.[2] The terrible result with negative selection, where 79 percent of the problems are actually reported as satisfiable, indicates a mismatch with our heuristics, since with pure resolution this heuristic is complete. Nevertheless, unsigned literal selection is appropriate because we wish to eliminate occurrences of elementary functions regardless of their sign.

---

[2] Tests were run on a 2.66 GHz Mac Pro allowing 10 seconds per problem.

A more significant change concerns the ordering itself. Metis follows most resolution theorem provers in providing the Knuth-Bendix ordering (KBO) [3, p. 124], whose advantages include computational efficiency and a tendency to prefer simpler terms. The latter property, however, can be a drawback. We are concerned with clauses such as the following:

$$\neg 0 < X \vee \neg[((2X^3 - 9X^2 + 18X - 11)/6) \leq Y] \vee \ln(X) \leq Y.$$

This combines the upper bound property $\ln(x) \leq (2x^3 - 9x^2 + 18x - 11)/6$ with transitivity, allowing $\ln(X)$ to be replaced by its bound. We would like resolution to select the literal $\ln(X) \leq Y$ in order to eliminate an occurrence of $\ln(t)$ from another clause. Unfortunately, standard KBO will want to select $\neg[((2X^3 - 9X^2 + 18X - 11)/6) \leq Y]$ because it is syntactically larger than $\ln(X) \leq Y$. We can attempt to force the issue by assigning ln a high *weight*. Weights (typically positive integers) can be assigned to all function symbols; the sum of the weights in a term is a key measure compared in the ordering. By choosing a sufficiently high weight, we can ensure that $\ln(X) \leq Y$ is selected. Unfortunately, the second literal will continue to be selected as well, because it contains several occurrences of the variable X. Both literals are maximal under KBO, for which the number of variable occurrences in terms is significant. The spurious selection needlessly expands the search space.

Ludwig and Waldmann [15] provide a solution to this difficulty. They give precise definitions of useful extensions to KBO, along with theory and implementation advice. We have modified Metis's built-in ordering so that a function can have not only a weight, but also a *subterm coefficient*. For example, if ln has a subterm coefficient of 10, then each occurrence of a variable in $\ln(t)$ is equivalent to 10 occurrences of that variable in t. By choosing subterm coefficients appropriately, we can ensure that a literal containing an elementary function is selected every time. This modification to Metis yields dramatic reductions in solution times for the great majority of problems.

3   Bounds for the Elementary Functions

We have devoted much effort to the choice of appropriate bounds. We first relied on Daumas et al. [9], who provide bounds for several elementary functions. Those bounds, however, were intended for a different application: to decide constant formulas using interval arithmetic. For each function, they supplied a family of increasingly precise bounds. Each bound included range reduction in order to ensure accuracy for all possible function arguments. In effect, each bound was an infinite family indexed in two dimensions. Resolution provers require a finite and preferably small axiom system.

We have simplified the bounds, largely eliminating the range reduction. We have used only a few members of the infinite family, preferring polynomials of modest degree (typically below 6, though in one case up to 15). These simplifications are adequate for our experiments, allowing us to focus on crucial issues such as the search space and the treatment of complex expressions. The original bounds were only claimed to hold over narrow intervals; these could often be relaxed. In other cases, we sought new bounds that were valid over wide intervals. Relaxing the range restrictions allows inequalities to be proved over infinite intervals. Our theorem prover can perform case analyses in order to join proofs involving bounds valid over different intervals, but of course it can only consider finitely many cases.

Remark: we use $\overline{\exp}(x)$ and $\underline{\exp}(x)$ to stand for various upper or lower bounds of $\exp(x)$, etc.; Daumas et al. [9] supply specific and fixed definitions of such functions.

3.1   The Logarithm Function

Daumas et al. [9] derive bounds for $\ln(x)$ from Taylor approximations,

$$\sum_{i=1}^{n} (-1)^{i+1} \frac{(x-1)^i}{i},$$

for the range $1 < x \leq 2$. With this series, even values of n yield lower bounds while odd values of n yield upper bounds. For $x > 2$, they perform range reduction by writing $x = 2^m y$ where $1 < y \leq 2$. When $0 < x < 1$, they use the identity $\ln(x) = -\ln(1/x)$.

Our upper bounds come from this Taylor series, using odd values of n up to 7. They are valid not merely for $1 < x \leq 2$, but for all $x > 0$.

Proposition 1. *If n is odd and $x > 0$ then*

$$\ln(x) \leq \sum_{i=1}^{n} (-1)^{i+1} \frac{(x-1)^i}{i},$$

with equality only if $x = 1$.

Remark: We would be grateful for a reference to a published proof of this fact, which has a straightforward proof using Rolle's theorem.

As for range reduction, note that if $\ln(x) \leq \overline{\ln}(x)$ for all $x > 0$ then

$$\ln(2^m y) = m \ln(2) + \ln(y) \leq m \overline{\ln}(2) + \overline{\ln}(y) = \overline{\ln}(2^m y)$$

for all $y > 0$. Therefore, the range reduction technique suggested by Daumas et al. [9] yields upper bounds that hold for all positive arguments, and not merely for the intervals they claim. At present, we use three of these, all with $m = 1$ and thus intended originally for the interval $(2, 4]$; the simplest of them is $\ln(x) < x/2$.

Our lower bounds are those that Daumas et al. [9] use for the interval $[1/2, 1)$. That is, they are given by the series

$$\sum_{i=1}^{2n+1} \frac{1}{i} \left(\frac{x-1}{x} \right)^i$$

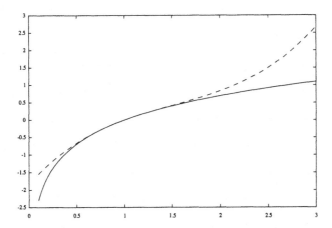

Fig. 2. An Upper Bound for the Logarithm Function

for $n = 0$, 1, 2. For $x > 1$, they are only slightly worse than the corresponding (and higher degree) lower bounds for $\ln(x)$ given by Daumas et al. We therefore use them for all $x > 0$.

The current lack of range reduction means that our upper bounds are poor when $x > 2$ and similarly our lower bounds are poor when $x < 1/2$. This point is evident in Fig. 2, which plots $\ln(x)$ against the upper bound $(2x^3 - 9x^2 + 18x - 11)/6$ as x ranges from 0.1 to 3. Bounds of higher degree are actually worse as x increases. We are considering alternative logarithm bounds that may exhibit less extreme behaviour.

3.2   The Exponential Function

Daumas et al. [9] derive bounds for $\exp(x)$ from its Taylor expansion, but only for $-1 \leq x < 0$. They use a complicated system of transformations, first covering the negative numbers in separate intervals of the form $[k-1, k)$ for integer $k < 0$. For $x > 0$, they use the identity

$$\exp(-x) = \frac{1}{\exp(x)}.$$

The rapid growth of the exponential function necessitates a degree of case analysis and range reduction, but we have managed to find simpler bounds valid over wide ranges.

We use a crucial fact about the Taylor expansion [7, p. 83].

Proposition 2. *If n is odd and $x \neq 0$ then*

$$\exp(x) > \sum_{i=0}^{n} \frac{x^i}{i!}.$$

If n is even then this inequality holds if $x > 0$, while the opposite inequality holds if $x < 0$. Obviously we have equality when $x = 0$.

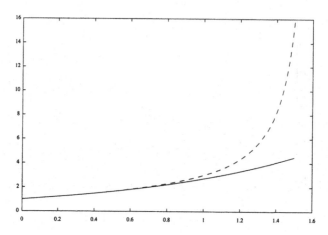

Fig. 3. An Upper Bound for the Exponential Function

Our upper bounds for $x \leq 0$ rely on this opposite inequality. The bound using $n = 4$ is already poor when $x < -2$, so they could benefit from range reduction, but they are valid for all $x \leq 0$. Our upper bounds for $x \geq 0$ rely on Prop. 2 with odd n. We define $\overline{\exp}(-x) = 1/\underline{\exp}(x)$: dividing by a positive lower bound yields an upper bound. Obviously the exponential function is not bounded by any rational function for $x > 0$, and one might imagine that the exponential function overtakes its bound after a certain point. In fact, our upper bounds are never overtaken, but reach a discontinuity as the denominator goes to zero. With $n = 3$, the upper bound is $6/(6 - 6x + 3x^2 - x^3)$; its denominator is a cubic equation with one real root at $x \approx 1.60$. The divergence of the upper bound function can be seen in Fig. 3, which plots $\exp(x)$ against the upper bound (dashed line) as x ranges from 0 to 1.5. We could get a much tighter fit by choosing $n = 5$, which diverges at $x \approx 2.18$.

Due to the limited range of these bounds, we also include a few versions with range reduction, via the identity $\exp(x) = \exp(x/k)^k$, for $k = 2, 3, 4$. With $k = 4$ and $n = 3$, the denominator of the upper bound becomes a 12th degree polynomial.

The treatment of lower bounds is simple, thanks to Prop. 2. The Taylor expansion of $\exp(x)$ for odd n is a valid lower bound over the entire real line. Figure 4 plots $\exp(x)$ against the lower bounds $1 + x$ (dashed line) and $1 + x + x^2/2 + x^3/6$ (dotted line) as x ranges from -4 to 3. It is clear that the lower bounds are poor when $x < -3$. We can again use $\exp(x) = \exp(x/k)^k$ for range reduction, but only for odd k. We want to deduce

$$\underline{\exp}(x/k)^k \leq \exp(x/k)^k = \exp(x)$$

from $\underline{\exp}(x/k) \leq \exp(x/k)$, for which we need k to be odd because $\underline{\exp}(x/k)$ could be negative.

No single lower bound is uniformly best, so we use the Taylor expansion with $n = 1, 3, 5$ and 7. For $n = 3$ and $n = 5$, we also perform range reduction as

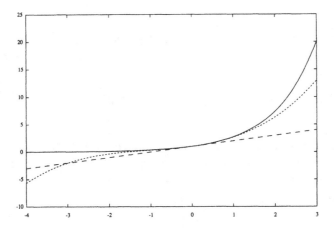

Fig. 4. Lower Bounds for the Exponential Function

described above with $k = 3$; the best of these bounds has degree 15 and is an excellent fit for $-8 \le x \le 8$.

3.3 Other Functions

For the functions $\sin(x)$, $\cos(x)$ and $\tan^{-1}(x)$, we use a selection of the bounds recommended by Daumas et al. [9]. They also suggest families of bounds that converge to π, but we cannot use families and have simply chosen two fractions based on the decimal expansion of π. We have improved the accuracy of their bounds for $\sqrt{x}$ while retaining their use of Newton's method.

We specify the absolute value function by a pair of clauses:

$$\neg(0 \le x) \vee |x| = x \qquad 0 \le x \vee |x| = -x$$

Strengthening the second clause to $0 < x \vee |x| = -x$ harms performance. The theorem prover uses these axioms to remove occurrences of the absolute value function while introducing new literals of the form $\neg(0 \le t)$ or $0 \le t$. Presumably, it helps to make these literals complementary.

3.4 Axioms

The guiding principle behind our axiom system is to avoid all use of general properties of orderings, such as transitivity, antisymmetry and monotonicity of addition and multiplication. Necessary instances of these properties are built into other axioms, built into simplification or left to the decision procedure. To limit problem size and search space, we include only axioms that are relevant to the functions that appear in the problem. It is often obvious by inspection whether upper or lower bounds are required. At present the user has to convert the problem to clause form and to include the required sets of axioms, although these steps would be trivial to automate.

A significant change from our previous paper [1] is that the less-than relation no longer exists. We have only one ordering relation, $\leq$. The equivalence $X < Y \iff \neg(Y \leq X)$, formerly a pair of clauses, is now built into the parser.

To illustrate our formalization of bounds, consider the fact $1 + x \leq \exp(x)$. We could combine it with transitivity for $\leq$ and $<$ by asserting two axioms:

$$\neg(Y \leq 1 + X) \vee Y \leq \exp(X)$$
$$\neg(Y < 1 + X) \vee Y < \exp(X)$$

However, writing each bound twice would be inconvenient. Instead we introduce a generalized less-than relation. Its first argument indicates which relation it designates. We express the following two equivalences using the obvious four axiom clauses:

$$\mathrm{lgen}(0, X, Y) \iff X \leq Y$$
$$\mathrm{lgen}(1, X, Y) \iff X < Y$$

Now, the lower bound axiom for $\leq$ and $<$ can be expressed by a single clause:

$$\neg(\mathrm{lgen}(R, Y, 1 + X)) \vee \mathrm{lgen}(R, Y \leq \exp(X)).$$

The theorem prover will then generate the two clauses shown above.

We use weights and subterm coefficients to ensure that the exp literals are selected. The lower bound clauses combine with literals of the form $\neg(t \leq \exp(u))$ or $\neg(t < \exp(u))$, respectively. These can substitute the lower bound in both conjectures and facts.

- They reduce a conjecture of the form $t \leq \exp(u)$ to $t \leq 1 + u$, and similarly for $<$.
- They can resolve with facts of the form $\exp(u) < t$ yielding the new fact $1 + u < t$, and similarly for $\leq$.

As before [1], we include axioms concerning division, which we give a high weight to encourage its replacement by multiplication.

$$\neg(X \leq Y \times Z) \vee X/Z \leq Y \vee Z \leq 0$$
$$\neg(X \leq Y/Z) \vee X \times Z \leq Y \vee Z \leq 0$$
$$\neg(X \times Z \leq Y) \vee X \leq Y/Z \vee Z \leq 0$$
$$\neg(X/Z \leq Y) \vee X \leq Y \times Z \vee Z \leq 0$$

Now that simplification pulls quotients to the outside of a term (Sect. 2.1), we can dispense with the other division axioms mentioned in that paper. These axioms concern the case $Z > 0$; we also need axioms for $Z < 0$.

4   Results

Our previous paper presented a table of results for about 30 simple problems. As of this writing, we have 179 problems, which 68 percent are proved in under

Table 1. Problems and Runtimes

problem	seconds
$-1 < x \implies 2|x|/(2+x) \leq 1 + |\ln(1+x)|$	0.373
$|x| < 1 \implies |\ln(1+x)| \leq -\ln(1-|x|)$	0.112
$|x| < 1 \implies |x|/(1+|x|) \leq |\ln(1+x)|$	1.478
$|x| < 1 \implies |\ln(1+x)| \leq |x|(1+|x|)/|1+x|$	2.052
$|x| < 1 \implies |x|/4 < |\exp(x) - 1|$	2.068
$0 < |x| < 1 \implies |\exp(x) - 1| < 7|x|/4$	1.071
$|\exp(x) - 1| \leq \exp(|x|) - 1$	1.760
$|\exp(x) - (1+x)| \leq |\exp(|x|) - (1+|x|)|$	3.617
$|\exp(x) - (1+x/2)^2| \leq |\exp(|x|) - (1+|x|/2)^2|$	27.802
$0 \leq x \implies 2x/(2+x) \leq \ln(1+x)$	0.147
$-1/3 \leq x \leq 0 \implies x/\sqrt{1+x} \leq \ln(1+x)$	0.241
$1/3 \leq x \implies \ln((1+x)/x) \leq 2/3 + (12x^2 + 12x + 1)/(12x^3 + 18x^2 + 6x)$	0.216
$1/3 \leq x \implies \ln((1+x)/x) \leq 1/3 + 1/\sqrt{x^2 + x}$	0.528
$0 \leq x \leq 1 \implies \exp(x - x^2) \leq 1 + x$	0.068
$x \leq 1/2 \implies \exp(-x/(1-x)) \leq 1 - x$	0.094
$|x| < 1 \implies |\sin(x)| \leq 6/5|x|$	0.342
$0 < x < 100/201 \implies \cos(\pi x) > 1 - 2x$	0.079
$\cos(x) - 1 + x^2/2 \geq 0$	0.004
$0 < x \leq \pi \implies \cos(x) \leq \sin(x)/x$	0.130
$0 < y < x \implies (1/2)\ln(x/y) > (x-y)/(x+y)$	1.176

60 seconds. For this paper, we present (Table 1) a small sample of the more interesting and difficult problems. The runtimes were measured on a 2.66 GHz Mac Pro running Poly/ML.

Limitations of our approach can be seen in the facts that cannot be proved. We cannot prove $2 > 2/(e(\ln 2)^2)$ or other problems in which functions are multiplied together. We can prove $\cos(\pi x) > 1 - 2x$ under the assumption $0 < x < 100/201$ but unfortunately not under the assumption $0 < x < 1/2$: our approximation to π is fixed and some precision is inevitably lost. Some problems involving square roots can be proved, but each square root $\sqrt{E}$ in the problems presented here has been manually replaced by a new variable y such that $y \geq 0$ and $y^2 = E$. This transformation encodes square roots as algebraic constraints and can easily be automated. It is only useful if E is algebraic.

5   Conclusions

MetiTarski, which combines a resolution theorem prover with specialized simplification and a decision procedure, can prove numerous facts about the elementary functions automatically. By further refining our techniques, particularly for products of functions, we expect to prove increasingly difficult theorems. The approach is flexible, and should work with any well-behaved functions.

Our objective is to perform proofs that could in principle be reconstructed or checked using other tools. Such reconstruction would be difficult however because QEPCAD performs lengthy computations using sophisticated algorithms. Our use of $x/0 = 0$ is well-known in interactive theorem proving. Although it is unconventional, it is logically consistent with traditional mathematics, which ascribes $x/0$ no value at all.

Resolution is traditionally regarded first as a formal calculus and only second as an implementation. A resolution calculus is first developed, then proved to be complete, before an implementation is contemplated. Our choice of Metis could also be questioned. On this point, however, we have the support of SPASS+T implementer Uwe Waldmann: "switching to SPASS would make sense for you only if you plan to deal with huge problems" and we would spend a lot of time "getting memory (de)allocation right."[3] ML's garbage collector certainly simplifies our task. Our results demonstrate that modifying an implementation can, at least, deliver proofs and insights. Our modifications are sympathetic to the overall architecture of resolution: we modify its notions of simplification and subsumption and its ordering. We ignore completeness because proving something is better than proving nothing. Nonetheless, we welcome suggestions for achieving completeness under particular circumstances.

Acknowledgements. The research was funded by the EPSRC grant EP/C013409/1, *Beyond Linear Arithmetic: Automatic Proof Procedures for the Reals.* Joe Hurd offered much help with his Metis prover. Christopher W. Brown and Ian Grant provided support for QEPCAD. John Harrison, David Lester, Cesar Muñoz and Uwe Waldmann answered various queries.

References

1. Akbarpour, B., Paulson, L.: Extending a resolution prover for inequalities on elementary functions. In: Logic for Programming, Artificial Intelligence, and Reasoning, pp. 47–61 (2007)
2. Akbarpour, B., Paulson, L.C.: Towards automatic proofs of inequalities involving elementary functions. In: Cook, B., Sebastiani, R. (eds.) PDPAR: Pragmatics of Decision Procedures in Automated Reasoning, pp. 27–37 (2006)
3. Baader, F., Nipkow, T.: Term Rewriting and All That. Cambridge University Press, Cambridge (1998)
4. Bachmair, L., Ganzinger, H.: Resolution theorem proving. In: Robinson, A., Voronkov, A. (eds.) Handbook of Automated Reasoning, ch. 2, vol. I, pp. 19–99. Elsevier Science, Amsterdam (2001)
5. Beeson, M.: Automatic generation of a proof of the irrationality of e. JSC 32(4), 333–349 (2001)
6. Brown, C.W.: QEPCAD B: a program for computing with semi-algebraic sets using CADs. SIGSAM Bulletin 37(4), 97–108 (2003)
7. Bullen, P.S.: A Dictionary of Inequalities, Longman (1998)
8. Clarke, E., Zhao, X.: Analytica: A theorem prover for Mathematica. Mathematica Journal 3(1), 56–71 (1993)

---

[3] E-mail dated 20 November 2007.

9. Daumas, M., Muñoz, C., Lester, D.: Verified real number calculations: A library for integer arithmetic, http://hal.archives-ouvertes.fr/hal-00168402/fr/
10. Dolzmann, A., Sturm, T., Weispfenning, V.: Real quantifier elimination in practice. Technical Report MIP-9720, Universität Passau, D-94030, Germany (1997)
11. Harrison, J.: Formalized mathematics. Technical Report 36, Turku Centre for Computer Science (TUCS), Lemminkäisenkatu 14 A, FIN-20520 Turku, Finland (1996), http://www.cl.cam.ac.uk/jrh13/papers/form-math3.html
12. Hong, H.: QEPCAD — quantifier elimination by partial cylindrical algebraic decomposition. Sources and documentation are on the Internet, http://www.cs.usna.edu/qepcad/B/QEPCAD.html
13. Hörmander, L.: The Analysis of Linear Partial Differential Operators II: Differential Operators with Constant Coefficient. Springer, Heidelberg (2006) (First published in 1983); cited by Mclaughlin and Harrison [17]
14. Hurd, J.: Metis first order prover (2007), http://gilith.com/software/metis/
15. Ludwig, M., Waldmann, U.: An extension of the Knuth-Bendix ordering with LPO-like properties. In: Dershowitz, N., Voronkov, A. (eds.) LPAR 2007. LNCS (LNAI), vol. 4790, pp. 348–362. Springer, Heidelberg (2007)
16. McCune, W., Wos, L.: Otter: The CADE-13 competition incarnations. Journal of Automated Reasoning 18(2), 211–220 (1997)
17. McLaughlin, S., Harrison, J.: A proof-producing decision procedure for real arithmetic. In: Nieuwenhuis, R. (ed.) CADE 2005. LNCS (LNAI), vol. 3632, pp. 295–314. Springer, Heidelberg (2005)
18. Prevosto, V., Waldmann, U.: SPASS+T. In: Sutcliffe, G., Schmidt, R., Schulz, S. (eds.) FLoC 2006 Workshop on Empirically Successful Computerized Reasoning. CEUR Workshop Proceedings, vol. 192, pp. 18–33 (2006)

High-Level Theories*

Jacques Carette and William M. Farmer

Department of Computing and Software
McMaster University
Hamilton, Ontario, Canada
{carette,wmfarmer}@mcmaster.ca

Abstract. We introduce high-level theories in analogy with high-level
programming languages. The basic point is that even though one can
define many theories via simple, low-level axiomatizations, that is neither
an effective nor a comfortable way to work with such theories. We present
an approach which is closer to what users of mathematics employ, while
still being based on formal structures.

1 Introduction

The mission of *mechanized mathematics* is to develop software systems that support the process people use to create, explore, connect, and apply mathematics. There are historically two main kinds of *mechanized mathematics systems (MMSs): theorem proving systems (TPSs)* and *computer algebra system (CASs).*
Both kinds of systems encapsulate a body of mathematical knowledge and a collection of tools for using this knowledge. The tools of TPSs tend to be primarily deductive, while those of CASs tend to be primarily computational.

MMSs have two kinds of users. *End users* use the tools of an MMS to help them do mathematics, whatever that involves. *Developers* use the tools of an MMS to produce new mathematical knowledge and new tools to facilitate the work of end users. Developers need to have a deep understanding of the logical and mathematical foundation of their MMS of choice. They are interested in the structure of mathematics, the problems involved in formalizing mathematics, and the MMS. End users need much less depth in their understanding of the MMS. They are computer scientists, engineers, and other scientists who are primarily interested in an MMS as a (mathematical) tool to solve a problem.

Mathematics is usually done in informal high-level reasoning environments that include a rich set of concepts and practical tools. The tools involve a mixture of computation and deduction and are highly integrated with each other. While the setting appears informal, enough rigor is applied that, in theory, the results *could* be made formal. MMS end users want to work in similar high-level environments, and MMS developers want to build such environments. But contemporary MMSs do not provide such environments, nor do they provide the tools to build them. We would like to provide something akin to the ease-of-use

* This research was supported by NSERC.

S. Autexier et al. (Eds.): AISC/Calculemus/MKM 2008, LNAI 5144, pp. 232–245, 2008.

of *Theorema* [4] with the computational correctness and efficiency provided by *Focal* [27] and the soundness of Coq [9].

Contemporary TPSs provide low-level reasoning environments based on axiomatic theories. An *axiomatic theory* consists of a set of formulas (called *axioms*) in a formal language. An axiomatic theory encodes a body of mathematical knowledge declaratively; the "truths" of the body of knowledge are the logical consequences of the axioms. In a TPS, new concepts are expressed by writing definitions, and the consequences of the concepts are explored by stating and proving conjectures. Making definitions and proving theorems is the primary emphasis; performing computations and introducing derived reasoning rules is secondary. As a result, doing mathematics in a TPS requires working at a very low conceptual level—much like programming in an assembly language. TPSs give developers the access they need to low-level details, but generally not the capability to build the kind of high-level reasoning tools that end users want. As a result, TPSs are almost useless for end users who are interested in "getting work done" and lack the necessary understanding of mathematics at the level of axiomatic theories.

Contemporary CASs provide high-level computational environments based on algorithmic theories. An *algorithmic theory* consists of a set of algorithms that perform symbolic computations over a formal language. An algorithmic theory encodes a body of mathematical knowledge procedurally; the "truths" of the body of knowledge are the results that can be obtained by running the algorithms on the range of inputs. In a CAS, new algorithms can be expressed in the theory by writing programs, usually in a special, system-supplied programming language. Reasoning is narrowly focused on computation; there is usually little support for deductive reasoning. Moreover, not only is the background theory of the algorithms largely hidden from the user, these background theories are (unfortunately) inconsistent from algorithm to algorithm, which makes the results obtained from such computations frequently untrustworthy. As a result, doing mathematics in a CAS is like programming in a high-level programming language with an inaccessible, untrustworthy compiler. CASs offer end users a high-level reasoning environment, but one that supports only computation and provides untrustworthy results. By not giving developers access to their logical foundations, CASs have very limited use for developers of mechanized mathematics.

We argue in this paper that, in order for mechanized mathematics to achieve its potential, MMSs must provide end users with high-level environments for reasoning and computation, similar to the informal environments they are used to, in which they can work in a sound and convenient fashion. MMSs must also provide developers with the capabilities to build high-level environments for end users that are derived from a solid logical foundation. Toward this goal, we introduce the notion of a *high-level theory*, a semi-formal high-level environment for reasoning and computation that is analogous to a high-level programming language.

The paper is organized as follows. Section 2 discusses what a high-level reasoning environment is. Section 3 introduces the notion of a high-level theory and explores the analogy between high-level theories and high-level programming languages. How mathematics is performed within a high-level theory is the subject of section 4. How high-level theories are created and connected is covered in section 5. Two examples of high-level theories are briefly discussed in section 6. We comment in section 7 on some related issues and conclude in section 8 with some remarks and a recommendation that MMS builders design and implement systems that offer high-level theories instead of just axiomatic or algorithmic theories.

2   High-Level Reasoning Environments

As we have mentioned above, mathematics practitioners work in high-level reasoning environments that offer integrated sets of concepts and deductive, computational, visual, and other kinds of tools. Working in them is more convenient and practical than working in an axiomatic theory or even in a network of axiomatic theories.

For example, consider the informal reasoning environment of *natural number arithmetic* (which is also called *number theory*). Even though an axiomatization of natural number arithmetic is relatively simple, the informal environment that people actually work in is quite sophisticated. For instance, it includes a set of algebraic operators, a linear order, several lattice structures, a collection of induction principles, a collection of algorithms for adding, multiplying, dividing (with remainder), etc., and various connections to set theory, analysis, and abstract algebra. An axiomatization of natural number arithmetic—even one augmented with many definitions and theorems—is an enfeebled reasoning environment in comparison to this standard informal high-level environment. Figure 1 gives a taste of what we mean. It is important to notice that the "signature" of Nat does not export the implementation details of the representation of Nat. We are thus free to provide "implementations" via Peano's axioms or via some other (more efficient) means.

Another good example of an informal high-level environment is what is often called *group theory*. It includes the basic definitions of the algebraic structure called a group, machinery connecting groups via homomorphisms, tools such as the Sylow theorems and the orbit-stabilizer theorem for analyzing the structure of groups, and standard applications of groups to various symmetric structures and problems in mathematics. Group theory cannot be naturally derived from an axiomatization of a group. It is based on a set of axiomatic theories that includes a theory of a single group, several copies of this theory for homomorphisms, a theory of a group action, a theory of natural number arithmetic, and a theory of sets and functions. In common use, group theory is more about its connections to other theories than about groups themselves.[1]

---

[1] There are exceptions, naturally.

```
Theory Nat:
  concepts 0,1 : Nat.
  transformers +, * : Nat -> Nat -> Nat. total, commutative, associative.
  transformers <, = : Nat -> Nat -> Bool.
  theorem: Nat is an ordered semi-ring.
  language AE(Nat) = {0,1,+,*}           // Arithmetic Expressions over Nat
  derive transformer: eval : AE -> Nat           // Evaluation, derived
  theorem: total(eval).
  derive transformer induction : Prop(AE) -> InductiveProof
  ...                               // Structural induction principle
```

Fig. 1. Nat as a sample high-level theory

---

1. Convenient, human-oriented, sound, and precise.
2. Supports deduction, computation, and mixtures of the two.
3. Allows the end user to work at a high conceptual level.
4. Includes a well-constructed, highly integrated set of tools.
5. Constructed modularly.
6. Efficiently implemented with respect to resources.
7. Enables multiple modes of interaction (e.g., graphical).

Fig. 2. The pragmatic properties of a high-level theory

Reasoning in one of these high-level environments is analogous to programming in a high-level programming language like Java or ML. The reasoning *can* be reduced to the level of axiomatic theories, but this is rarely necessary or even desirable.

3   High-Level Theories

So what exactly are high-level theories? Informally, they should be to mechanized mathematics what high-level programming languages are to programming. This analogy is quite rich, and deserves to be expanded upon. But first, we will explicitly list in Figure 2 the pragmatic properties that we want a high-level theory to have. While these might all sound quite desirable, each is a nontrivial constraint. Furthermore, if they are not *designed into* a system, they are rather unlikely to be emergent properties of an implementation.

3.1   The High-Level Programming Analogy

Taking a wide, top-down view of high-level programming languages, we first encounter programming *paradigms*, namely procedural, functional, object-oriented, and logical/relational. Mathematicians also have styles. Some like to prove purely existential theorems, others are engaged in giant computations, whilst others like to find relations between various theories; there are entire books and active conferences dedicated to studying these topics. While our goal is to support as many of these activities as possible, we will focus on deduction and computation.

If we look at most modern high-level programming languages, we get a large set of features, even though we know that we could achieve much the same by directly programming in an assembly language (or a Turing machine or the λ-calculus). The same reasons that drove programming languages to include such high-level features should guide our search. Furthermore we should not look merely at language advances or even programming systems (languages combined with a standard library). It is important to realize that some languages thrive because they inhabit a complete ecosystem, with rich IDEs offering non-ASCII based methods of interaction, project management features, etc. It is sobering to remember that N.G. de Bruijn had foreseen some of this 40 years ago [12].

At the simplest level, we want to combine oft-used chains of primitive deductions into new transformers which are meant to be used as units. For example, loops are so pervasive in programming that all languages offer high-level constructs for this, with a range of semantics. This varies from the one-size-fits-all `while` loop of early imperative languages to the semantically richer `foreach` loop, and the even richer `fmap` and `foldl` (from Haskell). In an MMS, once we have theorems that prove the correctness of algorithms for addition and multiplication over `Nat`, we should simply add these algorithms as new "fundamental" tools.

Between the two extremes of programming paradigm and low-level primitives, programming languages offer further tools, for "programming in the large", like classes, modules, or functors. The analogy extends: we want structuring mechanisms for our MMSs. We favor using *theories* and *parameterized theories* for that purpose. And even though our aim is to present to users rich high-level theories, we still firmly believe in the *little theories method* [19]. These can be assembled in a principled and modular fashion, and implemented atop a module system like Mei [29,30].

Perhaps the biggest difference is that a high-level theory needs to support more than just computation, it also needs to support deductive reasoning (and vice versa). These activities should not just co-exist: they should be tightly integrated with each other as they are in mathematical practice. Furthermore, reasoning and computation should not be restricted to objects of a particular theory: they should be applicable to theories and their interconnections [7].

3.2   Informal Definition

In this subsection, we give a preliminary, informal definition of a high-level theory, while in the next section we show how to effectively make this definition precise. In other words, we give an abstract specification now and then an implementation later. The aims of this subsection is to convey the intuition behind our ideas.

Definition 1. *A* high-level theory (HLT) *is a tuple* $(\mathcal{C}, \mathcal{T})$ *of* concepts *and* transformers *that possesses the pragmatic properties given in Figure 2. Concepts* $\mathcal{C}$ *are the basic objects of discourse and transformers* $\mathcal{T}$ *are n-ary functions on expressions concerning concepts.*

Implicit in the above definition is the notion of a *language* and a *base theory* over which everything is defined, which enters more directly in the next definition.

A *concept* is a pair (s, d) of a new symbol s and a definition d for s. In other words, a concept consists of a name and its meaning—defined in a formal language over a theory. This can be a basic object like 0 or 1, a particular group G, the definition of the fundamental group π_1 of a surface, the Gaussian elimination algorithm, an algorithm for integer factorization, the set theory NBG, or an abstract 2-category. The definition d of a concept can be given implicitly as a certain set of properties or explicitly as an expression that denotes an object.

A *transformer* is a function that maps expressions to expressions. It can embody computations or deductions. So a transformer can be as simple as the modus ponens deduction rule or integer arithmetic, up through induction or Gröbner basis computations, to proof rules for applying tactics, or an algorithm for solving PDEs symbolically. It is important to note that all these are maps from some pieces of syntax to other pieces of syntax, although our chief interest in all of them is what that syntax denotes. This point is worth emphasizing because this is a principal difference between theorem proving systems and computer algebra systems: both implement transformers, but they differ greatly in what *meaning* is a priori given to each transformer.

Another characteristic of mathematics which is important to model is the pervasive use of conceptual layers and abstraction. Concepts often appear in various guises: for example, (computable) functions can both be used directly as transformers and can also be a concept of study. Directed graphs with labeled edges and nodes can be studied directly or can be used to conveniently represent other concepts like commutative diagrams, which themselves represent equations in a theory. An HLT should give us the tools to draw a commutative diagram (as a labeled graph) and have the MMS properly interpret the result.

3.3   Semi-formal Definition

In the previous subsection we defined an HLT to be a collection of concepts and transformers that satisfy certain pragmatic requirements. Concepts are names representing mathematical ideas and objects, while transformers are functions mapping expressions to expressions that represent deduction and computation rules. In this section we will introduce a semi-formalization of an HLT, based on the notion of a biform theory [16,20].

We will begin by formalizing the notion of a transformer. Fix a set $\mathcal{E}$ of *expressions* that includes a set of *formulas*. For $n \geq 0$, an *n-ary transformer for* $\mathcal{E}$ is $\Pi = (\pi, \hat{\pi})$ where π is a symbol and $\hat{\pi}$ is an algorithm that implements a (possibly partial) function $f_{\hat{\pi}} : \mathcal{E}^n \to \mathcal{E}$. The symbol π serves as a name for the algorithm $\hat{\pi}$. There is no restriction on how the algorithm is presented. For example, it could be a lambda-expression in $\mathcal{E}$ or a program written in a high-level programming language.

Definition 2. *A biform theory T is a triple $(\mathcal{E}, \mathcal{T}, \Gamma)$ where $\mathcal{E}$ is a set of expressions, $\mathcal{T}$ is a set of transformers for $\mathcal{E}$, and Γ is a set of formulas in $\mathcal{E}$.*

The set $\mathcal{E}$ is generated from a set of symbols. Each symbol is either the name of a concept of T or is the name of a transformer of T. The members of Γ are the *axioms* of T. They specify the meanings of the concepts and transformers of T. Implicit in the above definition is the notion of a *background logic* that provides a semantic foundation for the meaning of a biform theory. We may define a biform theory $T = (\mathcal{E}, \mathcal{T}, \Gamma)$ to be an *axiomatic theory* if $\mathcal{T}$ is empty and an *algorithmic theory* if Γ is empty. Thus a biform theory is a generalization of both an axiomatic theory and an algorithmic theory.

A *rule for* $\mathcal{E}$ is a formula A in $\mathcal{E}$ of form

$$\forall e_1 : \mathcal{E}, \ldots, e_m : \mathcal{E} \, . \, B$$

where B contains one or more occurrences of an expression of the form $\pi(a_1, \ldots, a_n)$, which represents an application of the algorithm $\hat{\pi}$ to n expressions denoted by $a_1, \ldots, a_n$. The application of A to an input list $E_1, \ldots, E_m$ of expressions in $\mathcal{E}$ is the formula A' obtained by replacing each occurrence of the form $\pi(E'_1, \ldots, E'_n)$ in

$$B[e_1 \mapsto E_1, \ldots e_m \mapsto E_m]$$

with the result of applying $\hat{\pi}$ to the list $E'_1, \ldots, E'_n$ of expressions. A rule can be a rule of computation, deduction, or a mixture of the two.

Declaratively, a rule is a formula that specifies the set of transformers whose names occur in the formula. Functionally, a rule maps a list of expressions to a formula that relates the expressions as inputs to the expressions that are produced as outputs.

Definition 3. *A high-level theory (HLT) is a biform theory* $T = (\mathcal{E}, \mathcal{T}, \Gamma)$ *such that* Γ *includes a set of rules that have the pragmatic attributes listed in Figure 2.*

Since rules are statements about both the syntax of expressions and what the expressions mean, nontrivial biform theories are not easy to formalize in traditional logics such as first-order logic or simple type theory [16]. A logic is needed in which reasoning about the syntax of expressions (normally performed outside the logic) is integrated with reasoning about the semantics of expressions (normally performed in the logic itself). We have proposed a logic of this kind named *Chiron* [17,18] that we believe is exceptionally well-suited for formalizing biform theories. A derivative of von-Neumann-Bernays-Gödel (NBG) set theory, Chiron supports several reasoning paradigms by integrating set theory with elements of type theory, a scheme for handling undefinedness, and a facility for reasoning about the syntax of expressions.

3.4   Extended Example

To give a taste of what we would like to do, we present a "simple" example of what we can express in this setting.

Suppose that we define the concept *commutative* as the property $\forall a, b \, . \, a \circ b \simeq b \circ a$ of a (possibly) partial binary operation $\circ : A \times A \rightarrow A$. As expected, commutativity is a property of binary operators; less usual is the generalization to the partial setting.

Now consider the theory of Abelian groups. We would obtain a rather unusual theory if we used the above definition of commutativity. Yet, that is exactly the definition that we wish to use. Will we then need to rebuild all of group theory for a generalization that does not seem important? No, as we also have the following property:

Proposition 3.1. *Let* $* : A \times A \rightarrow A$ *be a total binary operation. Then* $*$ *is commutative if and only if* $\forall a, b : A . a * b = b * a$.

This property is really an immediate corollary of a more fundamental equivalence, namely

$$\forall a, b \in \mathcal{E} . (a\downarrow \wedge b\downarrow) \rightarrow (a \simeq b \leftrightarrow a = b).$$

($a\downarrow$ means a is defined.)

We recall that a binary operation $*$ is total if and only if $\forall a, b : A . (a * b)\downarrow$. Combining this with the property above allows us to create a *theory level* transformer which (in Chiron with syntactic sugar) reads

$$\lambda e : \mathcal{E} . \text{if}(\text{match}(e, e_1 \simeq e_2) \wedge [\![e_1]\!]\downarrow \wedge [\![e_2]\!]\downarrow, e_1 = e_2, e_1 \simeq e_2).$$

In other words, when we instantiate the concept of commutativity in the process of creating the theory of Abelian groups, our theory building operators can use the above as a *simplification* rule[2]. This rule belongs in the HLT dedicated to building theories, which is a computation on the syntactic representation of theories but relies on intermediate deductions (and computations) for its proper application. Another aspect is to note the different quantifications we are using above—universal for the definition of commutativity, over expressions for the equivalence of $\simeq$ and $=$, and over values for the definition of total.

4   Exploring High-Level Theories

The ultimate purpose of an HLT is to provide a convenient environment for end users to formulate mathematical problems and to explore possible solutions using deduction, computation, and other techniques such as visualization. An HLT is intended to be self-contained in the sense that everything the end user needs to reason within the HLT is available in the HLT. The end user should not have to introduce many new concepts, construct many new transformers, or look for many results in other HLTs. Not only does the end user have the tools he or she needs, the tools are designed to work together efficiently.

Since the tools of an HLT should usually involve both deduction and computation, an HLT-based MMS should provide a *derivation facility* for proving conjectures and performing computation in which proving and computing are

---

[2] It is important to note that we are implicitly using deep inference and inference "in context" in the statement of this rule.

```
> using ParnasTables
```

$$> f(x) = \frac{\boxed{x < 1 \mid x \geq 1 \wedge x \leq 5 \mid x > 5}}{\boxed{0 \qquad x \qquad 5}} \text{ denotes } \begin{cases} 0 & x < 1 \\ x & x \leq 5 \\ 5 & \texttt{otherwise} \end{cases}$$

```
> f(π)
```

$$\pi$$

```
> total? f
```

$$\textit{true} \quad \boxed{\text{proof}}$$

Fig. 3. (Mock up of) working in an HLT of Parnas tables

mixed. For example, in derivation a computation can involve the proof of side conditions and a proof can involve the computation of expressions. Figure 3 shows an idealized end user interaction with the HLT of Parnas tables [21] (as a particularly useful visual metaphor for piecewise functions [6]). A notable "feature" is the first-class proof object in the last interaction, while we retain the dubious feature of Parnas tables using only first-order logic to specify how to partition the domain of f. Note the difference with Figure 1 which gives the developer's view of the definition of an HLT, whereas Figure 3 is the end user's view. The vocabulary is the same, but an end user will typically interact via (the names of) transformers on expressions and see their results pretty-printed, while the developer gets a more theory-centric view.

5   Creating and Connecting High-Level Theories

The job of an MMS developer is to create HLTs. There are several approaches for doing this. First, an HLT can be created from scratch. This will be difficult and labor intensive if the subject matter of the HLT is complex or unfamiliar. On the other hand, this could be feasible if the HLT mirrors a well-understood and well-tested high-level reasoning environment such as elementary calculus. In either case, there is a significant danger that the concepts and transformers of the HLT may be inconsistent with each other. Another danger is that it could be extremely difficult to connect an HLT developed from scratch to another HLT so that results shared between the HLTs can be trusted, as witnessed by CASs.

A second approach is to construct an HLT incrementally starting from a set of very low-level axiomatic theories. Using concept and transformer definition techniques as well as module building techniques such as extension, union, renaming, and parameter instantiation [30], a network of interconnected biform theories can be build on top of the starting set of axiomatic theories. As one of the biform theories in the network, the HLT has a modular construction that is recorded in the structure of the theory network. The HLT is thus derived in a structured fashion from its underlying set of axiomatic theories.

The concepts and transformers of the HLT can be viewed as its *interface* and the theory network that records its construction can be viewed as its *implementation*. Just like the interface of a software module, the interface should not include everything in the implementation. Many low-level tools that are needed by developers to construct an HLT are either of no use to the end user or are subsumed by the high-level tools in the HLT's interface. Moreover, it is certainly possible that the same HLT can be derived from several different sets of axiomatic theories. That is, an HLT can have more than one implementation.

The implementation of an HLT is crucial for connecting one HLT to another HLT. Suppose we would like to connect an HLT T_1 to an HLT T_2. Let us assume that T_i is derived from a set S_i of axiomatic theories for $i = 1, 2$. We first construct translations from the axiomatic theories of S_1 to the axiomatic theories of S_2. Next we show that these translations are meaning preserving, i.e., are *theory interpretations* [14,15]. The last step is to use the constructions of T_1 and T_2 as a guide to lift and merge these axiomatic theory interpretations to a theory interpretation of T_1 in T_2. This last step would ideally be performed automatically. This has very natural categorical semantics in terms of limits of diagrams (in the category of biform theories).

A third approach is to construct an HLT, or least part of an HLT, automatically from an axiomatic theory. The work by R. McCasland on mechanical theorem discovery is an interesting step in this direction [24]. Two classical examples are the use of the Knuth-Bendix completion algorithm to automatically generate a terminating term rewrite system from a set of equational axioms [23] and the use of Buchberger's algorithm to construct a Gröbner basis for a system of polynomials [3]. For the algorithmic aspects, one can instantiate generic algorithms and still get efficient implementations [5,8].

We will finish this section with an important observation. Mathematical knowledge as a whole is a network of interconnected smaller bodies of mathematical knowledge. The interconnectivity of mathematics allows problems to be expressed and solved in a general context (e.g., metric spaces) and then their solutions to be applied in more specialized contexts (e.g., real analysis). The *little theories method* [19] models the network of interconnected bodies of mathematical knowledge as a network of separate, but interconnected axiomatic theories. The *big theory method* [19] models mathematical knowledge as one big theory. The little theories method is central in the creation and connection of HLTs, but the big theory method is followed in the exploration of HLTs.

6   Examples of High-Level Theories

We further examine how we might formalize the two examples of high-level reasoning environments given in section 2.

An HLT formalization of natural number arithmetic would be a biform theory T_1 containing several hundred concepts, transformers, and axioms. It would be carefully constructed in two ways.

First, T_1 would be constructed in a modular fashion from a well-understood, low-level axiomatization of natural number arithmetic, like Peano's axioms, and a set of supporting low-level theories about such things as sets and real numbers. Its construction would demonstrate that each of its axioms is a theorem derived from the low-level theories. How the low-level theories are axiomatized is not important as long as the axioms of T_1 can be derived from the theories.

Second, the concepts, transformers, and axioms of T_1 would be carefully chosen. They would not include every known concept, transformer, and theorem of natural number arithmetic. In particular, T_1 would not be simply the sum of the low-level theories from which it is constructed. Instead the constituents of T_1 would have a high level of coverage and a low level of redundancy. For example, the concepts of T_1 might well not include the successor function since it can be easily expressed using the addition function. And there would be no need for the recursive definitions of addition and multiplication if T_1 includes transformers for computing sums and products. The axioms of T_1 would be high-level theorems such as the fundamental theorem of arithmetic and the Chinese remainder theorem, high-level deduction rules such as various induction principles, and high-level computation rules such as those for computing greatest common divisors and factoring natural numbers into primes.

To end users, T_1 would look like a formalization of what mathematicians call "number theory"; the concepts, transformers, and axioms of T_1 would be basic ideas, tools, and assumptions of the theory. To developers, T_1 would look like the end result of a large, complicated theory development; the concepts, transformers, and axioms of T_1 would be the high-level ideas, tools, and theorems derived from the underlying low-level theories.

An HLT formalization of group theory would be a biform theory T_2 that contains, like T_1, several hundred concepts, transformers, and axioms, and constructed in the same careful manner to embody "group theory". In particular, "group theory" does not care whether an inverse function is provided axiomatically or as a derived property.

7   Related Work

As we have already mentioned, the working environment of a mainstream CAS (like Maple and Mathematica) gives the impression of working in a HLT, but in reality only achieves properties (3) and (7) (of those in Figure 2), and somewhat implements (5) and (6). Theorema [4] adds (2). But since they are unsound, it is unclear if that all amounts to much.

Current large TPSs (like Coq [9], Isabelle [26], and PVS [25]) also seem to be moving in this direction, with their strength being properties (4) and frequently (5), and slowly moving in the direction of (2). Mizar [28] is hobbled by being nonmodular. There is a lot of work being done to integrate computation into deduction [1,2,13]. This work will certainly have an effect on ours, but we feel that it is too asymmetric compared to the symmetry between computation and deduction in biform theories.

The most direct implementation of something akin to HLTs is in Focal [27]. Unfortunately, this system is only really comfortable for dedicated developers and does not yet enable multiple modes of interaction.

It is possible to build a safe computational system atop a theorem prover, as Kaliszyk and Wiedijk show [22]. While a definite achievement, this seems to embody (part of) one handbuilt HLT.

Lastly, we should note that it is possible to encode biform theories in the Calculus of Inductive Constructions [10,11], and thus in Coq. It is however our current feeling that Chiron is better suited for this task.

8 Conclusion

A high-level theory (HLT) is a model of the high-level reasoning environments employed in mathematical practice. Roughly speaking, it consists of a well-crafted set of concepts and transformers. More precisely, it is a biform theory with certain pragmatic properties. In particular, it includes high-level tools for deduction, computation, and a mixture of the two. An HLT is to a low-level axiomatic theory or algorithmic theory as a high-level programming language is to an assembly language. Working in an HLT is much more effective and convenient than working in a low-level theory.

We recommend that the ultimate goal of an MMS should be to provide a library of HLTs that are useful and accessible to a wide range of mathematics practitioners. The library's HLTs should include the best of the features currently found in the axiomatic and algorithmic theories of contemporary TPSs and CASs. Low-level axiomatic and algorithmic theories should be considered as part of the supporting infrastructure of the library, not as the end product of the system. To facilitate the development and expansion of the library, the MMS should include a facility with a powerful set of tools for developers to construct HLTs from low-level theories. We believe that an MMS that offers HLTs to end users and the tools for building HLTs to developers has the best chance of realizing the immense potential of mechanized mathematics.

References

1. Blanqui, F., Jouannaud, J.-P., Strub, P.-Y.: Building decision procedures in the calculus of inductive constructions. In: Duparc, J., Henzinger, T.A. (eds.) CSL 2007. LNCS, vol. 4646, pp. 328–342. Springer, Heidelberg (2007)
2. Blanqui, F., Jouannaud, J.-P., Strub, P.-Y.: From formal proofs to mathematical proofs: A safe, incremental way for building in first-order decision procedures. In: TCS 2008: 5th IFIP International Conference on Theoretical Computer Science. Springer, Heidelberg (2008)
3. Buchberger, B.: Theoretical basis for the reduction of polynomials to canonical forms. SIGSAM Bulletin 39, 19–24 (1976)
4. Buchberger, B., Craciun, A., Jebelean, T., Kovacs, L., Kutsia, T., Nakagawa, K., Piroi, F., Popov, N., Robu, J., Rosenkranz, M., Windsteiger, W.: Theorema: Towards computer-aided mathematical theory exploration. Journal of Applied Logic 4, 470–504 (2006)

5. Carette, J.: Gaussian Elimination: a case study in efficient genericity with MetaO-Caml. Science of Computer Programming 62(1), 3–24 (2004); Special Issue on the First MetaOCaml Workshop (2004)
6. Carette, J.: A canonical form for piecewise defined functions. In: Proceedings of the 2007 International Symposium on Symbolic and Algebraic Computation (ISSAC), pp. 77–84. ACM Press, New York (2007)
7. Carette, J., Farmer, W.M., Sorge, V.: A rational reconstruction of a system for experimental mathematics. In: Kauers, M., Kerber, M., Miner, R., Windsteiger, W. (eds.) MKM/CALCULEMUS 2007. LNCS (LNAI), vol. 4573, pp. 13–26. Springer, Heidelberg (2007)
8. Carette, J., Kiselyov, O.: Multi-stage programming with Functors and Monads: eliminating abstraction overhead from generic code (accepted, 2008); Special issue for GPCE 2004 and 2005
9. Coq Development Team. The Coq Proof Assistant Reference Manual, Version 7.4 (2003), http://pauillac.inria.fr/coq/doc/main.html
10. Coquand, T., Huet, G.: The calculus of constructions. Information and Computation 76, 95–120 (1988)
11. Coquand, T., Paulin-Mohring, C.: Inductively defined types. In: Martin-Löf, P., Mints, G. (eds.) COLOG 1988. LNCS, vol. 417, pp. 50–66. Springer, Heidelberg (1990)
12. de Bruijn, N.G.: Automath, a language for mathematics. In: Siekmann, J., Wrightson, G. (eds.) Automation of Reasoning 2: Classical Papers on Computational Logic 1967-1970, pp. 159–200. Springer, Heidelberg (1983)
13. Dowek, G., Hardin, T., Kirchner, C.: Theorem proving modulo. J. Autom. Reasoning 31(1), 33–72 (2003)
14. Enderton, H.B.: A Mathematical Introduction to Logic, 2nd edn. Academic Press, London (2000)
15. Farmer, W.M.: Theory interpretation in simple type theory. In: Heering, J., et al. (eds.) HOA 1993. LNCS, vol. 816, pp. 96–123. Springer, Heidelberg (1994)
16. Farmer, W.M.: Biform theories in Chiron. In: Kauers, M., Kerber, M., Miner, R., Windsteiger, W. (eds.) MKM/CALCULEMUS 2007. LNCS (LNAI), vol. 4573, pp. 66–79. Springer, Heidelberg (2007)
17. Farmer, W.M.: Chiron: A multi-paradigm logic. In: Matuszewski, R., Zalewska, A. (eds.) From Insight to Proof: Festschrift in Honour of Andrzej Trybulec, Studies in Logic, Grammar and Rhetoric, vol. 10(23), pp. 1–19, University of Białystok (2007)
18. Farmer, W.M.: Chiron: A set theory with types, undefinedness, quotation, and evaluation. SQRL Report No. 38, McMaster University (2007) (revised 2008)
19. Farmer, W.M., Guttman, J.D., Thayer, F.J.: Little theories. In: Kapur, D. (ed.) CADE 1992. LNCS, vol. 607, pp. 567–581. Springer, Heidelberg (1992)
20. Farmer, W.M., von Mohrenschildt, M.: An overview of a formal framework for managing mathematics. Annals of Mathematics and Artificial Intelligence 38, 165–191 (2003)
21. Janicki, R., Parnas, D.L., Zucker, J.: Tabular representations in relational documents. In: Brink, C., Kahl, W., Schmidt, G. (eds.) Relational Methods in Computer Science, pp. 184–196. Springer, Heidelberg (1997)
22. Kaliszyk, C., Wiedijk, F.: Certified computer algebra on top of an interactive theorem prover. In: Calculemus/MKM, pp. 94–105 (2007)
23. Knuth, D.E., Bendix, P.B.: Simple word problems in universal algebra. In: Leech, J. (ed.) Computational Problems in Abstract Algebra, pp. 263–297. Pergamon Press, Oxford (1970)

24. McCasland, R.L., Bundy, A., Smith, P.F.: Ascertaining mathematical theorems. Electronic Notes in Theoretical Computer Science 151, 21–38 (2006)
25. Owre, S., Rajan, S., Rushby, J.M., Shankar, N., Srivas, M.: PVS: Combining specification, proof checking, and model checking. In: Alur, R., Henzinger, T.A. (eds.) CAV 1996. LNCS, vol. 1102, pp. 411–414. Springer, Heidelberg (1996)
26. Paulson, L.C.: Isabelle: A Generic Theorem Prover. LNCS, vol. 828. Springer, Heidelberg (1994)
27. Prevosto, V.: Certified mathematical hierarchies: The FoCal system. In: Coquand, T., Lombardi, H., Roy, M.-F. (eds.) Dagstuhl Seminar Proceedings of Mathematics, Algorithms, Proofs, Dagstuhl, Germany. Internationales Begegnungs- und Forschungszentrum (IBFI), vol. 05021. Schloss Dagstuhl, Germany (2005)
28. Rudnicki, P.: An overview of the MIZAR project. Technical report, Department of Computing Science, University of Alberta (1992)
29. Xu, J.: Mei — A module system for mechanized mathematics systems. In: Programming Languages for Mechanized Mathematics Workshop, Hagenberg, Austria (2007)
30. Xu, J.: Mei — A Module System for Mechanized Mathematics Systems. PhD thesis, McMaster University (January 2008)

Parametric Linear Arithmetic over Ordered Fields in Isabelle/HOL

Amine Chaieb

Institut für Informatik
Technische Universität München

Abstract. We use higher-order logic to verify a quantifier elimination procedure for linear arithmetic over ordered fields, where the coefficients of variables are multivariate polynomials over *another* set of variables, we call parameters. The procedure generalizes Ferrante and Rackoff's algorithm for the non-parametric case. The formalization is based on axiomatic type classes and automatically carries over to e.g. the rational, real and non-standard real numbers. It is executable, can be applied to HOL formulae and performs well on practical examples.

1 Introduction

Most LCF-like theorem provers provide proof automation for several theories of arithmetic. These include universal real [14] and linear integer arithmetic, Presburger arithmetic [21,6] and the full first order theories of linear and non-linear real or complex arithmetic [11,20,12,7,5] using quantifier elimination (qe.). See also [4] for mixed real-integer linear arithmetic and [13,8,5] for ring problems. Present procedures able to deal with multiplication in quantified problems have non-elementary complexity and do not scale in practice. We are not aware of any proof-procedure with elementary complexity in an LCF-like theorem prover, but see [19] for very promising progress in verifying CAD in Coq.

In this paper we present a formally verified quantifier elimination procedure (qep.) for linear arithmetic over ordered fields, where the coefficients of variables are multivariate polynomials over *another* set of variables, we call parameters. The procedure is a generalization of Ferrante and Rackoff's algorithm [9] for the non-parametric case. Our formalization is based on axiomatic type classes [24] and hence automatically carries over to several interesting structures such as the rational, real and non-standard real numbers. The formalization is executable and can be applied to HOL formulae. Our procedure has doubly exponential space complexity and is hence optimal [23]. It is however, slightly less efficient than the procedure by Loos and Weispfenning [18], who gave the first qep. for parametric linear arithmetic. A description of the present work also appears in the author's thesis [5].

This paper is structured as follows. In § 2 we give the informal qep.. In § 3 we present a formalization of multivariate polynomials, which will serve as parameters. We formalize the qep. in § 4 and § 5 and integrate it in § 6.

S. Autexier et al. (Eds.): AISC/Calculemus/MKM 2008, LNAI 5144, pp. 246–260, 2008.
© Springer-Verlag Berlin Heidelberg 2008

2   The Main Algorithm

Let $\mathcal{F}_+^\rho$ be the first order theory of parametric linear arithmetic over ordered fields, i.e. coefficients of bound variables are multivariate polynomials over another set of variables. Note that $\mathcal{F}_+^\rho$ is more expressive than the non-parametric theory. Consider for instance $P(x) = a^2 = 2 \wedge x < a$ over the reals. Then $S(P) = \{x \mid P(x)\} =]-\infty, \sqrt{2}[$, which is not expressible in the non-parametric theory. Expressible sets in $\mathcal{F}_+^\rho$ are finite unions of intervals whose endpoints are infinite or algebraic (in the non-parametric case replace algebraic by rational), see [18].

The goal of this section is to give a qe. algorithm informally using the usual mathematical notation, except that under an ordered field F we understand an ordered field satisfying $0^{-1} = 0$ for all x, i.e. F is 0-totalized [2,15]. This assumption is also made in [18] and is satisfied in all practical instances in Isabelle/HOL for our formalization.

We consider an ordered field F and two disjoint infinite sets of variables: $\mathcal{V}$ and $\mathcal{V}_p$. A parameter is any multivariate polynomial over $\mathcal{V}_p$. Terms are built up from the field constants and variables in $\mathcal{V}$ by addition, subtraction and restricted multiplication *by parameters*. Atomic formulae are *True, False*, $s = t, s \neq t, s < t$ and $s \leq t$, where s and t are terms. The set of all formulae is then obtained by the closure under the usual boolean operations $\neg, \wedge, \vee, \rightarrow$ and $\leftrightarrow$, and by universal $\forall$ and existential $\exists$ quantification over variables in $\mathcal{V}$. Let L denote this language. It is a standard result of qe. that we only need to provide a qep. for $\exists x.P$, where P is qf. Furthermore, we can transform any qf. P into Q, which contains no negations and, by "gathering" the same terms, only contains atoms of the form $a \cdot x = t, a \cdot x \neq t, a \cdot x < t, a \cdot x \leq t$ or those not involving the existentially bound variable x. Thereby, a is a parameter and t is a term not involving x. Let $\mathcal{U}_Q$ denote the set of all pairs (t, a) such that $a \cdot x \bowtie t, \bowtie \in \{=, \neq, <, \leq\}$ occurs in Q. Furthermore let $Q_{-\infty}$ and $Q_{+\infty}$ be the formulae resulting after replacing the atoms in Q as in the table below, then Theorem 1 holds.

Atom	$Q_{-\infty}$	$Q_{+\infty}$
$a \cdot x = t$	$a = t = 0$	$a = t = 0$
$a \cdot x \neq t$	$a \neq 0 \vee t \neq 0$	$a \neq 0 \vee t \neq 0$
$a \cdot x < t$	$a > 0 \vee a = 0 \wedge t > 0$	$a < 0 \vee a = 0 \wedge t > 0$
$a \cdot x \leq t$	$a > 0 \vee a = 0 \wedge t \geq 0$	$a < 0 \vee a = 0 \wedge t \geq 0$

Theorem 1. *For x, P and Q as above, the formula $\exists x.P(x)$ holds if and only if $Q_{-\infty} \vee Q_{+\infty} \vee \bigvee_{(a,t) \in \mathcal{U}_Q, (a',t') \in \mathcal{U}_Q} Q((\frac{t}{a} + \frac{t'}{a'})/2)$ holds.*

Proof. First note that P and Q are equivalent. Moreover let $Q(\mathcal{U}_Q)$ abbreviate $\bigvee_{(a,t) \in \mathcal{U}_Q, (a',t') \in \mathcal{U}_Q} Q((\frac{t}{a} + \frac{t'}{a'})/2)$. It is trivial that if $Q(\mathcal{U}_Q)$ holds then $\exists x.Q(x)$ holds. To finish the proof of the *only if* direction we need to show that $Q_{-\infty}$ and $Q_{+\infty}$ respectively imply $\exists x.Q(x)$. This is a simple consequence of the fact that $Q_{-\infty}$ and $Q_{+\infty}$ are equivalent to Q for arguments that are arbitrarily large negative and positive respectively, i.e. they mimic Q near $-\infty$ and $+\infty$. Concretely the properties (a) and (b), easy to prove by induction on Q, hold.

$$\exists z. \forall x < z. Q(x) \leftrightarrow Q_{-\infty} \tag{a}$$

$$\exists z. \forall x > z. Q(x) \leftrightarrow Q_{+\infty} \tag{b}$$

Now for the *if* direction of Theorem 1, assume $Q(x)$ for some x and that neither $Q_{-\infty}$ nor $Q_{+\infty}$ hold, i.e. x is a neither too large nor too small witness for Q. From (a) and (b) it easy to see that $l \le x \le u$ holds for some l and u. In fact, a simple proof by induction on the structure of Q yields that l and u are $\frac{t}{c}$ and $\frac{t'}{c'}$ for some t, c, t', c' such that $(t, c) \in \mathcal{U}_Q$ and $(t, c) \in \mathcal{U}_Q$. Hence x is either equal to $\frac{s}{d}$, where $(s, d) \in \mathcal{U}_Q$, in which case we are done, or $l < x < u$ for l and u as above, but such that for no $(s, d) \in \mathcal{U}_Q$ we have $l < \frac{s}{d} < u$. The existence of such *smallest* interval is easy to show, since $\mathcal{U}_Q$ is finite and F is ordered. Now we show that Q actually holds over the whole interval $]l, u[$ and hence in particular for $\frac{l+u}{2}$. The proof is by induction on the structure of Q, and is interesting only for atoms. The case where Q is $a \cdot x = r$ is trivial since it contradicts the assumptions. Assume that Q is $a \cdot x < r$. If $a = 0$ the statement is trivial. Otherwise assume $a > 0$, then $x < \frac{r}{a}$. Consider an arbitrary y such that $l < y < y$, we must show that $a \cdot y < r$ holds. First note that $y \ne \frac{r}{a}$, since $(r, a) \in \mathcal{U}_Q$. Assume for contradiction that $y > \frac{r}{a}$, then $l < x < \frac{r}{a} < y < u$, i.e. $l < \frac{r}{a} < u$ which contradicts the assumptions, since $(r, a) \in \mathcal{U}_Q$. Hence Q holds for every $y \in]l, u[$. The case $a < 0$ is analogous. The cases $a \cdot x \ne r$ and $a \cdot x \le r$ are analogous to the $a \cdot x < r$ case.   □

In order to obtain qe. for $\mathcal{F}_+^\rho$, we just need to argue that we can encode the substitution of "informal fractions" $\frac{t}{c}$ into a formula using our simple language L of rings, i.e. $Q(\frac{t}{c})$ indeed represents a L-formula. This is easy using standard techniques [18,23,4]. We formalize two variants in detail in § 5.

Corollary 1. *The theory $\mathcal{F}_+^\rho$ admits qe. by terms. Moreover, for purely existential statements, the procedure above computes concrete witnesses for the existential quantifiers.*

This concludes the general description of the algorithm. In the following we present our formalization of polynomials in § 3, formulae and generic qe. in § 4 and finally the qep. in § 5.

3   Formalized Polynomials

Our formalizations of polynomials have a strong algorithmic flavor, since we want to compute and prove with them. In § 3.1 we formalize univariate polynomials as "functions": given a list of coefficients $c_0, \ldots, c_n$, they describe a function $x \mapsto \sum_{i=0}^n c_i \cdot x^i$. This approach was successfully used quite early in HOL, but due to the lack of classes or equivalent specification mechanisms, the formalizations are duplicated for $\mathbb{R}$ and $\mathbb{C}$. Our formalization is based on locales [17,1], and hence carries over to several instances including $\mathbb{Z}, \mathbb{Q}, \mathbb{R}, \mathbb{C}$ and the non-standard reals.

In § 3.2, we present a formalization of multivariate polynomials with rational coefficients using axiomatic type classes [24]. The theorems we present were proved in locales or classes with as few axioms as possible. We do not present the exact requirements for space and clarity reasons. The reader can simply assume that the theorems hold in any ordered field. See [5] for more details. An interesting alternative to our completely verified approach is presented in [3,22] using the FOC language.

Notation. The rest of the paper has a different status than the previous section: it has been formalized in Isabelle/HOL. In particular all statement *numbered* and all figures correspond to formal definitions or statements formally proved in Isabelle/HOL. The notation of HOL conforms to usual mathematical one. We use $\mathbb{B}, \mathbb{Z}, \mathbb{N}, \mathbb{Q}, \mathbb{R}$ to denote the *types* of truth values, integers, natural, rational and real numbers respectively. The space of total functions is denoted by $\Rightarrow$. The notation $t :: \tau$ means that the term t has type τ. Datatypes are defined using the keyword **datatype** and HOL allows the definition of recursive functions using pattern matching. Lists are built up from the empty list $[]$ and consing $x \cdot xs$. Multiplication and consing are both denoted by $\cdot$, but the meaning should be clear from the context. For a list xs we denote the *set* of its elements by $\{\!\{xs\}\!\}$.

3.1 Univariate Polynomials as Functions

We formalize *univariate* polynomials as functions. Given a list of coefficients $[c_0, \ldots, c_n]$, then $\overline{[c_0, \ldots, c_n]} = \lambda x. \sum_{i=0}^{n} c_i \cdot x^i$.

$$\overline{[]} \, x = 0 \qquad | \; \overline{c \cdot cs} \, x = (\overline{cs} \, x) \cdot x + c \tag{1}$$

This formalization is very appealing to switch views: a) view polynomials syntactically as a list of coefficients and b) view a list of coefficients p as a polynomial *function* $\overline{p}$. The zero polynomial is $\overline{[]}$. Note that the list representation is not unique, since we can always append zeros and $\overline{p} = \overline{p@[0]}$.

The first step is to implement algorithms for the usual operations on the syntax and prove them correct. The formalization includes addition $\overline{+}$, multiplication $\overline{\cdot}$, exponentiation $\lambda p, n.p^n$, subtraction and negation both denoted by $\overline{-}$ and normalization **norm** to remove superfluous zero coefficients. It also includes other notions we do not use here, such as the degree, the multiplicity of a root and square free conditions.

The definitions of addition and multiplication etc. over the syntax are straightforward. For instance addition is defined in (2). We prove all these operations correct, e.g. (3) for addition, multiplication and subtraction.

$$[] \overline{+} q = q \quad | \; p \overline{+} [] = p \quad | \; (c \cdot cs) \overline{+} (d \cdot ds) = (c + d) \cdot (cs \overline{+} ds) \tag{2}$$

$$\overline{p \bowtie q} \, x = (\overline{p} \, x) \bowtie (\overline{q} \, x) \quad \text{for } \bowtie \in \{+, \cdot, -\} \tag{3}$$

We present here only a few interesting theorems just to give a rough idea about the formalization. Like in algebra texts, stronger properties hold in classes with more axioms. The key property (4) for roots-factorization holds e.g. in commutative rings with unity. Moreover in an integral domain of characteristic zero, every polynomial has finitely many roots exactly when it is not the zero polynomial (cf. (5)) and the entirety property (6) holds. Here another strong and interesting property holds: a polynomial (*as a function*) is zero exactly when all its coefficients are zero (cf. (7)). This last property plays a central role in the implementation and also for the uniqueness property of multivariate polynomials in § 3.2. Let a polynomial p be normal (isorm p) if norm $p = p$ holds. Polynomials in normal form in integral domains are unique, cf. (8)

$$\overline{p}\, a = 0 \leftrightarrow p = [] \vee \exists q. p = [-a, 1] \dot{\cdot} q \tag{4}$$

$$\overline{p} \neq \overline{[]} \leftrightarrow \text{finite}\{x | \overline{p}\, x = 0\} \tag{5}$$

$$\overline{p \dot{\cdot} q} = \overline{[]} \leftrightarrow \overline{p} = \overline{[]} \vee \overline{q} = \overline{[]} \tag{6}$$

$$\overline{p} = \overline{[]} \leftrightarrow \forall c \in \{\!|p|\!\}.c = 0 \tag{7}$$

$$\text{isorm } p \wedge \text{isorm } q \rightarrow \overline{p} = \overline{q} \leftrightarrow p = q \tag{8}$$

The formalization in (1) is very suitable for abstract reasoning about *univariate* polynomials. We can also generate code if the underlying (semi)ring allows it. Note the ability of the code-generator framework [10] to deal with classes using dictionaries. The generated code depends on the implementation of the operations and hence *cannot* be used to compute abstractly. The main drawback of this formalization is that it does not "naturally" carry over to multivariate polynomials. The type corresponding to $R[x_1, \ldots, x_n]$ would be $[\ldots [\alpha] \ldots]$, nested n times, which can not be expressed in HOL.

3.2   Reflected Multivariate Polynomial Utilities

We present here an executable formalization of *multivariate* polynomials over fields. As in § 3.1 stronger theorems only hold in classes with more axioms. To generate code running for all instances, we restrict the coefficients to the sub-field $\{i/j \mid i, j \in \mathbb{Z}\}$, i.e. rational numbers.

Rational numbers. We implement rational numbers by pairs of integers $i : j$ interpreted as field elements by $(\!|i : j|\!)_r = i/j$. This representation is not unique in general, but it is for those in normal form (i.e. satisfying isorm$_r$, see (9)):

$$\text{isorm}_r\ (i : j) = \text{if } i = 0 \text{ then } j = 0 \text{ else } b > 0 \wedge \gcd i\, j = 1$$

$$\text{isorm}_r\ a \wedge \text{isorm}_r\ b \rightarrow (\!|a|\!)_r = (\!|b|\!)_r \leftrightarrow a = b \tag{9}$$

We implement the usual arithmetical operations $(+_r, \cdot_r, -_r, /_r)$ and ordering relations $(<_r, \leq_r)$ and prove them correct:

$$\text{isorm}_r\ a \wedge \text{isorm}_r\ b \rightarrow$$

$$(\!|a \circ b|\!)_r = (\!|a|\!)_r \circ (\!|b|\!)_r \wedge (a \bowtie b \leftrightarrow (\!|a|\!)_r \bowtie (\!|b|\!)_r) \wedge \text{isorm}_r\ (a \circ b), \tag{10}$$

$$\text{for } (\circ, \circ) \in \{(+_r, +), (\cdot_r, \cdot), (-_r, -), (/_r, /)\}$$

$$\text{and } (\bowtie, \bowtie) \in \{(<_r, <), (\leq_r, \leq)\}.$$

$$\text{datatype } \rho = \widehat{\mathbb{Z} : \mathbb{Z}} | v_\mathsf{N} | - \rho | \rho + \rho | \rho - \rho | \rho * \rho | \rho^\mathsf{N}$$

$$\begin{aligned}
(\!(\widehat{c})\!)^e_\rho &= (\!(\widehat{c})\!)_r & (\!(p+q)\!)^e_\rho &= (\!(p)\!)^e_\rho + (\!(q)\!)^e_\rho \\
(\!(v_n)\!)^e_\rho &= e_{[n]} & (\!(p-q)\!)^e_\rho &= (\!(p)\!)^e_\rho - (\!(q)\!)^e_\rho \\
(\!(-p)\!)^e_\rho &= -(\!(p)\!)^e_\rho & (\!(p*q)\!)^e_\rho &= (\!(p)\!)^e_\rho \cdot (\!(q)\!)^e_\rho \\
& & (\!(p^n)\!)^e_\rho &= ((\!(p)\!)^e_\rho)^n
\end{aligned}$$

Fig. 1. Syntax and semantics of polynomial expressions

The implementation does not raise an exception but correctly returns 0 as a result for $x/0$. This reflects Isabelle/HOL's behavior correctly. We also provide a function norm_r to normalize any $i:j$ and prove it correct. In the following i_r denotes $i:1$ for any integer $i \neq 0$ and $\mathbf{0}_r$ denotes $0:0$.

Polynomial expressions. The syntax (datatype ρ) and its semantics in Fig. 1 reflect multivariate polynomial expressions. The semantics $(\!(.)\!)_\rho$ is parameterized by an environment e (a list of "field-elements"). We represent variables by de Bruijn indices: v_n represents the bound variable with index $n :: \mathsf{N}$. Note that $(\!(v_n)\!)^e_\rho = e_{[n]}$ is the nth element of e. The bold symbols $+, *$ etc. are constructors and reflect their counterparts $+, \cdot$ etc. in the logic. We reflect a rational number i/j by $\widehat{i:j}$, i.e.$\widehat{}$ is a constructor of ρ.

The normal form defined by ishorn imposes a Horner scheme for multivariate polynomials where the ordering on variables is induced by their indices.

$$\begin{aligned}
\mathsf{ishornh}\ \widehat{c}\ n &= \mathit{True} \\
\mathsf{ishornh}\ (c + v_m * p)\ n &= p \neq \mathbf{0}_\rho \wedge m \geq n \wedge \mathsf{ishornh}\ c\ (m+1) \wedge \mathsf{ishornh}\ p\ m \\
\mathsf{ishorn}\ p &= \mathsf{ishornh}\ p\ 0
\end{aligned}$$

For example $5 \cdot y \cdot x^2 + y \cdot (x+2)$ is reflected by $(\!(q)\!)^{[x,y]}_\rho$ and $(\!(r)\!)^{[x,y]}_\rho$, for $q = \widehat{5} * v_1 * v_0{}^2 + v_1 * (v_0 + \widehat{2})$ and $r = (\widehat{0} + v_1 * \widehat{2}) + v_0 * ((\widehat{0} + v_1 * \widehat{1}) + v_0 * (\widehat{0} + v_1 * \widehat{5}))$. Only r is in normal form.

Ultimately we would like to compute the normal form of any ρ-polynomial. It is legitimate to write *the* normal form, since we show uniqueness in a certain sense: two ρ-polynomials in normal form are *syntactically* equal if and only if their interpretations are equal in *all* possible environments. We present a function norm_ρ to normalize any ρ-polynomial. For this, it applies algorithms for addition $\oplus$, multiplication $\circledast$, negation and subtraction $\ominus$ and power $\lambda p, n.p^{\downarrow n}$, with the additional property that they preserve the normal form. See Fig. 2 for the definitions. Subtraction and negation are straightforward. Taking p to the power of n repeatedly applies $\circledast$ depending on the binary scheme of n. All these operations preserve normal form, cf. (11), and semantics, cf. (12). Now the definition of norm_ρ is simple and mainly replaces the constructors with the definitions above. The main property of norm_ρ in (13) uses (11) and (12) and is proved by structural induction. For $\ominus$ there is a syntactical property, cf. (14).

$$
\begin{array}{ll}
\widehat{c} \oplus \widehat{d} & = \widehat{c +_r d} \\
\widehat{c} \oplus (d + v_k * q) & = (\widehat{c} \oplus d) + v_k * q \\
(c + v_n * p) \oplus \widehat{d} & = (c \oplus \widehat{d}) + v_n * p \\
(c + v_n * p) \oplus (d + v_k * q) & = \textbf{if } n < k \textbf{ then } (c \oplus (d + v_k * q)) + v_n * p \\
& \quad \textbf{else if } k < n \textbf{ then } ((c + v_n * p) \oplus d) + v_k * q \\
& \quad \textbf{else let } cd = c \oplus d; pq = p \oplus q \\
& \qquad \textbf{in if } pq = \mathbf{0}_\rho \textbf{ then } cd \textbf{ else } cd + v_n * pq \\
a \oplus b & = a + b
\end{array}
$$

$$
\begin{array}{ll}
\widehat{c} \circledast \widehat{d} & = \widehat{c \cdot_r d} \\
\widehat{c} \circledast (d + v_k * q) & = \textbf{if } c = 0 \textbf{ then } \mathbf{0}_\rho \textbf{ else } (\widehat{c} \circledast d) + v_k * (\widehat{c} \circledast q) \\
(c + v_n * p) \circledast \widehat{d} & = \textbf{if } d = 0 \textbf{ then } \mathbf{0}_\rho \textbf{ else } (\widehat{d} \circledast c) + v_n * (\widehat{d} \circledast p) \\
(c + v_n * p) \circledast (d + v_k * q) & = \textbf{if } n < k \textbf{ then} \\
& \qquad (c \circledast (d + v_k * q)) + v_n * (p \circledast (d + v_k * q)) \\
& \quad \textbf{else if } k < n \textbf{ then} \\
& \qquad ((c + v_n * p) \circledast d) + v_k * ((c + v_n * p) \circledast q) \\
& \quad \textbf{else } ((c + v_n * p) \circledast d) \oplus (\mathbf{0}_\rho + v_n * ((c + v_n * p) \circledast q)) \\
a \circledast b & = a * b
\end{array}
$$

$$
\begin{array}{ll}
p^{\downarrow 0} & = \mathbf{1}_\rho \\
p^{\downarrow n} & = \textbf{let } q = p^{\downarrow \frac{n}{2}}; d = q \circledast q \\
& \quad \textbf{in if even } n \textbf{ if } d \textbf{ else } p \circledast d
\end{array}
$$

Fig. 2. Addition, multiplication and power for ρ-polynomials

$$\mathsf{ishorn}\ p \rightarrow \mathsf{ishorn}\ p^{\downarrow n} \wedge (\mathsf{ishorn}\ q \rightarrow \mathsf{ishorn}\ p \oplus q \wedge \mathsf{ishorn}\ p \circledast q) \tag{11}$$

$$(\!|p \oplus q|\!)^e_\rho = (\!|p + q|\!)^e_\rho \wedge (\!|p \circledast q|\!)^e_\rho = (\!|p * q|\!)^e_\rho \wedge (\!|p^{\downarrow n}|\!)^e_\rho = (\!|p^n|\!)^e_\rho \tag{12}$$

$$\mathsf{ishorn}\ (\mathsf{norm}_\rho\ p) \wedge (\!|\mathsf{norm}_\rho\ p|\!)^e_\rho = (\!|p|\!)^e_\rho \tag{13}$$

$$\mathsf{ishorn}\ p \wedge \mathsf{ishorn}\ q \rightarrow p \ominus q = \mathbf{0}_\rho \leftrightarrow p = q \tag{14}$$

Using (13) we obtain an incomplete method to prove equality of two polynomials. We show completeness in the following.

The uniqueness property. We show in the following that ρ-polynomials in normal form are unique by using the uniqueness property of *univariate* polynomials in § 3.1. Given an environment e, then function $\lambda p.[p]^e$ in (15) connects ρ-polynomials to the univariate ones of § 3.1. Obviously it satisfies (16). This simple connection transfers properties from § 3.1 to ρ-polynomials.

$$[p]^e = \mathsf{map}\ (\lambda q.(\!|\mathsf{decr}_\rho\ q|\!)^e_\rho)\ (\mathsf{coeffs}\ p) \tag{15}$$

$$\mathsf{ishornh}\ p\ n_0 \rightarrow (\!|p|\!)^{x \cdot e}_\rho = \overline{[p]^e}\ x \tag{16}$$

Let max^v_p denote the maximal n, where v_n occurs in p. Property (17) is the analog of (7) and states that a normalized polynomial is $\mathbf{0}_\rho$ exactly when it evaluates to 0 for any reasonable environment e. The proof of (17) is by complete induction on max^v_p where (7) is applied to the coefficients of $[p]^e$. This corresponds to an induction over the number of variables n of the multivariate polynomial ring $R[x_1, \ldots, x_n]$. From (14) and (17) we derive the uniqueness property (18).

$$\text{ishorn } p \rightarrow (\forall e.|e| \geq \max_p^v \rightarrow (\!|p|\!)_\rho^e = 0) \leftrightarrow p = \mathbf{0}_\rho \tag{17}$$

$$\text{ishorn } p \wedge \text{ishorn } q \rightarrow (\forall e.(\!|p|\!)_\rho^e = (\!|q|\!)_\rho^e) \leftrightarrow p = q \tag{18}$$

An important impact of (18) is that all interesting algebraic properties about $+, \cdot$, etc. ultimately carry over to $\oplus, \circledast$, etc. *on the syntactical level*, e.g. distributivity and commutativity hold for $\oplus$ and $\circledast$. The only drawback we must accept is that the properties proved in this manner, although purely syntactical, hold only under the axioms of ordered fields.

4   Formalized Generic Quantifier Elimination

We formalize terms and formulae by datatypes τ and ϕ and their semantics in Fig. 3, which are parameterized with two environments: π is a list of field elements considered as parameters and used to interpret the polynomial coefficients, whereas e is a list of field elements interpreted as variables which can be bound by quantifiers. We use de Bruijn indices for variables and hence quantifiers need not carry names. When bound by a quantifier, the new bound variable is inserted into the environment e of variables, cf. Fig. 3, so we can refer to it by u_0.

Due to the bad complexity of the qe. problem [23], we must constantly care about efficiency in our implementation. In our experience, maintaining suitable normal forms for terms and formulae and using optimized versions of the constructors is invaluable. It is an easy exercise (see e.g. [6,7]) to implement and verify a normalizer for τ-terms to have the form $a_{i_1} * u_{i_1} + \ldots a_{i_k} * u_{i_k} + \widetilde{a : b}$, where for all $j \in \{1 \ldots k\}$ a_{i_j} is ρ-normalized, $a_{i_j} \neq \mathbf{0}_\rho$ and $i_1 < \cdots < i_k$. Moreover, we implement a simplifying version of every constructor of ϕ, e.g. for $\wedge$:

$$p \underline{\wedge} q = \text{if } p = F \vee q = F \text{ then } F$$
$$\text{else if } p = T \text{ then } q \text{ else if } q = T \text{ then } p \text{ else } p \wedge q$$

datatype $\tau = \widetilde{\rho} \mid u_N \mid - \tau \mid \tau + \tau \mid \tau - \tau \mid \rho * \tau$
datatype $\phi = T \mid F \mid \tau = \tau \mid \tau \neq \tau \mid \tau < \tau \mid \tau \leq \tau$
$\mid \neg \phi \mid \phi \wedge \phi \mid \phi \vee \phi \mid \phi \rightarrow \phi \mid \phi \leftrightarrow \phi \mid \exists \phi \mid \forall \phi$

$$
\begin{aligned}
(\!|c|\!)_\pi^e &= (\!|c|\!)_\rho^\pi & (\!|T|\!)_\pi^e &= \textit{True} \\
(\!|u_n|\!)_\pi^e &= e_{[n]} & (\!|F|\!)_\pi^e &= \textit{False} \\
(\!|-t|\!)_\pi^e &= -(\!|t|\!)_\pi^e & (\!|t \bowtie s|\!)_\pi^e &= ((\!|t|\!)_\pi^e \bowtie (\!|s|\!)_\pi^e) \\
(\!|t+s|\!)_\pi^e &= (\!|t|\!)_\pi^e + (\!|s|\!)_\pi^e & (\!|\neg p|\!)_\pi^e &= (\neg(\!|p|\!)_\pi^e) \\
(\!|t-s|\!)_\pi^e &= (\!|t|\!)_\pi^e - (\!|s|\!)_\pi^e & (\!|p \Diamond q|\!)_\pi^e &= ((\!|p|\!)_\pi^e \Diamond (\!|q|\!)_\pi^e) \\
(\!|c*t|\!)_\pi^e &= (\!|c|\!)_\rho^\pi \cdot (\!|t|\!)_\pi^e & (\!|\exists p|\!)_\pi^e &= (\exists x.(\!|p|\!)_\pi^{x \cdot e}) \\
& & (\!|\forall p|\!)_\pi^e &= (\forall x.(\!|p|\!)_\pi^{x \cdot e})
\end{aligned}
$$

Fig. 3. Semantics of parametric linear arithmetic formulae

$$
\begin{aligned}
\text{lift } qe \ (\forall p) &= \neg qe(\text{lift } qe \ (\neg p)) \\
\text{lift } qe \ (\exists p) &= qe(\text{lift } qe \ p) \\
\text{lift } qe \ (\neg p) &= \neg(\text{lift } qe \ p) \\
\text{lift } qe \ (p \bowtie q) &= (\text{lift } qe \ p) \bowtie (\text{lift } qe \ p) \text{ for } \bowtie \in \{\wedge, \vee, \rightarrow, \leftrightarrow\} \\
\text{lift } qe \ p &= \text{simp}_\phi \ p
\end{aligned}
$$

Fig. 4. Generic quantifier elimination for ϕ-formulae

To avoid cumbersome notation, every occurrence of any constructor in the rest of the paper, except in pattern matching, represents its optimized version. This means that e.g. $(\widetilde{4 : 3 < 0_\rho}) = \boldsymbol{F}$ and $(p \vee \boldsymbol{T}) = \boldsymbol{T}$ hold. Using the simplifying constructors we obtain a simple simplification procedure simp_ϕ (not shown).

Now that we have the syntax of formulae at hand, we define a recursive predicate qfree :: $\phi \Rightarrow \mathbb{B}$ which holds for p exactly when p contains neither $\exists$ nor $\forall$. The definition is very simple and omitted. Let $\text{isqe}_\phi \ qe$ for any $qe :: \phi \Rightarrow \phi$ be a shorthand for $\forall e, \pi, p.\text{qfree}(qe \ p) \wedge (\!(qe \ p)\!)^e_\pi \leftrightarrow (\!(p)\!)^e_\pi$, i.e. qe is a qep. for ϕ-formulae. Our goal is to define a function fr such that isqe_ϕ fr holds.

To formalize the argument that in order to eliminate all quantifiers, we only need qe a qep. for $\exists p$ where p is qf., we define lift in Fig. 4 to apply qe recursively to all quantifiers from innermost to outermost. Thereby, we reduce $\forall p$ to $\neg \exists \neg p$. Let $\text{isqe}_\exists \ qe$ formalize that qe is a qep. for $\exists p$, where p is qf., i.e. its is shorthand for $\forall e, \pi, p.\text{qfree} \ p \rightarrow \text{qfree}(qe \ p) \wedge (\!(qe \ p)\!)^e_\pi \leftrightarrow (\!(p)\!)^e_\pi$. We prove (19) automatically. In § 5 we formalize $\text{fr}_\exists$ and prove $\text{isqe}_\exists \ \text{fr}_\exists$.

$$
\text{isqe}_\exists \ qe \rightarrow \text{isqe}_\phi \ (\text{lift } qe) \tag{19}
$$

Given qe satisfying $\text{isqe}_\exists \ qe$, it is crucial for efficiency to apply it to formulae as small as possible, using the following rules for the existential quantifier:

$$
(\exists x.P \ x \vee Q \ x) \leftrightarrow ((\exists x.P \ x) \vee (\exists x.Q \ x)) \tag{20}
$$

$$
(\exists x.Q \wedge P \ x) \leftrightarrow (Q \wedge \exists x.P \ x) \tag{21}
$$

For this purpose, we introduce functions $\text{split}_\wedge$ and $\text{split}_\vee$ to return all conjuncts and disjuncts of a formula, respectively. Similarly, $\text{list}_\wedge$ and $\text{list}_\vee$ turn a list of formulae into their conjunction and disjunction, respectively. Given $f :: \phi \Rightarrow \phi$ and formulae $p_1, \ldots p_n$, then $\text{eval}_\vee \ f \ [p_1, \ldots, p_n]$ returns $f \ p_1 \vee \ldots f \ p_n$ evaluated lazily. If f is a qep. for one $\exists$, then $\lambda p.\text{eval}_\vee \ f \ (\text{split}_\vee \ p)$ distributes f over disjunctions, and is therefore a qep. for one $\exists$. Let $p :: \phi$ be qf. and $ps = \text{split}_\wedge \ p$. Function $\text{eval}_\wedge$ first partitions ps into two list: as consist of all formulae not involving the bound variable and bs consists of the rest. The call $\text{eval}_\wedge \ f \ p$ then returns $\text{decr}_\phi \ (\text{list}_\wedge \ as) \wedge f \ (\text{list}_\wedge \ bs)$. If f is a qep. as above, then $\text{eval}_\wedge \ f$ applies f only to the conjuncts not involving the bound variable. Summed up, given a qep. qe, then $\text{eval}_\vee \ (\text{eval}_\wedge \ qe)$ is an optimized qep. satisfying our purpose. By defining $\text{qelim} = \lambda qe.\text{lift} \ (\text{eval}_\vee \ (\text{eval}_\wedge \ qe))$ we obtain an optimized version of lift.

Concretely, we prove the following:

$$\mathsf{isqe}_\exists \; qe \to \mathsf{isqe}_\exists \; (\mathsf{eval}_\vee \; qe) \tag{22}$$

$$\mathsf{isqe}_\exists \; qe \to \mathsf{isqe}_\exists \; (\mathsf{eval}_\wedge \; qe) \tag{23}$$

$$\mathsf{isqe}_\exists \; qe \to \mathsf{isqe}_\phi \; (\mathsf{qelim} \; qe) \tag{24}$$

5 The Formalized Procedure

In this section we formalize the algorithm in § 2. In the following we write $\mathbf{0}_\tau$ as a shorthand for $\widetilde{\mathbf{0}_\rho}$. Let $\mathsf{unbound}_\tau \; t$ (resp. $\mathsf{unbound}_\phi \; p$) formalize that the τ-term t (resp. ϕ-formula p) does not contain $\mathbf{u}_0$ and $\mathsf{decr}_\phi \; p$ be p with all de Bruijn indices $\mathbf{u}_n$ decremented. These are simple recursive functions that satisfy:

$$\mathsf{unbound}_\tau \; t \to \forall x, y.(\!|t|\!)_\pi^{x \cdot e} = (\!|t|\!)_\pi^{y \cdot e} \tag{25}$$

$$\mathsf{unbound}_\phi \; p \to \forall x, y.(\!|p|\!)_\pi^{x \cdot e} \leftrightarrow (\!|p|\!)_\pi^{y \cdot e} \wedge \forall x.(\!|\mathsf{decr}_\phi \; p|\!)_\pi^e \leftrightarrow (\!|p|\!)_\pi^{x \cdot e} \tag{26}$$

Let a ϕ-formula p be linear ($\mathsf{islin}_\phi \; p$) if it does not involve $\mathbf{u}_0$ ($\mathsf{unbound}_\phi \; p$) or has the form $f \lozenge g$, for linear f and g and $\lozenge \in \{\wedge, \vee\}$, or $c * \mathbf{u}_0 + r \bowtie \mathbf{0}_\tau$, for $\bowtie \in \{<, \leq, =, \neq\}$, a normalized polynomial $c \neq \mathbf{0}_\rho$ and $r :: \tau$ not involving $\mathbf{u}_0$. Any qf. ϕ-formula is transformed into an equivalent linear ϕ-formula by lin_ϕ. We do not present this in detail, but see [4,7]. The important property is

$$\mathsf{qfree} \; p \to \mathsf{islin}_\phi \; (\mathsf{lin}_\phi \; p) \wedge (\!|\mathsf{lin}_\phi \; p|\!)_\pi^e \leftrightarrow (\!|p|\!)_\pi^e \tag{27}$$

Fig. 5 defines p_-, p_+, and U_p for a linear ϕ-formula p. They are analogs of $P_{-\infty}$, $P_{+\infty}$ and $\mathcal{U}_P$ in § 2 for $P = \lambda x.(\!|p|\!)_\pi^{x \cdot e}$, but encode the implicit dependency on the polynomial parameters into the resulting formula by explicit case distinction. Here we abuse notation and write $a \bowtie_1 b \bowtie_2 c$ for $a \bowtie_1 b \wedge b \bowtie_2 c$, for $\bowtie_i \in \{=, \neq, <, \leq\}$.

It is very easy to verify that p_- and p_+ do not depend on $\mathbf{u}_0$ and that they mimic p for values small (resp. large) enough in the underlying ordered field.

$$\mathsf{islin}_\phi \; p \to \mathsf{unbound}_\phi \; p_- \wedge \mathsf{unbound}_\phi \; p_+ \tag{28}$$

$$\mathsf{islin}_\phi \; p \to \exists z. \forall x < z.(\!|p_-|\!)_\pi^{x \cdot e} \leftrightarrow (\!|p|\!)_\pi^{x \cdot e} \tag{29}$$

$$\mathsf{islin}_\phi \; p \to \exists z. \forall x > z.(\!|p_+|\!)_\pi^{x \cdot e} \leftrightarrow (\!|p|\!)_\pi^{x \cdot e} \tag{30}$$

$$\mathsf{islin}_\phi \; p \to \forall (t, c) \in \{\!\{\mathsf{U}_p\}\!\}.\mathsf{unbound}_\tau \; t \wedge \mathsf{ishorn} \; c \wedge c \neq \mathbf{0}_\rho \tag{31}$$

p	U_p	p_-	p_+
$q \lozenge r$	$\mathsf{U}_q @ \mathsf{U}_r$	$q_- \lozenge r_-$	$q_+ \lozenge r_+$
$c * \mathbf{u}_0 + t = \mathbf{0}_\rho$	$[(t, c)]$	$c = t = \mathbf{0}_\rho$	$c = t = \mathbf{0}_\rho$
$c * \mathbf{u}_0 + t \neq \mathbf{0}_\rho$	$[(t, c)]$	$c = \mathbf{0}_\rho \neq t \vee c \neq \mathbf{0}_\rho$	$c = \mathbf{0}_\rho \neq t \vee c \neq \mathbf{0}_\rho$
$c * \mathbf{u}_0 + t < \mathbf{0}_\rho$	$[(t, c)]$	$t < \mathbf{0}_\rho = c \vee \mathbf{0}_\rho < c$	$t < \mathbf{0}_\rho = c \vee c < \mathbf{0}_\rho$
$c * \mathbf{u}_0 + t \leq \mathbf{0}_\rho$	$[(t, c)]$	$t \leq \mathbf{0}_\rho = c \vee \mathbf{0}_\rho \leq c$	$t \leq \mathbf{0}_\rho = c \vee c \leq \mathbf{0}_\rho$
-	$[\,]$	p	p

Fig. 5. Definition of U_p, p_- and p_+

A proof similar to § 2 yields the reflection of Theorem 1:

$$\text{islin}_\phi\ p \to \exists x. (\!(p)\!)^{x\cdot e}\pi \leftrightarrow (\!(p_- \lor p_+)\!)^{x\cdot e}_\pi \lor \exists ((t,c),(s,d)) \in \{\mathsf{U}_p\}^2. (\!(p)\!)^{(\frac{(\!(t)\!)^{y\cdot e}_\pi}{-2\cdot(\!(c)\!)^\pi_\rho} + \frac{(\!(s)\!)^{y\cdot e}_\pi}{-2\cdot(\!(d)\!)^\pi_\rho})\cdot e}_\pi \tag{32}$$

5.1   A First Naive Implementation

For a full implementation only a modified substitution $p[\frac{-t}{2\cdot c} + \frac{-s}{2\cdot d}]$ of the "expression" $\frac{t}{-2\cdot c} + \frac{s}{-2\cdot d}$ for u_0 in p satisfying (33) is missing. We use the same technique as for p_- and p_+ and encode case splits on parameters into the result, see Fig. 6. Recall that p_- and p_+ are modified substitutions of very large values.

$$\text{islin}_\phi\ p \land \text{unbound}_\tau\ t \land \text{unbound}_\tau\ s \to$$

$$(\!(p[\frac{-t}{2\cdot c} + \frac{-s}{2\cdot d}])\!)^{x\cdot e}_\pi \leftrightarrow (\!(p)\!)^{(\frac{(\!(t)\!)^{x\cdot e}_\pi}{-2\cdot(\!(c)\!)^\pi_\rho} + \frac{(\!(s)\!)^{x\cdot e}_\pi}{-2\cdot(\!(d)\!)^\pi_\rho})\cdot e}_\pi \land \text{unbound}_\phi\ p[\frac{-t}{2\cdot c} + \frac{-s}{2\cdot d}] \tag{33}$$

The proof of (33) is only interesting for atoms and we show only the case of $(a * u_0 + r < 0_\tau)[\frac{-t}{2\cdot c} + \frac{-s}{2\cdot d}]$. The $\le$-case is analogous and the $=$ and $\neq$ are even simpler. For this, fix x and environments e and π. Clearly there are 9 disjoint cases depending on the strict sign of $(\!(c)\!)^\pi_\rho$ and $(\!(d)\!)^\pi_\rho$. These are exactly the cases encoded in Fig. 6. Assume $(\!(c \circledast d)\!)^\pi_\rho > 0$, then $(\!(c)\!)^\pi_\rho \neq 0 \land (\!(d)\!)^\pi_\rho \neq 0$ and hence $\frac{(\!(t)\!)^{x\cdot e}_\pi}{-2\cdot(\!(c)\!)^\pi_\rho} + \frac{(\!(s)\!)^{x\cdot e}_\pi}{-2\cdot(\!(d)\!)^\pi_\rho} = -\frac{(\!(d*t+c*s)\!)^{x\cdot e}_\pi}{(\!(2_r \circledast c \circledast d)\!)^\pi_\rho}$. The claim now follows using the property $\forall a,b,c.b > 0 \to \frac{a}{b} < c \leftrightarrow a < c \cdot b$ and simple algebraic manipulations. The other cases are similar.

Finally, we implement $\text{fr}_\exists$ to eliminate one $\exists$, and fr the full qep. below. The function call $\text{eval}_\lor\ f\ [x_1, .., x_n]$ returns the disjunction $f\ x_1 \lor \ldots \lor f\ x_n$ lazily evaluated. We prove the main qe. theorem in (40)

$$\text{fr}_\exists\ q = \textbf{let}\ p = \text{lin}_\phi\ q\ ;\ U = \text{allpairs}(\text{remdups}\ (\mathsf{U}_p))$$
$$\textbf{in}\ \text{decr}_\phi (p_- \lor p_+ \lor \text{eval}_\lor\ (\lambda((t,c),(s,d)).q[\frac{-t}{2\cdot c} + \frac{-s}{2\cdot d}])\ U)$$

$$\text{fr}\quad = \text{qelim}\ \text{fr}_\exists$$

$$\text{qfree}\ (\text{fr}\ p) \land (\!(\text{fr}\ p)\!)^e_\pi \leftrightarrow (\!(p)\!)^e_\pi \tag{34}$$

5.2   Drawbacks and a Better Solution

By inspecting Fig. 6, we predict fr yields huge formulae and is not practical. Our tests corroborate this prediction. Fig. 6 shows *many* duplications of case splits. Keeping in mind that we substitute the same "fraction", the conditions on the coefficients involved in that fraction must not be encoded *at the atoms level* but rather globally. We present in the following an alternative substitution and procedure to achieve this goal.

Let us first reconsider the qe. theorem (32) and in particular the substitution of the fraction on the RHS. Let p be linear and let (t,c) and (s,d) be two

p	$p[\frac{-t}{2\cdot c} + \frac{-s}{2\cdot d}]$
$p \wedge q$	$p[\frac{-t}{2\cdot c} + \frac{-s}{2\cdot d}] \wedge q[\frac{-t}{2\cdot c} + \frac{-s}{2\cdot d}]$
$p \vee q$	$p[\frac{-t}{2\cdot c} + \frac{-s}{2\cdot d}] \vee q[\frac{-t}{2\cdot c} + \frac{-s}{2\cdot d}]$
$a * u_0 + r = 0_\tau$	$\tilde c = s = r = 0_\tau \vee$
	$\tilde c \neq 0_\tau = \tilde d \wedge a * t = (\widehat{2_\tau} \circledast c) * r \vee$
	$\tilde d \neq 0_\tau = \tilde c \wedge a * s = (\widehat{2_\tau} \circledast d) * r \vee$
	$\widetilde{c \circledast d} \neq 0_\tau \wedge a * (d * t + c * s) = (\widehat{2_\tau} \circledast c \circledast d) * r$
$a * u_0 + r < 0_\tau$	$\tilde c = \tilde d = r = 0_\tau \vee$
	$\tilde d < 0_\tau = c \wedge a * s < (\widehat{2_\tau} \circledast d) * r \vee$
	$c = 0_\tau < \tilde d \wedge (\widehat{2_\tau} \circledast d) * r < a * s \vee$
	$\tilde c < 0_\tau = d \wedge a * t < (\widehat{2_\tau} \circledast c) * r \vee$
	$d = 0_\tau < \tilde c \wedge (\widehat{2_\tau} \circledast c) * r < a * t \vee$
	$0_\tau < \widetilde{c \circledast d} \wedge (\widehat{2_\tau} \circledast c \circledast d) * r < a * (d * t + c * s) \vee$
	$\widetilde{c \circledast d} < 0_\tau \wedge a * (d * t + c * s) < (\widehat{2_\tau} \circledast c \circledast d) * r$
$a * u_0 + r \leq 0_\tau$	$\ldots$
$a * u_0 + r \neq 0_\tau$	$\ldots$
p	p

Fig. 6. Modified substitution in ϕ-formulae

elements of U_p. Furthermore fix environments e and π and let $P = \lambda x.(p)_\pi^{x \cdot e}$ and $\bar t, \bar s, \bar c$ and $\bar d$ denote $(t)_\pi^{y \cdot e}$, $(s)_\pi^{y \cdot e}$, $(c)_\rho^\pi$ and $(d)_\rho^\pi$ respectively. Our goal is to construct a ϕ-formula semantically equivalent to $P(\frac{\bar t}{-2\cdot \bar c} + \frac{\bar s}{-2\cdot \bar d})$ but without case splits on c and d at the atoms-level. For that consider all sign combinations of $\bar c$ and $\bar d$. If both are zero then we have $P(0)$. If exactly one is zero, say $\bar d$, then we have $P(\frac{\bar t}{-2\cdot \bar c})$ and the whole disjunction (for only such cases) reduces to $\exists (t, c) \in \{U_p\}.\bar c \neq 0 \wedge P(\frac{\bar t}{-2\cdot \bar c})$. For the last case we have $\bar c \neq 0 \neq \bar d$ and hence $\frac{\bar t}{-2\cdot \bar c} + \frac{\bar s}{-2\cdot \bar d} = \frac{\bar d \cdot \bar t + \bar c \cdot \bar s}{-2\cdot \bar c \cdot \bar d}$. Hence only two case splits on the strict sign of the denominator are necessary to obtain a simpler substitution. Summed up, we prove the following qe. theorem, where $p[a]_\phi$ denotes the "normal" substitution of term a for u_0 in p:

$$\text{islin}_\phi\ p \rightarrow (\exists p)_\pi^e \leftrightarrow (p_- \vee p_+ \vee p[0_\tau]_\phi)_\pi^{y \cdot e} \vee \exists (t, c) \in \{U_p\}. (c)_\rho^\pi \neq 0 \wedge (p)_\pi^{\frac{(t)_\pi^{y \cdot e}}{-2\cdot(c)_\rho^\pi} \cdot e} \vee$$

$$\exists ((t, c), (s, d)) \in \{U_p\}^2. (c)_\rho^\pi \neq 0 \wedge (d)_\rho^\pi \neq 0 \wedge (p)_\pi^{\frac{(d*t+c*s)_\pi^{y \cdot e}}{-2\cdot(c*d)_\rho^\pi} \cdot e}$$

$$(35)$$

Now we only need to find a substitution $p[\frac{t}{c}]_\phi^{\neq}$ of a "fraction" $\frac{t}{c}$ with "non-zero" denominator. Our substitution first splits over the strict sign of c and then performs two modified substitutions of $\frac{t}{c}$: the first $p[\frac{t}{c}]_\phi^{>}$ assumes $\bar c > 0$, and the second $p[\frac{t}{c}]_\phi^{<}$ assumes $\bar c < 0$. The definition of $p[\frac{t}{c}]_\phi^{\neq}$ (cf. (36)) and that of $p[\frac{t}{c}]_\phi^{>}$ for atoms (cf. (37)) are simple. The definition of $p[\frac{t}{c}]_\phi^{<}$ is analogous and omitted. It is not hard to prove the main properties in (38).

$$p[\tfrac{t}{c}]^{\neq}_{\phi} = c < \mathbf{0}_\tau \wedge p[\tfrac{t}{c}]^{<}_{\phi} \vee c > \mathbf{0}_\tau \wedge p[\tfrac{t}{c}]^{>}_{\phi} \qquad (36)$$

$$(a * \boldsymbol{u}_0 + b \bowtie \mathbf{0}_\tau)[\tfrac{t}{c}]^{>}_{\phi} = a * t + c * b \bowtie \mathbf{0}_\tau \quad \text{for } \bowtie \in \{=, <, \leq\} \qquad (37)$$

$$\mathsf{islin}_\phi\, p \wedge (\!(c)\!)^{\pi}_{\rho} \bowtie 0 \rightarrow (\!(p[\tfrac{t}{c}])\!)^{x\cdot e}_{\pi} \leftrightarrow (\!(p)\!)^{\frac{(\!(t)\!)^{x\cdot e}_{\pi}}{(\!(c)\!)^{\pi}_{\rho}}\cdot e}_{\pi} \quad \text{for } \bowtie \in \{>, <, \neq\}. \qquad (38)$$

Now we implement a new version of $\mathsf{fr}_\exists$ and define $\mathsf{fr} = \mathsf{qelim}\ \mathsf{fr}_\exists$. We prove (39) and (40) using the previous properties and (35).

$$\begin{aligned}
\mathsf{fr}_\exists\, q = \ &\mathbf{let}\ p = \mathsf{lin}_\phi\, q\ ; U = \mathsf{remdups}\ (U_p);\ U_2 = \mathsf{allpairs}\ U \\
&\mathbf{in}\ \mathsf{decr}_\phi(p_- \vee\ p_+ \vee p[\mathbf{0}_\tau]_\phi \vee \mathsf{eval}_\vee\ (\lambda(t,c).q[\tfrac{t}{-2_\rho \circledast c}]_\phi)\ U \\
&\vee \mathsf{eval}_\vee\ (\lambda((t,c),(s,d)).q[\tfrac{d*t+c*s}{-2_\rho \circledast c \circledast d}]_\phi)\ U_2)
\end{aligned}$$

$$\mathsf{qfree}\ q \rightarrow \mathsf{qfree}\ (\mathsf{fr}_\exists\, q) \wedge ((\!(\mathsf{fr}_\exists\, q)\!)^e_\pi \leftrightarrow (\!(\exists\, q)\!)^e_\pi) \qquad (39)$$

$$\mathsf{qfree}\ (\mathsf{fr}\, p) \wedge ((\!(\mathsf{fr}\, p)\!)^e_\pi \leftrightarrow (\!(p)\!)^e_\pi) \qquad (40)$$

6   Integration

So far we can apply fr to any $p :: \phi$ to obtain its qf.-equivalent. For the evaluation of $\mathsf{fr}\ p$ we can use Isabelle's rewriting facility, but thus is intractable in practice. Alternatively we can use fast evaluation techniques based on normalization by evaluation (cf. [16]) or extract an implementation into an external programming language. The framework in [10] allows extraction into SML, Haskell and OCaml. SML is more appealing for us, since it is the implementation language of Isabelle. Furthermore, we want to apply fr to HOL formulae, i.e. of type $\mathbb{B}$ and not ϕ. For this we must transform HOL formulae into their ϕ-representation. This is called *reification* and is performed in SML. Given $P :: \mathbb{B}$, we compute $p :: \phi$ and environments e and π and then *prove* that $P \leftrightarrow (\!(p)\!)^e_\pi$ holds. This ensures that we have guessed p properly. Using (40) we replace $(\!(p)\!)^e_\pi$ by $(\!(\mathsf{fr}\ p)\!)^e_\pi$ and evaluate $\mathsf{fr}\ p$ efficiently to say q. Now, and this is a non-trivial step, we assume that $\mathsf{fr}\ p = q$ holds in HOL, i.e. our evaluation have simulated a proof inside HOL. This is the only proof-step which is not performed by means of the logic but rather by meta-theory. In Coq for instance, such a reasoning could indeed be performed without leaving the logic, using e.g. the ι-reduction rule. Now that we have $\mathsf{fr}\ p = q$ its is easy to obtain a qf.-equivalent formula to P, i.e. $(\!(q)\!)^e_\pi$.

Consider for instance the HOL formula P which states $\forall x, y.(1 - t) \cdot x \leq (1 + t) \cdot y \wedge (1 - t) \cdot y \leq (1 + t) \cdot x \rightarrow 0 \leq y$. Reification finds that p, e and π should be $\forall \forall (\mathbf{1}_\rho - v_0) * \boldsymbol{u}_1 \leq (\mathbf{1}_\rho + v_0) * \boldsymbol{u}_0 \wedge (\mathbf{1}_\rho - v_0) \cdot \boldsymbol{u}_0 \leq (\mathbf{1}_\rho + v_0) * \boldsymbol{u}_1 \rightarrow \mathbf{0}_\tau \leq \boldsymbol{u}_0$, [] and $[t]$, respectively. We prove that $P \leftrightarrow (\!(p)\!)^{[]}_{[t]}$ holds and compute $\mathsf{fr}\ p$, which is quite large to be included here. This whole computation takes 0.17 seconds. The computation of qe. equivalent to the Collins/Jones problem (cf. (*)) takes 2.1 seconds. All timings are take on a PowerPC G4 1.67 GHz running OS X with 1.5 GB of memory.

$$\exists r.0 < r < 1 \wedge 0 < (2 - 3 \cdot r) \cdot (a^2 + b^2) + 2 \cdot a \cdot r$$
$$\wedge (2 - 3 \cdot r) \cdot (a^2 + b^2) + 4 \cdot a \cdot r - 2 \cdot a - r < 0 \tag{*}$$

7 Conclusion

We have presented a formalized qep. for $\mathcal{F}^{\rho}_{+}$ in Isabelle/HOL, based on type classes [24]. It is hence generic since it applies to e.g. $\mathbb{Q}, \mathbb{R}$ and the nonstandard reals without extra effort. The integration using reflection reduces deduction to computation. Our formalization in § 3.1 needed 1000 lines and that in § 3.2 1700 lines of Isabelle Proofs. The full qep. needed further 3000 lines of proofs and round 140 lines of SML code for the reification and the tactic.

Acknowledgment. I thank Jeremy Avigad, John Harrison and Tobias Nipkow for helpful discussions and suggestions. I also thank the anonymous referees for constructive suggestions.

References

1. Ballarin, C.: Locales and locale expressions in Isabelle/Isar. In: Berardi, S., et al. (eds.) TYPES 2003. LNCS, vol. 3085. Springer, Heidelberg (2004)
2. Bergstra, J.A., Tucker, J.V.: The rational numbers as an abstract data type. J. ACM 54(2) (2007)
3. Boulm, S., Hardin, T., Rioboo, R.: Some hints for polynomials in the Foc project. In: Linton, S., Sebastiani, R. (eds.) CALCUMEUS 2001, 9th Symposium on the Integration of Symbolic Computation and Mechanized Reasoning, Siena, Italy, pp. 142–154 (June 2001)
4. Chaieb, A.: Verifying mixed real-integer quantifier elimination. In: Furbach, U., Shankar, N. (eds.) IJCAR 2006. LNCS (LNAI), vol. 4130, pp. 528–540. Springer, Heidelberg (2006)
5. Chaieb, A.: Automated methods for formal proofs in simple arithmetics and algebra. PhD thesis, Technische Universität München, Germany (April 2008)
6. Chaieb, A., Nipkow, T.: Verifying and reflecting quantifier elimination for Presburger arithmetic. In: Sutcliffe, G., Voronkov, A. (eds.) LPAR 2005. LNCS (LNAI), vol. 3835, pp. 367–380. Springer, Heidelberg (2005)
7. Chaieb, A., Nipkow, T.: Proof Synthesis and Reflection for Linear Arithmetic. J. of Aut. Reasoning (to appear, 2008)
8. Chaieb, A., Wenzel, M.: Context aware calculation and deduction — ring equalities via Gröbner Bases in Isabelle. In: Kauers, M., Kerber, M., Miner, R., Windsteiger, W. (eds.) MKM/CALCULEMUS 2007. LNCS (LNAI), vol. 4573, pp. 27–39. Springer, Heidelberg (2007)
9. Ferrante, J., Rackoff, C.: A decision procedure for the first order theory of real addition with order. SIAM J. of Computing 4(1), 69–76 (1975)
10. Haftmann, F., Nipkow, T.: A code generator framework for Isabelle/HOL. In: Schneider, K., Brandt, J. (eds.) TPHOLs 2007. LNCS, vol. 4732. Springer, Heidelberg (2007)

11. Harrison, J.: Theorem proving with the real numbers. PhD thesis, University of Cambridge, Computer Laboratory (1996)
12. Harrison, J.: Complex quantifier elimination in HOL. In: Boulton, R.J., Jackson, P.B. (eds.) TPHOLs 2001. LNCS, vol. 2152, pp. 159–174. Springer, Heidelberg (2001)
13. Harrison, J.: Automating elementary number-theoretic proofs using Gröbner bases. In: Pfenning, F. (ed.) CADE 2007. LNCS (LNAI), vol. 4603, pp. 51–66. Springer, Heidelberg (2007)
14. Harrison, J.: Verifying nonlinear real formulas via sums of squares. In: Schneider, K., Brandt, J. (eds.) TPHOLs 2007. LNCS, vol. 4732, pp. 102–118. Springer, Heidelberg (2007)
15. Hodges, W.: Model Theory. Encyclopedia of Mathematics and its Applications, vol. 42. Cambridge University Press, Cambridge (1993)
16. Aehlig, K., Haftmann, F., Nipkow, T.: A compiled implementation of normalization by evaluation (submitted)
17. Kammüller, F., Wenzel, M., Paulson, L.C.: Locales: A sectioning concept for Isabelle. In: Bertot, Y., Dowek, G., Hirschowitz, A., Paulin, C., Théry, L. (eds.) TPHOLs 1999. LNCS, vol. 1690. Springer, Heidelberg (1999)
18. Loos, R., Weispfenning, V.: Applying linear quantifier elimination. Computer Journal 36(5), 450–462 (1993)
19. Mahboubi, A.: Contributions à la certification des calculs sur ℝ: théorie, preuves, programmation. PhD thesis, Univ. Nice Sophia-Antipolis (2006)
20. McLaughlin, S., Harrison, J.: A Proof-Producing Decision Procedure for Real Arithmetic. In: Nieuwenhuis, R. (ed.) CADE 2005. LNCS (LNAI), vol. 3632, pp. 295–314. Springer, Heidelberg (2005)
21. Norrish, M.: Complete integer decision procedures as derived rules in HOL. In: Basin, D., Wolff, B. (eds.) TPHOLs 2003. LNCS, vol. 2758, pp. 71–86. Springer, Heidelberg (2003)
22. Prevosto, V., Doligez, D.: Algorithms and proof inheritance in the foc language. Journal of Automated Reasoning 29(3-4), 337–363 (2002)
23. Weispfenning, V.: The complexity of linear problems in fields. J. of Symb. Comp. 5(1/2), 3–27 (1988)
24. Wenzel, M.: Type classes and overloading in higher-order logic. In: Gunter, E.L., Felty, A.P. (eds.) TPHOLs 1997. LNCS, vol. 1275. Springer, Heidelberg (1997)

A Global Workspace Framework for Combining Reasoning Systems

John Charnley and Simon Colton

Combined Reasoning Group, Department of Computing, Imperial College, London
jwc04@doc.ic.ac.uk, sgc@doc.ic.ac.uk
http://www.doc.ic.ac.uk/crg

1 Introduction

Stand-alone Artificial Intelligence systems for performing specific types of reasoning – such as automated theorem proving and symbolic manipulation in computer algebra systems – are numerous, highly capable and constantly improving. Moreover, systems which combine various forms of reasoning have repeatedly been shown to be more effective than stand-alone systems. For example, the ICARUS system for reformulating constraint satisfaction problems [1] and the HOMER system for conjecture making in number theory [2]. However, in general, such combinations have been ad-hoc in nature and designed with a specific task in mind. With little general design consideration or a suitable framework for combining reasoning, in general every new combination has to be built from scratch and the resulting system is often inflexible and difficult to manage. We believe it is imperative that generic frameworks are developed if the field of combining reasoning systems is to progress. Such generic frameworks would provide standardised rule sets and toolkits to simplify the development of combined systems.

We describe here a generic framework based on the cognitive science theory of the Global Workspace Architecture [3]. In our framework, the individual reasoning techniques are each encapsulated within specialist processes attached to a blackboard-style global workspace, which is visible to all processes. We achieve relative simplicity in the framework by requiring fairly severe restrictions upon the behaviour of the attached processes. In particular, there is no inter-process communication other than what is broadcast on the global workspace. These restrictions help ensure that the resulting system is simple to understand. Furthermore, the encapsulation of reasoning techniques within discrete individual processes adds clarity and flexibility. We explain our framework, and how it is used, in §2. To demonstrate the capability of the framework, we have implemented combined systems incorporating Prover9 [4], Maple [5] and SICStus Prolog. In §3, we describe applications to mathematical theorem discovery and conjecture making which produce results comparable to the ICARUS and HOMER systems, respectively. This demonstrates that while the framework is easy to use, it is as powerful as the ad-hoc systems.

2 A Framework for Combining Reasoning Systems

The architecture defined by our framework is inspired by the Global Workspace Architecture [3]. Each of the processes attached to the global workspace performs either

S. Autexier et al. (Eds.): AISC/Calculemus/MKM 2008, LNAI 5144, pp. 261–265, 2008.

some type of reasoning (e.g., by encapsulating a theorem prover or a computer algebra system) or a useful administrative task such as checking for redundancy in outputs. The framework defines how processing takes place on a round-by-round basis. In addition, it outlines rules which all attached processes must follow. A round starts with the broadcast of some reasoning *artefact* (e.g., a conjecture, proof, example, etc.) which each attached process may ignore or may react to in various ways. Specifically, a process may do one or more of the following:

- Construct a *proposal* for broadcast, consisting of a reasoning artefact and a numerical (heuristic) value of importance that the process ascribes to that artefact.
- Detach itself from the framework.
- Attach new processes to the framework.

At the end of each round, various processes will have been added to and removed from the global workspace, and a set of broadcast proposals will have been submitted to the framework. At the start of the next round, the framework chooses the proposal with the highest importance value, and broadcasts the reasoning artefact from that proposal. In the case where multiple proposals have equal heuristic value, one is chosen from them randomly. All non-broadcast proposals are discarded and will not be considered for broadcast later unless they are re-proposed.

To create a combined system, a developer must create a *configuration* of the framework, by defining:

- The reasoning artefacts that may be broadcast on the workspace.
- The processes that may be attached to the workspace and their behaviour, which must conform to the framework rules. In particular, how each process reacts to broadcasts, the processing or reasoning they perform, the proposals they can make and the method they use in determining the heuristic rating of importance.
- The starting state, i.e. the initially attached processes.

We have developed the GC toolkit which enables developers to easily configure combinations of reasoning systems for particular tasks within the framework. GC, which takes its name from global-workspace and combining, allows users to develop their configurations into full system implementations. It includes the core code for the round-by-round processing and a number of pre-coded processes which encapsulate specific reasoning tasks. For example, the toolkit currently provides a process which appeals to the Prover9 theorem prover in attempts to prove broadcast conjectures. Users can choose and adapt processes from GC's pre-coded selection for use in their configurations or they can develop their own processes by with the aid of libraries provided in the toolkit.

3   Applications and Results

Our first configuration demonstrates how the framework can combine machine learning, example construction and theorem proving processes to perform automated theory formation, similar to that performed by the HR system [6]. In overview, the configuration is required to invent new concepts (built from a set of user-supplied background concepts), make empirical conjectures which relate the concepts and then prove that some of the conjectures follow from a set of user-supplied axioms. We specified four types of broadcast artefacts, as follows:

1. Definition, in the form **def**(D), where D is a prolog-readable definition of a concept.

2. Concept, in the form **conc**($D|E$), with D as above and E being a list of examples which satisfy that concept definition.

3. Conjecture, in the form **conj**($[D_1,D_2]|K$), where D_1 and D_2 are concept definitions and K is a keyword indicating the type of conjecture; either *im*, which denotes that D_1 is conjectured to imply D_2; or *eq*, denoting D_1 is conjectured to be equivalent to D_2.

4. Explanation, in the form **exp**($[D_1,D_2]|K|P$), where D_1, D_2 and K represent a conjecture, as above, and P is a proof of that conjecture.

Our initial configuration uses the following processes:

1. DefinitionFormer processes propose new *Definitions*. They each encapsulate a different concept formation method, akin to production rules in HR. They react to *Concept* broadcasts, **conc**($D|E$). Some formation methods involve modifying a single concept definition, where they attempt to create a new definition from D. Others combine two definitions, in which case they remember D, by spawning a clone process that reacts to *Concept* broadcast, **conc**($D'|E'|C'$), by attempting to combine D and D'.

2. DefinitionReviewer, which reacts to *Definition* broadcasts, **def**(D), removes redundancy by checking whether D has been seen before. If not, it proposes for broadcast **conc**($D|\emptyset$), i.e. a concept with that definition and an empty example set.

3. ExampleFinder, encapsulates a Prolog database containing examples for the initial background concepts. All concept definitions are Prolog terms and *ExampleFinder* can generate example sets for new concepts by querying Prolog with the definition. *ExampleFinder* reacts to *Concept* broadcasts with empty example sets, **conc**($D|\emptyset$), by generating an example set E. If E is non-empty it proposes **conc**($D|E$).

4. ConjectureMaker, compares the example sets of two *Concept* broadcasts. It reacts to the first *Concept* broadcast, **conc**($D_1|E_1$), (where $E_1 \neq \emptyset$), by spawning a clone process, P, which itself reacts to future *Concept* broadcasts **conc**($D_2|E_2$). In particular, if P finds that $E_1 = E_2$ it proposes **conj**($[D_1, D_2]$,*eq*). Alternatively, if $E_1 \subset E_2$, it proposes **conj**($[D_1, D_2]$,*im*) (or **conj**($[D_2, D_1]$,*im*) if $E_2 \subset E_1$).

5. Prover processes encapsulate the Prover9 theorem prover with axioms for the domain under investigation. It attempts to prove conjectures in any *Conjecture* broadcast, **conj**($[D_1, D_2]$,K), and proposes **exp**($[D_1, D_2]$,K)$|P$), whenever a proof, P, is found.

In addition for this configuration, we specified a process which proposes the background concepts at the start of the session. Moreover, we specified a simple rating scheme which assigns a rating of 1 to *Definitions*, 2 to *Concepts*, 3 to *Conjectures* and 4 to *Explanations*. We enhanced this configuration by preventing the dual development of empirically equivalent concepts (i.e. if **conj**($[D_1, D_2]$,*eq*) is broadcast, then D_2 is no longer considered) which reduces duplication of effort. We implemented the configuration using GC and used it to find implied constraints about QG-quasigroups similar to those found by HR embedded in the ICARUS [1] combined system. Working with QG3, QG4 and QG5 quasigroups, the configuration generated the same theorems as ICARUS. For example, it found the same three theorems, $\forall a\, b\, (a * a = b \leftrightarrow b * b = a)$, $\forall a\, b\, ((a * b = b * a) \rightarrow a = b)$ and $\forall a\, b\, ((a * a = b * b) \rightarrow a = b)$; which were all used by ICARUS in reformulating QG3 constraint programs.

To demonstrate the flexibility of the framework, we extended this initial configuration to applications in number theory similar to those performed by the HOMER system [2]. We introduced new processes encapsulating Maple to provide background solutions to number theory functions and used several prover processes, each with different axiom sets, to perform conjecture filtering. We used the background functions $\sigma(n)$ (the sum of divisors of a number), $\tau(n)$ (the number of divisors) and $isprime(n)$ (a boolean predicate indicating whether a number is prime), together with the notion of equality. Our configuration achieved results very similar to those produced by HOMER, in terms of the concepts and conjectures discovered. Like HOMER, our system filtered out approximately 90% of all the conjectures it created, by showing them to be simple consequences of the definitions and hence uninteresting. Importantly, our system re-discovered the most interesting results from [2], including e.g., $isprime(\sigma(a)) \rightarrow isprime(\tau(a))$. Moreover, our system highlighted potential weaknesses in HR, by showing that concepts had been repeated due to variable ordering.

4   Conclusions and Future Work

Compared to building an ad-hoc combined system from scratch – which is currently the norm – it is relatively straightforward to construct systems using our GWA-based framework. Despite the framework's restrictions, it can be configured to achieve results equivalent to previous bespoke ad-hoc systems. We will continue to develop the GC toolkit, by adding additional reasoning processes for tasks such as model generation and SAT-solving. In addition, we aim to develop the user interface by providing graphical tools for selecting and tuning processes and for specifying new processes without having to write code explicitly. We will continue to improve our configurations in efforts to improve upon previous system results, for example, by introducing more sophisticated rating schemes. We intend to create new configurations for existing combined reasoning tasks, such as correcting false conjectures [7] and algebra classification [8], and to tackle new problems with the framework. Furthermore, the core-processing of the toolkit will be enhanced to take advantage of the distributed parallel nature of the underlying architecture, which should enhance performance. We hope to have demonstrated the potential of our GW-based generic framework for combining reasoning systems and we hope in future to add to the weight of evidence that combining reasoning systems is imperative for the advancement of Artificial Intelligence.

References

1. Charnley, J., Colton, S., Miguel, I.: Automatic generation of implied constraints. In: Proceedings of ECAI (2006)
2. Colton, S.: Automated conjecture making in number theory using HR, Otter and Maple. Journal of Symbolic Computation 39(5), 593–615 (2004)
3. Baars, B.: A cognitive theory of consciousness. Cambridge University Press, Cambridge (1988)
4. McCune, W.: Prover9, http://www.cs.unm.edu/mccune/prover9/
5. Waterloo Maple. Maple Manual, http://www.maplesoft.on.ca
6. Colton, S.: Automated Theory Formation in Pure Mathematics. Springer, Heidelberg (2002)
7. Colton, S., Pease, A.: The TM system for repairing non-theorems. In: Proceedings of the IJCAR 2004 Disproving workshop (2004)
8. Colton, S., Meier, A., Sorge, V., McCasland, R.: Automatic generation of classification theorems for finite algebras. In: Basin, D., Rusinowitch, M. (eds.) IJCAR 2004. LNCS (LNAI), vol. 3097, pp. 400–414. Springer, Heidelberg (2004)

Effective Set Membership in Computer Algebra and Beyond

James H. Davenport

Department of Computer Science
University of Bath, Bath BA2 7AY, United Kingdom
J.H.Davenport@bath.ac.uk

Abstract. In previous work, we showed the importance of distinguishing "I know that $X \neq Y$" from "I don't know that $X = Y$". In this paper we look at effective set membership, starting with Gröbner bases, where the issues are well-expressed in algebra systems, and going on to integration and other questions of 'computer calculus'.

In particular, we claim that a better recognition of the role of set membership would clarify some features of computer algebra systems, such as 'what does an integral mean as output'.

1 Introduction

In [7] we discussed the various ideas of equality that can be found in computer algebra, and showed the importance of distinguishing "I know that $X \neq Y$" from "I don't know that $X = Y$". In this paper (a fuller version of which is in [8]) we look at effective set membership. While sets can be defined in a variety of ways, we will be interested in sets defined as

$$S := \{x \in A \mid P(x)\} \tag{1}$$

where A is a set for which membership is "obvious", e.g. by construction, and P is some predicate, which will generally involve some existential quantifiers. The problem of **effective set membership**, then, is the following problem.

Problem 1. *Given some $x \in A$, produce*

either an effective *[5]* proof of $P(x)$
or *a proof of $\neg P(x)$.*

In general, it is the second part of the problem that is the hard one.

2 Ideals and Gröbner Bases

The classic definition of an *ideal* $(p_1, \ldots, p_m)$ in $k[x_1, \ldots, x_n]$ as

$$(p_1, \ldots, p_m) = \left\{ \sum_{i=1}^{m} f_i p_i : f_i \in k[x_1, \ldots, x_n] \right\} \tag{2}$$

means that exhibiting the f_i becomes a proof of **either**. But how to do so, and how to prove **or**? So in this case problem 1 becomes the following.

S. Autexier et al. (Eds.): AISC/Calculemus/MKM 2008, LNAI 5144, pp. 266–269, 2008.
© Springer-Verlag Berlin Heidelberg 2008

Problem 2. *For given* $p_1, \ldots, p_m$ *and given* f

either *exhibit* $f_i \in k[x_1, \ldots, x_n]$ *such that* $f = \sum_{i=1}^{m} f_i p_i$
or *demonstrate that none such exist.*

We have, of course, the process of polynomial reduction.

Algorithm 1 (Polynomial Reduction). *[1, Algorithm REDPOL]*
Input: $f, p_1, \ldots, p_m \in k[x_1, \ldots, x_n]$, *a monomial order* $>$
Output: $\hat{f}, f_1, \ldots, f_m \in k[x_1, \ldots, x_n]$:
$\hat{f} = f - \sum_{i=1}^{m} f_i p_i$: $\hat{f}$ *irreducible by the* p_i *(with respect to* $>$)

Clearly, *if* this process terminates with $\hat{f} = 0$, we have proved the **either**, and
we have the f_i.

Theorem 1 (Buchberger [4]). *If the* p_i *are a Gröbner basis, then Algorithm
1 precisely solves problem 2, i.e.* $\hat{f} \neq 0$ *is a proof that* $f \notin (p_1, \ldots, p_m)$.

Since being a Gröbner basis is algorithmically testable, we have a complete
process for solving problem 2 *if* we are given a Gröbner basis. Furthermore,
Buchberger's algorithm lets us compute a Gröbner basis for any polynomial
ideal starting from *any* finite set of generators. We have come to expect this of
computer algebra systems, and would be rather surprised to see a system take
a set of polynomial equations and just say "I can't solve these".

3 Integration in Elementary Terms

The problem of (indefinite) integration is not normally viewed as a set membership problem, but it can be. We refer the reader to [3] for the standard definitions,
and we let $\mathcal{I}$ be some class of functions (elementary, Liouvillian, $\mathcal{EL}$ [14] etc.).
When we say "given an $\mathcal{I}$ function", we mean that it is given *effectively*, i.e. it is
given as a member of an effective field of $\mathcal{I}$ functions. Then the set of $\mathcal{I}$-integrable
functions is

$$\{f \mid \exists g \in \mathcal{I} \quad g' = f\}$$

and the $\mathcal{I}$-integration problem (as perceived since [11,13]) becomes

Problem 3. *For given* f *(normally* $f \in \mathcal{I}$)

either *exhibit* $g \in \mathcal{I}$ *such that* $f = g'$
or *demonstrate that no such* g *exists.*

It was not always thus: [15] essentially perceived the problem as

Problem 4. *For given* f

either *exhibit* $g \in \mathcal{I}$ *such that* $f = g'$
or *return* `failed` *(g might exist, but hadn't been found),*

and a successful program was one which did not return `failed` when a freshman could see the answer.

The shift from problem 4 to problem 3 was essentially due to the rediscovery of Liouville's Theorem [10], which, in the case $\mathcal{I}=$"elementary", reduced problem 3 to the following.

Problem 5. *For given f in a differential field K*

either *exhibit f as* $v_0' + \sum_{i=1}^{n} c_i \frac{v_i'}{v_i}$, *with* $v_0 \in K$, $c_i \in C = \overline{\{g \in K \mid g' = 0\}}$,
$v_i \in CK$;
or *prove that no such decomposition exists.*

When K is purely transcendental over its field of constants, this problem is soluble [13] and generally implemented[1]. Hence, when such a system returns an unevaluated integral, this *should be* a proof that no such elementary integral exists. However, the documentation may not say so: for example Maple 11 says merely

> If Maple cannot find a closed form expression for the integral, the function call is returned.

When K is algebraic, the problem is solved in principle [2], but not completely implemented. Hence, when such a system returns an unevaluated integral, this can mean any of:

1. there is no elementary integral, i.e. the [**or**] of problem 5;
2. the implementation is fundamentally inadequate, e.g. Reduce's integrator uses [6], which only works for quadratic algebraic functions of x;
3. the implementation has attempted to address the question, but has failed, which may be reported as "implementation incomplete"; ([9] reports this of Axiom), or just as an unevaluated integral;
4. the implementation may be of some (theoretically[2]) weaker algorithm, without a proof of completeness.

In general, the user does not know which of these applies, and the standard notation of computer algebra provides no convenient way of telling the user, though a warning (on the lines of the error reported in case (3) above) would at least be useful.

4   Other Areas

There are other areas in which set membership problems are, at least in principle, decidable. One obvious example is the solution of differential equations in terms of Liouvilian functions [16]. Again, it is not clear how these decision procedures can be effectively 'sold' to the user.

---

[1] Subject to undecidability problems over constants [12]. This is an important caveat in principle, but less so in practice, and we shall ignore it from now on.

[2] It may be stronger in practice, however, as reported in [9].

5   Conclusions

In one area of computer algebra (polynomial ideals) we are now used to the fact that we have a decision procedure for set membership, and would be surprised if anything other than a clear-cut answer were obtained. Elsewhere, e.g. integration, we have decision procedures, but the user community is apparently willing to settle for not knowing whether a decision procedure has been applied or not. Put bluntly, the user, no matter how expert, has no way of knowing what an unevaluated integral means, and in many ways the situation has gone back to the user expectations of [15], where we are merely asking "can the software find any answer".

References

1. Becker, T., Weispfenning, V., Kredel, H.: Groebner Bases. A Computational Approach to Commutative Algebra. Springer, Heidelberg (1993)
2. Bronstein, M.: Integration of elementary function. J. Symbolic Comp. 9, 117–173 (1990)
3. Bronstein, M.: Symbolic Integration I, 2nd edn. Springer, Heidelberg (2005)
4. Buchberger, B.: Ein Algorithmus zum Auffinden des basiselemente des Restklassenringes nach einem nulldimensionalen Polynomideal. PhD thesis, Math. Inst. Universität Innsbruck (1965)
5. Davenport, J.H.: Effective Mathematics — the Computer Algebra viewpoint. In: Richman, F. (ed.) Proceedings Constructive Mathematics Conference 1980 [Springer Lecture Notes in Mathematics 873], pp. 31–43. Springer, Heidelberg (1981)
6. Davenport, J.H.: On the Integration of Algebraic Functions. Springer Lecture Notes in Computer Science vol.102 (1981)
7. Davenport, J.H.: Equality in computer algebra and beyond. J. Symbolic Comp. 34, 259–270 (2002)
8. Davenport, J.H.: Effective Set Membership in Computer Algebra and Beyond. Technical Report CSBU–2008–03, Dept. Computer Science, University of Bath (2008),
   http://www.cs.bath.ac.uk/department/technical-report-series/technical-report-series/index.php
9. Kauers, M.: Integration of Algebraic Functions: A Simple Heuristic for Finding the Logarithmic Part(2008),
   http://www.risc.uni-linz.ac.at/publications/download/risc_3390/main.pdf
10. Liouville, J.: Mémoire sur l'intégration d'une classe de fonctions transcendantes. Crelle's J., 13, 93–118 (1835)
11. Moses, J.: Symbolic Integration. PhD thesis M.I.T., & Project MAC TR-47 (1967)
12. Richardson, D.: Some Unsolvable Problems Involving Elementary Functions of a Real Variable. Journal of Symbolic Logic 33, 514–520 (1968)
13. Risch, R.H.: The Problem of Integration in Finite Terms. Trans. A.M.S. 139, 167–189 (1969)
14. Singer, M.F., Saunders, B.D., Caviness, B.F.: An Extension of Liouville's Theorem on Integration in Finite Terms. SIAM J. Comp. 14, 966–990 (1985)
15. Slagle, J.: A Heuristic Program that Solves Symbolic Integration Problems in Freshman Calculus. PhD thesis, Harvard U. (1961)
16. van Hoeij, M., Ragot, J.-F., Ulmer, F., Weil, J.-A.: Liouvillian Solutions of Linear Differential Equations of Order Three and Higher. J. Symbolic Comp. 28, 589–609 (1999)

Formalizing in Coq Hidden Algebras to Specify Symbolic Computation Systems*

César Domínguez

Departamento de Matemáticas y Computación, Universidad de La Rioja
Edificio Vives, Luis de Ulloa s/n, E-26004 Logroño (La Rioja, Spain)
cesar.dominguez@unirioja.es

Abstract. This work is an attempt to formalize, using the Coq proof assistant, the algebraic specification of the data structures appearing in two symbolic computation systems for algebraic topology called EAT and Kenzo. The specification of these structures have been obtained through an operation, called *imp* operation, between different specification frameworks as standard algebraic specifications and hidden specifications. Reusing previous Coq implementations of universal algebra and category theory we have proposed a Coq formalization of the *imp* operation, extending the representation to the particular hidden algebras which take part in this operation.

Keywords: Coq proof assistant, hidden algebras, symbolic computation.

1 Introduction

The formal description of a computation system can be tackled from different points of view depending on the various approaches in the formal methods area. Two of the most important methods on this area are algebraic specification and type theory. On the one hand, algebraic specification [14] models programs as many sorted algebras consisting in a collection of sets of data values together with functions over those sets. At this level of abstraction, mathematical theorems can be applied in order to obtain some mathematical properties of the programs. On the other hand, type theory [16] emphases the program syntax and some formal systems of rules are used in order to obtain theorems in this formal context.

Our interest focuses on the formalization of two symbolic computation systems called EAT and Kenzo [19,8]. These systems were designed by Sergeraert implementing his ideas on *Constructive Algebraic Topology* [20]. Different formal methods have been applied to the specification of these systems. On the one hand, techniques of algebraic specification have been used in order to obtain the formalization of the data structures of the systems [13,7,6]. On the other hand, a project for representing the systems in type theory is ongoing [5]. Moreover, important lemmas of algebraic topology used in the systems have been already formalized and checked with the help of proof assistants as Isabelle [2] and Coq [5] or theorem provers as ACL2 [1].

---

* Partially supported by Ministerio de Educación y Ciencia, project MTM2006-06513.

S. Autexier et al. (Eds.): AISC/Calculemus/MKM 2008, LNAI 5144, pp. 270–284, 2008.

This work deals with the relationship between the two previous issues. In particular, we try to apply the techniques of type theory to the mathematical theorems obtained in algebraic specification. This could be interesting because it will increase the reliability of the mathematical results obtained and open bridges between both formal methods.

More concretely, in [13] an operation between specification frameworks, called *imp* operation, is used to obtain the formalization of the data structures appearing in the systems under study. These frameworks include hidden specifications [10] which are used in the literature to model object oriented programming languages [10,11]. In this context the data structures can be nicely modeled as a final object in a particular hidden category. In our approach, this object is built in a very natural way, which could contrast with the "magical formula" used in a general hidden category [10]. It is worth noting that the first version of this final object [10] contained a subtle error which was corrected in subsequent versions [11,17].

In this work, a formalization of our particular category of hidden algebras and a final object in it are specified using the Coq proof assistant [15]. Appreciable work can be found in the literature related to this line. For instance [3] or [12] tried to represent in type theory universal algebra and category theory respectively (notions which are intensively used in algebraic specification). Both works implemented their developments in the Coq proof assistant. In this work we will try to reuse as far as possible these previous developments.

The paper is organized as follows. Section 2 introduces some preliminaries on algebraic specification and Section 3 describes the *imp* operation. In Section 4 a Coq implementation of notions in algebraic specification and category theory is briefly explained. This code is reused to obtain some first categorical results in algebraic specification in Section 5. Section 6 includes a Coq formalization of the *imp* construction in a standard algebraic specification. In Section 7, the formalization of the hidden context in which this operation should be placed is performed, and in Section 8, a description of the final object in this category is proposed. The paper ends with a conclusions and future work section. The files of the implementation are available at www.unirioja.es/cu/cedomin/fcha.html.

2   Preliminaries on Algebraic Specification

In this section, we will briefly introduce some basic notions on algebraic specification; see [14] for a systematic presentation.

In Mathematics, when dealing with an algebraic structure, such as for instance a group, we refer to a set G together with some operations on G, $+: G \times G \to G$, $-: G \to G$, $e: \to G$. This way of working is abstracted in the field of *universal algebra*, where structured-sets in this sense are studied in a generic way. Roughly speaking, algebraic specifications can be understood as universal algebra enriched with some syntactic constructs that establish a link between programming languages (through the notion of programming *type*) and mathematical structures.

More precisely, a *signature* Σ is a pair (S, Ω) of sets of symbols, whose elements are called *sorts* and *operations*, respectively. Each operation $\omega \in \Omega$ has associated a $(n + 1)$-tuple $(s_1, \ldots, s_n, s)$ of elements in S with $n \geq 0$, called the *arity* of the operation. The sorts $(s_1, \ldots, s_n)$ are called *argument* sorts of the operation and the sort s its *target* sort. An operation is often denoted by $\omega\colon s_1 \ldots s_n \to s$. In the case $n = 0$, the operation is called a *constant of sort s*. In the example of a group, the convenient signature, denoted by GRP, has one sort g and three operations $prd\colon g\, g \to g$, $inv\colon g \to g$, $unt\colon \to g$.

Then, the structures of universal algebra are retrieved by means of the notion of Σ-*algebra*. Let $\Sigma = (S, \Omega)$ be a signature. An *algebra for Σ* (or Σ-*algebra*) assigns a set A_s to each sort $s \in S$, called the *carrier set* of the sort s, and a function $\omega_A\colon A_{s_1} \times \ldots \times A_{s_n} \to A_s$ to each operation $\omega\colon s_1 \ldots s_n \to s \in \Omega$. In the example, we can define a Σ-algebra A taking $A_g = G$, $prd_A = +$, $inv_A = -$ and $unt_A = e$.

An important algebra for a signature is the *term algebra*. The carrier sets of such an algebra are the terms freely generated by the operation symbols of the signature. Then, the functions in the algebra are defined in a natural way.

The Σ-algebras can be organized as a category $Alg(\Sigma)$ using the following notion of morphism. Let A, B be two Σ-algebras, with $\Sigma = (S, \Omega)$. A Σ-*homomorphism* $h\colon A \to B$ *from A to B* is a family $\{h_s\colon A_s \to B_s\}_{s \in S}$ of functions which verifies the following homomorphism condition: $h_s(\omega_A(a_1, \ldots, a_n)) = \omega_B(h_{s_1}(a_1), \ldots, h_{s_n}(a_n))$ for any operation $\omega\colon s_1 \ldots s_n \to s \in \Omega$ and for all $a_i \in A_{s_i}$, $i = 1, \ldots, n$.

The category of algebras for a signature has an initial algebra: the term algebra associated to this signature.

In our previous work, we were interested in a particular case of algebraic specifications, known as *hidden specifications* (see [10] for details). These specifications have been useful in the formalization of some characteristics in the object oriented paradigm [10,11].

Let $V\Sigma = (VS, V\Omega)$ be a signature and let us fix a $V\Sigma$-algebra D. The elements of VS are called *visible sorts* and those of $V\Omega$ are called *visible operations*. The $V\Sigma$-algebra D is called *data domain*. Then a *hidden signature*, on $V\Sigma$ and D, is a signature $H\Sigma = (S, \Omega)$ such that:

- $S = HS \sqcup VS$; the elements of HS are called *hidden sorts* of $H\Sigma$.
- $\Omega = H\Omega \sqcup V\Omega$ and for each operation $\omega\colon s_1 \ldots s_n \to s$ in $H\Omega$ the following property holds: in $s_1, \ldots, s_n$ there is at most one hidden sort.

(The $\sqcup$ symbol denotes the disjoint union.)

A *hidden algebra* A for a hidden signature $H\Sigma$, on $V\Sigma$ and D, is a $H\Sigma$-algebra such that $A_{V\Sigma} = D$ (in other words, the restriction of A to the visible part is equal to the data domain D). A *hidden morphism* between two hidden algebras is a $H\Sigma$-homomorphism h such that h_{VS} is the identity on D.

The hidden algebras for a hidden signature $H\Sigma$, on $V\Sigma$ and D, together with the hidden morphisms, define a category denoted by $HAlg^D(H\Sigma)$. Besides, a forgetful functor can be defined between $HAlg^D(H\Sigma)$ and $Alg(H\Sigma)$.

In hidden categories, final semantics are more important than initial ones [10,11]. But in this context, a final object does not always exist and some additional restrictions must be imposed in the category in order to guarantee its definition. Under these conditions a final object is defined through a "magical formula" in [10]. It is worth noting that the definition of this object contained a subtle error which was corrected in [11,17,18]. In the following section, we are going to introduce a particular category of hidden algebras which allow us to define this object in a natural way. These hidden algebras are relevant for the specification of symbolic computation systems [13,6,7].

3   The *imp* Construction

In this section, we will explain an operation between specification frameworks which was used for modeling the structures of data implementing in two symbolic computation systems called Kenzo and EAT [8,19]. In these systems, and in general in any symbolic computation system, you do not work only with a unique data structure, a group for example; in contrast, you deal at runtime with families of this data structure. So, we can identify two *layers* of data structures. In the first layer, one finds the usual data structures as integer numbers or (finite) lists of symbols. In the second layer, one must deal with algebraic structures as groups or chain complexes whose elements are data belonging to the first layer. To model this situation, in [13] an operation between specification frames, called *imp* operation, was defined (this name was chosen because this kind of specifications are related to implementations of structures rather than to the structures themselves, i.e., the treatment at low level that EAT makes of them).

The following simple example is used to explain the syntactic aspects of this construction. Let GRP be the signature included in the previous section. This signature is obviously the basis of the algebraic specification for a group, whose underlying set is abstracted by the sort g. But if, as it is usual in symbolic computation systems, it is necessary to handle several groups on the same underlying data set, a new sort, which remains hidden in the signature GRP, has to be considered: the type of groups represented on g. If we make explicit this invisible (or hidden) type, we obtain a new signature denoted by GRP_{imp}, with a new sort imp_GRP, and operations: $imp\_prd\colon imp\_GRP\ g\ g \to g$, $imp\_inv\colon imp\_GRP\ g \to g$, $imp\_unt\colon imp\_GRP \to g$.

In general, given a signature $\Sigma = (S, \Omega)$, a new signature $\Sigma_{imp} = (S_{imp}, \Omega_{imp})$ can be defined as follows:

- $S_{imp} = S \cup \{imp\_\Sigma\}$ with $imp\_\Sigma \notin S$,
- for each operation $\omega\colon s_1 \ldots s_n \to s$ in Ω, an operation $imp\_\omega\colon imp\_\Sigma\ s_1 \ldots s_n \to s$ is included in Ω_{imp}.

The Σ_{imp}-algebras satisfy the property that each element of the carrier set for the distinguished sort allows to retrieve a Σ-algebra.

Proposition 1. *Let* $\Sigma = (S, \Omega)$ *be a signature and* $A = \langle A_{imp\_\Sigma}, (A_s)_{s\in S}, \{imp\_\omega_A : A_{imp\_\Sigma} \times A_{s_1} \times \ldots \times A_{s_n} \to A_s\}_{\omega\in\Sigma}\rangle$ *be* a Σ_{imp}-algebra, each element $a \in A_{imp\_\Sigma}$ defines a Σ-algebra A_a in the following way: $A_a = \langle (A_s)_{s \in S}, \{imp\_\omega_A(a, \_) : A_{s_1} \times \ldots \times A_{s_n} \to A_s\}_{\omega \in \Sigma}\rangle$.

In [13], the *imp* construction was studied considering that a signature and its corresponding *"imp signature"* belong to two different specification frameworks: the standard algebraic specification for the former and the hidden specification framework for the latter. From a programming point of view, when you are dealing with implementations of algebraic structures (groups, for instance), usually you are only interested in structures whose elements share the same syntactic pattern (it is useful to identify the ground elements of some sort with a unique programming *type*). This restriction leads, at the model level, to the need of fixing a data family $D = \{D_s\}_{s \in S}$ for a signature $\Sigma = (S, \Omega)$ and to consider only the Σ-algebras with carrier sets on D.

When you include this restriction in the *imp* construction, the signature Σ_{imp} can be considered as a hidden signature with one hidden sort: the new sort, and data domain D. Then, we are interested in the category of hidden algebras $HAlg^D(\Sigma_{imp})$. In that context, we can identify the two layers of data structures in the algebra. The visible sorts represent the first layer: fixed data that are used to build the algebraic structures of the second layer which is represented by the hidden sort. This hidden sort will be used as an index of these structures, i.e. each element of the distinguished sort can be understood as a parameter allowing to retrieve a structure.

Under these conditions, the category $HAlg^D(\Sigma_{imp})$ has a canonical object denoted by A^{can}. This object can be presented as a set of functional tuples in the carrier set for the hidden sort. More concretely, $A^{can}_{imp\_\Sigma} := \{(f_\omega)_{\omega \in \Omega} \mid \langle D, (f_\omega)_{\omega \in \Omega}\rangle \in Alg(\Sigma)\}$. Then, the functions of this algebra are defined by the application of the corresponding function in the tuple to the rest of the arguments: $imp\_\omega_{A^{can}}((f_\delta)_{\delta \in \Omega}, d_1, \ldots, d_n) := f_\omega(d_1, \ldots, d_n)$, for each $imp\_\omega : imp\_\Sigma\ s_1 \ldots s_n \to s \in \Omega$, each $(f_\delta)_{\delta \in \Omega} \in A^{can}_{imp\_\Sigma}$ and each tuple $(d_1, \ldots, d_n) \in (D_1 \ldots, D_n)$. Besides, this canonical object is a final object in $HAlg^D(\Sigma_{imp})$. This result is reflected in the following theorem.

Theorem 1. *The canonical object A^{can} is a final object in $HAlg^D(\Sigma_{imp})$.*

Proof. For each object $B \in HAlg^D(\Sigma_{imp})$, it is possible to define a hidden Σ_{imp}-homomorphism, h^{can}, which has as component for the distinguished sort the function $h^{can}_{imp\_\Sigma} : B_{imp\_\Sigma} \to A^{can}_{imp\_\Sigma}$, such that $h^{can}_{imp\_\Sigma}(a) := (\delta_{B_a})_{\delta \in \Omega}$ for each $a \in B_{imp\_\Sigma}$. The homomorphism condition can be directly proved knowing that $imp\_\omega_{A^{can}}((\delta_{B_a})_{\delta \in \Omega}, d_1, \ldots, d_n) = \omega_{B_a}(d_1, \ldots, d_n)$, for each operation $imp\_\omega : imp\_\Sigma\ s_1 \ldots s_n \to s \in \Omega$ and each $a \in B_{imp\_\Sigma}$, $d_i \in D_i$, $i = 1, \ldots, n$. Besides, it is easy to prove that this homomorphism is unique by definition of the canonical object.

The description of this final object nicely corresponds with the implementation of the data structures chosen in Kenzo and EAT (see [13] for a more detailed explanation).

4   Reusing Coq Code

An important objective of this work consists in reusing previous Coq developments instead of beginning from scratch. Obviously, we do not refer here to the Coq standard library [15] which is widely required by Coq users, but other useful codes included in the literature. In this line, we are going to cite two important works for us. The first one is an attempt to formalize concepts in universal algebra developed by Capretta [3]. The second one includes the representation of notions in category theory developed by Huet and Saïbi [12]. In the following subsections, we will briefly explain some fragments of these works that we are going to reuse.

4.1   Coq Formalization of Concepts in Universal Algebra

In [3], Capretta presented a description of concepts in universal algebra inside Type Theory that was formalized using the Coq proof assistant. The version 6.2.3 of Coq was used in that work. So, in order to reuse the proposed development, a previous conversion of the code to the current version 8.1 was required. This code translation was carried out with the help of semi-automatic transformation tools implemented by the Coq Development Team [15]. In this section, we will focus on the formalization of the concepts of signature, algebra and homomorphism included in this paper.

An essential idea incorporated into [3] consists in formalizing the intuitive notion of set as a *setoid*. A setoid is defined as a pair formed by a set and an equivalence relation over it. The formal definition of a setoid in Coq is:

```
Record Setoid: Type:= setoid
  {s_el:> Set; s_eq: s_el -> s_el -> Prop; s_proof: equiv s_eq}.
```

where (equiv s_eq) is the proposition stating that s_eq is an equivalence relation over s_el. The implicit coercion included in the definition allows to identify a setoid S with its first element (s_el S). We usually indicate the relation s_eq by the infix operator [=]. The setoid of functions from a setoid S_1 to a setoid S_2, S_1[->]S_2, is defined as the setoid on the type of functions S_1->S_2 with the extensional equality relation.

The notion of signature is formalized in the following way. A natural number n: nat is chosen to indicate the number of sorts of a signature (only signatures with a finite number of sorts are considered, but this is enough for our purposes). Then, the sorts are represented by elements of the finite set with n elements defined by a type Finite n. An operation is represented by its arity, which is formalized as a pair of a list of argument sorts and a target sort:

```
Definition Function_type: Set:= (list(Finite n) * (Finite n)).
```

Finally, a signature is a pair $< n, l >$ where n is the number of sorts and l is a list of function types:

```
Record Signature: Set:= signature
  {sorts_num: nat; function_types: list(Function_type sorts_num)}.
```

For instance, the group signature defined in Section 2 is represented by (`signature 1 [<[o,o],o>,<[o],o>,<[],o>]`), where o represents the unique element in `Finite 1`.

An algebra for a signature can be formalized as a structure that assigns a setoid to every sort in the signature and a setoid function to every operation. More precisely, given a signature `sigma: Signature` with `n:= sorts_num sigma`, the carrier sets of an algebra for this signature are represented as a family of n setoids `Sorts_interpretation:= Finite n -> Setoid`. Then, the interpretation of a function type consists in a setoid function with just one argument (all the argument sorts indexed on a finite type):

```
Definition Function_type_interpretation (f: Function_type n): Setoid:=
  Fun_arg_function _ (fun i: Finite(length(function_type_arguments f)) =>
              sorts(fin_proj(Finite n) _ i))
              (sorts(function_type_result f)).
```

where `sorts` is a variable of type `Sorts_interpretation`, `function_type_arguments` and `function_type_result` obtain the argument and target sorts of a function type, `fin_proj` is a projection function on finite sets and `Fun_arg_function` is an auxiliary definition for building a setoid function.

Finally, using a type for lists of function type interpretations defined in a natural way, we can obtain a type for algebras:

```
Record Algebra (sigma: Signature): Type:= algebra
  {sorts:> Sorts_interpretation (sorts_num sigma);
   functions: Function_list_interpretation sorts (function_types sigma)}.
```

The representation of a homomorphism is obtained as follows. Given a signature `sigma:Signature` and two algebras `A B:Algebra sigma`, let `n:= sorts_num sigma` and `m:= fun_num sigma` be the number of sorts and functions of `sigma`. First of all, a homomorphism is a family of setoid functions `phi: forall i: Finite n, (sorts A) i[->](sorts B) i`. Then, this family must verify the homomorphism condition. To this aim, fixed the i-th function of the signature `i: Finite m, fi:= nth_function i`, the following auxiliary components are defined: `r:= function_type_arity fi` is the number of arguments of `fi`, `a:= function_type_arg fi (j: Finite r)` a projection of the j-th argument of `fi`, and `h:= function_type_result fi` the target sort of this function. Assuming generic arguments for this i-th function in A, `argsA:(Fun_arg_arguments A i)` (where `Fun_arg_arguments` is the setoid extension of the function `Function_arguments_sorts A i (j: Finite r):= sorts (a j)`), the needed arguments in B are obtained by an application of phi, `argsB: Fun_arg_arguments B i:= fun j: Finite r => phi (a j) (argsA j)`. Then, the homomorphism condition can be defined by the property (`functions B i) argsB[=]phi h ((functions A i) argsA`), called `Is_homomorphism phi`. With this property, the type of homomorphism can be defined by the record:

```
Record Homomorphism: Set:= homomorphism
  {hom_function:> forall i:Finite n, (sorts A) i[->](sorts B) i;
   hom_proof: Is_homomorphism hom_function}.
```

The term algebra for a signature and a homomorphism from this algebra to any other algebra of the signature are defined as examples in [3].

4.2   Category Theory in Coq

Constructive type theory has been shown to be adequate for formalizing category reasoning [12,5]. For instance, we will remark the representation in Coq of notions in category theory by Huet and Saïbi [12]. The code produced is available on the Coq web [15] and is compatible with version 8.1 of Coq.

In this work, a *category* is formalized as a record type called `Category` which includes types for objects and morphisms `Ob:> Type`, `Hom: Ob -> Ob -> Setoid` together with slots for identity morphisms `Id`, composition maps `Op_comp` and category laws. In the same way, a *functor* between two categories `C D: Category` is formalized also as a record type with includes slots for the functions between objects `FOb:> C -> D` and morphisms `FMap: forall a b : C, Hom a b -> Hom (FOb a) (FOb b)` and for the functor laws.

Categorical notions that are useful for us are the *initial* or *final objects* of a category. For instance, given a category `C: Category`, an initial object of the category is represented by a record with an object of the category `Initial_ob:> C` and a morphism for each object in the category `MorI: forall b: C, Hom Initial_ob b` which verify the following initial law `Definition Initial (a: C) (h: forall b: C, Hom a b):= forall (b: C) (f: Hom a b), f [=] h b`.

Another categorical concept that we are going to use is the *full subcategory* of a category. Given a category `C: Category` and two variables `I: Type, a: I -> C`, it is possible to restrict the morphisms in `C: Category` to those obtained from objects of `I, FSC_mor_setoid (i j: I):= Hom (a i) (a j)`. Then, if the identity and composition morphisms are defined from the corresponding ones in the category, it is possible to prove the category laws which define a full subcategory.

5   First Categorical Results in Algebraic Specification

Using the formalization of notions in algebraic specification and category theory presented in the previous section, some first category results in algebraic specification can be proposed. For instance, algebras and homomorphisms for a signature build a category or the term algebra associated to this signature is initial in this category.

A problem arises when different Coq developments are brought together: a single concept can be formalized in different ways. In our case, this problem appears in the representation of the constructions related to setoids. Both developments are essentially equal but different labels are assigned to same concepts. In our case, this problem can be solved by simply renaming some components in one of the versions. We have chosen to rename the category theory version. A possible better alternative consists in using the setoid formalization proposed in the standard library of Coq [15]. This version includes advanced features for working with setoids which should be probably useful in our developments, but modifications beyond a simple renaming operation would be necessary in both files in order to take advantage of these features. This option needs further work.

Given a signature `sigma: Signature` with `n:= sorts_num sigma`, a category of algebras and homomorphisms for this signature can be defined in the following way. Firstly, for each `A: Algebra(sigma)` the identity morphism is built using the identity setoid function `id_sf(i: Finite n):=(s_id (sorts A i))`. This morphism is defined as an example in [3]. Secondly, given three algebras `A B C: Algebra sigma` and two homomorphisms `g: Homomorphism A B`, `h: Homomorphism B C`, the composition morphism is defined through the composition of the homomorphism setoid functions `comp_sf (i: Finite n):= (s_fun_comp (h i) (g i))`. Finally, in order to prove the category laws, it is necessary to establish when two homomorphism are equal. To this aim, a setoid for homomorphisms is defined using as equality the setoid equality of all the functions in `hom_function`. In this setoid it is not difficult to prove the category laws. This completes a `Category` structure called `Algebra_Category(sigma)` which formalizes $Alg(\Sigma)$ where `sigma` represents Σ.

In order to prove that the term algebra for `sigma` is initial in `Algebra_Category(sigma)`, it is only necessary to prove the initial law for the term evaluation homomorphism defined in [3]. This proof is easy by construction of this homomorphism.

6   Coq Formalization of the *imp* Construction

In this section we derive a Coq formalization for the *imp* construction explained in Section 3. The first step in this construction consists in building the *imp* signature for a given signature `sigma: Signature`. This signature is defined by adding one new sort to the sorts of `sigma` which must be included as the first argument in each operation of `sigma`:

```
Definition imp_sigma(sigma:Signature):=signature
(S (sorts_num sigma)) (imp_functions (function_types sigma)).
```

The `imp_functions` operation maps the following definition to all function types of the list:

```
Definition imp_function_type: Function_type (S n):=
(imp_sort::(trans_list_n_Sn (fst f)),(trans_n_Sn (snd f))).
```

where `imp_sort` renames the n+1-th element of `Finite(S n)`, `trans_n_Sn` is intended to transform the i-th element of `Finite n` in the i-th element of `Finite(S n)` and `trans_list_n_Sn` maps the previous function to a list of elements of `Finite n`.

It is worth noting that, due to the chosen formalization, the representation of each sort in `sigma` is different from the representation of the corresponding one in `imp_sigma(sigma)`. Indeed, they have a different type. The technique used to solve the type conflict between a component of a signature which is intended to be equal to a component in the another signature consists in the definition of *transformation type functions*. In the previous case, the transformation is guided by the property that establishes that both sorts are in the same position in different finite sets. This problem did not appear in [3]. Indeed, this work did not include any relation between signatures such as for instance signature morphisms [14].

Now, algebras for this *imp* signature are directly defined as Algebra(imp_sigma(sigma)).

Proposition 1 establishes that an *imp* algebra imp_A: Algebra(imp_sigma(sigma)) defines a family of algebras for sigma where the indices of this family are the elements of the distinguished sort. This result can be formalized in the following way in our context. When fixing an element e: (sorts imp_A imp_sort), it is possible to define an algebra, which will be called A, in Algebra(sigma). The interpretation of the sorts for A is defined by the corresponding interpretation in imp_A: sorts_A_interpretation(i: Finite n): Setoid:= (sorts imp_A (trans_n_Sn i)).

The interpretation of the operations for A corresponds to the interpretation of the operations for imp_A with the element e fixed. This definition needs a more elaborated process. Firstly, the following auxiliary function which appends the fixed element to a generic argument for an operation is needed. Let fi := nth_function i be the i-th operation of sigma (with m:= fun_num sigma and i: Finite m) and let B0 be the type of the arguments for this function in A. More concretely, B0 is the type of the setoid extension of the function fun c: Finite r => (sorts_A_interpretation (a c)) with r and a representing the number of arguments of fi and a projection of the j-th argument as above. Then, the required function is defined as:

```
Definition append_e(ai:B0):B0imp:=
fun i0:Finite r_imp => match zerop(fin_extr i0) with
 |left h1 => trans_0argument h1 e
 |right h2 => trans_Sargument h2 (ai(fin_pred h2))
 end.
```

where B0imp corresponds to the type of the setoid extension of the function fun c: Finite r_imp => (sorts imp_A (a_imp c)), being r_imp and a_imp the corresponding components for the i-th operation of imp_sigma(sigma). The function fin_extr defined in [3] extracts the natural number which corresponds to the position (minus one) of an element in a finite type, zerop has type forall n:nat, {n=0} + {0<n} and fin_pred is the predecessor function for finite sets. Besides, two transformation type functions are needed. The functions trans_0argument and trans_Sargument change the type of e and of ai(fin_pred h2) to the required type (sorts imp_A (a_imp i0)) respectively. It should be observed that the first argument sort of an *imp* operation is the new sort and another argument is the one situated in a predecessor position in the arguments of the initial operation.

Then, the definition which allows us to build the i-th function of the algebra A is the following:

```
Definition function_A (ai:BO):=
trans_result (functions imp_A (trans_pos_func i) (append_e ai)).
```

This function appends to `ai` the element `e` in the first position which defines an argument in BOimp for the i-th function of imp_A. Then, the corresponding function in the algebra imp_A is applied. In the process, some transformation type functions are needed again. Finally, it is necessary to prove that this function is indeed a setoid function. This completes the definition of an algebra A for sigma.

When the previous construction is generalized, we obtain a proof of Proposition 1 in Coq:

```
Definition family_algebra_A (e: (sorts imp_A imp_sort)):= (A e).
```

7   Coq Formalization of Hidden Algebras

In this section, we will propose a Coq formalization of the hidden algebras which appear in the hidden specification of the *imp* construction. As it is explained in Section 3, when we include our *imp* signatures in a hidden context, only the new sort is considered as hidden. In this case, a hidden algebra for this signature can be formalized in the following way. Given a signature sigma: Signature with n:= sorts_num sigma and a fixed data domain D: (Finite n) -> Setoid for the visible sorts, the sort interpretation of a hidden algebra for imp_sigma(sigma) only needs a setoid IMP for the distinguished sort. Then, an auxiliary function can be built which assigns the IMP setoid to the n+1-th sort of the signature and the corresponding D setoid to other visible sorts:

```
Definition sorts_interpretation_D (j: Finite (S n)): Setoid:=
  match (Finite_dec j) with
  (left h1 ) => (D (trans_Sn_n h1))
 |(right h2) => IMP
  end.
```

where `Finite_dec` establishes the following decidable property on finite sets: `forall (j: Finite (S n)), {n>fin_extr j} + {fin_extr j=n}` and `trans_Sn_n` defines the transformation function type between the elements in `Finite (S n)` different from the n+1-th element to elements in `Finite n` which is inverse to `trans_n_Sn`. With the help of this function, a hidden algebra for imp_sigma(sigma) can be formalized by:

```
Record Imp_Algebra_D:= imp_algebra_D
{IMP:>Setoid;
  functions_D: Function_list_interpretation(sorts_interpretation_D D IMP)
              (function_types(imp_sigma sigma))}.
```

It should be noted that an Imp_Algebra_D(sigma D) structure is not an Algebra(imp_sigma(sigma)) structure, but it can be directly transformed into one by considering as sorts slot the sorts_interpretation_D function. This operation, called inclusion_Algebra_D_in_Algebra, allows us to represent the Imp_Algebra_D(sigma D) structures as a full subcategory of the category of algebras Algebra_Category(imp_sigma(sigma)).

However, the category of hidden algebras is not a full subcategory because we are only are interested in the homomorphisms with identity components in the visible sorts. For our particular *imp* hidden algebras, only a setoid function for the distinguished sort is necessary for defining a hidden homomorphism. In this way, given a signature `sigma: Signature` with `n:= sorts_num sigma`, a data domain `D: (Finite n) -> Setoid` and two structures `IMP_A IMP_B: Imp_Algebra_D(sigma D)` with `As_IMP:= IMP IMP_A` and `imp_A:= (inclusion_Algebra_D_in_Algebra IMP_A)` (with similar components for `IMP_B`), it is possible to define a hidden homomorphism for `Imp_Algebra_D(sigma D)` structures in the following way:

```
Record Imp_Homomorphism_D:Type:=  imp_homomorphism
  {IMP_function: As_IMP [->] Bs_IMP;
   IMP_hom_proof: Is_homomorphism (A:=imp_A)(B:=imp_B)
                  (phi_imp IMP_function)}.
```

where `phi_imp`, defined in a way similar to `sorts_interpretation_D`, assigns the setoid function `IMP_function` to the n+1-th sort and the identity setoid function in the other case.

Again, an `Imp_Homomorphism_D(sigma D)` structure between `Imp_Algebra_D(sigma D)` structures is not a `Homomorphism(imp_sigma(sigma))` structure, but it can be easily transformed into one between the corresponding `Algebra(imp_sigma(sigma))` structures.

Finally, it is possible to prove that `Imp_Algebra_D(sigma D)` and `Imp_Homomorphism_D(sigma D)` structures defines a `Category` structure called `Imp_Algebra_Category_D(sigma D)` which formalizes $HAlg^D(\Sigma_{imp})$ where `imp_sigma(sigma)` and `D` represents Σ_{imp} and D respectively. Besides, the two transformations defined on objects and morphisms allow us to build a functor between this category and the category of algebras `Algebra_Category(imp_sigma(sigma))` which corresponds to the forgetful functor between hidden algebras and standard algebras.

8   Defining the Final Object in `Imp_Algebra_Category_D`

In Section 3 it has been explained that a canonical object can be defined in the category of hidden algebras for an *imp* signature with a fixed domain. Besides, Theorem 1 establishes that this object is final in the category. In this section we will propose a formalization of this object in `Imp_Algebra_Category_D`.

Basically, the canonical object has as carrier set for the distinguished sort a collection of tuples of functions, one function for each operation in the signature. So, given a signature `sigma: Signature` with `n:= sorts_num sigma` and a fixed data domain `D: (Finite n) -> Setoid`, a family of functions for `sigma` with this domain can be directly defined by `IMPSort_functions: Setoid:= Function_list_interpretation D (function_types sigma)`.

Then, for each operation, the assigned function in this canonical object is defined through the application of the corresponding function in the tuple to the rest of visible arguments. So, given `impai: B0imp` a generic argument for the i-th function of the canonical object (in this case `B0imp` represents the setoid corresponding to the function `fun c: Finite r_imp =>((sorts_interpretation_D D IMPSort_functions) (a_imp c))`, with `r_imp` and `a_imp` defined as above), it is possible to define auxiliary functions which extract the first element `extract_IMP_component` and the rest of elements `extract_D_component` of this argument. For instance, this second function is defined as follows:

```
Let extract_D_components (impai: B0imp): B0:=
fun i0:Finite r => trans_result_D(impai(trans_imp_arity(fin_succ i0))).
```

being `fin_succ` the successor function for finite sets. Again, in this function some transformation type functions are needed. Then, the following function applies the i-th function of the tuple in the distinguished sort (first argument) to the rest of the arguments:

```
Let apply_f(impai: B0imp):=
trans_result_imp((extract_IMP_component impai i)
      (extract_D_components impai)).
```

The proof that this function is indeed a setoid function allows to complete the definition of the canonical object called `Final_Algebra(sigma D)`.

In order to prove that this canonical object is final in `Imp_Algebra_Category_D(sigma D)`, a homomorphism `Imp_Homomorphism_D(sigma D)` must be defined for every object in `Imp_Algebra_Category_D(sigma D)` to `Final_Algebra(sigma D)`. The crucial point in the proof of the homomorphism condition (see Section 3) consists in the application of the appropriated function in the family of functions (which corresponds to the hidden argument) to some arguments (which correspond to the rest of visible arguments). So, essentially, it consists in one step of reduction in a simplification process.

We can illustrate the previous comments with a very simple example: a signature with one sort and one unary operation. In this case, a prefixed data domain consists in a setoid `D:Setoid`, and an algebra could be represented by a function setoid `f: D[->]D`. Then, a structure in `Imp_Algebra_D` is formalized by a setoid `imp: Setoid` and a setoid function `f_imp: imp[->]D[->]D`. Obviously, given an algebra `A_imp: Imp_Algebra_D`, each element of the distinguished sort `e:(imp A_imp)` defines an algebra for the initial signature through the setoid function `((f_imp A_imp) e)`.

Now, the final object has, as distinguished set, lists of setoid functions with one function `(D[->]D)`. The definition of this final object, named `final_object`, is completed by assigning the setoid function defined by `apply_function_D(f_D:D[->]D)(d:D):=(f_D d)` to the unique *imp* operation.

Now, given an algebra `A_imp: Imp_Algebra_D`, the setoid function of an `Imp_Homomorphism_D` between `A_imp` and `final_object`, is defined through the function which assigns, to every element of the distinguished sort, the algebra of the family defined by this element. More precisely, it can be defined in the following way:

```
Definition h_final:(imp A_imp)->(imp final_object):=
fun a: (imp A_imp)=> ((f_imp A_imp) a).
```

Finally, since the visible component in the homomorphism is an identity, the homomorphism condition is expressed as follows:

```
Theorem hom_proof_final: forall (a:imp A_imp)(d:D),
(((f_imp A_imp) a) d) [=] (((f_imp final_object) (h_final a)) d).
```

This theorem is easily proved in Coq. It is also not difficult to prove that this homomorphism is unique. To this aim, the equality defined between homomorphisms is needed.

The very natural proof in this simplified case illustrates the soundness of our approach. Nevertheless, the Coq proof in the general case is not finished yet, and we are reconsidering some of the definitions and decisions taken in early stages of our formalization. These aspects are commented in the following (and last section).

9   Conclusions and Further Work

In this paper we have formalized, using the Coq proof assistant, an algebraic specification of the data structures appearing in two symbolic computation systems for algebraic topology called EAT and Kenzo. In particular, hidden algebras play an important role in that specification. So, trying to reuse previous Coq developments in universal algebra and category theory, a Coq description for hidden algebras corresponding to these structures is proposed.

With respect to the continuation of the work, to finish in Coq the proof of Theorem 1 seems to us rather tedious, since a lot of technical lemmas would be needed in order to simplify the intricate syntactical constructions raised by our initial definitions. Thus, we are considering some alternative approaches. On the one hand, it would be possible to design other structures with an easier processing. For instance, the inclusion of the setoid formalization defined in the available standard library of Coq [15] could provide new helpful tools. It is also possible to try to generalize our transformation type functions which are defined in this work in an *ad hoc* manner. On the other hand, we could change our initial formalization of the algebraic concepts included in [3] using, for instance, a more abstract point of view. For example, our results can be set in a more general context provided by the concept of institution [6].

This idea of using institutions (a notion which tries to abstract the concept of specification framework; see [9]), could give another research line to continue our work. For that, it could be necessary to complete the Coq implementation of standard algebraic specifications (including equational reasoning and signature morphisms, a work that was partiality carried out in [4]), and hidden specifications (including behavioral reasoning).

References

1. Andrés, M., Lambán, L., Rubio, J.: Executing in Common Lisp, Proving in ACL2. In: Kauers, M., Kerber, M., Miner, R., Windsteiger, W. (eds.) MKM/CALCULEMUS 2007. LNCS (LNAI), vol. 4573, pp. 1–12. Springer, Heidelberg (2007)
2. Aransay, J., Ballarin, C., Rubio, J.: A Mechanized Proof of the Basic Perturbation Lemma. Journal of Automated Reasoning 40(4), 271–292 (2008)
3. Capretta, V.: Universal Algebra in Type Theory. In: Bertot, Y., Dowek, G., Hirschowitz, A., Paulin, C., Théry, L. (eds.) TPHOLs 1999. LNCS, vol. 1690, pp. 131–148. Springer, Heidelberg (1999)
4. Capretta, V.: Equational Reasoning in Type Theory (preprint)
5. Coquand, T., Spiwack, A.: Towards Constructive Homological Algebra in Type Theory. In: Kauers, M., Kerber, M., Miner, R., Windsteiger, W. (eds.) MKM/CALCULEMUS 2007. LNCS (LNAI), vol. 4573, pp. 40–54. Springer, Heidelberg (2007)
6. Domínguez, C., Lambán, L., Rubio, J.: Object-Oriented Institutions to Specify Symbolic Computation Systems. Rairo - Theoretical Informatics and Applications 41, 191–214 (2007)
7. Domínguez, C., Rubio, J., Sergeraert, F.: Modeling Inheritance as Coercion in the Kenzo System. Journal of Universal Computer Science 12(12), 1701–1730 (2006)
8. Dousson, X., Sergeraert, F., Siret, Y.: The Kenzo Program, Institut Fourier, Grenoble (1999), http://www-fourier.ujf-grenoble.fr/sergerar/Kenzo/
9. Goguen, J., Burstall, R.: Institutions: Abstract Model Theory for Specification and Programming. Journal of the Association for Computing Machinery 39(1), 95–146 (1992)
10. Goguen, J., Malcolm, G.: A Hidden Agenda. Theoretical Computer Science 245, 55–101 (2000)
11. Goguen, J., Malcolm, G., Kemp, T.: A Hidden Herbrand Theorem: Combining the Object, Logic and Functional Paradigms. Journal of Logic and Algebraic Programming 51, 1–41 (2002)
12. Huet, G., Saïbi, A.: Constructive Category Theory. In: Proof, language, and interaction: essays in honour of Robin Milner. Foundations of Computing series, pp. 239–275. MIT Press, Cambridge (2000)
13. Lambán, L., Pascual, V., Rubio, J.: An Object-Oriented Interpretation of the EAT System. Applicable Algebra in Engineering, Communication and Computing 14(3), 187–215 (2003)
14. Loeckx, J., Ehrich, H.D., Wolf, M.: Specification of Abstract Data Types. Wiley-Teubner, Chichester (1996)
15. LogiCal project. The Coq Proof Assistant (2008), http://coq.inria.fr/
16. Martin-Löf, P.: Twenty-five years of constructive type theory, Oxford Logic Guides, vol. 36, pp. 127–172 (1998)
17. Roşu, G.: Hidden Logic. PhD thesis. University of California at San Diego (2000)
18. Rubio, J.: Locally Effective Objects and Artificial Intelligence. In: Campbell, J.A., Roanes-Lozano, E. (eds.) AISC 2000. LNCS (LNAI), vol. 1930, pp. 223–226. Springer, Heidelberg (2001)
19. Rubio, J., Sergeraert, F., Siret, Y.: EAT: Symbolic Software for Effective Homology Computation, Institut Fourier, Grenoble (1997), ftp://ftp-fourier.ujf-grenoble.fr/pub/EAT/
20. Rubio, J., Sergeraert, F.: Constructive Algebraic Topology. Bulletin Sciences Mathématiques 126, 389–412 (2002)

Symbolic Computation Software Composability

Sebastian Freundt[1], Peter Horn[2], Alexander Konovalov[3],
Steve Linton[3] and Dan Roozemond[4]

[1] Fakultät II - Institut für Mathematik, Technische Universität Berlin,
Berlin, Germany
`freundt@math.tu-berlin.de`
[2] Fachbereich Mathematik, Universität Kassel, Kassel, Germany
`hornp@mathematik.uni-kassel.de`
[3] School of Computer Science, University of St Andrews, Scotland
`{alexk,sal}@mcs.st-and.ac.uk`
[4] Department of Mathematics and Computer Science, Technische Universiteit
Eindhoven, Netherlands
`d.a.roozemond@tue.nl`

Abstract. We present three examples of the composition of Computer
Algebra Systems to illustrate the progress on a composability infrastructure as part of the SCIEnce (Symbolic Computation Infrastructure
for Europe) project[1]. One of the major results of the project so far is
an OpenMath based protocol called SCSCP (Symbolic Computation
Software Composability Protocol). SCSCP enables the various software
packages for example to exchange mathematical objects, request calculations, and store and retrieve remote objects, either locally or accross the
internet. The three examples show the current state of the GAP, KANT,
and MuPAD software packages, and give a demonstration of exposing
Macaulay using a newly developed framework.

1 Introduction

The SCIEnce project (Symbolic Computation Infrastructure for Europe) [25]
brings together the developers of four powerful symbolic computation software
packages (GAP [8], KANT [14], Maple [17], and MuPAD [19]), a major symbolic
computation research institute (RISC-Linz [21]), and research groups expert in
essential underpinning technologies (CNRS Palaiseau (France) [4], TU Eindhoven (Netherlands) [6], IeAT (Romania) [13] and Heriot-Watt University (UK)
[5]). The aim is to unite the European community of researchers in, and users
of, symbolic computation.

In this paper we report on one of the activities the SCIEnce project consists of,
namely *NA3: Software Composability*. This activity focuses on the development
and implementation of standards in order for the various Computer Algebra
Systems (CASes) to communicate. The main goal of this activity is to allow
these systems to be efficiently composed to solve complex problems.

---

[1] The project 026133 "SCIEnce—Symbolic Computation Infrastructure for Europe"
is supported by the EU FP6 Programme.

S. Autexier et al. (Eds.): AISC/Calculemus/MKM 2008, LNAI 5144, pp. 285–295, 2008.
© Springer-Verlag Berlin Heidelberg 2008

This part of the project has some common concerns with the well known SAGE project [22], as both projects try to unite several mathematical software packages. There are, however, important differences. SAGE presents itself as an integrated system in which users interact with the SAGE frontend and the contributing CASes are used as backend servers. Our goal in the SCIEnce project is to create a framework that will allow services to be both provided and consumed by any CAS. An important technical difference is that we use an existing language for representing mathematical objects, namely OpenMath [20], whereas SAGE uses a custom internal representation. We expect the use of OpenMath to facilitate third party developers to expose their software using SCSCP – although the conversion to and from the XML-based OpenMath format is potentially more time-consuming, we obtain a stable system-independent representation.

We illustrate the progress made in this activity by means of three examples. First, we introduce the Computer Algebra Systems involved in Section 2, and the OpenMath standard in Section 3. We give an overview of the newly designed protocol for the composition of symbolic computation software, SCSCP, in Section 4. The first example is the factorization in KANT of polynomials created in MuPAD (Section 5). In the second example (Section 6) we show a Gröbner basis computation executed in Macaulay on polynomials created in GAP. The third example (Section 7) demonstrates cross-platform use of GAP using SCSCP. Comments on the current status and intended future research can be found in Section 8.

2 The Computer Algebra Systems Involved

In this section, we briefly describe the four computer algebra systems involved in the SCIEnce project.

GAP [8] is a free, open and extendable system for computational discrete algebra, with particular emphasis on Computational Group Theory. GAP provides a programming language, a library of thousands of functions implementing algebraic algorithms written in the GAP language as well as large data libraries of algebraic objects. GAP is developed by international cooperation of many contributors, and coordinated by the four GAP centers: Aachen (Germany), Braunschweig (Germany), Fort Collins (USA), and St Andrews (UK).

KANT [14] is a computer algebra system for sophisticated computations in algebraic number fields that has been developed at Technische Universität Berlin. The KANT functions are accessible through a user-friendly shell named KASH (KAnt SHell) that is freely available.

Maple [17] is the general purpose computer algebra system developed in Waterloo, Canada. Its latest features include an intuitive smart document environment and embedded GUI components such as buttons and sliders.

MuPAD Pro [19] is a general purpose computer algebra system for exact symbolic and numeric computing with arbitrary precision. It provides a Pascal-like programming language allowing imperative, functional, and object-oriented

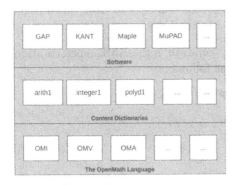

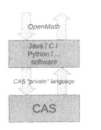

Fig. 1. OpenMath structure   **Fig. 2.** CAS implementations

programming. MuPAD is developed by the SciFace company, based in Pader-
born, Germany. The SCIEnce development with respect to MuPAD is performed
at the University of Kassel, Germany [26].

3   OpenMath

The OpenMath standard is made for the representation of mathematics in such
a way that mathematical objects can easily be exchanged between computer
programs by way of rich semantics. A rough overview of this standard can be
found in Figure 1. The 3 layers are explained as follows:

Language. The OpenMath language defines the grammar, i.e., notions such as
Variables, Errors, Applications, Integers, etc.

Content Dictionary. A Content Dictionary (CD) is a document describing
mathematical notions for some area of Mathematics. At the moment of writing about 180 content dictionaries are provided on the OpenMath homepage,
both official and experimental. They cover not only general areas, such as basic arithmetic (for example the 'arith1' CD describes 'minus', 'plus', 'power',
etc) or polynomials, but also more specific areas such as permutation groups,
planar geometry, finite fields, and much more.

Software. This third layer consists of all software built using the basic language
and content dictionaries, commonly referred to as "Phrasebooks". In Figure
2 the two most common approaches in our setting are described.

First the piece of translator software separate from the CAS: this software
(commonly written by a third party) takes care of the translation from Open-
Math into the CAS proprietary language, and back. Especially the translation back can be highly non-trivial, as the semantics of the CAS output
cannot always be read off from the output itself.

The second option is a piece of translator software that is built into the
CAS. The disadvantage is that one needs to have access to the source code,
which is not always possible. On the other hand, the major advantage of

this approach is that generally the translation can be done using the internal representation of mathematical objects in the CAS.

The most common representation for OpenMath is the XML-representation. For example, $3 - \frac{4}{5}$ could be represented as follows:

```
<OMA><OMS cd="arith1" name="minus"/>
  <OMI>3</OMI>
  <OMA><OMS cd="nums1" name="rational"/>
    <OMI>4</OMI>
    <OMI>5</OMI>
  </OMA>
</OMA>
```

4 SCSCP

To simplify the communication between the various CASes, we have developed a protocol called "Symbolic Computation Software Composability Protocol", abbreviated SCSCP. The protocol is XML-based; in particular, the protocol messages are in the OpenMath language, and its TCP-sockets-based implementation uses XML processing instructions to delimit these messages and indicate major failures that may arise during the processing of a request. Communication takes place using port 26133, reserved for SCSCP by the Internet Assigned Numbers Authority (IANA).

We have developed two Content Dictionaries for SCSCP, called scscp1 and scscp2. The protocol supports calling functions with mathematical objects as arguments, on either a local or remote system, and sending back successful results or failure reports. It also supports basic options such as limits on memory or CPU and information such as memory or CPU time used.

Moreover, SCSCP has support for remote objects. The client may indicate a preference for the reply to either contain a full mathematical object, or a reference to that same object. We envisage a scenario where a client can let almost all computations be performed remotely, possibly in another CAS, and where details of mathematical objects are not transmitted unless necessary.

An example of a call to a CAS and a response follows:

```
<OMOBJ>
  <OMATTR>
    <OMATP>
      <OMS cd="scscp1" name="call_ID"/>
      <OMSTR>a1d0c6e83f027327d8461063f4ac58a6</OMSTR>
      <OMS cd="scscp1" name="option_max_memory"/>
      <OMI>100000</OMI>
      <OMS cd="scscp1" name="option_return_object"/>
      <OMSTR/>
    </OMATP>
    <OMA>
      <OMS cd="scscp1" name="procedure_call"/>
```

```
<OMSTR>Evaluate </OMSTR>
<OMA>
    <OMS cd="arith1" name="plus"/>
    <OMI>16603777328095411 </OMI>
    <OMI>9529248804930722 </OMI>
</OMA>
</OMA>
</OMATTR>
</OMOBJ>
```

with response

```
<OMOBJ>
  <OMATTR>
    <OMATP>
      <OMS cd="scscp1" name="call_ID"/>
      <OMSTR>a1d0c6e83f027327d8461063f4ac58a6 </OMSTR>
      <OMS cd="scscp1" name="info_runtime "/>
      <OMI>3</OMI>
      <OMS cd="scscp1" name="info_memory "/>
      <OMI>2876</OMI>
    </OMATP>
    <OMA>
      <OMS cd="scscp1" name="procedure_completed"/>
      <OMI>26133026133026133 </OMI>
    </OMA>
  </OMATTR>
</OMOBJ>
```

More details and examples can be found in the specification [15] and the two Content Dictionaries [23], [24].

Basic SCSCP support, both as server and as client, is now available for development versions of GAP, KANT, and MuPAD. This means that a user of one of these software packages can invoke one of the other systems (or the same system on a different machine) without leaving the software packages he is working in himself. In particular, one uses GAP syntax to use SCSCP from within GAP, for example, even though one may be calling out to KANT.

We have also created a Java implementation of SCSCP, intended as a framework to enable third party developers to expose their software easily to e.g. users of one of the systems involved.

Furthermore, as SCSCP is a specialized protocol, it would have to be implemented in a each software package to allow access to the capabilities of the systems involved. In order to offer access by means of a more widely used protocol than SCSCP, we have developed a WebProxy that connects to an arbitrary number of SCSCP-compliant systems and offers a SOAP-interface as well as a simplistic html-interface to these systems.

One of the other parallel activities in the SCIEnce project is *JRA1: Symbolic Computing on the Grid*, which focuses on developing a suitable framework for

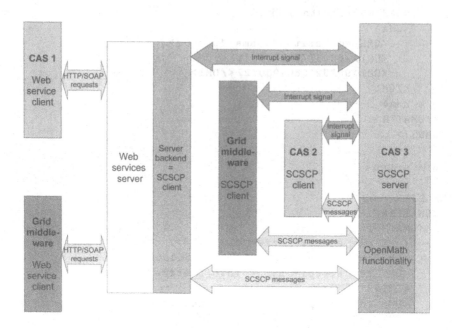

Fig. 3. SCSCP Overview

symbolic computing on computational grids. In this activity SCSCP is used for the communication between the server and the various systems.

5   Using KANT from MuPAD

As a first example, we consider factoring of polynomials defined similarly to Swinnerton-Dyer polynomials, but without the requirement that the primes are consecutive: for a set of distinct prime numbers $p_1, p_2, \ldots, p_n$, we define the polynomial $P_n(x)$ as

$$P_n(x) = \prod \left(x \pm \sqrt{p_1} \pm \sqrt{p_2} \ldots \pm \sqrt{p_n} \right),$$

where the product runs over all possible combinations of plus and minus signs, yielding 2^n factors in total. Hence the degree of such a polynomial is 2^n.

This polynomial is easily seen to be irreducible over $\mathbb{Z}$. On the other hand, suppose $\mathbb{F}$ is a finite field; then P_n splits into linear factors over a quadratic extension of $\mathbb{F}$, so it will only have linear and quadratic factors over $\mathbb{F}$ itself. In particular, P_n $(n > 1)$ is reducible over every finite field.

These polynomials are worst-case inputs for the Berlekamp-Zassenhaus algorithm for the factoring of polynomials over $\mathbb{Z}$. See [9, Section 15.3] for more information.

```
>> package("OpenMath"):
>> swindyer:=proc(plist) <some details omitted> :
>> R  := Dom::UnivariatePolynomial(x, Dom::Rational):
>> p1  := R(expand(swindyer([2,3,5,7,11]))):
>> p2  := R(expand(subs(swindyer([2,3,5,7,13,17])), x=3*x-2)):
>> p  := p1 * p2:
>> degree(p), nterms(p)
96, 49
```

So at this point we have constructed a univariate polynomial p, the product of two of these Swinnerton-Dyer like polynomials, with an affine transformation applied to the argument of the second one. It has 49 terms and degree $96 = 2^5 + 2^6$.

```
>> st  := time(): F1  := factor(p): time()-st
38431
```

So factoring it in MuPAD takes 38 seconds.

On the other hand, KANT has one of the fastest univariate polynomial factorizers available. If we convert the polynomial into OpenMath, transmit it to a machine running KANT almost 400 kilometers away, convert it to KANT syntax, factor it, convert it back into OpenMath, transmit it back to the original machine, and finally convert it back into MuPAD syntax:

```
>> kant  := SCSCP("scscp.math.tu-berlin.de", 26133):
>> st:=rtime(): F2:=kant::compute(hold(factor)(p)):rtime()-st
1221
```

So factoring in KANT only takes 1.2 seconds. To verify that the two results have the same factors:

```
>> FS1  := {op(Factored::factors(F1))}:
>> FS2  := {op(map(F2, X -> R(subs(expr((X[1])), '#1'=x))))}:
>> bool(FS1=FS2)
TRUE
```

The two-line conversion of the object KANT returns is necessary because one needs to explicitly state that this object $FS2$ is to be in the same polynomial ring that the original polynomial p was in.

The OpenMath objects are transmitted in uncompressed XML syntax, a few kilobytes for polynomials of this order of magnitude. Moreover, even though at this stage of the project no particular effort has been put into optimizing the conversions between CAS syntax and OpenMath, in our case these translations take only about 20 milliseconds each.

6   Using Macaulay from GAP

Using the framework mentioned earlier, we have created an SCSCP interface to Macaulay 2 [10]. We use this interface to perform a Gröbner basis computation on polynomials created in GAP. These polynomials can be used to create an automatic proof of the circle theorem of Apollonius.

```
gap> R := PolynomialRing(Rationals,
>           ["a","b","s","y","m1","m2","p1","p2"]);;
gap> a := R.1;; b := R.2;; s := R.3;; y := R.4;;
gap> m1 := R.5;; m2 := R.6;; p1 := R.7;; p2 := R.8;;
gap> pols := [
>     (m1-a)^2 + m2^2 - s^2,
>     m1^2 + (m2-b)^2 - s^2,
>     (m1-a)^2 + (m2-b)^2 - s^2,
>     -2*a*p1+2*b*p2,
>     -2*a*p2-2*b*p1+2*a*2*b,
>     a*b*y-1
> ];;
```

We have created a polynomial ring in 8 variables over $\mathbb{Q}$, and a list of 6 polynomials. We can try to compute a Gröbner Basis (with respect to the graded reverse lexicographic ordering) in GAP:

```
gap> B := GroebnerBasis(pols, MonomialGrevlexOrdering());
```

but the computation does not end within 30 minutes. We can also use the Macaulay 2 interface we created:

```
gap> I := Ideal(R, pols);;
gap> B2 := EvaluateBySCSCP("Macaulay2-Groebner", [I],
>               "scscp.win.tue.nl", 26133);;
#I  Creating a socket ...
#I  Connecting to a remote socket via TCP/IP ...
#I  Got connection initiation message
#I  Request sent ...
#I  Waiting for reply ...
gap> B2.object;
[ b-2*m2, a-2*m1, -4*m2*p2+p1^2+p2^2, m1*p1-m2*p2,
  4*m1*m2-m1*p2-m2*p1, s^2-m1^2-m2^2, y*m1*p2+y*m2*p1-1,
  4*y*m2^2*p2-p1, 4*y*m1^2*p2-4*m1+p1 ]
```

This calculation took only a few seconds. Moreover, the majority of the time is actually spent in the conversion of OpenMath to Macaulay syntax and back, so an even larger improvement could be obtained by optimization of this translation.

7   Further GAP Examples

The previous section demonstrated GAP as an SCSCP client. Here we would like to give one more example of the use of GAP as an SCSCP server with the development version of the GAP package SCSCP [16]. We will outline simple steps needed for the design and provision of the SCSCP service within the framework provided by SCSCP.

The GAP Small Groups Library [2] contains all groups of orders up to 2000, except groups of order 1024. The GAP command SmallGroup(n,i) returns the i-th group of order n. Moreover, for any group G of order $1 \leq |G| \leq 2000$ where $|G| \notin \{512, 1024\}$, GAP can determine its *library number*: the pair [n,i] such that G is isomorphic to SmallGroup(n,i). This is in particular the most efficient way to check whether two groups of "small" order are isomorphic or not.

For groups of order 512 the Small Groups Library contains all 10494213 non-isomorphic groups of this order and allows the user to retrieve any group by its library number, but it does not provide an identification facility. However, the GAP package ANUPQ [7] provides a function IdStandardPresented512Group that performs the latter task. Because the ANUPQ package only works in a UNIX environment it is useful to design an SCSCP service for identification of groups of order 512 that can be called from within GAP sessions running on other platforms (note that the client version of the SCSCP package for GAP does work under Windows).

First we need to decide how the client should transmit a group to the server: How should the group be encoded in OpenMath? Should it be converted into a permutation representation, which can be encoded using existing content dictionaries? Or should we develop new content dictionaries for other kinds of groups? Luckily, the SCSCP protocol provides enough freedom for the user to select his own data representation, and since we are interfacing between two copies of the GAP system, we are free to use a GAP-specific data format, namely the *pcgs code*: an integer, describing the *polycyclic generating sequence (pcgs)* of the group, to pass the data to the server. See the GAP manual and [1] for more details about the pcgs code.

First we create a function that takes the pcgs code of a group of order 512 and returns the number of this group in the GAP Small Groups library:

```
gap> IdGroup512ByCode := function( code )
>     local G, F, H;
>     G := PcGroupCode( code, 512 );
>     F := PqStandardPresentation( G );
>     H := PcGroupFpGroup ( F );
>     return IdStandardPresented512Group( H );
>     end;;
```

After such a function was created on the server, we need to make it "visible" as an SCSCP procedure under the name IdGroup512:

```
gap> InstallSCSCPprocedure ("IdGroup512", IdGroup512ByCode );
InstallSCSCPprocedure : procedure IdGroup512 installed.
```

For the convenience of the user, we provide the client's counterpart for this service, carrying all technical details (server, port) and also checking that the group is of order 512:

```
gap> IdGroup512:=function( G )
>     local code, result;
>     if Size( G ) <> 512 then
>       Error( "|G|<>512\n" );
>     fi;
>     code := CodePcGroup( G );
>     result := EvaluateBySCSCP ("IdGroup512ByCode", [ code ],
>                                "scscp.st-and.ac.uk", 26133);
```

```
>      return result.object;
>      end;;
```

Now when the client calls the function IdGroup512, it looks almost like the standard GAP function IdGroup (the user may switch off intermediate information messages):

```
gap> IdGroup512( DihedralGroup( 512 ) );
#I  Creating a socket ...
#I  Connecting to a remote socket via TCP/IP ...
#I  Got connection initiation message
#I  Request sent ...
#I  Waiting for reply ...
[ 512, 2042 ]
```

The GAP package SCSCP also offers functionality for parallel computations that may be used for example on a multi-core machine. It provides convenient functions for the user to parallelize computation by sending out two or more requests, and then collect either all results or (in the case when several methods are used for the same computation and it is not a priori clear which one will be fastest) get the first available result. More higher-level examples are contained in the package's manual which is available upon request and will be a part of the official release of the package.

8   Status and Future Research

At the moment of writing (March 2008) we implemented support for communications using SCSCP, both as servers and as clients, for the development versions of GAP, KANT, and MuPAD. Also, progress is being made with respect to OpenMath and SCSCP support in Maple. A part of ongoing work is to extend the range of mathematical objects that are understood by our systems, i.e., to enable translations from and to OpenMath for a wider set of OpenMath content dictionaries. We expect OpenMath support to be improved over the year 2008, greatly increasing the possibilities for exchanging mathematics between the various computer algebra systems.

As demonstrated, currently we expose the systems involved as SCSCP services. Future research includes using SCSCP to expose these systems as proper Web services, i.e. extending the WebProxy. We may also look into experience accumulated in the MONET project [18] and other existing technologies such as MathServe [11] and MathBroker II [12].

Furthermore, while developing this protocol we discovered that we have some need for representing mathematical objects in OpenMath that are not met by the current set of content dictionaries. This includes, for example, finitely presented groups, character tables of finite groups, and efficient representation of large matrices over finite fields. We plan to investigate these difficulties and create new content dictionaries where necessary.

Acknowledgement. The authors wish to thank the anonymous reviewers for their useful comments.

References

1. Besche, H.U., Eick, B.: Construction of finite groups. J. Symbolic Comput. 27(4), 387–404 (1999)
2. Besche, H.U., Eick, B., O'Brien, E.: The Small Groups Library, http://www-public.tu-bs.de:8080/~beick/soft/small/small.html
3. The Centre for Interdisciplinary Research in Computational Algebra (St Andrews, Scotland), http://www-circa.mcs.st-and.ac.uk/
4. CNRS, École Polytechnique (Palaiseau, France), http://www.polytechnique.fr/
5. The Dependable Systems Research Group at Heriot-Watt University, Edinburgh, Scotland, http://www.macs.hw.ac.uk/~dsg/content/public/home/home.php
6. The Discrete Algebra and Geometry group at the Technical University of Eindhoven, Netherlands, http://www.win.tue.nl/dw/dam/
7. Gamble, G., Nickel, W., O'Brien, E.: ANUPQ — ANU p-Quotient, GAP4 package, http://www.math.rwth-aachen.de/~Greg.Gamble/ANUPQ/
8. The GAP Group: GAP — Groups, Algorithms, and Programming, http://www.gap-system.org
9. Von zur Gathen, J., Gerhard, J.: Modern Computer Algebra. Cambridge University Press, Cambridge (1999)
10. Grayson, D.R., Stillman, M.E.: Macaulay 2, a software system for research in algebraic geometry, http://www.math.uiuc.edu/Macaulay2/
11. The MathServe Framework, http://www.ags.uni-sb.de/~jzimmer/mathserve.html
12. MathBroker II: Brokering Distributed Mathematical Services, http://www.risc.uni-linz.ac.at/research/parallel/projects/mathbroker2/
13. Institute e-Austria Timisoara, Romania, http://www.ieat.ro/
14. The KANT group at the Technical University of Berlin, Germany, http://www.math.tu-berlin.de/~kant/
15. Freundt, S., Horn, P., Konovalov, A., Linton, S., Roozemond, D.: Symbolic Computation Software Composability Protocol (SCSCP) Specification, Version 1.1. CIRCA (preprint, 2008), http://www.symbolic-computation.org/scscp/
16. Konovalov, A., Linton, S.: SCSCP — Symbolic Computation Software Composability Protocol. GAP 4 package
17. Maplesoft, Inc, Waterloo, Canada, http://www.maplesoft.com/
18. MONET, http://monet.nag.co.uk/
19. MuPAD, http://www.sciface.com
20. OpenMath, http://www.openmath.org
21. RISC-Linz, Austria, http://www.risc.uni-linz.ac.at/
22. SAGE: Open Source Mathematics Software, http://www.sagemath.org/
23. Roozemond, D.: OpenMath Content Dictionary: scscp1, http://www.win.tue.nl/SCIEnce/cds/scscp1.html
24. Roozemond, D.: OpenMath Content Dictionary: scscp2, http://www.win.tue.nl/SCIEnce/cds/scscp2.html
25. Symbolic Computation Infrastructure for Europe, http://www.symbolic-computation.org/
26. Research Group Computational Mathematics, Department of Mathematics, University of Kassel, Germany, http://www.mathematik.uni-kassel.de/compmath

Using Coq to Prove Properties of the Cache Level of a Functional Video-on-Demand Server*

J. Santiago Jorge, Victor M. Gulias, and Laura M. Castro

MADS group, Department of Computer Science, Universidade da Coruña
Campus de Elviña, s/n, 15071 A Coruña (Spain)
{sjorge,gulias,lcastro}@udc.es

Abstract. In this paper we describe our experiences applying formal software verification in a real-world distributed *Video-on-Demand* server. As the application of formal methods to large systems is extremely difficult, relevant properties of a particular subsystem have been identified and then verified separately. Conclusions on the whole system can be drawn later. The development consists of two parts: first, the definition of the algorithm in the CoQ proof assistant; second, codification of the theorems with the help of some new tactics derived from the abstraction of verification patterns common to different proofs.

Keywords: Formal methods, software verification, theorem provers, functional programming, real-world applications.

1 Introduction

Once a system has been developed, correcting remaining bugs represents one of the highest costs in software production [1]. In as much as testing does not guarantee software correction due to incompleteness in input testing data [2], formal verification of software properties is a mandatory alternative to assure quality of critical systems. This work proposes the verification of properties of a real-world application: a distributed video-on-demand server [3] which has been developed in the concurrent functional language ERLANG. Functional languages [4] have often been suggested as a suitable tool for writing programs which can later be formally analysed. This is due to their *referential transparency*, a powerful mathematical property of the functional paradigm that assures that equational reasoning makes sense.

As the application of formal methods to large systems is difficult, we propose a method [5,6] to separately verify relevant properties of part of the system, so that when other partial results are combined later on, conclusions on the whole system can be drawn. This method starts with building a model of the system in the CoQ proof assistant [7], and then checks the properties against the model.

This paper is structured as follows: first, an overview of the VoDKA project is shown. Section 3 highlights a relevant sample of code to be verified, and presents a CoQ model used to describe the algorithm; then, we sketch the certification process of the algorithm model. Finally, we present our conclusions.

* Supported by MEC TIN2005-08986 and Xunta de Galicia PGIDIT06PXIC105164PN.

S. Autexier et al. (Eds.): AISC/Calculemus/MKM 2008, LNAI 5144, pp. 296–299, 2008.

2   Building a Functional Video-on-Demand Server

A *Video-on-Demand* (VoD) server is a system that provides video services to multiple clients simultaneously. A VoD user can request a particular video at any time, with no pre-established temporal constraints. VoDKA (*VoD Kernel Architecture*, http://vodka.madsgroup.org) is a project supported by a local cable operator to provide video-on-demand services to its clients. A functional language, ERLANG, was chosen for the development.

ERLANG is a notable successful exception of a functional language applied to real-world problems [8]. Clusters built from cheap off-the-shelf components are proposed as an affordable solution for the huge amount of resources required by the VoD system. Cluster resources are organised as a hierarchical storage system with three levels: (a) *Repository level*, to store all the available media; (b) *Cache level*, responsible for storing videos requested from the repository level, before being streamed; and (c) *Streaming level*, a front end in charge of protocol adaption and media streaming to the final client.

3   Verification of Properties Using the Model

A sample piece of interesting software to be verified is part of the cache subsystem of the *Video-on-Demand* server: the block allocation algorithm. If a media object must be fetched to cache (because of a cache miss), enough space must be booked to load the object from the storage. If it was necessary to release some blocks, we have to assure that these blocks are not in use by other pending tasks. As locality in cache is important for performance, blocks of the same media object are in close proximity and we can speak of *block intervals*. A block interval is modeled as a triple (a, b, x): the interval between blocks a and b (inclusive) has x pending tasks. A list is used to store the whole sequence of block intervals.

```
Inductive interval: Set:= tuple: nat -> nat -> nat -> interval.
Inductive seq: Set:= Nil: seq | Cons: interval -> seq -> seq.
```

The function **add** sums up a request over a block. In order to certify the correctness of the algorithm implementation (or at least increase our confidence about it), we write a model of the implementation in the COQ proof assistant (figure 1). Due to the use of a functional language such ERLANG, the translation to COQ is quite straightforward.

At any moment each block has a particular number of ongoing requests. The key property that function **add** has to satisfy is that after adding up a request to a block, the number of requests over that particular block is incremented by one while all the other blocks remain unchanged. This property is split into two theorems provided that i equals n or not. So, **add** is certified by the proof of both laws. To certify the correctness of **add**, we need an auxiliary function nth that returns the ongoing requests for the nth block.

```
Theorem nth_add_1: (l:seq;n,i:nat) i=n -> nth n (add i l) = S (nth n l).
Theorem nth_add_2: (l:seq;n,i:nat) ~i=n -> nth n (add i l) = nth n l.
```

```
Fixpoint add [n:nat; l: seq]: seq := Cases l of
 | Nil => (Cons (tuple n n (S O)) Nil)
 | (Cons (tuple a b x) l') => Cases (le_lt_dec a n) of
   | (left _) => Cases (le_lt_dec n b) of
     | (left _) => Cases (eq_nat_dec a n) of
       | (left _) => Cases (eq_nat_dec n b) of
         | (left _) => (Cons (tuple a b (S x)) l')                          (*a=n, n=b*)
         | (right _) => (Cons (tuple a a (S x)) (Cons (tuple (S a) b x) l'))  (*a=n, n<b*)
         end
       | (right _) => Cases (eq_nat_dec n b) of
         | (left _) => (Cons (tuple a (pred b) x) (Cons (tuple b b (S x)) l'))  (*a<n, n=b*)
         | (right _) => (Cons (tuple a (pred n) x)                            (*a<n, n<b*)
                         (Cons (tuple n n (S x)) (Cons (tuple (S n) b x) l')))
         end
       end
   | (right _) => (Cons (tuple a b x) (add n l'))                            (*b<n*)
   end
 | (right _) => (Cons (tuple n n (S O)) l)                                   (*n<a*)
 end
end.
```

Fig. 1. Definition of add in the COQ system

The COQ proof system provides us with predefined tactics. Those can be extended with new ad-hoc tactics that define new proof patterns that prevent us from repeating steps when proving laws. The definition of new ad-hoc tactics, resulting from the abstraction of recurrent proof strategies applied at different proofs, helps us to avoid repetitive steps in the process.

3.1   Enforcing Canonical Representation of Interval Sequences

We enforce the use of a canonical representation of sequences, so that two different representations always correspond to two different objects. To improve efficiency, sequences $[(a_1, b_1, x_1), \ldots, (a_n, b_n, x_n)]$ are kept sorted $\forall i, a_i \leq b_i \wedge b_i < a_{i+1}$. What is more, to save space, the sequence is kept compact or *packed*, i.e. $\forall i, x_i \neq 0$, and $(x_i \neq x_{i+1}) \vee ((x_i = x_{i+1}) \wedge (b_i + 1 < a_{i+1}))$. The canonical form assumes that a sequence of intervals is both sorted and compact, thus achieving algorithms with better space and time behaviour.

First, we demonstrate that the application of add to a sequence of intervals in canonical representation delivers another sequence in canonical form. The output of add holds the predicate ascend but it does not hold packed; hence we need a new function pack which when applied to an *ascendant* sequence returns an equivalent *packed* one.

```
Lemma nth_pack: (l:seq;n:nat) ascend l -> nth n l = nth n (pack l).
```

Moreover, the result of add is a sequence sorted in ascendant order. And the result of pack on an ascendant sequence delivers a packed one.

```
Lemma ascend_add: (l:seq; i:nat) ascend l -> ascend (add i l).
Lemma packed_pack: (l: seq) ascend l -> packed (pack l).
```

Finally, we state and prove the law which establishes that given a canonical sequence, adding a request to any block with add, and applying pack after that, acquires a new canonical sequence.

```
Theorem can_pack_add:(l:seq;n:nat) canonical l->canonical (pack (add n l)).
```

4   Conclusions

In this work, we prove critical properties of software by modeling the relevant pieces of code using COQ, and then formally stating and proving those properties. A real-world system has been studied: a distributed *Video-on-Demand* server, built using the functional paradigm and used in several actual deployments with demanding real-world requirements.

The use of a functional language as implementation platform eases the construction of a COQ model and makes it conceivable to think of an (at least partially) automatic translation. Being a model, especially a hand-made one, it can not guarantee the full correctness of the actual implementation; however, the process helps to increase the reliability of the system.

Since we deal with a real-world case study, it is necessary to take into account the additional complexity involved in efficient (both in space and time) implementation of algorithms. In this case study, formalising the canonical representation of block interval sequences is not too complex, but the algorithms that handle those canonical intervals are harder and, thus, proving properties on their implementations becomes more complex and tedious. One of the key points to highlight is the identification, abstraction and reuse of verification patterns in the proof process.

Even though COQ has been used to address this case study, similar application of the method presented can be performed using other theorem provers.

References

1. The Risks Digest, http://catless.ncl.ac.uk/Risks
2. Ghezzi, C., Jazayeri, M., Mandrioli, D.: Fundamentals of Software Engineering. Prentice Hall, Englewood Cliffs (1991)
3. Gulías, V.M., Barreiro, M., Freire, J.L.: VODKA: Developing a video-on-demand server using distributed functional programming. Journal of Functional Programming 15, 403–430 (2005)
4. Hudak, P.: Conception, evolution, and application of functional programming languages. ACM Computing Surveys 21, 359–411 (1989)
5. Jorge, J.S.: Estudio de la verificación de propiedades de programas funcionales: de las pruebas manuales al uso de asistentes de pruebas. PhD thesis, University of A Coruña, Spain (2004)
6. Jorge, J.S., Gulías, V.M., Freire, J.L.: Certifying properties of an efficient functional program for computing Gröbner bases. Journal of Symbolic Computation (2008)
7. Bertot, Y., Casteran, P.: Interactive Theorem Proving and Program Development, Coq'Art: The Calculus of Inductive Constructions. Springer, Heidelberg (2004)
8. Wadler, P.: Functional programming: An angry half dozen. ACM Sigplan Notices 33, 25–30 (1998)

Automating Side Conditions in Formalized Partial Functions

Cezary Kaliszyk

Institute for Computing and Information Sciences,
Radboud University Nijmegen, The Netherlands
cek@cs.ru.nl

Abstract. Assumptions about the domains of partial functions are necessary in state-of-the-art proof assistants. On the other hand when mathematicians write about partial functions they tend not to explicitly write the side conditions. We present an approach to formalizing partiality in real and complex analysis in total frameworks that allows keeping the side conditions hidden from the user as long as they can be computed and simplified automatically. This framework simplifies defining and operating on partial functions in formalized real analysis in HOL LIGHT. Our framework allows simplifying expressions under partiality conditions in a proof assistant in a manner that resembles computer algebra systems.

1 Introduction

1.1 Motivation

When mathematicians write partial function they tend not to explicitly write assumptions about their domains. It is common for mathematical texts to include expressions like:

$$\ldots \frac{1}{x} \ldots$$

without specifying the type of the variable x and without giving any assumptions about it.

On the other hand these assumptions are necessary in proof assistants. Since most proof assistants are total frameworks, a similar formula expressed there looks like:

$$\forall x \in \mathbb{C}. x \neq 0 \Rightarrow \ldots \frac{1}{x} \ldots$$

The assumptions about the domain are obvious for any mathematician, in fact they can be generated by an algorithm. All names that have not been defined previously are considered to be universally quantified variables and all applications of partial functions give raise to conditions about their arguments. Inferring the types of variables is something that proof assistants are already good at. Giving the type of just one of the terms in an expression is often enough for a proof assistant to infer the types of the other. Mathematicians often work in a particular setting, where arithmetic operations and constants are assumed to be of

S. Autexier et al. (Eds.): AISC/Calculemus/MKM 2008, LNAI 5144, pp. 300–314, 2008.

particular types. Some proof assistants have mechanisms that allow to achieve a similar effect, eg. prioritizing a type in HOL LIGHT or using a local scope in COQ [5].

There are many examples of statements in libraries of theorems for proof assistants that include assumptions which are often omitted in mathematical practice. In particular the HOL LIGHT library part concerning real analysis includes statements like EXP_LN:

$$\forall x.0 < x \Rightarrow exp(\ln(x)) = x$$

Here the type of x is inferred automatically as real from the type of applied functions (the complex versions of the exponent and logarithm functions have different names in the library), but the domain conditions are not taken care of. The real logarithm is defined only for positive numbers, but the positivity assumption is not only in the statement of the theorems that include it, but also appears many times in the proofs that use this fact.

Computer algebra systems allow applying partial functions to terms and some of them have assumptions about variables computed automatically. This might be one of the reasons why for mathematicians computer algebra systems are usually more appealing than proof assistants. Unfortunately the way assumptions are handled in those systems is often approximate, and this is one of the reasons computer algebra systems sometimes give erroneous answers [2]. Therefore handling assumptions cannot be done in the same way in theorem proving.

In [11] we show, that it is possible to implement a prototype computer algebra system in HOL LIGHT and that proof assistants are already able to perform many simplification operations that one would expect from computer algebra. The prototype is able to perform many computations that involve total functions[1], but even simplest operations that require understanding partiality fail, since HOL LIGHT is a total framework:

```
In1 := diff (diff (\x. &3 * sin (&2 * x) + &7 + exp (exp x)))
Out1 := \x. exp x pow 2 * exp (exp x) + exp x * exp (exp x) +
        -- &12 * sin (&2 * x)
In2 := diff (\x. &1 / x)
Out2 := diff (\x. &1 / x)
```

The problem with the above example is that the function $\frac{1}{x}$ is mathematically a partial function that is not defined in zero. Still computer algebra systems asked for the derivative of it reply with $-\frac{1}{x^2}$. This answer is correct since the original function is differentiable on the whole domain where it is defined, and its derivative has the same domain. The proposed approach will let the framework correctly compute this kind of expressions.

We would also like to check whether approach for handling partiality in an automated way can be useful not only in formalizing partiality but might generalize to formalizing functions that operate on more complicated data structures, like when formalizing multivaluedness.

---

[1] The & operator is the coercion from natural numbers.

1.2 Approach

The domain of the function can be often inferred from the function itself. For example the domain of $1 + \frac{1}{x}$ can be computed to be $\lambda x.x \neq 0$. In such circumstances the domain can be represented by the function itself relieving the user from typing unnecessary expressions. This is not always the case. For example if the function $\lambda x.\frac{1}{x} - \frac{1}{x}$ is simplified to $\lambda x.0$ deriving the domain $\lambda x.x \neq 0$ is not possible. When a singularity point is not removed from the domain, the domain can be recomputed from the expression itself. Expressions in which singularity points are removed occur rarely in practical examples.

When we apply an operation in a CAS system[2] to a function f in a domain D, a function f' and its domain D' are returned. If the system can prove that D' represents the same domain as the one which we can compute from f' we can discard D'. We will be able to recompute it whenever it will be needed.

Our approach is to let the user input the partial functions as values from and to the `option` type and show them to the user as such, but to perform all operations on a total function of the underlying proof assistant with keeping the domain predicate alongside with the function. To do this we have two representations for functions and convert between them. The first representation is functions that operate on values in the option type and the second is pairs of total functions and domain predicates. We show how higher order functions (differentiation) can be defined in this framework and how terms involving it can be treated automatically.

1.3 Related Work

There are multiple approaches and frameworks for formalizing partial recursive functions. Ana Bove and Venanzio Capretta [4] introduce an approach to formalizing partial recursive functions and show how to apply it in the Coq proof assistant. Normally recursive functions are defined directly using `Fixpoint`, but that allows only primitive recursion. They propose to create an inductive definition that has a constructor for every recursive definition and create a `Fixpoint` that recurses over this definition. Alexander Krauss [12] has developed a framework for defining partial recursive functions in ISABELLE/HOL, that formally proves termination by searching for lexicographic combinations of size measures. William Farmer [9] proposes a scheme for defining partial recursive functions and implements it in IMPS. The main difference is that those approaches and frameworks compute the domains of partial recursive functions whereas we concentrate on functions in analysis which cannot be obtained by recursion and where the domain is limited because there are no values of the functions that would match their intuitive definition or that would allow properties like continuity.

The existing libraries for proof assistants contain formalized properties of functions in real and complex analysis. There are common approaches to partiality in existing libraries. It is common to define every function total. This is the case for the HOL LIGHT [10] library. Division is defined to return zero when

---

[2] We refer to the computer algebra functionality embedded in HOL as the CAS system.

dividing by zero. The resulting theory is consistent, but to make some standard theorems true assumptions are required. For example REAL_DIV_REFL:

$$\forall x . x \neq 0 \Rightarrow \frac{x}{x} = 1.$$

Another common approach is to require proofs that arguments applied to partial functions are in their domains. This is the case for the CoRN library [7] of formalized real and complex analysis for Coq. There division takes three arguments, the third one is a proof that the second argument is different from zero.

There are approaches to include partiality in the logic of the proof assistant. Those unfortunately complicate the logic and are already complicated for first order logics [16]. Some proof assistants are based on logics that support partial functions. An example is PVS [15] where partial functions are obtained by subtyping and IMPS [8] where there is a built-in notion of definedness of objects in the logic.

Olaf Müller and Konrad Slind [14] present an approach for lifting functions with the option monad that is closest to the one presented here. Their approach is aimed at partial recursive functions where computation of the domains of functions is not possible. Our approach is similar to applying the option monad to the real and complex values, but since particular functions need to have their domains reduced, we explicitly compute and keep the domains of functions and be able to transform these values back to original ones.

Finally computer algebra systems have their own approaches to partiality, eg "provisos" [6]. The main difference is they are intended to obtain maximum usability, sometimes at the cost of correctness. This is why those approaches cannot be used in a theorem proving environment.

1.4   Contents

This paper is organized as follows: in Section 2 we give the basic definitions of the two representations of partial functions and we define the operations used to convert between those representations. We also show a simplified example of a computation with partial functions. In Section 3 we present the design decisions and the details of our formalization. We show how does the automation work and show its limitations. Finally in Section 4 we present a conclusion and possible future work.

2   Proposed Approach

2.1   Basic Definitions

Our approach involves two representations of partial functions. The first representation is: as pair of a total extension of the original function and a domain predicate. The second representation is: a function from an option type to an option type. The first representation will be used in all automated calculations and the latter will be used in the user input and if possible in the output since it resembles better mathematical notation.

An option type is a type built on another type. The option type has two constructors. One denoting that the variable has a value and one used for no value. In proof assistants they are usually written as SOME α and NONE. We will denote those with $\overline{\alpha}$ and $-$. To simplify reading of the types, variables of the option type will be denoted as z and real variables as x both in the paper and in the shown examples from the system. The approach works for partial values of different types, but since HOL LIGHT does not have dependent types we cannot generalize over types, so we present our approach for a single type of partial values. We chose real numbers and not complex numbers since there are more decision procedures available in HOL LIGHT for real numbers and we make use of them.

We define two operations to convert between the two representations. Creating operations that work on the option type from the operations on the underlying proof assistant type is similar to applying the `option` monad operations *bind* composed with *return* to the functions in the proof assistant. In fact this is equivalent to the presented approach for functions that are really total. For functions that are undefined on a part of their original domain we additionally require the desired domain predicate so we create an operation that will additionally require the domain predicate and check it in the definition. We define @ that converts functions from the pair representation to the option representation (written as **papp** in the HOL LIGHT formalization) and $@^{-1}$ that converts a function in the option type to a pair (**punapp** in HOL LIGHT). The definition of @ is straightforward:

$$(f, D)@z = \begin{cases} \overline{fx} & \text{if } z = \overline{x} \land D(x) \\ - & \text{otherwise} \end{cases}$$

The inverse operation can be defined using the Hilbert operator (which we will denote as ϵ). This operator takes a property and returns an element that satisfies this property if such an element exists. It returns an undefined value when applied to a property that is not satisfied for any element. The inverse operation is defined as:

$$@^{-1}f = (\lambda x.\epsilon v.f(\overline{x}) = \overline{v}, \lambda x.\exists v.f(\overline{x}) = \overline{v})$$

The $@^{-1}$ function is the left inverse of @, (in fact we prove this in our formalization that for any F, D and z)[3]:

$$@^{-1}(\lambda z.(f, D)@z) = (f, D)$$

With the two operations definitions of the translations of the standard arithmetic operations are simple. The @ operator will check that the arguments applied to plus are defined. Note that in HOL LIGHT the syntax and semantics

---

[3] The $@^{-1}$ function is not the right inverse of @. This would require that for any function f and any z, $(@^{-1}f)@z = f(z)$. But this equality is not true for $z = $ NONE and $f(\text{NONE}) = \text{SOME}(0)$.

of expressions are very close, namely when syntactic expressions are applied to values they can be reduced to its syntax. This is why we will not distinguish between syntax and semantics in the paper:

$$a + b =_{\text{def}} @(\lambda xy.(x+y), \lambda xy.\top)$$

We can also define higher order functions that operate on partial functions by embedding the existing higher order operators from the proof assistant, first in the pair representation:

$$(f, D)' =_{\text{def}} (f', \lambda x.D(x) \wedge f \text{ is differentiable in } x)$$
$$f'(z) =_{\text{def}} (@^{-1}(f)')@z$$

2.2   Example in Mathematical Notation

With the definitions from the previous section it is possible to automatically simplify the side conditions in partial functions, we will first show it in the example and then show the full HOL LIGHT definitions and the algorithm for simplification in Section 3.2.

We will show a simplified example of automatically computing a derivative of a partial function in our framework. We will denote the derivative of a function $f(x)$ as $f(x)'$. The user types an expression:

$$(\lambda z.\pi z^2 + cz + \frac{2}{z})'$$

The expression that the user sees is written with standard mathematical operators. All the operator symbols are overloaded, and they are understood as the operations on partial functions, that is functions of type $(\mathbb{R})$option $\rightarrow$ $(\mathbb{R})$option $\rightarrow \ldots \rightarrow (\mathbb{R})$option. In the above expression z is the only variable of the $(\mathbb{R})$option type. All the other constants and expressions are their translations from the underlying total functions or constants. The only functions that are really partial (that is undefined on a part of the proof assistant domain) are division and differentiation and they are defined by providing their domain. The system unfolds the translation of all operators and constants, and computes a total function and its domain[4]:

$$(\lambda z.\langle \lambda x.\pi x^2 + cx + \frac{2}{x}, \lambda x.x \neq 0\rangle @z)'$$

We finally translate the derivative. For the obtained function we add the requirement that the derivative of the original function exists in the given point, otherwise a function defined in one point would always be differentiable there.

---

[4] In some proof assistants all computation is really simplification done by rewrite rules. This is the case in HOL LIGHT in which we will be formalizing this example, but we will refer to those simplifications as computation in the text.

This domain condition will be often combined with the assumptions about the domain of the original function:

$$\lambda z.(\langle(\lambda x.\pi x^2 + cx + \frac{2}{x})', \lambda x.x \neq 0 \wedge (\lambda x.\pi x^2 + cx + \frac{2}{x}) \text{ is differentiable in } x\rangle @z)$$

We can then apply the decision procedure for computing derivatives of total functions in the underlying proof assistant. It was not possible earlier, since the result of the procedure is a predicate with additional assumptions. It is possible with the use of the **papp** operator, since its definition ensures that the result does not depend on the function outside its desired domain. Since we also know the set on which the reciprocal is differentiable the domain can be simplified:

$$\lambda z.(\langle(\lambda x.2\pi x + c - \frac{2}{x^2}), \lambda x.x \neq 0\rangle @z)$$

Finally we try to return to the partial representation. This is done by reconstructing a partial function with the same symbols and recomputing its domain.

$$\lambda z.2\pi z + c - \frac{2}{z^2} = \lambda z.(\langle(\lambda x.2\pi x + c - \frac{2}{x^2}), \lambda x.x \neq 0\rangle @z)$$

Since the domains agree we can convert back and display the left hand side of the above equation as the final result to the user.

Returning from the representation of the function as a total function and its domain to the option type representation is not always possible, since a partial expression does not need to have an original form (can be expressed in the option representation). On the other hand the simplification is often possible and when it is possible it is desirable since it allows for greater readability. An example where it is not possible is:

$$\lambda z.\frac{1}{z} - \frac{1}{z} = \lambda z.(\langle\lambda x.0, \lambda x.x \neq 0\rangle @z)$$

The value is not equal to the constant function zero, since the expression does not have a value when x is zero. Furthermore for values of the option type even the term $z - z$ is not equal to zero if $z =$ NULL, therefore even after simplification to zero its value will depend on the variable z.

There are two approaches of treating this kind of terms. One can either simplify it to zero while leave the domain condition or not simplify the expression at all. We currently do not simplify expressions for which we cannot find a valid partial representation to return to. This is to avoid showing the user the complicated representation with the domain conditions. A possible approach that allows those simplifications and displays results in the option representation will be mentioned in Section 4.1.

3   The Implementation

3.1   Design Decisions

For our formalization we chose HOL LIGHT. The factors that influenced our choice were: a good library of real and complex analysis, as well as the possibility to write conversions in the same language as the language of the prover itself. HOL LIGHT is written in OCAML and is provided as an extension of it. This is very convenient for developing since it allows generating definitions and simplification rules by a programs and immediately using them in the prover.

In the representation with option types we use the vector type $\mathbb{R}^n \to \mathbb{R}$ instead of the curried types $\mathbb{R} \to \mathbb{R} \to \ldots \to \mathbb{R}$ to represent functions. One can convert between these two representations and the latter representation is often preferred since it allows partial application. The reason why we chose to work with the vector representation is that HOL LIGHT does not have general dependent types. Instead it has a bit less powerful mechanism that only allows proving theorems that reason about A^n for any n. We will use this to prove theorems about n-ary functions. With this approach some definitions (**papp** mentioned below and its properties) will have to be defined for multiple arities. On the other hand the theorems that are hard to prove will be only be needed to be proved once. Otherwise they would be needed for all versions of curried functions.

3.2 HOL LIGHT Implementation Details

In this section we will give the formalization details. To understand them knowledge of HOL LIGHT [10] is required. We will show an example of automatically computing the derivative of the partial function

$$f(z) = \pi z^2 + cz + \frac{2}{z}.$$

When the user inputs this function in the correct syntax in the main loop of the CAS, the system responds with the correct answer:

```
In1 := pdiff (\z. SOME pi * z * z + SOME c * z + & 2 / z)
Out1 := \z. & 2 * SOME pi * z + SOME c + --& 2 / (z * z)
```

The system computed this derivative automatically, but we will look at the conversions performed step by step. First lets examine the types in the entered expression. The variable z used in the function definitions is of the type `(real)option`. We overload all the standard arithmetic operators to their versions that take arguments of the `(real)option` type and produce results of this type. The coercion from naturals operator `&` creates values of this type. We decided not to overload the `&` operator to the coercion from real numbers (`SOME`), since this would lead to typing ambiguity and would require some types to be explicitly given in expressions.

The semantics of the standard arithmetic operations is to return a value if all arguments have a value and `NONE` if any of the arguments is `NONE`. For real partial functions we define an operation (called **papp**) that will create a partial function of type `(real)option` $\to$ `(real)option` $\to \ldots \to$ `(real)option` from a pair of a HOL LIGHT total function $\text{real}^n \to \text{real}$ and a predicate expressing its domain $\text{real}^n \to \text{bool}$. We show below the definitions of **papp** for one and two variables. In the formalization we see them as **papp1**, **papp2**, $\ldots$, but in the text we will refer to all those definitions together as **papp**:

```
new_definition '(papp1 (f, d) (SOME x) = if (d (lambda i.x)) then
  (SOME (f (lambda i.x))) else NONE) /\
  (papp1 ((f:A^1->A), (d:A^1->bool)) NONE = NONE)'
```

```
new_definition '(papp2 ((f:A^2->A), (d:A^2->bool)) (SOME x) (SOME y) =
  if (d (lambda i.if i = 1 then x else y)) then
    (SOME (f (lambda i.if i = 1 then x else y))) else NONE) /\
  (papp2 (f, d) NONE v = NONE) /\ (papp2 (f, d) v NONE = NONE)';;
```

In the above definitions we see the usage of **lambda** and below we see the usage of **$**. Those are used to create vectors and refer to vector elements. The reasons for using the vector types instead of curried type for functions was discussed in Section 3.1.

The total binary operations can be defined by applying a common operator, that defines binary operators in terms of **papp** for two variables. The types of all defined binary operations is (real)option → (real)option → (real)option. We show only the definition of addition on partial values:

```
new_definition 'pbinop (f:A->A->A) x y =
  papp2 ((\x:A^2. (f:A->A->A) (x$1) (x$2)),(\x:A^2.T)) x y';;
```

```
new_definition 'padd = pbinop real_add';;
```

The first partial function is division defined in terms of the reciprocal.

```
new_definition 'pinv = papp1 (partial ((\x:real^1. inv (x$1)),
  \x:real^1. ~((x$1) = &0)))';;
```

```
new_definition 'pdiv x y = pmul x (pinv y)';;
```

pdiff is the unary differentiation operator. It takes partial functions of the type (real)option→(real)option and returns functions of the same type. Since the derivative may not always exist it is defined using the Hilbert operator. Given a (partial) function it returns a partial function being a derivative of the given one on the intersection of its domain and the set on which it is differentiable. We will again define it in terms of **papp** applied to a total function and its domain. Since we are given a function and need to find its underlying total function and domain to apply the original differentiation predicate we will define **punapp** that returns this pair. For our definition it returns a pair of real→real and real→bool:

```
new_definition 'punapp1 f = ((\x:real^1. @v:real. (f(SOME (x$1))) =
  (SOME v)), (\x:real^1. ?v. (f (SOME (x$1))) = (SOME v)))';;
```

```
new_definition 'pdiff_proto (f:real^1->real, d:real^1->bool) =
  ( (\x:real^1. if d x /\ ?v. ((\x. f (lambda i. x)) diffl v) (x$1)
  then @v. ((\x. f (lambda i. x)) diffl v) (x$1) else &0) ,
  (\x:real^1. d x /\ ?v. ((\x. f (lambda i. x)) diffl v) (x$1)) )';;
```

```
new_definition 'pdiff f = papp1 (pdiff_proto (punapp1 f))';;
```

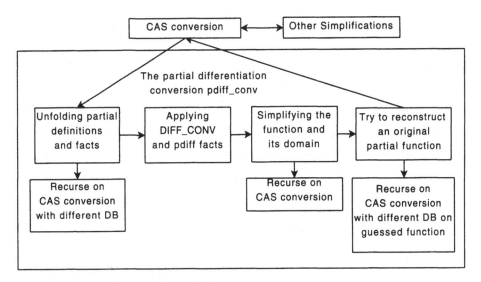

Fig. 1. A schematic view of the simplification performed by the partial differentiation conversion

The simplification of the term will be performed by a partial differentiation conversion `pdiff_conv` (Fig. 1). This conversion is a part of the knowledge base of the CAS and will be called by the CAS framework when the term has a `pdiff` term in it. To simplify the implementation of the partial differentiation conversion it will recursively call the CAS conversion with a modified database to simplify terms. The first step is a simplification performed by the main CAS conversion with the database of theorems for extended to include the above definitions of the partial operators and some basic facts, that will be described below. The conversion proves:

```
|- pdiff (\z. SOME pi * z * z + SOME c * z + & 2 / z) =
papp1 (((\x. @v. (((\x. x pow 2 * pi + c * x + &2 * inv x) diffl v) (x$1)),
(\x. ~(x$1 = &0) /\ (?v. (((\x. if ~(x = &0)
then x pow 2 * pi + c * x + &2 * inv x else @v. F) diffl v) (x$1)))))
```

All the partial operators and the `pdiff` operator were unfolded to their definitions. We notice that the partiality included in division (reciprocal) and differentiation have been propagated to the term. All occurrences of variables are pulled inside the `papp` terms and consecutive `papp` applications are combined by a set of reduction rules. This set includes a number of theorems, for the categories we give only single examples for one variable:

- rewrite rules that reduce the number of `papp` applications for SOME terms for arbitrary numbers of variables. An example for the second of two variables:

```
# papp2_beta_right;;
val it : thm = |- papp2 (f, d) (a:(A)option) (SOME b) =
  papp1 (((\x. f (lambda i. if i = 1 then x$1 else b)),
  (\x. d (lambda i. if i = 1 then x$1 else b))) a
```

– rewrite rules that combine multiple occurrences of the same variable:

```
# papp2_same;;
val it : thm =
|- papp2 (f, d) x x = papp1 ((\x:real^1. f ((lambda i. x$1):real^2)),
 \x:real^1. d ((lambda i. x$1):real^2)) x
```

– rewrite rules that combine consecutive applications of papp possibly with different numbers of abstracted variables:

```
# papp1_papp1;;
val it : thm = |- papp1 (f1, d1) (papp1 (f2, d2) (x:(A)option)) =
    papp1 ((\x. f1 (lambda i.(f2 x))),
    (\x. d2 x /\ d1 (lambda i.(f2 x)))) x
```

The next step performed by the partial differentiation conversion extracts the function to which the diff1 term is applied. The HOL LIGHT DIFF_CONV is applied to this term. For total functions it produces a diff1 theorem with no additional assumptions. For partial functions DIFF_CONV produces conditional theorems that have additional assumptions about the domain. For our example:

```
# DIFF_CONV '(\x. x pow 2 * pi + x * &c + &2 * inv x)';;
val it : thm = |- !x. ~(x = &0) ==>
  ((\x. x pow 2 * pi + x * &c + &2 * inv x) diff1
  ((((&2 * x pow (2 - 1)) * &1) * pi + &0 * x pow 2) +
  (&1 * &c + &0 * x) + &0 * inv x + --(&1 / x pow 2) * &2) x
```

Our formalization includes certain theorems about derivatives of partial functions where the derivative exists depending on some condition. For the example case the used theorem is about derivatives of functions that are not differentiable in a single point. We provide some similar theorems for inequalities which may arise in differentiating more complicated functions. The exact statement of the theorem used here is:

```
# pdiff_but_for_point;;
val it : thm = |- (!x. ~(x = w) ==> (f diff1 (g x)) x) ==>
    papp1((\(x:real^1). @v. ((\x:real. f x) diff1 v) (x$1)),
        (\(x:real^1). (~(x$1 = w) /\ d (x$1)) /\
    ?v. ((\x:real. if ~(x = w) then f x else @v. F) diff1 v) (x$1))) =
    papp1((\(x:real^1). g (x$1)), \(x:real^1). ~(x$1 = w) /\ d (x$1))
```

The partial differentiation conversion combines the above facts to prove:

```
|- pdiff (\z. SOME pi * z * z + SOME c * z + & 2 / z) =
papp1 ((\x. (((&2 * x$1 pow (2 - 1)) * &1) * pi + &0 * x$1 pow 2) +
    (&0 * x$1 + &1 * c) + &0 * inv (x$1) + --(&1 / x$1 pow 2) * &2),
    (\x. ~(x$1 = &0)))
```

The above function can be easily simplified, and this simplification is performed by recursively calling the CAS conversion both on the function and on the domain. For our example only the function can be reduced. For the recursive call to the CAS conversion we do not include the facts about partiality to prevent looping. The conversion proves:

```
|- pdiff (\z. SOME pi * z * z + SOME c * z + & 2 / z) =
   pappl ((\x. &2 * pi * x$1 + c + -- &2 * inv (x$1 * x$1)),
   (\x. ~(x$1 = &0)))
```

The last part of `pdiff_conv` tries to convert the term back to the original representation. As described is Section 3.1 this is not always possible, but it will be possible in our case. The algorithm for computing the original form examines the tree structure of the total function and reconstructs a partial function with the same structure. In our case:

```
# pconvert '(&2 * pi * (x:real^1)$1 + c + -- &2 * inv (x$1 * x$1))';;
val it : term = '& 2 * SOME pi * x + SOME c + --& 2 * pinv (x * x)'
```

We now check if the domain of the guessed partial function is the same as the original real one. To do this we apply the CAS conversion to the guessed term with the partial function definitions and facts about them again:

```
# cas_conv it;;
val it : thm = |- & 2 * SOME pi * z + SOME c + --& 2 * pinv (z * z) =
   pappl ((\x. &2 * pi * x$1 + c + -- &2 * inv (x$1 pow 2)),
   (\x. ~(x$1 pow 2 = &0))) z
```

The domain of the converted function is the same as the domain of the function we that was computed by differentiation[5]. Therefore we can compose this theorem with the previous result arriving at the final proved theorem:

```
|- pdiff (\z. SOME pi * z * z + SOME c * z + & 2 / z) =
   (\z. & 2 * SOME pi * z + SOME c + --& 2 / (z * z))
```

And the user is presented with the right hand side of the equation.

3.3   How to Extend the System

In this section we will show examples that the system cannot handle automatically. We will then show how the user can add theorems to the knowledge base to add automation for simplification of those terms. Consider adding a new partial function being the real square root:

```
new_definition 'psqrt = pappl ((\x. sqrt (x$1)), (\x. (x$1) >= &0))';;
```

---

[5] The two domains can be expressed in a slightly different way, thus there may be some theorem proving involved to show that they are equal. In our implementation the only thing performed is the CAS conversion, that internally tries HOL Light decision procedures for reals and tautology solving.

The original HOL LIGHT differentiation conversion `DIFF_CONV` is able to differentiate the real square root producing a differentiation predicate with a condition:

```
# DIFF_CONV '\x. sqrt x';;
val it : thm =
|- !x. &0 < x ==> ((\x. sqrt x) diffl inv (&2 * sqrt x) * &1) x
```

The partial differentiation conversion can not simplify the derivative of the partial square root automatically without additional facts in its knowledge base. This is because the result of the original differentiation conversion is only a condition for the function to be differentiable. It does not prove that the function is not differentiable elsewhere (namely in zero). To be able to simplify this function the user needs to prove an additional theorem that would show that the function is differentiable if and only if the variable is greater than zero. Namely:

```
|- (?v. ((\x. if x >= &0 then sqrt x else @v. F) diffl v) ((x:real^1)$1))
   = x$1 > &0
```

Adding this to the knowledge base allows the partial differentiation conversion to handle automatically the partial square root function.

4   Conclusion

The presented approach and formalized framework allow the automation of side-conditions. Simple expressions with partial functions can be simplified transparently to the user. More complicated partiality conditions still appear in the expressions.

The approach allows mathematical expressions in proof assistants to resemble those seen in computer algebra. The language for writing equations and for calculations (rewriting in HOL LIGHT) becomes simpler.

It can be useful for formalizing partial functions that we encounter in engineering books, for example in Abramowitz and Stegun [1] or in the NIST DLMF project [13].

4.1   Future Work

Our primary goal is to check how easily our approach can be extended to more complicated partial operations. For example with integration it is hard to check whether the objects are defined. Of course even then our approach gives a response, but the existential expression in the result may be hard to simplify.

It is important to note, that the standard HOL LIGHT equality does not take into account the option type, so any objects that do not exist will be equal. Defining an equality that is not true for `NONE` is possible, and this is what has been done in IMPS. On the other hand it leads to two separate notions of equality, which makes the expressions more complicated.

We would like to add more automation. All the simplifications that we perform can be done with functions of arbitrary number of variables. Those can be proved on the fly by special conversions. Our formalization currently has all simplifications rules proved for functions with at most two optional variables. Also the papp definitions for more variables and facts about them are analogous to their simpler version and their definitions can be created automatically by a ML function that calls HOL LIGHT's definition primitives.

We are looking for a policy for simplifying expressions. Currently when an expression is simplified in the total representation, but we cannot find an original partial representation, the whole conversion fails and the expression is returned unchanged. The same conversions would succeed with assumptions about the domains of variables present in the CAS environment. It would be therefore desirable to suggest assumptions about variables that would allow further simplification of terms [3].

It would be most interesting to see if the presented approach can be extended to address multivaluedness. Multivalued functions are rarely treated in proof assistants. On the other hand multivalued expressions tend to be one of the common sources of mistakes performed by computer algebra systems. There are not too many theorems in prover libraries that concern multivalued functions. The representation of multivalued functions could be done in a similar way as partiality is done in our approach.

References

1. Abramowitz, M., Stegun, I.A.(eds.): Handbook of Mathematical Functions With Formulas, Graphs, and Mathematical Tables, United States Department of Commerce, Washington, D.C. National Bureau of Standards Applied Mathematics Series, vol. 55 (June 1964); 9th Printing, November 1970 with corrections
2. Aslaksen, H.: Multiple-valued complex functions and computer algebra. SIGSAM Bulletin (ACM Special Interest Group on Symbolic and Algebraic Manipulation) 30(2), 12–20 (1996)
3. Beeson, M.: Using nonstandard analysis to ensure the correctness of symbolic computations. Int. J. Found. Comput. Sci. 6(3), 299–338 (1995)
4. Bove, A., Capretta, V.: Modelling general recursion in type theory. Mathematical Structures in Computer Science 15(4), 671–708 (2005)
5. Coq Development Team. The Coq Proof Assistant Reference Manual Version 8.1. INRIA-Rocquencourt (2006)
6. Corless, R.M., Jeffrey, D.J.: Well ... it isn't quite that simple. SIGSAM Bulletin (ACM Special Interest Group on Symbolic and Algebraic Manipulation) 26(3), 2–6 (1992)
7. Cruz-Filipe, L., Geuvers, H., Wiedijk, F.: C-CoRN, the constructive Coq repository at Nijmegen. In: Asperti, A., Bancerek, G., Trybulec, A. (eds.) MKM 2004. LNCS, vol. 3119, pp. 88–103. Springer, Heidelberg (2004)
8. Farmer, W.M., Guttman, J.D., Thayer, F.J.: IMPS: An Interactive Mathematical Proof System (system abstract). In: Stickel, M.E. (ed.) CADE 1990. LNCS, vol. 449, pp. 653–654. Springer, Heidelberg (1990)

9. Farmer, W.M.: A scheme for defining partial higher-order functions by recursion. In: Butterfield, A., Haegele, K. (eds.) IWFM, Workshops in Computing. BCS (1999)

10. Harrison, J.: HOL light: A tutorial introduction. In: Srivas, M., Camilleri, A. (eds.) FMCAD 1996. LNCS, vol. 1166, pp. 265–269. Springer, Heidelberg (1996)

11. Kaliszyk, C., Wiedijk, F.: Certified computer algebra on top of an interactive theorem prover. In: Kauers, M., Kerber, M., Miner, R., Windsteiger, W. (eds.) MKM/CALCULEMUS 2007. LNCS (LNAI), vol. 4573, pp. 94–105. Springer, Heidelberg (2007)

12. Krauss, A.: Partial recursive functions in higher-order logic. In: Furbach, U., Shankar, N. (eds.) IJCAR 2006. LNCS (LNAI), vol. 4130, pp. 589–603. Springer, Heidelberg (2006)

13. Lozier, D.W.: Nist digital library of mathematical functions. Ann. Math. Artif. Intell. 38(1-3), 105–119 (2003)

14. Müller, O., Slind, K.: Treating partiality in a logic of total functions. Comput. J. 40(10), 640–652 (1997)

15. Owre, S., Rushby, J., Shankar, N.: PVS: A prototype verification system. In: Kapur, D. (ed.) CADE 1992. LNCS, vol. 607, pp. 748–752. Springer, Heidelberg (1992)

16. Wiedijk, F., Zwanenburg, J.: First order logic with domain conditions. In: Basin, D., Wolff, B. (eds.) TPHOLs 2003. LNCS, vol. 2758, pp. 221–237. Springer, Heidelberg (2003)

Combining Isabelle and QEPCAD-B in the Prover's Palette

Laura I. Meikle and Jacques D. Fleuriot

School of Informatics, University of Edinburgh
Appleton Tower, Crichton Street, Edinburgh, EH8 9LE, UK
{lauram,jdf}@dai.ed.ac.uk

Abstract. We present the Prover's Palette, a framework for combining mathematical tools, and describe an integration of the theorem prover Isabelle with the computer algebra system QEPCAD-B following this approach. Examples are used to show how results from QEPCAD can be used in a variety of ways, with and without trust. We include new functionality for instantiating witnesses automatically and auto-running where applicable. We conclude that user-centric design yields systems integrations which are extremely versatile and easy to use.

1   Introduction

Reasoning about real nonlinear polynomials is generally difficult in interactive theorem provers. There is little automation, and the user must typically apply large numbers of inference rules manually. This is a tedious and time-consuming process. We had first hand experience of this while verifying computational geometry algorithms in the theorem prover Isabelle [22]. Of course, we are not the first or only ones to level this criticism: two major proof developments, the Flyspeck project [13] and the Prime Number Theorem [4], also make this complaint.

This class of problems in real algebra was shown to be decidable by Tarski [26], however, and several decision procedures exist. Most computer algebra systems (CAS) include at least one, with QEPCAD-B [6] and REDUCE [16] among the most sophisticated. Generally speaking though, the formal methods community has been less inclined to implement such algorithms directly inside theorem provers. Of course, there are exceptions: notably the Cohen-Hörmander algorithm has been implemented in HOL Light [14] and a related algorithm has been implemented in OCaml alongside the theorem prover Coq, with correctness proofs and proof producing capabilities [21]. The main benefit of implementing a decision procedure in a formal setting is the high degree of confidence in the results, but this comes at a price: systems can be limited in the types of problems they can address. Although the current CAS procedures also suffer from being impractical in some situations, as noted by Dolzmann et al. [11], their power greatly exceeds that of the theorem prover implementations. For this reason, we decided to extend the capabilities of the theorem prover Isabelle by integrating it with an existing CAS, namely QEPCAD-B.

S. Autexier et al. (Eds.): AISC/Calculemus/MKM 2008, LNAI 5144, pp. 315–330, 2008.

Our systems integration follows a user-centric framework that we call the Prover's Palette. The aim is to provide a flexible environment for formal verification, whereby the user is provided with the option to interact easily and intelligently with a suite of tools at the GUI level. The Prover's Palette attempts to satisfy the general criteria for an *Open Mechanised Mathematics Environment*, whereby the framework should be easily extensible and provide interfaces to existing CAS and theorem provers [5]. We have put these principles into action by adopting the recently developed Eclipse Proof General Kit (PG Kit) [3] as the foundations of the Prover's Palette. We believe the Prover's Palette architecture enables especially powerful and versatile integrations, where the user can customise how the external tool is used and how the result is applied in the theorem prover. We demonstrate this by combining Isabelle with QEPCAD: with a single click of a button, the external tool (QEPCAD) can be used to test the validity of a subgoal and to produce witnesses or counter-examples. This can all be done without introducing a proof dependency on the external tool. Where a user is willing to trust the external tool and the translations, after inspecting the output, the integration can be used in "oracle" mode to apply any simplified result to the current proof simply with another click. The framework offers support for both automated *and* interactive usage, and by adjusting other settings in the GUI, the framework can provide further proof guidance and even loop invariant discovery (see §6.3).

The paper is organised as follows: Section §2 gives brief descriptions of Isabelle, QEPCAD-B and Eclipse PG Kit. The architecture of our Prover's Palette is then introduced in Section §3, followed by a section on our QEPCAD widget. Sections §5 and 6 present some illustrative examples on how the Isabelle/QEPCAD framework can be used. An overview of related work is given in Section §7 and we conclude by describing our plans for future work.

2   Preliminaries

2.1   Isabelle

Isabelle is a generic, interactive theorem prover, written in ML, which can be used as a specification and verification system [25]. It is based on the LCF architecture, which means proofs are constructed form a small, trusted kernel that defines the basic inference rules. Soundness is enforced by using type-checking of the underlying programming language. Isabelle's built-in logic, the *meta-logic*, is intended only for the formalisation of other logics, known as the *object-logics*. There are a number of object logics in which Isabelle allows the user to encode particular problems. Of specific interest to our work is the capacity for proofs in higher order logic (HOL). This provides a framework powerful enough to reason about algorithms and sophisticated mathematical notions. Isabelle also provides an extensive library of theories and some automatic proof methods which combine simplification and classical reasoning. These tools greatly help mechanisation. Despite this, the power to reason automatically and efficiently about non-linear arithmetic is still lacking.

2.2   QEPCAD-B

QEPCAD-B is a tool for performing quantifier elimination over real closed fields [6]. It is based on the technique of partial cylindrical algebraic decomposition (CAD), which was originally implemented by Hong [17]. It is written in C and uses the SACLIB library [7]. QEPCAD[1] runs as a command-line tool, interactively prompting for the variables and bindings, the problem, and a wide array of configuration and operating parameters. It takes problems of the form

$$(Q_k x_k)(Q_{k+1} x_{k+1})\ldots(Q_r x_r)\phi(x_1,\ldots,x_r)$$

where $Q_k,\ldots,Q_r$ are quantifiers, $x_1,\ldots,x_r$ are real variables and ϕ is a quantifier-free formula consisting of a boolean combination of equalities and inequalities of polynomials with integer coefficients. It is not guaranteed to handle all such formulae, but in our experience it will handle most problems with the right configuration choices; when successful, it produces an equivalent formula in which no quantified variables occur.

2.3   Eclipse Proof General Kit

Eclipse Proof General Kit is a generic front-end for interactive theorem provers [3], based on the Eclipse integrated development environment [12]. It offers a user-friendy GUI, integrated with a powerful and extensible suite of editing, browsing and debugging tools. The two main widgets for theorem proving in the IDE are the proof script editor and the prover output view.

Eclipse PG Kit also provides a communications protocol and broker middleware which manages proofs-in-progress and mediates between components. The API follows the design principle of separating the prover from the user interface (UI) and uses the message-passing Proof General Interaction Protocol (PGIP) between the prover and the UI. The basic principle for representing proof scripts in the PG Kit is to use the prover's native language, and mark up the content with PGIP commands which give the proof script the structure needed by the interface. The format of the messages is defined by an XML schema. Any theorem prover which wants to be compatible with Eclipse PG Kit must implement certain XML markups to make the structure of the proof script explicit. To date, this has only been done for the theorem prover Isabelle, but it is expected that other provers will soon be compatible.

It should be noted that PG Kit is under active development, and Eclipse PG in particular has usability issues at the date of this writing; however, with patience we found it extremely powerful, and we expect a stable version imminently.

3   System Architecture

The Prover's Palette is an approach which aims to combine theorem provers with external tools at the level of the user interface. In the Eclipse PG

---

[1] For readability we use "QEPCAD" to mean Brown's QEPCAD-B v. 1.48.

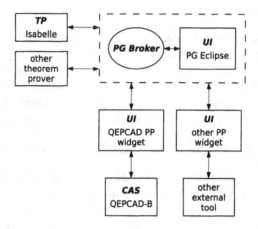

Fig. 1. Architecture of Prover's Palette Systems Integration

Kit, our Prover's Palette introduces a "View" (a graphical widget) for an external tool: this widget is backed by code which handles communication between the external tool and the PG Broker, and, more importantly, it presents details of the integration to the user interactively. The widget sits alongside the proof script editor and is easy to use when needed, but unobtrusive when not. It espouses the principle that, ultimately, the user knows how best to guide the interactive proof development, but should be freed of the burden of tedious or mundane details that might distract from the main proof effort.

Figure 1 shows the architecture of the integration. When a Prover's Palette widget starts, it registers a listener for changes to the proof state via the PG broker. When a new goal is observed, it is translated to a representation suitable for the external tool (QEPCAD, in this case), and presented to the user who may customise it. If any pre-processing is needed, the user is given the option to apply the appropriate prover commands (and the resulting proof state is reflected in the widget). The problem is then sent to the external tool (a new process is launched, in the case of QEPCAD, but other implementations could access network services) and the output from the tool is monitored and displayed to the user. When the final result is available, it is analysed and displayed along with options for how to apply it in the proof script (as a trusted oracle, as a subgoal, or any of the other modes presented in §6).

4   QEPCAD Widget

In Fig. 2, the "Start" tab in the QEPCAD widget shows the current subgoal reported by the PG broker (see bottom box). Clicking the "Next" button takes the user to the "Import" tab (Fig. 3). Here, the prover goal is dissected into assumptions and conclusion and it is made explicit which parts can be safely sent to QEPCAD. The variables and their bindings are also shown, and the user

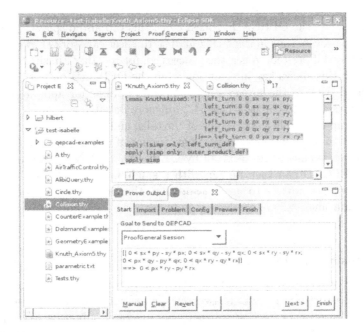

Fig. 2. Isabelle Eclipse Proof General with QEPCAD Widget

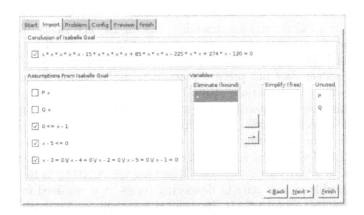

Fig. 3. QEPCAD Widget Import Tab

can choose whether to leave the variables bound or make them free. Clicking "Next" then takes the user to the "Problem" Tab. This describes the problem as it will be sent to QEPCAD and allows the user to edit any aspect of it. We note, in passing, that the "Problem" tab can in fact be usd as a convenient front-end to QEPCAD when it is running in stand-alone mode, i.e. without any associated theorem prover. Configuration settings, such as memory and projection type, can be adjusted in the "Config" tab.

Finally, the "Preview" tab displays to the user the script which will be sent to QEPCAD. As the QEPCAD process runs, this tab also displays QEPCAD's output. The process runs in the backgound so the user can work elsewhere and is easily cancelled if desired. When the process completes, the "Finish" tab (Fig. 4) is displayed: this shows key information from the QEPCAD output. The result is translated back to the prover's notation, and options for applying it to the proof are presented to the user (see §4.2 for more details).

It is of course possible to skip some of the tabs, and, where needed, the widget will attempt to make intelligent selections. In fact, if the original problem in the "Start" tab seems amenable to proof by QEPCAD, the "Finish" button will be enabled: QEPCAD can then be run with a single mouse click, making it very easy for complete novices to use. In addition, the user can also choose to run QEPCAD automatically in the background, alerting them only when a goal has been reduced.

4.1   Translation and Pre-processing

Translation between Isabelle and QEPCAD notation is done by the Prover's Palette infrastructure, in Java. The current prover state output is parsed to give a tree representation of the first subgoal[2]. For algebraic subterms — the parts of the goal we are interested in — this is relatively straightforward. Special symbols and uninterpreted functions are interpreted as prefix n-ary predicates. This may not always be correct, but for our purposes we can safely ignore such sub-terms: only the algebraic terms are relevant to QEPCAD in any case[3].

The Prover's Palette stresses flexibility, ease-of-use, and safety. To this end, where types other than "real" are used, the QEPCAD widget allows the goal to be sent to QEPCAD, but prevents the result from being used back in the Isabelle proof. The widget does not prevent potentially useful computations. Where a subterm is wholly incompatible with QEPCAD, the QEPCAD widget deselects it, by default, and presents the user with a number of options: transforming the subgoal to prenex normal form, expanding functions, reducing the subgoal without that term, or editting the term manually. Where a subgoal is fully compatible with QEPCAD, the widget will set all variable bindings (inferring them and their order where implicit or nested) and use QEPCAD to "solve" the problem. If a variable is used only in deselected terms, it is removed from the list which will be sent to QEPCAD. However if a variable occurs both in a deselected term and a selected term, the widget will initialize it as a free variable: the reason for this is that the desired action in this instance is more likely to be to attempt to "reduce" the problem in terms of that variable, i.e. using QEPCAD

---

[2] Although PG Kit uses XML, it does not currently include sufficient structural information about the goal for us always to disambiguate. We are in talks with the developers to provide this in future versions.

[3] As will be discussed in §7, this approach to translation is sub-optimal in theory, but expediant and reliable in practice. We have plans to adopt a more robust implementation as projects dedicated to this hard problem mature.

to eliminate all variables for which it has all known information and returning a result containing only those variables which have additional constraints.

4.2   Using the Result

The Prover's Palette attempts to give much flexibility when it comes to using the result of an external tool with a theorem prover. For QEPCAD, up to four modes may be applicable and, thus, be available for use:

> **Oracle:** the result *and* translation is to be believed by the prover (this makes use of an oracle method that we have written in Isabelle, for taking a statement and inserting it as a theorem)
>
> **Subgoal:** when the result is a simplified form (not completely reduced to "True" or "False"), a subgoal can be added asserting the equivalence of the input to the output
>
> **Instantiate:** if the result indicates an instantiation to use, this can automatically be applied
>
> **Thin:** if subterms of the subgoal are sent and the result indicates that certain assumptions and/or the conclusion are unnecessary, they can be "thinned out" of the proof state

These modes are illustrated in examples later in the paper. The widget also supports insertion of a comment in the proof script, containing details to re-run the computation in QEPCAD at a later time.

5   QEPCAD as Automated Oracle

As a demonstration of the oracle mode in the Isabelle/QEPCAD integration, we will focus on a theorem that we encountered while verifying Graham's Scan algorithm in Isabelle [22]. This algorithm is used to find the convex hull of a set of points in two dimensions and relies heavily on the notion of a *counter clockwise* (CC) system. In the book Axioms and Hulls, Knuth captures this CC system through a succinct axiomatisation which defines properties of *left turns* [20]. It would have been possible to adopt Knuth's axiomatic approach as a basis to formally represent Graham's Scan. However, we chose to follow Isabelle's methodology of maintaining consistency by developing new theories on top of old ones through conservative extensions only. In our case we built upon a theory of two dimensional vectors. The left turn property was then defined in terms of the outer product which itself is defined in terms of Cartesian coordinates. We write pqr to mean that the point r lies to the left of the directed line from p to q, which is represented algebraically as:

$$pqr \equiv (q_x - p_x)(r_y - p_y) - (q_y - p_y)(r_x - p_x) > 0$$

Along with this definition, we used a development of Floyd-Hoare logic to formalise Graham's Scan in Isabelle. This enabled the formal specification to closely resemble an implementation of the algorithm and also gave us a way of reasoning mathematically about imperative constructs. Under this approach, verification conditions (VCs) were automatically generated by Isabelle. By proving these VCs, the correctness of the algorithm is established. However, validating the VCs for Graham's Scan required proving many difficult subgoals, including:

$$KnuthsAxiom5 : tsp \wedge tsq \wedge tsr \wedge tpq \wedge tqr \longrightarrow tpr$$

This is presented as Axiom 5 in Knuth's book and is well known to be true. Despite this, it is not easy to prove in Isabelle alone: the proof breaks down into many case splits and took us several hours to complete, even when we set point t to be the origin. With the Isabelle/QEPCAD integration, we can use QEPCAD to solve this problem quickly.

In our system, the Isabelle problem appears automatically in the "Start" tab of the QEPCAD widget. The widget first recommends that the left turn definition should be expanded, which translates the problem to algebraic form. The "Finish" button is then enabled (see Fig. 2). Clicking the "Finish" button sends the problem off to QEPCAD in *solve* mode, which requires all implicit variable bindings to be made explicit. In this example, all bindings are automatically deduced to be universal. When a result is found—a matter of milliseconds in this case—the "Finish" tab is displayed (see Fig. 4). Selecting the "Oracle" button will generate the appropriate Isabelle command for the translated QEPCAD result to be trusted in Isabelle. In this example, the Isabelle lemma is then proved.

6   Guiding Fully Formal Proofs

In many domains, formal correctness requirements disallow reliance on tools such as QEPCAD. Nevertheless, there are many ways its results can be useful, from simplifying the subgoal to finding witnesses and missing assumptions.

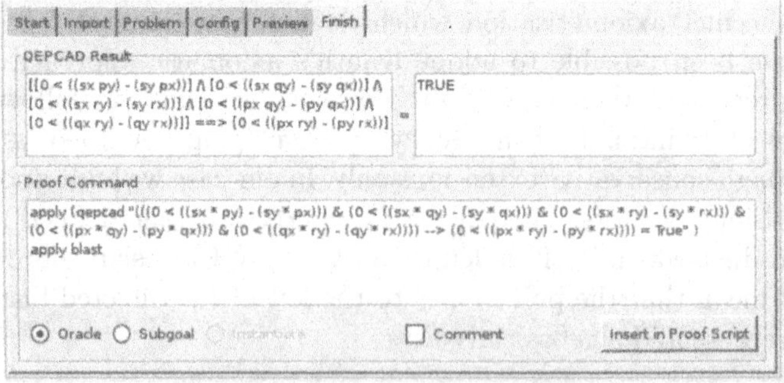

Fig. 4. QEPCAD Finish Tab

6.1   Simplifying Problems

Removing Superfluous Assumptions. Consider the following problem taken from our verification of Graham's Scan:

$$stq \wedge str \wedge stp \wedge trp \wedge srq \wedge spr \wedge tpq \wedge \neg psq \wedge tqr \longrightarrow pqs$$

In this goal there are superfluous assumptions which obscure the relevant facts needed in the proof. This is a common difficulty with interactive proof. However, the "Import" tab of the QEPCAD widget provides a way for the user to easily find a minimal set of assumptions which imply the conclusion. In this tab the user can remove assumptions and test whether the resulting statement still holds. If it does, the removed assumptions are not necessary in the proof. Following this basic strategy, we can discover (within two minutes) that the Isabelle subgoal above is equivalent to:

$$str \wedge tpq \wedge tqr \wedge srq \wedge spr \longrightarrow pqs$$

Whenever QEPCAD returns "True" with deselected assumptions, the widget presents an option to "Thin" in the "Finish" tab. Clicking this causes the redundant assumptions to be removed from the Isabelle goal automatically. Again, no proof dependency on QEPCAD is introduced.

Discovering Inconsistent Assumptions. Another way in which QEPCAD can simplify an Isabelle problem is by discovering if contradictory assumptions exist within the goal. In Isabelle, proving that assumptions are inconsistent is sometimes easier than trying to prove that the conclusion holds. In the verification of Graham's Scan, we found this to be a common situation, especially in proofs that involved case splits. As an example, consider the problem:

$$str \wedge tpq \wedge tqr \wedge srq \wedge spr \wedge tsr \longrightarrow pqs$$

Using a similar procedure to that described in the previous section, we can easily interact with the QEPCAD widget to discover if a contradiction is present in the assumptions. This is achieved by deselecting the conclusion in the "Import" tab so that only the assumptions of the goal are sent to QEPCAD. If QEPCAD returns "False" then we know there must be a contradiction present. A simple, interactive search can then be performed to deduce the minimal set of contradicting assumptions. For the above example we easily discover that the first and last assumptions contradict each other: $str \wedge tsr \longrightarrow False$. In this situation, the "Thin" button appears again and provides the user with a fully formal, automatic way to alter the original Isabelle goal to this new one. Proving the new goal in Isabelle is straightforward in this instance.

Reducing Number of Variables. Sending a formula to QEPCAD produces one of three results: True, False, or a simplified formula. For the latter case, this simplified formula is only in terms of the free variables. This can be useful for reducing the number of variables in an Isabelle problem, thereby easing the formal reasoning process. To utilise this feature, one has to over-ride the default Isabelle/QEPCAD "solve" mode (where all variables are quantified). This is achieved by going to the "Import" tab, which displays a table of variables and their associated bindings, and altering the bindings accordingly.

We motivate this functionality further by considering a collision problem described by Collins and Hong [17]. The problem is from robot motion planning and queries whether two moving objects will ever collide. Consider a moving circle (1) and a moving square (2), represented algebraically as:

$$
\begin{aligned}
(1) \quad & (x - t)^2 + y^2 \leq 1 \\
(2) \quad & -1 \leq x - \tfrac{17}{16}t \leq 1 \wedge -9 \leq y - \tfrac{17}{16}t \leq -7
\end{aligned}
$$

We want to know if these objects will ever collide. This is asking whether:

$$
\exists\, t\ x\ y.\ t \geq 0 \wedge (1) \wedge (2)
$$

QEPCAD quickly confirms that this is "True", there will be a time when the circle and square collide. In Isabelle alone, this is a difficult proof. If the user trusts the Isabelle/QEPCAD integration, of course, they can simply move on; but even when the user requires a fully formal proof, the integration can be of assistance. The parametric query can be transformed into the implicit representation of the problem, by keeping the variables x and y free and only binding t. Here, the "Subgoal" option is offered in addition to the "Oracle" option. This introduces a new subgoal in the proof which asserts that the original Isabelle subgoal is equivalent to the reduced form found by QEPCAD (with t eliminated, in this example). Proving this new subgoal formally in Isabelle allows the original goal to be transformed into the reduced problem. Proving the equivalence may still be challenging, but in many situations it is simpler than proving the original goal.

6.2   Producing Witnesses

After some experimentation, we realised that it would be useful if QEPCAD could give us witnesses to true formulae which are completely existentially quantified. We discussed this with Chris Brown, the developer of QEPCAD-B, and he has added this feature (in version 1.48). Our QEPCAD widget invokes this feature when applicable and shows the witness in the "Finish" tab. The "Instantiate" button is displayed in the "Finish" tab, and instantiations can be inserted and processed in the proof script.

Consider again the collision problem of the previous section. QEPCAD tells us that one possible solution is: $t = \tfrac{96}{17}$, $x = \tfrac{96}{17}$ and $y = -1$. Instantiating these variables in Isabelle completes the proof. In this example, the witness feature is more useful than our reduction to implicit form, but, of course, there are many instances where a reduced form is preferable (and it should be remembered that witnesses can only be found for problems which are fully existentially quantified).

6.3   Discovering Missing Assumptions

For complex proof developments it is a fairly common scenario to discover that
the first attempt at formally specifying a problem is incorrect. This is especially true when verifying algorithms using Hoare logic. In this setting, the user
has to provide the correct loop invariant (i.e. the facts that do not change on
each iteration of the loop) in order to prove the correctness of the algorithm.
Discovering the correct loop invariant is a challenging task, which is generally
accepted as non-trivial. From our own experience in verifying Graham's Scan,
we observed that our initial loop invariant needed refinement several times – a
process guided by failed proof attempts. Often, the root cause for failure was a
missing assumption, but identifying this was hard.

With our new Isabelle/QEPCAD integration, discovering the missing assumptions of invalid theorems is made easy. We provide two techniques: a counter-
example generator and a search strategy which explores the information gained
by altering the bindings of variables.

The counter-example feature is applicable when a false conjecture contains
only universally quantified variables. In these situations, the user can click a
button which automatically translates the negated original conjecture into the
equivalent existential form:

$$(\neg \forall x_1 \ldots x_n. \, \Psi(x_1,\ldots,x_n)) = (\exists x_1 \ldots x_n. \, \neg\Psi(x_1,\ldots,x_n))$$

where $\Psi(x_1, \ldots ,x_n)$ is a quantifier free formula. It then calls the witness function of QEPCAD to obtain a counterexample to the original conjecture. Let us
demonstrate this feature through an illustrative example taken from our verification of Graham's Scan:

$$bea \wedge abd \wedge cab \wedge ade \longrightarrow ace$$

Using our new framework, the QEPCAD widget quickly tells us this lemma is
false. In this particular case, we know from the context that a subgoal similar
to this is required. The counterexample generator tells us that the following
instantiations will falsify the original lemma:

$$a = (0,0) \quad b = (-1,-1) \quad c = (-1,-3) \quad d = (-1,-\tfrac{5}{2}) \quad e = (-1,-\tfrac{23}{8})$$

By drawing this particular case we gain an insight into why the conjecture is
false. It is clear to see that all the assumptions hold, but the conclusion does not.
From the diagram we can see that we want c to lie in the dark grey region. The
extra assumption $\neg adc$ would constrain the position of c and make the conjecture
true.

Despite the counter-example generator enabling us to understand why the
conjecture is false, it is a method that is not always applicable. Existential
quantifiers may be present in the conjecture or the number of variables and
assumptions may be so large that it would not be practical to draw the situation
and gain an appreciation of what is lacking in the assumptions.

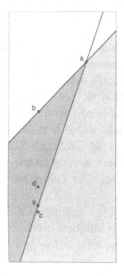

An alternative approach is to use the QEPCAD widget interactively to alter the bindings of variables: by keeping certain variables free and others bound we can identify what is missing. There is some art in selecting which variables should be free. In the previous example, variables a and b are used the most, so we translate a to be the origin and heuristically choose to keep b bound. With the other variables free, QEPCAD then returns:

$$d_y c_x - d_x c_y \geq 0 \vee d_x e_y - d_y e_x \leq 0 \vee e_y c_x - e_x c_y > 0$$

The second and third disjuncts are unenlightening (a negated assumption and the conclusion), but the first disjunct, however, is a hitherto missing condition to our Isabelle lemma. QEPCAD has told us that the lemma can be proven given $\neg adc$ with a at the origin. This is the same missing assumption we discovered using the counter-example generator. With this new fact, the framework has therefore led us to the discovery of a missing component in the loop invariant.

7   Related Work

The idea of integrating systems is certainly not new. The design of environments to combine several heterogeneous systems has been widely studied over the past decade and there is even a conference (FroCos) dedicated for research in this specific area. We do not attempt to survey the breadth of concepts and implementations here, but will instead focus on related work in the our domain, that of combining theorem provers and CAS.

Calmet et al. distinguish three main approaches to how logical and symbolic computations can be combined [9]: as an extension of a CAS to enable deduction (e.g. the Theorema project [8]), by implementing a CAS inside a theorem prover (e.g. the work of Harrison [14] and Kaliszyk et al. [18]), and finally through the combination of existing systems. The first two approaches have the attractive quality that they guarantee soundness, but also the major drawback that they require re-implementing substantial amounts of code. The second approach also tends to suffer from inefficiency, as mentioned in §1.

Our own work adopts the third methodology and takes advantage of the rich variety of existing systems. One of the biggest challenges with this endeavour is in translating between languages and representations. Our approach to translation, which uses syntactic conversions, ensures variables are of the correct type (see §4.1), but does not ensure the semantics of the operators are identical. This is a difficult problem and has been a concern of the computer algebra community for some time. There are several ambitious projects aimed at producing a standard for language and translation, notably OpenMath [24] and the OM-SCS framework [9]. We hope to take advantage of such work as more systems become compatible with standards. For now, we note that where systems vary and "Oracle" mode is used, results must be checked by the user of our Prover's Palette.

Systems integrations can be further classified on the basis of two questions:

1. How are different systems invoked?
2. How are results from different systems combined?

A common means of invoking systems is to introduce text commands in the proof script, for example the HOL/Maple [15] and PVS/QEPCAD [27] integrations. A drawback is that there can be a high barrier to entry as the user is required to learn new commands. This is especially difficult when the commands offer additional options to change operating parameters. Our approach differs in that it supports novices by allowing problems to be sent using the click of a mouse. The new Sledgehammer tool, which combines Isabelle with various automatic theorem provers, also works on this principle [23]: the user clicks an option from a scroll-down menu in the PG Emacs environment and, when a result is found, the user is alerted and the appropriate Isabelle commands are made available for cut-and-paste into the proof script. There are major differences between our work and that of Sledgehammer: most notably, we allow configuration parameters and translations to be adjusted, and we support multiple ways of using results (as well as single-click insertion and processing). When interfacing with a highly configurable tool such as QEPCAD, such flexibility is extremely useful, as we have shown; this in particular distinguishes our integration from [27].

The alternative to invoking a system interactively is to detect when it should be called automatically. Ballarin's integration of Isabelle and Maple follows this approach by extending Isabelle's simplification rules [5]. The Prover's Palette offers similar functionality, but instead of interfacing at the level of the prover, the widget is activated at the UI level: the user is then able to inspect the result and choose how best to use it.

When it comes to the integration of CAS results into theorem provers, the issue of trust becomes important. Having complete trust in a system and the translations between systems allows the results to be used easily within a proof (as shown with our "Oracle" mode). However, to maintain formally correct proofs is harder. Two methods which can been used to do this are:

(a) **Checking:** the CAS is used to compute a result which is then checked deductively (often in the prover)
(b) **Discovering:** a hierarchic *proof plan* is constructed using the results of the CAS and is then used to guide the formal proof

Our witness generator and "subgoal" option are examples of (a). In a similar vein, Harrison's so-called skeptic's approach to combining systems uses Maple to return simplifications and factorisations of polynomials which are then checked in HOL and used to guide fully formal proofs [15]. This approach is only useful when the verification of results is simpler than its discovery. We also employ an approach related to method (b) in our framework by aiding the discovery of counter-examples and missing assumptions, thereby formulating a strategy on how to fix failed proof attempts. Other work which has looked at how a CAS can formulated proof plans is that of Kerber et. al [19]. They view the CAS algorithms as *methods* which can give knowledge on how a result was found. This is similar to the work where CALVIN, an experimental CAS, is used to generate a trace along with the result [9]. This trace can then be used to reconstruct the main reasoning steps in the theorem prover IMPS.

The biggest drawback of our Prover's Palette approach, in our opinion, is the limited granularity at which the integration applies: it helps with a goal in the theorem prover; it does not combine proofs from different systems. Given the state of the art of IDEs and verifiers, this is realistic, but with ongoing developments in "Proof Engineering", we expect the range of possibilities that become available through integrations to be greatly extended. What is unique about our integration is the flexibility on how to use a result: we provide the functionality to trust, check and discover.

8   Future Work

Our main thrust of future work is to explore how our approach generalises. However, there are certain issues with the QEPCAD integration that we want to understand first, as we expect this will make our generalisation better. One major criticism of QEPCAD is that its double-exponential complexity makes it impractical in many situations. Performance can be improved by reducing the number of variables, decreasing the degree of polynomials, and using QEPCAD's specialised quantifiers such as "almost all". While users can transform problems manually, as we did with Knuth's Axiom 5, we believe there is scope for automating such transformations and we are working on ways of doing this in a formally correct manner.

We also plan to increase the class of Isabelle problems which can automatically be sent to QEPCAD. Some operators, including division, absolute value, and rational powers, can be included by adding suitable preprocessing rules. Permitting formulae which include exponentials, logarithms, and trigonometric functions can be achieved by replacing occurrences of them with polynomial upper and lower bounds. This approach is adopted by Akbarpour et al. [1]. They have integrated the theorem prover Metis with a formally proven decision procedure [14]. However, speed and coverage are issues with this particular integration: two examples are presented which their decision procedure cannot handle. QEPCAD returns false for these examples, indicating tighter bounds are needed. It could be attractive to combine their "bounds" approach, for reasoning about logarithms and exponentials, with our work, making QEPCAD available for use with the bounds they supply.

We believe our approach will work well for integrating other systems and are excited to test this hypothesis. We are currently exploring visualisation tools, model checkers, and other CAS for inclusion in the Prover's Palette. We note that, while QEPCAD is a good tool for the problems in §5 and §6, other tools could perform the same computations, and the results from those tools could be used in a proof integration in identical ways. The majority of the code for the QEPCAD integration is actually independent of QEPCAD: many of the GUI components, translation code, and other library routines are all applicable to other CAS, so we anticipate that incorporating other systems into the Prover's Palette will not be unduly difficult.

As more theorem provers are linked with PG Kit, we would also like to ensure they can be supported. In situations where theorem provers can communicate in a standard form, such as OpenMath or OMSCS, the integration becomes much easier. If the PGIP standard is extended to give richer structural details of proof states in a standard form, then a PGIP-compatible prover could directly be used with QEPCAD (or any Prover's Palette widget). Prover-specific customisation is needed only for the optional (though useful) preprocessing and result application steps.

Once we begin to expand the suite of tools in the Prover's Palette we will undoubtedly have to consider the complexity of the cooperations. A number of projects, including KOMET [9], PROSPER [10], Logic Broker [2], and PG Kit, could assist in such multi-system integrations: the Prover's Palette could offer a "meta tool" widget, recommending systems to use at crucial points in the proof development (and, where possible, automatically opening the corresponding widget).

Acknowledgements. We would like to thank the reviewers for their useful comments. This work was funded by the EPSRC grant EP/E005713/1.

References

1. Akbarpour, B., Paulson, L.C.: Extending a Resolution Prover for Inequalities on Elementary Functions. In: Dershowitz, N., Voronkov, A. (eds.) LPAR 2007. LNCS (LNAI), vol. 4790, pp. 47–61. Springer, Heidelberg (2007)
2. Armando, A., Zini, D.: Towards Interoperable Mechanized Reasoning Systems: the Logic Broker Architecture. AI*IA 3, 70–75 (2000)
3. Aspinall, D., Lüth, C., Winterstein, D.: A Framework for Interactive Proof. In: Kauers, M., Kerber, M., Miner, R., Windsteiger, W. (eds.) MKM/CALCULEMUS 2007. LNCS (LNAI), vol. 4573, pp. 161–175. Springer, Heidelberg (2007)
4. Avigad, J.: Notes on a formalization of the prime number theorem, http://www.andrew.cmu.edu/user/avigad/isabelle/pntnotes_a4.pdf
5. Ballarin, C., Homann, K., Calmet, J.: Theorems and Algorithms: An Interface between Isabelle and Maple. In: International Symposium on Symbolic and Algebraic Computation, pp. 150–157 (1995)
6. Brown, C.W.: QEPCAD B: a program for computing with semi-algebraic sets using CADs. SIGSAM Bulletin 37(4), 97–108 (2003)
7. Buchbergerger, B., et al.: SACLIB 1.1 user's guide. RISC-LINZ Report Services, Tech. Report No. 93-19 (1993)
8. Buchberger, B., et al.: The Theorema Project: A Progress Report. In: Symbolic Computation and Automated Reasoning. Proceedings of Calculemus (2001)

9. Calmet, J., Ballarin, C., Kullmann, P.: Integration of Deduction and Computation. In: Applications of Computer Algebra (2001)
10. Dennis, L.A., Collins, G., Norrish, M., Boulton, R., Slind, K., Melham, T.: The PROSPER Toolkit. Int. J. Software Tools for Technology Transfer 4(2) (2003)
11. Dolzmann, A., Sturm, T., Weispfenning, V.: Real Quantifier Elimination in Practice. Technical Report MIP-9720, Univrsit at Passau, D-94030 (1997)
12. The Eclipse Foundation, http://www.eclipse.org
13. Hales, T., et al.: The Flyspeck Project, http://code.google.com/p/flyspeck/
14. Harrison, J.: A Proof-Producing Decision Procedure for Real Arithmetic. In: Nieuwenhuis, R. (ed.) CADE 2005. LNCS (LNAI), vol. 3632. Springer, Heidelberg (2005)
15. Harrison, J., Théry, L.: A Skeptic's Approach to Combining HOL and Maple. Journal of Automated Reasoning (1997)
16. Hearn, A.: REDUCE User's Manual. RAND Publication (1995)
17. Collins, G.E., Hong, H.: Partial Cylindrical Algebraic Decomposition for Quantifier Elimination. Journal of Symbolic Computation 12(3) (1991)
18. Kaliszyk, C., Wiedijk, F.: Certified Computer Algebra on top of an Interactive Theorem Prover. In: Towards Mechanized Mathematical Assistants. Proceedings of Calculemus (2007)
19. Kerber, M., Kohlhase, M., Sorge, V.: Integrating Computer Algebra into Proof Planning. Journal of Automated Reasoning 21(3), 327–355 (1998)
20. Knuth, D.E.: Axioms and Hulls. LNCS, vol. 606. Springer, Heidelberg (1992)
21. Mahboubi, A., Pottier, L.: Élimination des quantificateurs sur les réels en Coq. Journées Francophones des Langages Applicatifs, Anglet (2002)
22. Meikle, L.I., Fleuriot, J.D.: Mechanical Theorem Proving in Computational Geometry. In: Hong, H., Wang, D. (eds.) ADG 2004. LNCS (LNAI), vol. 3763, pp. 1–18. Springer, Heidelberg (2006)
23. Meng, J., Quigley, C., Paulson, L.C.: Automation for interactive proof: first prototype. Inf. Comput. 204(10), 1575–1596 (2006)
24. The OpenMath Society, http://www.openmath.org
25. Paulson, L.C.: Isabelle: A Generic Theorem Prover. In: Paulson, L.C. (ed.) Isabelle. LNCS, vol. 828. Springer, Heidelberg (1994)
26. Tarski, A.: A Decision Method for Elementary Algebra and Geometry. University of California Press (1951)
27. Tiwari, A.: PVS-QEPCAD, http://www.csl.sri.com/users/tiwari/qepcad.html

Digital Mathematics Libraries:
The Good, the Bad, the Ugly

Thierry Bouche

Université de Grenoble I & CNRS,
Institut Fourier (UMR 5582) & Cellule Mathdoc (UMS 5638),
BP 74, 38402 St-Martin-d'Hères Cedex, France
`thierry.bouche@ujf-grenoble.fr`
`http://www-fourier.ujf-grenoble.fr/~bouche/`

Abstract. The mathematicians' Digital mathematics library (DML), which is not to be confused with libraries of mathematical objects represented in some digital format, is the generous idea that all mathematics ever published should end up in digital form so that it would be more easily referenced, accessible, usable. This concept was formulated at the very beginning of this century, and yielded a lot of international activity that culminated around years 2002–2005. While it is estimated that a substantial part of the existing math literature is already available in some digital format, nothing looking like *one* digital mathematics library has emerged, but a multiplicity of competing electronic offers, with unique standards, features, business models, access policies, etc.—even though the contents themselves overlap somewhat, while leaving wide areas untouched. The millenium's appealing idea has become a new Tower of Babel.

It is not obvious how much of the traditional library functions we should give up while going digital. The point of view shared by many mathematicians is that we should be able to find a reasonable archiving policy fitting all stakeholders, allowing to translate the essential features of the past library system—which is the central infrastructure of all math departments worldwide—in the digital paradigm, while enhancing overall performances thanks to dedicated information technology.

The vision of this library is rather straightforward: a third party to the academic publishing system, preserving, indexing, and keeping current its digital collections through a distributed network of partners curating the physical holdings, and a centralised access facility making use of innovative mining and interlinking techniques for easy navigation and discovery.

However, the fragmentation level is so high that the hope of a unique portal providing seamless access to everything relevant to mathematical research seems now completely out of reach. Nevertheless, we have lessons to learn from each one of the already numerous projects running. One of them is that there are too many items to deal with, and too many different initial choices over metadata sets and formats: it won't be possible to find a nontrivial greatest common divisor coping with everything already available, and manual upgrading is highly improbable.

S. Autexier et al. (Eds.): AISC/Calculemus/MKM 2008, LNAI 5144, pp. 331–332, 2008.

This is where future management techniques for loosely formalised mathematical knowledge could provide a new impetus by at last enabling a minimum set of features across projects borders through automated procedures. We can imagine e.g. math-aware OCR on scanned pages, concurrently with interpreters of electronic sources of born digital texts, both producing searchable full texts in a compatible semistructured format. The challenge is ultimately to take advantage of the high formalisation of mathematical texts rather than merely ignoring it!

With these considerations in mind, the talk will focus on achievements, limitations, and failures of existing digital mathematics libraries, taking the NUMDAM[1] and CEDRAM[2] programs as principal examples, hence the speaker himself as principal target...

---

[1] http://www.numdam.org
[2] http://www.cedram.org

Automating Signature Evolution in Logical Theories*

Alan Bundy

School of Informatics,
University of Edinburgh,
Edinburgh, UK
A.Bundy@ed.ac.uk

1 Introduction

The automation of reasoning as deduction in logical theories is well established. Such logical theories are usually inherited from the literature or are built manually for a particular reasoning task. They are then regarded as fixed. We will argue that they should be regarded as fluid.

1. As Pólya and others have argued, appropriate representation is the key to successful problem solving [Pólya, 1945]. It follows that a successful problem solver must be able to choose or construct the representation best suited to solving the current problem. Some of the most seminal episodes in human problem solving required radical representational change.
2. Automated agents use logical theories called *ontologies*. For different agents to communicate they must align their ontologies. When a large, diverse and evolving community of autonomous agents are continually engaged in online negotiations, it is not practical to manually pre-align the ontologies of all agent pairs – it must be done dynamically and automatically.
3. Persistent agents must be able to cope with a changing world and changing goals. This requires evolving their ontologies as their problem solving task evolves. The W3C call this *ontology evolution*[1].

Furthermore, in evolving a logical theory, it is not always enough just to add or delete axioms, definitions, rules, etc. — a process usually called *belief revision*. Sometimes it is necessary to change the underlying signature of the theory, e.g., to add, remove or alter the functions, predicates, types, etc. of the theory.

Below we present two projects to automate signature evolution in logic theories: one in the domain of online agents and one in the domain of theories of physics. Common themes emerge from these two projects that offer hope for a general theory of signature evolution.

* The research reported in this paper was supported by EPSRC grant EP/E005713/1. It will soon be supported by EPSRC grant EP/G000700/1 I would like to thank Michael Chan, Lucas Dixon and Fiona McNeill for their feedback on this paper and their contributions to the research referred to in it.
[1] http://www.w3.org/TR/webont-req/#goal-evolution

S. Autexier et al. (Eds.): AISC/Calculemus/MKM 2008, LNAI 5144, pp. 333–338, 2008.
© Springer-Verlag Berlin Heidelberg 2008

2   ORS: **Diagnosing and Repairing Agent Ontologies**

We first investigated the automation of signature evolution in ORS (Ontology Repair System): an automated system for repairing faulty ontologies in response to unexpected failures when executing multi-agent plans [McNeill & Bundy, 2007]. ORS forms plans to achieve its goals using the services provided by other agents. In forming these plans, ORS draws upon its knowledge base, which provides a representation of its world, including its beliefs about the abilities of other agents and under what circumstances they will perform various services. To request actions or ask questions of the other agents, ORS uses a simple performative language implemented in KIF[2], an ontology language based on first-order logic.

The representation of the world used by ORS may be faulty, not just in containing false beliefs, but also in using a signature that does not match that used by some of its collaborating agents. This mismatch will inhibit inter-agent communication, leading to faulty plans that will fail during execution. ORS analyses its failed plans, communicates with any agents that unexpectedly refused to perform a service, and proposes repairs to its ontology, including the signature of that ontology. Repairs can include: adding, removing or permuting arguments to predicates or functions, merging or splitting of predicates or functions and changing their types, as well as some belief revisions, such as adding or removing the precondition of an action.

Adding arguments to and splitting functions are examples of *refinement*, in which ontologies are enriched. Unfortunately, refinement operations are only partially defined. For instance, when an additional argument is added to a function it is not always clear what value each of its instances should take, or indeed whether any candidate values are available. When an old function is split into two or more new functions, each occurrence of the old function must be mapped to one of the new ones. It is not always clear how to perform this mapping.

The evaluation of ORS consisted of attempts to reproduce automatically the manual repairs we observed in KIF ontologies. Although this evaluation was successful, it was hampered by a lack of examples of before and after versions of ontologies, and of records of the fault in the before version, how it was diagnosed and how it was repaired to produce the after version. This led us to investigate domains in which ontological evolution was better documented. We picked the domains of physics and law. Our progress in the physics domain is the topic of the next section.

3   GALILEO: **Signature Evolution in Physics**

We are now applying and developing our techniques in the domain of physics [Bundy, 2007, Bundy & Chan, 2008]. This is an excellent domain because many of its most seminal advances can be seen as signature evolution, i.e., changing the way that physicists view the world. These changes are often triggered by a contradiction between existing theory and experimental observation. These

---

[2] http://logic.stanford.edu/kif/kif.html

contradictions, their diagnosis and the resulting repairs have usually been well documented by historians of science, providing us with a rich vein of case studies for the development and evaluation of our techniques, addressing the evaluation problem identified in the ORS project. The physics domain requires higher-order logic: both at the object-level, to describe things like planetary orbits and calculus, and at the meta-level, to describe the repair operations.

3.1   Repair Plans

We are developing a series of *repair plans* which operate simultaneously on a small set of small higher-order theories, e.g., one representing the current theory of physics, another representing a particular experimental set-up. Before the repair, these theories are individually consistent but collectively inconsistent. Afterwards the new theories are also collectively consistent. Each repair plan has a trigger formula and some actions: when the trigger is matched, the actions are performed. The actions modify the signatures and axioms of the old theories to produce new ones. Typical actions are similar to those described above for ORS. The repair plans have been implemented in the GALILEO system (Guided Analysis of Logical Inconsistencies Leads to Evolved Ontologies) using λProlog [Miller & Nadathur, 1988] as our implementation language, because it provides a polymorphic, higher-order logic.

This combination of repair plans and multiple interacting logic theories helps to solve several tough problems in automated signature evolution.

- The overall context of the plan completes the definition of the, otherwise only partially defined, refinement operations. For instance, it supplies the values of additional arguments and specifies which new function should replace which old one. Organising the theory as several interacting, small theories further guides the refinement, e.g., by enabling us to uniformly replace all the occurrences of an old function in one theory in one way, but all the occurrences in another theory in a different way.
- Grouping the operations into a predefined repair plan helps control search. This arises not only from inference, but also from repair choices. It also occurs not only within the evolving object-level theory but also in the meta-level theory required to diagnose and repair that object-level theory. This solution is adopted from our work on *proof plans* [Bundy, 1991].
- Having several theories helps us control inconsistency. A predictive theory and an observational one can be internally consistent, but inconsistent when combined. Since all sentences are theorems in an inconsistent theory, the triggers of all repair plans would be matched, creating a combinatorial explosion. This problem can be avoided when a trigger requires simultaneous matching across a small set of consistent ontologies.
- It is also enabling us to prove the minimality of our repair plans, i.e., to show that the repairs do not go beyond what is necessary to remove the inconsistency. We have extended the concept of *conservative extension* to signature evolution. We can now prove that the evolution of each separate

theory is conservative in this extended sense. Of course, we do not want the evolution of the combined theory to be conservative, since we want to turn an inconsistent combined theory into a consistent one.

3.2 Some Repair Plans and Their Evaluation

We have so far developed two repair plans, which we call *Where's my stuff?* (WMS) and *Inconstancy*. These roughly correspond to the refinement operations of splitting a function and adding an argument, respectively. We have found multiple examples of these repairs across the history of physics, but are always interested in additional ones.

The WMS repair plan aims at resolving contradictions arising when the predicted value returned by a function does not match the observed value. This is modelled by having two theories, corresponding to the prediction and the observation, with different values for this function. To break the inconsistency, the conflicting function is split into three new functions: *visible*, *invisible* and *total*. The conflicting function becomes the total function in the predictive theory and the visible function in the observation theory[3]. The invisible function is defined as the difference between them, and this new definition is added to the predictive theory. The intuition behind this repair is that the discrepancy arose because the function was not being applied to the same *stuff* in the predictive and the observational theories — the invisible stuff was not observed.

WMS has been successfully applied to conflicts between predictions of and observations of the following functions: the temperature of freezing water; the energy of a bouncing ball; the graphs relating orbital velocity of stars to distance from the galactic centre in spiral galaxies; and the precession of the perihelion of Mercury. In these examples the role of the invisible stuff is played by: the latent heat of fusion, elastic potential energy, dark matter and an additional planet, respectively.

The Inconstancy repair plan is triggered when there is a conflict between the predicted independence and the observed dependence of a function on some parameter, i.e., the observed value of a function unexpectedly varies when it is predicted to remain constant. This generally requires several observational theories, each with different observed values of the function, as opposed to the one observational theory in the WMS plan. To effect the repair, the parameter causing the unexpected variation is first identified and a new definition for the conflicting function is created that includes this new parameter. The nature of the dependence is induced from the observations using curve-fitting techniques.

Inconstancy has been successfully applied to the following conflicts between predictions and various observations: the ratio of pressure and volume of a gas; and again the graphs relating orbital velocity of stars to distance from the galactic centre in spiral galaxies. The unexpected parameter of the function is the temperature of the gas and the acceleration between the stars, respectively. The first of these repairs generalises Boyle's Law to the Ideal Gas Law and the second

---

[3] There are situations in which these roles are inverted [Bundy, 2007].

generalises the Gravitational Constant to Milgrom's MOND (MOdified Newtonian Dynamics). Interestingly, WMS and Inconstancy produce the two main rival theories on the spiral galaxy anomaly, namely dark matter and MOND. Since this is still an active controversy, its unfolding will help us develop mechanisms to choose between rival theory repairs. We are currently experimenting also with applying Inconstancy to the replacement of Aristotle's concept of instantaneous light travel with a finite (but fast) light speed, using conflicts between the predicted and observed times of eclipses by Jupiter's moon Io.

4   Conclusion

We have argued for the importance of automated evolution of logical theories to adapt to new circumstances, to recover from failure and to make them better suited to the current problem. We argue that this requires more than just belief revision — although this is part of the story. We also need *signature* revision, i.e., changes to the underlying syntax of the theory. We have begun the work of automating signature evolution in the ORS and GALILEO projects. We have developed repair plans over multiple theories, which address some of the tough problems of partial definedness, combinatorial explosion, coping with inconsistency and ensuring minimality, that beset this endeavour.

In the future, we plan to: develop additional repair plans, research additional case studies from the history of physics, refine our currently rather *ad hoc* logical theories, thoroughly evaluate our repair plans on a significant corpus of case studies, and explore notions of minimal repair and other aspects of a theory of signature evolution. Ideas for future repair plans include: the converses of WMS and Inconstancy; the use of analogy to create new theories; and the correction of faulty causal dependencies.

References

[Bundy & Chan, 2008]   Bundy, A., Chan, M.: Towards ontology evolution in physics. In: Hodges, W. (ed.) Procs. Wollic 2008. Springer, Heidelberg (2008)

[Bundy, 1991]   Bundy, A.: A science of reasoning. In: Lassez, J.-L., Plotkin, G. (eds.) Computational Logic: Essays in Honor of Alan Robinson, pp. 178–198. MIT Press, Cambridge (1991)

[Bundy, 2007]   Bundy, A.: Where's my stuff? An ontology repair plan. In: Ahrendt, W., Baumgartner, P., de Nivelle, H. (eds.) Proceedings of the Workshop on Disproving - Non-Theorems, Non-Validity, Non-Provability, Bremen, Germany, pp. 2–12 (July 2007), http://www.cs.chalmers.se/ ahrendt/ CADE07-ws-disproving/

[McNeill & Bundy, 2007]   McNeill, F., Bundy, A.: Dynamic, automatic, first-order ontology repair by diagnosis of failed plan execution. IJSWIS 3(3), 1–35 (2007); Special issue on ontology matching

[Miller & Nadathur, 1988]   Miller, D., Nadathur, G.: An overview of λProlog. In: Bowen, R. (ed.) Proceedings of the Fifth International Logic Programming Conference/ Fifth Symposium on Logic Programming. MIT Press, Cambridge (1988)

[Pólya, 1945]   Pólya, G.: How to Solve It. Princeton University Press, Princeton (1945)

A Tactic Language for Hiproofs

David Aspinall[1], Ewen Denney[2], and Christoph Lüth[3]

[1] LFCS, School of Informatics
University of Edinburgh
Edinburgh EH9 3JZ, Scotland
[2] RIACS, NASA Ames Research Center
Moffett Field, CA 94035, USA
[3] Deutsches Forschungszentrum für Künstliche Intelligenz
Bremen, Germany

Abstract. We introduce and study a tactic language, *Hitac*, for constructing hierarchical proofs, known as *hiproofs*. The idea of hiproofs is to superimpose a labelled hierarchical nesting on an ordinary proof tree. The labels and nesting are used to describe the organisation of the proof, typically relating to its construction process. This can be useful for understanding and navigating the proof. Tactics in our language construct hiproof structure together with an underlying proof tree. We provide both a big-step and a small-step operational semantics for evaluating tactic expressions. The big-step semantics captures the intended meaning, whereas the small-step semantics hints at possible implementations and provides a unified notion of proof state. We prove that these notions are equivalent and construct valid proofs.

1 Introduction

Interactive theorem proving is a challenging pursuit, made additionally challenging by the present state-of-the-art. Constructing significant sized computer checked proofs requires struggling with incomplete and partial automation, and grappling with many low-level system specific details. Once they have been written, understanding and maintaining such proofs is in some ways even harder: small changes often cause proofs to break completely, and debugging to find the failure point is seldom easy.

Moreover, notions of proof vary from one theorem prover to another, locking users into specific provers they have mastered. What is needed is a more abstract notion of proof which is independent of a particular prover or logic, but supports the relevant notions needed to interactively explore and construct proofs, thus improving the management of large proofs. In this paper, we study the notion of *hiproof* [1], which takes the *hierarchical* structure of proofs as primary. Although any system in which tactics can be defined from other tactics leads naturally to a notion of hierarchical proof, this is the first work to study the hierarchical nature of tactic languages in detail. By developing an idealised tactic language, therefore, we aim to work towards a generic proof representation language.

S. Autexier et al. (Eds.): AISC/Calculemus/MKM 2008, LNAI 5144, pp. 339–354, 2008.
© Springer-Verlag Berlin Heidelberg 2008

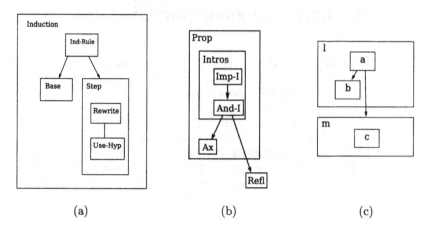

Fig. 1. Example Hiproofs

Figure 1 shows three example hiproofs which illustrate the basic ideas. Diagram (a) shows the structure of an induction procedure: it consists of the application of an induction rule, followed by a procedure for solving the base case and a procedure for solving the step case. The step case uses rewriting and the induction hypothesis to complete the proof. Diagram (b) shows a procedure for solving a positive propositional statement, using implication and then conjunction introduction, solving one subgoal using an axiom and leaving another subgoal unsolved, indicated by an arrow exiting the "Prop" box. The second subgoal is then solved using reflexivity. Diagram (c) shows two labelled hiproofs labelled l and m, respectively. l applies rule a, which produces two subgoals, the first of which is solved inside l with b, and the second of which is solved by the proof labelled m (consisting of a single rule c). We will use this third example as a test case later.

A hiproof is abstract: nodes are given names corresponding to basic proof rules, procedures or compositions thereof. Each node may be *labelled* with a name: navigation in the proof allows "zooming in" by opening boxes to reveal their content. Boxes which are not open are just visualised with their labels; details inside are suppressed. Hiproofs are an abstraction of a proof in an underlying logic or derivation system; we call a hiproof *valid* if it can be mapped on to an underlying proof tree, where nodes are given by derivable judgements.

The central topic, and novelty, introduced in this paper is a tactic programming language for constructing valid hiproofs. The language is general and not tied to a specific system. It is deliberately restrictive: at the moment we seek to understand the connection between hierarchical structure and some core constructs for tactic programming, namely, alternation, repetition and assertion; features such as meta-variable instantiation and binding are left to future work. Part of the value of our contribution is the semantically precise understanding of this core.

Outline. The next section introduces a syntax for hiproofs and explains the notion of validity. By extending this syntax we introduce *tactics* which can be used to construct programs. In Section 3 we study notions of *evaluation*: a tactic can be applied to a goal and, if successful, evaluated to produce a valid hiproof. We consider two operational semantics: a big-step relation which defines the meaning of our constructs, and a finer-grained small-step semantics with a notion of *proof state* that evolves while the proof is constructed. In Section 4 we demonstrate how our language can be used to define some familiar tactics. Section 5 concludes and relates our contribution to previous work.

2   Syntax for Hiproofs and Tactics

Hiproofs add structure to an underlying derivation system, introduced shortly. Hiproofs are ranged over by s and are given by terms in the following grammar:

$$
\begin{array}{lll}
s ::= & a & \text{atomic} \\
\mid & \text{id} & \text{identity} \\
\mid & [l]\, s & \text{labelling} \\
\mid & s\, ;\, s & \text{sequencing} \\
\mid & s \otimes s & \text{tensor} \\
\mid & \langle\rangle & \text{empty}
\end{array}
$$

We assume $a \in \mathcal{A}$ where $\mathcal{A}$ is the set of *atomic tactics* given by the underlying derivation system. The remaining constructors add the structure introduced above: labelling introduces named boxes (which can be nested arbitrarily deep); sequencing composes one hiproof after another, and tensor puts two hiproofs side-by-side, operating on two groups of goals. The identity hiproof has no effect, but is used for "wiring", to fill in structure. It can be applied to a single (atomic) goal. The empty hiproof is the vacuous proof for an empty set of goals. Hiproofs have a denotational semantics given in previous work [1]; the syntax above serves as an internal language for models in that semantics. The denotational semantics justifies certain equations between terms, in particular, empty is a unit for the tensor, and tensor and sequencing are associative.

When writing hiproof terms we use the following syntactic conventions: the scope of the label l in $[l]\, s$ extends as far as possible and tensor binds more tightly that sequencing.

Example 1. The hiproof in Fig. 1(c) is written as

$$([l]\, a\, ;\, b \otimes \text{id})\, ;\, [m]\, c.$$

Notice the role of id corresponding to the line exiting the box labelled l.

The underlying derivation system defines sets of *atomic tactics* $a \in \mathcal{A}$ and *atomic goals* $\gamma \in \mathcal{G}$. Typically, what we call a goal is given by a judgement form in the

$$\dfrac{\dfrac{\gamma_1 \cdots \gamma_n}{\gamma} \quad a_\gamma \in \mathcal{A}}{a \vdash \gamma \longrightarrow \gamma_1 \otimes \cdots \otimes \gamma_n} \qquad \text{(V-Atomic)}$$

$$\mathsf{id} \vdash \gamma \longrightarrow \gamma \qquad \text{(V-Identity)}$$

$$\dfrac{s \vdash \gamma \longrightarrow g}{[l]\, s \vdash \gamma \longrightarrow g} \qquad \text{(V-Label)}$$

$$\dfrac{s_1 \vdash g_1 \longrightarrow g \quad s_2 \vdash g \longrightarrow g_2}{s_1 \,;\, s_2 \vdash g_1 \longrightarrow g_2} \qquad \text{(V-Sequence)}$$

$$\dfrac{s_1 \vdash g_1 \longrightarrow g_1' \quad s_2 \vdash g_2 \longrightarrow g_2'}{s_1 \otimes s_2 \vdash g_1 \otimes g_2 \longrightarrow g_1' \otimes g_2'} \qquad \text{(V-Tensor)}$$

$$\langle\rangle \vdash \langle\rangle \longrightarrow \langle\rangle \qquad \text{(V-Empty)}$$

Fig. 2. Validation of Hiproofs

underlying derivation system. We work with lists of goals g written using the binary tensor:

$$g ::= \gamma \mid g \otimes g \mid \langle\rangle$$

The tensor is associative and unitary, with $\langle\rangle$ the unit (empty list). We write $g : n$ to indicate the length of g, i.e., when $g = \gamma_1 \otimes \cdots \otimes \gamma_n$ or $g = \langle\rangle$ for $n = 0$, called the *arity*. Elementary inference rules in the underlying derivation system can be seen as atomic tactics of the following form:

$$\dfrac{\gamma_1 \cdots \gamma_n}{\gamma} \, a$$

which given goals $\gamma_1, \ldots, \gamma_n$ produce a proof for γ. There are, of course, other atomic tactics possible. However, for a particular atomic tactic a, there may be a family of goals γ to which it applies; we write a_γ to make the instance of a precise. A restriction is that every instance of a must have the same arity, i.e., number of premises n. By composing atomic tactics, we can produce *proofs* in the derivation system. Thus, each hiproof s has a family of underlying proofs which consist of applications of the instances of the underlying atomic tactics. We say that s *validates* proofs from g_2 to g_1, written $s \vdash g_1 \longrightarrow g_2$. Validation is defined by the rules in Fig. 2.

Validation is a well-formedness check: it checks that atomic tactics are applied properly to construct a proof, and that the structural regime of hiproofs is obeyed. Notice that, although g_1 and g_2 are not determined by s, the arity restriction means that every underlying proof that s validates must have the same shape, i.e., the same underlying tree of atomic tactics. This underlying tree is known as the *skeleton* of the hiproof [1]. The (input) arity $s : n$ of a hiproof s with $s \vdash g_1 \longrightarrow g_2$ is n where $g_1 : n$; note that again by the restriction on atomic tactics, a hiproof has a unique arity, if it has any.

Example 2. Suppose we have a goal γ_1 which can be proved like this:

$$\frac{\dfrac{}{\gamma_2}\,\mathrm{b} \quad \dfrac{}{\gamma_3}\,\mathrm{c}}{\gamma_1}\,\mathrm{a}$$

Then $([l]\,a\;;\,b \otimes \mathsf{id})\;;\,[m]\,c \;\vdash\; \gamma_1 \longrightarrow \langle\rangle$, with arity 1.

To show how the abstract hiproofs may be used with a real underlying derivation system, we give two small examples with different sorts of underlying goals.

Example 3. Minimal propositional logic MPL has the formulae:

$$P ::= \mathsf{TT} \mid \mathsf{FF} \mid X \mid P \Longrightarrow P$$

where X is a propositional variable. Goals in MPL have the form $\Gamma \vdash P$, where Γ is a set of propositions. The atomic tactics correspond to the natural deduction rules:

$$\frac{}{\Gamma, P \vdash P}\,\mathrm{ax} \qquad \frac{\Gamma \cup \{P\} \vdash Q}{\Gamma \vdash P \Longrightarrow Q}\,\mathrm{impI} \qquad \frac{\Gamma \vdash P \Longrightarrow Q \quad \Gamma \vdash P}{\Gamma \vdash Q}\,\mathrm{impE}$$

$$\frac{\Gamma \vdash P}{\Gamma \cup \{Q\} \vdash P}\,\mathrm{wk}$$

Atomic tactics are rule instances (e.g., $\mathrm{ax}_{\{X\} \vdash X}$), which are viewed as being applied backwards to solve some given atomic goal; they have the obvious arities.

Example 4. Minimal equational logic MEL is specified by a signature Σ, giving a set of terms $\mathcal{T}_\Sigma(X)$ over a countably infinite set of variables X, together with a set of equations E of the form $a = b$ with a, b terms. Goals in this derivation system are equations of the same form. These can be established using the following atomic tactics (where $a, b, c, d \in \mathcal{T}_\Sigma(X)$):

$$\frac{}{a = a}\,\mathrm{refl} \qquad \frac{a = b \quad b = c}{a = c}\,\mathrm{trans} \qquad \frac{a = b}{b = a}\,\mathrm{sym} \qquad \frac{a = b \quad c = d}{a[c/x] = b[d/x]}\,\mathrm{subst}$$

Here, $a[c/x]$ denotes the term a with the variable x replaced by term c throughout. For example, the more usual substitutivity rule

$$\frac{a = b}{a[c/x] = b[c/x]}\,\mathrm{subst}'$$

can be derived with the hiproof $h_1 = subst\;;\,\mathsf{id} \otimes refl$ of arity 1, whereas the usual congruence rule say for a binary operation f

$$\frac{a_1 = b_1 \quad a_2 = b_2}{f(a_1, a_2) = f(b_1, b_2)}\,\mathrm{ctxt}$$

can be derived with the hiproof $h_2 = subst\;;\,(subst\;;\,refl \otimes \mathsf{id}) \otimes \mathsf{id}$ of arity 1.

2.1 Tactics and Programs

The hiproofs introduced above are static proof structures which represent the result of executing a tactic. We now present a language of tactics which can be evaluated to construct such hiproofs. These *tactic expressions* will be the main object of study. They are defined by extending the grammar for hiproofs with three new cases:

$$t ::= a \mid \mathsf{id} \mid [l]\, t \mid t\,;t \mid t \otimes t \mid \langle\rangle \qquad \text{as for hiproofs}$$
$$\mid\ t \mid t \qquad\qquad\qquad\qquad\qquad\ \text{alternation}$$
$$\mid\ \mathsf{assert}\ \gamma \qquad\qquad\qquad\qquad\ \text{goal assertion}$$
$$\mid\ T \qquad\qquad\qquad\qquad\qquad\quad \text{defined tactic}$$

Together the three new cases allow proof search: alternation allows alternatives, assertion allows controlling the control flow, and defined tactics allow us to build up a program of possibly mutually recursive definitions. Syntactic conventions for hiproofs are extended to tactic expressions, with alternation having precedence between sequencing (lowest) and tensor (highest). Alternation is also associative. Further, the arity $t : n$ of a tactic is defined as follows: for a hiproof s, it has been defined above; $[l]\,t : 1$; if $t_1 : n$, then $t_1\,;\,t_2 : n$; if $t_1 : n$ and $t_2 : m$, then $t_1 \otimes t_2 : n + m$; if $t_1 : n$ and $t_2 : n$, then $t_1 \mid t_2 : n$; $\mathsf{assert}\ \gamma : 1$; and if $T \stackrel{def}{=} t \in Prog$ and $t : n$, then $T : n$. This definition is partial, not all tactics can be given an arity.

Programs. A *tactic program Prog* is a set of definitions of the form $T_i \stackrel{def}{=} t_i$, together with a *goal matching* relation on atomic goals $\gamma \lesssim \gamma'$ which is used to define the meaning of the assertion expression. The definition set must not define any T more than once, and no label may appear more than once in all of the definition bodies t_i.

The uniqueness requirement on labels is so that we can map a label in a hiproof back to a unique origin within the program – although notice that, because of recursion, the same label may appear many times on the derivation.

We do not make restrictions on the goal matching relation. In some cases it may simply be an equivalence relation on goals. For MEL, a pre-order is more natural: the matching relation can be given by instantiations of variables, so a goal given by an equation $b_1 = b_2$ matches a goal $a_1 = a_2$ if there is an instantiation $\sigma : X \to T_\Sigma(X)$ such that $b_i = a_i\sigma$.

Example 5. We can give a tactic program for producing the hiproof shown in Fig. 1(c) by defining:

$$T_l \stackrel{def}{=} [l]\, a\,;\, b \otimes \mathsf{id}$$
$$T_m \stackrel{def}{=} [m]\, c$$
$$T_u \stackrel{def}{=} (\mathsf{assert}\ \gamma_3\,;\, T_m) \mid (T_l\,;\, T_u)$$

If we *evaluate* the tactic T_u applied to the goal γ_1, we get the hiproof shown earlier, $([l]\, a\,;\, b \otimes \mathsf{id})\,;\, [m]\, c$, with no left over goals.

The next section provides an operational semantics to define a notion of evaluation that has this effect.

3   Operational Semantics

To give a semantics to tactic expressions, we will consider an operational semantics as primary. This is in contrast to other approaches which model things as the original LCF-style tactic programming does, i.e., using a fixed-point semantics to explain recursion (e.g., ArcAngel [2]). We believe that an operational semantics is more desirable at this level, because we want to explain the steps used during tactic evaluation at an intensional level: this gives us a precise understanding of the internal proof state notion, and hope for providing better debugging tools (a similar motivation was given for Tinycals [3]).

There is a crucial difference between hiproofs and tactic expressions. Because of alternation and repetition, tactic evaluation is non-deterministic: a tactic expression can evaluate to many different hiproofs, each of which can validate different proofs, so we cannot extend the validity notion directly, even for (statically) checking arities – which is why our notion of arity is partial. This is not surprising because part of the point of tactic programming is to write tactics that can apply to varying numbers of subgoals using arbitrary recursive functions. It is one of the things that makes tactic programming difficult. In future work we plan to investigate richer static type systems for tactic programming, in the belief that there is a useful intermediate ground between ordinary unchecked tactics and what could be called "certified tactic programming" [4,5], where tactics are shown to construct correct proofs using dependent typing.

Here, we use untyped tactic terms, as is more usual. We begin by defining a big-step semantics that gives meaning to expressions without explicitly specifying the intermediate states. In Sect. 3.2 we give a small-step semantics which provides a notion of intermediate proof state.

3.1   Big-Step Semantics

The big-step semantics is shown in Fig. 3. It defines a relation $\langle g, t \rangle \Downarrow \langle s, g' \rangle$ inductively, which explains how applying a tactic t to the list of goals g results in the hiproof s and the remaining (unsolved) goals g'. A tactic t *proves* a goal, g, therefore, if $\langle g, t \rangle \Downarrow \langle s, \langle \rangle \rangle$, for some hiproof s. The relation is defined with respect to a program *Prog* containing a set of definitions.

The rules directly capture the intended meaning of tactic expressions. For example, (B-Label) evaluates a labelled tactic expression $[l] \, t$, by first evaluating the body t using the same goal γ, to get a hiproof s and remaining goals g. The result is then the labelled hiproof $[l] \, s$ and remaining goals g. Like (V-Label), this rule reflects key restrictions in the notion of hiproof (motivated in [1]), namely that a box has a unique entry point, its root, accepting a single (atomic) goal.

Notice that in (B-Assert), assertion terms evaluate to identity if the goal matches, or they do not evaluate at all. Similarly, (B-Atomic) only allows an

$$\frac{\gamma_1 \cdots \gamma_n}{\gamma} \; a_\gamma \in \mathcal{A}$$
$$\frac{}{\langle \gamma, a \rangle \Downarrow} \quad \text{(B-Atomic)}$$
$$\langle a, \gamma_1 \otimes \cdots \otimes \gamma_n \rangle$$

$$\frac{}{\langle \gamma, \mathsf{id} \rangle \Downarrow \langle \mathsf{id}, \gamma \rangle} \quad \text{(B-Id)}$$

$$\frac{\langle \gamma, t \rangle \Downarrow \langle s, g \rangle}{\langle \gamma, [l]\, t \rangle \Downarrow \langle [l]\, s, g \rangle} \quad \text{(B-Label)}$$

$$\frac{\langle g_1, t_1 \rangle \Downarrow \langle s_1, g_2 \rangle}{\langle g_2, t_2 \rangle \Downarrow \langle s_2, g_3 \rangle} \quad \text{(B-Seq)}$$
$$\overline{\langle g_1, t_1 \,;\, t_2 \rangle \Downarrow \langle s_1 \,;\, s_2, g_3 \rangle}$$

$$\frac{\langle g_1, t_1 \rangle \Downarrow \langle s_1, g_1' \rangle}{\langle g_2, t_2 \rangle \Downarrow \langle s_2, g_2' \rangle} \quad \text{(B-Tensor)}$$
$$\frac{\langle g_1 \otimes g_2, t_1 \otimes t_2 \rangle \Downarrow}{\langle s_1 \otimes s_2, g_1' \otimes g_2' \rangle}$$

$$\langle \langle \rangle, \langle \rangle \rangle \Downarrow \langle \langle \rangle, \langle \rangle \rangle \quad \text{(B-Empty)}$$

$$\frac{\langle g, t_1 \rangle \Downarrow \langle s, g' \rangle}{\langle g, t_1 \mid t_2 \rangle \Downarrow \langle s, g' \rangle} \quad \text{(B-Alt-L)}$$

$$\frac{\langle g, t_2 \rangle \Downarrow \langle s, g' \rangle}{\langle g, t_1 \mid t_2 \rangle \Downarrow \langle s, g' \rangle} \quad \text{(B-Alt-R)}$$

$$\frac{\gamma \lesssim \gamma'}{\langle \gamma', \mathsf{assert}\ \gamma \rangle \Downarrow \langle \mathsf{id}, \gamma' \rangle} \quad \text{(B-Assert)}$$

$$T \overset{def}{=} t \in Prog$$
$$\frac{\langle g, t \rangle \Downarrow \langle s, g' \rangle}{\langle g, T \rangle \Downarrow \langle s, g' \rangle} \quad \text{(B-Def)}$$

Fig. 3. Big-step semantics for *Hitac*

atomic tactic a to evaluate if it can be used to validate the given goal γ. Hence, failure is modelled implicitly by the lack of a target for the overall evaluation (i.e., there must be some subterm $\langle g, t \rangle$ for which there is no $\langle s, g' \rangle$ it evaluates to). The rules for alternation allow an angelic choice, as they allow us to pick the one of the two tactics which evaluate to a hiproof (if either of them does); if both alternatives evaluate, the alternation is non-deterministic.

While the obvious source of non-determinism is alternation, the tensor rule also allows the (possibly angelic) splitting of an input goal list into two halves $g_1 \otimes g_2$, including the possibility that g_1 or g_2 is the empty tensor $\langle \rangle$.

The crucial property is *correctness* of the semantics: if a hiproof is produced, it is a valid hiproof for the claimed input and output goals.

Theorem 1 (Correctness of big-step semantics)
If $\langle g, t \rangle \Downarrow \langle s, g' \rangle$ *then* $s \vdash g \longrightarrow g'$.

Proof By induction on the derivation of $\langle g, t \rangle \Downarrow \langle s, g' \rangle$.

Fact 1 (Label origin). *If t is label-free, $\langle g, t \rangle \Downarrow \langle s, g' \rangle$ and the label l appears in s, then l has a unique origin within some tactic definition x from Prog.*

The label origin property is immediate by the definition of program and the observation that evaluation only introduces labels from the program. It means that we can use labels as indexes into the program to find where a subproof was produced, which is the core motivation for labelling, and allows a source level debugging of tactical proofs.

3.2   Small-Step Semantics

Besides the big-step semantics given above, it is desirable to explain tactic evaluation using a small-step semantics. The typical reason for providing small-step semantics is to give meaning to non-terminating expressions. In principle we don't need to do this here (non-terminating tactics do not produce proofs), but in practice we are interested in debugging tactic expressions during their evaluation, including ones which may fail. A small step-semantics provides a notion of intermediate state which helps do this.

We now define an evaluation machine which evolves a *proof state* configuration in each step, eventually producing a hiproof. The reduction is again non-deterministic; some paths may get stuck or not terminate. Compared with the big-step semantics, the non-determinism in alternation does not need to be predicted wholly in advance, but the rules allow exploring both alternation branches of a tactic tree in parallel to find one which results in a proof.

Formulation of a small-step semantics is not as straightforward as the big-step semantics, because it needs to keep track of the intermediate stages of evaluation, which do not correspond to a well-formed hiproof. The difficulty is in recording which tactics have been evaluated and which not, and moving the goals which remain in subtrees out of their hierarchical boxes. It was surprisingly hard to find an elegant solution. The mechanism we eventually found is pleasantly simple; it was devised by visualising goals moving in the geometric representation of hiproofs; they move along lines, in and out of boxes, and are split when entering a tensor and rejoined afterwards.

This suggests a unified notion of proof state, where goals appear directly in the syntax with tactics. To this end, we define a compound term syntax for *proof states* which has hiproofs, tactic expressions and goal lists as sublanguages:

$$p ::= a \mid \mathsf{id} \mid [l]\, p \mid p \, ; p \mid p \otimes p \mid p \mid p \mid \mathsf{assert}\, \gamma \mid T \mid g$$

The general judgement form is $p \Rightarrow p'$, defined by the rules shown in Fig. 4. A proof state, p, consists of a mixture of open goals, g, active tactics, t, and successfully applied tactics, i.e., hiproofs, s. Composing proof states can be understood as connecting, or applying, the tactics of one state to the open goals of another. In particular, the application of tactic t to goal g has the form $g \, ; t$, and we treat the sequencing operator on proof states as associative. The notion of value is a fully reduced proof state with the form $s \, ; g'$. A complete reduction, therefore, has the form:

$$g \, ; t \Rightarrow^* s \, ; g'.$$

What happens is that the goals g move through the tactic t, being transformed by atomic tactics, until (if successful) the result is a simple hiproof s and remaining goals g'. Note that not all terms in this grammar are meaningful. In the rules, therefore, we will restrict attention to reductions of a meaningful form, and are careful to distinguish between syntactic categories for proof states, p, and the sublanguages of tactics t, hiproofs s and goals g, which can be embedded into the language of proof states. For example, in the rule (S-Alt), g has to be a goal and

$$\frac{\gamma_1 \cdots \gamma_n}{\gamma} \; a_\gamma \in \mathcal{A}$$
$$\frac{}{\gamma \; ; \; a \; \Rightarrow}$$
$$a \; ; \; \gamma_1 \cdots \otimes \cdots \gamma_n$$
(S-Atomic)

$$T \stackrel{def}{=} t \in Prog$$
$$\frac{}{g \; ; \; T \; \Rightarrow \; g \; ; \; t}$$
(S-Def)

$$\gamma \; ; \; \mathsf{id} \; \Rightarrow \; \mathsf{id} \; ; \; \gamma$$
(S-Id)

$$\frac{p \Rightarrow p'}{[l] \, p \; \Rightarrow \; [l] \, p'}$$
(S-Lab)

$$\gamma \; ; \; [l] \, t \; \Rightarrow \; [l] \, \gamma \; ; \; t$$
(S-Enter)

$$\frac{p_1 \Rightarrow p_1'}{p_1 \; ; \; p_2 \; \Rightarrow \; p_1' \; ; \; p_2}$$
(S-Seq-L)

$$[l] \, s \; ; \; g \; \Rightarrow \; ([l] \, s) \; ; \; g$$
(S-Exit)

$$g_1 \otimes g_2 \; ; \; p_1 \otimes p_2 \; \Rightarrow$$
$$(g_1 \; ; \; p_1) \otimes (g_2 \; ; \; p_2)$$
(S-Split)

$$\frac{p_2 \Rightarrow p_2'}{p_1 \; ; \; p_2 \; \Rightarrow \; p_1 \; ; \; p_2'}$$
(S-Seq-R)

$$(s_1 \; ; \; g_1) \otimes (s_2 \; ; \; g_2) \; \Rightarrow$$
$$s_1 \otimes s_2 \; ; \; g_1 \otimes g_2$$
(S-Join)

$$\frac{p_1 \Rightarrow p_1'}{p_1 \otimes p_2 \; \Rightarrow \; p_1' \otimes p_2}$$
(S-Tens-L)

$$g \; ; \; t_1 \mid t_2 \; \Rightarrow$$
$$(g \; ; \; t_1) \mid (g \; ; \; t_2)$$
(S-Alt)

$$\frac{p_2 \Rightarrow p_2'}{p_1 \otimes p_2 \; \Rightarrow \; p_1 \otimes p_2'}$$
(S-Tens-R)

$$(s_1 \; ; \; g) \mid p_2 \; \Rightarrow \; s_1 \; ; \; g$$
(S-Sel-L)

$$\frac{p_1 \Rightarrow p_1'}{p_1 \mid p_2 \; \Rightarrow \; p_1' \mid p_2}$$
(S-Alt-L)

$$p_1 \mid (s_2 \; ; \; g) \; \Rightarrow \; s_2 \; ; \; g$$
(S-Sel-R)

$$\frac{\gamma \lesssim \gamma'}{\gamma \; ; \; \mathsf{assert} \; \gamma' \; \Rightarrow \; \mathsf{id} \; ; \; \gamma}$$
(S-Assert)

$$\frac{p_2 \Rightarrow p_2'}{p_1 \mid p_2 \; \Rightarrow \; p_1 \mid p_2'}$$
(S-Alt-R)

Fig. 4. Small-step semantics for *Hitac*

t_1, t_2 must be tactics. Further, note that the empty goal and the empty hiproof and tactic are both denoted by $\langle\rangle$; this gives rise to the identity $\langle\rangle \; ; \; \langle\rangle = \langle\rangle \; ; \; \langle\rangle$ where on the left we have a tactic applied to an empty goal, and on the right a hiproof applied to an empty goal. The small-step semantics therefore does not need an explicit rule for the empty case.

The appearance of constrained subterms, and in particular, value forms $s \; ; \; g$, restricts the reduction relation and hints at evaluation order. Intuitively, joining tensors in (S-Join) only takes place after a sub-proof state has been fully evaluated. Similarly, in (S-Exit), when evaluation is complete inside a box, the remaining goals are passed out on to subsequent tactics. Alternatives need only be discarded in (S-Sel-L) or (S-Sel-R) after a successful proof has been found.

The theorems below show that restrictions do not limit the language, by relating it to the big-step semantics.

Example 6. Consider the tactic program from Ex. 5. We show the reduction of T_u applied to the goal γ_1. The steps name the major rule applied at each point.

$$\gamma_1 \; ; \; T_u \; \Rightarrow$$
$$\Rightarrow \gamma_1 \; ; \; (\mathsf{assert} \; \gamma_3 \; ; \; T_m) \mid (T_l \; ; \; T_u) \qquad \text{(S-Def)}$$
$$\Rightarrow (\gamma_1 \; ; \; \mathsf{assert} \; \gamma_3 \; ; \; T_m) \mid (\gamma_1 \; ; \; T_l \; ; \; T_u) \qquad \text{(S-Alt)}$$
$$\Rightarrow \ldots \gamma_1 \; ; \; ([l] \, b \; ; \; c \otimes \mathsf{id}) \; ; \; T_u \qquad \text{reduce on right, (S-Def)}$$
$$\Rightarrow \ldots ([l] \, \gamma_1 \; ; \; b \; ; \; c \otimes \mathsf{id}) \; ; \; T_u \qquad \text{(S-Enter)}$$

$$\Rightarrow \ldots ([l]\, b \,;\, \gamma_2 \otimes \gamma_3 \,;\, c \otimes \mathsf{id}) \,;\, T_u \qquad \text{(S-Atomic)}$$
$$\Rightarrow \ldots ([l]\, b \,;\, (\gamma_2 \,;\, c) \otimes (\gamma_3 \,;\, \mathsf{id})) \,;\, T_u \qquad \text{(S-Split)}$$
$$\Rightarrow \ldots ([l]\, b \,;\, (c \,;\, \langle\rangle) \otimes (\gamma_3 \,;\, \mathsf{id})) \,;\, T_u \qquad \text{(S-Atomic)}$$
$$\Rightarrow \ldots ([l]\, b \,;\, (c \,;\, \langle\rangle) \otimes (\mathsf{id} \,;\, \gamma_3)) \,;\, T_u \qquad \text{(S-Id)}$$
$$\Rightarrow \ldots ([l]\, b \,;\, (c \otimes \mathsf{id}) \,;\, \gamma_3) \,;\, T_u \qquad \text{(S-Join)},\ \langle\rangle \otimes \gamma_3 \equiv \gamma_3$$
$$\Rightarrow \ldots ([l]\, b \,;\, c \otimes \mathsf{id}) \,;\, \gamma_3 \,;\, T_u \qquad \text{(S-Exit)}$$
$$\Rightarrow \ldots ([l]\, b \,;\, c \otimes \mathsf{id}) \,;\, \gamma_3$$
$$ \,;\, (\mathsf{assert}\ \gamma_3 \,;\, T_m) \mid (T_l \,;\, T_u) \qquad \text{(S-Def)}$$
$$\Rightarrow \ldots \,;\, (\gamma_3 \,;\, \mathsf{assert}\ \gamma_3 \,;\, T_m) \mid (\gamma_3 \,;\, T_l \,;\, T_u)\ \text{(S-Alt)}$$
$$\Rightarrow \ldots \,;\, (\gamma_3 \,;\, T_m) \mid (\gamma_3 \,;\, T_l \,;\, T_u) \qquad \text{(S-Assert)}$$
$$\Rightarrow \ldots \,;\, (\gamma_3 \,;\, [m]\, c) \mid (\gamma_3 \,;\, T_l \,;\, T_u) \qquad \text{(S-Def)}$$
$$\Rightarrow \ldots \,;\, ([m]\, \gamma_3 \,;\, c) \mid (\gamma_3 \,;\, T_l \,;\, T_u) \qquad \text{(S-Enter)}$$
$$\Rightarrow \ldots \,;\, ([m]\, c \,;\, \langle\rangle) \mid (\gamma_3 \,;\, T_l \,;\, T_u) \qquad \text{(S-Atomic)}$$
$$\Rightarrow \ldots \,;\, ([m]\, c) \,;\, \langle\rangle \mid (\gamma_3 \,;\, T_l \,;\, T_u) \qquad \text{(S-Exit)}$$
$$\Rightarrow \ldots ([l]\, b \,;\, c \otimes \mathsf{id}) \,;\, ([m]\, c) \,;\, \langle\rangle \qquad \text{(S-Sel-L)}$$
$$\Rightarrow ([l]\, b \,;\, c \otimes \mathsf{id}) \,;\, ([m]\, c) \,;\, \langle\rangle \qquad \text{(S-Sel-R)}$$

The final value is as claimed in Ex. 5.

Our main result is that the two semantics we have given coincide. This shows that the small-step semantics is indeed an accurate way to step through the evaluation of tactic expressions.

Theorem 2 (Completeness of small-step semantics). *If* $\langle g, t \rangle \Downarrow \langle s, g' \rangle$, *then* $g \,;\, t \Rightarrow^* s \,;\, g'$

Proof. Straightforward induction on big-step derivation.

Theorem 3 (Soundness of small-step semantics). *If* $g \,;\, t \Rightarrow^* s \,;\, g'$ *then* $\langle g, t \rangle \Downarrow \langle s, g' \rangle$.

Proof. By induction on the length of the derivation, using Lemma 1.

Lemma 1 (Structure preservation)

1. If $[l]\, p \Rightarrow^* s \,;\, g$ then for some s', $s = [l]\, s'$ and there exists a reduction $p \Rightarrow^* s' \,;\, g$ with strictly shorter length.
2. If $p_1 \otimes p_2 \Rightarrow^* s \,;\, g$ and $p_1, p_2 \neq \langle\rangle$, then for some s_1, s_2, g_1 and g_2, we have $s = s_1 \otimes s_2$ and $g = g_1 \otimes g_2$ and there exist reductions $p_i \Rightarrow^* s_i \,;\, g_i$ with strictly shorter lengths.
3. If $p \,;\, t \Rightarrow^* s \,;\, g$ where p is not a goal, then for some s_1, s_2, g_1, we have $s = s_1 \,;\, s_2$ and there exist reductions $p \Rightarrow^* s_1 \,;\, g_1$, $g_1 \,;\, t \Rightarrow^* s_2 \,;\, g$ with strictly shorter length.
4. If $p_1 \mid p_2 \Rightarrow^* s \,;\, g$ then there exists a strictly shorter reduction of $p_1 \Rightarrow^* s \,;\, g$ or of $p_2 \Rightarrow^* s \,;\, g$.

Proof. Each part by induction on the lengths of sequences involved. For each constructor, a major rule is the base case and congruence rules are step cases.

Theorems 1 and 3 give correctness also for the small step semantics.

Corollary 1 (Correctness of small-step semantics). *If* $g \,;\, t \Rightarrow^* s \,;\, g'$ *then* $s \vdash g \longrightarrow g'$.

4   Tactic Programming

Tactics as above are procedures which produce hiproofs. To help with writing tactics, most theorem provers provide *tacticals* (tactic functionals or higher-order tactics), which combine existing tactics into new ones. The simplest examples of tacticals are the alternation and sequencing operations for tactics. Theorem provers like the original LCF, Isabelle, HOL or Coq provide more advanced patterns of applications; we concentrate on two illustrative cases here.

We will write tacticals as a meta-level notion, i.e., the following equations are meant as short-cuts defining one tactic for each argument tactic t:

$$\text{ALL } t \stackrel{def}{=} (t \otimes \text{ALL } t) \mid \langle \rangle$$

$$\text{ID} \stackrel{def}{=} \text{ALL id}$$

$$\text{REPEAT } t \stackrel{def}{=} (t \,;\, \text{REPEAT } t) \mid \text{ID}$$

ALL t applies t to as many atomic goals as available; in particular, ID is the 'polymorphic identity', which applied to any goal $g : n$ reduces to id^n, the n-fold tensor of id. REPEAT t applies t as often as possible. An application of this is a tactic to strip away all implications in the logic MPL (Example 3), defined as $stripImp \stackrel{def}{=} \text{REPEAT } impI$. To see this at work, here it is used in an actual proof:

$$\vdash A \Longrightarrow (B \Longrightarrow A) \,;\, stripImp$$
$$\Rightarrow \vdash A \Longrightarrow (B \Longrightarrow A) \,;\, (impI \,;\, \text{REPEAT } impI) \mid \text{ID}$$
$$\Rightarrow^* (\{A\} \vdash B \Longrightarrow A \,;\, \text{REPEAT } impI) \mid (\vdash A \Longrightarrow (B \Longrightarrow A) \,;\, \text{ID})$$
$$\Rightarrow^* (\text{id} \,;\, \{A, B\} \vdash A) \mid \ldots \mid \ldots$$
$$\Rightarrow^* \text{id} \,;\, \{A, B\} \vdash A$$

The last goal is easily proven with the atomic tactic ax.

4.1   Deterministic Semantics

The big- and small-step semantics given above are non-deterministic: a tactic t applied to a goal g may evaluate to more than one hiproof s (and remaining goals g'). This may result in many 'unwanted' reductions along with successful ones; e.g., **REPEAT** $t \,;\, g$ can always reduce to id $;\, g$. Non-determinism has its advantages: the tensor splitting allows a tactic such as ALL $b \otimes$ ALL c to solve the goal $\gamma_2 \otimes \gamma_2 \otimes \gamma_3$ by splitting the tensor judiciously: $\gamma_2 \otimes \gamma_2 \otimes \gamma_3 \,;\, \text{ALL } b \otimes$ ALL $c \Rightarrow (\gamma_2 \otimes \gamma_2 \,;\, \text{ALL } b) \otimes (\gamma_3 \,;\, \text{ALL } c)$. However, this behaviour is hard to implement (it requires keeping track of all possible reductions, and selecting the right ones after the fact), and it is in marked contrast to the usual alternation tactical (**ORELSE** in the LCF family) which always selects the first alternative if it is successful, and the second otherwise.

To give a deterministic behaviour for our language, we can define a restricted small-step semantics, which includes a strict subset of the reductions of the non-deterministic one. Since the principal sources of non-determinism are the alternation and the tensor splitting rules, the deterministic small-step semantics has the same rules as the small-step semantics from Sect. 3.2, but replaces rules (S-Sel-L), (S-Sel-R) and (S-Split) with the following:

$$(s_1 \; ; \; g) \mid p_2 \;\Rightarrow_D\; s_1 \; ; \; g \qquad \text{(TS}_D\text{-Alt-L)}$$

$$\frac{p_1 \;\not\Rightarrow_D\; s_1; h}{p_1 \mid (s_2 \; ; \; g) \;\Rightarrow_D\; s_2 \; ; \; g} \qquad \text{(TS}_D\text{-Alt-R)}$$

$$\frac{g_1 : n \qquad t_1 : n}{g_1 \otimes g_2 \; ; \; t_1 \otimes t_2 \;\Rightarrow_D\; (g_1 \; ; \; t_1) \otimes (g_2 \; ; \; t_2)} \qquad \text{(TS}_D\text{-Split)}$$

The right alternative is only chosen if the left alternative does not evaluate to any hiproof. Further, we only allow the argument goals to be split if the first component of the tactic has a fixed arity. This means that in the above example, $\gamma_2 \otimes \gamma_2 \otimes \gamma_3$; ALL $b \otimes$ ALL c does not reduce; to prove that goal under deterministic reduction, we need the tactic ALL $(b \mid c)$.

Theorem 4 (Deterministic small-step semantics). *The deterministic semantics is a restriction of the non-deterministic one:*

$$if\ t \; ; \; g \;\Rightarrow_D^*\; s \; ; \; g\ then\ t \; ; \; g \;\Rightarrow^*\; s \; ; \; g$$

Proof. A simple induction on the derivation of $\Rightarrow_D^*$. The rules (TS$_D$-Alt-L), (TS$_D$-Alt-R) and (TS$_D$-Split) are admissible in the non-deterministic small-step semantics.

Corollary 2. *(Soundness of the deterministic semantics) If $g \; ; \; t \;\Rightarrow_D^*\; s \; ; \; g$, then $\langle g, t \rangle \Downarrow \langle s, g' \rangle$.*

Theorem 4 implies that the deterministic semantics is weaker. In fact, it suffices for our simple example tacticals but some others (e.g., violating the arity restriction) are not covered. To recover more of the expressivity of the non-deterministic semantics, one could introduce a notion of backtracking, and treat the tensor split in a demand driven way to avoid the restriction of fixed arity. However, these extensions are beyond the scope of the present paper.

5   Related Work and Conclusions

This paper introduced a tactic language, *Hitac*, for constructing hierarchical proofs. We believe that hierarchical proofs offer the chance for better management of formal proofs, in particular, by making a connection between proofs and procedural methods of constructing them.

Our work with hierarchical structure in the form of hiproofs is unique, although there are many related developments on tactic language semantics and structured proofs elsewhere. A full survey is beyond our scope (more references can be found in [1]), but we highlight some recent and more closely connected developments.

Traditional LCF-style tactic programming uses a full-blown programming language for defining new tactics, as is also done in the modern HOL systems. The direct way of understanding such tactics is as the functions they define over proof states, suggesting a denotational fixed point semantics such as studied by Oliveira et al. [2]. Coq offers the power of OCaml for tactic programming, but also provides a dedicated functional language for writing tactics, $\mathcal{L}_{tac}$, designed by Delahaye [6]. This has the advantage of embedding directly in the Coq proof language, and offers powerful matching on the proof context. However, Delahaye did not formalise an evaluation semantics or describe a tactic tracing mechanism.

Kirchner [7] appears to have been the first to consider formally describing a small-step semantics for tactic evaluation, impressively attempting to capture the behaviour of both Coq and PVS within a similar semantic framework. He defines a judgement $e/\tau \to e'/\tau'$, which operates on a tactic expression e and proof context τ, to produce a simpler expression and updated context. So, unlike our simpler validation-based scheme, state based side-effecting of a whole proof is possible. However, the reduction notion is very general and the definitions for Coq and PVS are completely system-specific using semantically defined operations on proof contexts; there is a big gulf between providing these definitions and proving them correct.

Tinycals [3] is a recent small-step tactic language, implemented in the Matita system. The main motivation is to allow stepping inside tactics in order to extend Proof General-like interaction. In Coq and other systems of which we are aware, single-stepping defined tactics using their source is not possible, at best they can be *traced* by interrupting the tactic engine after a step and displaying the current state. Tinycals allows tracing linked back to the tactic expression, also showing the user information about remaining goals and backtracking points. The Tinycals language allows nested proof structure to be expressed in tactics, like hiproofs, and is also reminiscent of Isabelle's declarative proof language Isar [8], but there is no naming for the nested structure in either case.

One system that bears a closer structural resemblance to the hiproof notion is NuPrl's *tactic tree* [9], which extends LCF-style tactics by connecting them to proof trees, as would be done by combining our big-step semantics with the validation check which links a hiproof to an underlying tree. NuPrl allows navigating the tree and expanding or replacing tactics at each node.

Future work. Work still remains to fully describe the formal properties of the calculus and its extensions, including type systems for arity checking as hinted at in Sect. 3, and further deterministic evaluation relations. One important result for the small-step semantics is to characterise the normal forms. This requires a careful analysis of the "stuck" states (such as when an atomic tactic does not match a goal) that can be reached, and we have some preliminary results on this. Isolating failure points in stuck states will be important to help debugging.

The calculus we have presented here represents an idealised simple tactic language. We believe that this is a natural starting point for the formal study of tactic languages. We have kept examples concise on purpose to allow the reader to check them. Clearly, larger examples should be explored, but this will require mechanical assistance in some form. Moreover, there is clearly a diversity of concepts and constructs which are used to guide proof search in real systems. For example, many proof assistants allow goals to depend on each other via a meta-variable mechanism. More generally, we can envision interdependencies between each of tactics, goals, and proofs, and this leads us to speculate on the possibility of a "tactic cube" (analogous to Barendregt's Lambda Cube) of tactic languages.

On the practical side, the use of a generic tactic language offers hope that we will be able to write tactics that can be ported between different systems. We plan to investigate this and other issues with an implementation in Proof General. In associated work at Edinburgh, a graphical tool is being developed for displaying and navigating in hiproofs. Finally, one of us is developing a system which uses auto-generated formal proofs as evidence for the certification of safety-critical software. In this regard, the explicitly structured proofs which result from applications of tactics are likely to prove more useful than the unstructured proofs which are generated by most present theorem provers.

Acknowledgements. The authors would like to thank Gordon Plotkin, John Power, and Alan Bundy for useful discussions.

References

1. Denney, E., Power, J., Tourlas, K.: Hiproofs: A hierarchical notion of proof tree. In: Proceedings of Mathematical Foundations of Programing Semantics (MFPS). Electronic Notes in Theoretical Computer Science (ENTCS). Elsevier, Amsterdam (2005)
2. Oliveira, M.V.M., Cavalcanti, A.L.C., Woodcock, J.C.P.: ArcAngel: a tactic language for refinement. Formal Aspects of Computing 15(1), 28–47 (2003)
3. Coen, C.S., Tassi, E., Zacchiroli, S.: Tinycals: Step by step tacticals. Electr. Notes Theor. Comput. Sci. 174(2), 125–142 (2007)
4. Pollack, R.: On extensibility of proof checkers. In: Smith, J., Dybjer, P., Nordström, B. (eds.) TYPES 1994. LNCS, vol. 996, pp. 140–161. Springer, Heidelberg (1995)
5. Appel, A.W., Felty, A.P.: Dependent types ensure partial correctness of theorem provers. Journal of Functional Programming 14, 3–19 (2004)
6. Delahaye, D.: A tactic language for the system Coq. In: Parigot, M., Voronkov, A. (eds.) LPAR 2000. LNCS (LNAI), vol. 1955, pp. 85–95. Springer, Heidelberg (2000)
7. Kirchner, F.: Coq tacticals and PVS strategies: A small-step semantics. In: Archer, M., et al. (eds.) Design and Application of Strategies/Tactics in Higher Order Logics. NASA, pp. 69–83 (September 2003)
8. Wenzel, M.: Isar — a generic interpretative approach to readable formal proof documents. In: Bertot, Y., Dowek, G., Hirschowitz, A., Paulin, C., Théry, L. (eds.) TPHOLs 1999. LNCS, vol. 1690, pp. 167–184. Springer, Heidelberg (1999)
9. Griffin, T.G.: Notational Definition and Top-down Refinement for Interactive Proof Development Systems. PhD thesis, Cornell University (1988)

A   Example Reductions

Here are the reductions for the examples from Section 4 in detail. We first show how ALL id reduces applied to an empty goal list.

$\langle\rangle$; ALL id
$\Rightarrow \langle\rangle$; (id $\otimes$ ALL id) $| \langle\rangle$ (S-Def)
$\Rightarrow (\langle\rangle$; (id $\otimes$ ALL id) $| (\langle\rangle$; $\langle\rangle))$ (S-Alt)
$\Rightarrow \langle\rangle$; $\langle\rangle$ (S-Sel-R)

We can use this to show how ID reduces for a goal of arity 2.

$\gamma_1 \otimes \gamma_2$; ID
$\Rightarrow^* \gamma_1 \otimes \gamma_2$; (id $\otimes$ ALL id) $| \langle\rangle$ (S-Def)
$\Rightarrow (\gamma_1 \otimes \gamma_2$; id $\otimes$ ALL id) $| (\gamma_1 \otimes \gamma_2$; $\langle\rangle)$ (S-Alt)
$\Rightarrow ((\gamma_1$; id) $\otimes (\gamma_2$; ALL id)) $| \ldots$ (S-Split)
$\Rightarrow ((\text{id}$; $\gamma_1) \otimes (\gamma_2$; (id $\otimes$ ALL id) $| \langle\rangle)) | \ldots$ (S-Id),(S-Def)
$\Rightarrow^* ((\text{id}$; $\gamma_1) \otimes (\gamma_2 \otimes \langle\rangle$; id $\otimes$ ALL id) $| (\gamma_2$; $\langle\rangle)) | \ldots$ (S-Alt), $\gamma_2 \equiv \gamma_2 \otimes \langle\rangle$
$\Rightarrow^* ((\text{id}$; $\gamma_1) \otimes ((\gamma_2$; id) $\otimes (\langle\rangle$; ALL id) $| \ldots)) | \ldots$ (S-Split)
$\Rightarrow^* ((\text{id}$; $\gamma_1) \otimes ((\text{id}$; $\gamma_2) \otimes (\langle\rangle$; $\langle\rangle) | \ldots)) | \ldots$ (S-Id), see above
$\Rightarrow^* ((\text{id}$; $\gamma_1) \otimes (\text{id}$; $\gamma_2)) | \ldots$ (S-Join), $\gamma_2 \otimes \langle\rangle \equiv \gamma_2$, (S-Sel-R)
$\Rightarrow^* \text{id} \otimes \text{id}$; $\gamma_1 \otimes \gamma_2$ (S-Join), (S-Sel-L)

Finally, here is the full reduction from Section 4.

$\vdash A \Longrightarrow (B \Longrightarrow A)$; $stripImp$
$\Rightarrow \vdash A \Longrightarrow (B \Longrightarrow A)$; $(impI$; REPEAT $impI) | $ ID   (S-Def)
$\Rightarrow (\vdash A \Longrightarrow (B \Longrightarrow A)$; $impI$; REPEAT $impI) | (\vdash A \Longrightarrow (B \Longrightarrow A)$; ID) (S-Alt)
$\Rightarrow (\{A\} \vdash B \Longrightarrow A$; REPEAT $impI) | \ldots$ (S-Atomic), (S-Seq-L)
$\Rightarrow (\{A\} \vdash B \Longrightarrow A$; $(impI$; REPEAT $impI) | $ ID) $| \ldots$ (S-Def)
$\Rightarrow (\{A\} \vdash B \Longrightarrow A$; $impI$; REPEAT $impI) | (\{A\} \vdash B \Longrightarrow A$; ID) $| \ldots$ (S-Alt)
$\Rightarrow (\{A,B\} \vdash A$; REPEAT $impI) | \ldots | \ldots$ (S-Atomic), (S-Seq-L)
$\Rightarrow (\{A,B\} \vdash A$; $((impI$; REPEAT $impI) | $ ID)) $| \ldots | \ldots$ (S-Def)
$\Rightarrow (\{A,B\} \vdash A$; $impI$; REPEAT $impI) | (\{A,B\} \vdash A$; ID) $| \ldots | \ldots$ (S-Alt)
$\Rightarrow^* (\text{id}$; $\{A,B\} \vdash A) | \ldots | \ldots$ (S-Id), (S-Sel-R)
$\Rightarrow^* \text{id}$; $\{A,B\} \vdash A$ (S-Sel-L), (S-Sel-L)

The last goal is easily proven with the atomic tactic ax.

Logic-Free Reasoning in Isabelle/Isar

Stefan Berghofer and Makarius Wenzel*

Technische Universität München
Institut für Informatik, Boltzmannstraße 3, 85748 Garching, Germany
http://www.in.tum.de/~berghofe/
http://www.in.tum.de/~wenzelm/

Abstract. Traditionally a rigorous mathematical document consists of a sequence of definition – statement – proof. Taking this basic outline as starting point we investigate how these three categories of text can be represented adequately in the formal language of Isabelle/Isar.

Proofs represented in human-readable form have been the initial motivation of Isar language design 10 years ago. The principles developed here allow to turn deductions of the Isabelle logical framework into a format that transcends the raw logical calculus, with more direct description of reasoning using pseudo-natural language elements.

Statements describe the main result of a theorem in an open format as a reasoning scheme, saying that in the context of certain parameters and assumptions certain conclusions can be derived. This idea of turning Isar context elements into rule statements has been recently refined to support the dual form of elimination rules as well.

Definitions in their primitive form merely name existing elements of the logical environment, by stating a suitable equation or logical equivalence. Inductive definitions provide a convenient derived principle to describe a new predicate as the closure of given natural deduction rules. Again there is a direct connection to Isar principles, rules stemming from an inductive characterization are immediately available in structured reasoning.

All three categories benefit from replacing raw logical encodings by native Isar language elements. The overall formality in the presented mathematical text is reduced. Instead of manipulating auxiliary logical connectives and quantifiers, the mathematical concepts are emphasized.

1 Introduction

Isabelle/Isar [13, 14, 15] enables to produce formal mathematical documents with full proof checking. Similar in spirit to the Mizar system [12, 11], the user writes text in a formal language that is checked by the machine. As a side-effect of this, Isabelle/Isar produces high-quality documents using existing LATEX technology: the present paper is an example of such a formally processed document.

Rigorous mathematics is centered around *proofs*, and this view is taken to the extreme in Isabelle/Isar. The demands for human-readable proofs, which is the hardest part in formalized mathematics, are taken as guidelines for the design

* Supported by BMBF project "Verisoft" (grant 01 IS C38).

S. Autexier et al. (Eds.): AISC/Calculemus/MKM 2008, LNAI 5144, pp. 355–369, 2008.
© Springer-Verlag Berlin Heidelberg 2008

of the much simpler elements of *statements* and *definitions*. While the initial conception of the Isar proof language dates back almost 10 years, some more recent additions help to express structured statements and inductive definitions even more succinctly, in a "logic-free" style. This enables higher Isar idioms to focus on the mathematics of the application at hand, instead of demanding recurrent exercises in formal logic from the user. So *mathematical reasoning* is emphasized, and auxiliary logical constructions are left behind eventually.

Our basic approach works essentially in *bottom-up* manner, starting from primitive logical principles towards mathematical reasoning that is eventually free from the logic (which better serves in the background for foundational purposes only). As the art of human-readable formal reasoning evolves further, we hope to move towards a stage that meets with other approaches that are coming the *top-down* way from informal mathematics.

Overview. §2 reviews the original idea of "natural deduction" due to Gentzen, and its implementation in the Isabelle/Pure framework. §3 gives an overview of the Isar proof language as a linearized expression of structured proofs in the underlying logical framework. §4 introduces structured Isar statements, which enable to state and prove reasoning schemes conveniently, without going through the logical framework again. §5 covers a recent refinement of the well-known concept of inductive definitions, which enables to obtain natural deduction rules directly from basic definitions, without intermediate statements or proofs. §6 illustrates the benefits of the native "logic-free" style of Isar definitions, statements, and proofs by an example about well-founded multiset ordering.

2   Natural Deduction Revisited

About 75 years ago Gentzen introduced a logical calculus for "natural deduction" [3] that was intended to formalize the way mathematical reasoning actually works, unlike earlier calculi due to Hilbert and Russel. Since we share the motivation to approximate mathematical reasoning, we briefly review some aspects of traditional natural deduction as relevant for Isabelle/Isar.

Gentzen uses a two-dimensional diagrammatic representation of reasoning patterns, which may be composed to proof trees according to certain principles. Each logical connective is characterized by giving *introduction rules* and *elimination rules*. This is illustrated for $\longrightarrow$ and $\forall$ as follows (in our notation):

$$\frac{\begin{array}{c} [A] \\ \vdots \\ B \end{array}}{A \longrightarrow B} \ (\longrightarrow\! I) \qquad \frac{A \longrightarrow B \quad A}{B} \ (\longrightarrow\! E)$$

$$\frac{\begin{array}{c} [a] \\ \vdots \\ B(a) \end{array}}{\forall x.\ B(x)} \ (\forall I) \qquad \frac{\forall x.\ B(x)}{B(a)} \ (\forall E)$$

Inferences work by moving from assumptions (upper part) to conclusions (lower part). Nested inferences, as indicated by three dots and brackets, allow to refer to *local* assumptions or parameters, which are *discharged* when forming the final conclusion. Note that in $(\forall I)$ we have treated the locally "fresh" parameter a analogous to an assumption, which reflects the formal treatment in the Isabelle framework. Traditional logic texts often treat this important detail merely as a footnote ("eigenvariable condition").

The Isabelle/Pure framework [8, 9] implements a *generic* version of higher-order natural deduction, without presupposing any particular object-logic. Natural deduction rules are represented in Isabelle as propositions of the "meta-logic", which provides the framework connectives of implication $A \Longrightarrow B$ and quantification $\bigwedge x.\ B\ x$. This first-class representations of primitive and derived natural deduction rules is supported by two main operations: *resolution* for mixed forward-backward chaining of partial proof trees, and *assumption* for closing branches. Both may involve higher-order unification, which results in a very flexible rule-calculus that resembles higher-order logic programming [15, §2.2].

According to the initial "logical framework" idea of Isabelle [8, 9], the user may specify a new object-logic by declaring connectives as (higher-order) term constants, and rules as axioms. For example, the minimal logic of $\longrightarrow$ and $\forall$ could be declared as follows (using type i for individuals and o for propositions):

$$imp \quad :: o \Rightarrow o \Rightarrow o \qquad (\textbf{infix } \longrightarrow)$$
$$impI \ : \ \textstyle\bigwedge A\ B.\ (A \Longrightarrow B) \Longrightarrow A \longrightarrow B$$
$$impE \ : \ \textstyle\bigwedge A\ B.\ (A \longrightarrow B) \Longrightarrow A \Longrightarrow B$$
$$all \quad :: (i \Rightarrow o) \Rightarrow o \qquad (\textbf{binder } \forall)$$
$$allI \quad : \ \textstyle\bigwedge B.\ (\textstyle\bigwedge a.\ B\ a) \Longrightarrow \forall x.\ B\ x$$
$$allE \quad : \ \textstyle\bigwedge a\ B.\ (\forall x.\ B\ x) \Longrightarrow B\ a$$

Note that outermost $\bigwedge$ is usually left implicit. The above rules merely reflect the minimal logic of $\Longrightarrow$ and $\bigwedge$ of the framework. The idea of generic natural deduction becomes more apparent when the object-logic is enriched by further connectives, for example:

$$conj \quad :: o \Rightarrow o \Rightarrow o \qquad (\textbf{infix } \wedge)$$
$$conjI \ : \ A \Longrightarrow B \Longrightarrow A \wedge B$$
$$conjE \ : \ A \wedge B \Longrightarrow (A \Longrightarrow B \Longrightarrow C) \Longrightarrow C$$
$$disj \quad :: o \Rightarrow o \Rightarrow o \qquad (\textbf{infix } \vee)$$
$$disjI_1 \ : \ A \Longrightarrow A \vee B$$
$$disjI_2 \ : \ B \Longrightarrow A \vee B$$
$$disjE \ : \ A \vee B \Longrightarrow (A \Longrightarrow C) \Longrightarrow (B \Longrightarrow C) \Longrightarrow C$$
$$ex \quad :: (i \Rightarrow o) \Rightarrow o \qquad (\textbf{binder } \exists)$$
$$exI \quad : \ B\ a \Longrightarrow \exists x.\ B\ x$$
$$exE \quad : \ (\exists x.\ B\ x) \Longrightarrow (\textstyle\bigwedge a.\ B\ a \Longrightarrow C) \Longrightarrow C$$

These rules for predicate logic follow Gentzen [3], except for conjunction elimination. Instead of two projections $A \wedge B \Longrightarrow A$ and $A \wedge B \Longrightarrow B$, our *conjE* rule enables to assume local facts A and B, independently from the main goal. Other typical situations of elimination are represented by *disjE*, which splits the main goal into two cases with different local assumptions, and *exE*, which augments the main goal by a local parameter a such that $B\ a$ may be assumed.

This uniform presentation of eliminations is typical for Isabelle/Pure [8, 9], but often appears peculiar to users without a strong background in formal logic. Even in Gentzen's original article, the *disjE* and *exE* rules are explained with special care, while "the other rules should be easy to understand" [3]. In the Isar proof language (§3), we shall provide a refined view on elimination, that expresses directly the idea of being able to assume local assumptions over local parameters, potentially with a case-split involved.

The examples for natural deduction presented so far have referred to traditional connectives of predicate logic: $\longrightarrow$, $\forall$, $\wedge$, $\vee$, $\exists$ etc. There is nothing special about these in the generic framework of Isabelle/Pure. We may just as well reason directly with concepts of set theory, lattice theory etc. without going through predicate logic again. Here are natural deduction rules for $x \in A \cap B$:

$$interI : x \in A \implies x \in B \implies x \in A \cap B$$
$$interE : x \in A \cap B \implies (x \in A \implies x \in B \implies C) \implies C$$

In practice, such domain-specific rules are not axiomatized, but derived from the definitions of the underlying concepts. In fact, the majority of rules will be of the latter kind — after the initial object-logic axiomatization, regular users proceed in this strictly definitional manner. Thus the role of the logical framework as foundation for new logics is changed into that of a tool for plain mathematical reasoning with derived concepts. Then the main purpose of the special connectives $\implies$ and $\bigwedge$ is to outline *reasoning patterns* in a "declarative" fashion. Guided by the indicated structure of natural deduction rules, structured proofs are composed internally by means of the *resolution* and *assumption* principles.

The remaining question is how to obtain natural deduction rules conveniently. As we shall see later (§5), a refined version of the well-known concept of *inductive definitions* allows to produce elimination rules quite naturally from a "logic-free" specification of the introduction rules only. The system will derive a proper predicate definition internally, and derive the corresponding rules, which may then be turned immediately into Isar proof texts in the application.

3   Isar Proofs

The Isar proof language [13, 14, 15] enables to express formal natural deduction proofs in a linear form that approximates traditional mathematical reasoning. Gentzen [3] admits that his calculus looses information present in the "narrated" version of informal reasoning: it is unclear where to start reading two-dimensional proof trees. This linguistic structure is recovered in Isabelle/Isar: proof texts are written with pseudo-natural language elements, which are interpreted by the Isar proof processor in terms of the underlying logical framework of Isabelle/Pure, see [15, §3.3] for further details.

It is important to understand that Isar is not another calculus, but a *language* that is interpreted by imposing certain policies on the existing rule calculus of Isabelle/Pure. To this end, Isar introduces non-logical concepts to organize formal entities notably the *proof context*, the *goal state* (optional), and a register for the latest *result*. The overall proof configuration is arranged as a stack over these components, which enables block-structured reasoning within a flat logic.

An Isar proof body resembles a mathematical notepad: statements of various kinds may be written down, some refer to already established facts (**note**), some extend the context by new parameters and assumptions (**fix** and **assume**), some produce derived results (**have** and **show**, followed by a sub-proof). Moreover, there are several elements to indicate the information flow between facts and goals, notably **then, from, with, using, also, finally, moreover, ultimately**.

Previous facts may be referenced either by name, or by a literal proposition enclosed in special parentheses. For example, in the scope of **assume** a: A, both a and $\langle A \rangle$ refer to the same (hypothetical) result. In the subsequent examples, we mostly use the latter form for clarity. The labelled version is preferable in larger applications, when propositions are getting bigger. The special name *this* always refers to the result of the last statement.

From the perspective of the logical framework, the main purpose of Isar is to produce and compose natural deduction rules. The most elementary way to produce a rule works by concluding a result within the local scope of some extra hypotheses, which are discharged when leaving the scope. For example:

> {
>   **fix** x **and** y
>   **assume** A x **and** B y
>   **have** C x y $\langle proof \rangle$
> }
> **note** $\langle \bigwedge x\, y.\ A\, x \Longrightarrow B\, y \Longrightarrow C\, x\, y \rangle$

Within the body of a sub-proof, **fix**–**assume**–**show** yields a rule as above, but the result is used to refine a pending subgoal (matching both the assumptions and conclusion as indicated in the text). The structure of the goal tells which assumptions are admissible in the sub-proof, but there is some flexibility due to the way back-chaining works in the logical framework. For example:

> **have** $\bigwedge x\, y.\ A\, x \Longrightarrow B\, y \Longrightarrow C\, x\, y$ **have** $\bigwedge x\, y.\ A\, x \Longrightarrow B\, y \Longrightarrow C\, x\, y$
> **proof** −   **proof** −
>   **fix** x **and** y **fix** y **assume** B y
>   **assume** A x **and** B y **fix** x **assume** A x
>   **show** C x y $\langle proof \rangle$ **show** C x y $\langle proof \rangle$
> **qed**   **qed**

The **proof** and **qed** elements are not just delimiters, but admit initial and terminal refinements of pending goals. The default for **proof** is to apply a canonical elimination or introduction rule declared in the background context, using the "*rule*" method. The default for **qed** is to do nothing; in any case the final stage of concluding a sub-proof is to finish pending sub-goals trivially by assumption.

Further abbreviations for terminal proofs are "**by** *method*$_1$ *method*$_2$" for "**proof** *method*$_1$ **qed** *method*$_2$", and ".." for "**by** *rule*", and "." for "**by** *this*".

With standard introduction and elimination rules declared in the library, we can now rephrase natural deduction schemes (§2) as linear Isar text:

have $A \longrightarrow B$
proof
  assume A
  show B ⟨*proof*⟩
qed

assume $A \longrightarrow B$ **and** A
then have B ..

have $\forall x.\ B\ x$
proof
  fix a
  show $B\ a$ ⟨*proof*⟩
qed

assume $\forall x.\ B\ x$
then have $B\ a$ ..

Here we have mimicked Gentzen's diagrammatic reasoning, composing proof texts according to the structure of the underlying rules. Isar is much more flexible in arranging natural deduction proof outlines, though. Some of the rule premises may be established beforehand and pushed into the goal statement; the proof body will only cover the remaining premises. This allows mixed forward-backward reasoning according to the following general pattern:

from *facts*$_1$ **have** *props* **using** *facts*$_2$ **proof** (*method*$_1$) *body* **qed** (*method*$_2$)

For example, premise $A \longrightarrow B$ could be provided either as "**from** ⟨$A \longrightarrow B$⟩" before the goal, as "**using** ⟨$A \longrightarrow B$⟩" after the goal, or as "**show** $A \longrightarrow B$" in the body. It is up to the author of the proof to arrange facts adequately, to gain readability by the most natural flow of information. Sub-structured premises are usually addressed within a sub-proof, using **fix**–**assume**–**show** in backwards mode, as seen in the above introduction proofs of $\longrightarrow$ and $\forall$.

The other rules from §2 can be directly turned into Isar proof texts as well, but eliminations of the form $\ldots \Longrightarrow (\bigwedge a.\ B\ a \Longrightarrow C) \Longrightarrow C$ demand special attention. A naive rendering in Isar would require the main goal C given beforehand, and a sub-proof that proves the same C in a context that may be enriched by additional parameters and assumptions. Isar's **obtain** element [15, §3.1] supports this style of reasoning directly, in a logic-free fashion. For example:

{
  obtain x **and** y **where** $A\ x$ **and** $B\ y$ ⟨*proof*⟩
  have C ⟨*proof*⟩
}
note ⟨C⟩

The proof obligation of "**obtain** x **and** y **where** $A\ x$ **and** $B\ y$" corresponds to the rear-part of an eliminations rule: $(\bigwedge x\ y.\ A\ x \Longrightarrow B\ y \Longrightarrow thesis) \Longrightarrow thesis$ for a hypothetical *thesis* that is arbitrary, but fixed. Having finished that proof, the context is augmented by "**fix** x **and** y **assume** $A\ x$ **and** $B\ y$".

Results exported from that scope are unaffected by these additional assumptions, provided the auxiliary parameters are not mentioned in the conclusion!

We can now spell out the remaining natural deduction schemes of §2 adequately, only *disjE* requires explicit sub-proofs involving the main conclusion C, because **obtain** cannot split a proof text into several cases.

assume A **and** B
then have $A \wedge B$ ..

assume $A \wedge B$
then obtain A **and** B ..

assume A
then have $A \vee B$ ..

assume B
then have $A \vee B$ ..

assume $A \vee B$
then have C
proof
  assume A
  then show C $\langle proof \rangle$
next
  assume B
  then show C $\langle proof \rangle$
qed

assume B a
then have $\exists x.\ B\ x$ ..

assume $\exists x.\ B\ x$
then obtain a **where** $B\ a$ ..

4   Isar Statements

Isar proof composition is centered around natural deduction rules of the logical framework. Such rules may be established as regular theorems like this:

theorem r: $\bigwedge x\ y.\ A\ x \Longrightarrow B\ y \Longrightarrow C\ x\ y$
proof −
  fix x **and** y
  assume $A\ x$ **and** $B\ y$
  show $C\ x\ y$ $\langle proof \rangle$
qed

This is slightly unsatisfactory, because the structure of the result is specified redundantly in the main statement and the proof body, using framework connectives $\bigwedge / \Longrightarrow$ vs. Isar proof elements **fix–assume–show**, respectively. Moreover, exposing the Isabelle/Pure rendering of the intended reasoning scheme gives the head statement a rather technical appearance. This is even worse for elimination rules, due to extra rule nesting $\ldots \Longrightarrow (\bigwedge a.\ B\ a \Longrightarrow C) \Longrightarrow C$ etc.

Isar statements address these issues by introducing first-class notation for certain rule schemes. As seen in the initial example in §3, proof blocks allow to produce natural deduction rules on the spot, by discharging local parameters and assumptions, e.g. "**{ fix** x **assume** $A\ x$ **have** $B\ x$ $\langle proof \rangle$ **}**" for $\bigwedge x.\ A\ x \Longrightarrow B\ x$. Based on this idea we introduce three kinds of clausal Isar statements.

1. *Big clauses* have the form "**fixes** *vars* **assumes** *props* **shows** *props*" and specify the outermost structure of a natural deduction reasoning pattern.

The given **fixes** and **assumes** elements determine a local context, **shows** poses simultaneous local goals within that. The subsequent proof proceeds directly within the local scope; the final result emerges by discharging the context, producing corresponding $\bigwedge/\Longrightarrow$ rule structure behind the scenes.

2. *Dual clauses* have the form "**fixes** *vars* **assumes** *props* **obtains** *vars* **where** *props*" and abbreviate certain big clauses: "**obtains** a **where** $B\,a$" expands to "**fixes** *thesis* **assumes** $\bigwedge a.\ B\,a \Longrightarrow$ *thesis* **shows** *thesis*". Case-splits may be indicated by several clauses separated by "|", which corresponds to multiple branches of the form $\bigwedge a_i.\ B_i\,a_i \Longrightarrow$ *thesis*. According to the principles behind big clauses, the resulting rule will have exactly the elimination format described in §2. Within the proof body, each **obtains** case corresponds to a different hypothetical rule to conclude the main *thesis*; one of these possibilities has to be chosen eventually.

3. *Small clauses* are of the form "$B\,x$ **if** $A\,x$ **for** x" and indicate the second-level rule structure of framework propositions within big clauses. This corresponds directly to $\bigwedge x.\ A\,x \Longrightarrow B\,x$, but clausal notation may not be nested further.

The basic **fixes–assumes–shows** form of big clauses has been available in Isabelle/Isar for many years. The dual form is a recent addition, which has first appeared officially in Isabelle2007. Small clauses are not available in official Isabelle yet, but are an experimental addition for the present paper only.

Our initial proof of $\bigwedge x\,y.\ A\,x \Longrightarrow B\,y \Longrightarrow C\,x\,y$ is now rephrased as follows:

theorem r:
  fixes x **and** y
  assumes $A\,x$ **and** $B\,y$
  shows $C\,x\,y$ $\langle proof \rangle$

See also §6 for proofs involving **obtains**. To continue our running example of predicate logic, we rephrase the previous natural deduction rules from §2:

  theorem *impI*: **assumes** B **if** A **shows** $A \longrightarrow B$
  theorem *impE*: **assumes** $A \longrightarrow B$ **and** A **obtains** B

  theorem *allI*: **assumes** $B\,a$ **for** a **shows** $\forall x.\ B\,x$
  theorem *allE*: **assumes** $\forall x.\ B\,x$ **obtains** $B\,a$

  theorem *conjI*: **assumes** A **and** B **shows** $A \wedge B$
  theorem *conjE*: **assumes** $A \wedge B$ **obtains** A **and** B

  theorem $disjI_1$: **assumes** A **shows** $A \vee B$
  theorem $disjI_2$: **assumes** B **shows** $A \vee B$
  theorem *disjE*: **assumes** $A \vee B$ **obtains** A | B

  theorem *exI*: **assumes** $B\,a$ **shows** $\exists x.\ B\,x$
  theorem *exE*: **assumes** $\exists x.\ B\,x$ **obtains** a **where** $B\,a$

In other words, we have managed to express the inherent structure of reasoning schemes without demanding auxiliary logical connectives, not even those of the

Isabelle/Pure framework. Only concepts of the application, which happens to be predicate logic as an object-language here, and native Isar elements are involved. The same works for domain-specific rules, e.g. those for set theory seen before:

> **theorem** *interI*: **assumes** $x \in A$ **and** $x \in B$ **shows** $x \in A \cap B$
> **theorem** *interE*: **assumes** $x \in A \cap B$ **obtains** $x \in A$ **and** $x \in B$

5   Inductive Definitions

Inductive predicates provide a convenient way to define concepts by specifying a collection of characteristic *introduction rules*. Support for inductive definitions is available in many theorem provers. Melham [6] describes a version for the HOL system using an impredicative encoding, meaning that the definition involves universal quantification over predicate variables, whereas Harrison's inductive definition package for HOL [4] uses an encoding based on the Knaster-Tarski fixpoint theorem. The Coq system [2] is based on the *Calculus of Inductive Constructions* introduced by Paulin-Mohring, which contains inductive definitions as a primitive concept [7]. Inductive definitions in Isabelle were first introduced by Paulson [10], using fixpoints over the lattice of sets. Our refined version works on generic lattices, which subsume predicates in HOL.

Many well-known concepts of mathematics can be viewed as an inductive predicate. E.g. the transitive closure of a relation can be defined as follows:

inductive *trcl* **for** $R :: \alpha \Rightarrow \alpha \Rightarrow bool$
  where
    trcl $R\ x\ x$ **for** x
  | *trcl* $R\ x\ z$ **if** $R\ x\ y$ **and** *trcl* $R\ y\ z$ **for** $x\ y\ z$

The rules of **inductive** may be specified using the format of "small clauses" introduced in §4. Internally, the system derives further natural deduction rules that may be turned into Isar proofs as discussed in §3. By virtue of its definition as the least predicate closed under these rules, any inductive predicate admits an *induction* and an *inversion* principle (case analysis). For example:

assume *trcl* $R\ a\ b$
then have $P\ a\ b$
proof (*rule trcl.induct*)
  fix x
  show $P\ x\ x$ ⟨*proof*⟩ — induction base
next
  fix $x\ y\ z$
  assume $R\ x\ y$ **and** *trcl* $R\ y\ z$ **and** $P\ y\ z$
  then show $P\ x\ z$ ⟨*proof*⟩ — induction step
qed

This induction principle is a consequence of *trcl* being defined as the least fixpoint of a *predicate transformer* of type $(\alpha \Rightarrow \alpha \Rightarrow bool) \Rightarrow \alpha \Rightarrow \alpha \Rightarrow bool$:

$trcl \equiv$
$\lambda R.\ lfp\ (\lambda p\ x_1\ x_2.$
$\qquad (\exists x.\ x_1 = x \wedge x_2 = x)\ \vee$
$\qquad (\exists x\ y\ z.\ x_1 = x \wedge x_2 = z \wedge R\ x\ y \wedge p\ y\ z))$

The body of the function $(\lambda p\ x_1\ x_2.\ \ldots)$ is a disjunction, whose two parts correspond to the two introduction rules for $trcl$. Using the fact that the predicate transformer is monotonic, the induction principle follows from this definition using the Knaster-Tarski theorem for least fixpoints on complete lattices:

$$\frac{mono\ f \qquad f\ (lfp\ f \sqcap P) \sqsubseteq P}{lfp\ f \sqsubseteq P}$$

The ordering relation $\sqsubseteq$ and the infimum operator $\sqcap$ is defined on the complete lattice of n-ary predicates in a pointwise fashion:

$$P \sqsubseteq Q \equiv \forall x_1 \ldots x_n.\ P\ x_1 \ldots x_n \longrightarrow Q\ x_1 \ldots x_n$$
$$P \sqcap Q \equiv \lambda x_1 \ldots x_n.\ P\ x_1 \ldots x_n \wedge Q\ x_1 \ldots x_n$$

The premise $f\ (lfp\ f \sqcap P) \sqsubseteq P$ of the fixpoint theorem is established by the proofs of the induction base and the induction step in the above proof pattern.

Case analysis corresponds to the observation that if an inductive predicate holds, one of its introduction rules must have been used to derive it. This principle can be viewed as a degenerate form of induction, since there is no induction hypothesis. For the transitive closure, the case analysis scheme is:

```
assume trcl R a b
then have Q
proof (rule trcl.cases)
  fix x
  assume a = x and b = x
  then show Q ⟨proof⟩
next
  fix x y z
  assume a = x and b = z and R x y and trcl R y z
  then show Q ⟨proof⟩
qed
```

Although the case analysis rule could be derived from the above least fixpoint theorem as well, it is proved from the fixpoint unfolding theorem $mono\ f \implies lfp\ f = f\ (lfp\ f)$ which has the advantage that exactly the same proof technique can also be used in the case of *coinductive* predicates, using gfp in place of lfp.

Inductive predicates are very convenient to formalize mathematical concepts succinctly, even if there is no recursion involved. For example, the composition of two relations R and S can be defined as follows:

inductive $comp$ **for** $R :: \alpha \Rightarrow \beta \Rightarrow bool$ **and** $S :: \beta \Rightarrow \gamma \Rightarrow bool$
$\quad$ **where** $comp\ R\ S\ x\ z$ **if** $R\ x\ y$ **and** $S\ y\ z$ **for** $x\ y\ z$

For $comp$, the underlying primitive definition is $comp \equiv \lambda R\ S.\ lfp\ (\lambda p\ x_1\ x_2.$ $\exists x\ y\ z.\ x_1 = x \wedge x_2 = z \wedge R\ x\ y \wedge S\ y\ z)$. For fixpoints of constant functions like the above we have $lfp\ (\lambda x.\ t) = t$, which easily follows from the fixpoint unfolding theorem. Using the same principles, we can even characterize basic operators of predicate logic as inductive predicates with zero arguments. E.g.

inductive *and* **for** $A\ B :: bool$
 where *and* $A\ B$ **if** A **and** B

inductive *or* **for** $A\ B :: bool$
 where *or* $A\ B$ **if** A | *or* $A\ B$ **if** B

inductive *exists* **for** $B :: \alpha \Rightarrow bool$
 where *exists* B **if** $B\ a$ **for** a

Again, these operators are just examples. Real applications would introduce their genuine notions directly as inductive definitions.

6 Case-Study: Well-Founded Multiset Ordering

To illustrate the "logic-free" style of definitions, statements and proofs in Isar, we formalize some aspects of well-founded multiset ordering. A multiset is a finite "bag" of items, which can be modeled as a function from items to natural numbers that yields a non-zero value only on a finite domain. Multiset notation is reminiscent of plain sets: $\{a,\ a,\ b,\ b,\ b,\ c\}$ for enumeration, $a \in B$ for membership, $A \uplus B$ for union etc. The structure of multisets can also be characterized inductively, with base case $\{\}$ and step case $B \uplus \{a\}$ for a multiset B.

Given an ordering on items, multisets can be ordered by the following intuitive process: one item of the bag is removed and replaced by the content of another bag of strictly smaller items; this is repeated transitively. The main theorem states that the resulting relation on multisets is well-founded, provided that the item ordering is well-founded. Below we merely cover the basic definitions and a technical lemma required for the well-foundedness proof.[1]

Our development refers to a locally fixed *less* relation, which is introduced by commencing the following locale context (see also [1]).

locale *less-relation* = **fixes** *less* :: $\alpha \Rightarrow \alpha \Rightarrow bool$ (**infix** $\prec$ 50)
begin

The locale already contributes to the "logic-free" approach, since it avoids explicit abstraction or quantification over that parameter.

A bag of items is compared to a single item in point-wise manner as follows:

definition *lesser* (**infix** $\prec$ 50) **where** $B \prec a \leftrightarrow (\forall b.\ b \in B \longrightarrow b \prec a)$

lemma *lesserI*: **assumes** $b \prec a$ **for** b **shows** $B \prec a$
 using *assms* **unfolding** *lesser-def* **by** *auto*

---

[1] See http://isabelle.in.tum.de/dist/library/HOL/Library/Multiset.html for a rather old version of the complete formalization that mixes quite different styles; the main well-foundedness theorem is called *wf-mult* there.

lemma *lesserE*: **assumes** $B \prec a$ **and** $b \in B$ **obtains** $b \prec a$
  using *assms* **unfolding** *lesser-def* **by** *auto*

Obviously, the primitive predicate definition of $B \prec a$ is *not* logic-free, since it uses $\forall$ and $\longrightarrow$ connectives. The two extra "boiler plate" lemmas amend this by providing an alternative characterization in natural deduction style. (In the bits of proof shown below, we never need to analyze the *lesser* relation, though).

Next we define the main idea of the multiset ordering process. The subsequent inductive predicate $N \prec\!\!\prec M$ expresses a single step of splitting off an element from $M = B \uplus \{a\}$ and replacing it by a point-wise smaller multiset. (The full ordering emerges as the transitive closure of that relation.)

inductive *less-multiset* (**infix** $\prec\!\!\prec$ 50)
  where $B \uplus C \prec\!\!\prec B \uplus \{a\}$ **if** $C \prec a$ **for** a B C

This rather succinct logic-free definition characterizes the relation by a single clause — there are no other cases and no recursion either. Even this degenerate form of inductive definition is very convenient in formal reasoning. Here the decomposition of the two multisets is specified directly via pattern matching, with side-conditions and parameters expressed as native clauses of Isabelle/Isar.

In contrast, the original formulation from the Isabelle/HOL library uses an encoding that involves intermediate layers of predicate logic and set theory, with separate equations to express the decomposition.

definition *less-mult* $=$
  $\{(N, M). \exists a\, B\, C.\ M = B \uplus \{a\} \wedge N = B \uplus C \wedge C \prec a\}$

While this might look familiar to anybody trained in logic, manipulating such auxiliary structure in formal proof requires extra steps that do not contribute to the application. Nevertheless, even rather bulky encodings do often happen to work out in practice by means of reasonably strong "proof automation". We illustrate this by proving formally that both definitions are equivalent.

lemma $N \prec\!\!\prec M \leftrightarrow (N, M) \in$ *less-mult*
  unfolding *less-mult-def*
proof
  assume $N \prec\!\!\prec M$
  then obtain a B C **where** $M = B \uplus \{a\}$ **and** $N = B \uplus C$ **and** $C \prec a$
    by (*rule less-multiset.cases*)
  then show $(N, M) \in \{(N, M). \exists a\, B\, C.\ M = B \uplus \{a\} \wedge N = B \uplus C \wedge C \prec a\}$
    by *auto*
next
  assume $(N, M) \in \{(N, M). \exists a\, B\, C.\ M = B \uplus \{a\} \wedge N = B \uplus C \wedge C \prec a\}$
  then obtain a B C **where** $M = B \uplus \{a\}$ **and** $N = B \uplus C$ **and** $C \prec a$
    by *auto*
  from $\langle C \prec a \rangle$ **have** $B \uplus C \prec\!\!\prec B \uplus \{a\}$ **by** (*rule less-multiset.intros*)
  with $\langle M = B \uplus \{a\} \rangle$ **and** $\langle N = B \uplus C \rangle$ **show** $N \prec\!\!\prec M$ **by** *simp*
qed

This rather lengthy proof merely shuffles logical connectives back and forth, without being very informative. The *auto* method involved here is a fully-featured combination of classical proof search with equational normalization; it successfully bridges the gap between the intermediate statements given in the text. On the other hand, this extra overhead can be avoided by the logic-free characterization of the inductive definition from the very beginning. So we continue in that style now, working on the mathematics of multiset orderings instead of doing exercises in formal logic and automated reasoning.

The proof of the main theorem combines well-founded induction over the relation $\prec$ of items with structural induction over multisets. At some point in the induction step, the multiset ordering $N \prec\!\!\prec B \uplus \{a\}$ needs to be analyzed:

lemma *less-add-cases*:
  assumes $N \prec\!\!\prec B \uplus \{a\}$
  obtains
    (1) M **where** $M \prec\!\!\prec B$ **and** $N = M \uplus \{a\}$
  | (2) C **where** $C \prec a$ **and** $N = B \uplus C$

Ultimately, the rule resulting from this goal statement will split an arbitrary fact $N \prec\!\!\prec B \uplus \{a\}$ into two cases as specified above. In the present proof context, we are still in the course of establishing this claim. Here $N \prec\!\!\prec B \uplus \{a\}$ is available as a local fact, and there are two possibilities to finish the hypothetical main *thesis*, namely rule 1: *thesis* **if** $M \prec\!\!\prec B$ **and** $N = M \uplus \{a\}$ **for** M and rule 2: *thesis* **if** $C \prec a$ **and** $N = B \uplus C$ **for** C.

This means the subsequent proof already starts out in a nicely decomposed version of the idea of splitting cases and obtaining local parameters and assumptions, without having to work through auxiliary $\lor$, $\land$, $\exists$ connectives again:

proof −
  from $\langle N \prec\!\!\prec B \uplus \{a\} \rangle$
  obtain $a'\ B'\ C$ **where**
    $B \uplus \{a\} = B' \uplus \{a'\}$ **and**
    $N = B' \uplus C$ **and**
    $C \prec a'$
    by (*rule less-multiset.cases*) *simp-all*
  from $\langle B \uplus \{a\} = B' \uplus \{a'\} \rangle$ **show** *thesis*
  proof (*rule add-eq-cases*)
    assume $B = B'$ **and** $a = a'$
    with $\langle C \prec a' \rangle$ **and** $\langle N = B' \uplus C \rangle$
    have $C \prec a$ **and** $N = B \uplus C$ **by** *simp-all*
    then show *thesis* **by** (*rule 2*)
  next
    fix C' **assume** $B' = C' \uplus \{a\}$ **and** $B = C' \uplus \{a'\}$
    show *thesis*
    proof (*rule 1*)
      from $\langle C \prec a' \rangle$ **have** $C' \uplus C \prec\!\!\prec C' \uplus \{a'\}$ **by** (*rule less-multiset.intros*)
      with $\langle B = C' \uplus \{a'\} \rangle$ **show** $C' \uplus C \prec\!\!\prec B$ **by** *simp*
      from $\langle B' = C' \uplus \{a\} \rangle$ **and** $\langle N = B' \uplus C \rangle$
      show $N = C' \uplus C \uplus \{a\}$ **by** (*simp add: union-ac*)
    qed
  qed
qed

Above the initial **obtain** statement augments the local context by means of standard elimination of the $N \prec M$ relation, using the corresponding *cases* rule. The sub-proof via *add-eq-cases* involves another **obtains** rule proven in the background library; its statement is structurally similar to *less-add-cases*.

So our proof manages to maintain the logic-free style, no auxiliary connectives are involved, only some algebraic operators from the application domain. The old proof in the Isabelle/HOL library requires about two times more text, even though it uses many abbreviations for sub-terms and local facts. Moreover, it needs more automation to work through extraneous logical structure.

end

7   Conclusion and Related Work

Isabelle/Isar shares the mission of formal reasoning that approximates traditional mathematical style with the pioneering Mizar system [12, 11]. There are many similarities and dissimilarities, see also [17, 16] for some comparison.

Concerning the logical foundations, Isar uses the Isabelle/Pure framework [8, 9] which implements a generic higher-order version of Gentzen's natural deduction calculus [3]. In contrast, Mizar works specifically with classical first-order logic, and the style of reasoning is modeled after the "supposition calculus" due to Jaskowski [5]. The basic paradigm of structured proof composition in Mizar is quite different from Isar. Where Isar revolves around natural deduction rules that emerge from local proof bodies and refine pending goals eventually, Mizar allows to operate more directly on the logical structure of a claim in consecutive refinement steps: **let** to move past universal quantification, **assume** to move past an implication etc. In contrast, **fix** and **assume** in Isar do not operate on a goal structure, but construct a context that will impose a certain rule structure on the final **show** result. This can make a difference in practice: in proving an implication a Mizar proof needs to say **assume** A invariably, while in Isar the corresponding "**assume** A" is only required if that fact is actually used later.

Essentially, there are Mizar proof elements for each of the logical connectives of $\wedge, \vee, \longrightarrow, \forall, \exists$, but English words are used here both for the connectives and the corresponding proof elements. For example, the proposition **for x holds A[x]** can be established by **let x** and a proof of **A[x]** in that scope. Thus Mizar enables to produce a proof text according to principles from classical first-order logic, while Isar is more puristic in referring to generic natural deduction, where predicate logic is just one example. This different attitude is best illustrated by existential elimination, which works in Mizar via **consider a such that B[a]** and is closely tied to actual existential quantification **ex x st B[x]**. In Isar "**obtain** a **where** $B\ a$" merely espresses the more elementary idea of being able to augment the local scope by a hypothetical entity a with property $B\ a$. This might follow from a fact $\exists x.\ B\ x$, but the elimination is better performed by a domain-specific rule $\ldots \Longrightarrow (\bigwedge a.\ B\ a \Longrightarrow C) \Longrightarrow C$, or "**obtains** a **where** B a" as explained in the present paper. Our inductive definitions are particularly well suited to produce such rules.

This means certain aspects of Mizar are about predicate logic, rather than mathematics. In contrast, our "logic-free" style in Isar enables more direct expression of definitions, statements, and proofs — reducing the overall formality of the text.

References

[1] Ballarin, C.: Interpretation of locales in Isabelle: Theories and proof contexts. In: Borwein, J.M., Farmer, W.M. (eds.) MKM 2006. LNCS (LNAI), vol. 4108, pp. 31–43. Springer, Heidelberg (2006)

[2] Barras, B., et al.: The Coq Proof Assistant Reference Manual. INRIA (2006)

[3] Gentzen, G.: Untersuchungen über das logische Schließen (I + II). Mathematische Zeitschrift 39 (1935)

[4] Harrison, J.: Inductive definitions: automation and application. In: Schubert, E.T., Alves-Foss, J., Windley, P. (eds.) HUG 1995. LNCS, vol. 971. Springer, Heidelberg (1995)

[5] Jaskowski, S.: On the rules of suppositions. Studia Logica 1 (1934)

[6] Melham, T.F.: A package for inductive relation definitions in HOL. In: Archer, M., et al. (eds.) Higher Order Logic Theorem Proving and its Applications. IEEE Computer Society Press, Los Alamitos (1992)

[7] Paulin-Mohring, C.: Inductive Definitions in the System Coq – Rules and Properties. In: Bezem, M., Groote, J.F. (eds.) TLCA 1993. LNCS, vol. 664. Springer, Heidelberg (1993)

[8] Paulson, L.C.: The foundation of a generic theorem prover. Journal of Automated Reasoning 5(3) (1989)

[9] Paulson, L.C.: Isabelle: the next 700 theorem provers. In: Odifreddi, P. (ed.) Logic and Computer Science. Academic Press, London (1990)

[10] Paulson, L.C.: A fixedpoint approach to (co)inductive and (co)datatype definitions. In: Plotkin, G., Stirling, C., Tofte, M. (eds.) Proof, Language, and Interaction: Essays in Honour of Robin Milner. MIT Press, Cambridge (2000)

[11] Rudnicki, P.: An overview of the MIZAR project. In: 1992 Workshop on Types for Proofs and Programs, Chalmers University of Technology, Bastad (1992)

[12] Trybulec, A.: Some features of the Mizar language (1993) presented at Turin

[13] Wenzel, M.: Isar — a generic interpretative approach to readable formal proof documents. In: Bertot, Y., Dowek, G., Hirschowitz, A., Paulin, C., Théry, L. (eds.) TPHOLs 1999. LNCS, vol. 1690. Springer, Heidelberg (1999)

[14] Wenzel, M.: Isabelle/Isar — a versatile environment for human-readable formal proof documents. Ph.D. thesis, Institut für Informatik, TU München (2002)

[15] Wenzel, M.: Isabelle/Isar — a generic framework for human-readable proof documents. In: Matuszewski, R., Zalewska, A. (eds.) From Insight to Proof — Festschrift in Honour of Andrzej Trybulec, University of Białystok. Studies in Logic, Grammar, and Rhetoric, vol. 10(23) (2007), http://www.in.tum.de/~wenzelm/papers/isar-framework.pdf

[16] Wiedijk, F.(ed.): The Seventeen Provers of the World. LNCS (LNAI), vol. 3600. Springer, Heidelberg (2006)

[17] Wiedijk, F., Wenzel, M.: A comparison of the mathematical proof languages Mizar and Isar. Journal of Automated Reasoning 29(3-4) (2002)

A Mathematical Type for Physical Variables

Joseph B. Collins

Naval Research Laboratory
4555 Overlook Ave, SW
Washington, DC 20375-5337

Abstract. In identifying the requirements of a markup language for describing the mathematical semantics of physics-based models, we pose the question: "Is there a mathematical type for physical variables?" While this question has no a priori answer, since physics is fundamentally empirical, it appears that a large body of physics may be described with a single mathematical type. Briefly stated, that type is formed as the mathematical product of a physical unit, such as *meter* or *second*, and an element of a Clifford algebra. We discuss some of the properties of this mathematical type and its use in documentation of physics-based models.

1   Introduction

We are interested in creating a markup language for the representation of physical models, i.e., a physics markup language. Our primary requirement for a physics markup language is to represent the models that physicists and engineers create and so, necessarily, the components with which they build those models. The principal reason for creating such a language is to improve the communication of the semantics of models of the physical world in order to support interoperability of physics-based models with each other, such as with multiphysics simulation, and interoperability with other non-physics-based models. Physics-based models are used extensively in modeling and simulation (M&S) frameworks to support a wide array of predictive and decision making applications of practical importance. An open and standard way of documenting the physical and mathematical semantics of physics-based models, such as a markup language might provide, would go a long way towards lowering the costs of model development and validation. Additionally, since models form the basis of the theoretical development of physics, communication of research results and physics education would also be favorably impacted.

In approaching these goals, we ask "What information is it necessary to specify in order to transmit knowledge of a physical model and to make the transmission unambiguous?" In particular, we are interested in identifying the specific mathematical concepts necessary for expressing the physical semantics since, once identified, they may be dealt with somewhat independently. We observe that the typical computer code representing a physics-based model follows from a mathematical model derived from the application of mathematically phrased

S. Autexier et al. (Eds.): AISC/Calculemus/MKM 2008, LNAI 5144, pp. 370–381, 2008.
© Springer-Verlag Berlin Heidelberg 2008

physical laws to mathematical representations of physical objects. There is a rich array of mathematical concepts used in these mathematical representations. This raises the question as to how we approach the problem of representing all of these mathematical concepts. For example, we need to specify the dimensionality of the physical objects being modeled, their spatio-temporal extents, and the embedding space. We also note that: the physical quantities and corresponding units used to describe physical properties have a mathematical structure; the physical laws that are applied usually have a differential expression; and, invariance with respect to various transformations is a key concept. Each of these, while they carry physical semantics, must be mathematically expressed.

2   Mathematical Requirements of a Physics Markup Language

Physical semantics ultimately rests on mathematical phrasing. To be meaningful, scientific theories are required to provide predictions that are testable. In practice, this means we must be able to compare mathematically computed predictions to numerical measurement data. Accordingly, the first things we need to express in a physics markup language are the mathematical symbols that represent the properties of physical objects. To be useful, a physics-based model must represent a physical object with at least one of the object's measurable properties, which has physical dimension expressed in specified units. Very often these properties are modeled as variables and they are used to represent such things as the positions, velocities, and accelerations of a physical object, which typically vary as a function of time. A prediction results when, given the model, we can solve for a given variable. We refer to these variables as physical variables. We need to be able to express not only physical variables, but also the mathematical operations upon physical variables and the mathematical relationships between physical variables.

It is often said that most models in physics are ultimately partial differential equations with boundary conditions. In order to specify these relations between physical variables, we will require the ability to specify, in addition to the physical variables themselves, differential operators, such as gradient, divergence, and curl, acting on scalar and vector fields, as well as equality and inequality relationships. Note that this use of the term field is not the usual mathematical meaning as in, for example, "the real numbers form a field", but is specifically a physicists meaning. A physicists notion of a physical field (temperature field, gravitational field, etc.) is a scalar or vector quantity defined at each point in a space-time domain. Specifying a model in terms of differential equations is an implicit form of specification, since in order to express the variables as explicit functions of time we will require a solution to the equations.

There are many more mathematical concepts that we need to express to represent models with physical variables. We know, for example, that: classical mechanics makes use of scalars, vectors, and tensors defined in space-time; these vectors have length, giving metric properties to objects defined in space-time;

quantum physics makes use of Dirac spinors and Hilbert space vectors (bra-ket notation); general relativity requires transformation between covariant and contra-variant forms using a non-trivial metric tensor; and, models of physical objects possess spatial extent and often have defined boundary surfaces. Differential equations need to be expressed over definite volumes, and boundary conditions need to be expressed on the bounding surfaces of those volumes. We often want to specify a preferred geometric basis for the expressed geometry, such as rectangular, cylindrical, or spherical coordinates. Until we can express the semantics of these many mathematical concepts, we will not be able to express a large body of physical models.

Statements of invariance are also important relationships between physical variables. While the equations that make up a model may implicitly obey some invariance, and additional statement of such invariance may seem redundant, specific statements of known invariance are useful in understanding a particular model and in performing computational evaluations using the model. Invariance is, in general, specified with respect to operations performed on physical variables by particular operators. Such operators include Euclidean transformations (spatial rotations, translations, and reflections) and Lorentz transformations (space-time rotations, boosts, and reflections).

While in order to make specific predictions it is common to consider models as providing unique solutions for all of its physical variables as functions on space-time, this is not always necessary. There is value in using models to express incomplete knowledge of as well, which may result in sets of multiple possible solutions. We may, for example, only only be able to specify that two variables within a model have a functional dependence, i.e., X is a function of Y, without knowing more detail. We may want to specify that a variable has exclusive dependence on another or that it is independent of another. We often need to state physical principles as inequalities, for example, for which there are many solutions. It may be that we want to develop reusable models that can be used to predict many different variables, but not necessarily simultaneously, where each variable may have distinct dependencies, or lack thereof, on given initial conditions. In building these models, we may need to develop a more clear definition of what constitutes a model and under what conditions a model permits solutions to be determined.

To summarize, a markup language for physics must support the following mathematical concepts:

a) A physical attribute which has physical dimension and may be represented in defined units. It may be represented with a scalar magnitude, or, if it is a more complex property, by a vector, a tensor, or other object with the necessary algebraic properties.

b) A physical object has spatial presence and extent, properties that are represented as point-like, 1-dimensional, or arbitrary dimensional attributes. These properties may be described within the space-time reference frame of the physical object itself, or within the space-time reference frame of another physical object.

c) The attributes of a physical object may satisfy a specified set of differential equations or other mathematical relations.

Finally, it cannot be supposed that this is a complete tally of useful mathematical semantics. For example, statements of general physical laws, such as Newtonian universal gravitation, will be aided by the use of mathematical quantifiers to specify, for example, that gravitational forces are present between *all pairs* of massive physical objects within a model. In general, it seems desirable, if not necessary, to be able to express a full range of mathematical relations between variables in physical models.

3 A Type for Physical Variables

The fundamental components that physicists use to build models are physical variables, parameters that represent the physical quantity attributes of physical objects. A physical quantity is an observable, measurable property of an object in the physical world. A principal difficulty we have in representing physical variables in a markup language is that physical variables do not generally have a well-defined type, where we use the term type much as a computer scientist or mathematician would, i.e., a class of objects with a well-defined set of allowed operations. Physicists and engineers typically act as applied math practitioners with a well-schooled intuition, and they are not always fussy about mathematical formalism. The types of physical variables are rarely declared as part of a problem statement or model definition, and it is common to find abrupt transitions in usage, from one implied type to another. While one might well consider attempting to capture the reasoning abilities of these applied math practitioners as an exercise in artificial intelligence, that is a separate research topic of its own. We are undertaking here the problem of capturing as precisely as possible the mathematical description of such models, and describing as concise a set of clearly defined types as possible. The reason for looking more carefully at the formal mathematical representations of physical variables is to determine what is a sufficient amount of information to require for a semantic representation of physical models.

So, we begin this inquiry into developing a physics markup language by posing the following primary question: "What is the type necessary for representing physical variables?" Upon reflection, we may question why we should expect there to be a single type for representing physical variables. We state, somewhat axiomatically, that the objective of physics is to describe physical interactions mathematically. One may dispute the underlying axiomatic assumption that physical interactions may be described mathematically, but, pragmatically, we are only interested in those interactions that may be so described, since that is what affords us the ability to make predictions.

To answer the question as to why we should expect a single type for physical variables, consider the following. If for each interaction of two physical variables we were to be given a physical variable of a new type, it would not take very long for the resulting type proliferation to make it difficult, if not impossible, to describe the physical universe. Describing the physical universe is certainly easier if there is a countable, closed system of defined types, and easier still if there are but a finite number of defined types. More importantly, we should expect that if the physical universe is closed, so too in our mathematical description of the physical universe should the set of objects that represent physical variables be closed under those operations that represent physical interactions. The requirement of closure merely reflects the idea that physical interactions should be a function of the physical quantities of the interacting objects and should result in physical effects, where the effects may also be represented using physical variables. Without a requirement of closure for physical variables, we would allow non-physical results from the interaction of physical objects or we would allow physical effects to result from non-physical interactions. We therefore require the definition, from a formal perspective, of a type for representing physical variables, which has a mathematical description, being essentially a set that is closed under defined operations.

We also undertake this inquiry with the understanding that a practical solution today may well be improved upon later since it is impossible to anticipate all of the future developments of theoretical physics. This reality should not deter us, however, from attempting to answer our primary question, since there is significant challenge and great utility in handling only those representations of physical variables that have been described to date.

In summary, we need to represent the idea of physical variables, the mathematical symbols used to represent specific physical properties of physical objects. The physical variable may be thought of as having all of the mathematical properties that the applicable physical theories indicate that they should have, and also be capable of holding the corresponding measurable values. The values may be arrived at by measurement of the corresponding physical objects attributes, or by prediction arrived at by applying physical laws, e.g., equations, to other measured attributes of a system of physical objects.

4   The Physical Dimensional Properties of Physical Variables

The term "physical quantity" is a fundamental one in physics, narrowly defined by the International System of Units (SI) as the measurable properties of physical objects. Common usage often substitutes the phrase physical dimension for the SI defined phrase physical quantity, and uses the term physical quantity more loosely. A "physical dimension" in this sense should not be confused with the separate notion of spatial dimensions, e.g., those defined by three spatial basis vectors.

The SI has also defined base quantities: they are length, mass, time, electric current, thermodynamic temperature, amount of substance, and luminous intensity with corresponding dimensions represented by the symbols L, M, T, I, Θ, N, and J [1]. Derived quantities may be created by taking products, ratios, and powers of the base quantities. A measurement generally returns a positive, definite quantity and a zero value implies an immeasurably small amount of the quantity. The result of a simple (scalar) measurement of a physical quantity is represented as the product of a scalar real number and a physical unit, where the physical unit is a scale factor for a physical quantity or physical dimension. While there is debate within the physics research community as to how many physical dimensions are truly fundamental, standard practice is to use the seven SI base quantities mentioned above. The SI also provides corresponding standard base units for the seven base quantities: meter; second; kilogram; ampere; Kelvin; mole; and candela. Within the SI standard, many other units, called derived units, are defined in terms of these base units.

While the SI system is commonly used, it is not used exclusively. Other systems may have a different set of fundamental dimensions, base units, or both. A simple way to characterize the system used for a given model is to specify, for n fundamental dimensions, an n-tuple of defined units. This explicitly specifies the base units while implicitly specifying base dimensions and supports the expression of a model for any set of defined absolute quantities.

In its most comprehensible form, then, a physical variable represents a quantity, like a length, which is generally measured as a finite precision, real number of units, where the units are some reference or standard units. While an individual measurement is most easily thought of as a scalar quantity, physical variables may have multiple components which are more suitably represented as vectors or tensors. Measurement of these more complex objects is correspondingly complex.

4.1   The Mathematics of Units and Dimensions

As asserted earlier, the semantics of physics is largely contained within the mathematical properties of the components with which we describe physical models. We now begin to examine the mathematical properties of physical variables. The operation of taking the physical dimension of a physical variable, X, is usually written with square brackets, as $[X]$. This operation, which is idempotent, i.e., $[[X]] = [X]$, is like a projection, where the information about magnitude, units, and spatial directionality of the physical variable is all lost. We can enumerate some of the properties of physical variables under the physical dimension bracket operation:

All physical variables have physical dimension composed of the fundamental dimensions:

$$[X] = \mathsf{L}^\alpha \mathsf{M}^\beta \mathsf{T}^\gamma \mathsf{I}^\delta \Theta^\epsilon \mathsf{N}^\zeta \mathsf{J}^\eta, \tag{1}$$

where the exponents, α, β, γ, δ, ϵ, ζ, and η, are rational numbers.

Physical variables may be added if they are of the same dimension:

$$\text{If } [X] = [Y], \text{ then } [X + Y] = [X] = [Y]; \tag{2}$$

The physical dimension of a product of physical variables is the same as the commutative and associative product of the physical dimensions of the factor variables:

$$[X * Y] = [X] * [Y] = [Y] * [X] ; \tag{3}$$

$$[X * Y * Z] = [X * Y] * [Z] = [X] * [Y * Z] \tag{4}$$

The physical dimension of the reciprocal of a physical variable is the reciprocal of the physical dimension of the variable:

$$[X^{-1}] = [X]^{-1} \tag{5}$$

The physical dimension of a real number is defined to be 1. Formally, the physical dimensions of physical variables form a commutative, or abelian group. This group may be written multiplicatively, which corresponds to the usual way in which dimensional quantities are manipulated in most physical applications. Written multiplicatively, the group elements are the identity, 1, and, in the case of SI, $n = 7$ base quantities, L, M, T, I, Θ, N, and J, along with their powers and their products. Being abelian, this group may also be written additively, where the group element representation is as an n-tuple of exponents for the n base quantities. The additive representation of the group is useful in performing dimensional analysis. The exponents of the dimensions are often integers, although for convenience in some applications the exponents are extended to the rational numbers. When written additively, the physical dimensions of physical variables may be seen to form a vector space, where vector addition corresponds to multiplication of the underlying physical variables and scalar multiplication of the n-tuple of exponents corresponds to raising the physical variables to various powers.

By taking the physical dimension of a physical variable we have lost some essential pieces of information, which we now seek to recover. In particular, for what is commonly thought of as a scalar physical variable, we need to represent the combination of the units and magnitude of the physical property. In order to do so, we here introduce the following notation: $X = \{X\}_u * u$, where a physical variable is factored into two parts: the first part is $\{X\}_u$, while the second part is the unit, u, that the physical dimension is expressed in, i.e., $[X] = [u]$, The first part, $\{X\}_u$, which is properly scaled with respect to the unit, u, is the non-physically-dimensioned part of the physical variable, i.e., $[\{X\}_u] = 1$. We will call this part of the physical variable, $\{X\}_u$, the spatial part.

Units provide a scale factor for each of the base dimensions, giving a base unit for each base dimension. A unit is either a base unit or a unit derived by (commutative) products and ratios of base units. For example, a product of two units of length results in a unit having physical dimension $L * L = L^2$, a unit for area. A product or ratio of different units may be reduced if they have fundamental dimensions in common. A ratio of two different units having the same physical dimension, when reduced, results in a dimensionless real number called a *conversion factor*. For example, $[foot] = [meter] = L$ so $meter/foot \approx 3.28$.

We finally note that we can represent the logarithm of the physical variable as the formal sum

$$ln(X) = ln(\{X\}_u) + \alpha * ln(u_1) + ...\eta * ln(u_7) \tag{6}$$

where $u = \prod_{i=1}^{7} u_i$ in the case of seven base units. We can more simply represent this as

$$(z, \alpha, \beta, \gamma, \delta, \epsilon, \zeta, \eta) \tag{7}$$

where $z = ln(\{X\}_u)$, representing the measured quantity in units derived from base units. In this representation of the physical variable the operation of taking the physical dimension is seen to be a true projection operator, i.e.,

$$[(z, \alpha, \beta, \gamma, \delta, \epsilon, \zeta, \eta)] = (0, \alpha, \beta, \gamma, \delta, \epsilon, \zeta, \eta). \tag{8}$$

where the result is an element of the additive representation of the group of physical dimensions. The space of fundamental and derived physical dimensions so represented comprises a vector space, where the *vector addition* operation corresponds to multiplication of physically dimensioned quantities and the *scalar multiplication* operation corresponds to raising physical quantities to powers. A change of units is represented as a translation operation in the first (dimensionless) component of the $(1 + n)$-tuple that represents the physical variable when that component is an element of a scalar field.

While physicists routinely perform legitimate mathematical manipulations of physical dimensions, they do this intuitively and the formal mathematical structure of physical dimensions is rarely expressed.

5   A Type for the Spatial Part of Physical Variables

The spatial part of physical variables, i.e., $\{X\}_u$, has the following properties: we can multiply it by a scalar; we can add more than one together; we can multiply more than one together. The first two of these properties indicate that they form a vector space. The third property, multiplication of physical variables, is trivial when the spatial part of a physical variable is a scalar. After scalars, the most common object representing the spatial part of physical variables are *vectors*. When a physicist or engineer refers to a "vector", they usually mean a rank-1 tensor. Physicists and engineers also use higher rank tensors, most commonly rank-2 tensors.

Typically used vector multiplications are: the scalar, inner, or dot product; and, the vector cross product, or Gibbs' vector product. Well known, though less commonly used, is the dyadic, outer, or tensor product, where higher rank tensors may be constructed from lower rank tensors. Usually a metric is tacitly assumed, typically Euclidean. Other metrics are required for special and general relativistic mechanics.

The manner in which these vectors and tensors are manipulated by physicists is largely ad hoc, rather than uniform, and is usually derived from the work of prior physical scientists. Maxwell popularized Hamilton's quaternions, using them to express electrodynamics. Quaternions were superseded by the vector analysis of Gibbs [2] and Heaviside, which survives to this day, largely unaltered except by addition of new concepts, objects and operations. The mathematics used in quantum mechanics today follows the style of usage originated by the physicists that originally employed it. While we do not mean to suggest incorrectness in their treatment, much of the mathematics used by physicists is taught by physicists. A mathematician might find an absence of definition and uniformity in the mathematical properties of physical variables as they are most commonly used.

Considering these issues, a principle question that that we raise is: "What is the type of the spatial part of physical variables?" By asking this question we mean to proceed to understand the formal mathematical structure of these objects. Because physics is at root empirical, the best answer that can be provided is to propose a type of object that appears to meet the criteria of matching the known objects used by physicists as physical variables. Each time a new concept, object, or operation is added, it would be helpful to formally extend an axiomatic mathematical framework to incorporate the new in with the old. The purpose for doing this was stated previously: closure in the world of physical quantities and interactions should be reflected by mathematical closure in the physical variables used to represent the physical world. Happily, this question has been constructively considered and the best answer to date appears to be that the spatial part of physical variables may be described by Clifford algebras [3].

As usually encountered in the education of a physicist, physical variables, specifically the spatial part of physical variables, appear to consist of several types. Most commonly encountered are real scalars or vectors. Complex scalars and vectors are also commonly used in representing physical variables. Minkowski four-vector notation is well-known to students of physics to be a better notation than Gibbs' vector notation for electrodynamics, particularly the "Electrodynamics of Moving Bodies", [4] i.e., special relativity. General relativity introduces multi-ranked tensors; elementary quantum mechanics introduces Hilbert spaces and the non-commutative spinors. Finally, modern quantum particle theories make liberal use of elements of various Lie algebras. To the casual observer, there appears to be a multiplicity of types.

Clifford algebras are not commonly used by most physicists, though they are heavily used in some forefront research areas of theoretical physics. While there is currently some effort [5] to change this state of affairs, one may reasonably ask why we should introduce into a discussion of standards a construct that is not commonly used. The answer is based on two requirements. First, there is the important problem of being able to translate or otherwise relate models expressed in different notations. If there is one notational representation that can capture the semantics of a catchall of individual notations, then it is useful to have it present at least as an underlying representation, even if it is infrequently expressed explicitly in the specification of models. That is, since it represents the current understanding of the fundamental underlying mathematics for most, if not all, physical models, representing Clifford algebras is sufficient to represent physical variables in most known models. Secondly, since many models explicitly reference Clifford algebras, it is necessary to represent Clifford algebras in order to represent the semantics of those models.

5.1   Features of a Clifford Algebra

The objects of Clifford algebras are vectors, although they may not always seem as recognizable to physicists as the usual vectors that come from the Vector Analysis of Gibbs and Heaviside. The vectors of Clifford algebras are also referred to as *multivectors* and represent a richer set of objects than those in Gibbs' Vector Analysis. Some multivectors are the usual vectors of Gibbs' Vector Analysis, some are scalars and some are higher ranked tensors. Some of these multivectors represent formal sums of the usual scalars, vectors, and tensors. Some of these multivectors may be used to represent subspaces. Some of these multivectors are used to represent rotations, translations, spinors and other objects normally described by Lie groups. In summary, the principle mathematical objects of interest to physicists are all elements of Clifford algebras.

A key element of Clifford algebras is the Clifford product, an associative vector product with an inverse. The other vector products previously mentioned here do not have these properties. Since Clifford algebras also have an identity element and closure holds for the Clifford product, there is a resulting group structure for the vectors in a Clifford algebra. Of particular interest, Lie algebras are sub-algebras of Clifford algebras. A complete description of Clifford algebras is well beyond the scope of this paper and is well described elsewhere [3].

One may well ask "If Clifford algebras are as powerful as advertised, why did physicists ever commit to the standard Vector Analysis?" There may be several speculative answers possible [6]. Certainly the work of Grassman, which gave rise to Clifford algebras, may not have been as well publicized among physicists as Gibbs' work was, though Gibbs was certainly aware of it. Additionally, the standard Vector Analysis serves quite well for much of classical physics, so its continued use is a reasonable satisficing strategy. How, then, is Gibbs' Vector Analysis not the best fit for physics? It begins to be less comfortably used when vector objects of rank greater than one, i.e., tensors, are required, but, most certainly, spinors appear to be foreign objects within Vector Analysis. Perhaps one of the sorest points is that vector cross-product defined by Gibbs only exists in three dimensions. Modern physicists like to stretch well beyond three-dimensions. In Clifford algebras the cross product has been defined for spaces of any dimension. Outside of three dimensional space it is not a simple vector, and does not appear to be describable within Vector Analysis.

We note the following several points that may be of particular interest to mathematicians. Hestenes narrows the range of Clifford algebras of interest to physicists to *geometric algebras*. Geometric algebras are the subset of Clifford algebras defined over the reals and possessing a non-singular quadratic form [3] [7], so, from the mathematical perspective, expressing elements and operations of a Clifford algebra are sufficient for doing the same for elements of a geometric algebra. A concise axiomatic development of geometric algebra and its differential calculus, called geometric calculus, are provided by Hestenes [3]. Geometric calculus claims greater generality than Cartan's calculus of differential forms [3].

Of particular interest to physicists and other intuitive mathematicians, geometric algebras have natural and well developed geometric interpretations [11] which, interestingly, have been exploited in computer graphics rendering using the coordinate-free representations of rotations and translations. The work of reformulating physics in this coherent notation, not overwhelming, but no small task, has been underway for many years [8] [9] with the result that it appears to have great potential for unifying the mathematics of physics. Geometric calculus has even been successfully applied to gauge theory gravity [10], one of the more esoteric research frontiers in physics.

We are left to conclude that the standard Gibbs' Vector Analysis is by comparison just a convenient shorthand, derived from the ideas of Grassman which have reached a fuller and richer expression in Clifford algebras. In sum, Clifford algebras generally, and geometric algebra in particular, provide a coherent algebraic method for representing the spatial part of physical variables for most of classical and modern physics. It is certainly not the most commonly used notation, but other notations may be readily translated into it.

6   Summary

Our purpose here has been to sketch the essential mathematical properties of physical variables. One reason for doing this is to help clarify the mathematical semantics as separate from, though necessary to, the expression of physical semantics. Having made this separation, experts in representing mathematical semantics are now enabled to aid in the development of a physics markup language by independently expanding mathematical semantic representations. In particular, semantic representations of the mathematical properties of physical dimensions and units, and of Clifford algebras, which include geometric algebras, will greatly enable the expression of the physical semantics of physics-based models. We believe that the expression of Clifford algebras in this way will be significantly more straightforward from a mathematical perspective because it is mathematically better defined than the collection of notations used for different sub-theories within the physics community.

References

1. Comité International des Poids et Mesures: The International System of Units (SI), 8th edn. Bureau International des Poids et Mesures (2006)
2. Wilson, E.B.: Vector Analysis: Founded upon the Lectures of J. Yale University Press, Williard Gibbs (1901)
3. Hestenes, D., Sobczyk, G.: Clifford Algebra to Geometric Calculus: A Unified Language for Mathematics and Physics. D. Reidel Publishing (1984)
4. Einstein, A.: Zur Elektrodynamik bewegter Körper (On the Electrodynamics of Moving Objects). Annalen der Physik 17, 891–921 (1905)
5. Hestenes, D.: Reforming the Mathematical Language of Physics, Oersted Medal Lecture (2002). American Journal of Physics 71, 104 (2003)
6. Crowe, M.J.: A History of Vector Analysis: The Evolution of the Idea of a Vectorial System. Dover Publications (1994)
7. http://en.wikipedia.org/wiki/Geometric_algebra

8. Hestenes, D.: New Foundations for Classical Mechanics. Kluwer Academic Publishers, Dordrecht (1999)
9. Doran, C., Lasenby, A.: Geometric Algebra for Physicists. Cambridge University Press, Cambridge (2003)
10. Hestenes, D.: Gauge Theory Gravity with Geometric Calculus. D. Hestenes, Foundations of Physics 35(6), 903 (2005)
11. Dorst, L., Fontijne, D., Mann, S.: Geometric Algebra for Computer Science. Elsevier, Amsterdam (2007)

Unit Knowledge Management

Jonathan Stratford and James H. Davenport

Department of Computer Science
University of Bath, Bath BA2 7AY, United Kingdom
Jonathan.Stratford@alumni.bath.ac.uk,
J.H.Davenport@bath.ac.uk

Abstract. In [9], various observations on the handling of (physical) units in OpenMath were made. In this paper, we update those observations, and make some comments based on a working unit converter [21] that, because of its OpenMath-based design, is modular, extensible and reflective. We also note that some of the issues in an effective converter, such as the rules governing abbreviations, being more linguistic than mathematical, do not lend themselves to easy expression in Open-Math.

1 Introduction

For the purposes of this paper, we define a **unit** of measurement as *any determinate quantity, dimension, or magnitude adopted as a basis or standard of measurement for other quantities of the same kind and in terms of which their magnitude is calculated or expressed* [19, unit].

Units are generally thought of as "fairly easy", but, as this paper shows, there are some subtleties. One of the early design goals for Java was that it should be 'unit safe' as well as 'type safe', but this was dropped due to the difficulties [12].

There have been many famous examples where unit conversion was not undertaken, or where it was incorrectly calculated. The Gimli Glider [18,24], as it became known, was a (then) new Boeing 767 plane, which, during what should have been a routine flight in 1983, ran out of fuel just over halfway to its intended destination. The ensuing investigation established that an incorrect conversion had been performed, leading to a woefully insufficient fuel payload, because the aircraft was one of the first of its kind to use a metric measure of fuel, and the refuellers had used an imperial conversion instead of the correct metric one. In addition, although a second check was carried out between legs of the flight, the same incorrect conversion was used.

Large organisations such as NASA are not immune to such problems [17]. Software controlling the thrusters on the Mars Climate Orbiter was configured to use imperial units, while ground control, and the other parts of the space craft, interpreted values as if they were metric. This led to the orbiter entering an incorrect orbit too close to Mars, and ultimately to its being destroyed.

Again, even widespread systems such as Google can get this wrong — see the example in section 4 — as can attempts such as OntoWeb to "understand" MathML in terms of simple structures such as RDF [6] (section 4.3).

S. Autexier et al. (Eds.): AISC/Calculemus/MKM 2008, LNAI 5144, pp. 382–397, 2008.

2   Prior Work on Semantics of Units

2.1   OpenMath

OpenMath [3] is a standard for representing mathematical semantics. It differs from the existing versions[1] of Content MathML [4,5] in being *extensible*: new Content Dictionaries (CDs) can add new OpenMath symbols, known as OMS, and can prescribe their semantic, via Formal Mathematical Properties (FMPs). In contrast, OpenMath *variables*, known as OMVs, are purely names.

OpenMath is essentially agnostic with respect to type systems. However, one particular one, the Simple Type System [8] is used to provide arity and similar information that is mechanical, and also information that is human-readable, but not currently machine processable, such as stating that <OMS name="plus" cd ="arith1"/> takes its arguments from, and returns an answer in, the *same* Abelian semigroup, by having the following signature.

```
<OMA>
 <OMS name="mapsto" cd="sts"/>
 <OMA>
  <OMS name="nassoc" cd="sts"/>
  <OMV name="AbelianSemiGroup"/>
 </OMA>
 <OMV name="AbelianSemiGroup"/>
</OMA>
```

2.2   Prior Work on Units in OpenMath

The major previous work on the semantics of units on OpenMath is [9]. This proposes several Content Dictionaries of units: units_metric1, units_imperial1 and units_us1. These contain definitions of many common units covering a variety of dimensions (the dimensions themselves are defined in the Content Dictionary dimensions1) — metric (SI)[2] units are contained in units_metric1, for example. [9] suggests using the "usual" times operator (that stored in arith1) to represent a number in a particular unit — i.e. storing the value as the number

---

[1] Versions 1 and, to a lesser extent, 2. It is intended that OpenMath 3 and Content MathML 3 will have converged on this important point.

[2] The system in [9] actually differs in one respect from the SI system in [13,15]. [9] takes the fundamental unit of mass to be the gram, rather than the kilogram. This is necessary, as a slavish following of the general principles of [13] would lead to such absurdities as the millikilogram (see section 3.1 of this paper) rather than the gram. [13, section 3.2] explains the special rules for multiples of the kilogram, as follows.

Names and symbols for decimal multiples and submultiples of the unit of mass are formed by attaching prefix names to the unit name "gram", and prefix symbols to the unit symbol "g" (CIPM 1967, Recommendation 2; PV, 35, 29 and *Metrologia*, 1968, 4, 45).

multiplied by the unit, with the unit following the value to which it refers. The suggestions for unit "implementation" in OpenMath are stated as being based on those used by a complementary mathematics display language, MathML — although not blindly; where the authors believe MathML has some deficiencies, these have been corrected. This document also specifies a reasonable way of connecting a prefix to a unit (described in section 3.1), thus defining kilo as a separate concept, which can then be used to construct kilogram.

[9] uses STS in a novel (for OpenMath) manner. Rather than merely 'human-readable', as with <OMV name="AbelianSemiGroup"/> above, it uses formal OpenMath symbols as the type, thus the type of gram is

```
<OMS cd="dimensions1" name="mass"/>
```

More complicated dimensions can be expressed, e.g. Newton's type is

```
<OMS cd="dimensions1" name="force"/>
```

which has the formal property

```
<OMA>
  <OMS cd="relation1" name="eq"/>
  <OMS cd="dimensions1" name="force"/>
  <OMA>
    <OMS cd="arith1" name="times"/>
    <OMS cd="dimensions1" name="mass"/>
    <OMA>
      <OMS cd="arith1" name="divide"/>
      <OMS cd="dimensions1" name="length"/>
      <OMA>
        <OMS cd="arith1" name="power"/>
        <OMS cd="dimensions1" name="time"/>
        <OMI> 2 </OMI>
      </OMA>
    </OMA>
  </OMA>
</OMA>
```

Hence this system supports "dimensional analysis" (which should properly be called "dimensional algebra").

2.3 Unit Converters

There are a great many unit converters publicly available online. These have a range of units and features — see the analysis in [21, chapter 2]. However, in all cases, they are monolithic, in that new units cannot be added to them by the user. In some senses, this means that they go against modularity and incrementality, and are not reflective, in that they do not know that other units exist.

None of the converters surveyed seem to know about dimensions, and hence attitudes to the question "convert months into days", instead of being our (1), were generally (2), and surprisingly often (4).

1. They are both `time`, so the conversion is meaningful, but I don't have an exact conversion factor.
2. There are $30\frac{699}{1600} = 30.43687500$ days in a month, which is correct on average, but false for every month (for some reason, Google uses 30.4368499)!
3. There are 30 days in a month, which is "the nearest", but not the most common (e.g. `http://online.unitconverterpro.com/unit-conversion/ convert-alpha/time.html`), and which leads to absurdities such as "1 decade $= 121\frac{2}{3}$ months".
4. I don't know about months.

3   Abbreviations and Prefixes

Units have a variety of abbreviations and, particularly in the metric system, a range of prefixes. It is possible, as apparent in [14, section 5.3.5], to regard prefixed units as units in their own right, and introduce a unit `centimetre` with a formal property relating it to the `metre`, but this way lies, if not actual madness, vast repetition and the scope for error or inconsistency (who would remember to define the `yottapascal`?).

3.1   Prefixes

OpenMath therefore defines prefixes in the `units_siprefix1` CD, with FMPs to define the semantics, e.g. the following one for `peta`.

```
<OMA>
  <OMS name="eq" cd="relation1"/>
  <OMA>
    <OMS name="times" cd="arith1"/>
    <OMI> 1 </OMI>
    <OMA>
      <OMS name="prefix" cd="units_ops1"/>
      <OMS name="peta" cd="units_siprefix1"/>
      <OMV name="unit"/>
    </OMA>
  </OMA>
  <OMA>
    <OMS name="times" cd="arith1"/>
    <OMA>
      <OMS name="power" cd="arith1"/>
      <OMI> 10 </OMI>
      <OMI> 15 </OMI>
    </OMA>
```

```
        <OMV name="unit"/>
      </OMA>
    </OMA>
  </OMOBJ>
```

OpenMath uses a `prefix` operation (described as option 4 of [9, section 4]) to apply prefixes to OpenMath units. Its signature is given as follows.

```
<Signature name="prefix" >
<OMOBJ xmlns="http://www.openmath.org/OpenMath">
  <OMA>
    <OMS name="mapsto" cd="sts"/>
    <OMS cd="units_sts" name="prefix"/>
    <OMV name="dimension"/>
    <OMV name="dimension"/>
  </OMA>
</OMOBJ>
</Signature>
```

which can be seen as

$$\text{prefix} \times \text{unit} \to \text{unit}. \tag{1}$$

This has the slightly unfortunate property that it would allow, for example, 'millimicrometre', which is explicitly forbidden by [13, p. 122]. This could be solved by making the signature

$$\text{prefix} \times \text{unit} \to \text{prefixed unit}, \tag{2}$$

which should probably be done.

This construction also allows the use of prefixes with non-SI units, but this is in fact legitimate [13, p. 122].

3.2   Abbreviations

One issue not covered in [9] is that of abbreviations. Here we must confess to not having a completely worked-out and sensible solution yet. The following possibilities have been considered.

Alternative Definition in the same CD. This would mean that, for example, as well as `units_metric1` having the symbol `metre`, it would also have `m`. These would be linked via a FMP saying that the two were equal. Similarly, we would have prefixes `k` as well as `kilo`.

Pro. A small extension of [9].

Con. Allows "mixed" units such as `kilom` or `kmetre`, which are (implicitly) forbidden in [13].

Con. No built-in way of knowing which is the full name and which is the abbreviation.

Alternative Definition in different CDs. This would mean that units_
metric1 would have the symbol metre, and a new CD, say units_metricabbrev1,
would have the symbol m. Again, these would be linked via a FMP saying that the
two were equal. We would also have a new CD, say units_sipefixabbrev1, containing the abbreviations for the prefixes, and a *different* operation for combining
the two, say

```
<Signature name="prefixabbrev" >
<OMOBJ xmlns="http://www.openmath.org/OpenMath">
  <OMA>
    <OMS name="mapsto" cd="sts"/>
    <OMS cd="units_sts" name="prefixabbrev"/>
    <OMV name="dimensionabbrev"/>
    <OMV name="dimensionabbrev"/>
  </OMA>
</OMOBJ>
</Signature>
```

Pro. Prevents 'hybrid' units.

Pro. A converter such as [21] could output either full names or abbreviations
('symbols' in [13]) depending on which CDs were available on the output
side.

Con. Knowledge of which is the name and which is the abbreviation is still
implicit — merely moved from the name of the symbol to the name of the
CD. The linkage between the name of the CD and the fact that the symbol
should be regarded as <OMV name="dimensionabbrev"/> would be outside
the formal OpenMath system.

"This isn't an OpenMath problem". It could be argued that abbreviating
units and prefixes isn't an OpenMath problem at all, but a presentation one.
This is superficially tempting, but poses the question "Whose problem is it?"
Do we need a new layer of software to deal with it? One interesting sub-question
here is whether an ontology language such as OWL [7] would be better suited
to expressing such concepts.

3.3   Non-SI (but Metric) Units

The reader will have noticed that the CD is called units_metric1 rather than
units_si1. This is deliberate, as it includes the litre, which is explicitly *not*
an SI unit [13, Table 6, note (f)]. What of the other units in [13, Tables 6, 8]?

bar. This is 100kPa, and presumably is retained because of its convenience for
atmospheric pressure. Prefixes are valid with it [13, p. 127], though the only
common one is the millibar (which is also the mbar, since the bar, uniquely,
is its own abbreviation).

tonne. (alias 'metric ton') [13, Table 6] This is essentially an alias for the megagram, and as such does not take prefixes[3]. If the "different CDs" approach above were to be adopted, this could be in yet another CD, say units_metricmisc1, on which no prefixing[4] operated.

hectare. As 10^4m^2, this is in a very similar category to the tonne, and again does not take prefixes. The only question might be whether we ought to start with the **are** instead, but it is possible to argue that the **are** is obsolete, and conveys no advantage over the square decameter. If litre_pre1964 moves to a different CD, we could reasonably leave the **are** there as well.

ångström. Similarly.

nautical mile. (= 1852m) Similarly.

knot. ($=\frac{1852}{3600}$ m/s) Similarly. This is also an excellent argument for the representaton of definitional conversions (section 5.1) as exact fractions.

4   Not All Dimensions Are Monoids

[9] assumed, implicitly, that all physical dimensions could be regarded as (Abelian) monoids, in the sense that they could be added, and hence multiplied by integers. This is in fact not the case.

4.1   The Temperature Problem

One problem was not addressed in [9], but has been observed elsewhere [1, Celsius#_note-10], viz. that temperatures are not the same thing as temperature intervals. This confusion is widespread, as evidenced by the Google calculator's ability to produce computational absurdities such as

(1 degree Celsius) plus (1 degree Celsius) = 275.15 degrees Celsius

More subtly, the reader should compare the following Google outputs (generated from (-1C) in F and -(1C) in F respectively).

(-1) degree Celsius = 30.2 degrees Fahrenheit

with

-(1 degree Celsius) = -953.14 degrees Fahrenheit

Possibly the best explanation of the difference between relative and non-relative temperatures is in [22, Appendix B.9][5].

---

[3] The reader may ask "what about the megaton(ne)?" This is, of course, the mega[ton of TNT equivalent], and is not a unit of mass at all, but rather of energy, and is in fact 4.184 petajoules [22, Appendix B.8], where the figure 4.184 is definitional in the sense of section 5.1.

[4] Except that, in Belgium, the megaton (Nederlands) or megatonne (Français), and *certain* other multiples (kilo- to exa- only) are legal [2, Chap II, §4].

[5] http://physics.nist.gov/Pubs/SP811/appenB9.html#TEMPERATUREinterval

4.2   To Monoid or Not to Monoid

We can ask whether this is a peculiarity of temperature. The answer is in fact that it is not.

Most units form (Abelian) monoids, i.e. they can be added: 2 tonnes + 3 tonnes = 5 tonnes etc. *Non-relative*[6] temperatures are one obvious counter-example: $2°F + 3°F \neq 5°F$ or indeed any other temperature. The point is that *relative* temperatures, as in "*A* is ten degrees hotter than *B*", *are* additive, in the sense that if "*B* is twenty degrees hotter than *C*", then indeed "*A* is thirty degrees hotter than *C*", *are* additive, as are non-relative plus relative, but two absolute temperatures are *not* additive.

The same problem manifests itself with other scales such as decibels. Strictly speaking, these are purely relative, but in practice are also used in an absolute way, as in "the sound level exceeded 85dB". Again, the relative units form a monoid, but the absolute units do not.

This forces us to rethink the concept of "dimension". Though not using the word here (it is used in section 1.3), these are defined in [13, section 1.2] as follows.

> The base quantities used in the SI are length, mass, time, electric current, thermodynamic temperature, amount of substance, and luminous intensity.

This implies that all masses, for example, have the same dimension, and can be treated algebraically in the same way. But, as we have seen, not all temperatures are the same, and indeed have different algebraic properties. Two relative temperatures can be added, as in the example of *A*, *B* and *C* above. A relative temperature can be added to an absolute temperature, as in the following examples.

$$X \text{ was heated by } 10°C, \text{ from } 20°C \text{ to } 30°C. \tag{3}$$

$$X \text{ was heated by } 10K, \text{ from } 20°C \text{ to } 30°C. \tag{4}$$

$$X \text{ was heated by } 10°C, \text{ from } 293.15K \text{ to } 303.15K. \tag{5}$$

$$X \text{ was heated by } 10K, \text{ from } 293.15K \text{ to } 303.15K. \tag{6}$$

Equations (3) and (4) mean precisely the same thing (similarly for equations (5) and (6)), and this is obvious because, in the Content Dictionary defining relative temperatures, we state that the two are equal[7].

---

[6] Referring to 'absolute' temperatures would be likely to cause confusion, though that is what we mean in sense 10 of [19, absolute].

[7] This is the standard OpenMath way of doing so for units. It might make more sense simply to declare that the two symbols were precisely equal — see the discussion in section 7.2. It could be argued that, since the two symbols are equal, we do not actually need to have both — a minimalist view. This is similar to the discussion about `<OMS name="Landauin" cd="asymp1"/>` in [11], and our conclusion would be the same — convenience of rendering outweighs minimality.

```
<OMA>
  <OMS name="eq" cd="relation1"/>
  <OMA>
    <OMS name="times" cd="arith1"/>
    <OMI> 1 </OMI>
    <OMS name="relative_Kelvin" cd="units_metric1"/>
  </OMA>
  <OMA>
    <OMS name="times" cd="arith1"/>
    <OMI> 1 </OMI>
    <OMS name="relative_Celsius" cd="units_metric1"/>
  </OMA>
</OMA>
```

Equations (3) and (5) also mean precisely the same thing, but this time we need to rely on the definitions of non-relative temperatures, as in the following[8],

```
<OMA>
  <OMS name="eq" cd="relation1"/>
  <OMA>
    <OMS name="times" cd="arith1"/>
    <OMI> 1 </OMI>
    <OMS name="degree_Kelvin" cd="units_metric1"/>
  </OMA>
  <OMA>
    <OMS name="minus" cd="arith1"/>
    <OMA>
      <OMS name="times" cd="arith1"/>
      <OMI> 1 </OMI>
      <OMS name="degree_Celsius" cd="units_metric1"/>
    </OMA>
    <OMA>
      <OMS name="divide" cd="arith1"/>
      <OMI>   27315 </OMI>
      <OMI>     100 </OMI>
    </OMA>
  </OMA>
</OMA>
```

and need to do some actual arithmetic, as in [21].

The system of dimensions in [9] had, as the Simple Type System [8] signature of dimensions, the OpenMath <OMV name="PhysicalDimension"/>. As its format implies, this is a mere name with no formal semantic connotation. We therefore suggest that this be replaced by two objects: MonoidDimension

---

[8] This differs from the currently-published experimental CD units_metric1 in following the recommendation in section 5.1 that 273.15, as a defined number, should be represented as an element of **Q**.

for those cases where the dimension *does* represent an (additive) monoid, and `NonMonoidDimension`. While these *could* be OMVs as before, we now believe that it would make more sense for them to be OMSs, since there is a definite semantics being conveyed here *to software packages*, rather than to human beings. This also agrees with a growing feeling in the OpenMath community that STS could "do more".

4.3   The Confusion Is Widespread

It should be noted that the confusion between temperatures and relative temperatures manifests itself elsewhere in "web semantics". Consider an excerpt[9] from ontoworld's "approach to rewrite Content MathML so that it is expressable (sic) as RDF."

```
<owl:Class rdf:about="&phml;Temperature">
  <rdfs:subClassOf rdf:resource="&phml;PhysicalDimension"/>
</owl:Class>

<owl:Class rdf:about="&phml;TemperatureDifference">
  <rdfs:subClassOf rdf:resource="&phml;Temperature"/>
</owl:Class>
```

This could be argued to illustrate the difficulties of using a general-purpose language such as RDF beyond the semantics it is capable of handling, or, more simply perhaps, as an illustration of the fact that, since the *presentation* of temperatures and temperature intervals are the same, it is hard to distinguish the *semantics*, different though these may be.

5   Precision

5.1   Accuracy in the OpenMath

Conversion factors between units can be divided broadly into three categories.

Architected. There are those that, at least conceptually, arose when the unit(s) were defined. All the metric prefixes fall into this category, as do conversions such as "3 feet = 1 yard", or even "1 rod = $5\frac{1}{2}$ yards". These conversions, and their inverses, clearly ought to be stored as elements of $\mathbf{Q}$, i.e. as OpenMath integers (`OMI` objects) or fractions thereof.

Experimental. These are those that are truly determined by an experiment, such as the measurement of a standard of length in one system in terms of another such. An obvious example is those units that involve g, as in "1 slug $\approx$ 32.17405 pounds". These are probably best represented by means of floating-point numbers (`OMF` objects).

The reader might object that, since these items are only approximate, we should represent them by intervals, which are well-handled by OpenMath,

---

as in the CD `interval1`. This is a plausible point. We happen to disagree with it, for the reasons about to be given, but nevertheless it is fair to say that more usage of these factors is called for before a definitive decision can be made.

- Manipulation of intervals is not conservative unless it is done symbolically — [10]. Hence, if g were to be represented by an interval, say $[32, 33]$ (absurdly wide, but this makes the point better), one slug would be $[32, 33]$ pounds, which, on conversion back, would become $[\frac{32}{33} \approx 0.97, \frac{33}{32} \approx 1.03]$ slugs.
- Definitions in OpenMath are intended to be permanent, so an increase in precision would have to lead to a change in the formal definition.
- Experimentalists tend not to work in terms of intervals, but in terms of the standard accuracy [16]. It would be a fair argument, though, to say that there *ought to be* OpenMath interval types capable of representing these.

Definitional. These are those that started life as experimental, but have since been adopted as architected definitions. An obvious example is "1 yard = 0.9144 metre", which was adopted as a formal definition, replacing the previous experimental result[10] of "1 metre $\approx$ 39.370147 inches" [20] in 1959. Another example would be the value of "absolute zero", in the days of an independent celsius scale, which was about $-273.15°$C. Nowadays, this is fixed as precsiely this value, or, more accurately, the concept of $°$C is defined in terms of absolute zero and the number 273.15 [13, 2.1.1.5].

We now believe that all *definitional* numbers occurring in unit conversions, as well as those architected, should be expressed as elements of **Q**, i.e. as (fractions of) OMI. Hence the 0.9144 mentioned above should in fact be encoded as

```
<OMA>
  <OMS name="divide" cd="arith1"/>
  <OMI>  9144 </OMI>
  <OMI> 10000 </OMI>
</OMA>
```

This suggestion is well-characterised by the foot. Thus U.S. survey foot is defined[11] as $\frac{1200}{3937} \approx .3048006096$m, whereas the 'international' foot is defined as *precisely* 0.3048m. The difference is only just detectable in IEEE (single-precision) floating-point, and is best stored exactly.

5.2   Precision of Display

There is also the issue of how much precision to display in the result. In general terms, the result should not be more precise than the least precise value used in

---

[10] It is worth noting that [20] describes this as "1 yard = 0.91439841 metres", [1, Imperial_units] as 0.914398416 metres, and an accurate conversion of the headline figure in [20] is .9143984146. This illustrates the general point that a number and its reciprocal are unlikely both to be exact decimals.

[11] U.S. Metric Law of 1866.

the calculation. [21] currently supplies the entire result from the calculation, with a user-controllable "significant figures" level as part of the system's front-end. This was chosen on the basis that several alternatives considered appeared nonsensical or unreasonably difficult to implement or make firm decisions about. For example, internally, fractions (which in OpenMath are comprised of two infinite precision integers) are stored as finite-precision floating point numbers. It is impossible to tell, when presented with such a floating point number, whether it was made as such (again, OpenMath Floats, in decimal, can be of arbitrary precision) or whether it came from a fraction; in both these cases rounding would not be required, or if it was the result of a calculation, in which case rounding would be necessary. A nonsensical answer would clearly result if values were rounded during the calculation, and due to the aforementioned unknowable fact of where the value came from, it would also be impossible to maintain an internal counter of to how many significant figures the end result would be reasonable. With the chosen approach, a currently unanswered question regards the number of significant figures to display in a result such as `10 metre is 1 rod 5 yard 1 foot 3 inch 700.787401574787 mil`. Should the number of significant figures only cover the last part of the result?

6   The Two Meanings of "Obsolete"

According to the OpenMath standard [3], a content dictionary can be declared to be `obsolete`. This facility is needed so that, when an area of OpenMath gets rewritten in a (hopefully) better way, the semantics of existing OpenMath objects are preserved. However, there has been no need to deploy it yet. It is a feature of OpenMath[12] that this takes place at the content dictionary level, rather than the symbol level.

However, when we say that

```
<OMS name="litre_pre1964" cd="units_metric1"/>
```

is "obsolete", we do *not* mean that it is obsolete as an OpenMath symbol, rather that it is a current OpenMath symbol denoting an obsolete unit of measurement, and therefore that it *should* be in an `official` CD. Does this matter? There are two views.

No. This is the view of [9]. It is a unit, which may still be encountered as old texts/experiments etc. are analysed, so should be present.

Yes. In [21] we produced a unit converter that attempted to produce the "best" fit to a given input. Hence, as 175.98 `pints` converts to 100.002614775 `litres`, but also 99.99991478 `litre_pre1964`s, the latter conversion would, much to the user's surprise (and indeed ours on first encountering this issue), be preferred. In fact, we actually get

```
OpenMathConverter.exe --source_quantity 175.98 --source_unit
```

---

[12] At least at version 2. This may change in version 3.

```
pint --destination_unit metric
0.999998147801862 hectolitre_pre1964
```

Similarly, as 10 metres is 10.93613298 yards, but 1.988387814 rods, the latter will again be preferred[13].

From the point of view of 'user-friendliness', we are inclined to sympathise with [21], and state that obsolete units belong in separate CDs, in particular that litre_pre1964 should be moved from units_metric1 to, say, units_metricobs before units_metric1 becomes official.

7 Conclusion

We conclude that it is possible to use the OpenMath unit system (or ontology, as one might call it) of [9] to produce a serious and, unlike others, *extensible* unit converter, as in [21].

7.1 Recommendations for OpenMath Unit/Dimension CDs

The most important recommendation is a recognition that some (in our sense of the word) dimensions are (additive) monoids, and some are not, as outlined in section 4.

1. Move litre_pre1964 into a different CD, which is an official CD of "obsolete" units. Similar steps should be taken for "obsolete" imperial units.
2. Fix dimensions1 so as to have a definition for power.
3. Delete metre_squared from the units_metric1. It is anomalous (why isn't there metre_cubed, and why doesn't units_imperial1 have foot_squared?) and tempts a piece of software (such as earlier versions of [21]) into creating units such as

```
<OMA>
  <OMS name="prefix" cd="units_ops1"/>
  <OMS name="deci" cd="units_siprefix1"/>
  <OMs name="metre_squared" cd="units_metric1"/>
</OMA>
```

which is a deci(metre^2), as opposed to a (decimetre)^2, and is illegal [13, p. 121].
4. units_imperial1 is missing units such as inch, which need to be added.
5. Add a CD for U.S. units, where different (e.g. for volume). Move U.S. Survey units[14], currently in units_imperial1 into this.
6. Add a CD for E.U. units, where different. The only case known to the authors is the therm, which comes in both U.S. and E.U. variants. [22, footnote 25] states the following.

---

[13] Which will in fact come out as 10 metre is 1 rod 5 yard 1 foot 3 inch 700.787401574787 mil.

[14] See the discussion at the end of section 5.1 for the (small) difference.

Although the therm (EC), which is based on the International Table Btu, is frequently used by engineers in the United States, the therm (U.S.) is the legal unit used by the U.S. natural gas industry. The difference is about 0.02%.

7. Update all the semantics in the world of OpenMath units so as to adhere to the principles of section 5.1, in particular definitional numbers should be expressed as elements of **Q**, i.e. as (fractions of) OMI.

8. Sort out electrical energy definitions and other suggestions in [9].

9. Modify the signature of `prefix`, as described in section 3.1, from (1) to (2).

10. Update the definition of `pascal` to include an FMP: currently missing.

7.2 Further Considerations

We saw, in section 4.3, that the sort of semantics of RDF [6] are inadequate to convey the relationship between, for example, relative temperature and non-relative temperature. However, the OpenMath required to state that

```
<OMS name="relative_Kelvin" cd="units_metric1"/>
```

and

```
<OMS name="relative_Celsius" cd="units_metric1"/>
```

mean *precisely* the same thing is clumsy, and requires OpenMath-capable reasoning whereas all that is needed *in this case* is RDF-like, or OWL-like, reasoning.

We can also ask whether OWL would not be better at solving the abbreviations problem than OpenMath (see section 3.2).

Some units (`calendar_year` is the notable example) have multiple FMPs, whereas most of the other secondary units have only one, which is essentially a *defining* mathematical property in the sense [23, I, p. 11] that the definiens can be completely replaced by the definiendum. Making the distinction clear, as has been proposed elsewhere in the OpenMath community, would be a step forward.

7.3 Unsolved Problems

We see two currently unsolved problems.

1. The abbreviations issue — section 3.2, and the fact that legal prefixes can differ between countries, as in note [4]. This last might be a problem best solved outside OpenMath.

2. The `dimensions1` CD has symbols `length` and `displacement`, with the difference being explained (under `displacement`) as

   This symbol represents the spatial difference between two points. The direction of the displacement is taken into account as well as the distance between the points.

   It would be possible to read this as a non-monoid version of `length`, but more thought is necessary.

Acknowledgements. The authors are grateful to Dr Naylor for his explanations of [9], to Dr Donner for useful comments, and to Dr De Vos for her help with Belgian metrology.

References

1. Anonymous. Wikipedia, English (2008), http://www.wikipedia.org
2. The Kingdom of Belgium. Réglementation Métrologique, http://mineco.fgov. be/organization_market/metrology/showole_FR.asp?cParam=1150 Metrologische Reglementering, http://economie.fgov.be/organization_market/metrology/ showole_nl.asp?cParam=1152
3. Buswell, S., Caprotti, O., Carlisle, D.P., Dewar, M.C., Gaëtano, M., Kohlhase, M.: OpenMath Standard 2.0 (2004), http://www.openmath.org/standard/ om20-2004-06-30/omstd20.pdf
4. World-Wide Web Consortium. Mathematical Markup Language (MathML[tm]) 1.01 Specification (1999), http://www.w3.org/TR/MathML/
5. World-Wide Web Consortium. Mathematical Markup Language (MathML) Version 2.0 (Second Edition) (2003), http://www.w3.org/TR/MathML2/
6. World-Wide Web Consortium. RDF Semantics (2004), http://www.w3.org/TR/ 2004/REC-rdf-mt-20040210/
7. World-Wide Web Consortium. Web Ontology Language OWL (2004), http://www.w3.org/2004/OWL/
8. Davenport, J.H.: A Small OpenMath Type System. ACM SIGSAM Bulletin 34(2), 16–21 (2000), http://hdl.handle.net/10247/468
9. Davenport, J.H., Naylor, W.A.: Units and Dimensions in OpenMath. (2003), http://www.openmath.org/documents/Units.pdf
10. Davenport, J.H., Fischer, H.-C.: Manipulation of Expressions. In: Improving Floating-Point Programming, pp. 149–167 (1990)
11. Davenport, J.H., Libbrecht, P.: The Freedom to Extend OpenMath and its Utility. Mathematics in Computer Science (to appear, 2008)
12. Donner, M.: Private Communication (May 2008)
13. Organisation Intergouvernementale de la Convention du Mètre. The International System of Units (SI), 8th edn., http://www.bipm.org/utils/common/pdf/ si_brochure_8_en.pdf
14. World-Wide Web Consortium. Harder, D.W., Devitt, S.(eds.). Units in MathML (2003), http://www.w3.org/Math/Documents/Notes/units.xml
15. IEEE/ASTM. SI 10-1997 Standard for the Use of the International System of Units (SI): The Modern Metric System. IEEE Inc (1997)
16. International Standards Organisation. Guide to the Expression of Uncertainty in Measurement (1995)
17. Mars Climate Orbiter Mishap Investigation Board. Mars climate orbiter: Phase I report (November 1999), ftp://ftp.hq.nasa.gov/pub/pao/reports/1999/ MCO_report.pdf
18. Nelson, W.H.: The Gimli Glider. Soaring Magazine (1997)
19. Oxford University Press. Oxford English Dictionary (2008), http://dictionary. oed.com/entrance.dtl
20. Sears, J.E., Johnson, W.H., Jolly, H.L.P.: A New Determination of the Ratio of the Imperial Standard Yard to the International Prototype Metre. Phil. Trans. Roy. Soc. A 227, 281–315 (1928)

21. Stratford, J.: An OpenMath-based Units Converter. Dissertation for B.Sc. in Computer Science, University of Bath. Technical Report CSBU–2008–02 (2008), http://www.cs.bath.ac.uk/department/technical-report-series/technical-report-series/index.php
22. Taylor, B.N.: Guide for the Use of the International System of Units (SI) (May 2007), http://physics.nist.gov/Pubs/SP811
23. Whitehead, A.N., Russell, B.: Principia Mathematica. Cambridge University Press, Cambridge (1910)
24. Williams, M.: The 156-tonne Gimli Glider. Flight Safety Australia (2003)

Authoring Verified Documents by Interactive Proof Construction and Verification in Text-Editors

Dominik Dietrich, Ewaryst Schulz, and Marc Wagner

FR 6.2 Informatik, Saarland University, Saarbrücken, Germany
{dietrich,schulz,wagner}@ags.uni-sb.de

Abstract. Aiming at a document-centric approach to formalizing and verifying mathematics and software we integrated the proof assistance system ΩMEGA with the standard scientific text-editor TEXMACS. The author writes her mathematical document entirely inside the text-editor in a controlled language with formulas in LATEX style. The notation specified in such a document is used for both parsing and rendering formulas in the document. To make this approach effectively usable as a real-time application we present an efficient hybrid parsing technique that is able to deal with the scalability problem resulting from modifying or extending notation dynamically. Furthermore, we present incremental methods to quickly verify constructed or modified proof steps by ΩMEGA. If the system detects incomplete or underspecified proof steps, it tries to automatically repair them. For collaborative authoring we propose to manage partially or fully verified documents together with its justifications and notational information centrally in a mathematics repository using an extension of OMDOC.

1 Introduction

Unlike widely used computer-algebra systems, mathematical assistance systems have not yet achieved considerable recognition and relevance in mathematical practice. One significant shortcoming of the current systems is that they are not fully integrated into or accessible from standard tools that are already routinely employed in practice, like, for instance, standard mathematical text-editors. Integrating formal modeling and reasoning with tools that are routinely employed in specific areas is the key step in promoting the use of formal logic based techniques.

Therefore, in order to foster the use of proof assistance systems, we integrated the theorem prover ΩMEGA [7] into the scientific text-editor TEXMACS [15]. The goal is to assist the author inside the editor while preparing a TEXMACS document in a publishable format. The vision underlying this research is to enable a document-centric approach to formalizing and verifying mathematics and software. We tackle this vision by investigating two orthogonal approaches in parallel: On the one hand we start with mathematical documents written without any restrictions and try to extract the semantic content with natural language analysis techniques and accordingly generate or modify parts of the document using natural language generation. On the other hand we start with the semantic content and lift it to an abstract human-oriented representation without losing the benefits of machine processability. This paper describes our recent results for the second approach, resulting in a system in which the author can write a document

S. Autexier et al. (Eds.): AISC/Calculemus/MKM 2008, LNAI 5144, pp. 398–414, 2008.

in T$_E$X$_{MACS}$, which gets automatically proof-checked by ΩMEGA. When the document is changed, the dependent parts are automatically rechecked.

In our scenario the workflow for the author, who is preparing a mathematical document to be verified, consists of arbitrary combinations of the following operations: (1) writing theory with notation or citing theory and eventually redefining notation, (2) developing proofs by constructing or modifying proof steps that are continuously checked and possibly repaired by the proof assistance system if incomplete or underspecified, and (3) saving the current state of the document including the verification information in order to continue at a later date. This raises the following requirements for effectively supporting the author in real-time: (1) a fast parsing and rendering mechanism with efficient adaptation to changes in the notation rules, (2) quick incremental proof checking and repair techniques, and (3) an output format containing the formalized content of the document together with its justifications and notational information.

The paper is organized as follows: Section 2 presents in more detail our general architecture consisting of a mediator, a proof assistance system and a semantic repository. Section 3 introduces the hybrid parsing technique that efficiently deals with the controlled authoring language and notational extensions or modifications. The formal proof representation of the ΩMEGA system is defined in Section 4 including the notion of *proof view*. The techniques for management of change needed (1) to incrementally verify proof steps constructed or modified in the text-editor, and (2) to lift corrections or complete proofs from the internal format of the proof assistance system to the document are described in Section 5. In Section 6 we discuss how OMDOC [10] can be extended such that it can also store proof steps at different levels of granularity as well as parsing and rendering knowledge. We discuss the current situation for authoring verified mathematical documents as related work in Section 7 and summarize the paper in Section 8.

2 Architecture

Although this paper focuses on the interplay between a mediator and a proof assistance system, we propose to fill the authoring gap for semantic mathematics repositories with our complementary architecture. The envisioned architecture is a cooperation between a mediator, a proof assistance system and a semantic repository. The mediator parses and renders the informal content authored in a text-editor and propagates changes to the proof assistance system that provides services for verification and automatic proof construction. The semantic repository takes care of the management of formalized mathematics. Figure 1 illustrates the flow of mathematical knowledge. The big circles in the figure are abstract components of the architecture that can be instantiated by the concrete components attached to them, e.g. PLATΩ as mediator. The text between the arrows indicates the kind of knowledge that is exchanged. In detail, the roles and requirements of the components are:

Mediator. Following our document-centric philosophy, the document in the text-editor is both the human-oriented input and output representation for the proof assistance system, thus the central source of knowledge. The role of the mediator PLATΩ [16] is to preserve consistency between the text-editor and the proof assistance system

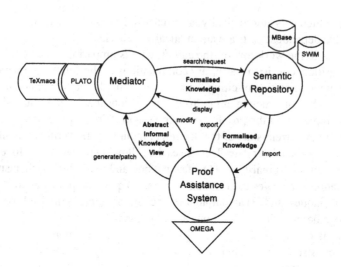

Fig. 1. Architecture for Authoring Mathematics Repositories

by incrementally propagating changes. Additionally, services and feedback of the proof assistance system are provided inside the text-editor through context-sensitive menus. The mediator allows to define, modify and overload the notation used in the document dynamically within the document.

Proof Assistance System. The role of the proof assistance system is to maintain the formal proof object in a way that it is verifiable and such that at the same time a human readable presentation can be extracted. It must be able to verify and integrate updates sent from the mediator, and provide means to automate parts of the proof. Finally, the system should be able to import from and export to standards such as OMDoc.

In our architecture we use the proof assistance system ΩMEGA, which is a representative of systems in the paradigm of *proof planning* and combines interactive and automated proof construction for domains with rich and well-structured mathematical knowledge. Proof planning is interesting for our architecture because it naturally supports to express proof plans and their expansion, i.e. verification.

Semantic Repository. The role of the semantic repository (e.g. in form of a database or a wiki) is to store and maintain the mathematical knowledge using structural semantic markup or scripting languages, including possibilities to search and retrieve knowledge and access control. The MBASE system [8] is for example a web-based, distributed mathematical knowledge base that allows for semantic-based retrieval. Semantic Wiki technologies like SWIM [11] are a current subject of research for collaboratively building, editing and browsing mathematical knowledge. Both types of semantic repositories are well-suited for our architecture because they store mathematical theories and statements in the OMDoc format and support dependency-aware semantic content retrieval.

Altogether, the proposed architecture allows for the incremental interactive development of verified mathematical documents at a high level of abstraction. By using the scientific WYSIWYG text-editor $T_{E}X_{MACS}$, the author additionally benefits from professional type-setting and powerful macro definition facilities like in LaTeX.

3   Hybrid Parsing

Let us first introduce an example of the kind of documents we want to support. Figure 2 shows a theory about *Simple Sets* written in the text-editor T$_E$X$_{MACS}$. This theory defines two base types and several set operators together with their axioms and notations. In general, theories are built on top of other theories and may contain definitions, notations, axioms, lemmas, theorems and proofs. Note that in the example set equality is written as an axiom because equality is already defined in the base theory.

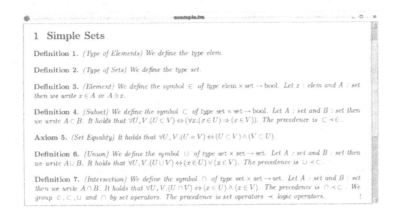

Fig. 2. Document in the text-editor T$_E$X$_{MACS}$

In our previous approach [16], the author had full freedom in writing her document but had to manually provide semantic annotations. We now use a controlled authoring language to skip the burden of providing annotations, thus increasing the overall usability by dealing with pure T$_E$X$_{MACS}$ documents. The grammar of the concrete syntax is given in Table 1. NAME and LABEL are unique string identifiers. URI is a resource location. SYM, VAR, TYPE, PAT and TERM represent symbol, variable, type, pattern and term respectively. Please note that T$_E$X$_{MACS}$ renders for example the macro "<definition|text>" into "**Definition 1.** text".

Dynamic Notation. This initial grammar can be extended on the fly by introducing new types and symbols as well as defining, extending or overloading their notations within a document. In [3] we presented a basic mechanism that allows the user to define notations by declaring local variables (e.g. *A,B*) and specifying notation patterns (e.g. $A \subset B, B \supset A$). The mechanism synthesizes automatically parsing rules from all patterns and the default rendering rule from the first pattern. The author can group operators, specify their associativity and define precedences as a partial ordering. Furthermore, if the notation is modified all affected formulas in the document are adapted efficiently, the right order of notation and formulas is checked, ambiguities are prevented using a family of theory-specific parsers and resolved by exploiting type information. The hierarchical structure of theories allows the reuse of concepts together with their notation even from other documents. Note that one can import theories together with their proofs and notations from other files by a reference in the document. Dynamic Notation

Table 1. Grammar for the Concrete Syntax of the Authoring Language

```
DOC    ::= THY*                                    AXM    ::= '<axiom|' '(' NAME ')' ALTC '>'
THY    ::= '<section|' NAME '>' CTX? THYC           LEM    ::= '<lemma|' '(' NAME ')' ALTC '>'
CTX    ::= 'We' 'use' CREFS '.'                     TEO    ::= '<theorem|' '(' NAME ')' ALTC '>'
CREFS ::= CREF (',' CREF)* ('and' CREF)?            ALTC   ::= 'It' 'holds' 'that' FORM '.'
CREF   ::= NAME|URI                                 PRF    ::= '<proof|' STEPS? '>'
THYC   ::= (DEF|AXM|LEM|TEO|PRF)*                   STEPS  ::= (OSTEP STEPS?)|CSTEP
DEF    ::= '<definition|' '(' NAME ')' DEFC '>'     OSTEP  ::= SET|ASS|FACT|GOAL|CGOAL
DEFC   ::= (DEFT|DEFS) NOTC? ALTC? SPEC*            CSTEP  ::= GOALS|CASES|CGOALS|TRIV
DEFT   ::= 'We' 'define' 'the' 'type' NAME '.'      TRIV   ::= 'Trivial' BY? FROM? '.'
DEFS   ::= 'We' 'define' 'the' 'symbol' NAME        SET    ::= 'We' 'define' FORMS '.'
           'of' 'type' TYPE '.'                     ASS    ::= 'We' 'assume' FORMS FROM? '.'
NOTC   ::= 'Let' TVARS 'then' 'we' 'write' PATS '.' FACT   ::= 'It' 'follows' 'that' FORMS BY? FROM? '.'
TVARS ::= VAR ':' TYPE (',' VAR ':' TYPE)*          GOAL   ::= 'We' 'have' 'to' 'prove' FORM BY? FROM? '.'
           ('and' VAR ':' TYPE)?                    GOALS  ::= 'We' 'have' 'to' 'show' FORMS BY? FROM? '.'
PATS   ::= PAT (',' PAT)* ('or' PAT)?                          SPRF*
PAT    ::= (VAR|STRING)+                            CASES  ::= 'We' 'have' 'the' 'cases' FORMS BY? FROM? '.'
SPEC   ::= GROUP|PREC|ASSOC                                    SPRF*
GROUP ::= 'We' 'group' SYM (',' SYM)* ('and' SYM)?  CGOAL  ::= 'We' 'have' 'to' 'prove' CFORM BY? FROM? '.'
           'by' NAME '.'                            CGOALS ::= 'We' 'have' 'to' 'show' CFORMS BY? FROM? '.'
PREC   ::= 'The' 'precedence' 'is' (SYM|NAME)                  SPRF*
           ('≺' (SYM|NAME))+ '.'                    SPRF   ::= 'We' 'prove' LABEL '.' STEPS?
ASSOC ::= 'The' 'operator' SYM 'is'                 BY     ::= 'by' NAME
           'right-associative' '.'                  FROM   ::= 'from' LABEL (',' LABEL)* ('and' LABEL)?
FORMS ::= FORM (',' FORM)* ('and' FORM)?            CFORMS ::= CFORM (',' CFORM)* ('and' CFORM)?
FORM   ::= ('(' LABEL ')')? TERM                    CFORM  ::= FORM 'assuming' FORMS
```

is aware of the positions of defining and using occurrences for notation but it does not take into account notational knowledge obtained by proven theorems yet; this is future work.

Speed Issues. Although we minimized the need for compiling parsers, the processing of the standard example in [3] took $\approx 1min$. The main reasons for inefficiency were (1) the parsers were compiled in interpreted mode in the text-editor, and (2) the scalability problem of LALR parser generators. Problem (1) has been solved by moving the parser generation to the mediator, but even in compiled mode the processing took $\approx 6sec$ which is still not sufficiently fast for real-time usage. The remaining issue (2) is severe because when notations are changed or extended all parsers for dependent theories in the hierarchy have to be recompiled. Therefore we integrated a directly-executable variant of an Earley parser [6] which is substantially faster than standard Earley parsers to the point where it is comparable with standard LALR(1) parsers. Although the time for parsing a single formula increases slightly, the overall processing of the example takes $\approx 0.1sec$ which is perfectly suitable for a real-time application.

Algorithm. First of all, the document is preprocessed and split into segments almost corresponding to sentences. Then the following steps are incrementally performed for each segment: (1) the static parts of the authoring language, i.e. the controlled phrase structure, are parsed using a precompiled LALR(1) parser; (2) the dynamic parts of the authoring language, i.e. the formulas and notations, are parsed using a theory-specific Earley parser; (3) the segment is propagated to the proof assistance system. Note that the dynamic parts are always strictly separated from the static parts in the document because they are written inside a math mode macro.

Normalization and Abstraction. The concrete syntax of the authoring language allows for variant kinds of syntax sugaring that has to be normalized for machine

Table 2. Grammar for the Abstract Syntax of the Proof Language

PROOF	::= STEPS	TRIVIAL	::= **trivial** BY FROM
STEPS	::= (OSTEP;STEPS)|CSTEP	SET	::= **set** FORMULA
OSTEP	::= SET|ASSUME|FACT	ASSUME	::= **assume** FORMULA FROM
CSTEP	::= GOALS|CASES|COMPLEX|TRIVIAL|ε	FACT	::= **fact** FORMULA BY FROM
FORMULA	::= (LABEL :)$^?$ TERM	GOALS	::= **subgoals** (FORMULA { PROOF })$^+$ BY FROM
BY	::= **by** NAME$^?$	CASES	::= **cases** (FORMULA { PROOF })$^+$ BY FROM
FROM	::= **from** (LABEL (, LABEL)*)$^?$	COMPLEX	::= **complex** COMP$^+$ BY FROM
		COMP	::= FORMULA **under** FORMULA { PROOF }

processing, e.g. formula aggregation ($x \in A, x \in B$ and $x \in C$) or the ordering of subproofs. Table 2 defines a normalized abstract syntax for the proof part of the authoring language. Aggregated formulas are composed to one formula by conjunction or disjunction depending on whether they are hypotheses or goals respectively. If an abstract proof step is generated by the system, the mediator tries to decompose the formula for aggregation accordingly. A proof is implicitly related to the last stated theorem previous to this proof in the document. Subproofs are grouped together with their subgoal or case they belong to. A single goal reduction is normalized to a subgoals step.

Management of Change. Using management of change we propagate incrementally arbitrary changes between the concrete and abstract representation. By additionally considering the semantics of the language we can optimize the differencing mechanism [12]. For example the reordering of subgoals or their subproofs in the text-editor is not propagated at all because it has no impact on the formal verification. The granularity of differencing is furthermore limited to the reasonable level of proof steps and formulas, s.t. deep changes in a formula are handled as a complete modification of the formula. The propagation of changes is essential for this real-time application because complete re-transformation and re-verification slows down the response time too much. Apart from that the differencing information allows for the local re-processing of affected segments instead of a global top-down re-processing using a replay mechanism. In order to recheck only dependent parts ΩMEGA uses Development Graphs [4] for the management of change for theories.

Let us now continue our example with a new theory in Figure 3 that refers to the previous one and that states a theorem the author already started to prove.

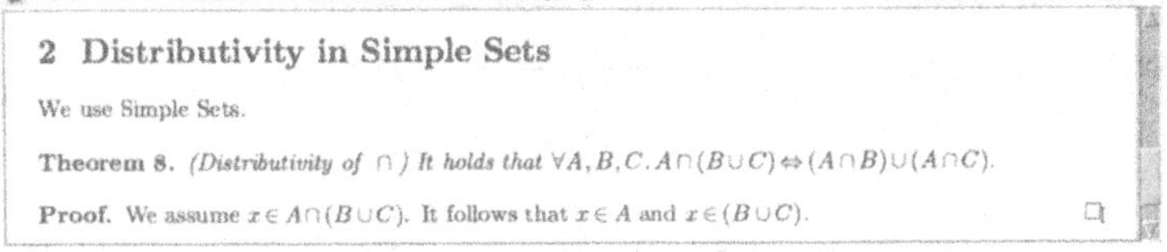

Fig. 3. Theorem with partial proof

This partial proof in concrete syntax is then abstracted as shown in Table 3, where we additionally emphasized underspecified parts with a dot.

Table 3. Proof in abstract syntax

> **Proof.**
>   **assume** $x \in A \cap (B \cup C)$ **from** . ;
>   **fact** $x \in A \wedge x \in (B \cup C)$ **by** . **from** . ; ε

4   Formal Proof Representation

In this section we describe how proofs are internally represented in ΩMEGA. This will allow us to describe how proof scripts are processed by the prover. In the ΩMEGA system proofs are constructed by the TASKLAYER that uses an instance of the generic proof data structure (PDS) [1] to represent proofs. One main feature of the PDS is the ability to maintain subproofs at different levels of granularity simultaneously, including a so-called *PDS-view* representing a complete proof at a specific granularity.

Task. At the TASKLAYER, the main entity is a task, a multi-conclusion sequent $F_1, \ldots, F_j \vdash G_1, \ldots, G_k$. Each formula can be named by assigning a label l to the formula. We denote the set of all term positions by **Pos**, the set of all admissible positions of a task T by **Pos**(T), and the position of the formula with label l by pos(l). Moreover, we write T_π to denote the subformula at position π and write $T_{\pi \leftarrow s}$ for the type compliant replacement of the subterm T_π by s. We use the notation $\Gamma \star \varphi$ to denote the set $\Gamma \cup \{\varphi\}$.

Agenda. The proof attempt is represented by an agenda. It maintains a set of tasks, that are the subproblems to be solved, and a global substitution which instantiates meta-variables. Formally an agenda is a triple $\mathcal{A} = \langle T_1, \ldots, T_n; \sigma; T_j \rangle$ where $T_1, \ldots, T_n$ are tasks, σ is a substitution, and T_j is the task the user is currently working on. We will use the notation $\langle T_1, \ldots, T_{j-1}, \underline{T_j}, T_{j+1} \ldots T_n; \sigma \rangle$ to denote that the task T_j is the current task. Note that the application of a substitution is a global operation. To reflect the evolutional structure of a proof, a substitution is applied to the open tasks of the agenda. Whenever a task is reduced to a list of subtasks, the substitution before the reduction step is stored within the PDS in the node for that task.

The Figure on the right shows the reconstruction of the first proof step of the partial proof of **Theorem 8**. Tasks are shown as oval boxes connected by justifications, where the squared boxes indicate which inference has been applied. The agenda to the shown PDS consists of the two leaf tasks. The global substitution σ is the identity *id*.

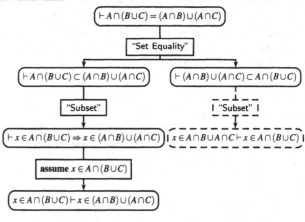

Fig. 4. Proof in ΩMEGA

5   Incremental Proof Step Verification and Correction

In this section we describe how information encoded as a proof script can be exchanged between the mediator and the prover. There are two possible information flows: First, a proof script sent by the mediator must be converted into the internal proof structure of the prover and thereby be checked. Second, given a proof generated by the prover, a corresponding proof script must be extracted, which can then be propagated to the mediator. Note that the last step is necessary as there is no guarantee that all parts of the proof have been constructed by a proof script. Before describing both directions we give an overview of the proof operators of the prover, which are *inferences* and *strategies*.

Inferences. Intuitively, an *inference* is a proof step with multiple premises and one conclusion augmented by (1) a possibly empty set of hypotheses for each premise, (2) a set of *application conditions* that must be fulfilled upon inference application, (3) a set of *completion functions* that compute the values of premises and conclusions from values of other premises

$$[H_1 : \varphi]$$
$$\vdots$$
$$\frac{P_1 : \psi}{C : \varphi \Rightarrow \psi} \; ImplIntro$$

and conclusions (see [2] for a formal definition). Each premise and conclusion consists of a unique name and a formula scheme. Consider for example the inference *ImplIntro* shown above. It consists of one conclusion with name C and formula scheme $\varphi \Rightarrow \psi$. Moreover, it has one premise with name P_1, formula scheme ψ, and hypothesis H_1, which has the formula scheme φ associated with it. In the sequel we write C to denote the conclusion of the inference and P_i to denote the ith premise of the inference.

Given a task, we can instantiate an inference with respect to the task by trying to find formulas in the task unifying with the formula scheme of a premise or conclusion. Technically, such an instantiation is represented by an *inference substitution* σ which binds premises and the conclusion to formulas or positions. We consider two parts of the substitution: σ_x contains instantiations of meta-variables of the task, and σ_c maps a premise or conclusion A to positions of the task (see [2] for details), which is $\perp$ in case that A is not matched. The instantiated formula scheme is denoted by $fs(A)$.

Strategies. As basis for automation ΩMEGA provides so-called strategies, which tackle a problem by some mathematical standard problem solving workflow that

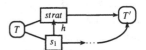

happens to be typical for this problem. To achieve a goal, a strategy performs a heuristically guided search using a dynamic set of inferences (as well as other strategies), and control rules. A strategy application either fails or constructs a subproof using the specified inferences and substrategies. In the latter case the constructed subproof is abstracted by inserting a justification labelled with the strategy application and connecting the task the strategy has been applied to with the nodes resulting from the strategy application. This has the advantage that the user can switch between the abstract step or the more detailed step. Note that from a technical point of view a strategy is similar to a tactic. However, by using explicit control knowledge its specification is declarative, whereas tactics are usually procedurally specified.

Consider a strategy *strat* which is applied to a task T, and constructs a subproof starting with the application of s_1 and finally leading to the task T'. In this case, a new justification labelled with *strat* is inserted connecting T and T'. To indicate that the strategy application is more abstract than the subproof a hierarchical edge h is inserted, defining an ordering on the outgoing edges of T. The resulting PDS is shown above.

5.1 Proof Checking and Repair

The verification of a single proof step can become time consuming if some information is underspecified. In the worst case a complete proof search has to be performed. To obtain adequate response times a given step is worked off in two phases. First, we perform a quickcheck, where we assume that the given step can be justified by the prover by a single inference application. This can be tested by a simple matching algorithm. If the test proceeds the step is sound, as the inference application is proved to be correct.

If it is not possible to justify the step with a single inference application, a more complex repair mechanism is started. This mechanism tries to find the missing information needed to justify the step by performing a heuristically guided resource bounded search. If we are not able to find a derivation within the given bound, a failure is reported and sent to the mediator, which can then initiate a specific reaction.

In the sequel we describe for each construct of our proof language its quickcheck and its repair mechanism, which are both modeled as a proof strategy in ΩMEGA. The quickcheck rules are summarized in Table 4. We use the notation $\langle \{\underline{\Gamma \vdash \Delta}, T_2, \ldots, T_n\}, \sigma \rangle : s ; S \to \langle \{\underline{\Gamma' \vdash \Delta'}, T_2, \ldots, T_n\}, \sigma \rangle : S'$ to indicate that under the agenda $\langle \{\underline{\Gamma \vdash \Delta}, T_2, \ldots, T_n\}, \sigma \rangle$ the proof step s in the sequence $s ; S$ can be checked, and that the checking results in a new agenda $\langle \{\underline{\Gamma' \vdash \Delta'}, T_2, \ldots, T_n\}, \sigma \rangle$ where the steps S' have to be checked.

Fact. The command **fact** derives a new formula φ with label l from the current proof context. The quickcheck tries to justify the new fact by the application of the inference or strategy *name* to term positions in the formulas with labels $l_1, \ldots, l_n$ in the current task. Although the specification of these formulas speeds up the matching process, both informations **by** and **from** can be underspecified in general. In this case, all inferences are matched against all admissible term positions, and the first one which delivers the desired formula φ is applied.

If the above check fails, the repair strategy for **fact** is started. It generates a new lemma, i.e. a new proof tree, containing the assumptions of the current task and the newly stated fact φ as goal. It then tries to automatically close the lemma by standard forward and backward reasoning. If the new lemma can be proved, the lemma is automatically transformed to an inference using the mechanism described in [2], which then justifies the step by a single inference application in the original proof.

Stating a new lemma has several advantages: Even if the lemma cannot automatically be checked, we can continue to check subsequent proof steps. The new lemma can then be proved with user interaction at a later time. Moreover, as the lemma is transformed into a single inference, it is globally available and can be used in similar situations by the quickcheck without performing any search.

Assume. The command **assume** introduces a new assumption φ on the left hand side of the current task. The quickcheck for **assume** checks whether one of the following situations occurs, each of which can be justified by a particular inference application:

- Δ contains $\varphi \Rightarrow \psi$. The implication is decomposed and $l : \varphi$ is added to the left side of the task.
- Δ contains $\neg\varphi$. Then $l : \varphi$ is added to the left side.
- Δ contains $\psi \Rightarrow \neg\varphi$. Then $l : \varphi$ is added to the left side and $\neg\psi$ to the right side of the task.

If the quickcheck fails we try to derive one of the above situations by applying inferences to the goal of the current task. The hypotheses of the task remain untouched.

Table 4. Quick checking rules

$\langle\{\underline{\Gamma \vdash \Delta}, T_2, \dots, T_m\}, \sigma\rangle : \mathbf{fact}\ l : \varphi\ \mathbf{by}\ name\ \mathbf{from}\ l_1, \dots, l_n; S \to \langle\{\underline{\Gamma \star l : \varphi \vdash \Delta}, T_2, \dots, T_m\}, \sigma_X^{name} \circ \sigma\rangle : S$
with $\sigma^{name}(C) = \bot$ and $\bigcup\sigma_C^{name}(P_i) = \bigcup\{\mathrm{pos}(l_i)\}$

$\langle\{\underline{\Gamma \vdash \varphi \Rightarrow \psi \star \Delta}, T_2, \dots, T_m\}, \sigma\rangle : \mathbf{assume}\ l : \varphi\ \mathbf{from}\ l_1; S \to \langle\{\underline{\Gamma \star l : \varphi \vdash \psi \star \Delta}, T_2, \dots, T_m\}, \sigma\rangle : S$
with $\sigma_C^{implintro}(C) = \mathrm{pos}(l_1)$ and $\sigma_C^{implintro}(P) = \bot$

$\langle\{\underline{\Gamma \vdash \neg\varphi \star \Delta}, T_2, \dots, T_m\}, \sigma\rangle : \mathbf{assume}\ l : \varphi\ \mathbf{from}\ l_1; S \to \langle\{\underline{\Gamma \star l : \varphi \vdash \Delta \star \bot}, T_2, \dots, T_m\}, \sigma\rangle : S$
with $\sigma_C^{contradiction}(C) = \mathrm{pos}(l_1)$ and $\sigma_C^{implintro}(P) = \bot$

$\langle\{\underline{\Gamma \vdash \psi \Rightarrow \neg\varphi \star \Delta}, T_2, \dots, T_m\}, \sigma\rangle : \mathbf{assume}\ l : \varphi\ \mathbf{from}\ l_1; S \to \langle\{\underline{\Gamma \star l : \varphi \vdash \neg\psi \star \Delta}, T_2, \dots, T_m\}, \sigma\rangle : S$
with $\sigma^{contrapositive}(C) = \mathrm{pos}(l_1)$ and $\sigma^{contrapositive}(P) = \bot$

$\langle\{\underline{\Gamma \vdash \psi \star \Delta}, T_2, \dots, T_m\}, \sigma\rangle : \mathbf{subgoals}\ l_1' : \varphi_1\{S_1\}, \dots, l_n' : \varphi_n\{S_n\}\ \mathbf{by}\ name\ \mathbf{from}\ l_1, \dots, l_n; S$
$\to \langle\{\underline{\Gamma \vdash \Delta \star l_1' : \sigma^{name}(P_1)}, \dots, \Gamma \vdash \Delta \star l_{n+k}' : \sigma^{name}(P_{n+k}), T_2, \dots, T_m\}, \sigma_X^{name} \circ \sigma\rangle : S_1; \dots; S_n; S$
with $T|_{\sigma^{name}(C)} = \psi$ and $(\bigcup\{\sigma_C^{name}(P_i)\} \cup \{\sigma_C^{name}(C)\}) = \bigcup\{\mathrm{pos}(l_i)\}$

$\langle\{\underline{\Gamma \vdash \Delta}, T_2, \dots, T_m\}, \sigma\rangle : \mathbf{cases}\ l_1' : \varphi_1\{S_1\}, \dots, l_n' : \varphi_n\{S_n\}\ \mathbf{from}\ l; S$
$\to \langle\{\underline{\Gamma \vdash \Delta|_{\mathrm{pos}(l) \leftarrow \varphi_1}}, \dots, \Gamma \vdash \Delta|_{\mathrm{pos}(l) \leftarrow \varphi_n}, T_2, \dots, T_m\}, \sigma\rangle : S_1; \dots; S_n; S$ with $T|_{\mathrm{pos}(l)} = \varphi_1 \vee \dots \vee \varphi_n$

$\langle\{\underline{\Gamma \vdash \Delta}, T_2, \dots, T_m\}, \sigma\rangle : \mathbf{set}\ x = t; S \to \langle\{\underline{\Gamma \vdash \Delta}, T_2, \dots, T_m\}, [x = t] \circ \sigma\rangle : S$ if x occurs in (Γ, Δ)

$\langle\{\underline{\Gamma \vdash \Delta}, T_2, \dots, T_m\}, \sigma\rangle : \mathbf{set}\ x = t; S \to \langle\{\underline{\Gamma \star x = t \vdash \Delta}, T_2, \dots, T_m\}, \sigma\rangle : S$ if x is new wrt. Γ, Δ

$\langle\{\underline{\Gamma \star \bot \vdash \Delta}, T_2, \dots, T_m\}, \sigma\rangle : \mathbf{trivial} \to \langle\{\underline{T_2}, \dots, T_m\}, \sigma\rangle :$

$\langle\{\underline{\Gamma \vdash \Delta \star \top}, T_2, \dots, T_m\}, \sigma\rangle : \mathbf{trivial} \to \langle\{\underline{T_2}, \dots, T_m\}, \sigma\rangle :$

$\langle\{\underline{\Gamma \star \varphi \vdash \Delta \star \varphi}, T_2, \dots, T_m\}, \sigma\rangle : \mathbf{trivial} \to \langle\{\underline{T_2}, \dots, T_m\}, \sigma\rangle :$

$\langle\{\underline{\Gamma \vdash \Delta}, T_2, \dots, T_m\}, \sigma\rangle : \mathbf{trivial}\ \mathbf{by}\ name\ \mathbf{from}\ l_1, \dots, l_n \to \langle\{\underline{T_2}, \dots, T_m\}, \sigma\rangle$
with $\bigcup\sigma_C^{name}(P_i) \cup \sigma_C^{name}(C) = \bigcup\{\mathrm{pos}(l_i)\}$

$\langle\{T_1, \dots, T_{j-1}, \underline{T_j}, T_{j+1}, \dots, T_m\}, \sigma\rangle : \varepsilon \to \langle\{T_1, \dots, T_{j-1}, T_j, \underline{T_{j+1}}, \dots, T_m\}, \sigma\rangle$

To increase the readability, subsequent steps which reduce a task to a single subtask are grouped together to a single step. Technically this is done by inserting a hierarchical edge. As default the most abstract proof step is propagated to the mediator.

Subgoals. The command **subgoals** reduces a goal of a given task to $n + m$ subgoals, each of which is represented as a new task, where n corresponds to the subgoals specified by the user and m denotes additional underspecified goals the user has omitted or forgotten. Each new task stems from a premise P_i of the applied inference, where the goal of the original task is replaced by the proof obligation for the premise, written as $pob(P_i)$. The quickcheck succeeds if the specified inference $name$ introduces at least the subgoals specified by the user.

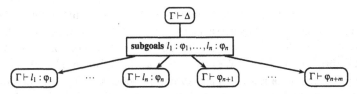

If there is no inference introducing the subgoals specified by the user within a single step the repair strategy tries to further reduce the goal in the current task, thus introducing further subgoals, until all specified subgoals are found. As in the assume case the antecedent of the sequent is untouched. If a subgoal matches a specified goal, it is not further refined. If all subgoals are found by a sequence of proof steps, these steps are abstracted to a single justification, which is by default propagated to the mediator.

Cases. The command **cases** reduces a task containing a disjunction on the left hand side of the task into $n + m$ subtasks where in each case an additional premise is added. As for the **subgoals** the user can leave out some of the cases. If the task does not contain a suitable disjunction, the repair strategy is executed, which tries to derive a desired disjunction by forward reasoning. The goal remains untouched. As for the subgoal case the sequence introducing the disjunction is abstracted to a single step.

Set. The command **set** is used to bind a meta-variable or to intro-duce an abbreviation for a term. If x is an unbound meta-variable in the proof state, **set** will instantiate this variable with the given term t. The substitution $x \to t$ is added to the proof state. If x is already bound, a failure is generated. If x does not occur in the proof state, the com-

mand **set** serves as a shortcut statement for the given term t. The formula $x = t$ will be added as a new premise to the task. The last case is shown on the right. Adding an equation $x = t$ with a fresh variable x as premise is conservative in the sense, that a proof using the new variable x can be converted in a proof without x by just substituting all occurrences of x by t. There is no repair strategy for the **set** command.

Trivial. The command **trivial** is used to indicate that a task is solved. This is the case if a formula φ occurs on both the left and the right hand side of the task, the symbol false occurs at top level on the left hand side of the task, or the symbol true occurs at top level on the right hand side of a task. A task can also be closed if the inference

name is applied and all its premises and conclusions are matched to term positions in the current task. In case that the quickcheck fails the repair strategy tries to close the task by a depth limited forward backward reasoning.

Complex. The command **complex** is an abstract command which subsumes an arbi-trary sequence of the previous commands. It is used to represent arbitrary abstract steps. Note that it is generally not possible to justify such a step with a single inference appli-cation, and without further information a blind search has to be performed to justify the step. Hence there is no quickcheck for **complex** . If however in the **by** slot a strategy is specified, this strategy needs to be executed and the result to be compared.

Example. Looking at our running example, the user wanted to show **Theorem 8** and stated already a partial proof (c.f. Table 3). As none of the proof checking rules for **assume** are applicable, the repair mode is started. The repair strategy tries to fur-ther refine the goal $A \cap (B \cup C) = (A \cap B) \cup (A \cap C)$ to construct a situation in which the **assume** step is applicable. Indeed, after three refinements the proof step becomes applicable (see Figure 4). Unnecessary derivations, indicated by a dotted line in the Figure, are deleted. The repair process brought a second subgoal out that now can be lifted to the abstract proof view. We offer the following lifting modes: (i) *fix* : repaired proof fragments are automatically patched into the document (ii) *inform* : the author is informed about repair patches and decides their execution. (iii) *silent* : the author is only informed about errors, no repair patches are offered.

5.2   Proof Lifting

Whenever a part of the proof is changed, it must be propagated back to the mediator. In principle the prover can insert arbitrary large parts and multiple hierarchies during the

Table 5. Proof lifting rules

$\{\mathrm{diff}(T,T_i)|T_i \in succ(T)\}$	proof step in abstract syntax
$\emptyset$	**trivial by** *name* **from** $\mathrm{lab}(name)$
$\{\langle\{\varphi\},\{\xi\}\rangle\}$	**assume** φ **by** *name* **from** $\mathrm{lab}(name)$
$\{\langle\emptyset,\emptyset\rangle\}$	**set** $\mathrm{diff}(\sigma,\sigma')$
$\{\langle\emptyset,\{\xi_1\}\rangle\}\dots\langle\emptyset,\{\xi_m\}\rangle\}\}$	**subgoals** $\xi_1,\dots,\xi_m$ **by** *name* **from** $\mathrm{lab}(name)$
$\{\langle\{x=t\},\emptyset\rangle\}$	**set** $x = t$
$\{\langle\{\varphi\},\emptyset\rangle\}$	**fact** φ **by** *name* **from** $\mathrm{lab}(name)$
$\{\langle\{\varphi_1\},\emptyset\rangle\dots\langle\{\varphi_n\},\emptyset\rangle\}$	**cases** $\varphi_1,\dots,\varphi_n$ **by** *name* **from** $\mathrm{lab}(name)$
$\{\langle\Gamma_1,\Delta_1\rangle,\dots,\langle\Gamma_m,\Delta_m\rangle\}$	**complex** Δ_1 **under** $\Gamma_1,\dots,\Delta_m$ **under** Γ_m

repair phase. As default the most abstract proof hierarchy is communicated as a proof script to the mediator. However, the mediator can ask for a more detailed or a more abstract version of the proof script. Given a selected proof hierarchy, each proof step has to be transformed into a command of the proof script language. This is done by a static analysis of the proof step.

Task Difference. Technically, a proof step is executed with respect to an agenda $\langle\{T_1,\dots,T_n\},\sigma\rangle$ and results in a new agenda $\langle\{T_1',\dots,T_k',T_2,\dots,T_n\},\sigma'\rangle$. The step has reduced the task T_1 to subtasks $succ(T) = \{T_1',\dots,T_k'\}$. In a first analysis, only the differences between the tasks are analyzed, defined as follows:

$$\mathrm{diff}(T,T') = \langle\{\varphi\in\Gamma'|\varphi\notin\Gamma\},\{\xi\in\Delta'|\xi\notin\Delta\}\rangle$$

If a task is reduced to several subtasks, we obtain a set of differences for each subtask. Moreover, we require that the *name* of the applied proof operator, i.e. the inference or strategy, for the reduction is given and we denote a substitution introduced by the proof step with σ. We assume a function *lab* which returns the set of those labels which are used in premises and conclusions of the proof operator and . if none of them has a label.

Lifting Rules. If $succ(T) = \emptyset$, the proof step is translated into a **trivial** step. If $succ(T) = \{T'\}$, there are the following possibilities: If the task T and its successor task T' are the same, we analyze the difference between σ and σ' to obtain the formula $x = t$ needed for the **set** case. If the difference between T and T' is only

Table 6. Proof repaired in abstract syntax

Proof.
  subgoals
    $A\cap(B\cup C)\subset(A\cap B)\cup(A\cap C):\{$
      assume $x\in A\cap(B\cup C)$ **from** .;
      fact $x\in A\wedge x\in(B\cup C)$ **by** . **from** .;$\varepsilon\}$
    $(A\cap B)\cup(A\cap C)\subset A\cap(B\cup C):\{\varepsilon\}$
    by Set Equality **from** .

one formula, and this formula has been added on the left hand side of the sequent, and is of the form $x = t$, where x is new, then the step is classified as a **set** case, otherwise as a **fact** step. If the new formula has been introduced as a goal, we classify the step to be a **subgoals** step. If several hypotheses are introduced, the step is classified to be a **cases** step. A formal definition of the proof lifting rules are given in

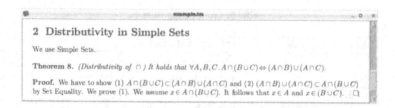

Fig. 5. Theorem with repaired partial proof

Table 5. Considering our running example, the abstract proof is repaired as shown in Figure 4, and presented to the author as illustrated in Figure 5.

6 Extending OMDoc for Authoring Verified Proofs

In ΩMEGA we use the OMDoc format already for theory repositories with acyclic theory dependencies, axioms, simple definitions and assertions. So far, we only support a subset of the OMDoc features and exclude e.g. complex theory morphisms and complex definitions. OMDoc's current proof module (PF) is designed for representing proofs given in a declarative or procedural proof language together with comments in natural language. Additionally, there is the possibility to store formal proofs as proof terms in `proofobject` elements. In the following we refer to the extension of the PF module proposed in [5] to store proofs with proofsteps on different levels of granularity. Moreover, as we use TEX$_{MACS}$ for authoring, we need to store additional *parsing* and *rendering* knowledge beyond the coverage of OMDoc's presentation module. For space reasons we use the compact form `<proof>...</>` for XML instead of `<proof>...</proof>`.

6.1 Hierarchical Proof Data Structure

As we require both, fast reconstruction of the PDS from OMDoc proofs and presentation of a given view of the proof, we need to store the proofs in OMDoc with all levels of granularity. As a simple example (with only one level of granularity) we show in Listing 1 the proof from Figure 4 in OMDoc format.

Each `proof` element represents a sequence of proofsteps. The proofstep consist of a `derive` element where the `type` attribute contains the proof command and the `method` element the **by** information, encoded in its `xref` attribute. The **from** information is stored in `premise` subelements which point to the corresponding labeled formulas. The labels are realized using the `id` attribute of the `OMOBJ` element.

An *assume* proofstep introduces always a new local hypothesis which is represented by the `hypothesis` element after the assume `derive` block. We represent a **subgoals** proofstep inside a `method` block by a sequence of `OMOBJ` - `proof` pairs corresponding to a subgoal followed by its subproof. We encode formulas in OMDoc using OPENMATH.

```
<proof xml:id="p1" for="#distr_inter">
  <derive xml:id="p1_d1" type="subgoals">
    <method xref="#definition_of_set_equality">
      <OMOBJ>A∩(B∪C)⊂A∩B∪A∩C</OMOBJ>
        <proof><derive xml:id="p1_d2" type="assume"><method xref="#definition_of_subset"/></>
            <hypothesis><FMP>x∈A∩(B∪C)</FMP></>
              <derive xml:id="p1_d3" type="fact"><FMP>x∈A∧x∈(B∪C)</FMP>
                <method xref="#definition_of_intersection"/></></>
      <OMOBJ>A∩B∪A∩C⊂A∩(B∪C)</OMOBJ></></></>
```

OMDoc Listing 1: Proof as XML tree

Finally, Listing 2 shows how we encode proof steps at different levels of granularity. The `alt` element contains as first `proof` element the strategy justification and then the expansion or refinement of this strategy. Table 7 gives an overview of the added attributes and content specifications.

Table 7. Extension of PF

```
<proof xml:id="p1" for="#distr_inter">
  <alt>
    <proof><derive xml:id="p1_dx" type="complex">
        <method xref="#strategy1"/></></>
    <proof> ... proof from Listing 1 ... </></></>
```

OMDoc Listing 2: Hierarchical Justifications

Element	Opt. Attrib.	Content
proof		alt
alt		proof+

Table 8. Extension of PRES

Element	Opt. Attrib.	Content
use	cop	symbol*, OMOBJ*

6.2 Parsing and Rendering Knowledge

The parsing and rendering facility of PLATΩ uses the following knowledge which we want to store for each theory and for each community of practice separately in OMDoc using the presentation module (PRES). Table 8 gives an overview of the extensions to the PRES module. Each knowledge item is encoded in a `presentation` block like

```
<presentation for="URI"><use format="texmacs" cop="name" attributes="type=TYPE">...</></>
```

Notations for mathematical symbols are given by NOTC. Typed variables are encoded as `symbol` elements and each pattern as an OMOBJ element[1]. While all patterns are allowed as parser input, the renderer uses by default the first pattern.

```
<presentation for="#union"><use ... attributes="type=symbol">&lt;cup&gt;</></>
<presentation for="#union"><use ... attributes="type=notation">
  <symbol xml:id="x"><type>set</type></symbol>
  <symbol xml:id="y"><type>set</type></symbol>
  <OMOBJ><OMA>
    <OMS cd="local" name="x"/><OMSTR>&lt;cup&gt;</OMSTR><OMS cd="local" name="y"/></></></></>
```

Symbolgroups are given by GROUP. We declare a symbolgroup by using a `symbol` element before specifying its elements in a `presentation` block.

```
<symbol xml:id="setops" role="symbolgroup"/>
<presentation for="#setops"><use ... attributes="type=symbolgroup">
  <OMOBJ><OMS cd="th1" name="union"/></OMOBJ>
  <OMOBJ><OMS cd="th1" name="intersection"/></OMOBJ></></>
```

---

[1] We encode a list of terms as argument list of an OMA element.

Associativity information for symbols, given by ASSOC, with values *right* or *left*.

```
<presentation for="#union"><use ... attributes="type=associativity">left</></>
```

Precedence constraints for two or more symbol(group)s, expressing that $s_1 \prec \ldots \prec s_n$, are given by PREC. The symbol(group) s_1 is referred to by the for attribute and the $s_2, \ldots, s_n$ are encoded as OMOBJ elements.

```
<presentation for="#intersection"><use ... attributes="type=precedence">
  <OMOBJ><OMS cd="th1" name="union"/></OMOBJ></></>
```

By storing this knowledge separately for each community of practice, our architecture supports for free the automatic notational translation of a document across communities of practice. The default notation for all symbols is prefix notation.

7   Related Work

The most prominent system for the publication of machine checked mathematics is MIZAR [14] with one of the largest libraries of formalized mathematics. The language of the library is a well-designed compromise between human-readability and machine-processability. Since the MIZAR system is not interactive, the usual workflow is to prepare an article, compile it and loop both steps until there is no error reported. In contrast to that, our architecture allows for both a human-oriented and a machine-oriented representation as well as techniques to lift or expand these representations respectively.

ISABELLE/ISAR [17] is a generic framework for human-readable formal proof documents, both like and unlike MIZAR. The ISAR proof language provides general principles that may be instantiated to particular object-logics and applications. ISABELLE tries to check an ISAR proof, shows the proof status but does not patch the proof script for corrections. We try to repair detected errors or underspecifications in proof steps.

A very promising representative of distributed systems for the publication of machine checked mathematics is LOGIWEB [9]. It allows the authoring of articles in a sophisticated customizable language but strictly separates the input from the output document, resulting in the usual LaTeX workflow. By using the WYSIWYG text-editor TeX_MACS we combine input and output representation in a document-centric approach.

Regarding parsing techniques the Matita system provides currently the best strategies for disambiguation [13]. Definitely, we plan to adapt these methods to our setting since they reduce efficiently the amount of alternative proofs to be verified.

8   Conclusion

In this paper we presented an architecture for authoring machine checked documents for mathematics repositories within a text-editor. To meet the real-time requirements of our scenario, we presented fast parsing and rendering mechanisms as well as incremental proof checking techniques. To increase the usability, the checking rules are enhanced by repair strategies trying to fix incomplete or underspecified steps. Finally, we presented an extension of the OMDOC format to store the formalized content of the document together with its justifications and notational information. Thus, the proof situations can be efficiently restored and verified at a later date and by other authors. With the approach presented in this paper we have a solid basis for further linguistic improvements. The plan is on the one hand to generate natural language with aggregation, topicalisation etc. from the controlled language and on the other hand to be able to understand that constantly increasing fragment of natural language to extract the controlled language. As future work we are going to investigate whether the proposed architecture is a well-suited foundation for collaborative authoring inside and across *communities of practice.*

References

1. Autexier, S., Benzmüller, C., Dietrich, D., Meier, A., Wirth, C.-P.: A generic modular data structure for proof attempts alternating on ideas and granularity. In: Kohlhase, M. (ed.) MKM 2005. LNCS (LNAI), vol. 3863, pp. 126–142. Springer, Heidelberg (2006)
2. Autexier, S., Dietrich, D.: Synthesizing proof planning methods and oants agents from mathematical knowledge. In: Borwein, J.M., Farmer, W.M. (eds.) MKM 2006. LNCS (LNAI), vol. 4108, pp. 94–109. Springer, Heidelberg (2006)
3. Autexier, S., Fiedler, A., Neumann, T., Wagner, M.: Supporting user-defined notations when integrating scientific text-editors with proof assistance systems. In: Kauers, M., Kerber, M., Miner, R., Windsteiger, W. (eds.) MKM/CALCULEMUS 2007. LNCS (LNAI), vol. 4573, Springer, Heidelberg (2007)
4. Autexier, S., Hutter, D., Mossakowski, T., Schairer, A.: The development graph manager MAYA. In: Kirchner, H., Ringeissen, C. (eds.) AMAST 2002. LNCS, vol. 2422. Springer, Heidelberg (2002)
5. Autexier, S., Sacerdoti-Coen, C.: A formal correspondence between OMDOC with alternative proofs and the $\bar{\lambda}\mu\tilde{\mu}$. In: Borwein, J.M., Farmer, W.M. (eds.) MKM 2006. LNCS (LNAI), vol. 4108, pp. 67–81. Springer, Heidelberg (2006)
6. Aycock, J., Horspool, N.: Directly-executable Earley parsing. In: Wilhelm, R. (ed.) CC 2001 and ETAPS 2001. LNCS, vol. 2027, p. 229. Springer, Heidelberg (2001)
7. Siekmann, J., et al.: Proof development with ΩMEGA: $\sqrt{2}$ is irrational. In: Baaz, M., Voronkov, A. (eds.) LPAR 2002. LNCS (LNAI), vol. 2514, pp. 367–387. Springer, Heidelberg (2002)
8. Franke, A., Kohlhase, M.: MBASE: Representing mathematical knowledge in a relational data base. In: Systems for Integrated Computation and Deduction. Elsevier, Amsterdam (1999)
9. Grue, K.: The Layers of Logiweb. In: Kauers, M., Kerber, M., Miner, R., Windsteiger, W. (eds.) MKM/CALCULEMUS 2007. LNCS (LNAI), vol. 4573, Springer, Heidelberg (2007)
10. Kohlhase, M.: OMDoc – An Open Markup Format for Mathematical Documents [version 1.2]. LNCS (LNAI), vol. 4180. Springer, Heidelberg (2006)
11. Lange, C.: A semantic wiki for mathematical knowledge management. Diploma thesis. IUB Bremen, Germany (2006)
12. Radzevich, S.: Semantic-based diff, patch and merge for XML-documents. Master thesis, Saarland University, Saarbrücken, Germany (April 2006)
13. Sacerdoti-Coen, C., Zacchiroli, S.: Spurious disambiguation error detection. In: Kauers, M., Kerber, M., Miner, R., Windsteiger, W. (eds.) MKM/CALCULEMUS 2007. LNCS (LNAI), vol. 4573, Springer, Heidelberg (2007)

14. Trybulec, A., Blair, H.: Computer assisted reasoning with MIZAR. In: Joshi, A. (ed.) Proceedings of the 9th Int. Joint Conference on Artifical Intelligence. M. Kaufmann, San Francisco (1985)

15. J.v.d. Hoeven.: GNU T$_{E}$X$_{MACS}$: A free, structured, WYSIWYG and technical text editor. Number 39-40 in Cahiers GUTenberg (May 2001)

16. Wagner, M., Autexier, S., Benzmüller, C.: PLATΩ: A mediator between text-editors and proof assistance systems. In: Benzmüller, C., Autexier, S. (eds.) 7th Workshop on User Interfaces for Theorem Provers (UITP 2006), Elsevier, Amsterdam (2006)

17. Wenzel, M.: Isabelle/Isar — a generic framework for human-readable proof documents. In: Matuszewski, R., Zalewska, A. (eds.) From Insight to Proof — Festschrift in Honour of Andrzej Trybulec. Studies in Logic, Grammar, and Rhetoric, University of Bialystok (2007)

Verification of Mathematical Formulae Based on a Combination of Context-Free Grammar and Tree Grammar

Akio Fujiyoshi[1], Masakazu Suzuki[2], and Seiichi Uchida[3]

[1] Department of Computer and Information Sciences, Ibaraki University
fujiyosi@mx.ibaraki.ac.jp
[2] Faculty of Mathematics, Kyushu University
suzuki@math.kyushu-u.ac.jp
[3] Faculty of Information Science and Electrical Engineering, Kyushu University
uchida@is.kyushu-u.ac.jp

Abstract. This paper proposes the use of a formal grammar for the verification of mathematical formulae for a practical mathematical OCR system. Like a C compiler detecting syntax errors in a source file, we want to have a verification mechanism to find errors in the output of mathematical OCR. Linear monadic context-free tree grammar (LM-CFTG) was employed as a formal framework to define "well-formed" mathematical formulae. For the purpose of practical evaluation, a verification system for mathematical OCR was developed, and the effectiveness of the system was demonstrated by using the ground-truthed mathematical document database INFTY CDB-1.

1 Introduction

Grammatical analysis is useful for many types of verification problems. For example, a C compiler grammatically analyzes a source file and returns error messages with the location and type of errors. For mathematical OCR [1], it is natural to think that such grammatical analysis helps to detect misrecognitions of characters and structures in mathematical formulae. This paper proposes a mathematical-formulae verification method for a practical mathematical OCR system based on a combination of context-free grammar [2] and tree grammar [3].

Grammatical analysis can be classified into two levels: syntactic analysis and semantic analysis. This paper concentrates only on the syntactic analysis of mathematical formulae because we wanted to build a very fast verification system. Needless to say, semantic analysis is also very important for the improvement of mathematical OCR. However, we will leave this task for another time. We use the term "well-formed" to mean syntactic correctness. Since syntactic correctness doesn't necessarily mean semantic correctness, we can consider unsatisfiable formulae, e.g., "$1 + 2 = 5$", and tautological formulae, e.g., "$x = x$", as "well-formed" formulae if they are syntactically correct.

The final aim of this study is to completely define "well-formed" mathematical formulae. In other words, we want to have a grammar to verify any mathematical

S. Autexier et al. (Eds.): AISC/Calculemus/MKM 2008, LNAI 5144, pp. 415–429, 2008.
© Springer-Verlag Berlin Heidelberg 2008

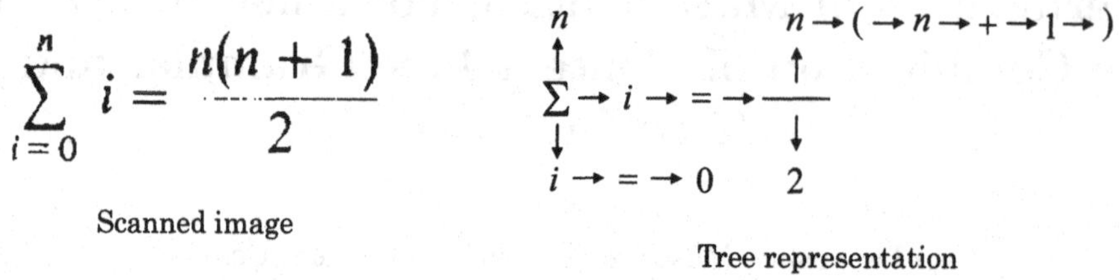

$$\sum_{i=0}^{n} i = \frac{n(n + 1)}{2}$$

Scanned image

$$\begin{array}{c} n \\ \uparrow \\ \sum \rightarrow i \rightarrow = \rightarrow \dfrac{n \rightarrow (\rightarrow n \rightarrow + \rightarrow 1 \rightarrow)}{2} \\ \downarrow \\ i \rightarrow = \rightarrow 0 \end{array}$$

Tree representation

Fig. 1. A result of structural analysis of a mathematical formula

$$Sum \rightarrow \begin{array}{c} Exp \\ | \\ \Sigma \\ | \\ Init \end{array} \qquad Frac \rightarrow \begin{array}{c} Exp \\ | \\ \rule{1cm}{0.4pt} \\ | \\ Exp \end{array}$$

Fan-out rules

$$Init \rightarrow Exp = Exp \qquad Term \rightarrow Factor$$
$$Exp \rightarrow Exp + Exp \qquad Factor \rightarrow 1$$
$$Exp \rightarrow Term \qquad Factor \rightarrow n$$
$$Term \rightarrow Term\ Term \qquad Factor \rightarrow (\ Exp\ )$$

Context-free rules

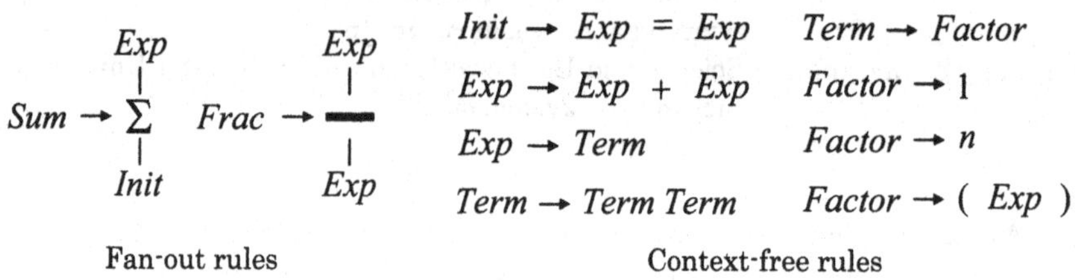

Fig. 2. Some rules of the grammar

formula that has appeared, or will appear, in a long-term build-up of mathematical documents. There were other grammatical approaches to the verification of mathematical formulae such as [4,5,6]. The proposed verification method will extend the coverage of those approaches.

In order to define "well-formed" mathematical formulae, we employed linear monadic context-free tree grammar (LM-CFTG) [3] as a formal framework. As shown in Fig. 1, a mathematical OCR system offers a tree representation of a mathematical formula from a scanned image. Therefore, we needed a grammar formalism to define a set of tree structures. An LM-CFTG defines a set of tree structures by arranging fan-out rules and context-free rules, where fan-out rules are used to describe the structural growth of a tree, and context-free rules are used to describe linear growth. For example, some fan-out rules and context-free rules of the grammar defining "well-formed" mathematical formulae are illustrated in Fig. 2.

The proposed verification method allows us to build a very fast verification system. Theoretically, the verification process of most mathematical formulae will be completed in linear time depending on the size of the input, though some exceptional mathematical formulae require cubic time. We need a very fast verification system because verification should be done for numerous recognition candidates to improve the reliability of mathematical OCR. We experimentally built a verification system and executed the system on the ground-truthed mathematical document database INFTY CDB-1 [7]. The verification of $21,967$ mathematical formulae (size: 48.1MB) was finished within 10 seconds by a PC (CPU: Pentium4 3.06GHz, RAM: 1GB).

The accomplishment of this very fast verification system mainly resulted from the following two features of the proposed verification method:

- Division of a mathematical formula into sub-formulae; and
- A grammar formalism with a fast recognition algorithm.

The idea of the division of a formula into sub-formulae is common to well-known algorithm design paradigms such as "Divide and Conquer" and "Dynamic Programming." The employment of LM-CFTG enables us to use not only parsing algorithms for LM-CFTG [8] but also well-established parsing techniques for context-free grammar (CFG) [2,9].

Although the proposed verification method may be useful in general, this paper mainly discusses the implementation of a verification system created to be used with InftyReader [10]. The information about InftyReader and other supporting software can be found on the Infty Project website [11].

This paper is organized as follows: In Section 2, the grammar defining "well-formed" mathematical formulae is explained; in Section 3, the outline of the proposed verification method is described; in Section 4, the results of the experiment are shown; in Section 5, LM-CFTG is introduced as a formal framework for the grammar defining "well-formed" mathematical formulae; and in Section 6, the conclusion is drawn and future work determined.

2 "Well-Formed" Mathematical Formulae

In order to define "well-formed" mathematical formulae, linear monadic context-free tree grammar (LM-CFTG) [3] was employed as a formal framework. The definition and the formal properties of LM-CFTG will be introduced in Section 5. To choose an appropriate grammar formalism, it was necessary for a grammar formalism to have sufficient descriptive power to process a diversity of mathematical formulae. In addition to descriptive power, we also required that a grammar formalism be accompanied by a very fast parser.

An LM-CFTG is defined by arranging *fan-out rules* and *context-free rules*. Fan-out rules are used to define possible structural configuration of mathematical formulae. We should arrange them for symbols which are possibly connected with adjunct symbols. Examples of those symbols are "capital sigma" for summation, "capital pi" for product, "radical sign" for square root, "long bar" for fraction, "integral sign" for definite integral, etc. Because any variable may have a subscript, we arranged a fan-out rule for all italic alphabet symbols. Context-free rules are used to define possible linear sequences of symbols of mathematical formulae. Context-free rules constitute a context-free grammar (CFG) [2], and thus we can use well-established parsing techniques for CFG [9].

We experimentally developed a grammar defining "well-formed" mathematical formulae. The grammar consists of 35 fan-out rules and 170 context-free rules. The number of rules will be increased with the refinement of the grammar. A representative sample of the grammar is illustrated in the appendix at the end of this paper.

Table 1. Grammatical categories

Category	Explanation and Example
Math	Acceptable mathematical formula "$u(a, b) = Int\ Frac$"
Range	Range of value of a variable "$1 \leq i \leq n$"
Init	Initialization of a variable "$i = 0$", "$i = k$"
Exp	Acceptable expression "$2 + 3$", "$n(n + 1)$"
ExpList	List of expressions connected with signs "$a < b < c < d$", "$z = x + y$"
Subscript	Subscript of a variable "2", "n", "$1, 2$", "$1, 2, 3, 4$"
Supscript	Supscript of a variable and expression "$\prime$", "$\prime\prime$", "2", "n", "$1, 2$"

On the development of the grammar, we tried to arrange context-free rules so that they constitute a deterministic context-free grammar (DCFG) [2] because we could take advantage of a linear-time parsing technique for DCFG [9]. Unfortunately, we needed to add some context-free rules, which break the condition of DCFG, and this is the reason why a verification process of some exceptional mathematical formulae requires cubic time. Most of those context-free rules are related to the vertical-bar symbol because the usage of vertical bar is too diverse: absolute value, divides, conditional probability, norm of a vector, etc.

Table 1 shows the grammatical categories defined by the context-free rules of the grammar.

3   Outline of the Verification Method

In this section, we describe the outline of the proposed verification method. We start with the input to the verification system, that is to say, the output of mathematical OCR.

The output of InftyReader is given in InftyCSV format. An example of an InftyCSV text expressing a mathematical formula is shown in Table 2. Each line corresponds to a symbol in the formula, where: "ID" is the number uniquely assigned to the symbol; "x_1, x_2, x_3, x_4" are the coordinates of the rectangular area; "Mode" is a flag showing if the symbol is a part of a mathematical formula; "Link" expresses the relationship with the parental symbol; "Parent" is the ID of the parental symbol; and "Code" is the internal character code of the symbol. The original image and the rectangular representation of the formula are shown at (1) and (2) in Fig. 3.

Table 2. An example of an InftyCSV text

ID	x_1	y_1	x_2	y_2	Mode	Link	Parent	Code
1,	1487,	708,	1535,	766,	1,	-1,	-1,	0x426C
2,	1542,	685,	1559,	758,	1,	0,	1,	0x1980
3,	1563,	704,	1603,	742,	1,	0,	2,	0x4161
4,	1610,	732,	1622,	753,	1,	0,	3,	0x142C
5,	1646,	683,	1679,	742,	1,	0,	4,	0x4162
6,	1686,	685,	1703,	758,	1,	0,	5,	0x1981
7,	1728,	708,	1780,	724,	1,	0,	6,	0x1D3D
8,	1801,	624,	1858,	812,	1,	0,	7,	0x33F0
9,	1853,	782,	1881,	810,	1,	2,	8,	0x4161
10,	1868,	622,	1891,	658,	1,	1,	8,	0x4162
11,	1909,	717,	2053,	722,	1,	0,	8,	0x33D1
12,	1945,	629,	1985,	689,	1,	5,	11,	0x4164
13,	1986,	650,	2016,	689,	1,	0,	12,	0x4163
14,	1911,	736,	1967,	800,	1,	6,	11,	0x0248
15,	1977,	740,	1994,	813,	1,	0,	14,	0x1980
16,	1999,	757,	2029,	796,	1,	0,	15,	0x4163
17,	2035,	740,	2051,	813,	1,	0,	16,	0x1981

3.1 Construction of a Tree Representation

First, the verification system converts an InftyCSV text into a linked list called
a *tree representation*. A node of a linked list is illustrated in Fig. 4. By preparing
nodes for all symbols and connecting them in accordance with "Link" and "Parent" information in an InftyCSV text, the tree representation of a mathematical
formula is constructed. The InftyCSV text in Table 2 is converted into the tree
representation shown at (3) in Fig. 3.

3.2 Division of a Mathematical Formula into Strings

Secondly, strings are extracted from a tree representation of a mathematical formula. Strings are obtained by concatenating symbols horizontally connected in a
tree representation. From the tree representation shown in Fig. 3, the following
five strings are extracted:

"u (a , b) $= Int\ Frac$",
"b",
"a",
"$d\ c$", and
"Θ (c)".

3.3 Grammatical Analysis

Grammatical analysis is executed in two stages: *linear sequence analysis* and
structural inspection. In the linear sequence analysis, a parser for a context-free

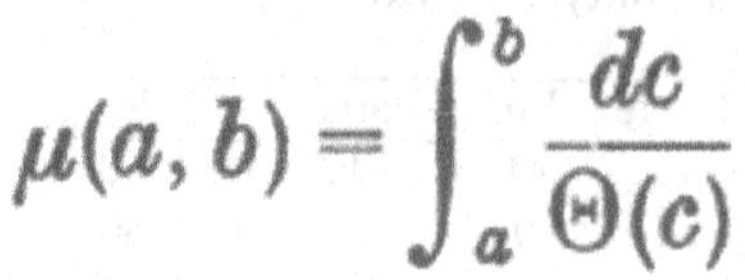

$$\mu(a, b) = \int_a^b \frac{dc}{\Theta(c)}$$

(1) Original image

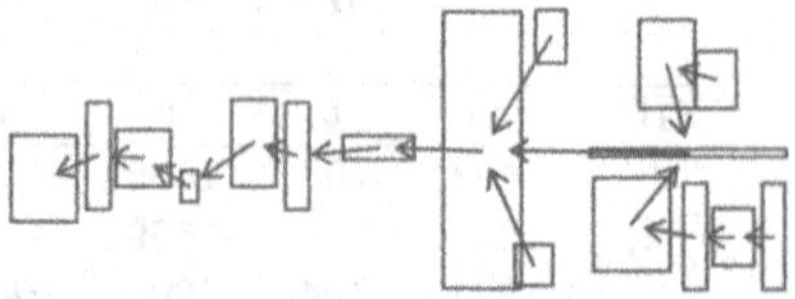

(2) Rectangular representation

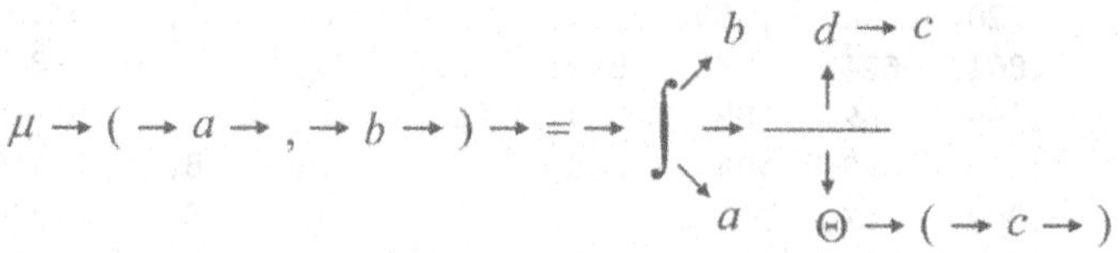

$$\mu \to (\to a \to , \to b \to) \to = \to \int \to \quad \begin{array}{c} b \quad d \to c \\ \nearrow \quad \uparrow \\ \\ \searrow \quad \downarrow \\ a \quad \Theta \to (\to c \to) \end{array}$$

(3) Tree representation

Fig. 3. The mathematical formula

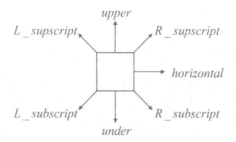

upper

L_supscript   *R_supscript*

horizontal

L_subscript   *R_subscript*

under

Fig. 4. A node

grammar (CFG) [2] is utilized, and, for each string extracted from a tree representation of a mathematical formula, the fitness to the grammatical categories is examined. In Table 3, the fitness to the grammatical categories for the strings is shown. The linear sequence analysis is the most time-consuming task in the proposed verification method, and may cost cubic time depending on the size of the input in the worst case, while the other tasks can be done in linear time.

After the linear sequence analysis, the structural inspection takes place. In the structural inspection, the connectivity of nodes is examined by searching for matching fan-out rules. The structural inspection process is illustrated in Fig. 5. The connection of the adjunct strings, "*b*" and "*a*", and the integral sign are inspected. The connection of the adjunct strings, "*d c*" and "Θ (*c*)", and the long bar are also inspected.

The mathematical formula in the example was successfully verified as a "well-formed" mathematical formula.

Table 3. Fitness to the grammatical categories

Strings	Math	Range	Init	Exp	ExpList	Subscript	Supscript
$u\,(\,a\,,\,b\,)\,=\,Int\ Frac$	yes	yes	yes	no	yes	no	no
b	yes	yes	no	yes	yes	yes	yes
a	yes	yes	no	yes	yes	yes	yes
$d\ c$	yes	yes	no	yes	yes	yes	yes
$\Theta\,(\,c\,)$	yes	yes	no	yes	yes	yes	yes

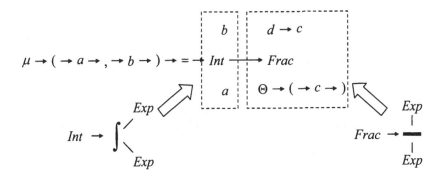

Fig. 5. Structural inspection

4 Experimental Results

We experimentally built a verification system in accordance with the proposed verification method. For implementation of the system, the program was written in C language, and GNU Bison [12], a parser generator for CFG, was utilized. For evaluation, we executed the system on the ground-truthed mathematical document database INFTY CDB-1 [7].

The verification of $21,967$ mathematical formulae in INFTY CDB-1 (size: 48.1MB) was finished within 10 seconds by a PC (CPU: Pentium4 3.06GHz; RAM: 1GB). The speed of the proposed verification method was experimentally confirmed. Theoretically, a verification process of most mathematical formulae will be finished in linear time depending on the size of the input, though some exceptional mathematical formulae require cubic time.

The verification system produces verification results in XHTML format with MathML inclusions and displays error messages on a web browser such as Mozilla Firefox [13]. An error message identifies the position and type of suspicious mathematical-formula error as enlarged and colored red.

(1), (2), (3) and (4) in Fig. 6 are error messages produced by the verification system. Original images corresponding to the error messages are shown in Fig. 7. As for (1), the verification system successfully detected the misrecognition of a comma before the letter 'b'. The comma was misrecognized as a period. With regard to (2), the verification system detected a faulty correspondence of parentheses. Looking at the original image, we noticed that this was an error from the

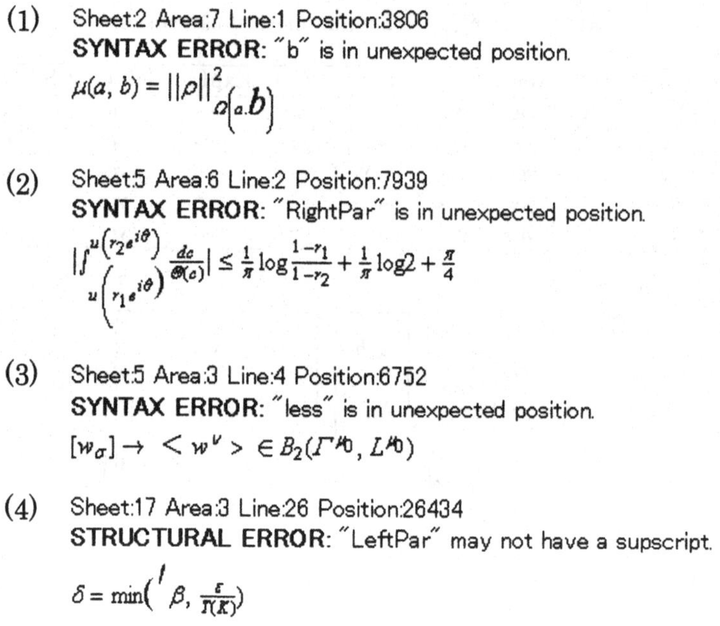

(1)   Sheet:2 Area:7 Line:1 Position:3806
      SYNTAX ERROR: "b" is in unexpected position.

$$\mu(a, b) = ||\rho||^2_{\Omega(a,b)}$$

(2)   Sheet:5 Area:6 Line:2 Position:7939
      SYNTAX ERROR: "RightPar" is in unexpected position.

$$\left| \int_{u\left(r_1 e^{i\theta}\right)}^{u\left(r_2 e^{i\theta}\right)} \frac{dc}{\Theta(c)} \right| \le \frac{1}{\pi} \log \frac{1-r_1}{1-r_2} + \frac{1}{\pi} \log 2 + \frac{\pi}{4}$$

(3)   Sheet:5 Area:3 Line:4 Position:6752
      SYNTAX ERROR: "less" is in unexpected position.

$$[w_\sigma] \to \ < w^\nu > \ \in B_2(\Gamma^\mu 0, L^\mu 0)$$

(4)   Sheet:17 Area:3 Line:26 Position:26434
      STRUCTURAL ERROR: "LeftPar" may not have a supscript.

$$\delta = \min\left({}' \beta, \frac{\varepsilon}{T(K)} \right)$$

Fig. 6. Error messages produced by the verification system

(1) $$\mu(a, b) = ||\varrho||^2_{\Omega(a,b)},$$

(2) $$\left| \int_{u(r_1 e^{i\theta})}^{u(r_2 e^{i\theta})} \frac{dc}{\Theta(c)} \right| \le \frac{1}{\pi} \log \frac{1-r_1}{1-r_2} + \frac{1}{\pi} \log 2 + \frac{\pi}{4}$$

(3) $$[w_\sigma] \to <w^\nu> \ \epsilon \ B_2(\Gamma^\mu 0, L^\mu 0)$$

(4) $$\delta = \min\left(\beta, \frac{\varepsilon}{T(K)} \right)$$

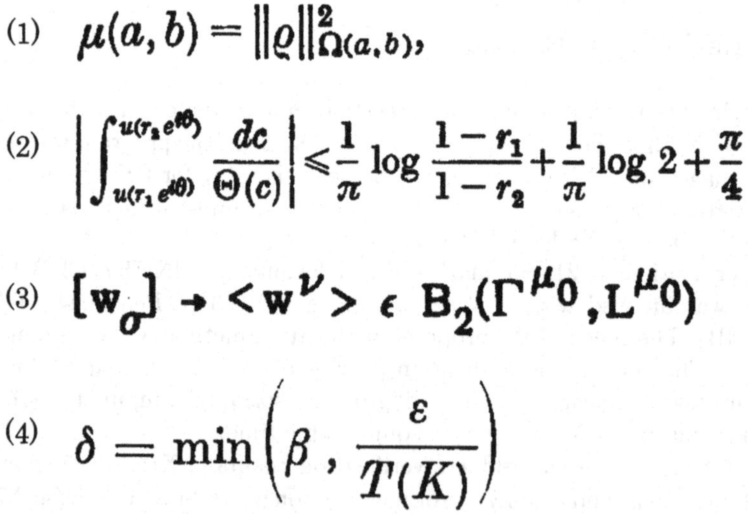

Fig. 7. Original images

original document. Concerning (3), the verification system successfully detected the misrecognition of the angle bracket. The angle bracket was misrecognized as a less-than sign. And about (4), a structural error was detected since a left parenthesis may not have a superscript. A portion of the left parenthesis was misrecognized as a prime symbol.

5   Formal Framework

In this section, we introduce the formal definitions of tree and linear monadic context-free tree grammar (LM-CFTG). LM-CFTG was employed as a formal framework to define "well-formed" mathematical formulae. We also introduce known results for LM-CFTG.

5.1   Tree

A *ranked alphabet* is a finite set of symbols in which each symbol is associated with a natural number, called the *arity* of a symbol. Let Σ be a ranked alphabet. For $n \geq 0$, let $\Sigma_n = \{a \in \Sigma \mid$ the arity of a is $n\}$. A ranked alphabet is *monadic* if the arity of each its element is at most 1.

The set of *trees* over Σ, denoted by T_Σ, is the smallest set of strings over elements of Σ, parentheses and commas defined inductively as follows:

(1) $\Sigma_0 \subseteq T_\Sigma$, and
(2) if $a \in \Sigma_n$ for some $n \geq 1$, and $t_1, t_2, \ldots, t_n \in T_\Sigma$, then $a(t_1, t_2, \ldots, t_n) \in T_\Sigma$.

Let x be a variable. $T_\Sigma(x)$ is defined as $T_{\Sigma \cup \{x\}}$ taking the rank of x to be 0. For $t, u \in T_\Sigma(x)$, $t[u]$ is defined as the result of substituting u for the occurrences of the variable x in t. A tree $t \in T_\Sigma(x)$ is *linear* if x occurs exactly once in t.

5.2   Linear Monadic Context-Free Tree Grammar

An *linear monadic context-free tree grammar* (LM-CFTG) is a four-tuple $\mathcal{G} = (N, \Sigma, P, S)$, where:

- N is a monadic ranked alphabet of *nonterminals*,
- Σ is a ranked alphabet of *terminals*, disjoint with N,
- $S \in N_0$ is the *initial nonterminal*, and
- P is a finite set of *production rules* of one of the following forms:

$$(1) \quad A \to u$$

with $A \in N_0$ and $u \in T_{N \cup \Sigma}$, or

$$(2) \quad A(x) \to u$$

with $A \in N_1$ and a linear tree $u \in T_{N \cup \Sigma}(x)$.

For an LM-CFTG $\mathcal{G}$, the *one-step derivation* $\underset{\mathcal{G}}{\Rightarrow}$ is the relation over $T_{N \cup \Sigma}(x)$ such that, for $t \in T_{N \cup \Sigma}(x)$, (1) if $A \to u$ is in P and $t = t'[A]$ for some linear tree $t' \in T_{N \cup \Sigma}(x)$, then $t \underset{\mathcal{G}}{\Rightarrow} t'[u]$, and (2) if $A(x) \to u$ is in P and $t = t'[A(t'')]$ for some linear trees $t', t'' \in T_{N \cup \Sigma}(x)$, then $t \underset{\mathcal{G}}{\Rightarrow} t'[u[t'']]$. See Fig. 8.

Let $\underset{\mathcal{G}}{\overset{*}{\Rightarrow}}$ denote the reflexive transitive closure of $\underset{\mathcal{G}}{\Rightarrow}$. The *tree language generated by* $\mathcal{G}$ is the set $L(\mathcal{G}) = \{t \in T_\Sigma \mid S \underset{\mathcal{G}}{\overset{*}{\Rightarrow}} t\}$.

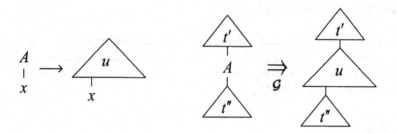

Fig. 8. One-step derivation

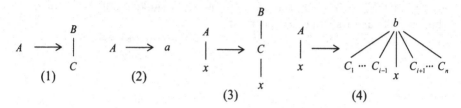

Fig. 9. Chomsky-like normal form

5.3   Known Results for LM-CFTG

First, we introduce normal forms for LM-CFTG. The reason fan-out rules and context-free rules are sufficient to define an LM-CFTG is based on Theorem 1.

Theorem 1. (Fujiyoshi [14]) [Chomsky-like normal form] Any LM-CFTG can be transformed into an equivalent one whose rules are in one of the following forms:

(1) $A \to B(C)$ with $A, C \in N_0$ and $B \in N_1$,
(2) $A \to a$ with $A \in N_0$ and $a \in \Sigma_0$,
(3) $A(x) \to B(C(x))$ with $A, B, C \in N_1$, or
(4) $A(x) \to b(C_1, \ldots, C_{i-1}, x, C_{i+1}, \ldots, C_n)$ with $A \in N_1$, $n \geq 1$, $b \in \Sigma_n$, $1 \leq i \leq n$ and $C_1, \ldots, C_{i-1}, C_{i+1}, \ldots, C_n \in N_0$.

See Fig. 9.

Theorem 2. (Fujiyoshi [14]) [Greibach-like normal form] Any LM-CFTG can be transformed into an equivalent one whose rules are in one of the following forms:

(1) $A \to a$ with $A \in N_0$ and $a \in \Sigma_0$,
(2) $A \to b(C_1, \ldots, C_{i-1}, u, C_{i+1}, \ldots, C_n)$ with $A \in N_0$, $n \geq 1$, $b \in \Sigma_n$, $1 \leq i \leq n$, $C_1, \ldots, C_{i-1}, C_{i+1}, \ldots, C_n \in N_0$ and $u \in T_N$, or
(3) $A(x) \to b(C_1, \ldots, C_{i-1}, u, C_{i+1}, \ldots, C_n)$ with $A \in N_1$, $n \geq 1$, $b \in \Sigma_n$, $1 \leq i \leq n$, $C_1, \ldots, C_{i-1}, C_{i+1}, \ldots, C_n \in N_0$, and $u \in T_{N_1}(x)$.

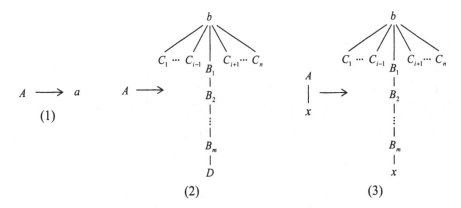

Fig. 10. Greibach-like normal form

See Fig. 10. Since N is monadic, all trees in T_N and $T_{N_1}(x)$ may be written as $B_1(B_2(\cdots(B_m(D))\cdots))$ and $B_1(B_2(\cdots(B_m(x))\cdots))$, respectively, for some $m \geq 0$, $B_1, B_2, \ldots, B_m \in N_1$ and $D \in N_0$. Note that m may be 0.

LM-CFTG is related to the tree adjoining grammar (TAG) [15,16,17], one of the most famous and well-studied mildly context-sensitive grammar formalisms. The definition of "weakly equivalent" is found in [3].

Theorem 3. (Fujiyoshi & Kasai [3]) LM-CFTG is weakly equivalent to TAG.

There exists an effective recognition algorithm for LM-CFTG.

Theorem 4. (Fujiyoshi [8]) There exists a recognition algorithm for LM-CFTG that runs in $O(n^4)$ time, where n is the number of nodes of an input tree.

It is known that a recognition algorithm that runs in $O(n^3)$ time can be obtained with some modifications to the $O(n^4)$-time algorithm in [8].

6   Conclusion and Future Work

We have proposed a verification method of mathematical formulae for a practical mathematical OCR system based on a combination of context-free grammar and tree grammar. Though we have recognized the usefulness of the proposed verification method by experimental results, we know the necessity of the improvement of the grammar defining "well-formed" mathematical formulae. Moreover, in order to avoid the ambiguity of the grammar, the inclusion of semantic analysis needs to be considered.

In the future, we plan to internalize a verification system within the recognition engine of a mathematical OCR system. Because the flexibility of the grammar is important, we want to allow users to manipulate the grammar. Therefore, we will reflect on ways users can update the grammar by themselves.

References

1. Chan, K.F., Yeung, D.Y.: Mathematical expression recognition: a survey. Int. J. Document Analysis and Recoginition 3(1), 3–15 (2000)
2. Hopcroft, J.E., Ullman, J.D.: Introduction to Automata Theory, Languages and Computation. Addison Wesley, Reading (1979)
3. Fujiyoshi, A., Kasai, T.: Spinal-formed context-free tree grammars. Theory of Computing Systems 33(1), 59–83 (2000)
4. Anderson, R.: Syntax-directed recognition of hand-printed two-dimensional mathematics. In: Interactive Systems for Experimental Applied Mathematics, pp. 436–459. Academic Press, London (1968)
5. Chou, P.A.: Recognition of equations using a two-dimensional stochastic context-free grammar. In: Proc. SPIE, vol. 1199, pp. 852–863 (1989)
6. Kanahori, T., Sexton, A., Sorge, V., Suzuki, M.: Capturing abstract matrices from paper. In: Borwein, J.M., Farmer, W.M. (eds.) MKM 2006. LNCS (LNAI), vol. 4108, pp. 124–138. Springer, Heidelberg (2006)
7. Suzuki, M., Uchida, S., Nomura, A.: A ground-truthed mathematical character and symbol image database. In: Proceedings of the 8th International Conference on Document Analysis and Recognition (ICDAR 2005), vol. 2, pp. 675–679 (2005)
8. Fujiyoshi, A.: Application of the CKY algorithm to recognition of tree structures for linear, monadic context-free tree grammars. IEICE Trans. Inf. & Syst. E90-D(2), 388–394 (2007)
9. Sikkel, K., Nijholt, A.: Parsing of Contex-Free Languages. In: Handbook of Formal Languages, vol. 2, pp. 61–100. Springer, Heidelberg (1997)
10. Suzuki, M., Tamari, F., Fukuda, R., Uchida, S., Kanahori, T.: Infty - an integrated OCR system for mathematical documents. In: Proceedings of ACM Symposium on Document Engineering 2003, pp. 95–104 (2003)
11. Infty Project, http://www.inftyproject.org/en/
12. Donnelly, C., Stallman, R.: Bison: The yacc-compatible parser generator (2006), http://www.gnu.org/software/bison/manual/
13. Mozilla Firefox, http://www.mozilla.com/firefox/
14. Fujiyoshi, A.: Analogical conception of chomsky normal form and greibach normal form for linear, monadic context-free tree grammars. IEICE Trans. Inf. & Syst. E89-D(12), 2933–2938 (2006)
15. Joshi, A.K., Levy, L.S., Takahashi, M.: Tree adjunct grammars. J. Computer & System Sciences 10(1), 136–163 (1975)
16. Joshi, A.K., Schabes, Y.: Tree-adjoining grammars. In: Handbook of Formal Languages, vol. 3, pp. 69–124. Springer, Berlin (1997)
17. Abeillé, A., Rambow, O. (eds.): Tree adjoining grammars: formalisms, linguistic analysis and processing. CSLI Publications, Stanford (2000)

Appendix: Representative Sample of the Grammar Defining "Well-Formed" Mathematical Formulae

Fan-Out Rules:

$$Int \rightarrow \int \begin{smallmatrix} Exp \\ \\ Exp \end{smallmatrix} \qquad Int \rightarrow \int \begin{smallmatrix} Exp \\ \\ Exp \end{smallmatrix} \qquad Int \rightarrow \int \begin{smallmatrix} \\ \\ Range \end{smallmatrix} \qquad Int \rightarrow \int \begin{smallmatrix} \\ \\ Range \end{smallmatrix}$$

$$Sum \rightarrow \sum \begin{smallmatrix} Exp \\ \\ Init \end{smallmatrix} \qquad Sum \rightarrow \sum \begin{smallmatrix} Exp \\ \\ Init \end{smallmatrix} \qquad Sum \rightarrow \sum \begin{smallmatrix} \\ \\ Range \end{smallmatrix} \qquad Sum \rightarrow \sum \begin{smallmatrix} \\ \\ Range \end{smallmatrix}$$

$$Prod \rightarrow \prod \begin{smallmatrix} Exp \\ \\ Init \end{smallmatrix} \qquad Prod \rightarrow \prod \begin{smallmatrix} Exp \\ \\ Init \end{smallmatrix} \qquad Prod \rightarrow \prod \begin{smallmatrix} \\ \\ Range \end{smallmatrix} \qquad Prod \rightarrow \prod \begin{smallmatrix} \\ \\ Range \end{smallmatrix}$$

$$Lim \rightarrow \lim \begin{smallmatrix} \\ Range \end{smallmatrix} \qquad Lim \rightarrow \lim \begin{smallmatrix} \\ Range \end{smallmatrix} \qquad Log \rightarrow \log \begin{smallmatrix} \\ Subscript \end{smallmatrix}$$

$$Letter \rightarrow a \begin{smallmatrix} Supscript \\ \\ Subscript \end{smallmatrix} \qquad Letter \rightarrow b \begin{smallmatrix} \\ Subscript \end{smallmatrix} \qquad Letter \rightarrow c \begin{smallmatrix} Supscript \\ \\ \end{smallmatrix}$$

$$Numeric \rightarrow 1 \begin{smallmatrix} Exp \\ \\ \end{smallmatrix} \qquad Frac \rightarrow \frac{Exp}{Exp} \qquad Sqrt \rightarrow \sqrt{} \begin{smallmatrix} \\ Exp \end{smallmatrix}$$

Context-Free Rules:

$Math$	$\rightarrow ExpList$	$ExpList$	$\rightarrow Exp$
	$\rightarrow Sign\ ExpList$		$\rightarrow ExpList\ Sign\ Exp$

Exp	$\rightarrow Term$	$Term$	$\rightarrow Factor$
	$\rightarrow UnaryOp\ Term$		$\rightarrow Term\ Factor$
			$\rightarrow Term\ BinaryOp\ Factor$
			$\rightarrow$ '∞'

$Range \rightarrow Exp$
$ \rightarrow Exp\ Sign\ Exp$
$ \rightarrow Exp\ Sign\ Factor\ Sign\ Exp$

$Init \rightarrow Exp$ '$=$' Exp

$SubScript$	$\rightarrow Exp$	$SupScript$	$\rightarrow Exp$
	$\rightarrow Sign$		$\rightarrow Sign$
	$\rightarrow SubScript$ ',' Exp		$\rightarrow$ '$\prime$' //prime
	$\rightarrow SubScript$ ',' $Sign$		$\rightarrow$ '$\prime\prime$' //doubleprime
	$\rightarrow Exp$ '$=$' Exp		$\rightarrow$ '$\prime\prime\prime$' //tripleprime

$Sign$	$\rightarrow$ '$=$'	$UnaryOp$	$\rightarrow$ '$+$'
	$\rightarrow$ '$\neq$'		$\rightarrow$ '$-$'
	$\rightarrow$ '$<$'		$\rightarrow$ '$\pm$'
	$\rightarrow$ '$\leq$'		$\rightarrow$ '$\mp$'
	$\rightarrow$ '$>$'		$\rightarrow$ '$\forall$'
	$\rightarrow$ '$\geq$'		$\rightarrow$ '$\exists$'
	$\rightarrow$ '$\in$'		
	$\rightarrow$ '$\ni$'		
	$\rightarrow$ '$\not\ni$'	$BinaryOp$	$\rightarrow$ '$+$'
	$\rightarrow$ '$\subset$'		$\rightarrow$ '$-$'
	$\rightarrow$ '$\not\subset$'		$\rightarrow$ '$\times$'
	$\rightarrow$ '$\supset$'		$\rightarrow$ '$/$'
	$\rightarrow$ '$\not\supset$'		$\rightarrow$ '$\cap$'
	$\rightarrow$ '$\equiv$'		$\rightarrow$ '$\cup$'
	$\rightarrow$ '$\cong$'		$\rightarrow$ '$\cdot$'
	$\rightarrow$ '$\sim$'		$\rightarrow$ '$\bullet$'
	$\rightarrow$ '$\rightarrow$'		$\rightarrow$ '$:$' //colon
	$\rightarrow$ '$\Rightarrow$'		
	$\rightarrow$ '$\leftrightarrow$'		
	$\rightarrow$ '$\Leftrightarrow$'		
	$\rightarrow$ '$\mapsto$'		
	$\rightarrow$ ',' //comma		
	$\rightarrow$ ';' //semicolon		
	$\rightarrow$ '$\vert$' //vert		

Factor	$\rightarrow$ *Variable*		*Variable*	$\rightarrow$ *Letter*
	$\rightarrow$ *Number*			
	$\rightarrow \emptyset$		*Letter*	$\rightarrow$ 'a'
	$\rightarrow *$			$\rightarrow$ 'b'
	$\rightarrow \triangle$			$\vdots$
	$\rightarrow \nabla$			$\rightarrow$ 'z'
	$\rightarrow \aleph$			
	$\rightarrow$ '(' ExpList ')'		*Number*	$\rightarrow$ *Integer*
	$\rightarrow$ '[' ExpList ']'			$\rightarrow$ *Integer* '.' Integer
	$\rightarrow$ '{' ExpList '}'			$\rightarrow$ '.' Integer
	$\rightarrow$ '(' ExpList ']'			
	$\rightarrow$ '[' ExpList ')'		*Integer*	$\rightarrow$ *Numeric*
	$\rightarrow$ '⟨' ExpList ')'			$\rightarrow$ *Integer Numeric*
	$\rightarrow$ *Frac*			
	$\rightarrow$ *Sqrt*		*Numeric*	$\rightarrow$ '0'
	$\rightarrow$ '|' Term '|'			$\rightarrow$ '1'
	$\rightarrow$ *TrigOp Factor*			$\vdots$
	$\rightarrow$ *SumOp Factor*			$\rightarrow$ '9'
	$\rightarrow$ *FuncOp Factor*			

TrigOp	$\rightarrow$ *Sin*		*Sin*	$\rightarrow$ 'sin'
	$\rightarrow$ *Cos*		*Cos*	$\rightarrow$ 'cos'
	$\rightarrow$ *Tan*		*Tan*	$\rightarrow$ 'tan'

SumOp	$\rightarrow$ *Int*		*Int*	$\rightarrow$ '$\int$'
	$\rightarrow$ *Sum*		*Sum*	$\rightarrow$ '$\sum$'
	$\rightarrow$ *Prod*		*Prod*	$\rightarrow$ '$\prod$'
	$\rightarrow$ *Bigcap*		*Bigcap*	$\rightarrow$ '$\bigcap$'
	$\rightarrow$ *Bigcup*		*Bigcup*	$\rightarrow$ '$\bigcup$'

FuncOp	$\rightarrow$ *Lim*		*Lim*	$\rightarrow$ 'lim'
	$\rightarrow$ *Log*		*Log*	$\rightarrow$ 'log'
	$\rightarrow$ *Min*		*Min*	$\rightarrow$ 'min'
	$\rightarrow$ *Max*		*Max*	$\rightarrow$ 'max'

Specifying Strategies for Exercises

Bastiaan Heeren[1], Johan Jeuring[1,2], Arthur van Leeuwen[2], and Alex Gerdes[1]

[1] School of Computer Science, Open Universiteit Nederland
P.O. Box 2960, 6401 DL Heerlen, The Netherlands
{bhr,jje,ale,age}@ou.nl
[2] Department of Information and Computing Sciences, Universiteit Utrecht

Abstract. The feedback given by e-learning tools that support incrementally solving problems in mathematics, logic, physics, etc. is limited, or laborious to specify. In this paper we introduce a language for specifying strategies for solving exercises. This language makes it easier to automatically calculate feedback when users make erroneous steps in a calculation. Although we need the power of a full programming language to specify strategies, we carefully distinguish between context-free and non-context-free sublanguages of our strategy language. This separation is the key to automatically calculating all kinds of desirable feedback.

1 Introduction

Tools like Aplusix [9], ActiveMath [14], MathPert [4], and our own tool for rewriting logic expressions [24] support solving mathematical exercises incrementally. Ideally a tool gives detailed feedback on several levels. For example, when a student rewrites $p \to (r \leftrightarrow p)$ into $\neg p \vee (r \leftrightarrow p$, our tool will tell the student that there is a missing parenthesis. If the same expression is rewritten into $\neg p \wedge (r \leftrightarrow p)$, it will tell the student that an error has been made when applying the definition of implication: correct application of this definition would give $\neg p \vee (r \leftrightarrow p)$. Finally, if the student rewrites $\neg(p \wedge (q \vee r))$ into $\neg((p \wedge q) \vee (p \wedge r))$, it will tell the student that although this step is not wrong, it is better to first eliminate occurrences of $\neg$ occurring at top-level, since this generally leads to fewer rewrite steps.

The first kind of error is a syntax error, and there exist good error-repairing parsers that suggest corrections to formulas with syntax errors. The second kind of error is a rewriting error: the student rewrites an expression using a non-existing or buggy rule. There already exist some interesting techniques for finding the most likely error when a student incorrectly rewrites an expression. The third kind of error is an error on the level of the procedural skill or strategy for solving this kind of exercises. This paper discusses how we can formulate and use strategies to construct the latter kind of feedback.

This paper. The main contribution of this paper is the formulation of a strategy language as a domain-specific embedded language, with a clear separation between a context-free and a non-context-free part. The strategy language can

S. Autexier et al. (Eds.): AISC/Calculemus/MKM 2008, LNAI 5144, pp. 430–445, 2008.

be used for any domain, and can be used to automatically calculate feedback on the level of strategies, given an exercise, the strategy for solving the exercise, and student input. Another contribution of our work is that the specification of a strategy and the calculation of feedback is separated: we can use the same strategy specification to calculate different kinds of feedback.

This paper is organized as follows. Section 2 introduces strategies, and discusses how they can help to improve feedback in e-learning systems or intelligent tutoring systems. We continue with some example strategies from the domain of logical expressions (Section 3). Then, we present our language for writing strategies in Section 4. We do so by defining a number of strategy combinators, and by showing how the various example strategies can be specified in our language. Section 5 discusses several possibilities for giving feedback or hints using our strategy language. The last section concludes and gives directions for future research.

2   Strategies and Feedback

Strategies. Whatever aspect of intelligence you attempt to model in a computer program, the same needs arise over and over again [8]:

- The need to have knowledge about the domain.
- The need to reason with that knowledge.
- The need for knowledge about how to direct or guide that reasoning.

In the case of exercises, the latter kind of knowledge is often captured by a so-called procedure or procedural skill. A procedure describes how basic steps may be combined to solve a particular problem. A procedure is often called a *strategy* (or meta-level reasoning, meta-level inference [8], procedural nets [6], plans, tactics, etc.), and we will use this term in the rest of this paper.

Many subjects require a student to learn strategies. At elementary school, students have to learn how to calculate a value of an expression, which may include fractions. At high school, students learn how to solve a system of linear equations, and at university, students learn how to apply Gaussian elimination to a matrix, or how to rewrite a logical expression to disjunctive normal form (DNF). Strategies are not only important for mathematics, logic, and computer science, but also for physics, biology (Mendel's laws), and many other subjects. Strategies are taught at any level, in almost any subject, and range from simple – for example the simplification of arithmetic expressions – to very complex – for example a complicated linear algebra procedure.

E-learning systems for learning strategies. Strategic skills are almost always acquired by practicing exercises, and indeed, students usually equate mathematics with solving exercises. In schools, the dominant practice still is a student performing a calculation using pen-and-paper, and the teacher correcting the calculation (the same day, in a couple of days, after a couple of weeks). There exist many software solutions that support practicing solving exercises on a computer. The simplest kinds of tools offer multiple-choice questions, possibly with an explanation of the error if a wrong choice is submitted. A second class of tools asks for an answer to a question, again, possibly with an analysis of the answer to give feedback when an error has been made. The class of tools we consider in the paper are tools that support the incremental, step-wise calculation of a solution to an exercise, thus mimicking the pen-and-paper approach more or less faithfully. Since e-learning tools for practicing procedural skills seem to offer many advantages, hundreds of tools that support practicing strategies in mathematics, logic, physics, etc. have been developed.

Should e-learning systems give feedback? In Rules of the Mind [1], Anderson discusses the ACT-R principles of tutoring, and the effectiveness of feedback in intelligent tutoring systems. One of the tutoring principles deals with student errors. If a student made a slip in performing a step (s)he should be allowed to correct it without further assistance. However, if a student needs to learn the correct rule, the system should give a series of hints with increasing detail, or show how to apply the correct rule. Finally, it should also be possible to give an explanation of an error made by the student. The question on whether or not to give immediate feedback is still debated. Anderson observed no positive effects in learning with deferred feedback, but observed a decline in learning rate instead. Erev et al. [13] also claim that immediate feedback is often to be preferred.

Feedback in e-learning systems supporting incrementally solving exercises. There are only very few tools that mimic the incremental pen-and-paper approach and that give feedback at intermediate steps different from correct/incorrect. Although the correct/incorrect feedback at intermediate steps is valuable, it is unfortunate that the full possibilities of e-learning tools are not used. There are several reasons why the feedback that is given is limited. The main reasons probably are that supporting detailed feedback for each exercise is very laborious, providing a comprehensive set of possible bugs for a particular domain requires a lot of research (see for example Hennecke's work [16] on student bugs in calculating fractions), and automatically calculating feedback for a given exercise, strategy, and student input is very difficult.

Feedback should be calculated automatically. We think specifying feedback together with every exercise that is solved incrementally is a dead-end: teachers will want to enter new exercises on a regular basis, and completely specifying feedback is just too laborious, error prone, and repetitive. Instead, feedback should in general be calculated automatically, given the exercise, the strategy for the exercise, buggy rules and strategies, and the input from the student. To automatically calculate feedback, we need information about the domain of the exercise, the rules for manipulating expressions in this domain, the strategy for solving the exercise, and common bugs. For example, for Gaussian elimination of a matrix, we have to know about matrices (which can be represented by a list of rows), the rules for manipulating matrices (the elementary matrix operations such as scaling a row, subtracting a row from another row, and swapping two rows), buggy rules and strategies for manipulating matrices (subtracting a row from itself), and the strategy for Gaussian elimination of a matrix, which can be found in the technical report corresponding to this paper [15].

Representing strategies. Representing the domain and the rules for manipulating an expression in the domain is often relatively straightforward. Specifying a strategy for an exercise is more challenging in many cases. To specify a strategy, we need the power of a full programming language: many strategies require computations of values. However, to calculate feedback based on a strategy, we need to know more than that it is a program. We need to know its structure and basic components, which we can use to report back on errors. Furthermore, we claim that if we ask a teacher to write a strategy as a program, instead of specifying feedback with every exercise, the automatic approach is not going to be very successful.

An embedded domain-specific language for specifying strategies. This paper discusses the design of a language for specifying strategies for exercises. The domains and rules vary for the different subjects, but the basic constructs for describing strategies are the same for different subjects ('first do this, then do that', 'either do this or that'). So the strategy language can be used for any domain (mathematics, logic, physics, etc). It consists of several basic constructs from which strategies can be built. These basic constructs are combined with program code in a programming language to be able to specify any strategy. The strategy language is formulated as an embedded domain-specific language (EDSL) in a programming language [19] to easily facilitate the combination of program code with a strategy. Here 'domain-specific' means specific for the domain of strategies, not specific for the domain of exercises. The separation into basic strategy constructs and program code offers us the possibility to analyse the basic constructs, from which we can derive several kinds of feedback.

What kind of feedback? We can automatically calculate the following kinds of feedback, many of which are part of the tutoring principles of Anderson [1].

- Is the student still on the right path towards a solution? Does the step made by the student follow the strategy for the exercise? What is the next step the student should take?
- We produce hints based on the strategy.
- Based on the position on the path from the starting point to the solution of an exercise we create a progress bar.
- If a student enters a wrong final answer, we ask the student to solve subproblems of the original problem.
- We allow the formulation of buggy strategies to explain common mistakes to students.

We do not build a model of the student to try to explain the error made by the student. According to Anderson, an informative error message is better than bug diagnosis.

How do we calculate feedback on strategies? The strategy language is defined as an embedded domain-specific language in Haskell [27]. Using the basic constructs from the strategy language, we can create something that looks like a context-free grammar. The sentences of this grammar are sequences of transformation steps (applications of rules). We can thus check whether or not a student follows a strategy by parsing the sequence of transformation steps, and checking that the sequence of transformation steps is a prefix of a correct sentence from the context-free grammar. Many steps require student input, for example when a student wants to multiply a row by a scalar, or when a student wants to subtract two rows. This part of the transformation cannot be checked by means of a context-free grammar, and here we make use of the fact that our language is embedded into a full programming language, to check input values supplied by the student. The separation of the strategy into a context-free part, using the basic strategy combinators, and a non-context-free part, using the power of the programming language, offers us the possibility to give the kinds of feedback mentioned above. Computer Science has almost 50 years of experience in parsing sentences of context-free languages, including error-repairing parsers, which we can use to improve feedback on the level of strategies.

Related work. Explaining syntax errors has been studied in several contexts, most notably in compiler construction [30], but also for e-learning tools [18]. Some work has been done on trying to explain errors made by students on the level of rewrite rules [16,21,25,5].

Already around 1980, but also later, VanLehn et al. [6,7,31], and Anderson and others from the Advanced Computer Tutoring research group at CMU [1,2] worked on representing procedures or procedural networks. VanLehn et al. already noticed that 'The representation of procedures has an impact on all parts of the theory.' Anderson et al. report that the technical accomplishment was 'no mean feat'. Both VanLehn et al. and Anderson et al. chose to deploy collections of condition-action rules, or production systems. In Mind bugs [31], VanLehn states several assumptions about languages for representing procedures. In Rules of the Mind [1], Anderson formulates similar assumptions. Their leading argument for selecting a language for representing procedures is that it should be psychologically plausible. We think our strategy language can be viewed as a production system. But our leading argument is that it should be easy to calculate feedback based on the strategy. Using an EDSL for specifying the context-free part of a strategy simplifies calculating feedback. Furthermore, our language satisfies the assumptions about representation languages given by VanLehn, such as the presence of variables in procedures, and the possibility to define recursive procedures. Neither VanLehn nor Anderson use parsing for the language for procedures to automatically calculate feedback.

Zinn [33] writes strategies as Prolog programs, in which rules and strategies ('task models') are intertwined.

3   Three Example Strategies

In this section we present three strategies for rewriting a classical logical expression to disjunctive normal form. Although the example strategies are relatively simple, they are sufficiently rich to demonstrate the main components of our strategy language.

The domain. Before we can define a strategy, we first have to introduce the domain of logical expressions and a collection of available rules. A logical expression is a logical variable, a constant *true* or *false* (written T and F), the negation of a logical expression, or the conjunction, disjunction, implication, or equivalence of two logical expressions. This results in the following grammar:

$$
\begin{array}{lll}
Logic & ::= & Var \mid T \mid F \mid \neg Logic \mid Logic \wedge Logic \\
 & \mid & Logic \vee Logic \mid Logic \rightarrow Logic \mid Logic \leftrightarrow Logic \\
Var & ::= & p \mid q \mid r \mid \dots
\end{array}
$$

If necessary, we write parentheses to resolve ambiguities. Examples of valid expressions are $\neg(p \vee (q \wedge r))$ and $\neg(\neg p \leftrightarrow p)$.

The rules. Logical expressions form a boolean algebra, and hence a number of rules for logical expressions can be formulated. Figure 1 presents a small collection of basic rules and some tautologies and contradictions. All variables in these rules are meta-variables and range over arbitrary logical expressions. The rules are expressed as equivalences, but are only applied from left to right. For most rules we assume to have a commutative variant, for instance, $T \wedge p = p$ for rule ANDTRUE. With these implicit rules, we can bring every logical expression to disjunctive normal form.

Every serious exercise assistant for this domain has to be aware of a much richer set of rules. In particular, we have not given rules for commutativity and associativity, several plausible rules for implications and equivalences are omitted, and the list of tautologies and contradictions is far from complete.

Strategy 1: apply rules exhaustively. The first strategy applies the basic rules from Figure 1 exhaustively: we proceed as long as we can apply *some* rule *somewhere*, and we will end up with a logical expression in disjunctive normal form. This is a special property of the chosen collection of basic rules, and this is not the case for a rule set in general. The strategy is very liberal, and approves every sequence of rules.

Strategy 2: four steps. Strategy 1 accepts sequences that are not very attractive, and that no expert would ever consider. We give two examples:

$$
\neg\neg(p \vee q) \stackrel{\text{DeMorganOr}}{\Longrightarrow} \neg(\neg p \wedge \neg q) \qquad T \vee (\neg\neg p) \stackrel{\text{NotNot}}{\Longrightarrow} T \vee p
$$

Basic Rules:

Constants:	ANDTRUE:	$p \wedge T = p$		ANDFALSE:	$p \wedge F = F$
	ORTRUE:	$p \vee T = T$		ORFALSE:	$p \vee F = p$
	NOTTRUE:	$\neg T = F$		NOTFALSE:	$\neg F = T$

Definitions: IMPLDEF: $p \rightarrow q = \neg p \vee q$
EQUIVDEF: $p \leftrightarrow q = (p \wedge q) \vee (\neg p \wedge \neg q)$

Negations: NOTNOT: $\neg\neg p = p$
DEMORGANAND: $\neg(p \wedge q) = \neg p \vee \neg q$
DEMORGANOR: $\neg(p \vee q) = \neg p \wedge \neg q$

Distribution: ANDOVEROR: $p \wedge (q \vee r) = (p \wedge q) \vee (p \wedge r)$

Additional Rules:

Tautologies: IMPLTAUT: $p \rightarrow p = T$ ORTAUT: $p \vee \neg p = T$
EQUIVTAUT: $p \leftrightarrow p = T$

Contradictions: ANDCONTR: $p \wedge \neg p = F$ EQUIVCONTR: $p \leftrightarrow \neg p = F$

Fig. 1. Rules for logical expressions

In both cases, it is more appealing to select a different rule (NOTNOT and ORTRUE, respectively). We define a new strategy that proceeds in four steps, and such that the above sequences are not permitted.

- **Step 1:** Remove constants from the logical expression with the rules for "constants" (see Figure 1), supplemented with constant rules for implications and equivalences. Apply the rules *top-down*, that is, at the highest possible position in the abstract syntax tree. After this step, all occurrences of T and F are removed.
- **Step 2:** Use IMPLDEF and EQUIVDEF to rewrite implications and equivalences in the formula. Proceed in a *bottom-up* order.
- **Step 3:** Push negations inside the expression using the rules for negations, and do so in a *top-down* fashion. After this step, all negations appear directly in front of a logical variable.
- **Step 4:** Use the distribution rule (ANDOVEROR) to move disjunctions to top-level. The order is irrelevant.

Strategy 3: tautologies and contradictions. Suppose that we want to extend Strategy 2, and use rules expressing tautologies and contradictions (for example, the additional rules in Figure 1). These rules introduce constants. Our last strategy is as follows:

- Follow the four steps of Strategy 2, however:
- *Whenever* possible, use the rules for tautologies and contradictions (*top-down*), *and*
- clean up the constants afterwards (step 1). Then continue with Strategy 2.

Buggy rules. In addition to the collection of rules and a strategy, we can formulate *buggy rules*. These rules capture mistakes that are often made, such as the following unsound variations on the two De Morgan rules:

BUGGYDM1 : $\neg(p \wedge q) \neq \neg p \wedge \neg q$ BUGGYDM2 : $\neg(p \vee q) \neq \neg p \vee \neg q$

The advantage of formulating buggy rules is that specialized feedback can be presented if the system detects that such a rule was applied. Note that these rules should not appear in strategies, since that would invalidate the strategy.

The idea of formulating buggy rules can easily be extended to buggy strategies. Such a strategy helps to recognize common procedural mistakes, in which case we can report a detailed message.

4   A Language for Strategies for Exercises

The previous section gives an intuition of strategies for exercises, such as the three DNF strategies. In this section we define a language for specifying such strategies. We explore a number of combinators to combine simple strategies into more complex ones. We start with a set of basic combinators, and gradually move on to more powerful combinators.

4.1   Basic Strategy Combinators

Strategies are built on top of basic rules, such as the logic rules from the previous section. Let r be a rule, and let a be some term. We write $r(a)$ to denote the application of r to a, which returns a set of terms. If this set is empty, we say that r is not applicable to a, and that the rule fails.

The basic combinators for building strategies are the same as the building blocks for context-free grammars. In fact, we can view a strategy as a grammar where the rules form the alphabet of the language.

- **Sequence.**   Two strategies can be composed and put in sequence. We write $s <\!\ast\!> t$ to denote the sequence of strategy s followed by strategy t.
- **Choice.**   We can choose between two strategies, for which we will write $s <\!|\!> t$. One of its argument strategies is applied.
- **Units.**   Two special strategies are introduced: *succeed* is the strategy that always succeeds, without doing anything, and *fail* is the strategy that always fails. These combinators are useful to have: *succeed* and *fail* are the unit elements of the $<\!\ast\!>$ and $<\!|\!>$ combinators.
- **Labels.**   With our final combinator we can label strategies. We write *label* ℓ s to label strategy s with some label ℓ. Labels are used to mark positions in a strategy, and allow us to attach content such as hints and messages to the strategy. Labeling does not change the language that is generated by the strategy.

We can apply a strategy s to a term a, written $s(a)$, just as we can apply some rule. We make the informal description of the presented combinators precise by giving a formal definition for each of the combinators:

$$(s <\!\!*\!\!> t)(a) = \{c \mid b \leftarrow s(a), c \leftarrow t(b)\}$$
$$(s <\!|\!> t)(a) = s(a) \cup t(a)$$
$$succeed(a) \quad = \{a\}$$
$$fail(a) \quad\quad = \varnothing$$
$$(label \; \ell \; s)(a) = s(a)$$

The rest of this section introduces more strategy combinators to conveniently specify strategies. All these combinators, however, can be defined in terms of the combinators that are given above.

4.2   Extensions

Extended Backus-Naur form (EBNF) extends the notation for grammars, and offers three new constructions that one often encounters in practice: zero or one occurrence (option), zero or more occurrences (closure), and one or more occurrences (positive closure). Along these lines, we introduce three new strategy combinators: *many s* means repeating strategy *s* zero or more times, with *many1* we have to apply *s* at least once, and *option s* may or may not apply strategy *s*. We define these combinators using the basic combinators:

$$many \;\; s = (s <\!\!*\!\!> many \; s) <\!|\!> succeed$$
$$many1 \; s = s <\!\!*\!\!> many \; s$$
$$option \; s = s <\!|\!> succeed$$

Observe the recursion in the definition of *many*. Depending on the implementation one prefers, the *many* combinator results in an infinite strategy (which is not at all a problem in a lazy programming language such as Haskell), or this combinator gets a special treatment and is implemented as a primitive. It is quite common for an EDSL to introduce a rich set of combinators on top of a (small) set of basic combinators.

4.3   Negation and Greedy Combinators

The next combinators we consider allow us to specify that a certain strategy is not applicable. Have a look at the definition of *not*, which only succeeds if the argument strategy *s* is not applicable to the current term *a*:

$$(not \; s)(a) = \textbf{if } s(a) = \varnothing \textbf{ then } \{a\} \textbf{ else } \varnothing$$

Observe that the *not* combinator can be specified as a single rule that either returns a singleton set or the empty set depending on the applicability of strategy *s*. A more general variant of this combinator is *check*, which receives a predicate as argument (instead of a strategy) for deciding what to return.

Having defined *not*, we now specify greedy variations of *many*, *many1*, and *option* (*repeat*, *repeat1*, and *try*, respectively). These combinators are greedy as they will apply their argument strategies whenever possible.

$$repeat \ \ s = many \ s \ <\!\!*\!\!> \ not \ s$$
$$repeat1 \ s = many1 \ s \ <\!\!*\!\!> \ not \ s$$
$$try \quad \ \ s = s <\!|\!> not \ s$$
$$s \triangleright t \quad \ \ = s <\!|\!> (not \ s <\!\!*\!\!> t)$$

The last combinator defined, $s \triangleright t$, is a left-biased choice: t is only considered when s is not applicable.

4.4   Traversal Combinators

In many domains, terms are constructed from smaller subterms. For instance, a logical expression may have several subexpressions. Because we do not only want to apply rules and strategies to the top-level term, we need some additional combinators to indicate that the strategy or rule at hand should be applied *somewhere*. For this kind of functionality, we need some support from the underlying domain. Let us assume that a function *once* has been defined on a certain domain, which applies a given strategy to exactly one of the term's immediate children. For the logic domain, this function would contain the following definitions:

$$once \ s \ (p \wedge q) = \{p' \wedge q \mid p' \leftarrow s(p)\} \cup \{p \wedge q' \mid q' \leftarrow s(q)\}$$
$$once \ s \ (\neg p) \quad = \{\neg p' \quad \mid p' \leftarrow s(p)\}$$
$$once \ s \ T \quad \quad = \emptyset$$

Using generic programming techniques [17], we can define this function once and for all, and use it for every domain.

With the *once* function, we can define some powerful traversal combinators. The strategy *somewhere* s applies s to one subterm (including the whole term itself).

$$somewhere \ s = s <\!|\!> once \ (somewhere \ s)$$

If we want to be more specific about where to apply a strategy, we can instead use *bottomUp* or *topDown*:

$$bottomUp \ s = once \ (bottomUp \ s) \triangleright s$$
$$topDown \ \ s = s \triangleright once \ (topDown \ s)$$

These combinators search for a suitable location to apply a certain strategy in a bottom-up or top-down fashion, without imposing an order in which the children are visited.

4.5   DNF Strategies Revisited

In Section 3, we presented three alternative strategies for turning a logical expression into disjunctive normal form. Having defined a set of strategy combinators, we can now give a precise definition of these strategies in terms of our combinators. We start with grouping the rules, as suggested by Figure 1:

$$basicRules = constants <|> definitions <|> negations <|> distribution$$
$$constants = \textsc{AndTrue} <|> \textsc{AndFalse} <|> \textsc{OrTrue} <|> \textsc{OrFalse}$$
$$<|> \textsc{NotTrue} <|> \textsc{NotFalse}$$

Definitions for the other groups are similar. The first two strategies can now conveniently be written as: The labels in the second strategy are not mandatory,

$$dnfStrategy1 = repeat\ (somewhere\ basicRules)$$
$$dnfStrategy2 = label\ \texttt{"step 1"}\ (repeat\ (topDown\ constants))$$
$$<*> label\ \texttt{"step 2"}\ (repeat\ (bottomUp\ definitions))$$
$$<*> label\ \texttt{"step 3"}\ (repeat\ (topDown\ negations))$$
$$<*> label\ \texttt{"step 4"}\ (repeat\ (somewhere\ distribution))$$

but they emphasize the structure of the strategy, and help to attach feedback to this strategy later on. The third strategy can be defined with the combinators introduced thus far, but we postpone the discussion and give a more elegant definition after the reflections.

4.6   Reflections

Is the set of strategy combinators complete? Not really, although we hope to have convinced the reader how easily the language can be extended with more combinators. In fact, this is probably the greatest advantage of using an EDSL instead of defining a new, stand-alone language. We believe that our combinators are sufficient for specifying the kind of strategies that are needed in interactive exercise assistants that aim at providing intelligent feedback. Our language is very similar to strategic programming languages such as Stratego [32,23], and very similar languages are used in parser combinator libraries [20,30], boiler-plate libraries [22], workflow applications [28], theorem proving (tacticals [26]) and data-conversion libraries [12], which suggests that our library could serve as a firm basis for strategy specifications.

One strategy combinator that we have not yet tackled is $s <\!\!|\!\!> t$, which applies the strategies s and t in parallel, i.e., interleaving steps from s with steps from t. Although this combinator would not allow us to define more strategies, it does help to specify certain strategies more concisely.

With parallel strategy combinators, we can give a concise definition for the third DNF strategy, in which we reuse our second strategy. We assume to have the left-biased variant of the parallel combinator at our disposal, for which we will write $s\ |\!\!> t$. Similar to the left-biased choice operator ($\triangleright$), this strategy applies a rule from s if possible. We first define a new and reusable combinator, followed by a definition for the strategy:

In the above definition, *step1* is equal to *repeat (topDown constants)*.

Our strategy language can be used to model strategies in multiple mathematical domains. In the technical report corresponding to this paper [15] we present a complete strategy that implements the Gaussian elimination algorithm in the

$whenever\ s\ t\ =\ repeat\ s\ |\!\!\!\triangleright\ t$

$dnfStrategy3 = whenever\ ((tautologies <\!|\!> contradictions) <\!\!*\!\!> step1)$
$\qquad\qquad\qquad dnfStrategy2$

linear algebra domain. This example is more involved than the strategies for DNF, and shows, amongst others, how rules can be parameterized, and how to maintain additional information in a context while running a strategy. The report also discusses a strategy for simplifying fractions in the domain of arithmetic expressions.

Producing a strategy is like programming, and might require quite some effort. We think that the effort is related to the complexity of the strategy. Gaussian elimination is an involved strategy, which probably should be written by an expert, but the basic strategy for DNF, *dnfStrategy1*, is rather simple, and could have been written by a teacher of a course in logic. Furthermore, due to compositionality of the strategy combinators, strategies are reusable. In the linear algebra domain, for example, many strategies we have written consist of Gaussian elimination, preceded and followed by some exercise-specific steps (e.g., to find the inverse of a matrix).

We can specify a strategy that is very strict in the order in which steps have to be applied (*dnfStrategy2* enforces a very strict order), or very flexible (*dnfStrategy1* doesn't care about which step is applied when). Furthermore, we can enforce a strategy strictly, or allow a student to deviate from a strategy, as long as the submitted expression is still equivalent to the previous expression, and the strategy can be restarted at that point (this is possible for most of the strategies we have encountered). If we want a student to take clever short-cuts through a strategy, then these shortcuts should either be explicitly specified in the strategy (which is always possible), or it should be possible to deviate from the given strategy.

5   Feedback on Strategies

This section briefly sketches how we use the strategy language, as introduced in the previous sections, to give feedback to users of our e-learning systems, or to users of other e-learning systems that make use of our feedback services. We have implemented several kinds of feedback. Most of these categories of feedback appear in the tutoring principles of Anderson [1], or in existing tools supporting the stepwise construction of a solution to an exercise. No existing tool implements more than just a few of these categories of feedback.

We do not try to tackle the problem of how feedback should be presented to a student in this paper. Here we look at the first step needed to provide feedback, namely to diagnose the problem, and relate the problem to the rules and the strategy for the exercise. We want users of our feedback services to determine how these findings are presented to the user. For example, we could generate a table from the strategy with the possible problems, and let teachers fill this table with the desired feedback messages.

Feedback after a step. After each step a student performs, we check whether or not this step is valid according to the strategy. Checking whether or not a step is valid amounts to checking whether or not the sequence of steps supplied by the student is a valid prefix of a sentence of the language specified by the context-free grammar corresponding to the strategy. Hence, this is essentially a parsing problem. As soon as we detect that a student no longer follows the strategy, we have several opportunities to react on this. We can force the student to undo the last step, and let the student strictly follow the strategy. Alternatively, we can warn the student that she has made a step that is invalid according to the strategy, but let the student proceed on her own path.

For steps involving argument- and variable-value computations we have to resort to generators, which calculate the correct values of these components, and check these values against the values supplied by the student. These generators are easily and naturally expressed in our framework.

Progress. Given an exercise and a strategy for solving the exercise, we determine the minimum number of steps necessary to solve the exercise, and show this information in a progress bar. Each time a student performs a correct step, the progress bar is updated.

Strategy unfolding. We have constructed an OpenMath binding with the Math-Dox system [11], in which a student enters a final answer to a question. If the answer is incorrect, we return a new exercise, which is part of the initial exercise. For example, if a student does not return a correct disjunctive normal form of a logical expression, we ask the student to solve the simpler problem of first eliminating all occurrences of *true* and *false* in the logical expression. After completing this part of the exercise, we ask to solve the remaining part of the exercise. This kind of feedback is also used by Cohen et al. [10] in an exercise assistant for calculus.

Hint. A student can ask for a hint. Given an exercise and a strategy for solving the exercise, we calculate the 'best' next step. The best next step is an element of the first set of the context-free grammar specified by the strategy. For the part of the strategy that is not context-free, we specify generators for generating the necessary variables and arguments. For example, when doing Gaussian elimination, our strategy specifies which rows have to be added or swapped when asked for a hint. An alternative possibility for hints is to also use strategy unfolding. Instead of giving the 'best' next step when asked for a hint, we tell the student which sub-problem should be solved first.

Completion problems. Sweller, based on his cognitive load theory, describes a series of effects and guidelines to create learning materials. The basic idea is that a student profits from having example solutions played for him or her, followed by an exercise in which the student fills out some missing steps in a solution [29]. We can use the strategy for a problem to play a solution for a student, and we can play all but the middle two (three, last two, etc.) steps, and ask the student to complete the exercise.

Buggy strategies. If a step supplied by a student is invalid with respect to the strategy specified, but can be explained by a buggy strategy for the problem, we give the error message belonging to the buggy strategy. Again, this amounts to parsing, not just with respect to the specified strategy, but also with respect to known buggy strategies.

6   Conclusions

We have introduced a strategy language with which we can specify strategies for exercises in many domains. A strategy is defined as a context-free grammar, extended with non-context-free constructs for, for example, manipulating variables and arguments. The formulation of a strategy as a context-free grammar allows us to automatically calculate several kinds of feedback to students incrementally solving exercises. Languages for specifying procedures or strategies for exercises have been developed before. Our language has the same expressive power and structure; our main contribution is the advanced feedback we can calculate automatically, and relatively easily. This is achieved by separating the strategy language into a context-free language, the strategy combinators, and a non-context-free language, the embedding as a domain-specific language.

We have several plans for the future. We hope to create bindings of our feedback service with more existing tools, such as ActiveMath [14]. For this purpose, we need to standardize the protocol for providing feedback. Also, we want to apply our ideas to domains with less structure, such as computer programming [3], software modelling, and maybe even serious games in which students have to cooperate to achieve a certain goal. Our tools are going to be used in several courses during 2008. A preliminary test with 9 students showed that they appreciated the strategic feedback within the domain of linear algebra. We will collect more data from the experiments, and analyze and report on the results.

Acknowledgements. This work was made possible by the support of the SURF Foundation, the higher education and research partnership organisation for Information and Communications Technology (ICT). For more information about SURF, please visit http://www.surf.nl. We thank the anonymous reviewers for their constructive comments. Discussions with Hans Cuypers, Josje Lodder, Wouter Pasman, Rick van der Meiden, and Erik Jansen are gratefully acknowledged.

References

1. Anderson, J.R.: Rules of the Mind. Lawrence Erlbaum Associates (1993)
2. Anderson, J.R., Corbett, A.T., Koedinger, K.R., Pelletier, R.: Cognitive tutors: lessons learned. The Journal of the Learning Sciences 4(2), 167–207 (1995)
3. Anderson, J.R., Skwarecki, E.: The automated tutoring of introductory computer programming. Communications of the ACM 29(9), 842–849 (1986)

4. Beeson, M.J.: Design principles of Mathpert: Software to support education in algebra and calculus. In: Kajler, N. (ed.) Computer-Human Interaction in Symbolic Computation, pp. 89–115. Springer, Heidelberg (1998)
5. Bouwers, E.: Improving automated feedback – a generic rule-feedback generator. Master's thesis, Utrecht University, department of Information and Computing Sciences (2007)
6. Brown, J.S., Burton, R.R.: Diagnostic models for procedural bugs in basic mathematical skills. Cognitive Science 2, 155–192 (1978)
7. Brown, J.S., VanLehn, K.: Repair theory: A generative theory of bugs in procedural skills. Cognitive Science 4, 379–426 (1980)
8. Bundy, A.: The Computer Modelling of Mathematical Reasoning. Academic Press, London (1983)
9. Chaachoua, H., et al.: Aplusix, a learning environment for algebra, actual use and benefits. In: ICME 10: 10th International Congress on Mathematical Education (2004) (retrieved May 2008), http://www.itd.cnr.it/telma/papers.php
10. Cohen, A., Cuypers, H., Jibetean, D., Spanbroek, M.: Interactive learning and mathematical calculus. In: Mathematical Knowledge Management (2005)
11. Cohen, A., Cuypers, H., Barreiro, E.R., Sterk, H.: Interactive mathematical documents on the web. In: Algebra, Geometry and Software Systems, pp. 289–306. Springer, Heidelberg (2003)
12. Cunha, A., Visser, J.: Strongly typed rewriting for coupled software transformation. Electron. Notes Theor. Comput. Sci. 174(1), 17–34 (2007)
13. Erev, I., Luria, A., Erev, A.: On the effect of immediate feedback (2006), http://telem-pub.openu.ac.il/users/chais/2006/05/pdf/e-chais-erev.pdf
14. Goguadze, G., González Palomo, A., Melis, E.: Interactivity of exercises in ActiveMath. In: International Conference on Computers in Education, ICCE 2005 (2005)
15. Heeren, B., Jeuring, J., van Leeuwen, A., Gerdes, A.: Specifying strategies for exercises. Technical Report UU-CS-2008-001, Utrecht University (2008)
16. Hennecke, M.: Online Diagnose in intelligenten mathematischen Lehr-Lern-Systemen (in German). PhD thesis, Hildesheim University, Fortschritt-Berichte VDI Reihe 10, Informatik / Kommunikationstechnik; 605. Düsseldorf: VDI-Verlag (1999)
17. Hinze, R., Jeuring, J., Löh, A.: Comparing approaches to generic programming in Haskell. In: Backhouse, R., Gibbons, J., Hinze, R., Jeuring, J. (eds.) SSDGP 2006. LNCS, vol. 4719, pp. 72–149. Springer, Heidelberg (2007)
18. Horacek, H., Wolska, M.: Handling errors in mathematical formulas. In: Ikeda, M., Ashley, K.D., Chan, T.-W. (eds.) ITS 2006. LNCS, vol. 4053, pp. 339–348. Springer, Heidelberg (2006)
19. Hudak, P.: Building domain-specific embedded languages. ACM Computing Surveys 28A(4) (December 1996)
20. Hutton, G.: Higher-order Functions for Parsing. Journal of Functional Programming 2(3), 323–343 (1992)
21. Issakova, M.: Solving of linear equations, linear inequalities and systems of linear equations in interactive learning environment. PhD thesis, University of Tartu (2007)
22. Lämmel, R., Jones, S.P.: Scrap your boilerplate: a practical approach to generic programming. ACM SIGPLAN Notices 38(3), 26–37 (2003); TLDI 2003
23. R. Lämmel, E. Visser, J. Visser.: The Essence of Strategic Programming. Draft, p. 18 (October15, 2002), http://www.cwi.nl/~ralf

24. Lodder, J., Jeuring, J., Passier, H.: An interactive tool for manipulating logical formulae. In: Manzano, M., Pérez Lancho, B., Gil, A. (eds.) Proceedings of the Second International Congress on Tools for Teaching Logic (2006)
25. Passier, H., Jeuring, J.: Feedback in an interactive equation solver. In: Seppälä, M., Xambo, S., Caprotti, O. (eds.) Proceedings of the Web Advanced Learning Conference and Exhibition, WebALT 2006, pp. 53–68. Oy WebALT (2006)
26. Paulson, L.C.: ML for the Working Programmer., 2nd edn. Cambridge University Press, Cambridge (1996)
27. Jones, S.P., et al.: Haskell 98, Language and Libraries. The Revised Report. Cambridge University Press, Cambridge (2003); A special issue of the Journal of Functional Programming, http://www.haskell.org/
28. Plasmeijer, R., Achten, P., Koopman, P.: iTasks: executable specifications of interactive work flow systems for the web. In: ICFP 2007, New York, NY, USA, pp. 141–152 (2007)
29. Sweller, J., van Merriënboer, J.J.G., Paas, F.: Cognitive architecture and instructional design. Educational Psychology Review 10, 251–295 (1998)
30. Swierstra, S.D., Duponcheel, L.: Deterministic, error-correcting combinator parsers. In: Launchbury, J., Sheard, T., Meijer, E. (eds.) AFP 1996. LNCS, vol. 1129, pp. 184–207. Springer, Heidelberg (1996)
31. VanLehn, K.: Mind Bugs – The Origins of Procedural Misconceptions. MIT Press, Cambridge (1990)
32. Visser, E., Benaissa, Z., Tolmach, A.: Building program optimizers with rewriting strategies. In: ICFP 1998, pp. 13–26 (1998)
33. Zinn, C.: Supporting tutorial feedback to student help requests and errors in symbolic differentiation. In: Ikeda, M., Ashley, K.D., Chan, T.-W. (eds.) ITS 2006. LNCS, vol. 4053, pp. 349–359. Springer, Heidelberg (2006)

Mediated Access to Symbolic Computation Systems*

Jónathan Heras, Vico Pascual, and Julio Rubio

Departamento de Matemáticas y Computación, Universidad de La Rioja,
Edificio Vives, Luis de Ulloa s/n, E-26004 Logroño (La Rioja, Spain)
{jonathan.heras, vico.pascual, julio.rubio}@unirioja.es

Abstract. Kenzo is a symbolic computation system devoted to Algebraic Topology. It has been developed by F. Sergeraert mainly as a research tool. The challenge is now to increase the number of users and to improve its usability. Instead of designing simply a friendly front-end, we have undertaken the task of devising a *mediated* access to the system, constraining its functionality, but providing guidance to the user in his navigation on the system. This objective is reached by constructing an *intermediary layer*, allowing us an *intelligent* access to some features of the system. This intermediary layer is supported by XML technology and interplays between a graphical user interface and the *pure* Common Lisp Kenzo system.

1 Introduction

Traditionally, symbolic computation systems have been oriented to research. This implies, in particular, that development efforts in the area of Computer Algebra systems have been centered in aspects such as the improvement of the efficiency (or the accuracy, in symbolic-numerical systems) or the extension of the scope of the applications. Things are a bit different in the case of widely spread commercial systems such as Mathematica or Maple, where some attention is also paid to connectivity issues or to special-purpose user interfaces (usually related to educational applications). But even in these cases the central focus is on the results of the calculations and not on the interaction with other kind of (software or human) agents.

The situation is, in a sense, similar in the area of interoperability among symbolic computation systems (including here both computer algebra systems and proof assistants). The emphasis has been put in the *universality* of the middleware (see, for instance, [5]). Even if important advances have been achieved, severe problems have appeared, too, such as difficulties in reusing previous proposals and the final obstacle of the speculative existence of a *definitive mathematical interlingua*. The irruption of XML technologies (and, in our context, of MathML [2] and OpenMath [4]) has allowed standard knowledge management,

---

* Partially supported by Comunidad Autónoma de La Rioja, project Colabora2007/16, and Ministerio de Educación y Ciencia, project MTM2006-06513.

S. Autexier et al. (Eds.): AISC/Calculemus/MKM 2008, LNAI 5144, pp. 446–461, 2008.

but they are located at the *infrastructure* level, depending always on higher-level abstraction devices to put together different systems. Interestingly enough, the initiative SAGE [28] producing an integrated environment seems to have no use for XML standards, intercommunication being supported by ad-hoc SAGE mechanisms.

In summary, in the symbolic computation area, we are always looking for *more powerful* systems (with more computation capacities or with more general expressiveness). However, it is the case that our systems became so powerful, that we can lose some interesting kinds of users or interactions. We have encountered this situation when designing and developing the *TutorMates project* [16]. TutorMates is aimed at linking an educational front-end with the Maxima system [26]. Since the final users were students (and teachers) at the high school level it was clear from the beginning of the project that Maxima should be *weakened* in a sense, in order to make its outputs meaningful for non mathematics-trained users. This approach is now transferred to the field of symbolic computation in Algebraic Topology, where the Kenzo system [12] provides a complete set of calculation tools, which can be considered difficult to use by a non-Common Lisp trained user (typically, an Algebraic Topology student, teacher or researcher). The key concept is that of *mediated access* by means of an *intermediary layer* aimed at providing an *intelligent middleware* between a user interface and the kernel Kenzo system.

The paper is organized as follows. In the next section antecedents are presented, reporting on the Kenzo and the TutorMates systems. Section 3 gives some insights on methodological and architectural issues, both in the development of the client interface and in the general organization of the software systems involved. The central part of the paper can be found in Section 4, where the basics on the intermediary layer are explained. The concrete state of our project to interface with Kenzo is the aim of Section 5. The paper ends with two sections devoted to open problems and conclusions, and finally the bibliography.

2   Antecedents

Kenzo [12] is a Common Lisp system, devoted to Symbolic Computation in Algebraic Topology. It was developed in 1997 under the direction of F. Sergeraert, and has been successful, in the sense that it has been capable of computing homology groups unreachable by any other means. Having detected accessibility and usability as two weak points in Kenzo (implying difficulties in increasing the number of users of the system), several proposals have been studied to interoperate with Kenzo (since the original user interface is Common Lisp itself, the search for other ways of interaction seems mandatory to extend the use of the system). The most elaborated approach was reported in [1]. There, we devised a remote access to Kenzo, using CORBA [25] technology. An XML extension of MathML played a role there too, but just to give genericity to the connection (avoiding the definition in the CORBA Interface Description Language [25] of a different specification for each Kenzo class and datatype). There was no intention of taking profit from the semantics possibilities of MathML. Being useful, this approach ended in a prototype, and its enhancement and maintenance were difficult, due both to the low level characteristics of CORBA and to the pretentious aspiration of providing *full* access to Kenzo functionalities. We could classify the work of [1] in the same line as [5] or in the initiative IAMC and its corresponding workshop series (see, for instance, [7,29]), where the emphasis is put into powerful and generic access to symbolic computation engines.

On the contrary, the TutorMates project [16] had, from its very beginning, a much more modest objective. The idea was to give access just to a part of Maxima, but guiding the user in his interaction. Since the purpose of TutorMates was educational (high school level), it was clear that many outputs given by Maxima were unsuitable for the final users, depending on the degree and the topic learned in each TutorMates session. To give just an example, an imaginary solution to a quadratic equation has meaning only in certain courses. In this way, a *mediated* access to Maxima was designed. The central concept is an intermediary layer that communicates, by means of an extension of XML, between the graphical user interface (Java based) and Maxima. The extension of MathML allows us to encode a *profile* for the interaction. A profile is composed of a role (student or teacher), a level and a lesson. In the case of a teacher (supposed to be preparing material for his students), full access to Maxima outputs is given, but a *warning* indicates to him whether the answer would be suitable inside the level and the lesson encoded in the profile. In this way, the intermediary layer allows the programmer to get an *intelligent* interaction, different from the "dummy" remote access obtained in [1].

Now, our objective is to emulate this TutorMates organization in the Kenzo context. The final users could be researchers in Algebraic Topology or students of this discipline. The problems to be tackled in the intermediary layer are different from those of TutorMates. The methodological and architectural aspects of this new product are presented in the following section.

3   Methodological and Architectural Issues

We have tried to guide our development with already proven methodologies and patterns. In the case of the design of the interaction with the user in our Graphical User Interface (GUI) front-end, we have followed the guidelines of the Noesis method [11,8]. In particular, our development has been supported by some Noesis models for control and navigation in user interfaces (see an example in Figure 1).

Even if graphical specification mechanisms have well-known problems (related with their scalability), Noesis models provide *modular tools*, allowing the designer to control the complexity due to the size of graphics. These models enable an exhaustive traversal of the interfaces, detecting errors, disconnected areas, lack of homogeneity, etc.

With respect to the general organization of the software system, we have been inspired by the *Microkernel* architectural pattern [3]. This pattern gives

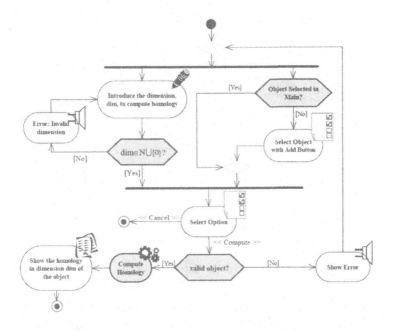

Fig. 1. A fragment of the control and navigation graph

a global view as a *platform*, in terminology of [3], which implements a virtual machine with applications running on top of it, namely a *framework* (in the same terminology). A high level perspective of the system as a whole is shown in Figure 2. Kenzo itself, wrapped with an interface based on XML-RPC [30], is acting as *internal server*. The *microkernel* acting as intermediary layer is based on an XML processor, allowing both a link with the standard XML-RPC used by Allegro Common Lisp [14], and intelligent processing. The view of the *external server* is again based on an XML processor, with a higher level of abstraction (since mathematical knowledge is included there) which can map expressions from and to the microkernel, and which is decorated with an *adapter* (the *Proxy* pattern, [15], is used to implement the adapter), establishing the final connection with the client, a Graphical User Interface in our case. A simplified version of the Microkernel pattern (without the external server) would suffice if our objective was to build a GUI for Kenzo. But we also pursue extending Kenzo by wrapping it in a framework which will link any possible client (other GUIs, web applications, web services, ...) with the Kenzo system. In this sense, our GUI is a client of our framework. The framework should provide each client with all necessary mathematical knowledge.

Which aspects of the intelligent processing must be dealt with in the external server or in the microkernel, is still controversial (in the current version, as we will explain later, we have managed the questions related to the input specifications in the external server and the most important mediations are done at the microkernel level). Moreover, the convenience of a double level of processing is

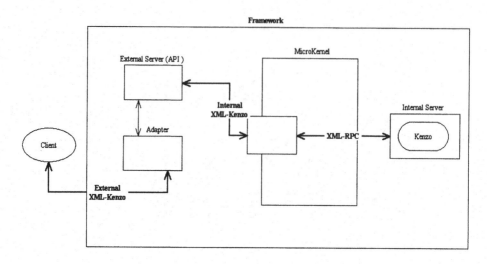

Fig. 2. Microkernel architecture of the system

clear, being based on, at least, two reasons. On the one hand the more concrete one (microkernel) is to be linked to Kenzo (via XML-RPC) and the more abstract one is aimed at being exported and imported, rendered by (extended) MathML engines, and so on. On the other hand, this double level of abstraction reflects the different languages in which the knowledge has to be expressed. The external one is near to Algebraic Topology, and it should offer a communication based on the concepts of this discipline to the final clients (this gives something as a type system; see Section 4). The internal part must communicate with Kenzo, and therefore a low level register of each session must be maintained (for instance, the unique identifier referring to each object, in order to avoid recalculations). There, a procedural language based on Kenzo conventions is needed.

As explained before, XML gives us the universal tool to transmit information along the different layers of the system. Besides the XML-RPC mechanism used by Allegro Common Lisp, two more XML formats (defined by means of XML schemas) are to be considered. The first one (used in the microkernel) is diagrammatically described in Figure 3, by using the Noesis method [11] again. The second format is used in the external server. Figure 4 shows a diagram corresponding to a part of this schema. The structure of this XML schema allows us to represent some knowledge on the process (for instance, it differentiates constructors from other kinds of algebraic manipulations); other more complex mathematical knowledge cannot be represented in the syntax of the schema (see Section 4). It is foreseen to adapt this XML schema to an MathML3 specification [2], by using the concept of MathML3 content dictionary (inspired by the corresponding OpenMath concept [4]). In Figure 5, we show how a Kenzo command (namely, the calculation of the third group of homology of the sphere of dimension 3) will be transformed from the user command on the GUI (top part of the figure) to the final XML-RPC format (the conventional Lisp call is shown,

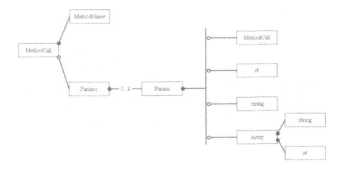

Fig. 3. Description of the Internal XML Kenzo Schema

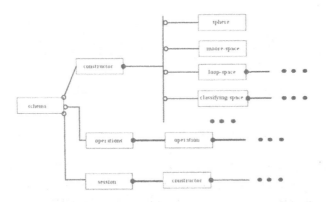

Fig. 4. Description of the External XML Kenzo Schema

too; however our internal server, Kenzo wrapped with an XML-RPC interface, will execute the command directly).

In the next section the behavior pursued with this architecture is explained.

4   Knowledge Management in the Intermediary Layer

The system as a whole will improve Kenzo including the following "intelligent" enhancements:

1. Controlling the input specifications on constructors.
2. Avoiding some operations on objects which will raise errors.
3. Chaining methods in order to provide the user with new tools.
4. Determining if a calculation can be done in a local computer or should be derived to a remote server.

The first aspect is attained, in an integrated manner, inside the Graphical User Interface. The three last ones are dealt with in the intermediary layer. From

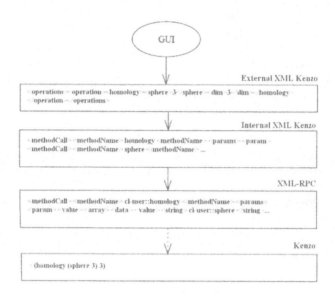

Fig. 5. Transforming XML representations

another point of view, the first three items are already partially programmed in the current version of the system; the last one is further work.

In order to explain the differences between points 1 and 2, it is worth noting that in Kenzo there are two *kinds* of data. The first one is representing *spaces* in Algebraic Topology (by *spaces* we mean here, any data structure having both behavior and elements belonging to it, such as a simplicial set, a simplicial group, a chain complex, and so on). The second kind of data is used to represent *elements* of the spaces. Thus, in a typical session with Kenzo, the users proceed in two steps: first, constructing some spaces, and second, applying some operators on the (elements of the) spaces previously built. This organization in two steps has been described by using Algebraic Specification methods in [17] and [10], for instance. Therefore, the first item in the enumeration refers to the inputs for the constructors of spaces, and the second item refers to some operations on *concrete* spaces. As we are going to explain, the first kind of control is naturally achieved in the GUI client (from the mathematical knowledge provided by the external XML format) but the second one, which needs some expert knowledge management, is better dealt with in the intermediary layer.

Kenzo is, in its pure mode, an untyped system (or rather, a dynamically typed system), inheriting its power and its weakness from Common Lisp. Thus, for instance, in Kenzo a user could apply a constructor to an object without satisfying its input specification. For instance, the method constructing the classifying space of a simplicial group could be called on a simplicial set without a group structure over it. Then, at runtime, Common Lisp would raise an error informing the user of this restriction. This is shown in the following fragment of a Kenzo session:

```
> (loop-space (sphere 4))
[K6 Simplicial-Group]
> (classifying-space (loop-space (sphere 4)))
[K18 Simplicial-Set]
> (sphere 4)
[K1 Simplicial-Set]
> (classifying-space (sphere 4))
;; Error: No method in generic function CLASSIFYING-SPACE
;; is applicable to arguments: [K1 Simplicial-Set]
```

With the first command, namely (loop-space (sphere 4)), we construct a simplicial group. Then, in the next step we are verifying that a simplicial group has a classifying space (which is, in general, just a simplicial set). In the third command, we check that the sphere of dimension 4 is constructed in Kenzo as a simplicial set. Thus, when in the last command we try to construct the classifying space of a simplicial set, the Common Lisp Object System (CLOS) raises an error.

In the current version of our system this kind of error is controlled, because the inputs for the operations between spaces can be only selected among the spaces with suitable characteristics. This same idea could be used to improve the reliability of internal processes, by controlling the outputs of the intermediary computations. The equivalence in our system of the example introduced before in pure Kenzo, is shown in Figure 6, where it can be seen that for the classifying operation just the spaces which are simplicial groups are candidates to be selected. This enriches Kenzo with something as a (semantical) type system which has been defined into the external XML schema.

With respect to the second item in the previous enumeration, the most important example in the current version is the management of the *connection degree* of spaces. Kenzo allows the user to construct, for instance, the loop space of a non simply connected space (as the sphere of dimension 1). The result is a simplicial set on which some operations (for instance, to compute the set of faces of a simplex) can be achieved without any problems. On the contrary, theoretical results ensure that the homology groups are not of finite type, and then they cannot be computed. In pure Kenzo, the user could ask for a homology group of such a space, catching a runtime error.

In our current version of the system, the intermediary layer includes a small expert system, computing, in a symbolic way (that is to say, working with the *description* of the spaces, and not with the spaces themselves considered as Common Lisp objects), the connection degree of a space. The set of rules gives a connection degree to each space builder (for instance, a sphere of dimension n has connection degree $n - 1$), and then a rule for each operation on spaces. For instance, loop space decreases the connection degree of its input in one unity, suspension increases it in one unity, a cartesian product has, as connection degree, the minimum of the connection degrees of its factors, and so on. From the

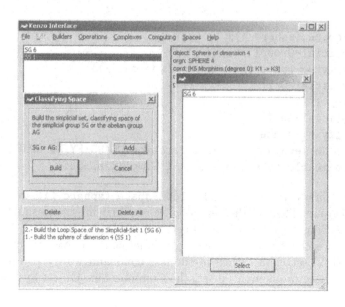

Fig. 6. Screen-shot of Kenzo Interface with a session related to classifying spaces

design point of view, a *Decorator* pattern [15] was used, decorating each space with an annotation of its connection degree in the intermediary layer. Then, when a computation (of a homology group, for instance) is demanded by a user, the intermediary layer monitors if the connection degree allows the transferring of the command to the Kenzo kernel, or a warning must be sent through the external server to the user.

As for item three, the best example is that of the computation of *homotopy groups*. In pure Kenzo, there is no final function allowing the user to compute them. Instead, there is a number of complex algorithms, allowing a user to chain them to get some homotopy groups. Our current user interface has an option to compute homotopy groups. The intermediary layer is in charge of chaining the different algorithms present in Kenzo to reach the final objective. In addition, Kenzo, in its current version, has limited capabilities to compute homotopy groups (depending on the homology of Eilenberg-Mac Lane spaces that are only partially implemented in Kenzo), so the *chaining* of algorithms cannot be *universal* (in this case, a possibility would be to *wire* the enhancement in the GUI, by means of the external XML schema, as in the case of item 1). Thus, the intermediary layer should process the call for a homotopy group, making some consultations to the Kenzo kernel (computing some intermediary homology groups, for instance) before deciding if the computation is possible or not (this is still work in progress).

Regarding point four, our system can be distributed, at present, in two manners: (a) as a stand-alone application, with a heavy client containing the Kenzo kernel to be run in the local host computer; (b) as a light client, containing just the user interface, and every operation and computation is done in a remote server. The second mode has obvious drawbacks related to the reliability of Internet connections, to the overhead of management where several concurrent users are allowed, etc. But option (a) is not fully satisfactory since interesting Kenzo computations used to be very time and space consuming (requiring, typically, several days of CPU time on powerful computing servers). Thus a mixed strategy should be convenient: the intermediary layer should decide if a concrete calculation can be done in the local computer or it deserves to be sent to a specialized remote server. (In this second case, as it is not sensible to maintain open an Internet connection for several days waiting for the end of a computation, some reactive mechanism should be implemented, allowing the client to disconnect and to be subscribed in some way, to the process of computation in the remote server). The difficulties of this point have two sources: (1) the knowledge here is not based on well-known theorems (as was the case in our discussion on the *connection degree* in the second item of the enumeration), since it is context-dependent (for instance, it depends on the computational power of a local computer), and so it should be based on *heuristics*; (2) the technical problems to obtain an optimal performance are complicated, due, in particular, to the necessity of maintaining a *shared state* between two different computers. These technical aspects are briefly commented in the Open Problems section.

With respect to the kind of heuristic knowledge to be managed into the intermediary level, there is some part of it that could be considered obvious: for instance, to ask for an homology group $H_n(X)$ where the degree n is big, should be considered harder than if n is small, and then one could wonder about a limit for n before sending the computation to a remote server. Nevertheless, this simplistic view is to be moderated by some expert knowledge: it is the case that in some kinds of spaces, difficulties decrease when the degree increases. The heuristics should consider each operation individually. For instance, it is true that in the computation of homology groups of iterated loop spaces, difficulties increase with the degree of iteration. Another measure of complexity is related to the number of times a computation needs to call the Eilenberg-Zilber algorithm (see [12]), where a double exponential complexity bound is reached. Further research is needed to exploit the expert knowledge in the area suitably, in order to devise a systematic heuristic approach to this problem.

5   State of the Project

The work done up to now has allowed us to reach one of the objectives: code reuse. This reusing has two aspects:

1. We have left the Kenzo kernel untouched. This was a goal since the team developing the framework and the user interface, and the team maintaining and extending Kenzo are different. Therefore, it is convenient to keep both systems as uncoupled as possible.
2. The intermediary level has been used, without changes, both in the standalone local version and in the light client with remote server version. A

first partial prototype, moving the view towards a web application client, seems to confirm that the degree of abstraction and genericity reached in our architecture (note that our framework includes several XML formats, each one with different abstraction level) is suitable.

In Figure 7, a screen-shot of our GUI is presented. The main toolbar is organized into 8 menus: *File, Edit, Builders, Operations, Complexes, Computing, Spaces* and *Help*. The rest of the screen is separated into three areas. On the left side, a list with the spaces already constructed during the current session is maintained. When a space is selected (the one denoted by SS 1 in Figure 7), a description of it is displayed in the right area. At the bottom of the screen, one finds a *history* description of the current session, which can be cleared or saved into a file. It is important to understand that a *history file* is different from a *session file*. The first one is just a plain text description of the commands selected by the user. The second kind of file is described in the next paragraph.

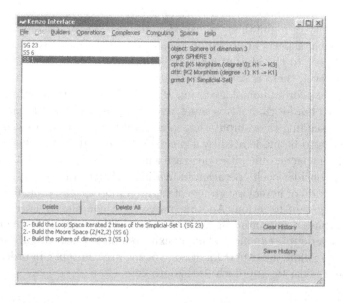

Fig. 7. Screen-shot of Kenzo Interface with an example of session

In the current version the *File* menu has just three options: *Exit, Save Session* and *Load Session*. When saving a session, a file is produced containing an XML description of the commands executed by the user in that session. In Figure 8 an example of session file can be found; this session file corresponds to the following Kenzo interaction:

```
> (sphere 3)
[K1 Simplicial-Set]
> (moore 4 2)
```

```
[K6 Simplicial-Set]
> (loop-space (sphere 3) 2)
[K23 Simplicial-Group]
> (crts-prdc (sphere 3) (moore 4 2))
[K35 Simplicial-Set]
```

These session files are stored using the external XML schema described in Section 3 (see Figure 4). Thus, the session files are exportable and, besides, an extensible stylesheet language (XSL) has been devised to render the sessions in standard displays (using mathematical notation).

```
<?xml version="1.0" encoding="UTF-8"?>

<?xml-stylesheet type="text/xsl" href="session.xsl"?>

<session xmlns:xsi="http://www.w3.org/2001/XMLSchema-instance"
xsi:noNamespaceSchemaLocation="url:http://www.unirioja.es/cu/joheras/esquemaxml2.xsd">
<constructor><sphere>3</sphere></constructor>
<constructor><moore-space>4 2</moore-space></constructor>
<constructor><loop-space><sphere>3</sphere><dim>2</dim></loop-space></constructor>
<constructor><crts-prdc><sphere>3</sphere><moore-space>4 2</moore-space></crts-prdc></constructor>
</session>
```

Fig. 8. Sample of a session file

The constructors of the spaces we have referred to the first point of Section 4, are collected by the menus *Builders, Operations* and *Complexes*. More specifically, the menu *Builders* includes the main ways of constructing new spaces from scratch in Kenzo as options: spheres, Moore spaces, Eilenberg-Mac Lane spaces, and so on (see [18] for Algebraic Topology notions). The menu *Operations* refers to the ways where Kenzo allows the construction of new simplicial spaces (see [23] for simplicial notions) from other ones: loop spaces, classifying spaces, Cartesian products, suspensions, etc. The menu *Complexes* is similar, but related to chain complexes instead of simplicial objects (here, for instance, the natural product is the tensorial product instead of the cartesian one).

The menus *Computing* and *Spaces* collect all the operations on concrete spaces (instead of *constructing spaces*, as in the previous cases). Both of them provide their items with all the necessary "intelligence" in order to avoid raising runtime errors. In *Computing* we concentrate on calculations *over* a space. We offer to compute homology groups, to compute the same but with explicit generators and to compute homotopy groups; in this last case we find the third kind of enhancement. In menu *Spaces* currently we only offer the possibility of showing the structure of a simplicial object (this is only applicable to *effective*, finite type spaces).

To consider a first complete (beta) version of the system, it is necessary to complete the questions already mentioned in the text relating to finishing the external XML schema definition and to controlling the cases in which homotopy groups can be effectively computed by Kenzo.

Moreover, we have planned to develop two more tools:

1. In the menu *Builders*, there is a still inactivated slot called *Build-finite-ss*, aimed at emulating, in our environment, the facility present in pure Kenzo which allows the user to construct step-by-step, in an interactive manner, a finite simplicial set (checking, in each step, whether faces are glued together in a coherent way). To this aim, we are thinking of designing a graphical tool.

2. In the menu *Spaces*, it is necessary to include the possibility of operating locally *inside* a selected space. For instance, given a simplex to compute one of its faces or given two simplexes in the same dimension we can compute its product in a selected simplicial group. One of the difficulties here is related to designing an editor for elements (data of the second kind, using the terminology in Section 4), which can be given as inputs to the local operations. This will give content to the *Edit* menu, in the main toolbar, which is now inactivated.

These extra functionalities are rather a matter of standard programming, and it is foreseen that no research problem will appear when tackling them. The questions discussed in the next section, on the contrary, could imply important challenges.

6   Open Problems

The most important issue to be tackled in the next versions of the system is how organizing the decision on when (and how) a calculation should be derived to a remote server. To understand the nature of the problem it is necessary to consider that there are two kinds of *state* in our context. Starting from the most simple, the *state of a session* can be described by means of the spaces that have been constructed so far. Then, to encode (and recover) such a state, a session file as explained in the previous section would be enough: an XML document containing a sequence of calls to different constructors and methods. In this case, when a calculation is considered too hard to be computed in a local computer, the whole session file could be transmitted to the remote server. There, executing step-by-step the session file, the program will re-find the same state of the local session, proceeding to compute the desired result and sending it to the client. Of course, as mentioned previously, some kind of subscription tool should be enabled, in such a way that the client could stop its running, and then to receive the result (or a notification indicating the result is already available somewhere), after some time (perhaps some days or weeks of computation on the remote server).

Even if this approach can be considered reasonable as a first step, it has turned out to be too simplistic to deal with the richness of Kenzo. A space in Kenzo consists in a number of methods describing its behavior (explaining, for instance, how to compute the faces of its elements). Due to the high complexity of the algorithms involved in Kenzo, a strategy of *memoization* has been systematically implemented. As a consequence, the *state of a space* evolves after it has been used in a computation (of a homology group, for instance). Thus, the time needed to compute, let us say, a face, depends on the concrete states of every space involved in the calculation (in the more explicit case, to re-calculate a face on a space could be negligible in time, even if in the first occasion this was very time consuming). This notion of *state of a space* is transmitted to the notion of *state of a session*. We could speak of two *states of a session*: the one *shallow* evoked before, that is essentially *static* and can be recovered by simply re-executing the top-level constructor calls; and the other *deep state* which is dynamic and depends on the computations performed on the spaces.

To analyse the consequences of this Kenzo organization, we should play with some scenarios. Imagine during a local session a very time consuming calculation appears; then we could simply send the *shallow state of the session* to the remote server, because even if some intermediary calculations have been stored in local memory, they can be re-computed in the remote server (finally, if they are cheap enough to be computed on the local computer, the price of re-computing them in the powerful remote server would be low). Once the calculation is remotely finished, there is no possibility of sending back the *deep state* of the remote session to the local computer because, usually, the memory used will exhaust the space in the local computer. Thus, it could seem that to transmit the *shallow state* would be enough. But, in this picture, we are losing the very reason why Kenzo uses the memoization (dynamic programming) style. Indeed, if after obtaining a difficult result (by means of the remote server) we resume the local session and ask for another related difficult calculation, then the remote server will initialize a new session from scratch, being obligated to re-calculate every previous difficult result, perhaps making the continuation of the session impossible. Therefore, in order to take advantages of all the possibilities Kenzo is offering now on powerful scientific servers, we are faced with some kind of *state sharing* among different computers (the local computers and the server), a problem known as difficult in the field of distributed object-oriented programming.

In short, even if our initial goal was not related to distributed computing, we found that in order to enable our intermediary layer as an *intelligent assistant* with respect to the classification of calculations as simple (runnable on a standard local computer) or complicated (to be sent to a remote server), we should solve problems of distributed systems. Thus, a larger perspective is necessary, and we are working with the Broker architectural pattern, see [3], in order to find a natural organization of our intermediary layer. From the symbolic computation literature, we will look for inspiration in different projects and frameworks such as the MathWeb software bus [22], its successor the MathServe Framework [21], the MoNET project [24,6,9] or the MathBroker [19] and MathBroker II [20] projects, as well as in other works as [27], [31] or [13].

7   Conclusions

The current state of our project can be considered solid enough to be a good point of continuation for all our objectives. We have showed how some *intelligent guidance* can be achieved in the field of Computational Algebraic Topology, without using standard Artificial Intelligence techniques. The idea is to build an intermediary layer, giving a mediated access to an already-written symbolic computation system. Putting together both Kenzo itself and the intermediary layer, we have produced a framework which is able to be connected to different clients (desktop GUIs, web applications and so on). In addition, with this framework, several profiles of interaction can be considered. In general, this can imply a restriction of the full capabilities of the kernel system, but the interaction with it is easier and enriched, contributing to the objective of increasing the number of users of the system.

References

1. Andrés, M., Pascual, V., Romero, A., Rubio, J.: Remote Access to a Symbolic Computation System for Algebraic Topology: A Client-Server Approach. In: Sunderam, V.S., van Albada, G.D., Sloot, P.M.A., Dongarra, J. (eds.) ICCS 2005. LNCS, vol. 3516, pp. 635–642. Springer, Heidelberg (2005)
2. Ausbrooks, R., et al.: Mathematical Markup Language (MathML) Version 3.0 (2 edn.) (2008), http://www.w3.org/TR/2008/WD-MathML3-20080409/
3. Buschmann, F., Meunier, R., Rohnert, H., Sommerland, P., Stal, M.: Pattern-oriented software architecture. A system of patterns, vol. 1. Wiley, Chichester (1996)
4. Buswell, S., Caprotti, O., Carlisle, D.P., Dewar, M.C., Gaëtano, M., Kohlhase, M.: OpenMath Version 2.0 (2004), http://www.openmath.org/
5. Calmet, J., Homann, K.: Towards the Mathematics Software Bus. Theoretical Computer Science 187, 221–230 (1997)
6. Caprotti, O., Davenport, J.H., Dewar, M., Padget, J.: Mathematics on the (Semantic) NET. In: Bussler, C.J., Davies, J., Fensel, D., Studer, R. (eds.) ESWS 2004. LNCS, vol. 3053, pp. 213–224. Springer, Heidelberg (2004)
7. Chiu, D., Zhou, Y., Zou, X., Wang, P.S.: Design, implementation, and processing support of MeML. In: Proceedings Internet Accessible Mathematical Computation (2003), http://www.symbolicnet.org/conferences/iamc03/meml.pdf
8. Cordero, C.C., De Miguel, A., Domínguez, E., Zapata, M.A.: Modelling Interactive Systems: an architecture guided by communication objects. In: HCI related papers of Interaccion 2004, pp. 345–357. Springer, Heidelberg (2006)
9. Dewar, M., Smirnova, E., Watt, S.M.: XML in Mathematical Web Services. In: Proceedings of XML 2005 Conference – Syntax to Semantics (XML 2005). IDEAlliance (2005)
10. Domínguez, C., Lambán, L., Rubio, J.: Object oriented institutions to specify symbolic computation systems. Rairo - Theoretical Informatics and Applications 41, 191–214 (2007)
11. Domínguez, E., Zapata, M.A.: Noesis: Towards a situational method engineering technique. Information Systems 32(2), 181–222 (2007)
12. Dousson, X., Sergeraert, F., Siret, Y.: The Kenzo program, Institut Fourier, Grenoble (1999), http://www-fourier.ujf-grenoble.fr/~sergerar/Kenzo/
13. Duscher, A.: Interaction patterns of Mathematical Services, Research Institute for Symbolic Computation (RISC), Johannes Kepler Univeristy, Linz, Technical report (2006)

14. Franz Inc. Allegro Common Lisp, http://www.franz.com/
15. Gamma, E., Helm, R., Johnson, R., Vlissides, J.: Design Patterns: Elements of Reusable Object-Oriented Software. Addison-Wesley, Reading (1994)
16. González-López, M.J., González-Vega, L., Pascual, A., Callejo, E., Recio, T., Rubio, J.: TutorMates, http://www.tutormates.es/
17. Lambán, L., Pascual, V., Rubio, J.: An object-oriented interpretation of the EAT system. Applicable Algebra in Engineering, Communication and Computing 14, 187–215 (2003)
18. MacLane, S.: Homology. Springer, Heidelberg (1963)
19. MathBroker: A Framework for Brokering Distributed Mathematical Services, http://www.risc.uni-linz.ac.at/projects/basic/mathbroker/
20. MathBroker II: Brokering Distributed Mathematical Services, http://www.risc.uni-linz.ac.at/projects/mathbroker2/
21. MathServe Framework, http://www.ags.uni-sb.de/~jzimmer/mathserve.html
22. MATHWEB-SB: A Software Bus for MathWeb, http://www.ags.uni-sb.de/jzimmer/mathweb-sb/
23. May, J.P.: Simplicial objects in Algebraic Topology. Van Nostrand Mathematical Studies (11) (1967)
24. MoNET: Mathematics on the Net, http://monet.nag.co.uk/cocoon/monet/index.html
25. Object Management Group. Common Object Request Broker Architecture (CORBA), http://www.omg.org
26. Schelter, W.: Maxima, http://maxima.sourceforge.net/index.shtml
27. Smirnova, E., So, C.M., Watt, S.M.: An architecture for distributed mathematical web services. In: Asperti, A., Bancerek, G., Trybulec, A. (eds.) MKM 2004. LNCS, vol. 3119, pp. 363–377. Springer, Heidelberg (2004)
28. Stein, W.: SAGE mathematical software system, http://www.sagemath.org/
29. Wang, P.S., Kajler, N., Zhou, Y., Zou, X.: Initial design of a Web-Based Mathematics Education Framework. In: Proceedings Internet Accessible Mathematical Computation (2002), http://www.symbolicnet.org/conferences/iamc02/wme.pdf
30. Winer, D.: Extensible Markup Language-Remote Procedure Call (XML-RPC), http://www.xmlrpc.com
31. Zhu, J., Wu, Y., Xie, F., Yang, G., Wang, Q., Mao, J., Shen, M.: A model for Distributed Computation over the Internet, In: Proceedings Internet Accessible Mathematical Computation (2003), http://www.symbolicnet.org/conferences/iamc03/zhu.pdf

Herbrand Sequent Extraction*

Stefan Hetzl[1], Alexander Leitsch[1],
Daniel Weller[1], and Bruno Woltzenlogel Paleo[1]

Institute of Computer Languages (E185),
Vienna University of Technology,
Favoritenstraße 9, 1040 Vienna, Austria
{hetzl, leitsch, weller, bruno}@logic.at

Abstract. Computer generated proofs of interesting mathematical theorems are usually too large and full of trivial structural information, and hence hard to understand for humans. Techniques to extract specific essential information from these proofs are needed. In this paper we describe an algorithm to extract Herbrand sequents from proofs written in Gentzen's sequent calculus **LK** for classical first-order logic. The extracted Herbrand sequent summarizes the creative information of the formal proof, which lies in the instantiations chosen for the quantifiers, and can be used to facilitate its analysis by humans. Furthermore, we also demonstrate the usage of the algorithm in the analysis of a proof of the equivalence of two different definitions for the mathematical concept of lattice, obtained with the proof transformation system **CERES**.

1   Introduction

Within mathematical knowledge management, the problem of analyzing and understanding computer generated proofs plays a fundamental role and its importance can be expected to grow, as automated and interactive deduction methods and computer processing power improve. Such computer generated proofs are *formal*, in the sense that they strictly follow *axioms* and *rules of inference* of formal logical calculi, as Hilbert calculi, natural deduction calculi or sequent calculi. The main advantages of formal proofs are:

- Formal proofs, when viewed and studied as a model and ideal for informal mathematical proofs, allow meta-mathematical investigations into the foundations of Mathematics.
- The correctness of formal proofs can be easily checked, by verifying whether the formal axioms and rules of the calculus were correctly employed.
- Formal proofs for formalized statements (formulas) can be constructed by computers executing automated or interactive theorem provers [15].
- Automated proof transformations can be employed to obtain new formal proofs from previously existing ones [2,3]. Subsequently, the analysis and

---

* Supported by the Austrian Science Fund (project P19875) and by the Programme Alban (project E05M054053BR).

S. Autexier et al. (Eds.): AISC/Calculemus/MKM 2008, LNAI 5144, pp. 462–477, 2008.

interpretation of the new formal proofs might lead to the discovery of new informal proofs of the original theorems containing interesting mathematical arguments.

However, formal representations of real mathematical proofs or computer generated proofs of real mathematical problems usually have some drawbacks that make them difficult to be analyzed and understood by mathematicians. Firstly, the size of a formal proof is usually huge ([1]), which makes it hard to be visualized as a whole. Secondly, many of its individual inferences are only structural, necessary not to carry some essential idea about the proof, but only to satisfy the formalities of the calculus. Thirdly, inference rules of proof calculi not always correspond easily to natural inferences in informal proofs. Together these drawbacks imply that, given a formal proof, it is not easy for humans to understand its essential idea, because it is hidden in a large data structure of repetitive, bureaucratic and non-intuitive formalities. Therefore there is a need for summarization of formal proofs or for extraction of its hidden crucial information, whenever these proofs are intended to be analyzed and understood by humans. This need has become especially clear to us during the development and use of our automated proof transformation system CERES[1] for the cut-elimination of real mathematical proofs in classical first-order logic [2].

This paper describes one possible technique that helps to overcome these difficulties in the particular case of first-order logic. Our technique relies on the concept of Herbrand sequent, a generalization of Herbrand disjunction [10], which can be used to summarize the creative content of first-order formal proofs [13], which lies in the instantiations chosen for quantified variables. Although we use sequent calculi, the idea described here could be adapted to other calculi, since it relies on a general property of first-order logic, as stated by Herbrand's theorem, and not on specific features of particular calculi.

After describing the technique, we demonstrate its use with the analysis of a computer generated proof of the equivalence of two different lattice definitions.

2 The Sequent Calculus LKDe

Our formal proofs are written in an extension of Gentzen's sequent calculus **LK**, which is called **LKDe** and has the following additional features:

- Arbitrary but pre-defined atomic formulas are allowed in the axioms. This has the advantage that typical mathematical axioms (e.g. symmetry and reflexivity of equality, associativity of addition, ...) do not need to be carried along all the formal proof in the antecedents of the sequents, but can instead appear simply as non-tautological axiom sequents in the leaf nodes of the proof. On the other hand, Gentzen's cut-elimination theorem [8] holds in this calculus only in a modified form, since atomic cuts are not necessarily eliminable.

---

[1] CERES Website: http://www.logic.at/ceres

- There are additional rules for equality and mathematical definitions, in order to make the calculus more comfortable to use in the formalization of real mathematical proofs, which use equality and definitions of concepts very often.

A partial description of the sequent calculus **LKDe** follows. A full description can be found in [3]. Additionally, by **LKe** we denote the **LKDe** calculus without definition rules.

Definition 1 (Sequent). *A sequent is a pair* $A_1, \ldots, A_n \vdash C_1, \ldots, C_m$ *of sequences of first-order logic formulas. The first sequence,* $A_1, \ldots, A_n$*, is the antecedent of the sequent and the second sequence,* $C_1, \ldots, C_m$*, is the consequent of the sequent. We use the symbols* Γ, Π, Λ *and* Δ*, possibly with subscripts, to denote sequences of formulas in the antecedent and consequent of sequents.*

1. **The Axioms:** We allow arbitrary atomic sequents as axioms. The logical axioms are of the form $A \vdash A$ for A atomic. For equality we use the reflexivity axiom scheme $\vdash t = t$ for all terms t.

2. **Propositional rules: LKDe** has rules for the propositional connectives: $\vee$, $\rightarrow$, $\neg$ and $\wedge$, as exemplified below:

$$\frac{\Gamma \vdash \Delta, A \quad \Pi \vdash \Lambda, B}{\Gamma, \Pi \vdash \Delta, \Lambda, A \wedge B} \wedge : r \qquad \frac{A, \Gamma \vdash \Delta}{A \wedge B, \Gamma \vdash \Delta} \wedge : l1 \qquad \frac{A, \Gamma \vdash \Delta}{B \wedge A, \Gamma \vdash \Delta} \wedge : l2$$

3. **First-order rules: LKDe** has rules for the existential ($\exists$) and universal ($\forall$) quantifiers.

$$\frac{A\{x \leftarrow t\}, \Gamma \vdash \Delta}{(\forall x)A, \Gamma \vdash \Delta} \ \forall : l \qquad \frac{\Gamma \vdash \Delta, A\{x \leftarrow \alpha\}}{\Gamma \vdash \Delta, (\forall x)A} \ \forall : r$$

$$\frac{A\{x \leftarrow \alpha\}, \Gamma \vdash \Delta}{(\exists x)A, \Gamma \vdash \Delta} \ \exists : l \qquad \frac{\Gamma \vdash \Delta, A\{x \leftarrow t\}}{\Gamma \vdash \Delta, (\exists x)A} \ \exists : r$$

The $\forall : r$ and $\exists : l$ rules must satisfy the eigenvariable condition: the variable α must not occur in Γ nor in Δ nor in A. Quantifiers introduced by them are called *strong quantifiers*. For the $\forall : l$ and the $\exists : r$ rules the term t must not contain a variable that is bound in A. Quantifiers introduced by them are called *weak quantifiers*.

4. **Equality rules:**

$$\frac{\Gamma \vdash \Delta, s = t \quad \Pi \vdash \Lambda, A[s]_\Xi}{\Gamma, \Pi \vdash \Delta, \Lambda, A[t]_\Xi} = (\Xi) : r1 \qquad \frac{\Gamma \vdash \Delta, s = t \quad A[s]_\Xi, \Pi \vdash \Lambda}{A[t]_\Xi, \Gamma, \Pi \vdash \Delta, \Lambda} = (\Xi) : l1$$

$$\frac{\Gamma \vdash \Delta, t = s \quad \Pi \vdash \Lambda, A[s]_\Xi}{\Gamma, \Pi \vdash \Delta, \Lambda, A[t]_\Xi} = (\Xi) : r2 \qquad \frac{\Gamma \vdash \Delta, t = s \quad A[s]_\Xi, \Pi \vdash \Lambda}{A[t]_\Xi, \Gamma, \Pi \vdash \Delta, \Lambda} = (\Xi) : l2$$

where Ξ is a set of positions in A and s and t do not contain variables that are bound in A.

5. **Structural rules:** weakening, contraction and permutation, as well as the following cut-rule:

$$\frac{\Gamma \vdash \Delta, A \quad A, \Pi \vdash \Lambda}{\Gamma, \Pi \vdash \Delta, \Lambda} \ cut$$

6. **Definition rules:** They correspond directly to the *extension principle* in predicate logic and introduce new predicate and function symbols as abbreviations for formulas and terms. Let A be a first-order formula with the free variables $x_1, \ldots, x_k$, denoted by $A(x_1, \ldots, x_k)$, and P be a *new* k-ary predicate symbol (corresponding to the formula A). Then the rules are:

$$\frac{A(t_1, \ldots, t_k), \Gamma \vdash \Delta}{P(t_1, \ldots, t_k), \Gamma \vdash \Delta} \; d:l \qquad \frac{\Gamma \vdash \Delta, A(t_1, \ldots, t_k)}{\Gamma \vdash \Delta, P(t_1, \ldots, t_k)} \; d:r$$

for arbitrary sequences of terms $t_1, \ldots, t_k$. Definition introduction is a simple and very powerful tool in mathematical practice, allowing the easy introduction of important concepts and notations (e.g. groups, lattices, ...) by the introduction of new symbols.

Definition 2 (Skolemization). *The skolemization of a sequent removes all its strong-quantifiers and substitutes the corresponding variables by skolem-terms in a validity preserving way (i.e. the skolemized sequent is valid iff the original sequent is valid).* **LKDe**-*proofs can also be skolemized, as described in [4], essentially by skolemizing the end-sequent and recursively propagating the skolemization to the corresponding formulas in the premises above. Such skolemized proofs can still contain strong quantifiers that go into cuts.*

Remark 1. There are many algorithms for skolemization. They can be classified as either *prenex*, which firstly transform formulas and sequents into a prenex form (i.e. with all quantifiers occurring in the beginning of formulas), or *structural*, which leave weak quantifiers in their places. It has been shown that prenex skolemization can result in a non-elementary increase in the Herbrand Complexity of an **LK**-Proof [4]. Moreover, prenexification impairs the readability of formulas. Therefore we use structural skolemization algorithms [5], whenever skolemization is necessary or desirable for our proof transformations.

Example 1. The sequent

$$(\forall x)((\exists z)P(x,z) \wedge (\forall y)(P(x,y) \rightarrow P(x,f(y)))) \vdash (\forall x)(\exists y)P(x,f^2(y))$$

can be structurally skolemized to

$$(\forall x)(P(x,g(x)) \wedge (\forall y)(P(x,y) \rightarrow P(x,f(y)))) \vdash (\exists y)P(a,f^2(y))$$

where a is a skolem-constant and g is a skolem-function.

3   The *CERES* Method

Our motivation to devise and implement Herbrand sequent extraction algorithms was the need to analyze and understand the result of proof transformations performed automatically by the CERES-system, among which the main one is Cut-Elimination by Resolution: the *CERES* method [6].

The method transforms any **LKDe**-proof with cuts into an *atomic-cut normal form* (ACNF) containing no non-atomic cuts. The remaining atomic cuts are, generally, non-eliminable, because **LKDe** admits non-tautological axiom sequents.

The ACNF is mathematically interesting, because cut-elimination in formal proofs corresponds to the elimination of lemmas in informal proofs. Hence the ACNF corresponds to an informal mathematical proof that is analytic in the sense that it does not use auxiliary notions that are not already explicit in the axioms or in the theorem itself.

The transformation to ACNF via Cut-Elimination by Resolution is done according to the following steps:

1. Construct the (always unsatisfiable [6]) *characteristic clause set* of the original proof by collecting, joining and merging sets of clauses defined by the ancestors of cut-formulas in the axioms of the proof.
2. Obtain from the characteristic clause set a grounded resolution refutation, which can be seen as an **LKe**-proof by exploiting the fact that the resolution rule is essentially a cut-rule restricted to atomic cut-formulas only.
3. For each clause of the characteristic clause set, construct a *projection* of the original proof with respect to the clause.
4. Construct the ACNF by plugging the projections into the leaves of the grounded resolution refutation tree (seen as an **LKe**-proof) and by adjusting the refutation accordingly. Since the projections do not contain cuts and the refutation contains atomic cuts only, the resulting **LKDe** proof will indeed be in atomic-cut normal form.

This method has been continuously improved and extended. The characteristic clause sets evolved to *proof profiles*, which are invariant under rule permutations and other simple transformations of proofs [12,11]. The resolution and the sequent calculi are now being extended to restricted second-order logics.

The CERES-system automates the method described above, using either Otter[2] or Prover9[3] as resolution-based first-order theorem provers to obtain the refutation of the characteristic clause set. However, current fully-automated resolution-based theorem provers have difficulties to refute some characteristic clause sets produced by CERES [1]. On the other hand, interactive theorem provers are typically not resolution-based. Therefore, we are currently developing our own flexible, interactive and resolution-based first-order theorem prover.

4   An Algorithm for Herbrand Sequent Extraction

Herbrand sequents are a generalization of Herbrand disjunctions [10] for the sequent calculus **LK**.

Definition 3 (Herbrand Sequents of a Sequent). *Let s be a closed sequent containing weak quantifiers only. We denote by s_0 the sequent s after removal of all its quantifiers. Any propositionally valid sequent in which the antecedent*

---

[2] Otter Website: http://www-unix.mcs.anl.gov/AR/otter/
[3] Prover9 Website: http://www.cs.unm.edu/ mccune/prover9/

(respectively, consequent) formulas are instances (i.e. their free variables are possibly instantiated by other terms) of the antecedent (respectively, consequent) formulas of s_0 is called a Herbrand sequent of s.

Let s *be an arbitrary sequent and* s' *a skolemization of* s. *Any Herbrand sequent of* s' *is a Herbrand sequent of* s.

Remark 2. In Gentzen's original sequent calculus **LK**, Herbrand sequents are tautologies. In a sequent calculus with arbitrary atomic axioms, as **LKDe**, a valid sequent is only valid with respect to the axioms used in the proof. Hence, the Herbrand sequent will not be a tautology, but only propositionally valid with respect to the axioms used in the proof.

Remark 3. It would be possible to define a Herbrand sequent of an arbitrary sequent s without using skolemization. This could be achieved by imposing eigenvariable conditions on the instantiations chosen for the originally strongly quantified variables. However, the use of skolemization is advantageous, because skolem symbols store information about how the originally strongly quantified variables depend on the weakly quantified variables. This information would be lost if, instead of skolem terms, we had eigenvariables. Hence, skolemization improves readability of the Herbrand sequent.

Example 2 (Herbrand Sequents). Consider the valid sequent

$$(\forall x)((\exists z)P(x,z) \land (\forall y)(P(x,y) \to P(x,f(y)))) \vdash (\forall x)(\exists y)P(x,f^2(y))$$

The following sequents are some of its Herbrand sequents, where g is a skolem-function and a is a skolem-constant produced by skolemization:

1. $P(a,g(a)) \land (P(a,g(a)) \to P(a,f(g(a))))$,
   $P(a,g(a)) \land (P(a,f(g(a))) \to P(a,f^2(g(a)))) \vdash P(a,f^2(g(a)))$
2. $P(g(a),g^2(a)) \land (P(g(a),g^2(a)) \to P(g(a),f(g^2(a))))$,
   $P(g(a),g^2(a)) \land (P(g(a),f(g^2(a))) \to P(g(a),f^2(g^2(a)))) \vdash P(g(a),f^2(g^2(a)))$
3. $P(b,g(b)) \land (P(b,c) \to P(a,f(c))), P(a,g(a)) \land (P(a,g(a)) \to P(a,f(g(a))))$,
   $P(a,g(a)) \land (P(a,f(g(a))) \to P(a,f^2(g(a)))) \vdash P(a,f^2(g(a))), P(a,f^2(d))$

The first two Herbrand sequents above are minimal in the number of formulas, while the third is not.

Apart from its usage as an analysis tool, as described in this paper, the concept of Herbrand disjunction (or Herbrand sequent) also plays an important role in the foundations of Logic and Mathematics, as expressed by Herbrand's Theorem. A concise historical and mathematical discussion of Herbrand's Theorem, as well as its relation to Gödel's Completeness Theorem, can be found in [7].

Theorem 1 (Herbrand's Theorem). *A sequent s is valid if and only if there exists a Herbrand sequent of s.*

Proof. Originally in [10], stated for Herbrand disjunctions. Also in [7] with more modern proof calculi.

Herbrand's theorem guarantees that we can always obtain a Herbrand sequent from a correct proof, a possibility that was realized and exploited by Gentzen in his Mid-Sequent Theorem for sequents consisting of prenex formulas only.

Theorem 2 (Mid-Sequent Theorem). *Let φ be a prenex* **LK***-Proof without non-atomic cuts. Then there is an* **LK***-Proof φ' of the same end-sequent such that no quantifier rule occurs above propositional and cut rules.*

Proof. The original proof, for Gentzen's original sequent calculus **LK** and cut-free **LK**-Proofs, defines rule permutations that shift quantifier rules downwards. The iterated application of the permutations eventually reduces the original proof to a normal form fulfilling the mid-sequent property [8]. The rule permutations can be easily extended to proofs containing atomic cuts.

Remark 4. φ' has a mid-sequent, located between its lower first-order part and its upper propositional part. This mid-sequent is a Herbrand sequent of the end-sequent of φ, after skolemization of φ'.

However, Gentzen's algorithm has one strong limitation: it is applicable only to proofs with end-sequents in prenex form. Although we could transform the end-sequents and the proofs to prenex form, this would compromise the readability of the formulas and require additional computational effort. Prenexification is therefore not desirable in our context, and hence, to overcome this and other limitations in Gentzen's algorithm, we developed three other algorithms. In this paper we describe one of them in detail [4], which was chosen to be implemented within **CERES**. The other two are described in [17,16]. Another different approach, based on functional interpretation and aiming at proofs with cuts, can be found in [9].

4.1   Extraction Via Transformation to Quantifier-Rule-Free LKe$_\mathbf{A}$

The algorithm requires a temporary transformation to an extension of the calculus **LKe**, called **LKe$_\mathbf{A}$** and obtained by the addition of the following two rules, which allow the formation of array formulas $\langle A_1, \ldots, A_n \rangle$ from arbitrary formulas A_j, $j \in \{1, 2, \ldots, n\}$. They will be used to replace some contraction rules in the original proof.

$$\frac{\Delta, A_1, \Gamma_1, \ldots, A_n, \Gamma_n, \Pi \vdash \Lambda}{\Delta, \langle A_1, \ldots, A_n \rangle, \Gamma_1, \ldots, \Gamma_n, \Pi \vdash \Lambda} \; \langle \rangle : l \qquad \frac{\Lambda \vdash \Delta, A_1, \Gamma_1, \ldots, A_n, \Gamma_n, \Pi}{\Lambda \vdash \Delta, \langle A_1, \ldots, A_n \rangle, \Gamma_1, \ldots, \Gamma_n, \Pi} \; \langle \rangle : r$$

To extract a Herbrand sequent of the end-sequent of a skolemized **LKDe**-proof φ such that cuts do not contain quantifiers (nor defined formulas that contain quantifiers), the algorithm executes two transformations:

1. Ψ (definition 4): produces a quantifier-rule-free **LKe$_\mathbf{A}$**-proof where quantified formulas are replaced by array-formulas containing their instances.

2. Φ (definition 5): transforms the end-sequent of the resulting $\mathbf{LKe_A}$-proof into an ordinary sequent containing no array-formulas.

Remark 5. The restriction to proofs such that cuts do not contain quantifiers (nor defined formulas that contain quantifiers) is necessary because otherwise the cut-formulas in each branch of the cut would be substituted by different arrays and the corresponding cut inference in the $\mathbf{LKe_A}$-proof would be incorrect. The extracted sequent would not be propositionally valid, and therefore not a Herbrand sequent. This restriction is not a problem, because we analyze proofs in atomic-cut normal form produced by the *CERES* method.

Let φ be a skolemized $\mathbf{LKDe}$-Proof such that cuts do not contain quantifiers (nor defined formulas that contain quantifiers). A Herbrand sequent of the end-sequent of φ can be obtained by computing:

$$H(\varphi) \doteq \Phi(\textit{end-sequent}(\Psi(\varphi)))$$

Definition 4 (Ψ: Transformation to Quantifier-rule-free $\mathbf{LKe_A}$). *The mapping Ψ transforms a skolemized $\mathbf{LKDe}$-Proof φ such that cuts do not contain quantifiers (nor defined formulas that contain quantifiers) into a quantifier-rule-free $\mathbf{LKe_A}$-Proof according to the inductive definition below:*

Base Case, Initial Axiom Sequents:

$$\Psi(A_1, \ldots, A_n \vdash B_1, \ldots, B_m) \doteq A_1, \ldots, A_n \vdash B_1, \ldots, B_m$$

Proofs Ending with Quantifier-Rules:

$$\Psi\left(\begin{array}{c} [\varphi'] \\ \dfrac{A(t), \Gamma \vdash \Delta}{(\forall x)A(x), \Gamma \vdash \Delta} \ \forall:l \end{array} \right) \doteq \Psi(\varphi')$$

$$\Psi\left(\begin{array}{c} [\varphi'] \\ \dfrac{\Gamma \vdash \Delta, A(t)}{\Gamma \vdash \Delta, (\exists x)A(x)} \ \exists:r \end{array} \right) \doteq \Psi(\varphi')$$

Proofs Ending with Definition-Rules:

$$\Psi\left(\begin{array}{c} [\varphi'] \\ \dfrac{A(t_1, \ldots, t_n), \Gamma \vdash \Delta}{P(t_1, \ldots, t_n), \Gamma \vdash \Delta} \ d:l \end{array} \right) \doteq \Psi(\varphi')$$

$$\Psi\left(\begin{array}{c} [\varphi'] \\ \dfrac{\Gamma \vdash \Delta, A(t_1, \ldots, t_n)}{\Gamma \vdash \Delta, P(t_1, \ldots, t_n)} \ d:r \end{array} \right) \doteq \Psi(\varphi')$$

Proofs Ending with Contractions:

$$\Psi\left(\begin{array}{c} [\varphi'] \\ \dfrac{\Gamma, A, \Lambda_1 \ldots, A, \Lambda_n \vdash \Delta}{\Gamma, A, \Lambda_1, \ldots, \Lambda_n \vdash \Delta} \ c:l \end{array} \right) \doteq \dfrac{\begin{array}{c} [\Psi(\varphi')] \\ \Gamma^*, A_1, \Lambda_1^*, \ldots, A_n, \Lambda_n^* \vdash \Delta^* \end{array}}{\Gamma^*, \langle A_1, \ldots, A_n \rangle, \Lambda_1^*, \ldots, \Lambda_n^* \vdash \Delta^*} \ \langle\rangle:l$$

$$\Psi\left(\begin{array}{c}[\varphi']\\ \dfrac{\Gamma \vdash \Delta, A, A, \Lambda_1 \ldots, A, \Lambda_n}{\Gamma \vdash \Delta, A, \Lambda_1, \ldots, \Lambda_n}\ c:r\end{array}\right) \doteq \dfrac{\begin{array}{c}[\Psi(\varphi')]\\ \Gamma^* \vdash \Delta^*, A_1, \Lambda_1^*, \ldots, A_n, \Lambda_n^*\end{array}}{\Gamma^* \vdash \Delta^*, \langle A_1, \ldots, A_n\rangle, \Lambda_1^*, \ldots, \Lambda_n^*}\ \langle\rangle:r$$

Proofs Ending with other Unary Rules:

$$\Psi\left(\begin{array}{c}[\varphi']\\ \dfrac{A, \Gamma \vdash \Delta}{A \wedge B, \Gamma \vdash \Delta}\ \wedge:l1\end{array}\right) \doteq \dfrac{\begin{array}{c}[\Psi(\varphi')]\\ A^*, \Gamma^* \vdash \Delta^*\end{array}}{A^* \wedge B^*, \Gamma^* \vdash \Delta^*}\ \wedge:l1$$

For other unary rules, Ψ is defined analogously to the case for $\wedge:l1$ above.

Proofs Ending with other Binary Rules:

$$\Psi\left(\begin{array}{c}[\varphi_1']\quad\ [\varphi_2']\\ \dfrac{A, \Gamma_1 \vdash \Delta_1 \quad B, \Gamma_2 \vdash \Delta_2}{A \vee B, \Gamma_1, \Gamma_2 \vdash \Delta_1, \Delta_2}\ \vee:l\end{array}\right) \doteq \dfrac{\begin{array}{c}[\Psi(\varphi_1')]\qquad [\Psi(\varphi_2')]\\ A^*, \Gamma_1^* \vdash \Delta_1^* \quad B^*, \Gamma_2^* \vdash \Delta_2^*\end{array}}{A^* \vee B^*, \Gamma_1^*, \Gamma_2^* \vdash \Delta_1^*, \Delta_2^*}\ \vee:l$$

For other binary rules, Ψ is defined analogously to the case for $\vee:l$ above. Additionally, the introduction of array-formulas requires an adaption of the positions in equality rules, so that replacements are executed in all formulas of the array.

Note that the definition of Ψ is sound, as for every proof φ and for all formula occurrences F in the end-sequent of φ, we can associate a corresponding formula occurrence F^* in the end-sequent of $\Psi(\varphi)$.

The transformation Ψ essentially omits quantifier-rules and definition-rules. Then it replaces contractions by array formations, because the auxiliary formulas of contractions will not be generally equal to each other anymore after the omission of quantifier-rules.

Example 3. Let φ be the following **LKDe**-proof:

$$\dfrac{\begin{array}{c}[\varphi']\\ \dfrac{\dfrac{\dfrac{P(0), P(0) \to P(s(0)), P(s(0)) \to P(s^2(0)) \vdash P(s^2(0))}{P(0), P(0) \to P(s(0)), (\forall x)(P(x) \to P(s(x)) \vdash P(s^2(0))}\ \forall:l}{P(0), \forall x(P(x) \to P(s(x)), (\forall x)(P(x) \to P(s(x)) \vdash P(s^2(0))}\ \forall:l}{P(0), \forall x(P(x) \to P(s(x)) \vdash P(s^2(0))}\ c:l\end{array}}{P(0) \wedge (\forall x)(P(x) \to P(s(x)) \vdash P(s^2(0))}\ \wedge:l$$

Let φ_c be the subproof of φ with the contraction as the last rule. Firstly we compute $\Psi(\varphi_c)$:

$$\dfrac{\begin{array}{c}[\Psi(\varphi')]\\ P(0), P(0) \to P(s(0)), P(s(0)) \to P(s^2(0)) \vdash P(s^2(0))\end{array}}{P(0), \langle P(0) \to P(s(0)), P(s(0)) \to P(s^2(0))\rangle \vdash P(s^2(0))}\ \langle\rangle:l$$

The instances of the auxiliary occurrences of the contraction, $P(0) \to P(s(0))$ and $P(s(0)) \to P(s^2(0))$, are not equal to each other anymore. A contraction is therefore generally not possible, and this is why Ψ replaces contractions by array formations. The array-formulas store all the instances that would have been contracted.

We proceed with the computation of $\Psi(\varphi)$:

$$[\Psi(\varphi')]$$

$$\frac{\dfrac{P(0), P(0) \to P(s(0)), P(s(0)) \to P(s^2(0)) \vdash P(s^2(0))}{P(0), \langle P(0) \to P(s(0)), P(s(0)) \to P(s^2(0))\rangle \vdash P(s^2(0))} \; \langle\rangle : l}{P(0) \wedge \langle P(0) \to P(s(0)), P(s(0)) \to P(s^2(0))\rangle \vdash P(s^2(0))} \; \wedge : l$$

Definition 5 (Φ: Expansion of Array Formulas). *The mapping Φ transforms array formulas and sequents into first-order logic formulas and sequents. In other words, Φ eliminates $\langle \ldots \rangle$ and can be defined inductively by:*

1. *If A is a first-order logic formula, then $\Phi(A) \doteq A$*
2. *$\Phi(\langle A_1, \ldots, A_n \rangle) \doteq \Phi(A_1), \ldots, \Phi(A_n)$*
3. *If $\Phi(A) = A_1, \ldots, A_n$, then $\Phi(\neg A) \doteq \neg A_1, \ldots, \neg A_n$*
4. *If $\Phi(A) = A_1, \ldots, A_n$ and $\Phi(B) = B_1, \ldots, B_m$, then $\Phi(A \circ B) \doteq A_1 \circ B_1, \ldots, A_1 \circ B_m, \ldots, A_n \circ B_1, \ldots, A_n \circ B_m$, for $\circ \in \{\wedge, \vee, \to\}$*
5. *$\Phi(A_1, \ldots, A_n \vdash B_1, \ldots, B_m) \doteq \Phi(A_1), \ldots, \Phi(A_n) \vdash \Phi(B_1), \ldots, \Phi(B_m)$*

*Φ has not been defined over formulas that contain array formulas in the scope of quantifiers. This is not necessary, because Ψ transforms **LKDe**-proofs into **LKe$_A$**-proofs where this situation never occurs.*

Example 4. Let $\Psi(\varphi)$ be the **LKe$_A$**-Proof in Example 3. Then, its end-sequent, after mapping array formulas to sequences of formulas, is:

$$\Phi(\text{end-sequent}(\Psi(\varphi))) = \begin{pmatrix} P(0) \wedge (P(0) \to P(s(0))), \\ P(0) \wedge (P(s(0)) \to P(s^2(0))) \end{pmatrix} \vdash P(s^2(0))$$

Theorem 3 (Soundness). *Let φ be a skolemized **LKDe**-Proof in atomic cut normal form. Then the sequent $H(\varphi)$, extracted by the algorithm defined above, is a Herbrand sequent of the end-sequent of φ.*

Proof. A proof (fully available in [17]) can be sketched in the following way. We have to show that:

1. The formulas of $H(\varphi)$ are substitution instances of the formulas of the end-sequent of φ without their quantifiers.
2. $H(\varphi)$ is a valid sequent.

Item 1 follows clearly from the definitions of Φ and Ψ, because Ψ substitutes quantified sub-formulas of the end-sequent by array-formulas containing only substitution instances of the respective sub-formulas, and Φ expands the the array-formulas maintaining the structure of the formulas where they are located.

Item 2 can be proved by devising a transformation Φ_P that maps the intermediary **LKDe$_A$**-proof $\Psi(\varphi)$ to an **LKDe**-proof $\Phi_P(\Psi(\varphi))$ by substituting $\langle\rangle : l$ rules by sequences of $\wedge : l$ rules and $\langle\rangle : r$ rules by sequences of $\vee : r$ rules. $\Psi(\varphi)$ does not contain quantifier rules. Therefore $\Phi_P(\Psi(\varphi))$ is essentially a propositional **LKe**-proof, in which the arrays of the end-sequent were substituted by either nested $\wedge$ connectives or nested $\vee$ connectives. The end-sequent of $\Phi_P(\Psi(\varphi))$ can be shown, by structural induction, to be logically equivalent to the extracted sequent $\Phi(\text{end-sequent}(\Psi(\varphi)))$. Therefore the extracted sequent is valid.

5   Analysis of the Lattice Proof

In this section, the usefulness of a Herbrand sequent for understanding a formal proof will be demonstrated on a simple example from lattice theory . There are several different, but equivalent, definitions of *lattice*. Usually, the equivalence of several statements is shown by proving a cycle of implications. While reducing the size of the proof, this practice has the drawback of not providing *direct* proofs between the statements. Cut-elimination can be used to automatically generate a direct proof between any two of the equivalent statements. In this section, we will demonstrate how to apply cut-elimination with the CERES-system followed by Herbrand sequent extraction for this purpose.

5.1   The Lattice Proof

Definitions 7, 8 and 10 list different sets of properties that a 3-tuple $\langle L, \cap, \cup \rangle$ or a partially ordered set $\langle S, \leq \rangle$ must have in order to be considered a lattice.

Definition 6 (Semi-Lattice). *A semi-lattice is a set L together with an operation $\circ$ which is*

- *commutative:* $(\forall x)(\forall y)\ x \circ y = y \circ x,$
- *associative:* $(\forall x)(\forall y)(\forall z)\ (x \circ y) \circ z = x \circ (y \circ z)$ *and*
- *idempotent:* $(\forall x)\ x \circ x = x.$

Definition 7 (Lattice: definition 1). *A L1-lattice is a set L together with operations $\cap$ (meet) and $\cup$ (join) s.t. both $\langle L, \cap \rangle$ and $\langle L, \cup \rangle$ are semi-lattices and $\cap$ and $\cup$ are "inverse" in the sense that*

$$(\forall x)(\forall y)\ x \cap y = x \leftrightarrow x \cup y = y.$$

Definition 8 (Lattice: definition 2). *A L2-lattice is a set L together with operations $\cap$ and $\cup$ s.t. both $\langle L, \cap \rangle$ and $\langle L, \cup \rangle$ are semi-lattices and the absorption laws*

$$(\forall x)(\forall y)\ (x \cap y) \cup x = x \quad and \quad (\forall x)(\forall y)\ (x \cup y) \cap x = x$$

hold.

Definition 9 (Partial Order). *A binary relation $\leq$ on a set S is called* partial order *if it is*

- *reflexive (R):* $(\forall x)\ x \leq x,$
- *anti-symmetric (AS):* $(\forall x)(\forall y)\ ((x \leq y \wedge y \leq x) \rightarrow x = y)$ *and*
- *transitive (T):* $(\forall x)(\forall y)(\forall z)\ ((x \leq y \wedge y \leq z) \rightarrow x \leq z).$

Definition 10 (Lattice: definition 3). *A L3-lattice is a partially ordered set $\langle S, \leq \rangle$ s.t. for each two elements of S there exist*

- *greatest lower bound (GLB)* $\cap$*, i.e.*

$$(\forall x)(\forall y)(x \cap y \leq x \wedge x \cap y \leq y \wedge (\forall z)((z \leq x \wedge z \leq y) \rightarrow z \leq x \cap y)),$$

– *least upper bound (LUB)* $\cup$, i.e.

$$(\forall x)(\forall y)(x \leq x \cup y \wedge y \leq x \cup y \wedge (\forall z)((x \leq z \wedge y \leq z) \rightarrow x \cup y \leq z)).$$

The above three definitions of lattice are equivalent. We will formalize the following proofs of $L1 \rightarrow L3$ and $L3 \rightarrow L2$ in order to extract a direct proof of $L1 \rightarrow L2$, i.e. one which does not use the notion of partial order.

Proposition 1. *L1-lattices are L3-lattices.*

Proof. Given $\langle L, \cap, \cup \rangle$, define $x \leq y$ as $x \cap y = x$. By idempotence of $\cap$, $\leq$ is reflexive. Anti-symmetry of $\leq$ follows from commutativity of $\cap$ as $(x \cap y = x \wedge y \cap x = y) \rightarrow x = y$. To see that $\leq$ is transitive, assume (a) $x \cap y = x$ and (b) $y \cap z = y$ to derive

$$x \cap z =^{(a)} (x \cap y) \cap z =^{(\text{assoc.})} x \cap (y \cap z) =^{(b)} x \cap y =^{(a)} x$$

So $\leq$ is a partial order on L.

By associativity, commutativity and idempotence of $\cap$, we have $(x \cap y) \cap x = x \cap y$, i.e. $x \cap y \leq x$ and similarly $x \cap y \leq y$, so $\cap$ is a lower bound for $\leq$. To see that $\cap$ is also the greatest lower bound, assume there is a z with $z \leq x$ and $z \leq y$, i.e. $z \cap x = z$ and $z \cap y = z$. Then, by combining these two equations, $(z \cap y) \cap x = z$, and therefore, $z \leq x \cap y$.

To show that $\cup$ is an upper bound, derive from the axioms of semi-lattices that $x \cup (x \cup y) = x \cup y$ which, by the "inverse" condition of L1 gives $x \cap (x \cup y) = x$, i.e. $x \leq x \cup y$ and similarly for $y \leq x \cup y$. Now assume there is a z with $x \leq z$ and $y \leq z$, i.e. $x \cap z = x$ and $y \cap z = z$ and by the "inverse" condition of L1: $x \cup z = z$ and $y \cup z = z$. From these two equations and the axioms of semi-lattices, derive $(x \cup y) \cup z = z$ which, by the "inverse" condition of L1, gives $(x \cup y) \cap z = x \cup y$, i.e. $x \cup y \leq z$.

Proposition 2. *L3-lattices are L2-lattices.*

Proof. We want to show the absorption law $(x \cap y) \cup x = x$. That $x \leq (x \cap y) \cup x$ follows immediately from $\cup$ being an upper bound. But $x \cap y \leq x$ because $\cap$ is a lower bound. Furthermore also $x \leq x$, so x is an upper bound of $x \cap y$ and x. But as $\cup$ is the lowest upper bound, we have $(x \cap y) \cup x \leq x$ which by anti-symmetry of $\leq$ proves $(x \cap y) \cup x = x$. For proving the other absorption law $(x \cup y) \cap x = x$, proceed symmetrically.

By concatenation, the above two proofs show that all L1-lattices are L2-lattices. However, this proof is not a direct one, it uses the notion of partially ordered set which occurs neither in L1 nor in L2. By cut-elimination we will generate a direct formal proof automatically.

5.2   Overview of the Analysis

The analysis of the lattice proof followed the steps below:

1. *Formalization of the lattice proof in the sequent calculus* **LKDe**: semi-auto-mated by HLK[4]. Firstly the proof was written in the language HandyLK, which can be seen as an intermediary language between informal mathematics and **LKDe**. Subsequently, HLK compiled it to **LKDe**.
2. *Cut-Elimination of the formalized lattice proof*: fully automated by CERES, employing the cut-elimination procedure based on resolution, sketched in Section 3, to obtain an **LKDe**-proof in Atomic-Cut Normal Form (ACNF), i.e. a proof in which cut-formulas are atoms.
3. *Extraction of the Herbrand sequent of the ACNF*: fully automated by CERES, employing the algorithm described in Section 4.
4. *Use of the Herbrand sequent* to interpret and understand the ACNF, in order to obtain a new direct informal proof.

5.3   Formalization of the Lattice Proof

The full formal proof has 260 rules (214 rules, if structural rules (except cut) are not counted). It is too large to be displayed here. Below we show only a part of it, which is close to the end-sequent and depicts the main structure of the proof, based on the cut-rule with $L3$ as the cut-formula. This cut divides the proof into two subproofs corresponding to propositions 1 and 2. The full proofs, conveniently viewable with ProofTool[5], are available in the website of CERES.

$$
\cfrac{
\cfrac{
[p_R] \quad
\cfrac{
[p_{AS}] \quad [p_T]
}{
\vdash AS \quad \vdash T
}
}{
\cfrac{\vdash R \quad \vdash AS \wedge T}{\cfrac{\vdash R \wedge (AS \wedge T)}{\vdash POSET} \; d:r} \wedge : r
} \wedge : r
\qquad
\cfrac{
\cfrac{[p_{GLB}] \quad [p_{LUB}]}{L1 \vdash GLB \wedge LUB} \wedge : r
}{\; \wedge : r}
}{
\cfrac{
\cfrac{L1 \vdash POSET \wedge (GLB \wedge LUB)}{L1 \vdash L3} \; d:r
\qquad
\cfrac{[p_3^2]}{L3 \vdash L2}
}{L1 \vdash L2} \; cut
}
$$

- $L1 \equiv \forall x \forall y ((x \cap y) = x \supset (x \cup y) = y) \wedge ((x \cup y) = y \supset (x \cap y) = x)$
- $L2 \equiv \forall x \forall y (x \cap y) \cup x = x \wedge \forall x \forall y (x \cup y) \cap x = x$
- $L3 \equiv POSET \wedge (GLB \wedge LUB)$
- p_{AS}, p_T, p_R are proofs of, respectively, anti-symmetry (AS), transitivity (T) and reflexivity (R) of $\leq$ from the axioms of semi-lattices.
- p_3^2 is a proof that $L3$-lattices are $L2$-lattices, from the axioms of semi-lattices.

5.4   Cut-Elimination of the Lattice Proof

Prior to cut-elimination, the formalized proof is skolemized by CERES, resulting in a proof of the skolemized end-sequent $L1 \vdash (s_1 \cap s_2) \cup s_1 = s_1 \wedge (s_3 \cup s_4) \cap s_3 = s_3$, where s_1, s_2, s_3 and s_4 are skolem constants for the strongly quantified variables of $L2$. Then CERES eliminates cuts, producing a proof in atomic-cut normal (also available for visualization with ProofTool in the website of CERES).

---

[4] HLK Website: http://www.logic.at/hlk/
[5] ProofTool Website: http://www.logic.at/prooftool/

The ACNF is still quite large (214 rules; 72 rules not counting structural rules (except cut)). It is interesting to note, however, that the ACNF is smaller than the original proof in this case, even though in the worst case cut-elimination can produce a non-elementary increase in the size of proofs [14].

Although the ACNF of the Lattice proof is still large (214 rules), the extracted Herbrand sequent contains only 6 formulas, as shown in Subsection 5.5. Therefore, the Herbrand sequent significantly reduces the amount of information that has to be analyzed in order to extract the direct mathematical argument contained in the ACNF.

5.5   Herbrand Sequent Extraction of the ACNF of the Lattice Proof

The Herbrand sequent of the ACNF, after set-normalization and removal of remaining sub-formulas introduced by weakening (or as the non-auxiliary formula of $\vee$ and $\wedge$ rules) in the ACNF, is:

$(A1)$ $s_1 \cup (s_1 \cup (s_1 \cap s_2)) = s_1 \cup (s_1 \cap s_2) \rightarrow s_1 \cap (s_1 \cup (s_1 \cap s_2)) = s_1$,

$(A2)$ $s_1 \cap s_1 = s_1 \rightarrow s_1 \cup s_1 = s_1$,

$(A3)$ $\underbrace{(s_1 \cap s_2) \cap s_1 = s_1 \cap s_2}_{(A3i)} \rightarrow (s_1 \cap s_2) \cup s_1 = s_1$,

$(A4)$ $(s_1 \cup (s_1 \cap s_2)) \cup s_1 = s_1 \rightarrow (s_1 \cup (s_1 \cap s_2)) \cap s_1 = s_1 \cup (s_1 \cap s_2)$,

$(A5)$ $\underbrace{s_3 \cup (s_3 \cup s_4) = s_3 \cup s_4}_{(A5i)} \rightarrow s_3 \cap (s_3 \cup s_4) = s_3$

$(C1)$ $\vdash \underbrace{(s_1 \cap s_2) \cup s_1 = s_1}_{(C1i)} \wedge \underbrace{(s_3 \cup s_4) \cap s_3 = s_3}_{(C1ii)}$

5.6   Construction of the Informal Proof

After extracting a Herbrand sequent from the ACNF, the next step is to construct an informal, analytic proof of the theorem, based on the ACNF, but using only the information about the variable instantiations contained in its extracted Herbrand sequent. We want to stress that in the following, we are not performing syntactic manipulations of formulas of first-order logic, but instead we use the formulas from the Herbrand sequent of the ACNF as a *guide* to construct an analytical mathematical proof.

Theorem 4. *All L1-lattices $\langle L, \cap, \cup \rangle$ are L2-lattices.*

Proof. As both lattice definitions have associativity, commutativity and idempotence in common, it remains to show that the absorption laws hold for $\langle L, \cap, \cup \rangle$. We notice that, as expected, these properties coincide with the conjunction $(C1)$ for arbitrary $s_1, \ldots, s_4$ on the right hand side of the Herbrand sequent and so we proceed by proving each conjunct for arbitrary $s_1, \ldots, s_4 \in L$:

1. We notice that $(A3i) + (A3)$ imply $(C1i)$. So we prove these properties:

(a) *First we prove* (A3i):

$$s_1 \cap s_2 =^{\text{(idem.)}} (s_1 \cap s_1) \cap s_2 =^{\text{(assoc.)}} s_1 \cap (s_1 \cap s_2) =^{\text{(comm.)}}$$
$$s_1 \cap (s_2 \cap s_1) =^{\text{(assoc.)}} (s_1 \cap s_2) \cap s_1$$

(b) *Assume* $(s_1 \cap s_2) \cap s_1 = s_1 \cap s_2$. *By definition of* L1-*lattices,* $(s_1 \cap s_2) \cup s_1 = s_1$. *Thus, we have proved* (A3).

2. *Again, we notice that* (A5i) + (A5) + *commutativity imply* (C1ii) *and use this fact:*

(a) $s_3 \cup s_4 =^{\text{(idem.)}} (s_3 \cup s_3) \cup s_4 =^{\text{(assoc.)}} s_3 \cup (s_3 \cup s_4)$. *We have proved* (A5i).

(b) *Assume* $s_3 \cup (s_3 \cup s_4) = s_3 \cup s_4$. *By definition of* L1-*lattices,* $s_3 \cap (s_3 \cup s_4) = s_3$. *This proves* (A5).

So we have shown that for arbitrary $s_1, \ldots, s_4 \in L$, *we have* $(s_1 \cap s_2) \cup s_1 = s_1$ *and* $(s_3 \cup s_4) \cap s_3 = s_3$, *which completes the proof.*

Contrary to the proof in Section 5.1, we can now directly see the algebraic construction used to prove the theorem. This information was hidden in the synthetic argument that used the notion of partially ordered sets and was revealed by cut-elimination.

This example shows that the Herbrand sequent indeed contains the essential information of the ACNF, since an informal direct proof corresponding to the ACNF could be constructed by analyzing the extracted Herbrand sequent only.

6   Conclusion

We have described a new algorithm for Herbrand sequent extraction, which is better than Gentzen's mid-sequent reduction because it can be applied to proofs that are not in prenex form. Its use as a tool for the analysis of computer generated proofs was successfully demonstrated with a simple proof about lattices, which was automatically transformed to atomic-cut normal form by the CERES system. The Herbrand sequent significantly reduced the amount of information that had to be analyzed in order to understand the atomic-cut normal form produced by CERES:

Our algorithm still lacks support for definition rules, because they are removed by the transformation to **LKe$_\mathbf{A}$**. We are planning to improve on this and to be able to reinsert defined formulas in the extracted Herbrand sequent, in order to further improve its readability.

The technique described here is not limited to sequent calculi, since it relies on Herbrand's theorem, which is applicable to first-order logic in general. Many computer-generated proofs are obtained, for example, by automated theorem provers that do not work with sequent calculi, but with resolution calculi. Such resolution proofs are usually even harder for humans to understand, and therefore we are planning to extend the general idea behind Herbrand sequent extraction to an algorithm applicable to resolution proofs as well. This could be done by firstly translating a resolution refutation into a corresponding proof in sequent calculus.

References

1. Baaz, M., Hetzl, S., Leitsch, A., Richter, C., Spohr, H.: Ceres: An Analysis of Fürstenberg's Proof of the Infinity of Primes. Theoretical Computer Science (to appear)
2. Baaz, M., Hetzl, S., Leitsch, A., Richter, C., Spohr, H.: Cut-Elimination: Experiments with CERES. In: Baader, F., Voronkov, A. (eds.) LPAR 2004. LNCS (LNAI), vol. 3452, pp. 481–495. Springer, Heidelberg (2005)
3. Baaz, M., Hetzl, S., Leitsch, A., Richter, C., Spohr, H.: Proof Transformation by CERES. In: Borwein, J.M., Farmer, W.M. (eds.) MKM 2006. LNCS (LNAI), vol. 4108, pp. 82–93. Springer, Heidelberg (2006)
4. Baaz, M., Leitsch, A.: On skolemization and proof complexity. Fundamenta Matematicae 20, 353–379 (1994)
5. Baaz, M., Leitsch, A.: Cut normal forms and proof complexity. Annals of Pure and Applied Logic 97, 127–177 (1999)
6. Baaz, M., Leitsch, A.: Cut-elimination and Redundancy-elimination by Resolution. Journal of Symbolic Computation 29(2), 149–176 (2000)
7. Buss, S.R.: On Herbrand's theorem. In: Leivant, D. (ed.) LCC 1994. LNCS, vol. 960, p. 195. Springer, Heidelberg (1995)
8. Gentzen, G.: Untersuchungen über das logische Schließen. In: Szabo, M.E. (ed.) The Collected Papers of Gerhard Gentzen, pp. 68–131. North-Holland Publishing Company, Amsterdam (1969)
9. Gerhardy, P., Kohlenbach, U.: Extracting herbrand disjunctions by functional interpretation. Archive for Mathematical Logic 44(5) (2005)
10. Herbrand, J.: Recherches sur la théorie de la démonstration. Travaux de la Societé des Sciences et des Lettres de Varsovie, Class III, Sciences Mathématiques et Physiques 33 (1930)
11. Hetzl, S.: Characteristic Clause Sets and Proof Transformations. PhD thesis, Vienna University of Technology (2007)
12. Hetzl, S., Leitsch, A.: Proof transformations and structural invariance. In: Aguzzoli, S., Ciabattoni, A., Gerla, B., Manara, C., Marra, V. (eds.) ManyVal 2006. LNCS (LNAI), vol. 4460, pp. 201–230. Springer, Heidelberg (2007)
13. Luckhardt, H.: Herbrand-Analysen zweier Beweise des Satzes von Roth: polynomiale Anzahlschranken. Journal of Symbolic Logic 54, 234–263 (1989)
14. Statman, R.: Lower bounds on Herbrand's theorem. Proceedings of the American Mathematical Society 75, 104–107 (1979)
15. Sutcliffe, G., Suttner, C.: The State of CASC. AI Communications 19(1), 35–48 (2006)
16. Paleo, B.W.: Herbrand sequent extraction. Master's thesis, Technische Universitaet Dresden; Technische Universitaet Wien, Dresden, Germany, Wien, Austria (2007)
17. Paleo, B.W.: Herbrand Sequent Extraction. VDM-Verlag. Saarbruecken, Germany (2008)

Visual Mathematics: Diagrammatic Formalization and Proof

John Howse and Gem Stapleton

The Visual Modelling Group
University of Brighton, Brighton, UK
{john.howse,g.e.stapleton}@brighton.ac.uk
www.cmis.brighton.ac.uk/research/vmg

Abstract. Diagrams have been used for centuries in the visualization of mathematical concepts and to aid the exploration and formalization of ideas. This is hardly surprising given the intuitive appeal of visual languages. Thus it seems very natural to establish how diagrams can play an integral part of mathematical formalization and reasoning, giving them the same status as the symbolic languages that they are used alongside. Indeed, recently we have seen the emergence of diagrammatic reasoning systems that are defined with sufficient mathematical rigour to allow them to be used as formal tools in their own right. Some of these systems have been designed with particular application areas in mind, such as number theory and real analysis, or formal logics. This paper focuses on the use of diagrammatic logics to formalize mathematical theories with the same level of rigour that is present in their corresponding predicate logic axiomatizations. In particular, extensions to the constraint diagram logic are proposed to make it more suitable for use in mathematics. This extended logic is illustrated via the diagrammatic formalization of some commonly occurring mathematical concepts. Subsequently, we demonstrate its use in the proofs of some simple theorems.

1 Introduction

The demonstrable popularity of diagrammatic communication in mathematics lies in the widespread use of diagrams to aid intuition and visualize concepts. In all probability the pages of an arbitrarily chosen mathematics book will contain illustrative diagrams; examples that emphasize the role of diagrams include [1,16,21]. Indeed, [7] argues "visual thinking in mathematics is rarely just a superfluous aid; it usually has epistemological value, often as a means of discovery."

The prevalent usage of diagrams to illustrate mathematical concepts and their role in discovery motivates the development of formal diagrammatic languages that are suitable for writing definitions, formulating theorems and constructing proofs. After all, the fact that there are benefits of using diagrams yields the obvious question as to whether diagrams can be used on an equal footing with symbolic notations. Recent results show that this is the case, with many different diagrammatic reasoning systems emerging [3,11,12,19,20,24,27,28]. In the

S. Autexier et al. (Eds.): AISC/Calculemus/MKM 2008, LNAI 5144, pp. 478–493, 2008.

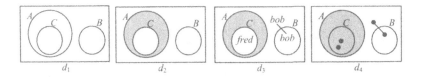

Fig. 1. Logics based on Euler diagrams

context of metric space analysis it has been shown that a diagrammatic logic specifically designed with this topic in mind allows students to outperform others who are taught using a more typical algebraic formalism [29], for example.

Most closely related to the contribution made in this paper is the development of diagrammatic logics based on Euler diagrams which are often mistakenly called Venn diagrams[1]. The Euler diagram d_1, figure 1, asserts that A and B are disjoint and C is a subset of A. The relative placement of the curves give, for free, that C is disjoint from B. This 'free ride' is one of the areas where diagrams are thought to be superior to symbolic languages [18]. This example also illustrates the concept of 'well-matchedness' [8] since the syntactic properties that the diagram uses to make assertions mirror those at the semantic level: the containment of one curve by another mirrors the interpretation that the enclosed curve, C, represents a subset of the set represented by the enclosing curve, A.

The expressiveness of Euler diagrams can be increased by incorporating traditional logical connectives, such as $\wedge$ and $\vee$, and the negation operator, $\neg$; this results in a language equivalent to monadic first order logic (MFOL) [25]. Also equivalent to MFOL is Shin's Venn-II language [19], which is based on Venn diagrams and uses shading to assert emptiness and $\otimes$ symbols to assert non-emptiness. In Venn-II, the only connective used is $\vee$. The use of shading to assert emptiness dates back to Venn; in Euler diagrams, emptiness can be asserted by the absence of a region. The intuitiveness of, and advantages of using, shading are discussed in [23].

Various extensions to Euler diagrams have been proposed, such as including syntax to represent named individuals [27], or assert the existence of arbitrary finite numbers of elements [11]. The Euler diagram d_2 in figure 1 is augmented with shading which asserts the emptiness of the set $A - C$ and the Euler/Venn diagram d_3 tells us, in addition, that *fred* is in the set C and *bob* is not in the set A. The spider diagram d_4 asserts the existence of two elements in the set C and at least one in the set $\overline{A}$; this is done through the use of the trees which are called existential spiders. The shading in d_3 is used to place an upper bound on the cardinality of A, limiting it to two: in the set represented by a shaded region, all elements must be denoted by spiders.

These extensions to Euler diagrams result in monadic languages. Constraint diagrams [13] further extend Euler diagrams and use *universal spiders*, which are represented by asterisks, to make universally quantified statements and arrows

---

[1] In a Venn diagram, all intersections between the closed curves must be present; this condition is removed in the case of Euler diagrams.

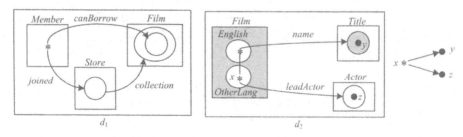

Fig. 2. Two constraint diagrams

to talk about properties of binary relations. Constraint diagrams were designed for formal software specification and, in figure 2, d_1 asserts that every member (using the asterisk) can only borrow films that are in the collections of the stores they have joined; this might be a typical constraint one may place on a film rental system.

Euler diagrams are frequently used in the teaching of set theory to illustrate concepts as they are introduced. More broadly, Euler diagrams are widely used in mathematics. As a consequence, logics that build on Euler diagrams may well enhance the teaching of logic and appeal to mathematicians. In this paper, we investigate the use of constraint diagrams in mathematics, as opposed to software engineering. Since constraint diagrams form a logic in their own right, it seems natural to ask whether they are capable of axiomatizing common theories.

Section 2 gives a brief introduction to constraint diagrams and some recent generalizations of them. We discuss the use of these generalized constraint diagrams to formalize mathematical concepts and subsequently present some extensions to the syntax that allow mathematical expressions to be more easily formulated. As two simple case studies, we define axioms for the theory of partially ordered sets and equivalence relations in our extended notation, presented in section 3. Further we illustrate the use of our extended constraint diagrams through the proof of some theorems in section 4. Section 5 presents further extensions, allowing a variety of second order statements to be made.

2   Constraint Diagrams

The constraint diagram language was introduced specifically for software specification [13] and in this section we overview its syntax and semantics; a formal treatment can be found in [4].

The so-called *unitary* diagram d_1 in figure 2 contains three *given contours*; these are closed curves labelled *Member, Film* and *Store* respectively. In a unitary diagram, no two distinct contours have the same label. Given contours represent sets, and their spatial relationships make statements about containment and disjointness; in d_1, figure 2, the given contours assert that *Member, Film* and *Store* are pairwise disjoint because no pair of contours overlap. Also in the diagram are three *derived contours* which assert the existence of a set; these are the closed curves that are not labelled and happen to be targeted by *arrows*.

The asterisk is called a *universal spider* and its *habitat* is the *region* in which it is placed. In other words, its habitat is the region inside the contour labelled *Member*. In this example, this region is also called the *domain* of the universal spider; the domain is the region which represents the set that the spider quantifies over, in this case the set *Member*. It is not necessarily the case that the habitat equals the domain; see [4] for full details. A region is a set of *zones*. A zone is a maximal set of points in the plane, described as being inside certain contours (possibly no contours) and outside the rest of the contours. The diagram d_1 in figure 2 contains seven zones, three of which are inside *Films*.

In a unitary diagram, every arrow is *sourced* on, and targets, a spider or a contour (given or derived). Every arrow has a label and, unlike given contours, two (or more) arrows can have the same label. The actual set represented by a derived contour is determined by any targeting arrow(s). In d_1, figure 2, given any member, m, the derived contour inside the given contour labelled *Store* represents the image of the relation *joined* when the domain is restricted to m (the derived contour represents the set of stores that m has joined).

Unitary diagrams can also contain *existential spiders* and *shading* and, strictly speaking, every unitary diagram is augmented with a *reading tree* [4] (the diagram d_1 in figure 2 does not have a reading tree). Existential spiders are denoted by trees whose nodes are round and filled (as opposed to asterisks in the universal case) and they represent the existence of elements. For example in d_2, figure 2, there are two existential spiders; these are the dots labelled y and z and each of these two spiders has a single zone habitat. By contrast, the universal spider labelled x has a two zone habitat and quantifies over the set $English \cup OtherLang$.

Shading allows us to place upper bounds on set cardinality: in a shaded region, all of the elements are represented by spiders. So, in figure 2, d_2 expresses that the set $Film - (English \cup OtherLang)$ is empty since there are no spiders placed in the corresponding shaded zone. The meaning of the diagram is determined by the order in which the quantifiers (i.e. the spiders) are interpreted. The reading tree on the right of the diagram tells us the order in which to read the quantifiers, thus resolving ambiguities: we start with the universal spider and then we can read the existential spiders in either order and independently of each other. Moreover, the reading tree also provides quantifier scoping and bracketing information (the details can be found in [4]). The diagram expresses:

1. *Film*, *Title* and *Actor* are pairwise disjoint,
2. *English* and *OtherLang* form a partition of *Film* and
3. every film, x, has a unique name which is a title and, in addition, x has at least one lead actor.

The first two statements capture the information provided by the underlying Euler diagram. The uniqueness of each film's name is asserted by the use of an existential spider (giving the existence of the name) and shading (which asserts that the only elements in the set of names are represented by spiders); likewise, the existence of a lead actor is denoted by the use of a spider but there can be more than one lead actor due to the absence of shading. Some examples of specifications using constraint diagrams can be seen in [10,14].

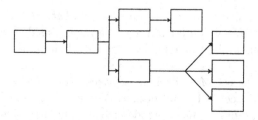

Fig. 3. Diagrams as partialorders

There are some issues regarding the well-matchedness of constraint diagrams to their meaning [22,23] although much of the time they work well for their intended application area; see [9] for discussions on well-matchedness in the context of software engineering diagrams. Specifically for constraint diagrams [22,23] reports on some improvements to the notation that allow well-matched statements to be made more often, presenting generalized constraint diagrams which do not require reading trees. Figure 3 illustrates the type of structure that might be present in a generalized constraint diagram, where the boxes are unitary diagrams. In general, each non-root unitary diagram in such a partial order is a copy of its immediate ancestor with some changes to the syntax. Arrows connect the boxes, with those of the form $-\!\!\!\prec$ indicating conjunction whereas those of the form $-\!\!\!\vdash$ indicate disjunction. As a trivial change, in this paper we simply use juxtaposition for $\wedge$ and write OR to indicate a disjunction. Bounding boxes are used for brackets. A key difference between generalized constraint diagrams and constraint diagrams augmented with reading trees is the way in which quantification works: in augmented constraint diagrams, quantification scopes only over parts of the unitary diagram in which it occurs but with generalized diagrams, the scope is, informally, the unitary diagram, d, in which it occurs and all of the ancestors of d.

There is a simple generalized constraint diagram in figure 4. The first (root) unitary diagram contains syntax which expresses

$$\forall x \in Member \{x\}.canBorrow \subseteq Film$$

where $\{x\}.canBorrow$ denotes the image of the relation $canBorrow$ when the domain is restricted to $\{x\}$. The second rectangle includes additional syntax, expressing that every element of $\{x\}.canBorrow$ is available to x.

However, this generalized version of constraint diagrams was designed with software specification in mind. When attempting to use the notation for defining mathematical concepts further modifications are beneficial.

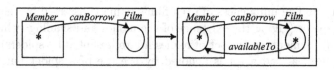

Fig. 4. A generalized constraint diagram

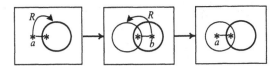

Fig. 5. Formalizing the symmetric property with generalized constraint diagrams

3   Formalizing Mathematical Concepts

In mathematics, we often wish to talk about properties of relations in a global sense, making assertions such as 'R is an equivalence relation' or 'the relation S is a total order'. Frequently, such assertions do not require us to assert the distinctness of elements and require us to leave open the possibility that an element may be related to itself. In these types of constructions constraint diagrams can become cluttered. In this section, we take an example based approach to demonstrating some improvements to generalized constraint diagrams that make them more amenable to use in mathematics.

A formalization of the relation R being symmetric using generalized constraint diagrams can be seen in figure 5, which is not particularly intuitive or visually clear. In part, this is because one must construct the set of elements to which a is related in some unitary diagram in which a appears, but taking care not to specify whether a is in that set; the lefthand diagram constructs the image of a, namely $a.R$, which is then used in the middle diagram but there is a requirement on the user to recall the contour with greater weight is $a.R$.

Constraint diagrams are very good at making strong statements and, in the context of object-oriented modelling this does not necessarily lead to visually cluttered diagrams. For example, associations (binary relations) often hold between disjoint classes (sets) of objects. This means that one often wants to talk about distinct objects, rather than having to worry about not specifying distinctness, or is able to assume that an element in the domain of a binary relation is not related to itself (since binary relations are often irreflexive).

Our first extension to the generalized constraint diagram syntax is to allow arrows to connect components placed in different unitary diagrams, allowing the images of relations to be constructed where we wish to make some statement about that image, rather than necessarily in the unitary diagram where the arrow source occurs. To illustrate, the diagram in figure 6 asserts that R is symmetric in a much more elegant manner than the generalized constraint diagram in figure 5. The left-hand diagram includes just a universal spider, allowing us to say 'for all a'. The arrow sourced on this spider targets a derived contour in the middle

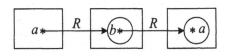

Fig. 6. The relation R is symmetric

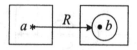

Fig. 7. Everything is related to something

diagram; this derived contour represents the set of elements to which a is related. Then, placing a universal spider, b, inside this derived contour allows us to talk about each element to which a is related and, using a further arrow between unitary parts, we assert that b is related to a (the right-hand diagram).

A particular feature of constraint diagrams is that arrows are used to construct the set of objects to which something is related. By contrast, typical symbolic constructs tend to just assert whether two elements are related. To illustrate, $\forall x \exists y R(x, y)$ says everything is related to something whereas a constraint diagram would assert that everything is related to a non-empty set of elements as in figure 7. Further useful extensions to the syntax are presented in the subsections below, in the context of formalizing commonly occurring mathematical concepts.

3.1 Equivalence Relations

A fundamental concept in mathematics is that of an equivalence relation. We have already shown how to formalize the symmetric property using our extended constraint diagrams (figure 6). The reflexive and transitive properties are captured by the lefthand and righthand diagrams respectively in figure 8. For the transitive property, the relevant diagram in figure 8 presents a formalization that is rather different in style to the usual $\forall x \forall y \forall z ((R(x, y) \wedge R(y, z)) \Rightarrow R(x, z))$. We believe that these two different presentations provide alternative perspectives on the same concept and could, if used in together, help students gain a deeper appreciation of transitivity.

3.2 Ordered Sets

Another commonly occurring construct is that of a partially ordered set, that is a set with a reflexive, antisymmetric and transitive relation on it. To formalize antisymmetry it is convenient to introduce the $\Rightarrow$ operator into the language, as well as allowing $=$ to be written between syntactic elements when we wish to make statements involving equality. We note that adding $\Rightarrow$ and $=$ to the syntax does not increase the first-order expressive power of the language but it does facilitate

Fig. 8. The relation R is reflexive and R is transitive

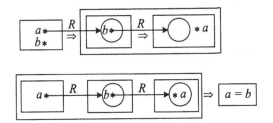

Fig. 9. Two approaches to formalizing antisymmetry

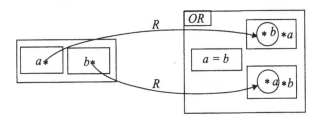

Fig. 10. Trichotomy

ease of construction of certain statements. Two syntactically different ways of formalizing antisymmetry can be seen in figure 9, with the bottom diagram closely
resembling the typical symbolic definition $\forall a \forall b((R(a,b) \wedge R(b,a)) \Rightarrow a = b)$
whereas the top diagram is closer in spirt to $\forall a \forall b(a \neq b \Rightarrow (R(a,b) \Rightarrow \neg R(b,a)))$.
Thus, the diagrams in figures 8 and 9 axiomatize the theory of partially ordered
sets. Further, we may want to consider strict total orders which require us to
specify the trichotomy property. This is done in figure 10.

4   Diagrammatic Proofs

Below we show a very simple proof constructed using these mathematical constraint diagrams. It establishes that, for an equivalence relation, R, any two
unrelated elements are not related to any common element. This is part of the
assertion that R partitions the universal set into equivalence classes. The remainder of this assertion is captured by the statements of theorem 2, whose proof
nicely demonstrates that related elements have the same image, and theorem 3
which asserts that the image of R is U; a first order predicate logic formalization
of this is given by the two sentences

$$\forall a \forall b (R(a,b) \Rightarrow \forall c(R(a,c) \Leftrightarrow R(b,c)))   \quad (*)$$

and $\forall a \exists b R(b,a)$.

Theorem 1. *Let R be an equivalence relation. Then*

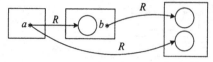

Proof. Assume for a contradiction that

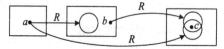

Then, by symmetry,

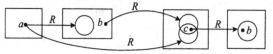

By transitivity we obtain a contradiction,

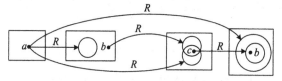

Theorem 2. *Let R be an equivalence relation. Then*

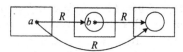

Proof. Suppose that

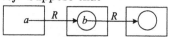

Then, by transitivity,

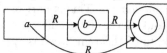

By symmetry,

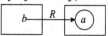

By transitivity,

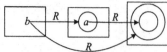

Hence

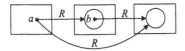

Theorem 3. *Let R be an equivalence relation. Then*

Proof. Assume for a contradiction that

Then, by reflexivity,

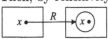

Hence we obtain a contradiction,

Some comments on the three presented proofs are in order. The proof of theorem 1 is by contradiction, starting by assuming the negation of the result; note that this does not require any explicit negation. It then builds up further information by using the symmetric and transitive properties. The final diagram is a contradiction: the second box indicates that a is not related to b, whereas the final box asserts that a is related to b. The proof of theorem 2 uses free variables (a concept not seen previously in constraint diagrams) visually represented by the labels a and b without annotations (i.e. $*$ for 'for all' or $\bullet$ for 'there exists'). The proof builds diagrams using symmetry and transitivity, that the image of a contains the image of b and also that the image of b contains the image of a and hence that these two images are equal. Applying generalization completes the last step in the proof. The proof of theorem 3 generates an obvious contradiction using only reflexivity.

We believe that an area in which constraint diagrams excel is where statements are being made about all of the elements related to some element; this is because derived contours and arrows naturally allow us to construct such the set of all such elements. Further desirable features include the presence of some implicit implication (which is achieved in theorem 1 by placing b inside the derived contour which represents the set over which it quantifies); implication is often hard for those learning about logic to understand. Moreover, the use of arrows and the placement of contours bring with them implicit universal quantification which could also be an advantage. To draw contrast, theorem 2 contains two universal spiders and makes no explicit use of $\Rightarrow$ whereas (*) includes three universal quantifiers and uses both $\Rightarrow$ and $\Leftrightarrow$.

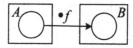

Fig. 11. Second order quantification: there exists a surjection

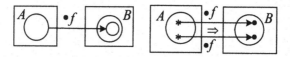

Fig. 12. Second order quantification: there exists an injection

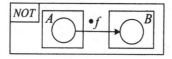

Fig. 13. There is not a surjection from A to B

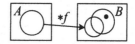

Fig. 14. An alternative formalization

5   Second Order Constraint Diagrams

To be of wider applicability in mathematics, constraint diagrams need to be able to make second order statements. For example, when stating the Schröder-Bernstein theorem, one needs to assert the existence of functions with certain properties. Thus, we generalized the use of derived contours, asterisks and dots to make second-order quantification possible. To illustrate, the diagram in figure 11 asserts the existence of a surjection from A to B, by placing • next to the function symbol f; thus, f is acting as a second order variable. The diagram in figure 12 asserts the existence of a function from A to B that is injective. From these two figures one can see how to formally state the Schröder-Bernstein theorem.

We might wish to specify that the cardinality of A is less than the cardinality of B. Formally, we could do this by asserting that there does not exist a surjection from A to B, as in figure 13. Alternatively, we can avoid the use of the 'negation box' and use second order universal quantification, as in figure 14.

To further demonstrate our extensions to the syntax, we consider defining a well-order. We have already defined ordered sets, so we concentrate on the part that specifies each non-empty subset has a least element. This can be seen in figure 15, where we start off by talking about all subsets, Z, and assumes that R is an order relation.

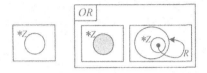

Fig. 15. Well-orders

6 Free Rides and Other Reasoning Advantages

It has been observed that one of the advantages diagrams have over symbolic notations is related to reasoning: sometimes one must perform reasoning to make a deduction in the symbolic world whereas the deduction may come 'for free' in the diagrammatic case. Euler diagrams have been well-studied in terms of their free rides, such as with regard to subset relationships [18]. These free rides also, therefore, apply to the images of relations under domain restrictions in our generalized constraint diagrams. A further example of a free ride can be seen in d_1, figure 16, where the placement of an existential spider, a, inside A and the relative positioning of A and B gives the information that the element represented by a is not in B for free. This type of free ride is very similar to those exhibited by Euler diagrams.

Other examples of free rides also occur. In figure 16, d_2 asserts that $x.f = A$, $A.g = y$ and $y.h = x$ for some elements x and y. We get, for free, that $x.f.g = y$, $A.g.h = x$, $y.h.f = A$, $x.f.g.h = x$ and so forth. In terms of reasoning from a diagram interpretation perspective, we point the reader to [23] which discusses features of constraint diagrams that are well-matched to meaning, highlighting other areas where constraint diagrams might outperform symbolic notations.

In the context of proof writing, it is an interesting and open problem as to what these diagrams buy you over symbolic proofs. This question will require extensive investigation to establish. Before we can begin to answer this question, further work is required to establish what constitutes a sound reasoning step in our mathematical constraint diagrams. We strongly anticipate that diagrammatic proofs and symbolic proofs can provide different perspectives on proof construction and neither approach is likely to be always better than the other. This belief is not restricted to the type of diagrams that we have presented here. It is an important challenge for the diagrammatic reasoning community to identify the relative effectiveness of diagrammatic and symbolic proofs. Perhaps

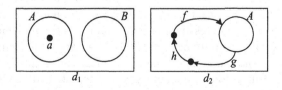

Fig. 16. Free rides involving spiders and arrows

a first step is to somehow classify proofs using patterns and identify which proofs are best within each realm (diagrammatic and symbolic), before attempting to identify when each notation outperforms the other.

7   Tool Support

The development of tools to support diagrammatic reasoning is well underway, and recent advances provide a basis for automated support for mathematical constraint diagrams. Such tools require varied functionality and the research challenges can be viewed as more broad than for symbolic logics. There are at least two major differences: first, it is more difficult to parse a 2D diagram than a 1D symbolic sentence; more significantly, when automatically generating proofs, the diagrams must be laid out in order for the user to read the proof. In respect of the second difference, possibly the hardest aspect of mathematical constraint diagram layout is in the initial generation of the underlying Euler diagram. There have been many recent efforts in this regard, including [2,5,17]. Further syntax can be automatically added later, as demonstrated in [15].

In terms of automated reasoning, this has been investigated for unitary Euler diagrams [26] and, to some extent, for spider diagrams, for example [6]. The approaches used rely on a heuristic search, guided by a function that provides a lower bound on proof length. Roughly speaking, the better this lower bound, the more efficiently the theorem prover finds proofs. An Euler diagram theorem prover, called EDITH, is freely available for download from www.cmis.brighton.ac.uk/research/vmg/autoreas.htm. We note that the main goals of automated reasoning in diagrammatic systems need not include outperforming symbolic theorem provers in terms of speed; of paramount importance is the production of proofs that are accessible to the reader and it may be that this readability constraint has a big impact on the time taken to find a proof.

8   Conclusion

In this paper, we have explored the use of constraint diagrams in mathematics, which were originally designed for software specification. We presented several extensions to the notation to make it more fit-for-purpose in a mathematical setting. In particular, we allowed arrows to be used between unitary parts of diagrams, equality to be asserted within unitary diagrams, and incorporated the explicit use of $\Rightarrow$ and $\neg$ (albeit written using English, but this is a trivial point).

The presented mathematical constraint diagrams, unlike augmented constraint diagrams, are capable of expressing proper second-order statements through the use of derived contours that are not the target of an arrow. Perhaps the simplest extension we have proposed is the explicit use of function symbols alongside relation symbols. To ensure wider applicability in mathematical domains, where there is often a need for second order quantification, we generalized derived contours so that they could range over all subsets of any given set and, moreover, we have introduced syntax that allows us to talk about the existence of functions and relations.

A fruitful avenue of future work will be to establish what is best rendered diagrammatically and what is best rendered symbolically. Insight into the relative strengths of each notation will allow their effective integration, resulting in hybrid notations that incorporate aspects from both paradigms. Moreover, we conjecture that it will be beneficial to use both symbolic and diagrammatic logics in a common context, providing different perspectives on the same problem. For example, one could construct two proofs of a theorem (one diagrammatic, one symbolic) which may provide different insights into the validity of the theorem, aiding understanding.

In addition to the points in the previous paragraph, the intuitive nature of diagrams has obvious benefits. The act of formalization brings with it greater understanding of the problem in hand; by using notations (such as mathematical constraint diagrams) that provide a different perspective on the construction of statements, mathematicians and their students may gain a deeper insight into the problem domain. Further, we argue that diagrammatic notations may provide more accessible languages to those who prefer visual approaches to problem solving over symbolic approaches. Thus, the integrated use of diagrams in formal mathematics is likely to bring with it a greater understanding and insight.

The next stage in this work is to identify a necessary and sufficient set of constraints that identify well-formed diagrams, so that we only prevent ambiguous diagrams from being created. This will be an integral part of defining the syntax and semantics, which in all likelihood will follow the style of those for generalized constraint diagrams [23]. We believe that it is important to restrict the syntax only where necessary when defining well-formed diagrams so that, when reasoning, one can make intuitive deductions which might have otherwise resulted in non-wellformed diagrams. We also intend to more fully explore the expression of second order statements using this language.

Future work also includes a more thorough exploration of the ways in which mathematical concepts can be defined and how proofs can be written using constraint diagrams. By producing a large variety of cases studies, we will be able to extract a set of reasoning (inference) rules that allow the construction of formal diagrammatic proofs that correspond well to those one would write in a rigorous setting. Such an approach to defining a reasoning system for these diagrams will have obvious benefits, since it will bring the way in which a mathematician typically works (i.e. by constructing rigorous proofs) closer to formal mathematics. An obvious challenge lies in promoting the uptake of diagrammatic formalization and reasoning in mathematics; this is challenging because of the overwhelmingly prevalent use of symbolic notations throughout mathematics.

Acknowledgements. This research is supported by EPSRC grant EP/E011160 for the Visualization with Euler Diagrams project. Thanks to the anonymous reviewers for their helpful comments.

References

1. Barwise, J., Etchemendy, J.: Language Proof and Logic. CSLI (1999)
2. Chow, S.: Generating and Drawing Area-Proportional Euler and Venn Diagrams. PhD thesis. University of Victoria (2007)
3. Dau, F.: http://www.dr-dau.net/eg_readings.shtml (2006)
4. Fish, A., Flower, J., Howse, J.: The semantics of augmented constraint diagrams. Journal of Visual Languages and Computing 16, 541–573 (2005)
5. Flower, J., Howse, J.: Generating Euler diagrams. In: Proceedings of 2nd International Conference on the Theory and Application of Diagrams, Georgia, USA, April 2002, pp. 61–75. Springer, Heidelberg (2002)
6. Flower, J., Masthoff, J., Stapleton, G.: Generating readable proofs: A heuristic approach to theorem proving with spider diagrams. In: Blackwell, A.F., Marriott, K., Shimojima, A. (eds.) Diagrams 2004. LNCS (LNAI), vol. 2980, pp. 166–181. Springer, Heidelberg (2004)
7. Giaquinto, M.: Visual Thinking in Mathematics. Clarendon Press (2007)
8. Gurr, C.: Aligning syntax and semantics in formalisations of visual languages. In: Proceedings of IEEE Symposia on Human-Centric Computing Languages and Environments, pp. 60–61. IEEE Computer Society Press, Los Alamitos (2001)
9. Gurr, C., Tourlas, K.: Towards the principled design of software engineering diagrams. In: Proceedings of 22nd International Conference on Software Engineering, pp. 509–518. ACM Press, New York (2000)
10. Howse, J., Schuman, S.: Precise visual modelling. Journal of Software and Systems Modeling 4, 310–325 (2005)
11. Howse, J., Stapleton, G., Taylor, J.: Spider diagrams. LMS Journal of Computation and Mathematics 8, 145–194 (2005)
12. Jamnik, M.: Mathematical Reasoning with Diagrams. CSLI (2001)
13. Kent, S.: Constraint diagrams: Visualizing invariants in object oriented modelling. In: Proceedings of OOPSLA 1997, pp. 327–341. ACM Press, New York (1997)
14. Kim, S.-K., Carrington, D.: Visualization of formal specifications. In: 6th Asia Pacific Software Engineering Conference, pp. 102–109. IEEE Computer Society Press, Los Alamitos (1999)
15. Mutton, P., Rodgers, P., Flower, J.: Drawing graphs in Euler diagrams. In: Blackwell, A.F., Marriott, K., Shimojima, A. (eds.) Diagrams 2004. LNCS (LNAI), vol. 2980, pp. 66–81. Springer, Heidelberg (2004)
16. Nelson, R.: Proofs without Words: Exercises in Visual Thinking, vol. 1. The Mathematical Association of America (1997)
17. Rodgers, P., Zhang, L., Stapleton, G., Fish, A.: Embedding wellformed euler diagrams. In: 12th International Conference on Information Visualization. IEEE, Los Alamitos (2008)
18. Shimojima, A.: Inferential and expressive capacities of graphical representations: Survey and some generalizations. In: Blackwell, A.F., Marriott, K., Shimojima, A. (eds.) Diagrams 2004. LNCS (LNAI), vol. 2980, pp. 18–21. Springer, Heidelberg (2004)
19. Shin, S.-J.: The Logical Status of Diagrams. Cambridge University Press, Cambridge (1994)
20. Shin, S.-J.: The Iconic Logic of Peirce's Graphs. Bradford Book (2002)
21. Sobanski, J.: Visual Math: See How Math Makes Sense. Learning Express (2002)
22. Stapleton, G., Delaney, A.: Towards overcoming deficiencies in constraint diagrams. In: IEEE Symposium on Visual Languages and Human-Centric Computing, pp. 33–40 (2007)

23. Stapleton, G., Delaney, A.: Evaluating and generalizing constraint diagrams. Visual Languages and Computing (available online) (accepted, 2008)
24. Stapleton, G., Howse, J., Taylor, J.: A decidable constraint diagram reasoning system. Journal of Logic and Computation 15(6), 975–1008 (2005)
25. Stapleton, G., Masthoff, J.: Incorporating negation into visual logics: A case study using Euler diagrams. In: Visual Languages and Computing 2007, pp. 187–194. Knowledge Systems Institute (2007)
26. Stapleton, G., Masthoff, J., Flower, J., Fish, A., Southern, J.: Automated theorem proving in Euler diagrams systems. Journal of Automated Reasoning 39, 431–470 (2007)
27. Swoboda, N., Allwein, G.: Using DAG transformations to verify Euler/Venn homogeneous and Euler/Venn FOL heterogeneous rules of inference. Journal on Software and System Modeling 3(2), 136–149 (2004)
28. Winterstein, D., Bundy, A., Gurr, C.: Dr Doodle: A diagrammatic theorem prover. In: Basin, D., Rusinowitch, M. (eds.) IJCAR 2004. LNCS (LNAI), vol. 3097, pp. 331–335. Springer, Heidelberg (2004)
29. Winterstein, D., Bundy, A., Gurr, C., Jamnik, M.: An experimental comparison of diagrammatic and algebraic logics. In: International Conference on the Theory and Application of Diagrams, pp. 432–434 (2004)

Normalization Issues in Mathematical Representations*

Manfred Kerber

School of Computer Science
University of Birmingham
Birmingham B15 2TT, England
http://www.cs.bham.ac.uk/~mmk

Abstract. Typically it is considered a strength of a language that the same situation can be described in different ways. However, when a human or a program is to check whether two representations are essentially the same it is much easier to deal with normal forms. For instance, (infinitely many) different sets of formulae may normalize to the same clause set. In the case of propositional logic formulae with a fixed number of boolean variables, the class of all clause sets is finite. Since it grows doubly exponentially, it is not feasible to construct the complete class even for small numbers of boolean variables. Hence further normalizations are necessary and will be studied. Furthermore some potential applications will be discussed.

1 Introduction

There are many different ways to represent the same situation and for particular problem classes particular logics have been developed. Even within a particular logic it is possible to state the same fact in different ways. We can ask, for instance, how many different ways there are to express a particular fact in a particular logic and typically there are infinitely many ways. Even a logic as simple as propositional logic allows a lot of variation and typically infinitely many ways exist to express the same situation. Let us look at this situation in more detail for very simple propositional logics with few boolean variables.

If we do not have any vocabulary (propositional logic with no boolean variables) then we have only one single option, namely to remain silent, we cannot say anything. With one boolean variable, things do not get much more interesting. However, it is already possible to state infinitely many consistent facts, e.g., X_1, $X_1 \wedge X_1$, $X_1 \wedge (X_1 \wedge X_1)$, $X_1 \wedge (X_1 \wedge (X_1 \wedge X_1))$ and so on, which mean all the same; as well as inconsistent facts $X_1 \wedge \neg X_1$; and tautological statements such as $X_1 \vee \neg X_1$; and obviously we still can remain silent. While there are infinitely many different formulae, they all are (logically) equivalent to only four different situations, namely the clause sets:

---

* I would like to thank Riccardo Poli for stimulating discussions which inspired this work.

S. Autexier et al. (Eds.): AISC/Calculemus/MKM 2008, LNAI 5144, pp. 494–503, 2008.
© Springer-Verlag Berlin Heidelberg 2008

- $\{X_1\}$, that is, X_1 must be true,
- $\{\neg X_1\}$, that is, X_1 must be false,
- $\{X_1, \neg X_1\}$, that is, an inconsistent statement, and
- $\emptyset$, that is, X_1 may be true or may be false.

That is, the infinite class of possible formulae can be reduced to a finite class of four different clause sets. For this reason it is much better to ask how many different clause sets there are. Formulae which reduce to the same clause set are logically equivalent.

Note that we do not have any tautologies in the clause sets above, and actually if we allowed them in we would have infinitely many clause sets. Deleting tautologies from a clause set is a cheap procedure, since it is local, that is, it needs to look at a single clause only at a time, that is, the complexity of the simplification is linear in the size of the clause set. It is typically applied in theorem proving; it does not change the meaning since tautologies do not carry any information. For instance, the clause sets $\{X_1 \vee X_2, X_2 \vee \neg X_2\}$ and $\{X_1 \vee X_2\}$ carry the same information. For this reason we assume in the following that a clause set does not contain any tautologies. As a consequence, given n boolean variables there are only finitely many clause sets. How many? We will look at this next.

After the removal of all tautologies, any clause can be characterized as a string over the alphabet $\{0, 1, \#\}$: Assumed we have n boolean variables $X_1, X_2, \ldots, X_n$ then we can represent a (non-tautological) clause C by a string $c_1 c_2 \cdots c_n$ as follows: for every i

$$c_i = \begin{cases} 0 & \text{if } X_i \text{ occurs negatively in } C \\ 1 & \text{if } X_i \text{ occurs positively in } C \\ \# & \text{if } X_i \text{ does not occur in } C \end{cases}$$

For instance, for $n = 6$, the string $0\#111\#$ represents $\neg X_1 \vee X_3 \vee X_4 \vee X_5$ and $1\#\#10\#$ represents $X_1 \vee X_4 \vee \neg X_5$.[2]

With this string representation of clauses we see that there are 3^n different clauses altogether. Since the clause sets which contain the empty clause, represented as $\#\# \cdots \#$, are all unsatisfiable anyway and can not only be identified semantically, but also recognized as such computationally very cheaply, we can take all these clause sets out of our consideration. That is, we get $3^n - 1$ different (non-empty) clauses with n boolean variables. Hence there are all in all $f(n) = 2^{3^n - 1}$ different clause sets (without tautologies and without clauses which do contain the empty clause). Although finite, this set grows doubly exponentially in the number of boolean variables. Normalization beyond clause normal form should be applied or a study becomes infeasible even for very small n.[3]

---

[2] The terminology is taken from the area of classifier systems. In SAT solvers like zChaff one would rather number the clauses and mention them only if they occur, either without prefix if positive, or a "-" if negative. For instance, the clause set $\{\neg X_1 \vee X_3 \vee X_4 \vee X_5,\ X_1 \vee X_4 \vee \neg X_5\}$ can be represented by the two lines -1 3 4 5 0 and 1 4 -5 0. The 0 indicates the end of a clause.

[3] Because of the doubly exponential growth, just taking out the clause sets which contain the empty clause halves the size of all clause sets under consideration.

n	$f(n)$
0	1
1	4
2	256
3	67108864
4	120892581961462917470617
5	7067388259113537318333190002971674063309935587502475832486424805170479104

For $n = 0$ there is only the one possibility to remain silent. For $n = 1$, we have the clauses X_1, $\neg X_1$, and the four clause sets: $\emptyset$, $\{X_1\}$, $\{\neg X_1\}$, and $\{X_1, \neg X_1\}$. For $n = 3$ there are already more than sixty million different clause sets.

In the following we will consider how to categorize the clause sets for $n = 2$ and $n = 3$ so that we have only relatively few classes to consider for a complete analysis of these cases. First, however, we will look in the next section at the reductions we want to apply and then – in the section after that – at an important property which a reduction may or may not have.

2   Reductions

Normally we select the names of boolean variables very carefully and associate some meaning with these names. It makes a difference to us as humans whether we say Loves_john_mary $\lor$ $\neg$Loves_mary_john or Mortal_socrates $\lor$ $\neg$Mortal_aristotle. On an abstract level, however, the two different representations are structurally equivalent, and if we use variable names X_i only, the two formulae are represented as $X_1 \lor \neg X_2$. In the same line, however, the clauses $X_1 \lor \neg X_2$ and $X_2 \lor \neg X_1$ are structurally equivalent as well[4], since they can be transformed into each other by swapping the names of X_1 and X_2. That is, if we are interested in the characterization of fundamentally different representations, then we can identify clause sets which can be transformed into each other by swapping variable names.

In this paper, we will look at four different normalizations. They are different in type; the one type consists of classical reductions as found in traditional theorem proving textbooks [1]: purity and subsumption. The other type is about symmetries: the renaming one above, and flipping the polarity of variables. Some of them are *stable* under the construction of the full class of all clause sets, one not. In the following it will be made precise what is meant by stable and which reductions do and do not have this property. Roughly spoken stable means that when creating the powerset we may apply stable reductions before applying the recursive step, but not instable ones (without changing the result). Before we go into details about stability, let us take a closer look at the four reductions we want to apply:

---

[4] Note that the second clause is represented as $0\,1$ and as a clause considered identical to $\neg X_1 \lor X_2$. Note furthermore that the two clauses $X_1 \lor \neg X_2$ and $X_2 \lor \neg X_1$ are *not* logically equivalent. In order not to get confused what is meant by equivalent, we use the term "structurally equivalent".

permutation. When considering a class of clause sets then we can partition the whole class into equivalence classes of those clause sets which can be obtained from each other by permuting the boolean variables. For instance, $\{X_1 \vee X_3 \vee X_4,\ \neg X_1 \vee X_2 \vee X_3 \vee \neg X_4\}$ (represented as $\{1\,\#11\,,0\,1\,1\,0\,\}$) and $\{X_2 \vee X_3 \vee X_4,\ X_1 \vee \neg X_2 \vee X_3 \vee \neg X_4\}$ (represented as $\{\#1\,1\,1\,,1\,0\,1\,0\,\}$) are structurally equivalent, since we can transform them into each other by permuting the first and the second variable.

flip. Flip the polarity of some variables. For instance, in $\{X_2 \vee X_3 \vee X_4,\ X_1 \vee \neg X_2 \vee X_3 \vee \neg X_4\}$ (represented as $\{\#1\,1\,1\,,1\,0\,1\,0\,\}$), we can flip the polarity of the first variable and get the structurally equivalent clause set $\{X_2 \vee X_3 \vee X_4,\ \neg X_1 \vee \neg X_2 \vee X_3 \vee \neg X_4\}$ (represented as $\{\#1\,1\,1\,,0\,0\,1\,0\,\}$).

pure. Remove pure literals. We say that a literal is *pure* if for this literal there is no corresponding literal of opposite sign in the whole clause set, that is, if the clause contains X_i, but there is no $\neg X_i$ in the clause set (or conversely if it contains a literal $\neg X_i$, but there is no X_i in the clause set). If we have in a clause a pure literal then this clause can be trivially satisfied by assigning to it the corresponding truth value, that is, true for a positive and false for a negative literal. Then the whole clause, in which it occurs, can be deleted from the clause set, and essentially a problem which originally contained n propositional logic variables is reduced to one with $n-1$ variables. Note that the deletion of clauses containing pure literals may make further literals pure so that the process is to be repeated until no more pure literals are left. The fully reduced clause set is considered.

subsumed. Remove all subsumed clauses, that is, remove less specific information in the presence of more specific information. For instance, $\{X_1 \vee \neg X_3,\ X_1 \vee X_2 \vee \neg X_3,\ \neg X_1 \vee \neg X_2 \vee X_3\}$ (represented as $\{1\,\#0\,,1\,1\,0\,,0\,0\,1\,\}$) can be reduced to $\{X_1 \vee \neg X_3,\ \neg X_1 \vee \neg X_2 \vee X_3\}$ (represented as $\{1\,\#0\,,0\,0\,1\,\}$) since $X_1 \vee \neg X_3$ (represented as $1\,\#0\,$) subsumes $X_1 \vee X_2 \vee \neg X_3$ (represented as $1\,1\,0\,$).

When a class of clause sets is given, we want to reduce it by the reductions so that for every equivalence class only one representative is left in the clause set after exhaustive application of all reductions.

Although it is theoretically possible to follow a naive approach and to first generate the whole class of all clause sets and then reduce it, this is not feasible for doubly exponential problems even for small n. The set of all clauses can be effectively generated for $n \leq 10$ (or modestly bigger n; for $n = 10$ there are $3^n - 1 = 59049$ different clauses). Even for $n = 3$ there are, however, more than 60 million different clause sets. For any bigger n it is certainly infeasible to compute the full class of all clause sets. For this reason it is necessary to keep the set on construction as small as possible, that is, we want to apply the reductions on construction as much as possible. But may we? Or will we change the result this way? We call the reductions which may be applied during construction *stable*. We will establish that the properties **permutation**, **flip**, and **subsumed** are stable, and that **pure** is not. Let us make this more precise in the next section.

3   Stability of Reductions in Generating All Clause Sets

Definition 1 (Reduction). *Let S be a class of clause sets and $\mathcal{R}$ be a set of reduction rules. Let $S_{\mathcal{R}}$ be the set of equivalence classes produced from S by completely reducing the clause sets in S by reductions from $\mathcal{R}$. Two clause sets S_1 and S_2 are said to be structurally equivalent if there are r_1 and r_2 in $\mathcal{R}^*$ such that $r_1(S_1) = r_2(S_2)$ (that is, the clause sets can be made equal by applying any number of reductions to S_1 and any number of reductions to S_2).*

Let us now clarify what we mean by stable with respect to the powerset construction.[5]

One way to compute the powerset (in Lisp code) is

```
(defun powerset(set)
  (if (null set)
      (list nil)
      (let* ((prev-level (powerset (rest set)))
             (first (first set)))
        (append (mapcar #'(lambda(el) (cons first el)) prev-level)
                prev-level)))))
```

The corresponding version in which reductions are applied at each recursive step is:

```
(defun powersetR(set R)
  (if (null set)
      (list nil)
      (let* ((prev-level
              (apply-reductions (powersetR (rest set) R) R))
             (first (first set)))
        (append (mapcar #'(lambda(el) (cons first el)) prev-level)
                prev-level)))))
```

where `apply-reductions` applies the reductions exhaustively until the set is reduced as much as possible. We say that the powerset algorithm is *stable* with respect to the set of reductions R if and only if for all sets holds (again the reductions R are applied exhaustively):

$$R(\texttt{powersetR}(\texttt{set}, R)) = R(\texttt{powerset}(\texttt{set})).$$

Remember, the class of all clause sets is generated as the powerset of the set of all clauses.

Property 1. *Powerset generation is not stable with respect to purity reduction, that is, it is not possible in the recursive call of powerset to first reduce by recursively deleting all pure clauses.*

---

[5] The notion can be easily generalized to arbitrary recursive procedures and any reductions, but we keep it at a concrete level here, since we are interested only in the powerset construction.

Proof: It is sufficient to show that when we delete pure clauses in the class of clauses sets with one boolean variable that then we do not get the full class when we create the class of clause sets with two boolean variables.

The set of all clause sets with boolean variable X_2 is $\{\emptyset, \{\neg X_1\}, \{X_1\}, \{\square\}\}$. If we apply purity reduction the class is reduced to the two elements $\{\emptyset, \{\square\}\}$. If we now add in the recursive call either $\neg X_1$, X_1, or nothing, we get as class of all reduced sets $\{\emptyset, \{\neg X_1\}, \{X_1\}, \{\square\}\}$, which reduces under purity reduction to $\{\emptyset, \{\square\}\}$.[6] However, the full class must contain more elements (there are 164 different clause sets which do not contain pure literals or the empty clause), for instance, the clause set $\{\neg X_1 \vee X_2, X_1 \vee \neg X_2\}$, which does not contain any pure literals. $\square$

Property 2. *Powerset generation is stable with respect to reductions which identify clauses generated by component-wise negation.*

Proof: Clause sets such as $\{X_1 \vee \neg X_2 \vee \neg X_3, X_1 \vee X_3, \neg X_1 \vee X_2, X_2 \vee X_3\}$ (represented as $\{100, 1\#1, 01\#, \#11\}$) and $\{\neg X_1 \vee \neg X_2 \vee \neg X_3, \neg X_1 \vee X_3, X_1 \vee X_2, X_2 \vee X_3\}$ (represented as $\{000, 0\#1, 11\#, \#11\}$), which can be transformed into each other by component-wise negation are identified since they can be transformed into each other by flipping the sign of some variables (of the first variable in the example). In the process of generating the powerset we have the commutative relationship that it does not matter whether we first flip and then add another component or the other way around first add another component and then flip. $\square$

Property 3. *Powerset generation is stable with respect to reductions which identify clauses which are generated by permutations.*

Proof: If one clause of length n is a permutation of another then by adding the same $n + 1$st component to each of them the resulting two clauses are permutations of each other. $\square$

Property 4. *Powerset generation is stable with respect to subsumption.*

Proof: If one clause of length n subsumes another then by adding the same $n + 1$st component to each of them the first resulting one subsumes the second resulting one. $\square$

Property 5. *The combination of the application of reductions by component-wise negation and permutation is more powerful than the two reductions applied consecutively.*

Proof: With the reduction by negation, clause sets such as $\{X_1 \vee \neg X_2, X_1 \vee X_2 \vee X_3\}$ (represented as $\{10\#, 111\}$) and $\{\neg X_1 \vee \neg X_2, \neg X_1 \vee X_2 \vee X_3\}$ (represented as $\{00\#, 011\}$) are identified (via negation of the first component). In the case of permutation clause sets such as $\{X_1 \vee \neg X_2, X_1 \vee X_2 \vee X_3\}$ (represented as

---

[6] Likewise all sets for bigger n would contain always only two elements.

$\{10\#,111\}$) and $\{\neg X_1 \lor X_2,\ X_1 \lor X_2 \lor X_3\}$ (represented as $\{01\#,111\}$) are identified (by permuting the first and second elements). However the second two clause sets, $\{\neg X_1 \lor \neg X_2,\ \neg X_1 \lor X_2 \lor X_3\}$ (represented as $\{00\#,011\}$) and $\{\neg X_1 \lor X_2,\ X_1 \lor X_2 \lor X_3\}$ (represented as $\{01\#,111\}$) are not related by either permutation or negation alone.   □

Property 6. *Purity reduction and subsumption do not commute.*

Proof: Purity reduction and subsumption each change the structure of clauses and after the application of subsumption further clauses may become pure. However, by the deletion of pure clauses no further clauses may become subsumed. For instance for $n = 2$, the clause set $\{\neg X_1 \lor X_2,\ X_1,\ X_1 \lor \neg X_2,\ X_1 \lor X_2\}$ (represented as $\{01,1\#,10,11\}$) does not reduce under purity. Application of subsumption results in $\{\neg X_1 \lor X_2,\ X_1\}$ (represented as $\{01,1\#\}$), in which the first clause is pure in the second literal.   □

It is advisable to apply the reductions in an order, namely apply subsumption and the combined negation/permutation reduction on construction and apply purity reduction last, since purity firstly cannot be used on construction and after the application of the other reductions has only a relatively small impact. Purity reduction should be applied after subsumption. Likewise permutation reduction should be applied after subsumption.

4   The Cases for $n \leq 3$

For $n = 3$, the total class with a size of 67108864 elements can be reduced by the application of subsumption to 15935 elements. This can be reduced further by the combined application of permutation/negation reduction to 522 elements. Finally purity reduction reduces the set to 410 elements.

Summarizing we get the following numbers of different clause sets after the cumulative application of reductions:

n	no red.	subsum	subsum+neg/perm	subsum+neg/perm+purity
1	4	4	3	2
2	256	47	14	8
3	67108864	15935	522	410

E.g., for $n = 2$ the 8 elements after all reductions are:

```
NIL

("00" "11")
("#1" "#0")

("00" "01" "10")
("#0" "01" "11")
("#0" "0#" "11")
```

```
("00" "01" "10" "11")
("#0" "#1" "0#" "1#")
```

These eight cases are clearly non-isomorphic since they can be classified by the number of clauses contained (0, 2, 3, or 4), and within these classes by the total number of # symbols contained in the clause sets. It is under current investigation to find a similar classification for the 410 elements for $n = 3$. If we had a general classification principle, this could be used in order to come up with an effective mechanism to create the reduced class of clause sets for $n = 4$ without having to go back to the powerset construction. Note, however, that it cannot be expected that the growth in number of the reduced clause set classes is in a better complexity class compared to the full class. That is, they are expected to be doubly exponential as well. However, the much reduced numbers make them more accessible to complete investigation for small n, in particular for $n = 4$. Because of the special situation of two-clauses, the case $n = 3$ is the first really interesting case and it would be good to have at least the case $n = 4$ at hand as well to come up with generalizable conjectures.

5   Unsatisfiable Clauses for $n = 3$

In the reduced class of clause sets we can investigate properties of the whole class. One of them is to ask: Given an unsatisfiable clause set how many resolution steps are necessary at most for a node to reduce it to the empty clause, and how many at least for the same node?

Let us add the empty clause to the class of reduced clause sets. Of the 411 different cases, 207 are not satisfiable and 204 are satisfiable. The non-satisfiable ones can be ordered as a partial order with respect to a relation of clause sets in which the empty clause comes lowest. A clause set is immediately below another one if the first can be generated from the second by a single application of the binary resolution rule. The resulting acyclic directed graph[7] is displayed in Fig. 1, however, it would be necessary to zoom in order to get detailed information from it.

The top most clause set is
$\{ \neg X_1 \lor \neg X_2 \lor \neg X_3, \ \neg X_1 \lor \neg X_2 \lor X_3, \ \neg X_1 \lor X_2 \lor \neg X_3, \ \neg X_1 \lor X_2 \lor X_3,$
$X_1 \lor \neg X_2 \lor \neg X_3, \ X_1 \lor \neg X_2 \lor X_3, \ X_1 \lor X_2 \lor \neg X_3, \ X_1 \lor X_2 \lor X_3 \ \}$
represented as {"000" "001" "010" "011" "100" "101" "110" "111"};
the bottom most one is the clause set consisting just of the empty clause. In order to get from the top most by binary resolution to the empty clause it is necessary to apply binary resolution at least 7 times and at most 18 times. That means that any heuristic to reduce proof search can at best reduce the search from 18 steps to 7 steps for this particular example. A long term goal of this work to theorem proving is to better understand the impact of theorem proving heuristics on the class of *all* problems rather than on a set of challenge problems.

---

[7] The figure is created with the help of the "dot (Graphviz)" graph visualization tool.

Fig. 1. Relationship of unsatisfiable clause sets with 3 boolean variables

This way this work may contribute to a better understanding of heuristics and help to speed up theorem provers.[8] The hope would be to generalize from cases with few propositional logic variables (e.g., $n = 2$, $n = 3$, and $n = 4$) to the arbitrary propositional logic case and to lift results to first order logic.

Another application of this work is in the search of similar expressions. The work can be used in order to detect structurally equivalent formulae. This is relevant when different formulations of the same problem are given; in this case these formulations are typically not logically equivalent. This becomes more relevant for logics which are more powerful than the ones studied here.

6   Related Work

The reduction of clause sets has been studied in theorem proving since the invention of the resolution principle by Robinson [5] in 1965 and it is well described in the text books on automated theorem proving such as [1]. Symmetry has been studied in the area of constraint satisfaction problems, see in particular the work by Frisch et al.[2,3], where permutations and changes of polarity of boolean variables play a major role. Representations and studies in the complexity of classifier systems [4] are related insofar as the same problem can be represented – using the string representation of clauses described in this paper – in different ways resulting in different complexities.

7   Conclusion

In this paper we looked at normalizations of very simple logical systems. The reduction was achieved by symmetry reductions as well as the traditional reductions of purity and subsumption. By these reductions it is possible to consider

---

[8] Note that two reductions used in this research (subsumption and purity) are routinely employed in theorem proving, while the others (permutation and flip) would not change the complexity of a particular theorem proving task.

instead of more than 60 million different clause sets only 410 structurally different ones. This reduction allows to answer, for instance, the question how many resolution steps may be necessary at most in order to show that a clause set with three different boolean variables is unsatisfiable.

The classification of the different cases is ongoing work and will lead hopefully to a computationally cheaper way to generate representative classes. In this context it is also interesting to see whether all 410 clause sets are non-isomorphic and whether the reductions used in this work are in some precise way best possible, that is, that any further identification would identify problems which should not be identified. Having a cheaper way to generate the set of all equivalence classes of reduced clause sets would allow to investigate the corresponding class for $n = 4$ and generate some conjectures by generalizing from the cases $n = 2, 3$, and 4.

The reduction to equivalence classes can be used to recognize and classify problems of a particular type. However, in the current paper only first steps have been made and a deeper understanding is necessary.

References

1. Chang, C.-L., Lee, R.C.-T.: Symbolic Logic and Mechanical Theorem Proving. Academic Press, New York (1973)
2. Frisch, A.M., Jefferson, C., Hernandez, B.M., Miguel, I.: The rules of constraint modelling. In: Proceedings of the 19th IJCAI, pp. 109–116 (2005), www.cs.york.ac.uk/aig/constraints/AutoModel/conjure-ijcai05.pdf
3. Frisch, A.M., Jefferson, C., Hernandez, B.M., Miguel, I.: Symmetry in the generation of constraint models. In: Proceedings of the International Symmetry Conference (2007), www.cs.york.ac.uk/aig/constraints/AutoModel/ISC07Conjure.pdf
4. Kovacs, T., Kerber, M.: A study of structural and parametric learning in XCS. Evolutionary Computation Journal 14(1), 1–19 (2006)
5. Robinson, J.A.: A machine oriented logic based on the resolution principle. Journal of the ACM 12, 23–41 (1965)

Notations for Living Mathematical Documents

Michael Kohlhase, Christine Müller, and Florian Rabe

Computer Science, Jacobs University Bremen
{m.kohlhase,c.mueller,f.rabe}@jacobs-university.de

Abstract. Notations are central for understanding mathematical discourse. Readers would like to read notations that transport the meaning well and prefer notations that are familiar to them. Therefore, authors optimize the choice of notations with respect to these two criteria, while at the same time trying to remain consistent over the document and their own prior publications. In print media where notations are fixed at publication time, this is an over-constrained problem. In living documents notations can be adapted at reading time, taking reader preferences into account.

We present a representational infrastructure for notations in living mathematical documents. Mathematical notations can be defined declaratively. Author and reader can extensionally define the set of available notation definitions at arbitrary document levels, and they can guide the notation selection function via intensional annotations.

We give an abstract specification of notation definitions and the flexible rendering algorithms and show their coverage on paradigmatic examples. We show how to use this framework to render OpenMath and Content-MathML to Presentation-MathML, but the approach extends to arbitrary content and presentation formats. We discuss prototypical implementations of all aspects of the rendering pipeline.

1 Introduction

Over the last three millennia, mathematics has developed a complicated two-dimensional format for communicating formulae (see e.g., [Caj93,Wol00] for details). Structural properties of operators often result in special presentations, e.g., the scope of a radical expression is visualized by the length of its bar. Their mathematical properties give rise to placement (e.g., associative arithmetic operators are written infix), and their relative importance is expressed in terms of binding strength conventions for brackets. Changes in notation have been influential in shaping the way we calculate and think about mathematical concepts, and understanding mathematical notations is an essential part of any mathematics education. All of these make it difficult to determine the functional structure of an expression from its presentation.

Content Markup formats for mathematics such as OpenMath [BCC+04] and content MathML [ABC+03] concentrate on the functional structure of mathematical formulae, thus allowing mathematical software systems to exchange mathematical objects. For communication with humans, these formats rely on a

S. Autexier et al. (Eds.): AISC/Calculemus/MKM 2008, LNAI 5144, pp. 504–519, 2008.

"presentation process" (usually based on XSLT style sheets) that transforms the content objects into the usual two-dimensional form used in mathematical books and articles. Many such presentation processes have been proposed, and all have their strengths and weaknesses. In this paper, we conceptualize the presentation of mathematical formulae as consisting of two components: the two-dimensional **composition** of visual sub-presentations to larger ones and the **elision** of formula parts that can be deduced from context.

Most current presentation processes concentrate on the relatively well-understood composition aspect and implement only rather simple bracket elision algorithms. But the visual renderings of formulae in mathematical practice are not simple direct compositions of the concepts involved: mathematicians gloss over parts of the formulae, e.g., leaving out arguments, iff they are non-essential, conventionalized or can be deduced from the context. Indeed this is part of what makes mathematics so hard to read for beginners, but also what makes mathematical language so efficient for the initiates. A common example is the use of $\log(x)$ or even $\log x$ for $\log_{10}(x)$ or similarly $[\![t]\!]$ for $[\![t]\!]_{\mathcal{M}}^{\varphi}$, if there is only one model $\mathcal{M}$ in the context and φ is the most salient variable assignment.

Another example are the bracket elision rules in arithmetical expressions: $ax + y$ is actually $(ax) + y$, since multiplication "binds stronger" than addition. Note that we would not consider the "invisible times" operation as another elision, but as an alternative presentation.

In this situation we propose to encode the presentational characteristics of symbols (for composition *and* elision) declaratively in **notation definitions**, which are part of the representational infrastructure and consist of "prototypes" (patterns that are matched against content representation trees) and "renderings" (that are used to construct the corresponding presentational trees). Note that since we have reified the notations, we can now devise flexible management process for notations. For example, we can capture the notation preferences of authors, aggregators and readers and adapt documents to these. We propose an elaborated mechanism to collect notations from various sources and specify notation preferences. This brings the separation of *function* from *form* in mathematical objects and assertions in MKM formats to fruition on the document level. This is especially pronounced in the context of dynamic presentation media (e.g., on the screen), we can now realize "*active documents*", where we can interact with a document directly, e.g., instantiating a formula with concrete values or graphing a function to explore it or "*living/evolving documents*" which monitor the change of knowledge about a topic and adapt to a user's notation preferences consistently.

Before we present our system, let us review the state of the art. Naylor, Smirnova, and Watt [NW01a,SW06b,SW06a] present an approach based on meta stylesheets that utilizes a MATHML-based markup of arbitrary notations in terms of their content and presentation and, based on the manual selection of users, generates user-specific XSLT style sheets [Kay06] for the adaptation of documents. Naylor and Watt [NW01a] introduce a one-dimensional context annotation of content expressions to intensionally select an appropriate

notation specification. The authors claim that users also want to delegate the
styling decision to some defaulting mechanism and propose the following hierarchy of default notation specification (from high to low): command line control,
input documents defaults, meta stylesheets defaults, and content dictionary defaults.

In [MLUM05], Manzoor et al. emphasize the need for maintaining uniform
and appropriate notations in collaborative environments, in which various authors contribute mathematical material. They address the problem by providing
authors with respective tools for editing notations as well as by developing a
framework for a consistent presentation of symbols. In particular, they extend
the approach of Naylor and Watt by an explicit language markup of the content
expression. Moreover, the authors propose the following prioritization of different notation styles (from high to low): individual style, group, book, author or
collection, and system defaults.

In [KLR07] we have revised and improved the presentation specification of
OMDOC1.2. [Koh06] by allowing a static well-formedness, i.e., the well-formedness of presentation specifications can be verified when writing the presentations
rather than when presenting a document. We also addressed the issue of flexible
elision. However, the approach does not facilitate to specify notations, which are
not local tree transformations of the semantic markup.

In [KMM07] we initiated the redefinition of *documents* towards a more *dynamic* and *living* view. We explicated the narrative and content layer and extended the *document model* by a third dimension, i.e., the *presentation layer*. We
proposed the *extensional* markup of the *notation context* of a document, which
facilitates users to explicitly select suitable notations for document fragments.
These extensional collection of notations can be inherited, extended, reused,
and shared among users. For the system presented in this paper, we have re-
engineered and extended the latter two proposals.

In Sect. 2, we introduce abstract syntax for notation definitions, which is used
for the internal representation of our notation objects. (We use a straightforward
XML encoding as concrete syntax.) In Sect. 3, we describe how a given notation
definition is used to translate an OPENMATH object into its presentation. After
this local view of notation definitions, the remainder of the paper takes a more
global perspective by introducing markup that permits users to control which
notation definitions are used to present which document fragment. There are
two conflicting ways how to define this set of available notation definitions:
extensionally by pointing to a notation container; or intensionally by attaching
properties to notation definitions and using them to select between them. These
ways are handled in Sect. 4 and 5, respectively.

2   Syntax of Notation Definitions

We will now present an abstract version of the presentation starting from the
observation that in content markup formalisms for mathematics formulae are
represented as "formula trees". Concretely, we will concentrate on OPENMATH

objects, the conceptual data model of OPENMATH representations, since it is sufficiently general, and work is currently under way to re-engineer content MATHML representations based on this model. Furthermore, we observe that the target of the presentation process is also a tree expression: a layout tree made of layout primitives and glyphs, e.g., a presentation MATHML or LaTeX expression.

To specify notation definitions, we use the one given by the abstract grammar from Fig. 1. Here $|$, $[-]$, $-^*$, and $-^+$ denote alternative, bracketing, and non-empty and possibly empty repetition, respectively. The non-terminal symbol ω is used for patterns φ that do not contain jokers. Throughout this paper, we will use the non-terminal symbols of the grammar as meta-variables for objects of the respective syntactic class.

Intuitions. The intuitive meaning of a notation definition $ntn = \varphi_1, \ldots, \varphi_r \vdash (\lambda_1 : \rho_1)^{p_1}, \ldots, (\lambda_s : \rho_s)^{p_s}$ is the following: If an object matches one of the patterns φ_i, it is rendered by one of the renderings ρ_i. Which rendering is chosen,

Notation declarations	ntn	::=	$\varphi^+ \vdash [(\lambda : \rho)^p]^+$
Patterns	φ	::=	
Symbols			$\sigma(n, n, n)$
Variables		$\|$	$v(n)$
Applications		$\|$	$@(\varphi[, \varphi]^+)$
Binders		$\|$	$\beta(\varphi, \Upsilon, \varphi)$
Attributions		$\|$	$\alpha(\varphi, \sigma(n, n, n) \mapsto \varphi)$
Symbol/Variable/Object/List jokers		$\|$	$\underline{s} \mid \underline{v} \mid \underline{o} \mid \underline{l}(\varphi)$
Variable contexts	Υ	::=	φ^+
Match contexts	M	::=	$[q \mapsto X]^*$
Matches	X	::=	$\omega^* \mid S^* \mid (X)$
Empty match contexts	μ	::=	$[q \mapsto H]^*$
Holes	H	::=	$\_ \mid {}^{\text{“”}} \mid (H)$
Context annotation	λ	::=	$(S = S)^*$
Renderings	ρ	::=	
XML elements			$\langle S \rangle \rho^* \langle / \rangle$
XML attributes		$\|$	$S = {}^{\text{"}}\rho^{*\text{"}}$
Texts		$\|$	S
Symbol or variable names		$\|$	$\underline{q}$
Matched objects		$\|$	$\underline{q}^p$
Matched lists		$\|$	$\text{for}(q, I, \rho^*)\{\rho^*\}$
Precedences	p	::=	$-\infty \mid I \mid \infty$
Names	n, s, v, l, o	::=	C^+
Integers	I	::=	integer
Qualified joker names	q	::=	$l/q \mid s \mid v \mid o \mid l$
Strings	S	::=	C^*
Characters	C	::=	character except /

Fig. 1. The Grammar for Notation Definitions

depends on the active rendering context, which is matched against the context annotations λ_i (see Sect. 5). Each context annotation is a key-value list designating the intended rendering context. The integer values p_i give the output precedences of the renderings.

The patterns φ_i are formed from a formal grammar for a subset of OPEN-MATH objects extended with named jokers. The jokers $\underline{o}$ and $\underline{l}(\varphi)$ correspond to \(.\) and \(φ\)$^+$ in Posix regular expression syntax ([POS88]) – except that our patterns are matched against the list of children of an OPENMATH object instead of against a list of characters. We need two special jokers $\underline{s}$ and $\underline{v}$, which only match OPENMATH symbols and variables, respectively. The renderings ρ_i are formed by a formal syntax for simplified XML extended with means to refer to the jokers used in the patterns. When referring to object jokers, input precedences are given that are used, together with the output precedences, to determine the placement of brackets.

Match contexts are used to store the result of matching a pattern against an object. Due to list jokers, jokers may be nested;therefore, we use qualified joker names in the match contexts (which are transparent to the user). Empty match contexts are used to store the structure of a match context induced by a pattern: They contain holes that are filled by matching the pattern against an object.

Example. We will use a multiple integral as an example that shows all aspects of our approach in action.

$$\int_{a_1}^{b_1} \ldots \int_{a_n}^{b_n} \sin x_1 + x_2 \, dx_n \ldots dx_1.$$

Let *int*, *iv*, *lam*, *plus*, and *sin* abbreviate symbols for integration, closed real intervals, lambda abstraction, addition, and sine. We intend *int*, *lam*, and *plus* to be flexary symbols, i.e., symbols that take an arbitrary finite number of arguments. Furthermore, we assume symbols *color* and *red* from a content dictionary for style attributions. We want to render into LaTeX the OPENMATH object

$$@\big(int, @(iv, a_1, b_1), \ldots, @(iv, a_n, b_n),$$
$$\beta\big(lam, v(x_1), \ldots, v(x_n), \; \alpha(@(plus, @(sin, v(x_1)), v(x_2)), color \mapsto red))\big)$$

as \int_{a_1}^{b_1}...\int_{a_n}^{b_n}\color{red}{\sinx_1+x_2}dx_n... dx_1
  We can do that with the following notations:

$@(int, \underline{\mathbf{ranges}}(@(iv, \underline{a}, \underline{b})), \beta(lam, \underline{\mathbf{vars}}(\underline{x}), \underline{f}))$
$\vdash ((format = latex) :$
   for($\underline{\mathbf{ranges}}$)\{\int_\{ $\underline{a}^\infty$ \}^\{ $\underline{b}^\infty$ \}\} $\underline{f}^\infty$ for($\underline{\mathbf{vars}}$, -1)\{d $\underline{x}^\infty$\})$^{-\infty}$

$\alpha(\underline{a}, color \mapsto \underline{\mathbf{col}}) \vdash ((format = latex) : \{\color\{ \underline{\mathbf{col}} \} \underline{a}^\infty \})^{-\infty}$

$@(plus, \underline{\mathbf{args}}(\underline{\mathbf{arg}})) \vdash ((format = latex) : \; \mathbf{for}(\underline{\mathbf{args}}, +)\{\underline{\mathbf{arg}}\})^{10}$

$@(sin, \underline{\mathbf{arg}}) \vdash ((format = latex) : \; \sin \underline{\mathbf{arg}})^0$

The first notation matches the application of the symbol *int* to a list of ranges and a lambda abstraction binding a list of variables. The rendering iterates first over the ranges rendering them as integral signs with bounds, then recurses into the function body $\underline{f}$, then iterates over the variables rendering them in reverse order prefixed with d. The second notation is used when $\underline{f}$ recurses into the presentation of the function body $a(@(plus, @(sin, v(x_1)), v(x_2)), color \mapsto red)$. It matches an attribution of *color*, which is rendered using the LaTeX color package. The third notation is used when $\underline{a}$ recurses into the attributed object $@(plus, @(sin, v(x_1)), v(x_2))$. It matches any application of *plus*, and the rendering iterates over all arguments placing the separator $+$ in between. Finally, *sin* is rendered in a straightforward way. We omit the notation that renders variables by their name.

The output precedence $-\infty$ of *int* makes sure that the integral as a whole is never bracketed. And the input precedences ∞ make sure that the arguments of *int* are never bracketed. Both are reasonable because the integral notation provides its own fencing symbols, namely $\int$ and d. The output precedences of *plus* and *sin* are 10 and 0, which means that *sin* binds stronger; therefore, the expression $\sin x$ is not bracketed either. However, an inexperienced user may wish to display these brackets: Therefore, our rendering does not suppress them. Rather, we annotate them with an elision level, which is computed as the difference of the two precedences. Dynamic output formats that can change their appearance, such as XHTML with JavaScript, can use the elision level to determine the visibility of symbols based on user-provided elision thresholds: the higher its elision level, the less important a bracket.

Well-formed Notations. A notation definition $\varphi_1, \ldots, \varphi_r \vdash (\lambda_1 : \rho_1)^{p_1}, \ldots, (\lambda_s : \rho_s)^{p_s}$ is well-formed if all φ_i are well-formed patterns that induce the same empty match contexts, and all ρ_i are well-formed renderings with respect to that empty match context.

Every pattern φ generates an *empty match context* $\mu(\varphi)$ as follows:

- For an object joker $\underline{o}$ occurring in φ but not within a list joker, $\mu(\varphi)$ contains $o \mapsto \_$.
- For a symbol or variable with name n occurring in φ but not within a list joker, $\mu(\varphi)$ contains $n \mapsto$ "".
- For a list joker $\underline{l}(\varphi')$ occurring in φ, $\mu(\varphi)$ contains
  - $l \mapsto (\_)$, and
  - $l/n \mapsto (H)$ for every $n \mapsto H$ in $\mu(\varphi')$.

In an empty match context, a hole $\_$ is a placeholder for an object, "" for a string, $(\_)$ for a list of objects, $((\_))$ for a list of lists of objects, and so on. Thus, symbol, variable, or object joker in φ produce a single named hole, and every list joker and every joker within a list joker produces a named list of holes (H). For example, the empty match context induced by the pattern in the notation for *int* above is

$$\mathbf{ranges} \mapsto (\_), \ \mathbf{ranges/a} \mapsto (\_), \ \mathbf{ranges/b} \mapsto (\_), \ \mathbf{f} \mapsto \_,$$

$$\mathbf{vars} \mapsto (\_), \ \mathbf{vars/x} \mapsto ("")$$

A pattern φ is well-formed if it satisfies the following conditions:

- There are no duplicate names in $\mu(\varphi)$.
- List jokers may not occur as direct children of binders or attributions.
- At most one list joker may occur as a child of the same application, and it may not be the first child.
- At most one list joker may occur in the same variable context.

These restrictions guarantee that matching an OpenMath object against a pattern is possible in at most one way. In particular, no backtracking is needed in the matching algorithm.

Assume an empty match context μ. We define well-formed renderings with respect to μ as follows:

- $\langle S \rangle \rho_1, \ldots, \rho_r \langle / \rangle$ is well-formed if all ρ_i are well-formed.
- $S = "\rho_1, \ldots, \rho_r"$ is well-formed if all ρ_i are well-formed and are of the form S' or $\underline{n}$. Furthermore, $S = "\rho_1, \ldots, \rho_r"$ may only occur as a child of an XML element rendering.
- S is well-formed.
- $\underline{n}$ is well-formed if $n \mapsto$ "" is in μ.
- $\underline{o}^p$ is well-formed if $o \mapsto \_$ is in μ.
- $\texttt{for}(\underline{l}, I, sep)\{body\}$ is well-formed if $l \mapsto (\_)$ or $l \mapsto ("")$ is in μ, all renderings in sep are well-formed with respect to μ, and all renderings in $body$ are well-formed with respect to μ^l. The step size I and the separator sep are optional, and default to 1 and the empty string, respectively, if omitted.

Here μ^l is the empty match context arising from μ if every $l/q \mapsto (H)$ is replaced with $q \mapsto H$ and every previously existing hole named q is removed. Replacing $l/q \mapsto (H)$ means that jokers occurring within the list joker l are only accessible within a corresponding rendering $\texttt{for}(\underline{l}, I, \rho^*)\{\rho^*\}$. And removing the previously existing holes means that in $@(\underline{o}, \underline{l}(\underline{o}))$, the inner object joker shadows the outer one.

3   Semantics of Notation Definitions

The rendering algorithm takes as input a notation context Π (a list of notation definitions, computed as described in Sect. 4), a rendering context Λ (a list of context annotations, computed as described in Sect. 5), an OpenMath object ω, and an input precedence p. If the algorithm is invoked from top level (as opposed to a recursive call), p should be set to ∞ to suppress top level brackets.

It returns as output either text or an XML element. There are two output types for the rendering algorithm: text and sequences of XML elements. We will use $O + O'$ to denote the concatenation of two outputs O and O'. By that, we mean a concatenation of sequences of XML elements or of strings if O and O' have the same type. Otherwise, $O + O'$ is a sequence of XML elements treating text as an XML text node. This operation is associative if we agree that consecutive text nodes are always merged. The algorithm inserts brackets if necessary. And to give the user full control over the appearance of brackets, we obtain the brackets by the rendering of two symbols for left and right bracket from a special fixed content dictionary. The algorithm consists of the following three steps.

1. ω is matched against the patterns in the notation definitions in Π (in the listed order) until a matching pattern φ is found. The notation definition in which φ occurs induces a list $(\lambda_1 : \rho_1)^{p_1}, \ldots, (\lambda_n : \rho_n)^{p_n}$ of context-annotations, renderings, and output precedences.
2. The rendering context Λ is matched against the context annotations λ_i in order. The pair (ρ_j, p_j) with the best matching context-annotation λ_j is selected (see Section 5.2 for details).
3. The output is $\rho_j{}^{M(\varphi,\omega)}$, the rendering of ρ_j in context $M(\varphi, \omega)$ as defined below. Additionally, if $p_j > p$, the output is enclosed in brackets.

Semantics of Patterns. The semantics of patterns is that they are matched against OPENMATH objects. Naturally, every OPENMATH object matches against itself. Symbol, variable, and object jokers match in the obvious way. A list joker $\underline{l}(\varphi)$ matches against a non-empty list of objects all matching φ.

Let φ be a pattern and ω a matching OPENMATH object. We define a match context $M(\varphi, \omega)$ as follows.

- For a symbol or variable joker with name n that matched against the sub-object ω' of ω, $M(\varphi, \omega)$ contains $n \mapsto S$ where S is the name of ω'.
- For an object joker $\underline{o}$ that matched against the sub-object ω' of ω, $M(\varphi, \omega)$ contains $o \mapsto \omega$.
- If a list joker $\underline{l}(\varphi')$ matched a list $\omega_1, \ldots, \omega_r$, then $M(\varphi, \omega)$ contains
  - $l \mapsto (\omega_1, \ldots, \omega_r)$, and
  - for every l/q in $\mu(\varphi)$: $l/q \mapsto (X_1, \ldots, X_r)$ where $q \mapsto X_i$ in $M(\varphi', \omega_i)$.

We omit the precise definition of what it means for a pattern to match against an object. It is, in principle, well-known from regular expressions. Since no back-tracking is needed, the computation of $M(\varphi, \omega)$ is straightforward. We denote by $M(q)$, the lookup of the match bound to q in a match context M.

Semantics of Renderings If φ matches against ω and the rendering ρ is well formed with respect to $\mu(\varphi)$, the intuition of $\rho^{M(\varphi,\omega)}$ is that the joker references in ρ are replaced according to $M(\varphi, \omega) =: M$. Formally, ρ^M is defined as follows.

- $\langle S \rangle \rho_1 \ldots \rho_r \langle / \rangle$ is rendered as an XML element with name S. The attributes are those $\rho_i{}^M$ that are rendered as attributes. The children are the concatenation of the remaining $\rho_i{}^M$ preserving their order.
- $S = "\rho_1 \ldots \rho_r"$ is rendered as an attribute with label S and value $\rho_1{}^M + \ldots + \rho_n{}^M$ (which has type text due to the well-formedness).
- S is rendered as the text S.
- $\underline{s}$ and $\underline{v}$ are rendered as the text $M(s)$ or $M(v)$, respectively.
- $\underline{o}^p$ is rendered by applying the rendering algorithm recursively to $M(o)$ and p.
- $\text{for}(\underline{l}, I, \rho_1 \ldots \rho_r)\{\rho'_1 \ldots \rho'_s\}$ is rendered by the following algorithm:
  1. Let $sep := \rho_1{}^M + \ldots + \rho_r{}^M$ and t be the length of $M(l)$.
  2. For $i = 1, \ldots, t$, let $R_i := \rho'_1{}^{M_i^l} + \ldots + \rho'_s{}^{M_i^l}$.
  3. If $I = 0$, return nothing and stop. If I is negative, reverse the list R, and invert the sign of I.

4. Return $R_I + sep + R_{2*I} \ldots + sep + R_T$ where T is the greatest multiple of I smaller than or equal to t.

Here the match context M_i^l arises from M as follows

- replace $l \mapsto (X_1 \ldots X_t)$ with $l \mapsto X_i$,
- for every $l/q \mapsto (X_1 \ldots X_t)$ in M: replace it with $q \mapsto X_i$, and remove a possible previously defined match for q.

Example. Consider the example introduced in Sect. 2. There we have

$$\omega = @\big(int, @(iv, a_1, b_1), \ldots, @(iv, a_n, b_n),$$
$$\beta\big(lam, v(x_1), \ldots, v(x_n), \alpha(@(plus, @(sin, v(x_1)), v(x_2)), color \mapsto red)\big)\big)$$

And Π is the given list of notation definitions. Let $\Lambda = (format = latex)$. Matching ω against the patterns in Π succeeds for the first notation definitions and yields the following match context M:

$\mathtt{ranges} \mapsto (@(iv, a_1, b_1), \ldots, @(iv, a_n, b_n))$, $\mathtt{ranges/a} \mapsto (a_1, \ldots, a_n)$,

$\mathtt{ranges/b} \mapsto (b_1, \ldots, b_n)$, $\mathtt{f} \mapsto \alpha(@(plus, @(sin, v(x_1)), v(x_2)), color \mapsto red)$,

$\mathtt{vars} \mapsto (v(x_1), \ldots, v(x_n))$, $\mathtt{vars/x} \mapsto (x_1, \ldots, x_n)$

In the second step, a specific rendering is chosen. In our case, there is only one rendering, which matches the required rendering context Λ, namely

$$\rho = \mathtt{for}(\underline{\mathtt{ranges}})\{\backslash\mathtt{int\_}\{\ \underline{\mathtt{a}}^\infty\ \}\hat{\ }\{\ \underline{\mathtt{b}}^\infty\ \}\}\ \underline{\mathtt{f}}^\infty\ \mathtt{for}(\underline{\mathtt{vars}}, -1)\{\mathtt{d}\ \underline{\mathtt{x}}^\infty\})^{-\infty}$$

To render ρ in match context M, we have to render the three components and concatenate the results. Only the iterations are interesting. In both iterations, the separator sep is empty; in the second case, the step size I is -1 to render the variables in reverse order.

4   Choosing Notation Definitions Extensionally

In the last sections we have seen how collections of notation definitions induce rendering functions. Now we permit users to define the set Π of available notation definitions extensionally. In the following, we discuss the collection of notation definitions from various sources and the construction of Π_ω for a concrete mathematical object ω.

4.1   Collecting Notation Definitions

The algorithm for the collection of notation definitions takes as input a tree-structured document, e.g., an XML document, an object ω within this document, and a totally ordered set S^N of source names. Based on the hierarchy proposed in [NW01b], we use the source names EC, F, Doc, CD, and SD explained below. The user can change their priorities by ordering them.

The collection algorithm consists of two steps: The collection of notation definitions and their reorganization. In the first step the notation definitions are collected from the input sources according to the order in $\mathcal{S}^N$. The respective input sources are treated as follows:

- *EC* denotes the **extensional context**, which associates a list of notation definitions or containers of notation definitions to every node of the input document. The effective extensional context is computed according to the position of ω in the input document (see a concrete example below). *EC* is used by authors to reference their individual notation context.
- *F* denotes an **external notation document** from which notation definitions are collected. *F* can be used to overwrite the author's extensional context declarations.
- *Doc* denotes the **input document**. As an alternative to *EC*, *Doc* permits authors to embed notation definitions into the input document.
- *CD* denotes the **content dictionaries** of the symbols occurring in ω. These are searched in the order in which the symbols occur in ω. Content dictionaries may include or reference default notation definitions for their symbols.
- *SD* denotes the **system default** notation document, which typically occurs last in $\mathcal{S}^N$ as a fallback if no other notation definitions are given.

In the second step the obtained notation context Π is reorganized: All occurrences of a pattern φ in notation definitions in Π are merged into a single notation definition preserving the order of the $(\lambda \dot{} \rho)^p$ (see a concrete example below).

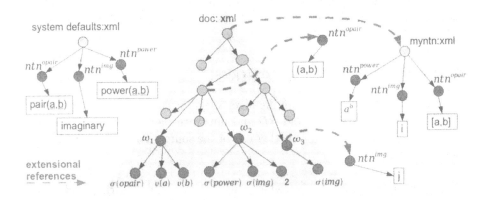

Fig. 2. Collection Example

We base our further illustration on the input document in Fig. 2 figure above, which includes three mathematical objects. For simplicity, we omit the `cdbase` and `cd` attributes of symbols.

$$\omega_1 : @(\sigma(opair), v(a), v(b)) \rightsquigarrow (a, b) \qquad \omega_2 : @(\sigma(power), \sigma(img), 2) \rightsquigarrow i^2 \qquad \omega_3 : \sigma(img) \rightsquigarrow j$$

The dashed arrows in the figure represent extensional references: For example, the ec attribute of the document root *doc* references the notation document "*myntn*", which is interpreted as a container of notation definitions.

We apply the algorithm above with the input object ω_3 and $\mathcal{S}^N = (EC, SD)$ and receive Π_{ω_3} in return. For simplicity, we do not display context annotations and precedences.

---

1. We collect all notation definitions yielding Π_{ω_3}
1.1 We collect notation definitions from EC
1.1.1 We compute the effective extensional context based on the position of ω_3 in the input document: $ec(\omega_3) = (ntn^{img}, myntn)$
1.1.2 We collect all notation definition based on the references in $ec(\omega_3)$:
$\Pi_{\omega_3} = (ntn^{img}, ntn^{power}, ntn^{img}, ntn^{opair})$
1.2. We collect notation definitions from SD and append them to Π_{ω_3}
$\Pi_{\omega_3} = (ntn^{img}, ntn^{power}, ntn^{img}, ntn^{opair}, ntn^{opair}, ntn^{img}, ntn^{power})$
1.3. The collected notation definition form the notation context Π_{ω_3}
$\Pi_{\omega_3} = (\ \varphi_1 \vdash j, \varphi_2 \vdash \underline{a}^{\underline{b}}, \varphi_1 \vdash i, \varphi_3 \vdash [\underline{a}, \underline{b}], \varphi_3 \vdash pair(\underline{a}, \underline{b}), \varphi_1 \vdash imaginary,$
  $\varphi_2 \vdash power(\underline{a}, \underline{b})\ )$
2. We reorganize Π_{ω_3} yielding Π'_{ω_3}
$\Pi'_{\omega_3} = (\ \varphi_1 \vdash j, i, imaginary; \varphi_2 \vdash \underline{a}^{\underline{b}}, power(\underline{a}, \underline{b}); \varphi_3 \vdash [\underline{a}, \underline{b}], pair(\underline{a}, \underline{b})\ )$

To implement EC in arbitrary XML-based document formats, we propose an ec attribute in a namespace for notation definitions, which may occur on any element. The value of the ec attribute is a whitespace-separated list of URIs of either notation definitions or any other document. The latter is interpreted as a container, from which notation definitions are collected. The ec attribute is empty by default. When computing the effective extensional context of an element, the values of the ec attributes of itself and all parents are concatenated, starting with the inner-most.

4.2   Discussion of Collection Strategies

In [KLM+08], we provide the specific algorithms for collecting notation definitions from EC, F, Doc, CD and SD and illustrate the advantages and drawbacks of basing the rendering on either one of the sources. We conclude with the following findings:

1. Authors can write documents which only include content markup and do not need to provide any notation definitions. The notation definitions are then collected from CD and SD.
2. The external document F permits authors to store their notation definitions centrally, facilitating the maintenance of notational preferences. However, authors may not specify alternative notations for the same symbol on granular document levels.
3. Authors may use the content dictionary defaults or overwrite them by providing F or Doc.

4. Authors may embed notation definitions inside their documents. However, this causes redundancy inside the document and complicates the maintenance of notation definitions.
5. Users can overwrite the specification inside the document with F. However, that can destroy the meaning of the text, since the granular notation contexts of the authors are replaced by only one alternative declaration in F.
6. Collecting notation definitions from F or Doc has benefits and drawbacks. Since users want to easily maintain and change notation definitions but also use alternative notations on granular document levels, we provide EC. This permits a more controlled and more granular specification of notations.

5   Choosing Renderings Intensionally

The extensional notation context declarations do not support authors to select between alternative renderings inside one notation definition. Consequently, if relying only on this mechanism, authors have to take extreme care about which notation definition they reference or embed. Moreover, other users cannot change the granular extensional declarations in EC without modifying the input document. They can only overwrite the author's granular specifications with their individual styles F, which may reduce the understandability of the document.

Consequently, we need a more *intelligent, context-sensitive* selection of renderings, which lets users guide the selection of alternative renderings. We use an intensional rendering context Λ, which is matched against the context annotations in the notation definitions. In the following, we discuss the collection of contextual information from various sources and the construction of Λ_ω for a concrete mathematical object ω.

5.1   Collecting Contextual Information

We represent contextual information by contextual key-value pairs, denoted by $(d_i = v_i)$. The key represents a *context dimension*, such as *language*, *level of expertise*, *area of application*, or *individual preference*. The value represents a *context value* for a specific context dimension. The algorithm for the context-sensitive selection takes as input an object ω, a list L of elements of the form $(\lambda : \rho)^p$, and a totally ordered set $\mathcal{S}^C$ of source names. We allow the names GC, CCF, IC, and MD. The algorithm returns a pair (ρ, p).

The selection algorithm consists of two steps: The collection of contextual information Λ_ω and the selection of a rendering. In the first step Λ_ω is computed by processing the input sources in the order given by $\mathcal{S}^C$. The respective input sources are treated as follows:

- GC denotes the **global context** which provides contextual information during *rendering time* and overwrites the author's intensional context declarations. The respective $(d_i = v_i)$ can be collected from a user model or are explicitly entered. GC typically occurs first in $\mathcal{S}^C$.
- CCF denotes the **cascading context files**, which permit the contextualization analogous to cascading stylesheets[Cas99].

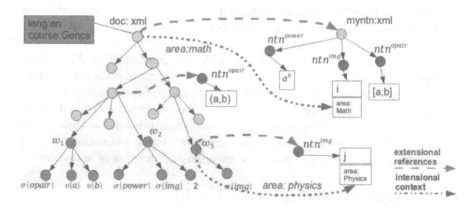

Fig. 3. Rendering Example

- IC denotes the **intensional context**, which associates a list of contextual key-value pairs $(d_i = v_i)$ to any node of the input document. These express the author's intensional context. For the implementation in XML formats, we use an `ic` attribute similar to the `ec` attribute above, i.e., the effective intensional context depends on the position of ω in the input document (see a concrete example below).
- MD denotes **metadata**, which typically occurs last in $\mathcal{S}^C$.

In the second step, the rendering context Λ_ω is matched against the context annotations in L. We select the pair (ρ, p) whose corresponding context annotation satisfies the intensional declaration best (see [KLM$^+$08] for more details).

In Fig. 3, we continue our illustration on the given input document. The dashed arrows represent extensional references, the dashed-dotted arrows represent intensional references, i.e., implicit relations between the `ic` attributes of the input document and the context-annotations in the notation document. A global context is declared, which specifies the language and course dimension of the document. We apply the algorithm above with the input object ω_3, $\mathcal{S}^C = (GC, IC)$, and a list of context annotations and rendering pairs based on the formerly created notation context Π_{ω_3}. For convenience, we do not display the system's default notation document and the precedences.

---

1. We compute the intensional rendering context 1.1. We collect contextual information from GC
$\Lambda_{\omega_3} = (lang = en, course = GenCS)$ 1.2. We collect contextual information from IC an append them to Λ_{ω_3}
$\Lambda_{\omega_3} = (lang = en, course = GenCS, area = physics, area = math)$

2. We match the rendering context against the context annotations of the input list L and return the rendering with the best matching context annotation: $L = [\,(\lambda_0 = j), (\lambda_1 = i), (\lambda_2 = imaginary)\,]$
$\lambda_0 = (area = physics)$,
$\lambda_1 = (area = maths)$, and $\lambda_2 = \emptyset$

For simplicity, we compute the similarity between Λ_{ω_3} and λ_i based on the number of similar $(d_i = v_i)$: λ_0 includes $(area = physics)$. λ_1 includes $(area = math)$. λ_2 is empty. λ_0 and λ_1 both satisfy one $(d_i = v_i)$ of Λ_{ω_3}. However, since $(\lambda_0 = \rho_0)$ occurs first in Π_{ω_3}, the algorithm returns ρ_0.

5.2 Discussion of Context-Sensitive Selection Strategies

In [KLM$^+$08], we illustrate and evaluate the collection of contextual information from GC, CCF, IC and MD in detail. We conclude with the following findings:

1. The declaration of a *global context* provides a more intelligent intensional selection between alternative $(\lambda : \rho)^p$ triples inside one notation definition: The globally defined $(d_i = v_i)$ are matched against the context-annotations λ to select an appropriate rendering ρ. However, the approach does not let users specify intensional contexts on granular levels.
2. Considering *metadata* is a more granular approach than the global context declaration. However, metadata may not be associated to any node in the input document and cannot be overwritten without modifying the input document. Moreover, the available context dimensions and values are limited by the respective metadata format.
3. The *intensional context* supports a granular selection of renderings by associating an intensional context to any node of the input document. However, the intensional references cannot be overwritten on granular document levels.
4. *Cascading Context Files* permit a granular overwriting of contexts.

6 Conclusion and Future Work

We introduced a representational infrastructure for notations in living mathematical documents by providing a flexible declarative specification language for notation definitions together with a rendering algorithm. We described how authors and users can extensionally extend the set of available notation definitions on granular document levels, and how they can guide the notation selection via intensional context declarations. Moreover, we discussed different approaches for collecting notation definitions and contextual information.

To substantiate our approach, we have developed prototypical implementations of all aspects of the rendering pipeline:

- The Java toolkit `mmlkit` [MMK07] implements the conversion of OPENMATH and Content-MATHML expressions to Presentation-MATHML. It supports the collection of *notation definitions* from various sources, constructs rendering contexts based on contextual annotations of the rendered object, identifies proper renderings for the conversion.
- The semantic wiki SWiM [Lan08] supports the collaborative browsing and editing of notation definitions in OPENMATH content dictionaries.
- The *panta rhei* [Mül07] reader integrates mmlkit to present mathematical documents, provides facilities to categorize and describe notations, and uses these context annotations to adapt documents.

We will invest further work into our implementations as well as the evaluation of our approach. In particular, we want to address the following challenges:

- Write Protection: In some cases, users should be prevented to overwrite the author's declaration. On the contrary, static notations reduce the flexibility and adaptability of a document (see [KLM+08] for more details).
- Consistency: The flexible adaptation of notations can destroy the meaning of documents, in particular, if we use the same notation to denote different mathematical concepts.
- Elision: In [KLM+08], we have already adapted the elision of arbitrary parts of formulae from [KLR07].
- Notation Management: Users want to reuse, adapt, extend, and categorize notation definitions (see [KLM+08] for more details).
- Advanced notational forms: Ellipses and Andrews' dot are examples of advanced notations that we cannot express yet.

Acknowledgments. Our special thanks go to Normen Müller for the initial implementation of the presentation pipeline. We would also like to thank Alberto Gonzáles Palomo and Paul Libbrecht for the discussions on their work. This work was supported by JEM-Thematic-Network ECP-038208.

References

ABC+03. Ausbrooks, R., Buswell, S., Carlisle, D., Dalmas, S., Devitt, S., Diaz, A., Froumentin, M., Hunter, R., Ion, P., Kohlhase, M., Miner, R., Poppelier, N., Smith, B., Soiffer, N., Sutor, R., Watt, S.: Mathematical Markup Language (MathML) version 2.0, 2nd edn. W3C recommendation, World Wide Web Consortium (2003), http://www.w3.org/TR/MathML2

BCC+04. Buswell, S., Caprotti, O., Carlisle, D.P., Dewar, M.C., Gaetano, M., Kohlhase, M.: The Open Math standard, version 2.0. Technical report, The Open Math Society (2004), http://www.openmath.org/standard/om20

Caj93.   Cajori, F.: A History of Mathematical Notations. Courier Dover Publications(1993) (Originally published in 1929)

Cas99.   Cascading Style Sheets (1999), http://www.w3.org/Style/CSS/

Kay06.   Kay, M.: XSL Transformations (XSLT) Version 2.0. W3C Candidate Recommendation, World Wide Web Consortium (W3C) (June 2006), http://www.w3.org/TR/2006/CR-xslt20-20060608/

KLM+08. Kohlhase, M., Lange, C., Müller, C., Müller, N., Rabe, F.: Adaptation of notations in living mathematical documents. KWARC report, Jacobs University Bremen (2008)

KLR07.   Kohlhase, M., Lange, C., Rabe, F.: Presenting mathematical content with flexible elisions. In: Caprotti, O., Kohlhase, M., Libbrecht, P. (eds.) Open-Math/ JEM Workshop 2007 (2007)

KMM07. Kohlhase, M., Müller, C., Müller, N.: Documents with flexible notation contexts as interfaces to mathematical knowledge. In: Libbrecht, P. (ed.) Mathematical User Interfaces Workshop 2007 (2007)

Koh06.   Kohlhase, M.: OMDoc – An open markup format for mathematical documents [Version 1.2]. LNCS (LNAI), vol. 4180. Springer, Heidelberg (2006)

Lan08.   Lange, C.: Mathematical Semantic Markup in a Wiki: The Roles of Symbols
            and Notations. In: Lange, C., Schaffert, S., Skaf-Molli, H., Völkel, M. (eds.)
            Proceedings of the 3rd Workshop on Semantic Wikis, European Semantic
            Web Conference 2008, Costa Adeje, Tenerife, Spain (June 2008)

MLUM05.   Manzoor, S., Libbrecht, P., Ullrich, C., Melis, E.: Authoring Presentation
            for OPENMATH. In: Kohlhase, M. (ed.) MKM 2005. LNCS (LNAI), vol. 3863,
            pp. 33–48. Springer, Heidelberg (2006)

MMK07.   Müller, C., Müller, N., Kohlhase, M.: mmlkit - a toolkit for handling
            mathematical documents and MathML3 notation definitions. mmlkit v0.1
            (November 2007) http://kwarc.info/projects/mmlkit

Mül07.   Müller, C., Rhei, P.: http://kwarc.info/projects/ panta-rhei/ (August
            2007)

NW01a.   Naylor, B., Watt, S.M.: Meta-Stylesheets for the Conversion of Mathemat-
            ical Documents into Multiple Forms. In: Proceedings of the International
            Workshop on Mathematical Knowledge Management (2001)

NW01b.   Naylor, B., Watt, S.M.: Meta-Stylesheets for the Conversion of Mathemat-
            ical Documents into Multiple Forms. In: Proceedings of the International
            Workshop on Mathematical Knowledge Management [NW01a]

POS88.   IEEE POSIX, ISO/IEC 9945 (1988)

SW06a.   Smirnova, E., Watt, S.M.: Generating TeX from Mathematical Content with
            Respect to Notational Settings. In: Proceedings International Conference
            on Digital Typography and Electronic Publishing: Localization and Inter-
            nationalization (TUG 2006), Marrakech, Morocco, pp. 96–105 (2006)

SW06b.   Smirnova, E., Watt, S.M.: Notation Selection in Mathematical Computing
            Environments. In: Proceedings Transgressive Computing 2006: A conference
            in honor of Jean Della Dora (TC 2006), Granada, Spain, pp. 339–355 (2006)

Wol00.   Wolfram, S.: Mathematical notation: Past and future. In: MathML and
            Math on the Web: MathML International Conference, Urbana Champaign,
            USA (October 2000), http://www.stephenwolfram.com/publications/
            talks/mathml/

Cross-Curriculum Search for Intergeo

Paul Libbrecht[1], Cyrille Desmoulins[2], Christian Mercat[3],
Colette Laborde[2], Michael Dietrich[1], and Maxim Hendriks[4]

[1] DFKI GmbH, Saarbrücken, Germany
[2] LIG, Université Joseph Fourier, Grenoble, France
[3] I3M, Université Montpellier 2, France
[4] Technische Universiteit Eindhoven, The Netherlands

Abstract. Intergeo is a European project dedicated to the sharing of interactive geometry constructions. This project is setting up an annotation
and search web platform which will offer and provide access to thousands of
interactive geometry constructions and resources using them. The search
platform should cross the boundaries of the curriculum standards of Europe. A topics and competency based approach to retrieval for interactive geometry with designation of the semantic entities has been adopted:
it requests the contributor of an interactive geometry resource to input
the competencies and topics involved in a construction, and allows the
searcher to find it by the input of competencies and topics close to them;
both rely on plain-text-input.

This paper describes the current prototypes, the input-methods, the
workflows used, and the integration into the Intergeo platform.

1 Introduction

The last decade has seen a bloom in tools that allow teachers to enrich their
teaching with interactive data, whether in face-to-face or distant mode. This
wealth has its drawbacks and teachers need support to navigate through this
diversity: which software should I use, where can I find resources, will this
resource work for my class? Indeed, apart from pioneer work by dedicated
teachers, the actual practices in the classroom have not evolved much. The reasons are manifold. Here are the three main ones:

- All the communities that have grown around the different technical solutions
  and software available have produced resources that they share in one way
  or another. They have all thought about their practice and produced different approaches. Currently these cannot be merged, because the data they
  produce is scattered, both physically and semantically. The resources need
  to be *centrally visible and exchangeable*.
- As well as being difficult to find and analyze, the resources are usually diverse
  in quality and relevance to a specific need. Teachers are unsure in which
  situation a given resource, even if apparently interesting, could actually be
  used, and whether it adds pedagogical value to the learning experience [1,2].
  They wait for a bolder colleague to report on her attempt. The *resources
  need to be tested*, and *published reports need to reflect these tests*.

S. Autexier et al. (Eds.): AISC/Calculemus/MKM 2008, LNAI 5144, pp. 520–535, 2008.
© Springer-Verlag Berlin Heidelberg 2008

– Mastering a piece of software is time-consuming, and very few teachers grow to become power-users of their tool. The resources need to be *easy to use, share and adapt*, in spite of software choices.

In order to solve these issues at least for one specific subject, interactive geometry, we propose to centralize educational resources from this field on the Intergeo web platform. All resources will have clear Intellectual Property Rights, promoting open licences. And they will be there in an interoperable file format we are going to create, based on OpenMath [3]. This format will be supported by the most common software programs for interactive geometry, so teachers can keep on using their own. In this article we will detail the way in which resources are annotated and how our search tool works with the competencies of many curriculum standards.

1.1 Outline

Sub-section 1.2 provides a very short description of what interactive geometry software is. Section 2 is a survey of learning object repositories comparable to what the Intergeo platform should be and presents some of the rationale behind the choice of platform. Section 3 deals with the preliminary phase of the development of the search tool while analyzing the workflow for the search and annotation tasks. Section 4 describes the design process of the ontology used for both representing the various European curricula and the resources we want to see shared among all users. Section 5 describes the two methods of inputting queries, by typing and explicitly selecting competencies and topics or by pointing in a curriculum or a textbook. The search process, from the query to a list of resources ordered by matching scores, is described in section 6. The paper ends with a vision towards the dynamic evolution of the ranking algorithm (section 8).

1.2 What Is Interactive Geometry?

The Intergeo project is driven by European leaders in interactive geometry software. We are going to explain what is understood by *interactive* or *dynamic geometry*, a way of doing geometry which is required of math and science teachers more and more often. Interactive geometry allows for the manipulation and the visualization of a construction (a figure) on a computer. The construction depends on some free parameters, like the position of one or several control points. The user manipulates the figure through the keyboard, the mouse or a tracking device, by changing one or more of these free parameters. The construction then changes accordingly.

Of course, the main entities and relations in interactive geometry are of geometrical type. You will find triangles, circles, lines and points, barycentres, tangents, secants with given angles and distances [4]. But it is much more general than antique Greek geometry – you can have functions, derivatives, colors, random variables, all sorts of constructs that allow you to visualize and manipulate concepts that arise in all sorts of contexts, inside mathematics as well as outside [5,6,7].

2   Survey of Current Repositories

In order to approach the realization of the intergeo platform for sharing inter-active geometric constructions across curriculum boundaries, we give a brief survey of the state of learning object repositories which are closest to what the intergeo platform should be.

2.1   Annotations and Retrieval in Learning Object Repositories

As far as we could observe, learning object repositories all classify learning *objects* of a highly variable nature using a certain amount of bibliographic information augmented by pedagogical and topical information. Unfortunately, there is rarely enough information to allow fine-grained search. Topical information is, at most, encoded in broad taxonomies such as the Mathematical Science Classification (MSC)[8]. The most fine-grained taxonomy for mathematics seems to be the WebALT repository [9] which attempts to refine the MSC to a level close to a curriculum standard but seems to stick to a single organization.

Other approaches that tend to be fine-grained are the tag-based approaches, where any tag can be attributed freely by any person providing content. While this approach works fine for statistical similarity and in communities that share a single language, it could only offer translation capabilities if mostly used by multilingual users and users that bridge several communities; we have not found, yet, such users to be common.

A learning object repository that provides topical information directly within the curriculum is GNU Edu [10]: this platform catalogues learning objects accor-ding to the skills described in a curriculum, split into years and chapters. GNU Edu allows the skills to be annotated with keywords which can be used to access the skills directly. The keywords are translated and this is how GNU Edu achieves cross-curriculum search: a query matches a set of keywords, each matching skills from each curriculum. GNU Edu does not, however, rank the results or generalize a query so that related keywords also matched.

The emergent repository TELOS from the LORNET research network, and its associated competency framework [11] have been considered, but rejected for their too generic approach. We are not concerned with the design and organiza-tion of coherent courses or evaluations; on the contrary, Intergeo resources will be aimed at being used as building blocks by more elaborate Learning Content Management Systems.

Several approaches to link resources to curricula are available. England's Cur-riculum Online [12], a concerted effort between the Education Board of England and several publishers to present the curriculum standard of England associated with resources that schools may purchase. Microsoft Lesson Connection is a joint of effort of Microsoft and a publisher to do the same for the curricula of the USA [13]. Most of these approaches seem to be based on directly and manually asso-ciating resources to lines in curricula, something which is clearly not an avenue for us, since we want the resources to *cross the curriculum barriers*, even being available for a freshly encoded curriculum.

The American commercial project ExploreLearning [14] has a similar view on cross-curriculum and textbook search: They propose interactive figures that are associated to both curricula of the different states of the United States, and to standard textbooks. We don't know whether this association relies on skills and topics ontology or (more probably) is performed manually.

The analysis above leads us to the belief that text search engines, based on information retrieval principles, still tend to be the most used approach for learning object identification. Information retrieval, the science of search engines, is a very mature field with pioneer works such as [15]. Software tools such as Apache Lucene [16] provide a sturdy basis to apply the theories of this field with good performance expectations. Indeed, we shall exploit partial search queries as often as 100 times a second for the purpose of *designating* the topics. Information retrieval is mostly for word matches. It has taught us the fundamental approach to quantify the *relevance* of a document matching a query: this yields search results that are ranked from most to least relevant and expects users to read only the most relevant results.

One way to generalize a query is to make it tolerant to typos or to match phonetically. Another way is to generalize the search by including *semantically close* words. An example is the Compass tool [17], which uses an ontology of all concepts to generalize queries using concepts related to the query words. But even the Compass approach needs to be complemented for cross-curriculum search of interactive geometry, since we wish that a search in French for the topic *théorème de Thalès* should match (at least mildly) a construction contributed by an English speaker who has annotated it with the competency of *recognizing an enlargement*. As a result, the Intergeo project needed an approach that imitates the query-expansion mechanism found in Compass and others but that performs this expansion with the mathematical relationships. Hence we need to tackle the work of encoding the geometric parts of curriculum standards of Europe in a way that identifies the common topics and their relationships. In particular, the search engine that associates topics and competencies to queries will be able to help annotate forthcoming curricula quickly and resources matching its entries will appear instantly.

2.2   Choice of Repository Platform

Learning object repositories can be compared by the services they offer. We shall tackle services which are relevant for interactive geometry constructions, on the authoring side, in order to upload, version, preview, convert, encapsulate into easily edited web-content, deliver and annotate the resources, and on the user's side communicate and report within a chosen community, especially in order to promote enhancement quality cycles through quality evaluation of resources and more casual forums.

In order to obtain all these objectives, we settled on building on the foundation of the Curriki learning object portal [18], an open-source extension of the XWiki platform, which provides textschtml-editing and communication services, and appeared easy enough to be developed further to accomodate needed extensions such as the search tool.

Having documented the general problematics of applicable technologies, we now turn to a more precise description of the user workflows, and afterwards shall cover the literature relevant to curriculum encoding.

3   User Workflow for Searching and Annotating

The Intergeo platform's main goal is to allow sharing of interactive geometric constructions and related materials. This material can take on the form of interactive geometric constructions, with or without concrete learner tasks attached to them, as well as web-based materials that encompass these. We shall use the term *resource* here, as has been done often on the web, to denote any of these data types. What does the sharing mean? Overall, it is the execution of the following *roles*:

- the *annotator role*: provision of authoring, licensing, topical, and pedagogical information about a resource contributed to the Intergeo platform;
- the *searcher role*: navigation and search through the platform's database to find relevant resources to use in teaching, to edit, or to evaluate.

The roles described here shall be complemented with the *curriculum encoder* role (described below) and the *quality evaluator* role described in section 7.

A crucial condition for the annotator's and searcher's roles to work is that together, they use a similar vocabulary to input the information about the resources and to search for the resources. A fundamental aspect of Intergeo is to solve this in a cross curriculum fashion, so that the annotator and searcher can express themselves in vocabularies that may be in different human languages and in different environments.

A simple example of a matching that crosses curriculum boundaries is the construction of the division of a segment in n equal parts. This should be matched by queries using strings such as "divide in equal parts", "diviser en parties de même longueur", etc. Curriculum standards, however, do not all speak about this topic in the very same way. The English curriculum only mentions the operation of *enlargement*, whereas the French national program of study mentions "connaître et utiliser dans

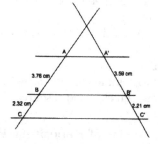

Fig. 1.   Théorème   de   Thalès. $AB/BC = A'B'/B'C' = 1.62$.

une situation donnée les deux théorèmes suivants" and provides the formulation of the "Théorème de Thalès" and its converse [19].

A simple example of a mismatching across some of the curriculum boundaries is the name "Thales' Theorem". In French (*théorème de Thalès*) and Spanish (*teorema de Tales*) it indicates the intercepting lines theorem, concluding proportionalities of segments, as in figure 1. However, Thales' Theorem in English or in German (Satz des Thales) refers to the theorem that if one takes a point on a circle and draws segments to the two endpoints of a diameter of the circle, these segments will be perpendicular, as in figure 2.

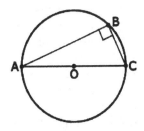

Fig. 2. Satz des Thales

Thus the role of the annotator is to provide sufficiently detailed topical and educational context information so that users in other curricula can find resources using the language of their curriculum as well as using everyday language. For this to work, we have added a third role to this workflow, that of curriculum encoder. This person makes sure that every competency and topic in a curriculum standard he or she is responsible for, is encoded in our ontology. The curriculum standards come from different sources, mainly official ones such as ministries of education publications, and often go by a different name. But everyday practices of teachers lead us to consider more practical implementations of curricula: textbooks that teachers ordinarily use in the classroom. We will ask editors to provide at least the table of contents of their textbooks and manuals to annotate them similarly.

4   GeoSkills: a Cross-Curriculum Ontology for Geometric Constructions

4.1   Ontologies and Sub-ontologies

The Intergeo project defines an ontology called GeoSkills [20], consisting of several sub-ontologies (to which we also simply refer as ontologies in themselves). These contain classes for *competencies, topics* and *educational level*, respectively. The first two reflect an agreement of the community as to what is actually being taught. The third one will be discussed a little more in the next section. The ontology mostly describes mathematics learned at the secondary school level, but could of course be extended to cover much more. Let's make clear what we mean by competencies, topics and educational levels:

- a topic is an object of knowledge such as *isosceles triangle* or *Thales theorem*;
- a competency is the compound of an ability (a verb) and a topic such as *identify parallel lines*;
- an educational level is a stage in the development of a learner, in the context of a specific educational region and educational pathway (school type). For example "Eerste klas" of the pathway "secundair onderwijs" in "Vlaanderen", an educational region within Belgium.

The competency ontology makes it possible to represent that the competency "use of scale" taken from the English national program of study [21] is related to "intercept theorem", itself linked to triangle, enlargement, similarity of triangles, measuring segments and measuring angles. And it enables us to capture the fact that the resource depicted in figure 2 refers to this competency.

One thing the ontology aims at is providing European curriculum experts the means to encode localized geometry curricula with a common semantics. Another is to enable searching, which will be discussed in the next section. The goal of this section is to describe the approach and the design decisions made in order to provide the ontology for curriculum encoding by experts. It presents the tool we used to create the ontology collaboratively. The methodology we followed is to rely on mature and widespread tools and practices both at the theoretical and practical level. On the theoretical level, the approach is to rely on well-defined semantics, decidable knowledge representation and widely interoperable languages. On the practical level, the idea is to use tools providing enough affordance for non computer scientists like curriculum experts from several countries, and to ask them to collaboratively construct the ontology and benchmark it with instances.

4.2   Other Projects

To design the competency ontology, we first surveyed tools providing curriculum mappings, encodings, and cross-curriculum search, especially in Europe.

Dragan Gasevic and Marek Hatala [22] developed a curriculum mapping using SKOS between the ACM (Association for Computing Machinery) classification [23] and the IEEE Information Management Course Curriculum Recommendation [24].

The CALIBRATE EU project has been working on curriculum encoding of competencies for curriculum-based resource browsing [25]. Their ontology is composed of an Action Verb Taxonomy and a Topic Taxonomy. A text fragment is tagged with a specific competency described with an Action Verb and a set of Topics taken from the ontology, together called a Tuple. The Tuple approach is well-suited for curriculum indexing. They also developed TopicMapper, a tool enabling curriculum expert to encode curriculum texts in HTML format into this ontology. It is a tool based on the XTM language providing an easy-to-use Graphical User Interface. However, Topic Mapper is a standalone application. It does not work on the web and uses local files.

We chose to use a similar approach to CALIBRATE's one (a verb plus a set of topics) to design our competency ontology.

4.3   Editing GeoSkills with Protégé OWL

The Protégé tool [26] has been chosen, both to design and edit the curriculum ontology and to provide an ontology-based curriculum encoding facility for the national experts. It corresponds to our need of a widespread tool. Protégé is the most widely used ontology editor at this moment. It also provides a simple Graphical User Interface for designing the concepts and properties of the ontology and for encoding curriculum competencies as instances of this ontology.

Protégé offers two major ontological representations (and a corresponding interface): frame based language [27] or OWL. We chose OWL, because it is an interoperable format provided by the W3C [28], and because it has a well-defined semantics. OWL-DL has been proven to be decidable, which was therefore our final choice. Additionally, several inference engines are available that could help searching [29,30,31]. This contrasts with the previously mentioned topic maps.

There exists a standardised language [32] for them and an editing tool [33]. But this editor is less widely used than Protégé and, more importantly, there are no results about the decidability of algorithms on topic maps.

In order to collaboratively design and populate the ontology with our *sample* curriculum, we first tried to use the WEB versioning system (LibreSource so6 synchronizer [34]). After some files were corrupted, we switched to the use of Protégé Server. It saves changes of each concurrent co-author in real time, thanks to JAVA RMI, thus providing a truly collaborative tool. The limits we encountered are twofold. Firstly, as RMI saves each change on the server, the network bandwidth is critical and sometimes not enough. Secondly, Protégé's user interface is not editable in the server version, despite it really being important to let curriculum experts work. It was solved by stopping the server when performing changes to the user interface, which was quite rarely.

At this moment, a first version of the curriculum and resource ontology have been designed and the parts of curricula around the intercepting lines theorem have been encoded for four countries (Great Britain, France, Germany, Spain) without major difficulties.

4.4   Design of the Ontology

The competency ontology contains two main concepts. The first is the class **Competency**, a hierarchy of action verbs divided in two main subclasses : **TransversalCompetency** (such as **Apply, Calculate, Explore**) and **GeometricCompetency** (such as **Construct, Infer, ToMeasure**). An instance of one of the Competency class is described with:

- a set of instances of the topic ontology (at least one);
- a set of curricula it belongs to;
- names (strings) that can be common names, uncommon names, rare names or false friends.

The various **names** properties provide an easy way to qualify the type of names related to a competency and consequently serve as a basis to implement a fuzzy search among names (a common name is more probable to be matched than any other type). They also provide a way to manage localized names in a simple way, as Protégé OWL provides a user-interface requesting the language of each string.

The second main class is **Topic**, a hierarchy of geometry topics mainly divided into the following sub-classes : **Object, Operation, Proof, Theorem** and **Tool**. Part of the topic ontology is shown in figure 3.

Educational levels are encoded following [35]. Similar to competencies and topics, levels are named. The levels branch of the ontology encodes the pathways of learning in the countries of the EU, and the particular country and age that are associated with a certain context. The latter two pieces of information then allow the system to find resources of the appropriate educational level with ordering and distance provided by the pathway and age range.

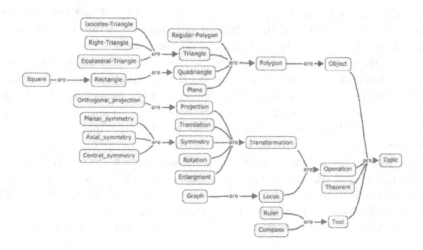

Fig. 3. An extract of the **Topics** branch of the GeoSkills ontology

To test the ontology with concrete resources, a second simple ontology has been developed representing them. It contains a three class hierarchy: Resource, LearningSubject and SchoolLevel. A resource is described by a title and an URI. It is linked to school levels, learning subjects, and competencies (taken from the competency ontology). This represents the objects which shall be indexed well.

5   Using the Ontology in Searching

Search engines are a crucial part of everyday internet usage, they are the applications that power information retrieval (see [15]). Both the comprehensive nature of the major search engines on the web and their simple query mechanism are extremely attractive. This simplicity is created on the one hand by simple text input and on the other hand by the responsiveness of results. These stimulate numerous search attempts and refinements to attain the right set of documents.

But because search engines are generally text-based, they are improper to search for conceptual entities such as described in the previous section, which can be made of several (overlapping) words. Therefore we designed two means to let the users easily designate nodes of the ontology.

5.1   Designating by Typing: SkillsTextBox

To let users designate a node of the ontology, we extend the familiar auto-completion: they can type a few words in the search field, these are matched to the terms of the names of the individual nodes; the auto-completion pop-up presents, as the user types, a list of matching nodes similar to figure 4. This list presents, for each candidate ontology node, the full name of the node, the number of related resources, an icon of the type, and a link to browse the ontology around that node. When chosen using either a click, or a few presses of the down key followed by the return key, the sequence of words is replaced by the name of the node, surrounded by square brackets to indicate an exact reference to a conceptual entity in our ontology.

This process is used not only to search but also when annotating a resource: individual competencies, topics, and educational usage are then provided.

SkillsTextBox uses a simple HTML form equipped with a GWT script [36]. SkillsTextBox also uses the Rocket GWT library. This script submits the fragments typed to the index on the server which uses all the retrieval matching capabilities (stemming, fuzziness through

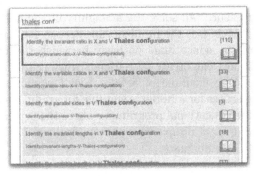

Fig. 4. Choosing among competencies about "Thales conf"

edit distance or phonetic matching) to provide an object description of the best matching 20 nodes of the ontology, which the script renders as an auto-completion list. This process is depicted in figure 5. More information about it is at http://www.activemath.org/projects/SkillsTextBox/.

5.2 Which Names to Match?

For SkillsTextBox to come up with the right resources, it is also vital that it knows the educational context in which a query is submitted. For one, it is a basic necessity that the system works transparently for the user: when typing a query, the user should be able to use his or her own language.

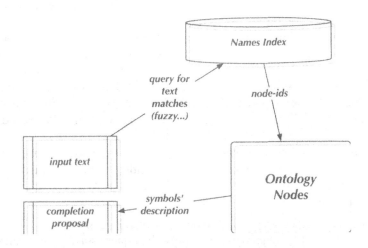

Fig. 5. Token designation using SkillsTextBox

The resulting suggestions for competencies and topics should also be in this language, and the returned resources probably suitable for students native to that language should be higher ranked. So letting the person who inputs a query select their language is a first step (done once in the user's preferences). This measure will solve the ambiguity of the name "Thales" either in German or in French. But the problem runs deeper, for there is a bigger cultural context, that of the educational pathway in which the user is working. Not regarding this aspect can create false friends when searching.

One example of a false friend would be when a French-speaking teacher searches for a resource suitable for grade *5ème*. The intergeo system could match the educational programme where the typical age 10–11 of Western Switzerland or that of France where the typical age 12–13. The context of use can exploited to disambiguate.

5.3   Designating by Pointing in a Book

Supplementary to letting users search for resources by explicitly selecting competencies and topics, we will offer the possibility to do this implicitly by letting the system infer these automatically from specific sections in curriculum standards or in text books they know well and that include geometry. Although we shall mostly not be able to offer whole text books to browse through, we expect it to be unproblematic to display tables of contents.

A user can then browse through a table of content and click on sections of interest. This click will trigger the selection of the competencies and topics associated to these sections, adding the necessary queries in the search field.

6   The Search Tool at Work

We have described, in the previous section how a set of words is used to identify interactively a node in the ontology, such as a competency or an educational level. In this section we turn to the actual search, from a query as simple-text to a list of documents, ordered by matching score.

Once a query is launched by the user, it is decomposed into a boolean combination of search terms. Fragments of texts between square brackets indicate queries to individual node names in the ontology whereas isolated words indicate the generic word-query. Consider the example [Identify parallel lines] [Enlargement] keystage 3, which includes a reference to a competency [Identify...], a topic [Enlargement] and two words keystage and 3 that do not yet designate an educational level.

First, the plain words appearing in the query are matched with names of the nodes of the ontology. The query is then expanded to include queries for the competencies that include these words, with a low boosting. Next, the query is transformed as follows:

- query competencies and topics pointed to with high boost;
- for each competency queried, expand with a query for a competency with the parent competency-verb with lower-boost;

- for each competency and topics queried, add weakly boosted queries for the ingredient nodes.

This last expansion step is where a query for the word *Thalès*, not identified with a topic by the auto-completion mechanism, is expanded to a query for Thales-theorem (with high boost) or for enlargement (with lower boost), or for the task of dividing a segment in equal parts or for parallel lines (both with low boost). The isolated words appearing in this query are also matched outside of the ontology (in order of preference):

- in the title of the construction;
- in the author names;
- in the names (along their varying degree of commonality) of the nodes associated to each resource;
- in the plain-text content of the resources.

The expanded query for the example above might look like:

comp:identify_parallel_lines $\cdot$ 100 + top:Identify $\cdot$ 30 + top:Parallel_r $\cdot$ 30

+ top:Enlargement_r $\cdot$ 100 + top:Amplification_r $\cdot$ 100

+ top:Reduction_r $\cdot$ 100 + top:Operation_r $\cdot$ 100 + txt-en:keystage $\cdot$ 20

+ txt:keystage $\cdot$ 5 + txt-en:3 $\cdot$ 20 + txt:3 $\cdot$ 20 + lvl:keystage_3 $\cdot$ 5

This expanded query has now taken full advantage of the ontology, it is passed to the resources' index, also a Lucene index, which returns the first few matching documents with the highest overall score. A presentation similar to the prototype of figure 6 is being implemented.

The combination described yields a search engine with the following characteristics:

- Most importantly, the nodes of the GeoSkills ontology, encoded in the query and in the annotation, are matched against each other. The queries are generalized using the relationships in the ontology. This is multilingual and multicultural, e.g., through the use of topics and competencies, but is expressed using a language-dependent and culture-dependent vocabulary.
- Less importantly, the query words are matched to the resources information and contents. This match is mostly single-language: e.g. queries in english search contents in english and maybe contents in another language

7   Enhancing Quality of Retrieval and Resources

In this section we will present our vision towards a dynamic evolution of the ranking algorithm based on social network behavior and quality evaluation by peers.

First, the quality asserted by peer review is going to play an important role [37]: the eLearning objects that we gather on our website will be used in the

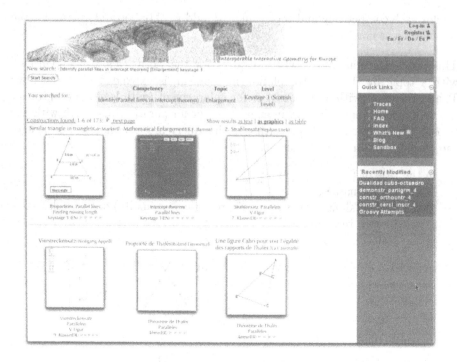

Fig. 6. Prototype of the result of searching for [Identify parallel lines in intercept theorem] [Enlargement] keystage 3

classroom, and we will organize and collect data about this use. We will collect automatic server data like the number of downloads of the resource but also user reports on the usage: on informal forums and chats where user's opinions will be expressed in their own words, but as well through a short questionnaire regarding the adequacy of the resource for the advertised purpose. This questionnaire will be available online to every identified user, for a priori quality assessment for a teacher enrolling for a teaching experience, and a posteriori quality assessment after the experimentation in the classroom has taken place.

Therefore a resource with many positive users feedback will be ranked before a resource with fewer or absent or negative feedback. To achieve this, we shall enrich the index, during nightly updates, with the results of the quality statements, and queries will be expanded to take it in account. A forum will be attached to resources to promote the evolution of the resource that should not always be used "as is" but should be periodically revised, adapted and improved. This quality assessment as well as the sense of community (see below) will promote responsibility and we expect that authors will syndicate around subjects to organize the evolution and production of quality educational content.

Second, the pedagogical and personal context has to be taken into account, with items such as the country and the language in which the teaching is going to happen, the circle of friends a user has or pioneer experts she tries to emulate; we might want that a resource deemed relevant with regard to the internationalized query to be ranked before another one, despite its lower query-matching score, because it was previously used and appreciated by fellow teachers belonging in the same real or virtual community. This shall be realized by enriching quality results in the index with contextual informations (per educational context and per named-user), the query expansion will then favour resources validated in a similar educational level or by a user tagged as a *friend* [38].

The paradigm behind this is philosophical as well as practical: *the users know better*. Therefore the social interactions and the actual use of the resources should dictate the distances and the scoring, not the reverse.

8   Conclusion

8.1   Implementation Status

The GeoSkills ontology is reaching completeness in structure, it can be seen at http://i2geo.net/ontologies/dev/GeoSkills.owl and a rendering can be seen at http://i2geo.net/ontologies/dev/index.html. The ontology is being completed for the most learning pathways of Germany, France, UK, The Netherlands and Spain, before October 2008. Curriculum-encoding will then be done by contributing curriculum-experts during the remaining two years of the intergeo project.

The SkillsTextBox GWT project can be enjoyed and downloaded from its project page http://ls.activemath.org/projects/SkillsTextBox. It is made available under the Apache Public License. The Search Tool is under active development and will be made available to the public in the summer.

The intergeo platform is under a first harvesting phase where interested parties report about intent to contribute interactive geometry constructions, with license. Since its launch several hundreds of reports have been submitted. The platform is accessible on http://i2geo.net/. The second phase will be activated at the end of Spring, based on Curriki, the annotations system will be incorporated in August. Finally, the quality framework will be embedded in the platform and in the search engine at the end of the year.

8.2   Summary

In this paper we have presented an approach to cross-curriculum search relying on a multinational and domain-aware ontology. The ontology basis is the major ingredient for both helping annotating and searching the constructions, of which we expect to receive several thousands. It will be the result of the coordinated work of curriculum encoders, which we expect to be done by curriculum experts in their community aware of the language of others cultures. This ontology is the key to enable the multinationality of the seach and annotation process tool.

Acknowledgements

We wish to thank Odile Bénassy and Albert Creus-Mir for their participation and contribution to this research.

References

1. Guin, D., Trouche, L.: Intégration des tice: concevoir, expérimenter et mutualiser des ressources pédagogiques. Repéres (55), 81–100 (2004)
2. Ruthven, K., Hennessy, S., Brindley, S.: Teacher representations of the successful use of computer based and resources in secondary school english, mathematics and science. Teaching and Teacher Education 20(3), 259–275 (2004)
3. Buswell, S., Caprotti, O., Carlisle, D., Dewar, M., Gaëtano, M., Kohlhase, M.: The openmath standard, version 2.0. Technical report, The OpenMath Society (2004), http://www.openmath.org/
4. Philippe, J.: Exploiter les logiciels de géométrie dynamique. 4 constructions géométriques avec Géoplan. Les Dossiers de l'ingénierie éducative (54), 35–37 (2006)
5. Ait Ouassarah, A.: Cabri-géométre et systémes dynamiques. Bulletin de l'APMEP (433), 223–232 (2001)
6. Falcade, R., Laborde, C., Mariotti., A.: Approaching functions: Cabri tools as instruments of semiotic mediation. Educational Studies in Mathematics (66.3), 317–333 (2007)
7. Hohenwarter, M.: GeoGebra: Dynamische Geometrie, Algebra und Analysis für die Schule. Computeralgebra-Rundbrief (35), 16–20 (2004)
8. American Mathematical Society: Mathematical Subject Classfication (2000), http://www.ams.org/msc/
9. Karhima, J., Nurmonen, J., Pauna, M.: WebALT Metadata = LOM + CCD. In: Proceedings of the WebALT 2006 Conference, The WebALT project (2006)
10. OFSET: GNU Edu. (2008), http://gnuedu.ofset.org/
11. Paquette, G.: An ontology and a software framework for competency modeling and management. Educational Technology and Society 10(3), 1–21 (2007)
12. British Educational Communication and Technology Agency: Curriculum online (2008), http://www.curriculumonline.gov.uk/
13. Microsoft: Microsoft Lesson Connection Launched At Technology + Learning Conference (1999), http://www.microsoft.com/presspass/ press/1999/nov99/lessonpr.mspx
14. ExploreLearning: Correlation of gizmos by state and textbooks (2005), http://www.explorelearning.com
15. van Rijsbergen, C.: Information Retrieval. Butterworths (1979), http://www.dcs.gla.ac.uk/~iain/keith/
16. Hatcher, E., Gosnopedic, O.: Lucene in Action. Manning (2004)
17. Graupmann, J., Biwer, M., Zimmer, C., Zimmer, P., Bender, M., Theobald, M., Weikum, G.: COMPASS: A Concept-based Web Search Engine for HTML, XML, and Deep Web Data. In: Proceedings of the Thirtieth International Conference on Very Large Data Bases, pp. 1313–1316 (2004), http://citeseer.ist.psu.edu/graupmann04compass.html
18. The Global Education Learning Community: Curriki (2008), http://www.curriki.org/

19. Ministére de l'Éducation Nationale: Programmes des classes de troisieme des colleges. Bulletin Officiel de l'Education Nationale (10), 108 (1998)
20. The Intergeo Consortium: D2.1 internationalized ontology (2008), http://www.inter2geo.eu/en/deliverables
21. Qualifications and Curriculum Authority: National Curriculum Mathematics Keystage 3. Technical report, Qualifications and Curriculum Authority (2007), http://curriculum.qca.org.uk/subjects/mathematics/keystage3/
22. Gasevic, D., Hatala, M.: Ontology mappings to improve learning resource search. British Journal of Educational Technology 37(3), 375–389 (2006)
23. Association of Computer Machinery: ACM Classification (2008), http://www.acm.org/class/1998/
24. IEEE: ACM/IEEE Computer Society Computing Curricula (2008), http://www.computer.org/education/cc2001/final/
25. Van Asche, F.: Linking learning resources to curricula by using competencies. In: First International Workshop on Learning Object Discovery and Exchange, Crete (2007)
26. Stanford Medical Informatics: Protégé ontology editor version 3.3.1 (2007), http://protege.standford.edu
27. Kifer, M., Lausen, G., Wu, J.: Logical foundations of object-oriented and frame-based languages. Journal of the ACM 42(4), 741–843 (1995)
28. McGuinness, D.L., van Harmelen, F.: OWL Web Ontology Language Overview (2004), http://www.w3.org/TR/owl-features/
29. Racer Systems GmbH and Co. KG: Racerpro owl reasoner version 1.9.2 (2007), http://www.racer-systems.com/
30. HP Labs: Jena - a semantic web framework for java version 2.5.5 (2007), http://jena.sourceforge.net/
31. Parsia, C.: Pellet owl dl reasoner version 1.5.1 (2007), http://pellet.owldl.com/
32. Pepper, S., Moore, G.: Xml topic maps (xtm) 1.0 – topicmaps.org specification (2001), http://www.topicmaps.org/xtm/1.0/
33. Dicheva, D.: Towards reusable and shareable courseware: Topic maps-based digital libraries (2005), http://compsci.wssu.edu/iis/nsdl/
34. Skaf-Molli, H., Molli, P., Marjanovic, O., Godart, C.: Libresource: Web based platform for supporting collaborative activities. In: 2nd IEEE Conference on Information and Communication Technologies: from Theory to Applications (2006)
35. Eurydice European Unit: Key data on education in Europe 2005, Eurydice European Unit (2005), http://www.eurydice.org/portal/page/portal/Eurydice/showPresentation?pubid=052EN
36. Google Inc.: Google Web Toolkit (GWT), a java to javascript compiler and toolkit (2008), http://code.google.com/webtoolkit/
37. The Intergeo Consortium: D6.1 quality assessment plan (2008), http://www.inter2geo.eu/en/deliverables
38. Xerox Research: Knowledge pump (2006), http://www.xrce.xerox.com/competencies/past-projects/km/kp/home.html

Augmenting Presentation MathML for Search

Bruce R. Miller[1] and Abdou Youssef[2]

[1] Information Technology Laboratory, National Institute of Standards and Technology, Gaithersburg, MD
bruce.miller@nist.gov
[2] Department of Computer Science, George Washington University, Washington, DC 20052
ayoussef@gwu.edu

Abstract. The ubiquity of text search is both a boon and bane for the quest for math search. A bane in that user's expectations are high regarding accuracy, in-context highlighting and similar features. Yet also a boon with the availability of highly evolved search engine libraries; Youssef has previously shown how an appropriate 'textualization' of mathematics into an indexable form allows standard text search engines to be applied.

Furthermore, given sufficiently semantic source forms for the math, such as LaTeX or Content MathML, the indexed form can be enhanced by co-locating synonyms, aliases and other metadata, thus increasing the accuracy and richness of expression.

Unfortunately, Content MathML is not always available, and the conversion from LaTeX to Presentation MathML (pMML) is too complex to carry out on the fly. Thus, one loses the ability to provide query-specific, fine-grained highlighting within the pMML displayed in search results to the user.

Where semantic information is available, however, such as for pMML generated from a richer representation, we propose augmenting the generated pMML with those semantics from which synonyms and other metadata can be reintroduced. Thus, in this paper, we aim to have both the high accuracy introduced by semantics while still obtaining fine-grained highlighting.

1   Introduction

The achievements of modern text search on the web have raised standards and user expectations. The relevance of the top ranked results to the query are often astounding. Concise summaries of the search results with matching terms highlighted allows users to quickly scan to find what they are looking for. Search has become, for better, sometimes worse, one of the first tools used to solve many information problems. These high expectations are carried over to math search; anecdotally, we see users uninterested in its unique challenges — the chess playing dog rationalization[1] carries little weight.

---

[1] He doesn't play well, but that he plays at all is impressive.

S. Autexier et al. (Eds.): AISC/Calculemus/MKM 2008, LNAI 5144, pp. 536–542, 2008.
© Springer-Verlag Berlin Heidelberg 2008

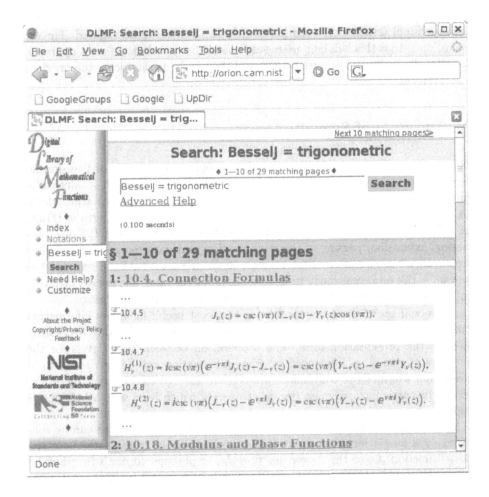

Fig. 1. Search results for `BesselJ = trigonometric` with coarse-grained math highlighting

In previous work [1], we described a strategy for math-aware search in which the mathematics, in whatever representation, is encoded into a linear textual form that can be processed by a conventional text search engine. This allows us to leverage the advances and tools in that field[2]. Recently, we have reported progress in improving the accuracy and ranking of math-aware search through the embedded semantics [4,5], and work in these areas continue.

In this paper, we describe on-going work in which we will add fine-grained highlighting to our math search engine. Our test corpus is the Digital Library of Mathematical Functions². In the search results 'hit list', we show a summary of each matching document containing the fragments that match the query. Rather than merely highlighting complete math expressions that have matches

---

² http://dlmf.nist.gov/

(See Figure 1 for an example), we intend to refine the presentation so that only the individual terms within the MathML[3] that match are highlighted. The expectation is that this will help users scan the hit list to select the most appropriate ones. Of course, this must be accomplished without sacrificing the previous improvements.

2   The Role of Semantics

Document processing typically proceeds through several levels of transformation beginning with the 'authored' form possibly passing through semantically-enhanced forms (e.g. some XML variant), and ending up in a presentational form for display to the user (e.g. HTML). In our case, the authored form is LaTeX; we use LaTeXML [3] to carry out the conversion which involves TeX-like processing, expansion, parsing of the mathematics, and data reorganization, while preserving as much authored information (both presentational and semantic) as we can — the process is fairly time-consuming for the large documents involved, and is clearly better suited for batch processing than server-side usage.

As we will show, the semantic form is the most useful for many information extraction tasks, particularly indexing for search. The highlighting necessarily involves the presentation form. The cooperation between these levels for the purposes of search, is influenced by what information can be carried between levels, and whether the conversions are efficient enough to do on-the-fly, in the server. In particular, we must decide at exactly which level search indexing is applied.

Mathematical notation is extremely symbolic; much information can be conveyed by a single letter or slight change in position. A J may stand for the Bessel function J_ν, or the Anger function $\boldsymbol{J_\nu}$, or perhaps Jordan's function J_k. Of course, any content-oriented markup worth its salt will make these distinctions clear.

Yet, a user may search for a specific function, say `BesselJ`, expecting only matches to the first function, or may search more generically for J and reasonably expect to match all of the above functions. Thus, for good accuracy and expressiveness, the indexed form of math needs to reflect both semantic and presentational aspects. A technique we call token *co-location* allows storing multiple forms of a term in the search index. Aliases, synonyms and other metadata can also be handled in this way, such as associating 'trigonometric' with sin and cos. The query in Figure 1 demonstrates these aspects. Furthermore, by co-locating the aliases, they are stored as if they appeared at the same document location, preserving information which will be essential for highlighting.

It is important to note at this point that, given the smaller corpus of a digital library like the DLMF[1], such techniques as latent semantic indexing are harder to apply, but may be worth investigating in the future to turn latent semantics into explicitly stated semantics for richer search capabilities.

---

[3] http://www.w3.org/Math/

3   The Role of Presentation

If the application needs only to find which documents match a query, one has much flexibility regarding at which level of processing the document is indexed. However, it is very useful to a reader to see the portions of the matching documents that match the query, along with some context. Given that a search query is at best a crude approximation to what they really wanted, such summaries help them select the most documents from the results.

In the case of text documents, a common approach is to break each matching document into fragments (typically sentences) at search time. The query is then reapplied to the fragments to select the relevant ones. Finally, the matching tokens within each fragment are then marked for highlighting.

This approach needs to be refined to deal with richly structured XML documents. In such cases, reasonable summary fragments can include not only sentences, but equations, figures and portions of tables. Moreover, the reconstructed summary must itself also be a valid XML document. Furthermore, one would like to store the fragments as XML document fragments, and thus the indexing must 'skip over' the XML markup. In fact, this is easily achieved by manipulating term positions during indexing, and thus the eventual highlighting preserves the XML structures.

This additional complexity is a strong inducement to carry out the fragmenting in batch mode, rather than during search, and to generate a separate fragment index. At search time, the query is applied to the document index to find the relevant documents, as before. But, to summarize each document, the fragment index is searched to find matching fragments for that document.

Moreover, this argues for indexing the documents at the presentation level, in contrast to the arguments of the previous section. This greatly simplifies the processing needed at search time — one never gives up the hubris that one's site will be The Next Big Thing. However, token co-location again can be used to associate any desired semantics with the presentational tokens, provided they are made available.

Turning to the mathematics specific aspects of this argument, we note that in our previous work, each math entity was treated as a unit by converting it to a textual form corresponding to the LaTeX markup originally used to author it. Since we have adopted a semantically enhanced LaTeX markup (e.g. macros like \BesselJ)[3], we gain the benefits discussed in the previous section. However, converting from LaTeX form to presentation on-the-fly is not feasible, and so we lose the direct correspondence with the structure of the Presentation MathML. Consequently, we are unable to highlight individual mathematical terms or variables within the summary, but are only able to highlight the entire formula, as can be seen in Figure 1.

4   A Strategy

We propose the following strategy to work around the limitations and contradictions described in the previous sections. Firstly, we will convert from the

1: 10.4. Connection Formulas

. . .

▶ 10.4.5
$$J_\nu(z) = \csc(\nu\pi)(Y_{-\nu}(z) - Y_\nu(z)\cos(\nu\pi)).$$

. . .

▶ 10.4.7
$$H_\nu^{(1)}(z) = i\csc(\nu\pi)\left(e^{-\nu\pi i}J_\nu(z) - J_{-\nu}(z)\right) = \csc(\nu\pi)\left(Y_{-\nu}(z) - e^{-\nu\pi i}Y_\nu(z)\right),$$

▶ 10.4.8
$$H_\nu^{(2)}(z) = i\csc(\nu\pi)\left(J_{-\nu}(z) - e^{\nu\pi i}J_\nu(z)\right) = \csc(\nu\pi)\left(Y_{-\nu}(z) - e^{\nu\pi i}Y_\nu(z)\right).$$

. . .

Fig. 2. The query of Figure 1 with fine-grained highlighting

content-oriented level of representation to the presentation-oriented level while augmenting the presentation with any semantic information available. In our particular case, we can simply carry the 'meaning' attribute of LaTeXML's internal representations over to the generated Presentation MathML token elements in a special attribute, say `ltxml:meaning`. These 'meanings' form a rough ontology, and could be (but have not yet been) defined by OpenMath[4] Content Dictionaries. Associated with each such meaning is a set of keywords and aliases. Thus, associated with the `sin`, generated by `\sin`, are the keywords: trigonometric, sine, function, elementary and periodic. A similar approach could be used when converting Content MathML or other representations.

A careful treewalk of the augmented MathML generates the same textualization as was produced by textualizing the LaTeX, as described above. In this case, however, we take care to manipulate the positions of the textualized tokens as they are being indexed so that each corresponds to the appropriate MathML fragment with the expression. This assures that the eventual highlighting will be both possible, and that the result will be well-formed XML. Whenever a token with assigned meaning is encountered, the additional aliases are co-located with the principal MathML token.

5 Preliminary Results

A preliminary implementation of these ideas has been carried out, with the result as shown in Figure 2. Although no detailed benchmarking has been carried out, there appears to be no execution time penalty for either indexing or search.

Two kinds of highlighting inaccuracies could be expected from this relatively simple approach.

---

[4] http://www.openmath.org/

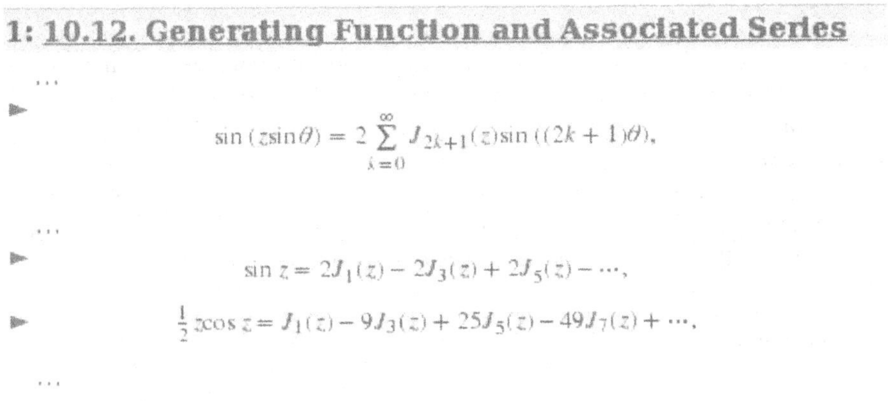

1: 10.12. Generating Function and Associated Series

. . .

$$\sin(z\sin\theta) = 2 \sum_{k=0}^{\infty} J_{2k+1}(z)\sin((2k+1)\theta),$$

. . .

$$\sin z = 2J_1(z) - 2J_3(z) + 2J_5(z) - \cdots,$$

$$\tfrac{1}{2}z\cos z = J_1(z) - 9J_3(z) + 25J_5(z) - 49J_7(z) + \cdots.$$

. . .

Fig. 3. First hit from the query J_1(z)

- tokens present in the query but not matching the full query may be highlighted when they should not;
- non-indexed tokens or markings (e.g. parentheses, surds) may fail to be highlighted when they should be.

And in fact, the first is seen in Figure 2 and both are demonstrated in Figure 3.

Although subjectively exploring the DLMF corpus with fine-grained highlighting is more pleasant and helpful than the coarse-grained whole-formula highlighting, the errors, especially the first type can be quite distracting; the second problem is more an aesthetic issue.

In fact, the over-highlighting problem is commonly seen in text search engines, but it simply is not as disturbing as within math. For example, in Figure 3, every occurrence of z is highlighted, not just when it is the argument of J_1. Users may find this confusing, and it begins to defeat the purpose of the fine-grained highlighting, namely to make easy the selection of good matches.

The solution will be to apply more of the restrictive 'logic' of the query (ands, ors, proximity) to the highlighting phase than is currently done. The search engine we are using, Lucene[5], apparently only makes token positions available when searching using primitive token queries. If that is indeed the case, we will be forced to reimplement or approximate some of that logic during highlighting. Tedious, but doable.

The second problem seems less serious but may yield to one of two approaches. One would be to extend the textualization language to allow indexing MathML structural schema rather than just token elements. Another would be to broaden the highlighting to a parent when many children of a node are highlighted.

---

[5] http://lucene.apache.org/

6   Conclusion

We have described techniques that should give math-aware search engines the capabilities of both high accuracy and relevance, while still presenting the user with a concise and appropriate summary of each result that indicates how the document relates to the query.

This paper describes work that is in progress. However, initial implementation gives satisfying results, with some flaws. We anticipate being able to remedy some or all of those flaws and demonstrate the results at the conference.

References

1. Miller, B., Youssef, A.: Technical Aspects of the Digital Library of Mathematical Functions. Annals of Mathematics and Artificial Intelligence 38, 121–136 (2003)
2. Baeza-Yates, R.A., Riberto-Neta, B.(eds.): Modern Information Retrieval. Addison Wesley, Reading (1999)
3. Miller, B.: Creating Webs of Math Using LaTeX. In: Proceedings of the 6th International Congress on Industrial and Applied Mathematics, Zürich, Switzerland, July 17 (2007)
4. Youssef, A.: Information Search And Retrieval of Mathematical Contents: Issues And Methods. In: The ISCA 14th International Conference on Intelligent and Adaptive Systems and Software Engineering (IASSE-2005), Toronto, Canada, July 20-22 (2005)
5. Youssef, A.: Methods of Relevance Ranking and Hit-content Generation in Math Search. In: The 6th Mathematical Knowledge Management Conference, Hagenberg, Austria, June 27-30, pp. 393–406 (2007)

Disclaimer: Certain products, commercial or otherwise, are mentioned for informational purposes only, and do not imply recommendation or endorsement by NIST or GWU.

Automated Classification and Categorization of Mathematical Knowledge

Radim Řehůřek and Petr Sojka

Masaryk University, Faculty of Informatics, Brno, Czech Republic
xrehurek@fi.muni.cz,sojka@fi.muni.cz

Abstract. There is a common Mathematics Subject Classification (MSC) System used for categorizing mathematical papers and knowledge. We present results of machine learning of the MSC on full texts of papers in the mathematical digital libraries DML-CZ and NUMDAM. The F_1-measure achieved on classification task of top-level MSC categories exceeds 89%. We describe and evaluate our methods for measuring the similarity of papers in the digital library based on paper full texts.

1 Introduction

> We thrive in information-thick worlds because of our marvelous and everyday capacity to select, edit, single out, structure, highlight, group, pair, merge, harmonize, synthesize, focus, organize, condense, reduce, boil down, choose, *categorize*, catalog, *classify*, list, abstract, scan, look into, idealize, isolate, discriminate, distinguish, screen, pigeonhole, pick over, sort, integrate, blend, inspect, filter, lump, skip, smooth, chunk, average, approximate, cluster, aggregate, outline, summarize, itemize, review, dip into, flip through, browse, glance into, leaf through, skim, refine, enumerate, glean, synopsize, winnow the wheat from the chaff and separate the sheep from the goats. (Edward R. Tufte)

Mathematicians are used to classifying their papers. One of the first mathematical classification schemes appeared in the subject index for *Pure Mathematics* of 19 volumes of the *Catalogue of Scientific Papers 1800–1900* [1]. This attempt was continued but not completed by the *International Catalogue of Scientific Literature (1901–1914)*. About two hundred classes were used. Headings in the *Jahrbuch* [2] may be considered as another classification scheme.

The Library of Congress classification system has 939 subheadings under the heading QA–Mathematics. Another schemes used in many libraries around the world are the Dewey Decimal system and the Referativnyi Zhurnal System used in the Soviet Union. To add to this wide variety of schemes, we may mention systems used by NSF Mathematics Programs, by various encyclopaedia projects such as Wikipedia, or by the ARXIV Preprint project. However, the most commonly used classification system today is the Mathematics Subject Classification (MSC) scheme (http://www.ams.org/msc/), developed and supported jointly by reviewing databases *Zentralblatt MATH* (ZBL) and *Mathematical Reviews* (MR).

S. Autexier et al. (Eds.): AISC/Calculemus/MKM 2008, LNAI 5144, pp. 543–557, 2008.
© Springer-Verlag Berlin Heidelberg 2008

In order to classify papers some system of paper categorization has to be chosen. We may pick up some established system developed by human experts or we may try to induce one from digital library of papers by clever document clustering and machine learning techniques.

1.1 Mathematics Subject Classification

It is clear that no fixed classification scheme can survive longer time period, since new areas of mathematics appear every year. Mathematicians entered the new millennium with the MSC version 2000, migrating from MSC of 1991. Draft version of MSC 2010 has already been prepared and published at msc2010.org recently. The primary and secondary keys of MSC 2000, requested today by most mathematical journals are used for indexing and categorizing a vast amount of new papers—see the exponential growth of publications in Figure 1.

We believe that automated classification system, good article similarity measures and robust math paper classifiers allowing more focused math searching capabilities will help to tackle the future information explosion as predicted by Stephen Hawking:

> "If [in 2600] you stacked all the new books being published next to each other, you would have to move at ninety miles an hour just to keep up with the end of the line. Of course, by 2600 new artistic and scientific work will come in electronic forms, rather than as physical books and paper. Nevertheless, if the exponential growth continued, there would be ten papers a second in my kind of theoretical physics, and no time to read them."

Editors of mathematical journals usually require the authors themselves to include the MSC codes in manuscripts submitted for publication. However, most retrodigitized papers published before the adoption of MSC are not classified yet.

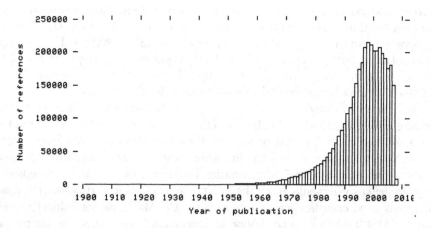

Fig. 1. Number of references in The Collection of Computer Science bibliographies (http://liinwww.ira.uka.de/bibliography/) as of March 2008

Some projects, e.g. JAHRBUCH, use MSC 2000 even for the retroclassification of papers. Human classification needs significant resources of qualified mathematicians and reviewers. A similar situation is in the other retrodigitization projects such as NUMDAM [3] (http://www.numdam.org), or DML-CZ [4,5] (http://www.dml.cz): classifying digitized papers with MSC 2000 manually is expensive.

As there are already many papers properly classified (by authors and reviewers) in recent publications, methods of machine learning may be used to train an automated classifier based on the full texts author- and/or reviewer-classified papers.

This paper is organized as follows. In Section 2 we start by describing our data sources and preprocessing needed for our experiments. Section 3 on page 547 discusses results of automated document classification. In Section 4 on page 550 we show what results we obtained while computing mathematical document similarity. Finally, we discuss possible future experiments and work in Section 5 on page 555.

2   Data Preprocessing

> We run carelessly to the precipice, after we have put something before us to prevent
> us seeing it. (Blaise Pascal)

The data available for experiments are metadata and full texts of mathematical journals covered by the DML-CZ and NUMDAM projects.

2.1   Primary Data

During the first three years of the DML-CZ project, we digitized and collected data in the digital library, accessible via a web tool called Metadata editor (editor.dml.cz). To date (March 2008), in the digitized part there are 369 volumes of 14 journals and book collections: 1,493 issues, 11,742 articles on 177,615 pages.

From NUMDAM, we got another 15,767 full texts of articles (in simple XML format) for our research. We converted them into DML-CZ format as utf8 encoded text and excluded 134 articles due to inconsistencies such as having the same ID for parts of paper, invalid MSC etc. There were 5,697 papers tagged as English, 4,587 as French, 384 as Italian, 84 as German and there was no language tag for the remaining 4,881 papers available—language can be reliably detected by established statistical methods [6].

For experiments, we have used two types of data:

1. Texts from scanned pages of digitized journals (usually before 1990, where no electronic data are available). There are of course errors in full text, especially in mathematical formulae, as these were not recognized by OCR.
2. Texts from 'digital-born' papers, written in TeX, as papers of the journal *Archivum Mathematicum* (http://www.emis.de/journals/AM/) from years

1992–2007, where we had access to TeX source files. The workflow of the paper publishing process in some journals was modified somewhat so that all fine-grained metadata including the full text are exported for the digital library for long-term storage (CEDRAM project).

We started our experiments with retrodigitized articles, where texts were obtained by the OCR process [7].

After excluding papers with no MSC code we were left with 21,431 papers. From those, we only used papers tagged as English and with only a primary MSC classification (no secondary MSC) for our current experiments. This left us with 5,040 articles.

We started with our experiments with the task of classification of top-level (the first 2 digits) MSC categories. To ensure meaningful results, we used only a part of the text corpus: only top-level categories with more than 30/40/60 papers in them were considered. Without this pruning step, we could not expect the automated classifiers to learn well: given tiny classes comprising only a few papers, generalizing well is not straightforward. In this way, we were left with 31, 27 and 20 top-level MSC classes for the minimum 30, 40 and 60 papers per class limit, respectively. The total amount of articles after this pruning step is 4,618, 4,481 and 4,127 articles, respectively.

2.2 Preprocessing and Methods Used

It is widely known that the design of the learning architecture is very important, as is preprocessing, learning methods and their parameters [8].

For the purpose of building an automated MSC classification system, we chose the standard Vector Space Model (VSM) together with statistical Machine Learning (ML) methods. In order to convert the text in the natural language to vectors of features, several preprocessing steps must be taken—for a more thorough explanation, see e.g. [8]. A detailed description of all ML methods and IR notions is beyond the scope of this paper; the reader is referred to the overviews [9,10,11] for exact definitions and notation used.

The setup of the experiments is such that we run a vast array of training attempts in multidimensional learning space of tokenizers, feature selectors, term weighting types, classifiers and learning methods' parameters:

tokenization and lemmatization: the first part of the preprocessing relates to how the text is split into tokens (words)—alphabetic, lowercase, Krovetz stemmer [12], lemmatization, bi-gram tokenization (collocations chosen by MI-score);

feature selectors: how to choose the tokens that discriminate best—χ^2, mutual information (MI-score) [13,14,15];

feature amount: how many features are needed to classify best—500, 2,000 or 20,000 features [14];

term weighting: how the features will be weighted (*tfidf* variants [16] or [11, Fig. 6.15]) and smart weights normalizations (*atc* (augmented term frequency), *bnn* and *nnn*) [17];

classifiers: Naïve Bayes (NB), k-Nearest Neighbours (kNN), Support Vector Machines (SVM), decision trees, Artificial Neural Nets (ANN), K-star algorithm, Hyperpipes;

threshold estimators: how to choose the category status of the classifier based on a threshold—*fixed* or *s-cut* strategy for threshold setting [18];

evaluation and confidence estimation: how results are measured and how the confidence is estimated in them—Receiver Operating Characteristic (ROC), Normalized Cross Entropy (NCE) [19].

To give an example, evaluating one particular combination might mean that we tokenize the corpus using an alphabetic tokenizer, convert the tokens to lower case, select the best 2,000 tokens (words aka features aka terms) using χ^2 and weigh them using an *atc* scheme. One part of the corpus is then used for training the binary classifiers and the rest is evaluated to see whether the predicted MSC equals the expected MSC. Each binary classifier is responsible for one category (MSC class), and given a full text on input, returns whether the input belongs to the category or not. Each article may thus be predicted to belong to any number of categories, including none or all.

Out of the seven classifiers listed above, only the first three were used in the final experiments. The other four were discarded on the ground of poor performance in preliminary experiments not reported here. On the other hand, there are several recent hierarchical classification algorithms [20] that we did not have time to explore yet.

In order to evaluate the quality of each learned classifier, we compute an average of ten cross-validation runs. We measure micro/macro F_1, accuracy, precision, recall, correlation coefficient, break-even point and their standard deviations [11,8]. Since the popular accuracy measure is highly unsuitable for our task (extremely unbalanced ratio of positive/negative test examples), we will report results using the even more popular F_1 measure in this paper.

All these results are then compared to see which 'points' in the parameter space perform best. Our framework allows easy comparison of the evaluated parameters with visualization of the whole result space methods chosen—see multidimensional data visualization on Figure 2.

As the number of different learning setup combinations grows exponentially, methods that performed poorly in preliminary tests were excluded in the full testing.

3   MSC 2000 Automated Classification

> The classification here adopted has been the subject of more or less unfavourable criticism; the principal objection to it, however, seems to be that it is different from any of those previously employed, and is therefore to this extent inconvenient without any obvious advantage in the innovations. (J. A. Allen, [21])

A detailed evaluation of classification accuracy shows that, while automatically classifying the first two letters of primary MSC, we can easily reach an 80% F_1

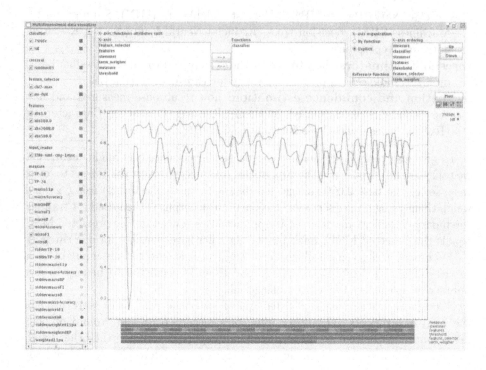

Fig. 2. Framework for comparing learning methods [8]. The two differently colored curves correspond to the chosen learning methods (k-NN, Naïve Bayes in the legend on the right). From the colors below chosen function values, one immediately sees which combination (at the bottom) of preprocessing methods leads to which particular value.

classification score with almost any combination of methods. With fine-tuning the best method (Support Vector Machines with a large number of features seems to be the winner) we can increase the F_1 score to 89% or more. The microaveraged accuracy measure is above 99%, but is uninteresting as the baseline score, which can be achieved by a trivial reject-all classifier, is as high as $30/31 = 97\%$. The same difficulty does not arise with microaveraged F_1, where trivial classifiers score under 6%. In this light, our best result of nearly 90% F_1 score is quite encouraging.

In Figure 3 on the facing page there is a side-by-side plot of three different corpora which result from setting the $30/40/60$ minimum articles per category threshold. The intuition is that, given less training examples, the task of learning a classifier would become harder and classification accuracy would drop. This can indeed be observed here. On the other hand, the drops are not dramatic but rather graceful (about two F_1 percentage points for going from a minimum of 60 to 40, and another 1% for going from 40 to 30). Also to be noted is another factor contributing to this drop—with a lower article per category threshold, we are in fact classifying into more classes (recall that the number of classes for the 60, 40 and 30 threshold is 20, 27 and 31 classes, respectively). Again, this makes more room for error and lowers the score.

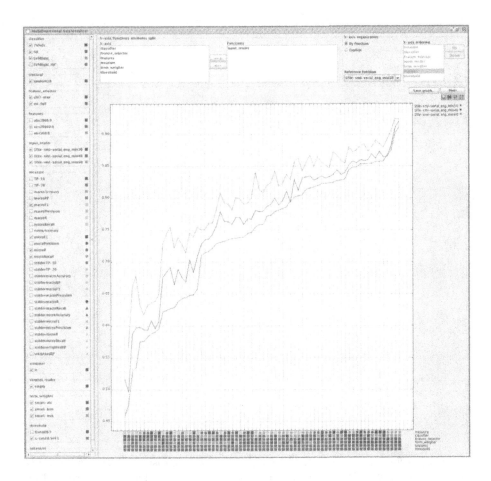

Fig. 3. Dependency of performance on the number of examples per class limit. From the three curves one can see that by increasing the threshold of minimum category size one gets better results in every aspect (color square combination at the bottom).

Figure 4 on the next page enables us to examine the best performing combinations of methods and parameters. It may be observed that the best classifiers are exclusively SVM and kNN; the performance of NB depends heavily on term weighting. Also the aggressive feature selection of only 500 features performed poorly. The best result of the micro-averaged F_1 score of 89.03% was achieved with SVM with linear kernel, χ^2 feature selection of 20,000 features, *atc* term weighting and decision threshold selected dynamically by *s-cut*. In the light of the previous comment, it is unsurprising that this maximum occurred in the dataset selected with a minimum article per category threshold of 60. F_1 scores at the very same configuration, but with a threshold of 40 and 30 articles per category read 86.28% and 85.72%, respectively.

Similarly, we measure and can visualize training times (computation expense) for every method tried. Many of these are computationally expensive—it takes

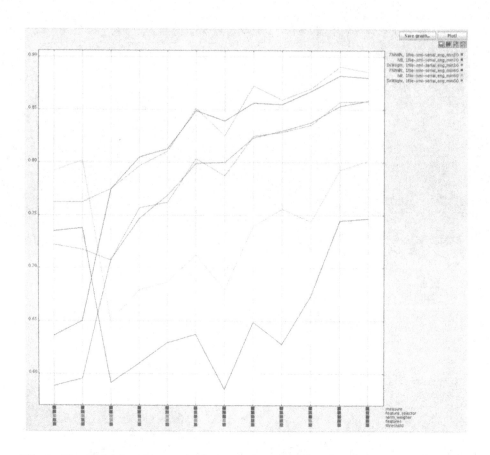

Fig. 4. Classifiers' learning methods comparison by F_1 measure. SVM and kNN run hand in hand while NB lags behind. The major influence is due to the threshold on minimum category size (see Figure 3 on the preceding page).

days to weeks on a server with four multithreaded processors to compute all the results to visualize and analyze.

4   Mathematical Document Similarity

It's false to assume that mathematics consists of discrete subfields, it's false to assume that there is an objective way to gather those subfields into main divisions, and it's false to assume that there is an accurate two-dimensional positioning of the parts. (Dave Rusin [22])

Recall that one of the purposes of the automated MSC classification detailed above is to enable a similarity search. Given MSC categories, the user may browse articles with similar MSCs and thus (hopefully) with similarly relevant content.

But we have also been intrigued by similarity searches based on raw full text, and not on metadata such as MSC codes. This differs in that there is no predefined class taxonomy that the articles ought to follow (such as MSC). The similarity of two articles is gauged directly based on the articles' content, with no reference to human-entered or human-revised metadata.

Because fine linguistic analysis tools would be ineffective (recall that our texts come from OCR, with errors appearing as early as at the character level), we opted for 'a brute' Information Retrieval approach. Namely, we tried computing paper similarities using *tfidf* [16] and Latent Semantic Analysis (LSA) [23] methods. Again, both use a Vector Space Model, first converting articles to vectors and then using the cosine of the angle between the two document vectors to assess their similarity. [11] The difference between them is that while *tfidf* works directly over tokens, LSA first extracts concepts, then projects the vectors into this conceptual space where it only computes similarity. For LSA we chose the 200 top latent dimensions (concepts) to represent the vectors, in accordance with standard Information Retrieval practise [23].

Evaluating the effectiveness of our similarity schemes is not as straightforward as in the classification task. This is due to the fact that, as far as we know, there exists no corpus with an explicitly evaluated similarity between each pair of papers. In this way, we are left with two options: either constructing such corpus ourselves, or approximating it. As the first option appears too costly, we decided to assume that MSC equality implies content similarity. Accordingly, we evaluated how closely the computed similarity between two papers corresponds to the similarity implied by them sharing the same MSC.

Again, to avoid data sparseness, we only took note of the top MSC categories (first two letters of the MSC codes). In Figures 5 and 6 on page 553 there are *tfidf* and LSA plots of similarities between all English papers in our database that are tagged only with a primary MSC code.

Two things can be seen immediately from the plot:

- articles within one top MSC group are usually very similar (lighter squares along the diagonal);
- the similarities of articles from different MSC groups are low (dark rectangles off diagonal).

There are also exceptions, such as patches of light colour off the diagonal as well as dark patches within the MSC group squares. This is however to be expected from noisy real-world data and cannot be fixed nor explained without actually inspecting the articles by hand. Clear small square areas in matrix detail on Figure 7 on page 554 show that papers exhibit similarity of MSC even when sharing MSC code prefix of length 3 or higher.

4.1 Experiments with Latent Semantic Analysis

Next experiment we tried with Latent Semantic Analysis [23] was to see which concepts are the most relevant ones.

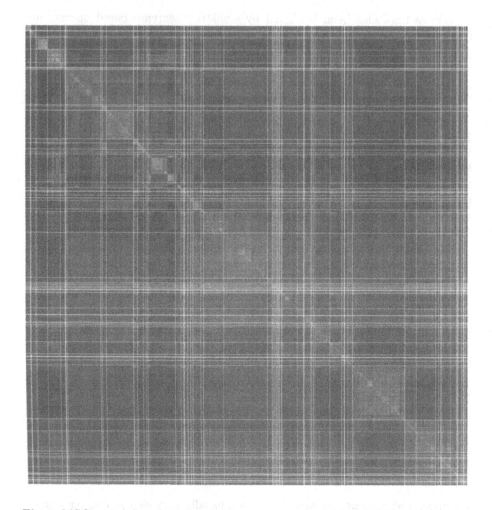

Fig. 5. MSC-sorted documents' similarity matrix computed by *tfidf*. The axes of this 5104×5104 matrix are articles, grouped together by their MSC code (white vertical and horizontal lines separate different top-level MSC categories) and sorted lexicographically by full five letter MSC code. The intensity of the plot shows similarity, with white being the most similar and black being completely dissimilar. Note that because the ordering of articles along both axes is identical, all diagonal elements must necessarily be white (completely similar), as each article is always fully similar to itself.

There were papers in several different languages in the *Czechoslovak Mathematical Journal* (CMJ). When we listed the top concepts in LSA of the CMJ corpus, it was clear that the first thing the method was going to decide was its language, as the first terms of top concepts are:

1. 0.3 "the" +0.19 "and" +0.19 "is" +0.18 "that" +0.15 "of" +0.14 "we" +0.14 "for" +0.11 "ε" +0.11 "let" +0.11 "then" +...
2. −0.41 "ist" −0.40 "die" −0.28 "und" −0.26 "der" −0.23 "wir" −0.21 "für" −0.17 "eine" −0.17 "von" −0.14 "mit" −0.13 "dann" +...

Fig. 6. MSC-sorted documents' similarity matrix computed by LSA. Interpretation is identical to Figure 5.

3. −0.31 "de" −0.30 "est" −0.29 "que" −0.27 "la" −0.26 "les" −0.2 "une"
    −0.2 "pour" −0.20 "et" −0.18 "dans" −0.18 "nous" +...

4. −0.36 "что" −0.29 "для" −0.23 "пусть" −0.19 "из" −0.19 "если" −0.16 "так"
    −0.16 "то" −0.14 "на" −0.14 "тогда" −0.131169 "мы" +...

5. −0.33 "semigroup" −0.25 "ideal" −0.19 "group" −0.18 "lattice" +0.18 "solution"
    +0.16 "equation" −0.16 "ordered" −0.15 "ideals" −0.15 "semigroups" +...

6. 0.46 "graph" +0.40 "vertices" +0.36 "vertex" +0.23 "graphs" +0.2 "edge"
    +0.19 "edges" −0.18 "ε" −0.15 "semigroup" −0.13 "ideal" +...

7. 0.81 "ε" −0.25 "semigroup" −0.16 "ideal" +0.12 "lattice" −0.11 "semigroups"
    +0.10 "i" −0.1 "ideals" +0.09 "ordered" +0.09 "ř" −0.08 "idempotent" +...

8. 0.29 "semigroup" −0.22 "space" +0.2 "ε" +0.19 "solution" +0.19 "ideal"
    +0.18 "equation" +0.16 "oscillatory" −0.15 "spaces" −0.16 "compact" +...

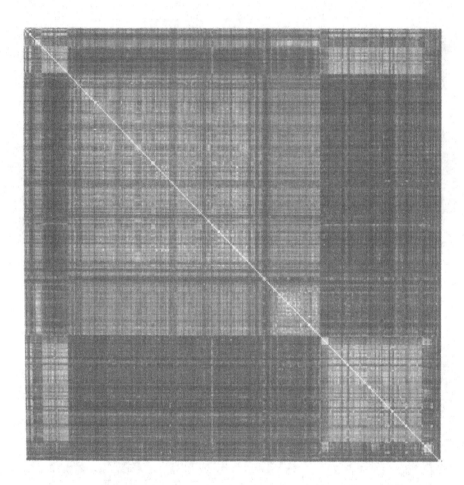

Fig. 7. Detail of MSC-sorted documents' similarity matrix computed by LSA for top-level MSC code 20-xx *Group theory and generalizations*. The white lower right square corresponds to the 20Mxx *Semigroups* subject papers. We can see strong similarity of 20Mxx to 20.92 *Semigroups, general theory* and 20.93 *Semigroups, structure and classification* (white lower left and upper right rectangles).

The first concepts clearly capture the language of the paper (EN, DE, FR, RU), and only then topical term-sets start to be grabbed. It is not surprising—the classifiers then have to be trained either for every language, or the document features have to be chosen language-independently by mapping words to some common topic ontology. To the best of our knowledge, nothing like EuroWordNet for mathematical subject classification terms or mathematics exists.

Given the amount of training data—papers of given MSC code for given language—we face the sparsity problem for languages such as Czech, Italian, German and even French presented in the digital library.

When we trained LSA on the monolingual corpora of *Archivum Mathematicum*, where mathematics formulae were used during tokenization (subcorpus created from original TEX files), we saw that even in the first concepts, there was significant proportion of mathematical terms with high weights in concepts created by LSA:

1. -0.32 "t" -0.24 "ds" -0.17 "u" -0.17 "_" -0.17 "x" -0.15 "solution" -0.12 "equation" -0.11 "q" -0.11 "x_" -0.11 "oscillatory" $+\dots$
2. 0.28 "ds" $+0.28$ "t" -0.22 "bundle" -0.16 "natural" $+0.15$ "oscillatory" -0.15 "vector" $+0.13$ "solution" -0.13 "connection" -0.13 "manifold" $+0.11$ "t_0" $+\dots$
3. -0.22 "bundle" $+0.19$ "ring" -0.17 "natural" -0.16 "oscillatory" $+0.15$ "fuzzy" -0.15 "ds" $+0.12$ "ideal" -0.11 "t" -0.11 "r_0" -0.11 "nonoscillatory" $+\dots$

It supports the idea that mathematical formulae have to be taken into account—having robust math OCR and finding its good discriminative feature representation we may get much better similarity and classification results in the future.

5   Conclusions and Future Work

> Words differently arranged have a different meaning,
> and meanings differently arranged have different effects.
> (Blaise Pascal)

Our results convincingly demonstrated the feasibility of a machine learning approach to the classification of mathematical papers. Although we compared and reported the results according to the F_1 measure, our approach can easily be tweaked to favour a different trade-off between higher recall and/or precision. Results in the form of guessed MSC and similarity lists are going to be directly used in the DML-CZ project.

Given enough data, when we extrapolate the best results of preliminary experiments done on our limited data, with linear machine learning methods (creating separable convex spaces in multidimensional feature space) we were able to approach a very high precision of 96% and recall of 92.5%, which are the current bests, for a combined F_1 score of well over 90%. Future research thus extends to evaluating the classification on all 64 top MSC categories, and using hierarchical classifiers to cover the full MSC taxonomy. With ambitions for even higher recall, there are several approaches, namely to either improve the preprocessing for vectors representing the documents by NLP techniques (characteristic words, bi-words, etc.) or use higher order models (deep networks). Mainstream machine learning research was concentrated on using "convex", shallow methods (SVM, shallow neural networks with back-propagation training) so far. State-of-the-art fine tuned methods allow very high accuracy even on large scale classification problems. However, the training of these methods is exceptionally high and the models are big. Using the ensembles of classifiers makes the situation even less satisfactory (size even bigger), and the final models need to be regularized. In future, we plan to try new algorithms for a hierarchical text classification [20] and training large models with non-convex optimization [24] that may give classifications that does not exhibit overfitting.

Further studies will encompass a fine-grained classification trained on bigger collections (using MSC tagged mathematical papers from (`ArXiv.org`), growing NUMDAM and DML-CZ libraries etc.), and a rigorous measure confidence evaluation [19].

For final large scale applications scaling issues, and fine-tuning the best performance by choosing the best set of preprocessing parameters and machine learning methods remains to be done. We will watch Apache Lucene Mahout project's code when scalability of machine learning will arise as a serious issue.

Acknowledgement. This study has been partially supported by the grants 1ET200190513 and 1ET208050401 of the Academy of Sciences of the Czech Republic and 2C06009 and LC536 of MŠMT ČR.

References

1. Royal Society of London: Catalogue of scientific papers 1800–1900 vol. 1–19 and Subject Index in 4 vols (published, 1867–1925) (1908), free electronic version available by project Gallica `http://gallica.bnf.fr/`
2. Ohrtmann, C., Müller, F., (eds.): Jahrbuch über die Fortschritte der Mathematik, vol. 1–68 (1868–1942) Druck und Verlag von Georg Reimer, Berlin (1871–1942); electronic version available by project ERAM, `http://www.emis.de/projects/JFM/`
3. Bouche, T.: Towards a Digital Mathematics Library? In: Rocha, E.M. (ed.) CMDE 2006: Communicating Mathematics in the Digital Era, pp. 43–68. A.K. Peters, MA, USA (2008)
4. Sojka, P.: From Scanned Image to Knowledge Sharing. In: Tochtermann, K., Maurer, H. (eds.) Proceedings of I-KNOW 2005: Fifth International Conference on Knowledge Management, Graz, Austria, Know-Center in coop, Graz Uni, pp. 664–672. Joanneum Research and Springer Pub. Co. (2005)
5. Bartošek, M., Lhoták, M., Rákosník, J., Sojka, P., Šárfy, M.: DML-CZ: The Objectives and the First Steps. In: Borwein, J., Rocha, E.M., Rodrigues, J.F. (eds.) CMDE 2006: Communicating Mathematics in the Digital Era, pp. 69–79. A.K. Peters, MA, USA (2008)
6. Dunning, T.: Statistical identification of language. Technical Report MCCS 94-273, New Mexico State University, Computing Research Lab (1994)
7. Sojka, P., Panák, R., Mudrák, T.: Optical Character Recognition of Mathematical Texts in the DML-CZ Project. Technical report, Masaryk University, Brno. CMDE 2006 conference in Aveiro, Portugal (presented, 2006)
8. Pomikálek, J., Řehůřek, R.: The Influence of Preprocessing Parameters on Text Categorization. International Journal of Applied Science, Engineering and Technology 1, 430–434 (2007)
9. Sebastiani, F.: Machine learning in automated text categorization. ACM Computing Surveys 34, 1–47 (2002)
10. Yang, Y., Joachims, T.: Text categorization. Scholarpedia (2008), `http://www.scholarpedia.org/article/Text_categorization`
11. Manning, C.D., Raghavan, P., Schütze, H.: Introduction to Information Retrieval. Cambridge University Press, Cambridge (2008)

12. Krovetz, R.: Viewing morphology as an inference process. In: Proceedings of the Sixteenth Annual International ACM SIGIR Conference on Research and Development in Information Retrieval. Linguistic Analysis, pp. 191–202 (1993)
13. Yang, Y., Pedersen, J.O.: A comparative study on feature selection in text categorization. In: Fisher, D.H. (ed.) Proceedings of ICML 1997, 14th International Conference on Machine Learning, pp. 412–420. Morgan Kaufmann, San Francisco (1997)
14. Galavotti, L., Sebastiani, F., Simi, M.: Experiments on the use of feature selection and negative evidence in automated text categorization. In: Borbinha, J.L., Baker, T. (eds.) ECDL 2000. LNCS, vol. 1923, pp. 59–68. Springer, Heidelberg (2000)
15. Forman, G.: An extensive empirical study of feature selection metrics for text classification. Journal of Machine Learning Research 3, 1289–1305 (2003)
16. Salton, G., Buckley, C.: Term-weighting approaches in automatic text retrieval. Information Processing and Management 24, 513–523 (1988)
17. Lee, J.H.: Analyses of multiple evidence combination. In: Proceedings of the 20th Annual International ACM SIGIR Conference on Research and Development in Information Retrieval. Combination Techniques, pp. 267–276 (1997)
18. Yang, Y.: A Study on Thresholding Strategies for Text Categorization. In: Croft, W.B., Harper, D.J., Kraft, D.H., Zobel, J. (eds.) Proceedings of the 24th Annual International ACM SIGIR Conference on Research and Development in Information Retrieval (SIGIR 2001), pp. 137–145. ACM Press, New York (2001)
19. Gandrabur, S., Foster, G., Lapalme, G.: Confidence Estimation for NLP Applications. ACM Transactions on Speech and Language Processing 3, 1–29 (2006)
20. Esuli, A., Fagni, T., Sebastiani, F.: Boosting multi-label hierarchical text categorization. Information Retrieval 11 (2008)
21. Allen, J.A.: The international catalogue of scientific literature. The Auk. 21, 494–501 (1904)
22. Rusin, D.: The Mathematical Atlas—A Gateway to Modern Mathematics (2002), http://www.math-atlas.org/welcome.html
23. Deerwester, S.C., Dumais, S.T., Landauer, T.K., Furnas, G.W., Harshman, R.A.: Indexing by latent semantic analysis. Journal of the American Society of Information Science 41, 391–407 (1990)
24. Bengio, Y., Lamblin, P., Popovici, D., Larochelle, H.: Greedy layer-wise training of deep networks. In: Schölkopf, B., Platt, J., Hoffman, T. (eds.) Advances in Neural Information Processing Systems 19, pp. 153–160. MIT Press, Cambridge (2007)

Kantian Philosophy of Mathematics and Young Robots

Aaron Sloman

School of Computer Science, University of Birmingham, UK
A.Sloman@cs.bham.ac.uk
http://www.cs.bham.ac.uk/~axs/

Abstract. A child, or young human-like robot of the future, needs to develop an information-processing architecture, forms of representation, and mechanisms to support perceiving, manipulating, and thinking about the world, especially perceiving and thinking about actual and possible structures and processes in a 3-D environment. The mechanisms for extending those representations and mechanisms, are also the core mechanisms required for developing mathematical competences, especially geometric and topological reasoning competences. Understanding both the natural processes and the requirements for future human-like robots requires AI designers to develop new forms of representation and mechanisms for geometric and topological reasoning to explain a child's (or robot's) development of understanding of affordances, and the proto-affordances that underlie them. A suitable multi-functional self-extending architecture will enable those competences to be developed. Within such a machine, human-like mathematical learning will be possible. It is argued that this can support Kant's philosophy of mathematics, as against Humean philosophies. It also exposes serious limitations in studies of mathematical development by psychologists.

Keywords: learning mathematics, philosophy of mathematics, robot 3-D vision, self-extending architecture, epigenetic robotics.

1   Introduction: Approaches to Mathematics

Some people have a central interest in mathematics, e.g.: mathematicians, whose job is to extend mathematical knowledge and to teach it; scientists, who routinely use mathematics to express data, analyse data, formulate theories, make predictions, construct explanations, etc.; and engineers, who use it to derive requirements for their designs and to check consequences of designs.

Others study aspects of mathematics: e.g. philosophers who discuss the nature of mathematical concepts and knowledge; psychologists who study how and when people acquire various mathematical concepts and kinds of mathematical competence; biologists interested in which animals have any mathematical competence, what genetic capabilities make that possible, how it evolved, and how that is expressed in a genome; and AI researchers who investigate ways of

S. Autexier et al. (Eds.): AISC/Calculemus/MKM 2008, LNAI 5144, pp. 558–573, 2008.

giving machines mathematical capabilities. Finally there are the children of all ages who are required to learn mathematics, including a subset who love playing with and learning about mathematical structures and processes, and many who hate mathematics and make little progress learning it.

My claim is that there are connections between these groups that have not been noticed. In particular, if we can understand how children and other animals learn about, perceive and manipulate objects in the environment and learn to think about what they are doing, we shall discover that the competences they need are closely related to requirements for learning about mathematics and making mathematical discoveries. Moreover, if we make robots that interact with and learn from the environment in the same way, they too will be able to be mathematical learners – a new kind of biologically inspired robot. The insights that we can gain from this link can shed light on old problems in philosophy of mathematics, and psychology of mathematics.

And finally, if we really make progress in this area, we may be able to revolutionise mathematical education for humans, on the basis of a much deeper understanding of what it is to be an intelligent learner of mathematics.[1]

2   Philosophies of Mathematics

Over many centuries, different views of the nature of mathematical knowledge and discovery have been developed. Those include differing philosophical views about the nature of numbers, for example (simplifying enormously):

1. Number concepts and laws are abstractions from operations on perceived groups of objects. (J.S.Mill [1] and some developmental psychologists. See [2] for discussion.)

2. Numbers are mental objects, created by human mental processes. Facts about numbers are discovered by performing mental experiments. (Intuitionist logicians, e.g. Brouwer. Heyting [3], Kant? [4])

3. Numbers and their properties are things we can discover by thinking about them in the right way (Kant, and many mathematicians, e.g [5]).

4. Numbers are sets of sets, or predicates of predicates, definable in purely logical terms. E.g. the number one is the set of all sets capable of being mapped bi-uniquely onto the set containing nothing but the empty set. (Frege [6], Russell [7], and other logicists).

5. Numerals are meaningless symbols manipulated according to arbitrary rules. Mathematical discoveries are merely discoveries about the properties of such games with symbols. (Formalists, e.g. Hilbert.)

6. Numbers are implicitly defined by a collection of axioms, such as Peano's axioms. Any collection of things satisfying these axioms can be called a set of numbers. The nature of the elements of the set is irrelevant. Mathematical

---

[1] See also http://www.cs.bham.ac.uk/research/projects/cogaff/talks/#math-robot (PDF presentation on whether a baby robot can grow up to be a mathematician).

discoveries about numbers are merely discoveries of logical consequences of the axioms. (Many mathematicians)

7. It doesn't matter what numbers are: we are only interested in which statements about them follow from which others (Russell, [8]).

8. There is no one correct answer to the question 'what are numbers?' People play a motley of 'games' using number words and other symbols, and a full account of the nature of numbers would simply be an analysis of these games (including the activity of mathematicians) and the roles they play in our lives. (Wittgenstein: *Remarks on the Foundations of Mathematics*)

J.S. Mill claimed that mathematical knowledge was empirical, abstracted from experiences of actions like counting or matching sets, and capable of being falsified by experience. Most thinkers regard mathematical knowledge as non-empirical, and not refutable by experiments on the environment, though interacting with the environment, including making drawings, or doing calculations on paper, help us notice mathematical truths, or help us find counter-examples to mathematical claims (Lakatos [9]).

Some philosophers who regard mathematical knowledge as non-empirical think it is all essentially empty of content, because it merely expresses definitions or "relations between our ideas" – i.e. such knowledge is "analytic" (defined in [10]). Hume had this sort of view of mathematics. Kant (1781) reacted by arguing for a third kind of knowledge, which is neither empirical nor analytic but "synthetic": these significantly extend our knowledge.

As a graduate student in Oxford around 1960 I found that something like Hume's view was common among the philosophers I encountered, so I tried, in my DPhil thesis [11], to explain and defend Kant's view, that mathematical knowledge is synthetic and *a priori* (non-empirical), which clearly accorded much better with the experience of doing mathematics.

"*A priori*" does not imply "innate". Discovering or understanding a mathematical proof can be a difficult achievement. Although mathematicians can make mistakes and may have to debug their proofs, their definitions, their algorithms, their axiom-systems, and even their examples, as shown by Lakatos in [9], mathematical knowledge is not empirical in the sense in which geological or chemical knowledge is, namely subject to refutation by new physical occurrences. Both flaws and the fixes in mathematics can be discovered merely by thinking.

3   Psychological Theories about Number Concepts

It is often supposed that the visual or auditory ability to distinguish groups with different numbers of elements (subitizing) displays an understanding of number. However this is simply a perceptual capability. A deeper understanding of numbers requires a wider range of abilities, discussed further below.

Rips *et al.* [2] give a useful survey of psychological theories about number concepts. They rightly criticise theories that treat number concepts as abstracted from perception of groups of objects, and discuss alternative requirements for a child to have a concept of number, concluding that having a concept of number involves having some understanding (not necessarily consciously) of a logical schema something like Peano's five axioms. They claim that that is what enables a child to work out various properties of numbers, e.g. the commutativity of addition, and the existence of indefinitely larger numbers. This implies that such children have the logical capabilities required to draw conclusions from the axioms, though not necessarily consciously. That immediately raises the question how children can acquire such competences. They conclude that somehow the Peano schema and the logical competences are innately built into the child's "background architecture" (but do not specify how that could work).

They do not consider an alternative possibility presented in [12,13] according to which such competences may be meta-configured, i.e. not determined in the genome, but produced through interactions with the environment that generate layers of meta-competences (competences that enable new competences to be acquired). Some hints about how that might occur are presented below.

Many psychologists and researchers in animal cognition misguidedly search for experimental tests for whether a child or animal does or does not understand what numbers are.[2] Rips *et al.* are not so committed to specifying an experimental test, but they do require a definition that makes a clear distinction between understanding and not understanding what numbers are.

4   Towards an AI Model of Learning about Numbers

As far as I know, no developmental psychologists have considered the alternative view, presented 30 years ago in [14], chapter 8, that there is no single distinction between having and not having a concept of number, because learning about numbers involves a never-ending process that starts from relatively primitive and general competences that are not specifically mathematical and gradually adds more and more sophistication, in parallel with the development of other competences. In particular, [14] suggests that learning about numbers involves developing capabilities of the following sorts:

1. performing a repetitive action;
2. memorising an ordered sequence of arbitrary names;
3. performing two repetitive actions together and keeping them in synchrony;
4. initiating such a process and then being able to use different stopping conditions, depending on the task;
5. doing all this when one of the actions is uttering a learnt sequence of names;
6. learning rules for extending the sequence of names indefinitely;
7. observing various patterns in such processes and storing information about them, e.g. information about successors and predecessors of numerals, or results of counting onwards various amounts from particular numerals;

---

[2] Compare the mistake of striving for a definitive test for whether animals of some species understand causation, criticised here in presentation 3:
http://www.cs.bham.ac.uk/research/projects/cogaff/talks/wonac

8. noticing commonalities between static mappings and process mappings (e.g. paired objects vs paired events);

9. finding mappings between components of static structures as well as the temporal mappings between process-elements;

10. noticing that such mappings have features that are independent of their order (e.g. counting a set of objects in two different orders must give the same result);

11. noticing that numbers themselves can be counted, e.g. the numbers between two specified numbers;

12. noticing possibilities of and constraints on rearrangements of groups of objects – e.g. some can be arranged as rectangular arrays, but not all;

13. learning to compare continuous quantities by dividing them into small components of a standard size and counting.

Such competences and knowledge can be extended indefinitely. Some can be internalised, e.g. counting silently. Documenting all the things that can be discovered about such structures and processes in the first few years of life could fill many pages. (Compare [15,16].) The sub-abilities involved in these processes are useful in achieving practical goals by manipulating objects in the environment and learning good ways to plan and control such achievements. An example might be fetching enough cups to give one each to a group of people, or matching heights of two columns made of bricks, to support a horizontal beam, or ensuring that enough water is in a big jug to fill all the glasses on the table.

Gifted teachers understand that any deep mathematical domain is something that has to be explored from multiple directions, gaining structural insights and developing a variety of perceptual and thinking skills of ever increasing power. That includes learning new constructs, new reasoning procedures, learning to detect partial or erroneous understanding, and finding out how to remedy such deficiencies. [14] presented some conjectures about some of the information-processing mechanisms involved. As far as I know, nobody has tried giving a robot such capabilities. It should be feasible in a suitably simplified context. I had hoped to do this in a robot project, but other objectives were favoured.[3]

5   Internal Construction Competences

The processes described above require the ability to create (a) new internal information structures, including, for example, structures recording predecessors of numbers, so that it is not necessary always to count up to N to find the predecessor of N, and (b) new algorithms for operating on those structures. As these internal information-structures grow, and algorithms for manipulating them are developed, there are increasingly many opportunities to discover more properties of numbers. The more you know, the more you can learn.

---

[3] See http://www.cs.bham.ac.uk/research/projects/cosy/PlayMate-start.html

Moreover those constructions do not happen instantaneously or in an error-free process. Many steps are required including much self-debugging, as illustrated in [17]. This depends on self-observation during performance of various tasks, including observations of external actions and of thinking. One form of debugging is what Sussman called detecting the need to create new critics that run in parallel with other activities and interrupt if some pattern is matched, for instance if disguised division by zero occurs.

The ongoing discovery of new invariant patterns in structures and processes produced when counting, arranging, sorting, or aligning sets of objects, leads to successive extensions of the learner's understanding of numbers. Initially this is just empirical exploration, but later a child may realise that the result of counting a fixed set of objects cannot depend on the order of counting. That invariance is intrinsic to the nature of one-to-one mappings and does not depend on properties of the things being counted, or on how fast or how loud one counts, etc. However, some learners may never notice this non-empirical character of mathematical discoveries until they take a philosophy class!

One of the non-empirical discoveries is that the natural numbers form an infinite set. Kant suggested that this requires grasping that a rule can go on being applied indefinitely. This contrasts with the suggestion by Rips *et al.* [2] that a child somehow acquires logical axioms which state that every natural number has exactly one successor and at most one predecessor, and that the first number has no predecessor, from which it follows logically that there is no final number and the sequence of numbers never loops. Instead, a child could learn that there are repetitive processes of two kinds: those that start off with a determinate stopping condition that limits the number of repetitions and those that do not, though they can be stopped by an external process. Tapping a surface, walking, making the same noise repeatedly, swaying from side to side, repeatedly lifting an object and dropping it, are all examples of the latter type.

The general notion of something not occurring is clearly required for intelligent action in an environment. E.g. failure of an action to achieve its goal needs to be detectable. So if the learner has the concept of a repetitive process leading to an event that terminates the process, then the general notion of something not happening can be applied to that to generate the notion of something going on indefinitely. From there, depending on the information processing architecture and the forms of representation available, it may be a small step to the representation of two synchronised processes going on indefinitely, one of which is a counting process.

What is more sophisticated is acquiring a notion of a sequence of sounds or marks that can be generated indefinitely without ever repeating a previous mark. An obvious way to do that is to think of marks made up of one or more dots or strokes. Then the sequence could start with a single stroke, followed by two strokes, followed by three strokes, etc., e.g.

| || ||| |||| ||||| etc.

That has the disadvantage that the patterns grow large very quickly. That can motivate far more compact notations, like arabic numerals, though any infinitely generative notation will ultimately become physically unmanageable.

6   Extending Simple Number Concepts

A different sort of extension is involved in adding *zero* to the natural numbers, which introduces "anomalies", such as that there is no difference between *adding* zero apples and *subtracting* zero apples from a set of apples.

Negative integers add further confusions. This extension is rarely taught properly, and as a result most people cannot give a coherent explanation of why multiplying two negative numbers should give a positive number. It cannot be *proved* on the basis of previous knowledge because what multiplying by a negative number means is undefined initially. For mathematicians, it is defined by the rules for multiplying negative numbers, and the *simplest* way to extend multiplication rules to negative numbers without disruption of previous generalisations, is to stipulate that multiplying two negatives produces a positive. (Similarly with defining what 3^{-1} and 3^0 should mean.)

Some teachers use demonstrations based on the so-called "number line" to introduce notions of negative integers, but this can lead to serious muddles (e.g. about multiplication). Pamela Liebeck [18] developed a game called "scores and forfeits" where players have two sets of tokens: addition of a red token is treated as equivalent to removal of a black token, and vice versa. (Multiplication was not included.) The person with the biggest surplus of black over red wins. Giving a player both a red and a black token, or removing both a red and a black token makes no difference to the total status of the player. Playing, and especially discussing, the game seemed to give children a deeper understanding of negative numbers and subtraction than standard ways of teaching, presumably because the set of pairs of natural numbers can be used to model accurately the set of positive and negative integers.

Cardinality and orderings are properties of *discrete* sets. Extending the notion of number to include *measures* that are continuously variable, e.g. lengths, areas, volumes and time intervals, requires sophisticated extensions to the learner's ontology and forms of representation – leading to deep mathematical and philosophical problems. In humans, an understanding of Euclidean geometry and topology seems to build on reasoning/planning competences combined with visual competences, as illustrated in [16]. This requires different forms of representation from counting and matching groups of entities. Some of these competences are apparently shared with some other animals – those that are capable of planning and executing novel spatial actions.

7   Doing Philosophy by Doing AI

After completing my D.Phil defending Kant in 1962, I gradually realised something was lacking. My arguments consisted mostly of illustrative examples.

Something deeper was required, to explain what goes on (a) when people acquire mathematical *concepts* (e.g. number, infinitely thin line, perfectly straight line, infinite set, etc.), and (b) when they acquire mathematical *knowledge* expressed using those concepts. In 1969, Max Clowes introduced me to programming and AI, and I soon realised that by building a working human-like mind (or suitable fragments of one) we could demonstrate the different modes of development of knowledge discussed by Kant. Many mathematical proofs, especially in Euclidean geometry and topology, but also in number theory, seemed to rely on our ability to perceive structures and structural relationships, so I concluded that explaining how mathematical discoveries were made, depended, in part, on showing how *visual* capabilities, or more generally, *spatial perception and reasoning* capabilities, were related to some kinds of mathematical reasoning.

At that time the dominant view in AI, represented by McCarthy and Hayes (1969) was that logical modes of representation and reasoning were all that an intelligent robot would need. In 1971, I submitted a paper to IJCAI [20], distinguishing "Fregean" from "analogical" forms of representation and arguing that spatial analogical forms of representation and reasoning could be used in *valid* derivations and could also in some cases help with the organisation of search. Fregean representations are those whose mode of syntactic composition and semantic interpretation use only application of functions to arguments, whereas analogical representations allow properties and relations of parts of a representation to refer to properties and relations of parts of a complex whole, though not necessarily the same properties and relations: in general analogical representations are not isomorphic with what they represent. (E.g. a 2D picture can represent a 3D scene without being isomorphic with it.) The paper was accepted, and subsequently reprinted twice. But it was clear that a lot more work needed to be done to demonstrate how the human spatial reasoning capabilities described therein could be replicated in a machine.

Many others (e.g. [21]) have pointed out the need to provide intelligent machines with spatial forms of representation and reasoning, but progress in replicating human abilities has been very slow: we have not yet developed visual mechanisms that come close to matching those produced by evolution. In part, this is because the requirements for human-like visual systems have not been analysed in sufficient depth (as illustrated in [22,23]). E.g. there is a vast amount of research on object recognition that contributes nothing to our understanding of how 3-D spatial structures and processes are seen or how information about spatial structures and processes is used, for instance in reasoning and acting.[4]

8 Requirements for a Mathematician's Visual System

To address this problem, I have been collecting requirements for visual mechanisms since 1971, in parallel with more general explorations of requirements for a complete human-like architecture (e.g., [14,24,25,26,27,23]). Full understanding

---

[4] Some of the differences between recognition and perception of 3-D structure are illustrated in http://www.cs.bham.ac.uk/research/projects/cogaff/challenge.pdf

of the issues requires us to investigate: (a) trade-offs between alternative sets of requirements and designs, including different biological examples [28,29]; (b) different kinds of developmental trajectories [30,12]; and (c) requirements for *internal* languages supporting structural variability and compositional semantics in other animals, pre-verbal humans, and future robots [31,13]. I have also tried to show how that analysis can lead to a new view of the evolution of language (http://www.cs.bham.ac.uk/research/projects/cogaff/talks/#glang). In particular, it allows internal languages with compositional semantics to include *analogical* forms of representation, whose manipulation can play an important role both in visual perception and in reasoning. This underlies human spatial reasoning abilities that are often used in mathematics. Human-like mathematical machines (e.g. robots that reason as humans do) will also need such competences.

9   Affordances, Visual Servoing, and Beyond

Analysis of biological requirements for vision (including human vision) enlarges our view of the functions of vision, requiring goals of AI vision researchers to be substantially expanded. An example is the role of vision in servo-control, including control of continuous motion described in [33] and Berthoz [34], as well as discrete condition-checking.

Gibson's work on affordances in [35] showed that animal vision provides information not merely about geometrical and physical features of entities in the environment, as in [36,37], nor about recognising or categorising objects (the focus of much recent AI 'vision' research), but about what the perceiver can and cannot do, given its physical capabilities and its goals. I.e. vision needs to provide information not only about actual objects, structures and motion, but also what processes *can and cannot occur in the environment* [38]. In order to do this, the visual system must use an ontology that is only very indirectly related to retinal arrays. But Gibson did not go far enough, as we shall see.

All this shows that a human-like visual sub-architecture must be multi-functional, with sub-systems operating concurrently at different levels of abstraction and engaging concurrently with different parts of the rest of the architecture, including central and motor subsystems. For example, painting a curved stripe requires continuous visual control of the movement of the brush, which needs to be done in parallel with checking whether mistakes have been made (e.g. bits not painted, or the wrong bits painted) and whether the task has been completed, or whether the brush needs to be replenished. For these reasons, in [24], I contrasted (a) "modular" visual architectures, with information flowing from input images (or the optic array) through a pipeline of distinct processing units, as proposed by Marr and others, with (b) "labyrinthine" visual architectures reflecting the multiplicity of functions of vision and the rich connectivity between subsystems of many kinds.

10   Perception of Actual and Possible Processes

Work on an EU-funded cognitive robotics project, CoSy, begun in 2004,[5] included analysis of requirements for a robot capable of manipulating 3-D objects, e.g. grasping them, moving them, and constructing assemblages, possibly while other things were happening. Analysis of the requirements revealed (a) the need for representing scene objects with parts and relationships (as everyone already knew), (b) the need for several ontological layers in scene structures (as in chapter 9 of [14]), (c) the need to represent "multi-strand relationships" because not only whole objects but also parts of different objects are related in various ways, (d) the need to represent "multi-strand processes", because when things move the multi-strand relationships change, e.g. with metrical, topological, causal, functional, continuous, and discrete changes occurring concurrently, and (e) the need to represent *possible* processes, and constraints on possible processes. I call the latter positive and negative "proto-affordances", because they are the substratum of affordances, but more general.

Not all perceived changes are produced or can be produced by the perceiver. Likewise seeing that a process is *possible*, e.g. an apple falling, or that possibilities are *constrained*, e.g. because a table is below the apple, does not presuppose that the perceiver desires to or can produce the process. So perception of proto-affordances and perception of processes in the environment makes it possible to take account of far more than one's own actions, their consequences and their constraints. As explained in [22,23], that requires an *amodal, exosomatic* form of representation of processes; one that is not tied to the agent's sensorimotor processes. That possibly is ignored by researchers who focus only on sensorimotor learning and representation, and base all semantics on "symbol-grounding".[6]

The ability to perceive a multi-strand process requires the ability to have internal representations of the various concurrently changing relationships. Some will be continuous changes, including those needed for servo-control of actions, while others may be discrete changes as topological relations change or goals become satisfied. Mechanisms used for perceiving multi-strand processes can also be used both to predict outcomes of possible processes that are not currently occurring (e.g. when planning), and to explain how a perceived situation came about. Both may use a partial simulation of the processes.[7] (Cf. Grush in [39].)

11   The Importance of Kinds of Matter

A child, or robot, learning about kinds of process that can occur in the environment needs to be able to extend the ontology she uses indefinitely, and not

---

[5] Described in http://www.cs.bham.ac.uk/research/projects/cosy/
[6] Reasons for preferring "symbol-tethering" to symbol-grounding theory are given in: http://www.cs.bham.ac.uk/research/projects/cogaff/talks/#models
[7] Examples are given in the presentation in Note 1, and in this discussion paper on predicting changes in action affordances and epistemic affordances: http://www.cs.bham.ac.uk/research/projects/cosy/papers/#dp0702

merely by defining new concepts in terms of old ones: there are also *substantive* ontology extensions (as in the history of physics and other sciences). For example, whereas many perceived processes involve objects that preserve all their metrical relationships, there are also many deviations from such rigidity, and concepts of different kinds of matter are required to explain those deviations: string and wire are flexible, but wire retains its shape after being deformed; an elastic band returns to its original length after being stretched, but does not restore its shape after bending. Some kinds of stuff easily separate into chunks in various ways, if pulled, e.g. mud, porridge, plasticine and paper. A subset of those allow restoration to a single object if separated parts are pressed together. There are also objects that are marks on other objects, like lines on paper, and there are some objects that can be used to produce such marks, like pencils and crayons. Marks produced in different ways and on different materials can have similar structures.

As demonstrated by Sauvy and Sauvy in [16], children, and presumably future robots, can learn to play with and explore strings, elastic bands, pieces of wire, marks on paper and movable objects, thereby learning about many different sorts of process patterns. Some of those are concerned with rigid motions some not. Some examples use patterns in non-rigid motions that can lead to development of topological concepts, e.g. a cup being continuously deformed into a toroid. Robot vision is nowhere near this capability at present.

12   Perception and Mathematical Discovery

I have argued that many mathematical discoveries involve noticing an invariant in a class of processes. For example, Mary Ensor, a mathematics teacher, once told me she had found a good way to teach children that the internal angles of a triangle add up to a straight line, demonstrated in the figure.

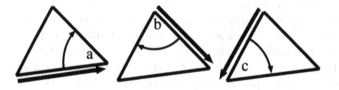

Consider any triangle. Imagine an arrow starting at one corner, pointing along one side. It can be rotated and translated as indicated, going through positions shown in the three successive figures. The successive rotations *a*, *b* and *c* go through the interior angles of the triangle, and because of the final effect they produce, they must add up to a straight line. This discovery may initially be made through empirical exploration with physical objects, but the pattern involved clearly does not depend on what the objects are made of and changing conditions such as colours used, lengths of lines, particular angles in the triangle, temperature, strength of gravitational or magnetic field cannot affect the property of the process. A robot learner should notice that it is not an empirically falsifiable discovery.

However, such discoveries can have "bugs" as Lakatos [9] demonstrated using Euler's theorem about polyhedra. That is sometimes wrongly taken to imply that mathematical knowledge is empirical in the same way as knowledge about the physical properties of matter. The discovery of bugs in proofs and good ways to deal with them is an important feature of mathematical learning. For example, the rotating arrow proof breaks down if the triangle is on the surface of a sphere. Noticing this can lead a learner to investigate properties that distinguish planar and non-planar surfaces. But that exploration does not *require* experiments in a physical laboratory, though it may benefit from them. Kant claimed that such discoveries are about the perceiver's forms of perception, but they are not restricted to any particular perceivers.

13   Humean and Kantian Causation

Adding properties of matter, such as rigidity and impenetrability, to representations of shape and topology allows additional reasoning about and prediction of results of processes. An example is the ability to use the fact that two meshed gear wheels are rigid and impenetrable to work out how rotating one will cause the other to rotate. That kind of reasoning is not always available.

If the wheels are not meshed, but there are hidden connections, then the only basis for predicting the consequence of rotating the wheels is to use a Humean notion of causation: basing predictions of results of actions or events solely on observed correlations. In contrast, where the relevant structure and constraints are known, and mathematical proofs (using geometry and topology) are possible, Kant's notion of causation, which is structure-based and deterministic, can be used. Causal relationships represented in Bayesian nets are essentially generalisations of Humean causation and based only on statistical evidence. However, a significant subset of the causal understanding of the environment that a child acquires is Kantian because it allows the consequences of novel processes to be *worked out* on the basis of geometric and topological relationships, and kinds of matter involved. For more on Humean vs Kantian causation in robots and animals see the presentations by Sloman and Chappell here: http://www.cs.bham.ac.uk/research/projects/cogaff/talks/wonac

Many kinds of learning involve strings. If an inelastic but flexible string is attached to a remote movable object, then if the end is pulled away from the object a process can result with two distinct phases: (1) curves in the string are gradually eliminated (as long as there are no knots), and (2) when the string is fully straightened the remote object will start moving in the direction of the pulled end. However, if the string is looped round a fixed pillar, the first sub-process does not produce a single straight string but two straight portions and a portion going round the pillar, and in the second phase the attached object moves toward the pillar, not toward the pulled end.[8]

14   Russell *vs* Feynman on Mathematics

At the beginning of the last century Russell and Whitehead [40] attempted to demonstrate that all of mathematics could be reduced to logic (Frege had attempted this only for Arithmetic). Despite the logical paradoxes and the difficulty of avoiding them, Russell thought the goals of the project could be or had been achieved, and concluded that mathematics was just *the investigation of implications that are valid in virtue of their logical form, independently of any non-logical subject matter.* He wrote: "Mathematics may be defined as the subject in which we never know what we are talking about, nor whether what we are saying is true" (*Mysticism and Logic* [8]). In some ways Russell seems to have been a philosophical descendant of David Hume, who had claimed that non-empirical propositions were in some sense trivial, e.g. mere statements of the relations between our ideas.

In contrast, the physicist Richard Feynman described mathematics as "the language nature speaks in". He wrote: "To those who do not know Mathematics it is difficult to get across a real feeling as to the beauty, the deepest beauty of nature. ... If you want to learn about nature, to appreciate nature, it is necessary to understand the language that she speaks in" (in [41]). I believe that Feynman's description is closely related to what I am saying about how a child (or a future robot) can develop powerful, reusable concepts and techniques related to patterns of perception and patterns of thinking that are learnt through interacting with a complex environment, part of which is the information-processing system within the learner. Despite the role of experience in such learning, the results of such learning are not empirical generalisations. Feynman seems to agree with Kant that mathematical knowledge is both non-empirical and deeply significant.

15   Conclusion

We need further investigation of architectures and forms of representation that allow playful exploration by a robot to produce discoveries of patterns in structures and processes that are the basis of deep mathematical concepts and mathematical forms of reasoning. The robot should be able to go through the following stages:

---

[8] More examples and their implications are discussed in the presentation in Note 1 and in http://www.cs.bham.ac.uk/research/projects/cosy/papers#dp0601: "Orthogonal recombinable competences acquired by altricial species".

1. Acquiring familiarity with some domain, e.g. through playful exploration;
2. Noticing (empirically) certain generalisations;
3. Discovering a way of thinking about them that shows they are not empirical;
4. Generalising, diversifying, debugging, deploying, that knowledge;
5. Formalising the knowledge, possibly in more than one way.

This paper merely reports on a subset of the requirements for working designs. Some more detailed requirements are in [22]. It is clear that AI still has a long way to go before the visual and cognitive mechanisms of robots can match the development of a typical human child going through the earlier steps. There is still a great deal more to be done, and meeting all the requirements will not be easy. If others are interested in this project, perhaps it would be useful to set up an informal network for collaboration on refining the requirements and then producing a working prototype system as a proof of concept, using a simulated robot, perhaps one that manipulates 2-D shapes in a plane surface, discovering properties of various kinds of interactions, involving objects with different shapes made of substances with various properties that determine the consequences of the interactions, e.g. impenetrability, rigidity, elasticity, etc.

Perhaps one day, a team of robot mathematicians, philosophers of mathematics and designers will also be able implement such systems. First we need a deeper understanding of the requirements.

Acknowledgements

Some of the ideas reported in this paper were developed as part of the requirements analysis activities in the EU-funded CoSy project http://www.cognitivesystems.org, and discussions with Jackie Chappell. Work by Jim Franklin referenced in the presentation in Note 1 is also relevant. Some related ideas are in early work by Doug Lenat, e.g. AM and Eurisko, [42]

References

1. Mill, J.S.: A System of Logic, Ratiocinative and Inductive. John W. Parker, London (1843)
2. Rips, L.J., Bloomfield, A., Asmuth, J.: From Numerical Concepts to Concepts of Number. The Behavioral and Brain Sciences (in press)
3. Heyting, J.: Intuitionism, an Introduction. North Holland, Amsterdam (1956)
4. Kant, I.: Critique of Pure Reason. Macmillan, London (1781); (translated by N.K. Smith, 1929)
5. Penrose, R.: The Emperor's New Mind: Concerning Computers Minds and the Laws of Physics. Oxford University Press, Oxford (1989)
6. Frege, G.: The Foundations of Arithmetic: a logico-mathematical enquiry into the concept of number. B.H. Blackwell, Oxford (1950); (original, 1884)
7. Russell, B.: The Principles of Mathematics. CUP, Cambridge (1903)
8. Russell, B.: Mysticism and Logic and Other Essays. Allen & Unwin, London (1917)

9. Lakatos, I.: Proofs and Refutations. CUP, Cambridge (1976)
10. Sloman, A.: Necessary, A Priori and Analytic. Analysis 26(1), 12–16 (1965), http://www.cs.bham.ac.uk/research/projects/cogaff/07.html#701
11. Sloman, A.: Knowing and Understanding: Relations between meaning and truth, meaning and necessary truth, meaning and synthetic necessary truth. PhD thesis, Oxford University (1962), http://www.cs.bham.ac.uk/research/projects/cogaff/07.html#706
12. Chappell, J., Sloman, A.: Natural and artificial meta-configured altricial information-processing systems. International Journal of Unconventional Computing 3(3), 211–239 (2007), http://www.cs.bham.ac.uk/research/projects/cosy/papers/#tr0609
13. Sloman, A., Chappell, J.: Computational Cognitive Epigenetics (Commentary on [32]). Behavioral and Brain Sciences 30(4), 375–386 (2007), http://www.cs.bham.ac.uk/research/projects/cosy/papers/#tr0703
14. Sloman, A.: The Computer Revolution in Philosophy. Harvester Press (and Humanities Press), Hassocks, Sussex (1978), http://www.cs.bham.ac.uk/research/cogaff/crp
15. Liebeck, P.: How Children Learn Mathematics: A Guide for Parents and Teachers. Penguin Books, Harmondsworth (1984)
16. Sauvy, J., Suavy, S.: The Child's Discovery of Space: From hopscotch to mazes – an introduction to intuitive topology. Penguin Education, Harmondsworth (1974) (Translated from the French by Pam Wells)
17. Sussman, G.: A computational model of skill acquisition. Elsevier, Amsterdam (1975)
18. Liebeck, P.: Scores and Forfeits: An Intuitive Model for Integer Arithmetic. Educational Studies in Mathematics 21(3), 221–239 (1990)
19. McCarthy, J., Hayes, P.: Some philosophical problems from the standpoint of AI. In: Meltzer, B., Michie, D. (eds.) Machine Intelligence 4, pp. 463–502. Edinburgh University Press, Edinburgh (1969), http://www-formal.stanford.edu/jmc/mcchay69/mcchay69.html
20. Sloman, A.: Interactions between philosophy and AI: The role of intuition and non-logical reasoning in intelligence. In: Proc 2nd IJCAI, pp. 209–226. William Kaufmann, London (1971), http://www.cs.bham.ac.uk/research/cogaff/04.html#200407
21. Glasgow, J., Narayanan, H., Chandrasekaran, B. (eds.): Diagrammatic Reasoning: Computational and Cognitive Perspectives. MIT Press, Cambridge (1995)
22. Sloman, A.: Architectural and representational requirements for seeing processes and affordances. Research paper, for Workshop Proceedings COSY-TR-0801, University of Birmingham, UK. School of Computer Science (March 2008), http://www.cs.bham.ac.uk/research/projects/cosy/papers#tr0801
23. Sloman, A.: Putting the Pieces Together Again. In: Sun, R. (ed.) Cambridge Handbook on Computational Psychology. CUP, New York (2008), http://www.cs.bham.ac.uk/research/projects/cogaff/07.html#710
24. Sloman, A.: On designing a visual system (towards a gibsonian computational model of vision). Journal of Experimental and Theoretical AI 1(4), 289–337 (1989), http://www.cs.bham.ac.uk/research/projects/cogaff/81-95.html#7
25. Sloman, A.: Architecture-based conceptions of mind. In: The Scope of Logic, Methodology, and Philosophy of Science (Vol II). Synthese Library, vol. 316, pp. 403–427. Kluwer, Dordrecht (2002), http://www.cs.bham.ac.uk/research/projects/cogaff/00-02.html#57

26. Sloman, A.: Beyond shallow models of emotion. Cognitive Processing: International Quarterly of Cognitive Science 2(1), 177–198 (2001)

27. Sloman, A.: Evolvable biologically plausible visual architectures. In: Cootes, T., Taylor, C. (eds.) Proceedings of British Machine Vision Conference, Manchester, BMVA, pp. 313–322 (2001)

28. Sloman, A.: Interacting trajectories in design space and niche space: A philosopher speculates about evolution. In: Deb, K., Rudolph, G., Lutton, E., Merelo, J.J., Schoenauer, M., Schwefel, H.-P., Yao, X. (eds.) PPSN 2000. LNCS, vol. 1917, pp. 3–16. Springer, Heidelberg (2000)

29. Sloman, A.: Diversity of Developmental Trajectories in Natural and Artificial Intelligence. In: Morrison, C.T., Oates, T.T., (eds.) Computational Approaches to Representation Change during Learning and Development, AAAI Fall Symposium 2007. Technical Report FS-07-03, Menlo Park, CA, pp. 70–79. AAAI Press (2007), http://www.cs.bham.ac.uk/research/projects/cosy/papers/#tr0704

30. Sloman, A., Chappell, J.: The Altricial-Precocial Spectrum for Robots. In: Proceedings IJCAI 2005. Edinburgh, IJCAI, pp. 1187–1192 (2005), http://www.cs.bham.ac.uk/research/cogaff/05.html#200502

31. Sloman, A.: The primacy of non-communicative language. In: MacCafferty, M., Gray, K. (eds.) The analysis of Meaning: Informatics 5 Proceedings ASLIB/BCS Conference, March 1979, pp. 1–15. Oxford, London (1979), http://www.cs.bham.ac.uk/research/projects/cogaff/81-95.html#43

32. Jablonka, E., Lamb, M.J.: Evolution in Four Dimensions: Genetic, Epigenetic, Behavioral, and Symbolic Variation in the History of Life. MIT Press, Cambridge (2005)

33. Sloman, A.: Image interpretation: The way ahead? In: Braddick, O., Sleigh, A. (eds.) Physical and Biological Processing of Images (Proceedings of an international symposium organised by The Rank Prize Funds, London, 1982.), pp. 380–401. Springer, Berlin (1982), http://www.cs.bham.ac.uk/research/projects/cogaff/06.html#0604

34. Berthoz, A.: The Brain's sense of movement. Perspectives in Cognitive Science. Harvard University Press, London (2000)

35. Gibson, J.J.: The Ecological Approach to Visual Perception. Houghton Mifflin, Boston (1979)

36. Barrow, H., Tenenbaum, J.: Recovering intrinsic scene characteristics from images. In: Hanson, A., Riseman, E. (eds.) Computer Vision Systems. Academic Press, New York (1978)

37. Marr, D.: Vision. Freeman, San Francisco (1982)

38. Sloman, A.: Actual possibilities. In: Aiello, L., Shapiro, S. (eds.) Principles of Knowledge Representation and Reasoning: Proceedings of the Fifth International Conference (KR 1996), pp. 627–638. Morgan Kaufmann Publishers, Boston (1996)

39. Grush, R.: The emulation theory of representation: Motor control, imagery, and perception. Behavioral and Brain Sciences 27, 377–442 (2004)

40. Whitehead, A.N., Russell, B.: Principia Mathematica, vol. I – III. CUP, Cambridge (1910–1913)

41. Feynman, R.: The Character of Physical Law. The 1964 Messenger Lectures. MIT Press, Cambridge (1964)

42. Lenat, D.B., Brown, J.S.: Why AM and EURISKO appear to work. Artificial Intelligence 23(3), 269–294 (1984)

Transforming the arχiv to XML

Heinrich Stamerjohanns and Michael Kohlhase

Computer Science, Jacobs University Bremen
{h.stamerjohanns,m.kohlhase}@jacobs-university.de

Abstract. We describe an experiment of transforming large collections of LaTeX documents to more machine-understandable representations. Concretely, we are translating the collection of scientific publications of the Cornell e-Print Archive (arXiv) using the LaTeX to XML converter which is currently under development.

The main technical task of our arXMLiv project is to supply LaTeXML bindings for the (thousands of) LaTeX classes and packages used in the arXiv collection. For this we have developed a distributed build system that reiteratively runs LaTeXML over the arXiv collection and collects statistics about e.g. the most sorely missing LaTeXML bindings and clusters common error events. This creates valuable feedback to both the developers of the LaTeXML package and to binding implementers. We have now processed the complete arXiv collection of more than 400,000 documents from 1993 until 2006 (one run is a processor-year-size undertaking) and have continuously improved our success rate to more than 56% (i.e. over 56% of the documents that are LaTeX have been converted by LaTeXML without noticing an error and are available as XHTML+MathML documents).

1 Introduction

The last few years have seen the emergence of various XML-based, content-oriented markup languages for mathematics and natural sciences on the web, e.g. OpenMath, Content MathML, or our own OMDoc. The promise of these content-oriented approaches is that various tasks involved in "doing mathematics" (e.g. search, navigation, cross-referencing, quality control, user-adaptive presentation, proving, simulation) can be machine-supported, and thus the working mathematician can concentrate in doing what humans can still do infinitely better than machines.

On the other hand LaTeX is and has been the preferred document source format for thousands of scientists who publish results that include mathematical formulas. Millions of scientific articles have been written and published using this document format. Unfortunately the LaTeX language mixes content and presentation and also allows to create additional macro definitions. Therefore machines have great difficulties to parse and analyze LaTeX documents and to extract enough information to represent the written formulas in a XML representation.

S. Autexier et al. (Eds.): AISC/Calculemus/MKM 2008, LNAI 5144, pp. 574–582, 2008.

In this paper, we will present an experiment of translating a large corpus of mathematical knowledge to a form that is more suitable for machine processing. The sheer size of the ARXIV [ArX07] poses a totally new set of problems for MKM technologies, if we want to handle (and in the future manage) corpora of this size. In the next section we will review the translation technology we build on and then present the corpus-level build system which is the main contribution of this paper.

2   TeX/LaTeX to XML Conversion

The need for translating LaTeX documents into other formats has been long realized and there are various tools that attempt this at different levels of sophistication. We will disregard simple approaches like the venerable `latex2html` translator that cannot deal with user macro definitions, since these are essential for semantic preloading. The remaining ones fall into two categories that differ in the approach towards parsing the TeX/LaTeX documents.

Romeo Anghelache's HERMES [Ang07] and Eitan Gurari's TeX4HT systems use special TeX macros to seed the `dvi` file generated by TeX with semantic information. The `dvi` file is then parsed by a custom parser to recover the text and semantic traces which are then combined to form the output XML document. While HERMES attempts to recover as much of the mathematical formulae as Content-MATHML, it has to revert to Presentation-MATHML where it does not have semantic information. TeX4HT directly aims for Presentation-MATHML.

The systems rely on the TeX parser for dealing with the intricacies of the TeX macro language (e.g. TeX allows to change the tokenization (via "catcodes") and the grammar at run-time). In contrast to this, Bruce Miller's `LaTeXML` [Mil07] system and the SGLR/ELAN4 system [vdBS03] re-implement a parser for a large fragment of the TeX language. This has the distinct advantage that we can fully control the parsing process: We want to expand abbreviative macros and recursively work on the resulting token sequence, while we want to directly translate semantic macros[1], since they directly correspond to the content representations we want to obtain. The LaTeXML and SGLR/ELAN4 systems allow us to do just this. In our conversion experiment we have chosen the `LaTeXML` system, whose LaTeX parser seems to have largest coverage.

The LaTeXML system consists of a TeX parser, an XML emitter, and a postprocessor. To cope with LaTeX documents, the system needs to supply `LaTeXML` **bindings** (i.e. special directives for the XML emitter) for the semantic macros in LaTeX packages. Concretely, every LaTeX package and class must be accompanied by a `LaTeXML` binding file, a PERL file which contains `LaTeXML` constructor-, abbreviation-, and environment definitions, e.g.

```
DefConstructor ("\Reals","<XMTok name='Reals'/>");
DefConstructor ("\SmoothFunctionsOn{}",
        "<XMApp><XMTok name='SmoothFunctionsOn'/>#1</XMApp>");
DefMacro("\SmoothFunctionsOnReals","\SmoothFunctionsOn\Reals");
```

---

[1] See [Koh08] for a discussion of semantic and abbreviative macros.

DefConstructor is used for semantic macros, whereas DefMacro is used for abbreviative ones. The latter is used, since the latexml program does not read the package or class file and needs to be told, which sequence of tokens to recurse on. The LaTeXML distribution contains LaTeXML bindings for the most common base LATEX packages.

For the XML conversion, the latexml program is run, say on a file doc.tex. latexml loads the LaTeXML bindings for the LATEX packages used in doc.tex and generates a temporary LTXML document, which closely mimics the structure of the parse tree of the LATEX source. The LTXML format provides XML counterparts of all core TEX/LATEX concepts, serves as a target format for LaTeXML, and thus legitimizes the XML fragments in the LaTeXML bindings.

In the semantic post-processing phase, the LATEX-near representation is transformed into the target format by the latexmlpost program. This program applies a pipeline of intelligent filters to its input. The LaTeXML program supplies various filters, e.g. for processing HTML tables, including graphics, or converting formulae to Presentation-MATHML. Other filters like transformation to OPENMATH and Content-MATHML are currently under development. The filters can also consist of regular XML-to-XML transformation process, e.g. an XSLT style sheet. Eventually, post-processing will include semantic disambiguation information like types, part-of-speech analysis, etc. to alleviate the semantic markup density for authors.

3   The Build System

To test and give feedback to improve LaTeXML, and to extend our collection of valid XHTML+MATHML documents which are being used for other projects such as our MathWebSearch [Mat07], we have chosen to use the articles that have been published in the ARXIV. This large heterogenous collection of scientific articles is a perfect source for experiments with scientific documents that have been written in LATEX.

The huge number of more than 400.000 documents (each one may include figures and its own style files and is located in its own subdirectory) in this collection made simple manual handling of conversion runs impossible. To handle the conversion process itself (invocation of ttlatexml and latexmlpost) Makefiles have been automatically created by scripts. But the usage of make has also some limitations: It does not easily allow to run distributed jobs on several hosts, a feature that is essential to be able to massively convert thousands of documents in one day. While distributed make utilities (such as dmake) or other grid tools may support distributed builds, all these tools are of limited use when only a restricted set of specific documents with certain error characteristics should be converted again, which would require complex and continuous rewriting of makefiles.

To overcome these limitations we have developed an ARXMLIV build system which allows make jobs to be distributed among several hosts, extracts and analyzes the conversion process of each document and stores results in its own SQL

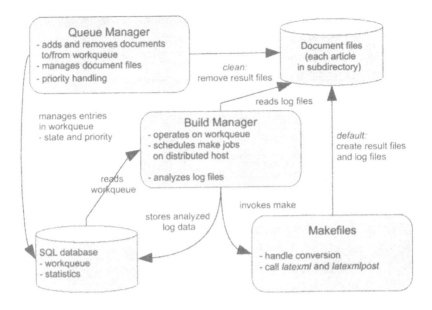

Fig. 1. Schematic overview of the ARXMLIV build system

database. The database allows to easily cluster documents which include a macro that is only partially supported or to gather statistics about the build process.

The ARXMLIV build system consists of a file system which is shared among all hosts, a queue manager, a build manager, and a relational database, which stores a workqueue and results statistics about each single converted file. The file system contains all the documents ($\approx$ 150 Gigabytes), classified by topic and each one located in its own subdirectory. The file system is exported via NFS to all hosts which take part in the build process.

To schedule conversion jobs, we operate the queue manager via the command line. A command like `php workqueue.php default cond-mat` will add all documents inside the `cond-mat` subdirectory — the ARXIV section for papers concerning *condensed matter* — to the current work queue, which is stored in the relational database.

The build manager is implemented in PHP, where SQL databases as well as process control functions can be easily used. It keeps an internal list of available hosts, reads the files to be converted next from the workqueue and distributes jobs to remote hosts. For each document that is to be converted the build manager forks off a new child process on the local machine. The child sets a timer to enable a limiting timeout of 180 s for the conversion process and then creates another child (the grandchild) which then calls the `make` on a host via remote `ssh` execution. The `make` process will then invoke `latexml` and `latexmlpost` to convert a TEX file to XML and XHTML. LaTeXML logs the inclusion of style files and also reports problems while converting from `latex` to `xml` to special log files.

After a timeout or the completion of the conversion process the build manager is notified via typical Unix signal handling. The build manager then parses and

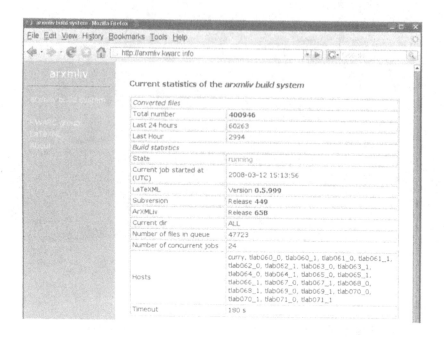

Fig. 2. The web interface shows the current state of the build system

analyzes log files, extracts information about the result state of conversion and collects the names of missing macros that are not yet supported. If the conversion has failed, the error message given by LaTeXML is also extracted. For each processed document the analyzed result data is then stored into the database for later use. With the stored result data it is also possible to instruct the queue manager to rerun specific documents which use a certain macro or to rerun all documents which resulted in a fatal error in the conversion process. The queue manager will then take care of removing the appropriate files and reset the status and add these files to the workqueue again. This has been especially useful when changes to the binding files have been applied or when an improved version of latexml becomes available.

Developers are able to retrieve the results via a web-interface which is available at http://arxmliv.kwarc.info. The main page describes the number of documents, the number of converted files in the last 24 hours and the current state of the build system. Since the binding files as well as LaTeXML are being developed in distributed environments, subversion is being used as a version control system. The current release version of the binding files and LaTeXML are also shown. Furthermore the active hosts that are currently being used for the conversion process is also displayed. For our experiment we have used 13 different hosts on 24 processors.

A further table gives detailed information about the results of the conversion process. The most important states are *success* where latexml has only issued some minor warnings, *missing macros*, where the conversion has been successfully completed, but some macro definitions could not be resolved. In this case the rendered layout may contain unexpected elements or not properly displayed elements. The status *fatal_error* is returned from the conversion process if there are too many unresolved macros or if some internal error condition during the LaTeXML conversion process has been triggered.

Conversion tex->xml

return value	count	%	marked for rerun
unknown	134	0.04	
no_latex	29718	n/a	1
missing_errlog	1128	n/a	
fatal_error	32501	8.78	87
timeout	23035	6.22	1
missing_macros	107244	28.98	13472
success	207186	55.98	

Fig. 3. The result status of converted documents

All these states are clickable and lead to a list of recently converted files with the specified status. The clickable file name leads to the source directory of the document where the document can be investigated in all its different representations, such as the TEX source, as an intermediate XML file that LaTeXML produces or as the XHTML+MATHML form. Also the full log file

Files that have given fatal error: *Unbalanced $ or } while ending group for @@close@inner@column*

No.	Date	Files	Errmsg
1	2008-03-11 07:17:53	/math/papers/0102185	Unbalanced $ or } while ending group for @@close@inner@column
2	2008-03-11 07:17:41	/cond-mat/papers/9909445	Unbalanced $ or } while ending group for @@close@inner@column
3	2008-03-11 07:17:27	/alg-geom/papers/9303003	Unbalanced $ or } while ending group for @@close@inner@column
4	2008-03-11 07:16:04	/alg-geom/papers/9508002	Unbalanced $ or } while ending group for @@close@inner@column
5	2008-03-11 07:15:39	/math/papers/0012161	Unbalanced $ or } while ending group for @@close@inner@column

Fig. 4. Documents that could not be successfully converted

containing detailed error messages can be easily be retrieved via the web browser.

The backend behind the web interface is also able to analyze the database content and create cumulated statistics. It applies some regular expressions to the error messages and clusters and cumulates these. By creating this information the backend is able to gather statistics such as a list of *Top Fatal Errors* and *Top Missing Macros* on-the-fly. Especially these two lists have proven to give valuable information not only the developer of the LaTeXML system but also to the implementers of binding files that are needed to support the conversion from LATEX to XML. With this information one can easily determine the most severe bugs in the still evolving conversion tool as well as determine the macros that are being used by many documents and that need further support.

The ARXIV articles use a total of more than 6000 different LATEX packages. Some of these style files are well known ones which are widely used, while other

Top Fatal Errors

No.	Count	Error Message
1	19873	Too many errors!
2	2576	Can't call method "currentColumn" on an undefined value at /soft/arXMLiv/dan/rep
3	2230	Missing $
4	1312	Unbalanced $ or } while ending mode inline_math for @@ENDINLINEMATH
5	1188	[Internal] T_PARAM[#] should never reach $tomach!
6	619	Can't locate object method "setAttribute" via package "XML::LibXML::DocumentFrag
7	401	Unbalanced $ or } while ending group for End
8	273	Missing { in sub/super-script argument
9	246	Unbalanced $ or } while ending mode text for endpicture
10	245	Unbalanced $ or } while ending mode display_math for end{equation}
11	233	Can't call method "newRow" on an undefined value at /soft/arXMLiv/dan/repos/arXiM
12	183	Unbalanced $ or } while ending group for @@close@inner@column
13	176	Input file appears to be binary:

Occurences of missing macros

No.	Macro	Count	In files
1	keywords	28254	found in files.
2	acknowledgements	13592	found in files.
3	institute	13067	found in files...
4	affil	10631	found in files..
5	inst	9568	found in files...
6	apj	9188	found in files..
7	altaffilmark	8506	found in files..
8	email	8355	found in files..
9	offprints	8345	found in files..
10	references	8325	found in files..
11	acknowledgments	7157	found in files..
12	titlerunning	6526	found in files...
13	mnras	6422	found in files...
14	altaffiltext	6386	found in files..
15	thesection@ID	6302	found in files..

Fig. 5. Lists of Top fatal errors and of macros that are currently not supported

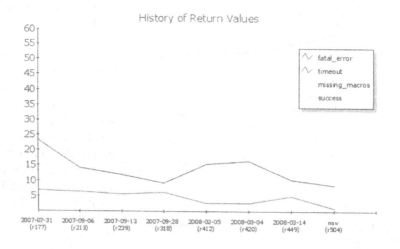

Fig. 6. The history of return values in our conversion experiment

are private enhancements which are used only once or very few times. While same macro names for different things are not a problem, since the binding files are created for each LaTeX package, there might be a problem that authors add private additions to well-known style files. We have chosen to ignore this problem since it is statistically insignificant.

With these statistics we have been able to focus on the most important macros and have been able to improve the success rate to now more than 58%. Although another 29% of the documents have also been successfully converted and are available as XHTML+MathML, we do not currently count them as full successes since support for some macros is still lacking and the layout that is rendered in a web browser might not be fully appropriate.

4   Conclusion and Outlook

By using the LaTeXML tool with our ARXMLIV build system to support the conversion process of large document collections, we have been able to successfully convert more than half of the more than 400,000 scientific articles of the ARXIV written in LaTeX to a semantically enriched XHTML+MATHML representation. We have been able to expand our collection of scientific MATHML documents which we need for further studies by more than 200,000 (real-world) documents. The build system has enabled us to cope with the conversion process of this huge collection of documents and helped us to improve our binding files that are needed to support various style files. The statistics the build system gathers have also been valuable contribution to the developer of LaTeXML since they clearly point to bugs and give hints where to enhance the software.

Although still under development, LaTeXML has shown to be a very promising tool to convert LaTeX documents to XML and hence XHTML+MATHML. Many existing LaTeX documents can already be fully converted to an XHTML+MATHML representation which may then be nicely rendered inside a browser. The possibility to render and fully integrate these documents inside a browser will enable us to add features on top of existing articles and offer added-value services which we might not even think of yet.

Up to now, the work in the ARXMLIV project has focused on driving up the coverage of the translation process and build a tool set that allows us to handle large corpora. With a conversion rate of over 80% we consider this phase as complete. We are currently working to acquire additional corpora, e.g. Zentralblatt Math[2] [ZBM07].

The next steps in the analysis of the ARXIV corpus will be to improve the LaTeXML post-processing, and in particular the OPENMATH/MATHML generation. Note that most of the contents in LaTeX documents are presentational in nature, so that content markup generation must be heuristic or based on linguistic and semantic analysis. Rather than relying on a single tool like the latexmlpost processor for this task we plan to open up the build system and compute farm to competing analysis tools. These get access to our corpus and we collect the results in a analysis database, which will be open to external tools for higher-level analysis tasks or end-user MKM services like semantic search [Mat07]. For this, we will need to generalize many of the build system features that are currently hard-wired to the translation task and the LaTeXML system. We also plan to introduce facilities for *ground-truthing* (i.e. for establishing the intended semantics of parts of the corpus, so that linguistic analysis can be trained on this). For the ARXIV corpus this will mean that we add feedback features to the generated XHTML+MATHML that allow authors to comment on the generation and thus the ARXMLIV developers to correct their LaTeXML bindings.

---

[2] First tests show that due to the careful editorial structure of this collection and the limited set of macros that need to be supported, our system can reach nearly perfect translation rates.

To allow manual tests for the developers, the build system also includes an additional interface (available at `http://tex2xml.kwarc.info`) where LaTeX files can be manually uploaded and then converted. It allows to test the conversion without the need to install `LaTeXML` and also makes use of the many additional binding files that we have created to support additional style files and are not part of the standard `LaTeXML` distribution. This interface may also be used to convert private LaTeX files to XHTML+MATHML, but because of limited resources only few users can concurrently use this system.

The build system itself is open source software and can be obtained from the authors upon request.

References

[Ang07]   Anghelache, R.: Hermes - a semantic xml+mathml+unicode e-publishing/self-archiving tool for latex authored scientific articles (2007), `http://hermes.roua.org/`

[ArX07]   arXiv.org e-Print archive (December, 2007), `http://www.arxiv.org`

[Koh08]   Kohlhase, M.: sTeX: Using TeX/LaTeX as a semantic markup format. Mathematics in Computer Science; Special Issue on Management of Mathematical Knowledge (accepted, 2008)

[Mat07]   Math Web Search (June 2007), `http://kwarc.info/projects/mws/`

[Mil07]   Miller, B.: LaTeXML: A LaTeXto xml converter. Web Manual (September 2007), `http://dlmf.nist.gov/LaTeXML/`

[vdBS03] van den Brand, M., Stuber, J.: Extracting mathematical semantics from latex documents. In: Bry, F., Henze, N., Małuszyński, J. (eds.) PPSWR 2003. LNCS, vol. 2901, pp. 160–173. Springer, Heidelberg (2003)

[ZBM07]   Zentralblatt MATH (December 2007), `http://www.zentralblatt-math.org`

On Correctness of Mathematical Texts from a Logical and Practical Point of View

Konstantin Verchinine[1], Alexander Lyaletski[2],
Andrei Paskevich[2], and Anatoly Anisimov[2]

[1] Université Paris 12, IUT Sénart/Fontainebleau,
77300 Fontainebleau, France
[2] Kyiv National Taras Shevchenko University, Faculty of Cybernetics,
03680 Kyiv, Ukraine*

Abstract. Formalizing mathematical argument is a fascinating activity in itself and (we hope!) also bears important practical applications. While traditional proof theory investigates deducibility of an individual statement from a collection of premises, a mathematical proof, with its structure and continuity, can hardly be presented as a single sequent or a set of logical formulas. What is called "mathematical text", as used in mathematical practice through the ages, seems to be more appropriate. However, no commonly adopted formal notion of mathematical text has emerged so far.

In this paper, we propose a formalism which aims to reflect natural (human) style and structure of mathematical argument, yet to be appropriate for automated processing: principally, verification of its correctness (we consciously use the word rather than "soundness" or "validity").

We consider mathematical texts that are formalized in the ForTheL language (brief description of which is also given) and we formulate a point of view on what a correct mathematical text might be. Logical notion of correctness is formalized with the help of a calculus. Practically, these ideas, methods and algorithms are implemented in a proof assistant called SAD. We give a short description of SAD and a series of examples showing what can be done with it.

1 Introduction

The question in the title of the paper is one to which we would like to get an answer formally. What we need to this aim is: a formal language to write down texts, a formal notion of correctness, a formal reasoning facility. And no matter what the content of the text in question is.

The idea to use a formal language along with formal symbolic manipulations to solve complex "common" problems, already appeared in G.W. Leibniz's writings (1685). The idea seemed to obtain more realistic status only in the early sixties

---

* This research is being supported by the INTAS project 05-1000008-8144. Some of its parts were performed within the scope of the project M/108-2007 in the framework of the Ukrainian-French Programme "Dnipro".

S. Autexier et al. (Eds.): AISC/Calculemus/MKM 2008, LNAI 5144, pp. 583–598, 2008.
© Springer-Verlag Berlin Heidelberg 2008

of the last century when first theorem proving programs were created [1]. It is worth noting how ambitious was the title of Wang's article! Numerous attempts to "mechanize" mathematics led to less ambitious and more realistic idea of "computer aided" mathematics as well as to the notion of "proof assistant" — a piece of software that is able to do some more or less complex deductions for you. Usually one has in mind either long but routine inferences or a kind of case analysis with enormously large number of possible cases. Both situations are embarrassing and "fault intolerant" for humans.

Mathematical text is not a simple sequence of statements, neither a linear representation of a sequent tree, nor a λ-term coding a proof. It is a complex object that contains axioms, definitions, theorems, and proofs of various kinds (by contradiction, by induction, by case analysis, etc). What its "correctness" might stand for? The formal semantics of a text can be given by packing the whole text in a single statement and considering the corresponding logical formula (which we may call the *formula image* of the text). Then the text is declared correct whenever its formula image is deducible in the underlying logic. The approach is simple, theoretically transparent but absolutely impracticable, e.g. the precise notion of correctness obtained in this way can hardly be considered as a formal specification of a proof assistant we would like to implement. That's why we develop a specific notion of text correctness that, though being less straightforward, can be formalized with the help of a logical calculus on one hand and can serve as a formal specification on the other.

Our approach to mathematical text correctness is implemented in the proof assistant called SAD (System for Automated Deduction). The SAD project is the continuation of a project initiated by academician V. Glushkov at the Institute for Cybernetics in Kiev more than 30 years ago [2]. The title of the original project was "Evidence Algorithm" and its goal was to help a working mathematician to verify long and tiresome but routine reasonings. To implement that idea, three main components had to be developed: an inference engine (we call it *prover* below) that implements the basic level of evidence, an extensible collection of tools (we call it *reasoner*) to reinforce the basic engine, and a formal input language which must be close to natural mathematical language and easy to use. Today, a working version of the SAD system exists [3,4,5] and is available online at http://nevidal.org.ua.

What is the place of SAD in the world of proof assistants w.r.t. proof representation style? Actually, we observe four major approaches to formal presentation of a mathematical proof (see also [6] for an interesting and detailed comparison).

Interactive proof assistants, such as Coq [7], Isabelle [8], PVS [9], or HOL derivatives [10], work with "imperative" proofs, series of tactic invocations.

Systems based on the Curry-Howard isomorphism, such as de Bruijn's Automath [11] or Coq, consider a proof of a statement as a lambda term inhabiting a type that corresponds to the statement. Since writing such a proof directly is difficult and time-consuming, modern systems of the kind let user build a proof in an interactive tactic-based fashion and construct the final proof term automatically.

The third branch deals with "declarative" proofs, which are structured collections of hypotheses, conjectures and claims expressed in the same language as the axioms and theorems themselves. The Mizar system [12] is the oldest and most known proof assistant working with proofs of declarative style. Isabelle, with introduction of Isar [13], accepts declarative proofs, too. Declarative proof presentation is employed and thoroughly studied in the works on Mathematical Vernacular started by N. de Bruijn [11] and later extended to Weak Type Theory [14,15] and MathLang [16]. In particular, MathLang and the ForTheL language, presented below, share a lot of similar traits owed to the common striving for a natural-like formal mathematical language.

Finally, there are systems that do not use user-given proofs and rely instead on proof generation methods: planning, rewriting, or inference search facilities to deduce each claim from premises and previously proved statements. The systems ACL2 (successor to Nqthm) [17], λCLAM [18], Theorema [19] and any classical automated prover (e.g. Otter) can be considered as proof assistants of the kind.

In a general setting, SAD may be positioned as a declarative style proof verifier that accepts input texts written in the special formal language ForTheL [20,4], uses an automated first-order prover as the basic inference engine and possesses an original reasoner (which includes, in particular, a powerful method of definition expansion).

The rest of the paper is organized as follows. In Section 2, we briefly describe the ForTheL language and write a ForTheL proof of Tarski's fixed point theorem. We define correctness of a ForTheL text with the help of a logical calculus in Section 3. We illustrate this calculus by verifying a simple text in Section 4. We conclude with a brief list of experiments on formalization performed in SAD.

2   ForTheL Language

Like any usual mathematical text, a ForTheL text consists of definitions, assumptions, affirmations, theorems, proofs, etc. Figure 1 gives an idea of what a ForTheL text looks like.

The syntax of a ForTheL sentence follows the rules of English grammar. Sentences are built of units: statements, predicates, notions (that denote classes of objects) and terms (that denote individual entities). Units are composed of syntactical primitives: nouns which form notions (e.g. "subset of") or terms ("closure of"), verbs and adjectives which form predicates ("belongs to", "compact"), symbolic primitives that use a concise symbolic notation for predicates and functions and allow to consider usual quantifier-free first-order formulas as ForTheL statements. Of course, just a little fragment of English is formalized in the syntax of ForTheL.

There are three kinds of sentences in the ForTheL language: assumptions, selections, and affirmations. Assumptions serve to declare variables or to provide some hypotheses for the following text. For example, the following sentences are typical assumptions: "Let S be a finite set.", "Assume that m is greater than n.". Selections state the existence of representatives of notions and can be used to declare variables, too. Here follows an example of a selection: "Take an even prime number X.". Finally, affirmations are simply statements: "If p divides n - p then p divides n.". The semantics of a sentence is determined by a series of transformations that convert a ForTheL statement to a first-order formula, so called *formula image*. For example, the formula image of the statement "all closed subsets of any compact set are compact" is:

∀A ((A is a set ∧ A is compact) ⊃

∀B ((B is a subset of A ∧ B is closed) ⊃ B is compact))

The sections of ForTheL are: sentences, sentences with proofs, cases, and top-level sections: axioms, definitions, signature extensions, lemmas, and theorems. A top-level section is a sequence of assumptions concluded by an affirmation. Proofs attached to affirmations and selections are simply sequences of low-level sections. A case section begins with a special assumption called *case hypothesis* which is followed by a sequence of low-level sections (the "proof" of a case).

Any section A or sequence of sections Δ has a formula image, denoted $|A|$ or, respectively, $|\Delta|$. The image of a sentence with a proof is the same as the image of that sentence taken without proof. The image of a case section is the implication ($H \supset$ thesis), where H is the formula image of the case hypothesis and thesis is a placeholder for the statement being proved (see Section 3). The formula image of a top-level section is simply the image of the corresponding sequence of sentences.

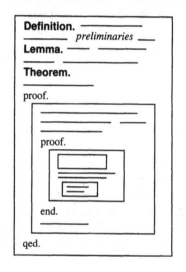

Fig. 1. ForTheL text's structure

The formula image of a sequence of sections A, Δ is an existentially quantified conjunction $\exists x_A(|A| \wedge |\Delta|)$, whenever A is a conclusion (affirmation, selection, case section, lemma, theorem); or a universally quantified implication $\forall x_A(|A| \supset |\Delta|)$, whenever A is a hypothesis (assumption, axiom, definition, signature extension). Here, x_A denotes the set of variables declared in A and can only be non-empty when A is an assumption or a selection. This set depends on the logical context of A, since any variable which is declared above A can not be redeclared in A. The formula image of the empty sequence is ⊤, the truth.

In this syntax, we can express various proof schemes like proof by contradiction, by case analysis, and by general induction. The last scheme merits special consideration. Whenever an affirmation is marked to be proved by induction, the system constructs an appropriate induction hypothesis and inserts it into the statement to be verified. The induction hypothesis mentions a binary relation which is declared to be a well-founded ordering, hence, suitable for induction proofs. Note that we cannot express the very property of well-foundness in ForTheL (since it is essentially a first-order language), so that the correctness of this declaration is unverifiable and we take it for granted. After that transformation, the proof and the transformed statement can be verified in a first-order setting, and the reasoner of SAD has no need in specific means to build induction proofs. The semantics of this and other proof schemes is considered in more detail in the next section.

Is ForTheL practical as a formalization language? Our numerous experiments show that rather often a ForTheL text is sufficiently close to its hand-made prototype. Consider for example an excerpt of a verified formalization of the Tarski's fixed point theorem:

```
Definition DefCLat. A complete lattice is a set S such that
    every subset of S has an infimum in S and a supremum in S.

Definition DefMono. f is monotone iff for all x,y << Dom f
    x <= y  =>  f(x) <= f(y).

Theorem Tarski.
    Let U be a complete lattice and f be an monotone function on U.
    Let S be the set of fixed points of f.
    S is a complete lattice.
Proof.
    Let T be a subset of S.
    Let us show that T has a supremum in S.
      Take P = { x << U | f(x) <= x and x is an upper bound of T in U }.
      Take an infimum p of P in U.
      f(p) is a lower bound of P in U and an upper bound of T in U.
      Hence p is a fixed point of f and a supremum of T in S.
    end.
    Let us show that T has an infimum in S.
      Take Q = { x << U | x <= f(x) and x is a lower bound of T in U }.
      Take a supremum q of Q in U.
      f(q) is an upper bound of Q in U and a lower bound of T in U.
      Hence q is a fixed point of f and an infimum of T in S.
    end.
  qed.
```

3   Text Correctness

We distinguish two types of correctness of a well-formed ForTheL text: ontological and logical.

Ontological correctness means that the text in question contains no occurrence of a symbol (constant, function, notion or relation) that comes from nowhere. First, every symbol must be either a signature symbol or be introduced by a definition. Second, in every occurrence of a symbol, the arguments, if any, must satisfy the guards of the corresponding definition or signature extension. Since ForTheL is a one-sorted and untyped language, these guards can be arbitrary logical formulas. Therefore, the latter condition cannot be checked by purely syntactical means nor by type inference of any kind. Instead, it requires proving statements about terms inside complex formulas, possibly, under quantifiers. Such reasoning can be performed in a sound way using the notion of local images [21].

Ontological correctness is to ForTheL what type correctness is to typed languages. It allows early detection of formalization errors which otherwise could hardly be detected. Indeed, accidental ontological incorrectness most often implies logical incorrectness. However, it is much harder to trace a failure log of a prover back to an invalid occurrence than to discover it in the first place. Also, during ontological verification we obtain some important knowledge about the text which will be used later in logical verification.

$$\frac{x = \mathcal{DV}_\Gamma(F) \qquad \Gamma \vdash \forall x\,(F \supset G') \supset G \qquad \Gamma, (\texttt{assume}\ F) \triangleright_{G'} \Delta}{\Gamma \ \triangleright_G \ (\texttt{assume}\ \Theta_G(F)), \Delta}$$

$$\frac{\mathcal{DV}_\Gamma(F) = \varnothing \qquad \Gamma \triangleright_F \Lambda \qquad \Gamma \vdash (F \wedge G') \supset G \qquad \Gamma, (\texttt{affirm}\ F\ [\Lambda]) \triangleright_{G'} \Delta}{\Gamma \ \triangleright_G \ (\texttt{affirm}\ \Theta_G(F)\ [\Lambda]), \Delta}$$

$$\frac{x = \mathcal{DV}_\Gamma(F) \qquad \Gamma \triangleright_{\exists x\,F} \Lambda \qquad \Gamma \vdash \exists x\,(F \wedge G') \supset G \qquad \Gamma, (\texttt{select}\ F\ [\Lambda]) \triangleright_{G'} \Delta}{\Gamma \ \triangleright_G \ (\texttt{select}\ \Theta_G(F)\ [\Lambda]), \Delta}$$

$$\frac{\mathcal{DV}_\Gamma(F) = \varnothing \qquad \Gamma, (\texttt{assume}\ F) \triangleright_G \Lambda \qquad \Gamma, (\texttt{case}\ (F \supset G)\ [\Lambda]) \triangleright_{G \vee F} \Delta}{\Gamma \ \triangleright_G \ (\texttt{case}\ (F \supset \texttt{thesis})\ [\Lambda]), \Delta}$$

$$\frac{\Gamma \triangleright_{\mathrm{IT}_t^\prec(G)} \Delta}{\Gamma \ \triangleright_G \ \Delta} \qquad\qquad \frac{\mathcal{DV}_{\Gamma,\Lambda}(\mathrm{IH}_t^\prec(G)) = \varnothing \qquad \Gamma \triangleright_{\mathrm{IT}_t^\prec(G)} \Lambda, (\texttt{assume}\ \mathrm{IH}_t^\prec(G)), \Delta}{\Gamma \ \triangleright_G \ \Lambda, \Delta}$$

$$\frac{\Gamma \vdash G}{\Gamma \ \triangleright_G} \qquad \frac{\Gamma \triangleright_\top \Lambda \qquad \Gamma, (\texttt{toplevel}\ |\Lambda|\ [\Lambda]) \triangleright_\top \Delta}{\Gamma \ \triangleright_\top \ (\texttt{toplevel}\ |\Lambda|\ [\Lambda]), \Delta} \qquad \frac{\Gamma, (\texttt{posit}\ F) \triangleright_\top \Delta}{\Gamma \ \triangleright_\top \ (\texttt{posit}\ F), \Delta}$$

Fig. 2. Calculus of Correctness **CTC**

Logical correctness is imposed on particular affirmations in the text: theorems, lemmas, intermediate statements in proofs. Any such affirmation must be deducible from its logical predecessors.

In what follows, a ForTheL section A will be considered as a triple $(T\,|A|\,[\Lambda])$, where T denotes the section type, $|A|$ is the formula image of A, and Λ is the sequence of subsections of A, if any. The type of A can be of the following: **toplevel** for any top-level section (axiom, definition, signature extension, theorem, lemma), **case** for a case section, **assume** for an assumption, **select** for a selection, **affirm** for an affirmation, **posit** for a postulate. Several remarks should be made here. A sentence with a supplied proof is considered to be of the same type as the same sentence without proof. They differ only in the third component of the triple, the list of subsections, which is empty for a sentence without proof. Recall that the formula image of a sentence does not depend on presence of a proof, too. In a case section, the case hypothesis belongs to the formula image of the section and does not appear among its subsections. A postulate is an affirmation at the end of an axiom, definition, or signature extension; in other words, an affirmation which is not meant to be proved.

A formula F is *logically correct* in view of a sequence of sections Γ (the logical context of F), denoted $\Gamma \vdash F$, whenever F can be deduced in the classical first-order predicate calculus from the formula images of sections from Γ.

Logical correctness of a ForTheL text is deduced from logical correctness of particular formulas in view of appropriate sets of premises according to the Calculus of Text Correctness, or **CTC**, given in Figure 2.

In **CTC**, we infer sequents of the form $\Gamma \triangleright_G \Delta$. Only those sequents are allowed where every free variable of G occurs free in Γ ($\mathcal{FV}(G) \subseteq \mathcal{FV}(\Gamma)$), and neither Γ nor G contain occurrences of thesis.

In such a sequent, Δ is a sequence of sections whose correctness is being verified and Γ is a sequence of sections that logically precede Δ. The formula G is a *current thesis*: a formula which we want to deduce from Γ with the help of auxiliary reasoning in Δ (note the rule $\frac{\Gamma \vdash G}{\Gamma \triangleright_G}$). Verification consists in counter-applying the rules of the calculus, reducing the sequent $\Gamma \triangleright_G \Delta$ to $\vdash$-premises which are to be checked directly.

A ForTheL text Δ is said to be *logically correct* whenever $\triangleright_\top \Delta$ can be inferred.

The expression $\Theta_G(F)$ denotes the formula F where some occurrences of G are replaced with thesis. There may be several $\Theta_G(F)$ for given F and G. One can consider $\Theta_G(F)$ as an abbreviated form of F; when we counter-apply the rules of **CTC** during verification, we pass from $\Theta_G(F)$ back to F, i.e. we expand the abbreviation and eliminate the placeholder thesis.

The expression $\mathcal{DV}_\Gamma(F)$ stands for the set of variables which are declared in the formula F in view of Γ. Basically, that means that x does not occur freely in Γ and F "says" that x belongs to some class described by a notion (like in "x is a fixed point of f"). In a similar fashion, we define $\mathcal{DV}_\Gamma(\mathbb{A})$ and $\mathcal{DV}_\Gamma(\Delta)$ (with the proviso that only assumptions and selections can declare variables). Recall that the formula image of a sequence of sections, $|\Delta|$ actually depends on $\mathcal{DV}_\Gamma(\Delta)$ and, hence, on Γ. Note that any free variable in a well-formed text must be declared either in that very sentence or somewhere above, so that $\mathcal{DV}_\Gamma(\mathbb{A})$ contains those and only those free variables of $\mathbb{A}$ which do not occur free in Γ: $\mathcal{DV}_\Gamma(\mathbb{A}) = \mathcal{FV}(\mathbb{A}) \backslash \mathcal{FV}(\Gamma)$.

The expressions $\mathrm{IT}_t^\prec(G)$ and $\mathrm{IH}_t^\prec(G)$ stand for the *induction thesis* and *induction hypothesis*, respectively. They are defined as follows. For a given formula G of the form $\forall \boldsymbol{x}_1 (H_1 \supset \forall \boldsymbol{x}_2 (H_2 \supset \ldots \forall \boldsymbol{x}_n (H_n \supset F) \ldots))$, an arbitrary term t, and a binary relation symbol $\prec$:

$$\mathrm{IH}_t^\prec(G) = \forall \boldsymbol{x}_1' (H_1 \sigma \supset \forall \boldsymbol{x}_2' (H_2 \sigma \supset \ldots \forall \boldsymbol{x}_n' (H_n \sigma \supset (t\sigma \prec t \supset F\sigma)) \ldots))$$
$$\mathrm{IT}_t^\prec(G) = \forall \boldsymbol{x}_1 (H_1 \supset \forall \boldsymbol{x}_2 (H_2 \supset \ldots \forall \boldsymbol{x}_n (H_n \supset (\mathrm{IH}_t^\prec(G) \supset F)) \ldots))$$

where σ is the renaming substitution $[\boldsymbol{x}_1'/\boldsymbol{x}_1, \boldsymbol{x}_2'/\boldsymbol{x}_2, \ldots, \boldsymbol{x}_n'/\boldsymbol{x}_n]$ and $\boldsymbol{x}_1', \boldsymbol{x}_2', \ldots,$ $\boldsymbol{x}_n'$ are some fresh variables.

The induction thesis $IT_t^\prec(G)$ is equivalent to the original thesis G on condition that $\prec$ denotes a well-founded ordering. Note that the well-foundness of $\prec$ cannot be finitely expressed in a first-order language and the calculus **CTC** takes it on trust. In other words, correctness of a ForTheL text is verified assuming the following axiom scheme of general induction: $\mathcal{I}nd = (IT_t^\prec(G) \supset G)$, where **G** and **t** are placeholders for a formula and a term, respectively.

Note that $IT_t^\prec(G)$ is equivalent to G if $\prec$ is the always false relation. Therefore the extension of first-order logic with the symbol $\prec$ and the axiom scheme $\mathcal{I}nd$ is conservative.

The first induction handling inference rule of **CTC** says that by proving the induction thesis, we automatically prove the initial one. In counter-application, it means that the verifier has the right to substitute the appropriate induction thesis for the initial thesis, when verifying a proof by induction. The second induction handling rule says additionally that the induction hypothesis need not to be put explicitly in the proof, but can be silently inserted there by the verifier. However, the induction hypothesis can not appear in the proof before all the free variables in it are declared.

Case section handling is another rule where an implicit logical predecessor, namely, the case hypothesis, is added by the verifier (in counter-application). Note that the Θ operation is not applied to a case hypothesis in the conclusion of the rule. That means that the word **thesis** can not appear in a ForTheL case hypothesis sentence, or it will not be verified.

Thesis handling. In the rules of **CTC** for assumptions, selections, and affirmations, we see $\vdash$-premises that relate a current thesis G to a new thesis G' (by "new", we mean that G' is used as the thesis for subsequent ForTheL proof sequence). Such a transformation of thesis reflects our perception of a proof development when a complex formula is being demonstrated.

For instance, whenever we want to prove a conjunction $F \wedge G$ and succeed to derive one part of it, say F, and write down the affirmation of F in the text, then the thesis can be reduced just to G. Furthermore, if we prove a universal statement about sets $\forall x\,(x$ **is a set** $\supset F)$ then we can begin by an assumption declaring x a set, thus reducing the thesis to F.

When we see such a connection between the current thesis and a sentence under consideration, we call that sentence *motivated*. Motivated affirmations, selections, assumptions allow to reduce the thesis to a new formula which would be probably simpler to prove. Sometimes, the connection is evident from the syntax, e.g. when the thesis is an implication and the assumption is the antecedent formula. Sometimes, the connection depends on several reasoning steps: for example, if variables S, T have been declared as sets and the current thesis is "S **is a subset of** T", then the assumption "**let x be an element of** S" is motivated and reduces the current thesis to "x **is an element of** T".

There is nothing special in non-motivated affirmations or selections. Whenever we meet such a sentence, we simply do not change the thesis. On the contrary, in a well-written mathematical proof, assumptions should be always motivated, i.e. "suggested" by a current thesis. A non-motivated assumption is an unjustified narrowing of the search space. Also, it may happen that our reasoning capabilities are too weak to discover the justification.

Now, how can we infer $\Gamma \rhd_G$ (assume F), Δ, if F is in no visible relation with G? Though the calculus **CTC** admits various solutions, in our implementation, the choice of the new thesis is guided by the form of Δ. Whenever the formula images of sentences in Δ contain occurrences of thesis (e.g. when there are case sections), we suppose that the proof of G continues under the non-motivated assumption and leave the thesis unchanged:

$$\frac{x = \mathcal{DV}_\Gamma(F) \qquad \Gamma \vdash \forall x \, (F \supset G) \supset G \qquad \Gamma, (\text{assume } F) \rhd_G \Delta}{\Gamma \, \rhd_G \, (\text{assume } \Theta_G(F)), \, \Delta}$$

The premise $\Gamma \vdash \forall x \, (F \supset G) \supset G$ is nontrivial: its is equivalent to the disjunction $G \vee (\exists x \, F)$. Recall that the free variables of G are all declared in Γ and thus cannot be among x.

If thesis does not occur in the formula images in Δ, we suppose that the rest of the proof is a sort of independent argument which should be considered by itself. Therefore we take for the new thesis the image of the whole rest of the proof sequence. The inference is as follows:

$$\frac{x = \mathcal{DV}_\Gamma(F) \qquad \Gamma \vdash \forall x \, (F \supset |\Delta|) \supset G \qquad \Gamma, (\text{assume } F) \rhd_{|\Delta|} \Delta}{\Gamma \, \rhd_G \, (\text{assume } \Theta_G(F)), \, \Delta}$$

Note that the new variables in Δ, not known from Γ or F, are all bound in $|\Delta|$.

Each assumption in Δ is an antecedent in the new thesis $|\Delta|$ and therefore will be considered as motivated. Each affirmation or selection in Δ will reduce the thesis, too, so that at the end of Δ the thesis will be simply $\top$. The premise $\Gamma \vdash \forall x \, (F \supset |\Delta|) \supset G$ finishes the demonstration by deducing G from the formula image of the proof sequence (assume F), Δ.

Altogether, the following theorem can be seen as the statement of soundness of **CTC**:

Theorem 1. *Let Γ and Δ be arbitrary sequences of ForTheL sections and G, an arbitrary formula. If $\Gamma \rhd_G \Delta$ can be inferred in* **CTC** *then $\mathcal{I}nd, \Gamma \vdash G$.*

Proof. The claim can be proved by induction on the number of steps in the inference of $\Gamma \rhd_G \Delta$. Let us consider the last inference step. If it is made by a rule with $\rhd_\top$ in conclusion, then G is $\top$, and the claim is trivial. Otherwise, we have seven cases to consider. We will denote the cases by the form of the conclusion of the corresponding inference rule.

Case $\Gamma \rhd_G$. The premise of this rule is $\Gamma \vdash G$, hence the claim.

Case $\Gamma \rhd_G$ (assume $\Theta_G(F)$), Δ. By the premises of the rule, we have $\Gamma \vdash \forall x \, (F \supset G') \supset G$ and $\Gamma, (\text{assume} F) \rhd_{G'} \Delta$, where $x = \mathcal{DV}_\Gamma(F) = \mathcal{FV}(F) \backslash \mathcal{FV}(\Gamma)$. By the induction hypothesis, the latter implies $\mathcal{I}nd, \Gamma, F \vdash G'$. Also, $\mathcal{FV}(G) \subseteq \mathcal{FV}(\Gamma)$ and $\mathcal{FV}(G') \subseteq \mathcal{FV}(\Gamma) \cup \mathcal{FV}(F)$. Therefore, $\mathcal{FV}(F \supset G') \backslash \mathcal{FV}(\Gamma) = x$. Hence $\mathcal{I}nd, \Gamma \vdash \forall x \, (F \supset G')$ and we have the claim.

Case $\Gamma \rhd_G$ (affirm $\Theta_G(F) \, [\Lambda]$), Δ. This is subsumed by the next case.

Case $\Gamma \rhd_G (\mathtt{select}\, \Theta_G(F)\,[\Lambda]), \Delta$. We have $\Gamma \rhd_{\exists \boldsymbol{x}\, F} \Lambda$ and $\Gamma \vdash \exists \boldsymbol{x}\, (F \wedge G') \supset G$, and $\Gamma, (\mathtt{select}\, F\, [\Lambda]) \rhd_{G'} \Delta$, where $\boldsymbol{x} = \mathcal{FV}(F) \backslash \mathcal{FV}(\Gamma)$. By the induction hypothesis, we have $\mathcal{I}nd, \Gamma \vdash \exists \boldsymbol{x}\, F$ and $\mathcal{I}nd, \Gamma, F \vdash G'$. Also, $\mathcal{FV}(G) \subseteq \mathcal{FV}(\Gamma)$ and $\mathcal{FV}(G') \subseteq \mathcal{FV}(\Gamma) \cup \mathcal{FV}(F)$. Hence $\mathcal{FV}(F \wedge G') \backslash \mathcal{FV}(\Gamma) = \boldsymbol{x}$ and $\mathcal{I}nd, \Gamma \vdash \forall \boldsymbol{x}\, (F \supset G')$. This implies $\mathcal{I}nd, \Gamma \vdash \exists \boldsymbol{x}\, (F \wedge G')$ and we have the claim.

Case $\Gamma\ \rhd_G\ (\mathtt{case}\ (F \supset \mathtt{thesis})\ [\Lambda]), \Delta$. From the premises, we obtain $\Gamma, (\mathtt{assume}\ F) \rhd_G \Lambda$ and $\Gamma, (\mathtt{case}\ (F \supset G)\ [\Lambda]) \rhd_{G \vee F} \Delta$. Also, $\mathcal{DV}_\Gamma(F) = \varnothing$ which means that $\mathcal{FV}(F), \mathcal{FV}(G) \subseteq \mathcal{FV}(\Gamma)$. By the induction hypothesis, we have $\mathcal{I}nd, \Gamma, F \vdash G$ and $\mathcal{I}nd, \Gamma, (F \supset G) \vdash (G \vee F)$. The former gives $\mathcal{I}nd, \Gamma \vdash (F \supset G)$. The latter gives $\mathcal{I}nd, \Gamma, (F \supset G) \vdash G$ and we have the claim.

Cases $\Gamma \rhd_G \Delta$ and $\Gamma \rhd_G \Delta, \Lambda$ (induction handling rules). By the premise of the rule and the induction hypothesis, we have $\mathcal{I}nd, \Gamma \vdash \mathrm{IT}_t^{\rightharpoonup}(G)$. By definition of $\mathcal{I}nd$, we have the claim. $\square$

4   Verification Example

Let us consider an example of a well-formed ForTheL text which, while being simple, contains proofs by case analysis and by induction (the symbol -<- below denotes a well-founded binary relation).

```
[number/numbers]        # let the parser know it is the same word

Signature Nat.                                          # 0
    A natural number is a notion.                       # 0.0
Signature Zer.                                          # 1
    0 is a natural number.                              # 1.0
Signature Suc.                                          # 2
    Let i be a natural number.                          # 2.0
    succ i is a natural number.                         # 2.1
Signature Add.                                          # 3
    Let i,j be natural numbers.                         # 3.0
    i + j is a natural number.                          # 3.1
Signature Ord.                                          # 4
    Let i,j be natural numbers.                         # 4.0
    i -<- j is an atom.                                 # 4.1

Axiom ZerSuc.                                           # 5
    For any natural number i if i != 0 then             # 5.0
    there exists a natural number j such that succ j = i.
Axiom AddZer.                                           # 6
    For any natural number i (i + 0 = i).               # 6.0
Axiom AddSuc.                                           # 7
    For all natural numbers i,j (i + succ j = succ (i+j)). # 7.0
Axiom OrdSuc.                                           # 8
    For any natural number i (i -<- succ i).            # 8.0

Lemma ZerAdd.                                           # 9
    For any natural number i (0 + i = i).               # 9.0
```

```
Proof by induction.
   Let i be a natural number.                        # 9.0.0
   Case i = 0.                                        # 9.0.1
   obvious.
   Case i != 0.                                       # 9.0.2
      Take a natural number j such that succ j = i.   # 9.0.2.0
      We have j -<- i.                                # 9.0.2.1
      Hence 0 + j = j.                                # 9.0.2.2
      Then we have the thesis.                        # 9.0.2.3
   end.
qed.
```

Note the numerical indexes in the comments. Each index denotes a position of a particular ForTheL section in the text. For example, 9.0 is the position of the main affirmation in the lemma ZerAdd, 9.0.0 is the position of the starting assumption in the proof, 9.0.1 and 9.0.2 point at the case sections.

Let us reconsider this text with the formula images in place of ForTheL sentences. In what follows, $t \, \varepsilon \, \text{NatNum}$ stands for "t is a natural number".

```
toplevel Nat
   posit  ∀x (x ε NatNum ⊃ ⊤)
toplevel Zer
   posit  ∀x (x ≈ 0 ⊃ x ε NatNum)
toplevel Suc
   assume  i ε NatNum
   posit  ∀x (x ≈ succ i ⊃ x ε NatNum)
toplevel Add
   assume  i ε NatNum ∧ j ε NatNum
   posit  ∀x (x ≈ i + j ⊃ x ε NatNum)
toplevel Ord
   assume  i ε NatNum ∧ j ε NatNum
   posit  i ≺ j ⊃ ⊤
toplevel ZerSuc
   posit  ∀i (i ε NatNum ⊃ i ≠ 0 ⊃ ∃j (j ε NatNum ∧ succ j ≈ i))
toplevel AddZer
   posit  ∀i (i ε NatNum ⊃ i + 0 ≈ i)
toplevel AddSuc
   posit  ∀i (i ε NatNum ⊃ ∀j (j ε NatNum ⊃ i + (succ j) ≈ succ (i + j)))
toplevel OrdSuc
   posit  ∀i (i ε NatNum ⊃ i ≺ succ i)
toplevel ZerAdd
   affirm  ∀i (i ε NatNum ⊃ 0 + i ≈ i)
      assume  i ε NatNum
      case  i = 0 ⊃ thesis
      case  i ≠ 0 ⊃ thesis
         select  j ε NatNum ∧ (succ j) ≈ i
         affirm  j ≺ i
         affirm  0 + j ≈ j
         affirm  thesis
```

We are going to study the inference steps which prove correctness of the main lemma (the top level of the text is pretty trivial). In order to fit into the page width we will write position indexes in parentheses in place of the corresponding sections. We proceed in a bottom-top manner, moving from the desired conclusion to axioms.

$$\cfrac{\cfrac{\Gamma \triangleright_{|(9.0)|} (9.0.0), (9.0.1), (9.0.2)}{\Gamma \triangleright_G (9.0.0), \mathbb{A}, (9.0.1), (9.0.2)} \qquad \cfrac{\Gamma \vdash (|(9.0)| \wedge \mathsf{T}) \supset \mathsf{T} \qquad \Gamma, (9.0) \triangleright_\mathsf{T}}{\Gamma, (9.0) \vdash \mathsf{T}}}{\Gamma \triangleright_\mathsf{T} (9.0)}$$

where

$$\Gamma = (0), \ldots, (8)$$
$$\mathbb{A} = (\textbf{assume } H)$$
$$|(9.0)| = \forall i\, (i \, \varepsilon \, \mathrm{NatNum} \supset 0 + i \approx i)$$
$$G = \mathrm{IT}_i^{\prec}(|(9.0)|) = \forall i\, (i \, \varepsilon \, \mathrm{NatNum} \supset (H \supset 0 + i \approx i))$$
$$H = \mathrm{IH}_i^{\prec}(|(9.0)|) = \forall i'\, (i' \, \varepsilon \, \mathrm{NatNum} \supset (i' \prec i \supset 0 + i' \approx i'))$$

Note the fragment of inference where we apply the induction rule. Instead of proving the statement of the affirmation (9.0) as is, we descend into the proof with a weakened current thesis G having the additional induction hypothesis H.

We begin by inserting that induction hypothesis H into the proof. Note that the variable i which is free in H is declared in (9.0.0) and therefore known at the position of added hypothesis. Also note how the two assumptions reduce the current thesis from G to G' and then to G''.

$$\cfrac{\Gamma \triangleright_G (9.0.0), \mathbb{A}, (9.0.1), (9.0.2)}{\Gamma \vdash \forall i\, (i \, \varepsilon \, \mathrm{NatNum} \supset G') \supset G \qquad \Gamma, (9.0.0) \triangleright_{G'} \mathbb{A}, (9.0.1), (9.0.2)}$$

$$\cfrac{\Gamma, (9.0.0) \triangleright_{G'} \mathbb{A}, (9.0.1), (9.0.2)}{\Gamma, (9.0.0) \vdash (H \supset G'') \supset G' \qquad \Gamma, (9.0.0), \mathbb{A} \triangleright_{G''} (9.0.1), (9.0.2)}$$

where

$$G' = (H \supset 0 + i \approx i) \qquad\qquad G'' = (0 + i \approx i)$$

The first case section is very short. Note that **thesis** in the formula image of (9.0.1) is replaced with the actual thesis in (9.0.1)':

$$\cfrac{\cfrac{\Gamma, (9.0.0), \mathbb{A}, \mathbb{C}_1 \triangleright_{G''}}{\Gamma, (9.0.0), \mathbb{A}, \mathbb{C}_1 \vdash G''}}{\Gamma, i \, \varepsilon \, \mathrm{NatNum}, i \approx 0 \vdash 0 + i \approx i} \qquad \cfrac{\Gamma, (9.0.0), \mathbb{A} \triangleright_{G''} (9.0.1), (9.0.2)}{\Gamma, (9.0.0), \mathbb{A}, (9.0.1)' \triangleright_{G'''} (9.0.2)}$$

where

$$\mathbb{C}_1 = (\textbf{assume } i \approx 0) \quad (9.0.1)' = (\textbf{case } (i \approx 0 \supset G'')) \quad G''' = G'' \vee i \approx 0$$

The second case section is longer but no more complex:

$$\frac{\Delta_0 \rhd_{G'''} (\tau)}{\Delta_0, \mathbb{C}_2 \rhd_{G'''} (\tau.0), (\tau.1), (\tau.2), (\tau.3) \qquad \dfrac{\Delta_0, (\tau)' \; \rhd_{G'''\vee i \not\approx 0}}{\dfrac{\Delta_0, (\tau)' \vdash G''' \vee i \not\approx 0}{\vdash G'' \vee i \approx 0 \vee i \not\approx 0}}}$$

$$\frac{\Delta_1 \rhd_{G'''} (\tau.0), (\tau.1), (\tau.2), (\tau.3)}{\dfrac{\Delta_1 \rhd_{\exists j F_1}}{\Delta_1 \vdash \exists j F_1} \qquad \Delta_1 \vdash \exists j \, (F_1 \wedge G''') \supset G''' \qquad \Delta_1, (\tau.0) \rhd_{G'''} (\tau.1), (\tau.2), (\tau.3)}$$

$$\frac{\Delta_2 \rhd_{G'''} (\tau.1), (\tau.2), (\tau.3)}{\dfrac{\Delta_2 \rhd_{j \prec i}}{\Delta_2 \vdash j \prec i} \qquad \Delta_2 \vdash (j \prec i \wedge G''') \supset G''' \qquad \Delta_2, (\tau.1) \rhd_{G'''} (\tau.2), (\tau.3)}$$

$$\frac{\Delta_3 \rhd_{G'''} (\tau.2), (\tau.3)}{\dfrac{\Delta_3 \rhd_{0+j \approx j}}{\Delta_3 \vdash 0+j \approx j} \qquad \Delta_3 \vdash (0+j \approx j \wedge G''') \supset G''' \qquad \Delta_3, (\tau.2) \rhd_{G'''} (\tau.3)}$$

$$\frac{\Delta_4 \rhd_{G'''} (\tau.3)}{\dfrac{\dfrac{\Delta_4 \rhd_{G'''}}{\Delta_4 \vdash G'''}}{\Delta_4 \vdash 0+i \approx i} \qquad \Delta_4 \vdash (G''' \wedge \top) \supset G''' \qquad \dfrac{\Delta_4, (\texttt{affirm } G''' \; [\,]) \rhd \top}{\Delta_4, (\texttt{affirm } G''' \; [\,]) \vdash \top}}$$

where

$$\tau = 9.0.2 \qquad\qquad \Delta_0 = \Gamma, (9.0.0), \mathbb{A}, (9.0.1)'$$
$$\mathbb{C}_2 = (\texttt{assume } (i \not\approx 0)) \qquad \Delta_1 = \Delta_0, \mathbb{C}_2$$
$$\Lambda = (\tau.0), (\tau.1), (\tau.2), (\tau.3) \qquad \Delta_2 = \Delta_1, (\tau.0)$$
$$(\tau)' = (\texttt{case } (i \not\approx 0 \supset G''') \; [\Lambda]) \qquad \Delta_3 = \Delta_2, (\tau.1)$$
$$F_1 = j \, \varepsilon \, \text{NatNum} \wedge \text{succ } j \approx i \qquad \Delta_4 = \Delta_3, (\tau.2)$$

A few comments should be made here. First, note the right-hand branch in the first inference fragment, where the goal $G'' \vee i \approx 0 \vee i \not\approx 0$ is proved. According to the rules of our calculus, each additional case section weakens the current thesis by putting it into a disjunction with the case's hypothesis. At the end of case analysis we have to prove the formula $G \vee H_1 \vee \cdots \vee H_n$, where G is the original thesis and $H_1, \dots, H_n$ are explored cases. Yet, it is a good style to make case analyses exhaustive so that just the disjunction $H_1 \vee \cdots \vee H_n$ would hold at the end.

Second, a selection sentence is valid whenever we can prove the existence of named objects, i.e. the non-emptiness of the classes corresponding to the listed notions. While in the ForTheL text in question the selection (9.0.2.0) does not change the current thesis, that may happen when the current thesis is a statement of existence.

Third, pay attention that the affirmation (9.0.2.2) is a direct consequence of the induction hypothesis H and the previous affirmation (9.0.2.1). If the assumption A (whose formula image is H) were not inserted in the proof, the affirmation (9.0.2.2) could not be proved. However, one can write a proof where no sentence requires the induction hypothesis in order to be verified. For example, the whole proof of the affirmation (9.0) could be simply omitted. Then the system would try to prove just the induction thesis G, which is not difficult and does not require any induction reasoning capabilities.

Fourth, let us consider the last inference fragment. The formula image of the affirmation (9.0.2.3) is just the atomic formula **thesis**, which stands for the current thesis, G'''. Once having this affirmation proved, we have no pending obligations so that the new thesis is simply $\top$. Recall that G''' is the formula $0 + i \approx i \vee i \approx 0$, that is, we must either prove the initial thesis (G'') or reduce the task to the previous case.

Now, assuming the validity of all the first-order leaves ($\vdash$-sequents) in our derivation, we have demonstrated the *logical correctness* of the ForTheL text under consideration.

5 Experiments

In the course of development of the SAD system, we have conducted a number of essays on formalization and verification of non-trivial mathematical results:

- Ramsey's Finite and Infinite theorems.
- Cauchy-Bouniakowsky-Schwarz inequality.
- Newman's lemma about term-rewriting systems [21].
- The square root of a prime number is irrational: 30 statements in preliminaries (integer numbers), 5 definitions, 7 lemmas, about 50 sentences in the proof of the main lemma (any prime dividing a product divides one of the factors), 10 sentences in the proof of the theorem (see [4] for details).
- Chinese remainder theorem and Bezout's identity in terms of abstract rings: 25 statements in preliminaries (ring axioms, operations on sets), 7 definitions (ideal, principal ideal, greatest common divisor, etc), 3 lemmas, 8 sentences in the proof of CRT, about 30 sentences in the proof of Bezout's identity.
- Tarski's fixed point theorem (cited above): 11 statements in preliminaries (ordered sets), 7 definitions (upper and lower bounds, supremum, infimum, complete lattice, isotone function, fixed point), 2 lemmas, 18 sentences in the proof of the theorem.

The texts listed above were written in ForTheL and automatically verified in SAD (using different background provers). This work have taught us many important lessons. To mention some:

- Formalization style is critical: the choice of symbols to introduce in definitions, the choice of preliminary facts, and even the way a proof is structured may decide whether the text will be verified or not.

- It is very desirable to comprehend the proofs before writing them in ForTheL. The SAD system may succeed to fulfil the gaps in a well thought-out reasoning, but it will not invent one for you.
- In most cases, the background prover finds the proof in three seconds — or does not find it at all.

6   Conclusion

While working on the development of the SAD project, we always felt that we strongly need an abstract, clear and transparent sight on the whole "formalization-and-verification" process. The calculus **CTC** proposed here is the first step in this direction. The next one will be a formal description of definition expansion and other reasoner's routines which is obviously missing now. Moreover, such a description must provide users with a kind of specific language to create their own reasoning strategies. Further on, something like a guide to problem formalization in a strong mathematical manner would be extremely useful, too.

Our abstract considerations of text verification resulted in the SAD system. Certainly, we could not give here a detailed description of all nice features of SAD. SAD is a powerful system and its power lies in its reasoning facility. Experiments show that, for example, the specific strategy of definition processing contributes a lot to the success of the whole verification process. If we use definitions straightforwardly — convert them into formula images and add the corresponding premises to the sequent that goes into a prover — we cannot verify the proof of Tarski fixed point theorem as it is formulated above, even when a winner of CASC competitions is chosen as the background prover.

SAD is not a perfect system (if any!). One can easily see how it may be improved and developed. Our research and implementation plans with respect to SAD are: extend ForTheL and SAD with some means to talk and reason about second-order objects (functions, vectors, sequences) and operations on them; develop and implement a mathematical library of SAD to accumulate verified portions of mathematical knowledge and to support further (deeper) advances in formalization.

References

1. Wang, H.: Towards mechanical mathematics. IBM J. of Research and Development 4, 2–22 (1960)
2. Glushkov, V.M.: Some problems of automata theory and artificial intelligence (in Russian). Kibernetika 2, 3–13 (1970)
3. Lyaletski, A., Paskevich, A., Verchinine, K.: Theorem proving and proof verification in the system SAD. In: Asperti, A., Bancerek, G., Trybulec, A. (eds.) MKM 2004. LNCS, vol. 3119, pp. 236–250. Springer, Heidelberg (2004)
4. Lyaletski, A., Paskevich, A., Verchinine, K.: SAD as a mathematical assistant — how should we go from here to there? Journal of Applied Logic 4(4), 560–591 (2006)
5. Verchinine, K., Lyaletski, A., Paskevich, A.: System for Automated Deduction (SAD): a tool for proof verification. In: Pfenning, F. (ed.) CADE 2007. LNCS (LNAI), vol. 4603, pp. 398–403. Springer, Heidelberg (2007)

6. Wiedijk, F. (ed.): The Seventeen Provers of the World. LNCS (LNAI), vol. 3600. Springer, Heidelberg (2006)
7. Bertot, Y., Castéran, P.: Interactive Theorem Proving and Program Development. Coq'Art: The Calculus of Inductive Constructions. Texts in Theoretical Computer Science (EATCS), vol. XXV. Springer, Heidelberg (2004)
8. Nipkow, T., Paulson, L.C., Wenzel, M.: Isabelle/HOL: A Proof Assistant for Higher-Order Logic. LNCS, vol. 2283. Springer, Heidelberg (2002)
9. Owre, S., Rushby, J.M., Shankar, N.: PVS: a prototype verification system. In: Kapur, D. (ed.) CADE 1992. LNCS, vol. 607, pp. 748–752. Springer, Heidelberg (1992)
10. Gordon, M.J.C., Melham, T.F.: Introduction to HOL: a theorem proving environment for higher order logic. Cambridge University Press, Cambridge (1993)
11. Nederpelt, R.P., Geuvers, J.H., de Vrijer, R.C.(eds.): Selected Papers on Automath. Studies in Logic and the Foundations of Mathematics, vol. 133. North-Holland, Amsterdam (1994)
12. Trybulec, A., Blair, H.: Computer assisted reasoning with Mizar. In: Proc. 9th International Joint Conference on Artificial Intelligence, IJCAI 1985, pp. 26–28. Morgan Kaufmann, San Francisco (1985)
13. Wenzel, M.: Isar — a generic interpretative approach to readable formal proof documents. In: Bertot, Y., Dowek, G., Hirschowitz, A., Paulin, C., Théry, L. (eds.) TPHOLs 1999. LNCS, vol. 1690, pp. 167–184. Springer, Heidelberg (1999)
14. Kamareddine, F., Nederpelt, R.P.: A Refinement of de Bruijn's Formal Language of Mathematics. Journal of Logic, Language and Information 13(3), 287–340 (2004)
15. Jojgov, G., Nederpelt, R.: A path to faithful formalizations of mathematics. In: Asperti, A., Bancerek, G., Trybulec, A. (eds.) MKM 2004. LNCS, vol. 3119, pp. 145–159. Springer, Heidelberg (2004)
16. Kamareddine, F., Maarek, M., Wells, J.B.: Flexible encoding of mathematics on the computer. In: Asperti, A., Bancerek, G., Trybulec, A. (eds.) MKM 2004. LNCS, vol. 3119, pp. 145–159. Springer, Heidelberg (2004)
17. Kaufmann, M., Manolios, P., Moore, J.S.: Computer-Aided Reasoning: An Approach. Kluwer, Dordrecht (2000)
18. Richardson, J., Smaill, A., Green, I.: System description: Proof planning in higher-order logic with Lambda-Clam. In: Kirchner, C., Kirchner, H. (eds.) CADE 1998. LNCS (LNAI), vol. 1421, pp. 129–133. Springer, Heidelberg (1998)
19. Buchberger, B., Crăciun, A., Jebelean, T., Kovács, L., Kutsia, T., Nakagawa, K., Piroi, F., Popov, N., Robu, J., Rosenkranz, M., Windsteiger, W.: Theorema: Towards computer-aided mathematical theory exploration. Journal of Applied Logic 4(4), 470–504 (2006)
20. Vershinin, K., Paskevich, A.: ForTheL — the language of formal theories. International Journal of Information Theories and Applications 7(3), 120–126 (2000)
21. Paskevich, A., Verchinine, K., Lyaletski, A., Anisimov, A.: Reasoning inside a formula and ontological correctness of a formal mathematical text. In: Calculemus/MKM 2007 — Work in Progress, University of Linz, Austria. Number 07-06 in RISC-Linz Report Series, pp. 77–91 (2007)

Author Index

Lecture Notes in Artificial Intelligence (LNAI)